# XIIth EUROPEAN CONFERENCE ON ANIMAL BLOOD GROUPS AND BIOCHEMICAL POLYMORPHISM

EUROPEAN SOCIETY FOR ANIMAL BLOOD GROUP RESEARCH

# XIIth EUROPEAN CONFERENCE ON ANIMAL BLOOD GROUPS AND BIOCHEMICAL POLYMORPHISM

Edited by

G. KOVÁCS and M. PAPP

1972

SPRINGER-SCIENCE+BUSINESS MEDIA, B.V.

XIIth European Conference on Animal Blood Groups and Biochemical Polymorphism
held in Budapest, July 6–11, 1970

*Honorary Members*

Professor I. Dimény, Minister of Agriculture and Food
Academician G. Láng
Academician R. Manninger
Academician J. Mócsy

*Executive Committee of the European Society for Animal Blood Group Research*

President: M. Braend, Norway
Secretary: J. Bouw, The Netherlands
Members: B. Larsen, Denmark
Gunvar Lindström, Finland
M. McDermid, Great Britain
J. Matoušek, Czechoslovakia
Ex-officio member: P. Soós, Hungary

*Organization Committee*

Chairman: Professor A. B. Kovács
Vice-chairman: Professor A. Kemény
Secretaries: M. Papp, G. Kovács and J. Stukovszky

*Technical Committee*

Gy. Lencsés
F. Ludrovszky

*Social Committee*

P. Bánfalvi
L. Fodor

ISBN 978-94-017-5458-3
DOI 10.1007/978-94-017-5456-9
ISBN 978-94-017-5456-9 (eBook)


Originally published by Dr. W. Junk N.V. in 1972

# CONTENTS

III. BLOOD GROUPS AND BIOCHEMICAL POLYMORPHISM IN PIGS

IV. BLOOD GROUPS AND BIOCHEMICAL POLYMORPHISM IN POULTRY

V. BLOOD GROUPS AND BIOCHEMICAL POLYMORPHISM IN HORSES

## IX. BLOOD GROUPS AND BIOCHEMICAL POLYMORPHISM IN OTHER SPECIES

# LIST OF PARTICIPANTS

ANDRESEN, E., Department of Animal Genetics, The Royal Veterinary and Agricultural University, 1870. Copenhagen V., 13 Bülowsvej, Denmark

ASHTON, G. C., Department of Genetics, University of Hawaii, Honolulu, Hawaii 96822, USA

BACKHAUSZ, R., "Human" Institute for Serobacteriological Production and Research, Szállás u. 5—7, Budapest, X., Hungary

BALBIERZ, H., Department of Immunopathology, Chair of Obstetrics and Reproduction Pathology, Faculty of Veterinary Science, School of Agriculture, Wrocław, Poland

BARBIER, Yolaine, 28me des la République, 78 Saint Germain-en-Laye, France

BARON, J. C., Office de la Recherche Scientifique et Technique, Outre-Mer, 114 R. Pasteur, 94 Bonneuil, France

BENDA, V., Institute of Experimental Biology and Genetics, Czechoslovak Academy of Sciences, Prague 6, Flemingovo 2, Czechoslovakia

BENESOVÁ, Jana, Institute of Experimental Biology and Genetics, Czechoslovak Academy of Sciences, Prague, Budejovická 1083, Czechoslovakia

BEZENKO, S. P., All-Union Research Institute of Animal Breeding, Moskovskaya oblast', Podolskii region, Dubrowitsy, WIZ, USSR

BÖHM, O., The Veterinary Institute of Slovenia, P.O.B. 257, Ljubljana, Yugoslavia

BOUQUET, Y., Faculty of Veterinary Medicine, University of Ghent, Heidestraat 19, 9220 Merelbeke, Belgium

BOUW, J., Laboratorium voor Bloedgroepen Onderzoek, Landbouwhogeschool, Duivendaal 5, Wageningen, The Netherlands

BRAEND, M., Faculty of Veterinary Science, P. O. Kabete, Kenya

BRILES, W. E., DeKalb AgResearch, Inc., DeKalb, Illinois 60115, USA

BRUNO, R., Istituto di Fisiologia Veterinaria, Università di Torino, Italy

BUIS, R. C., Laboratorium voor Bloedgroepen Onderzoek, Landbouwhogeschool, Duivendaal 5, Wageningen, The Netherlands

CARENZI, M., Istituto di Zootecnia Generale, Università di Milano, Via Celoria, 10, Milano 20133, Italy

ČERNE, I., The Veterinary Institute of Slovenia, Ljubljana, Yugoslavia

CERUTTI, F., Istituto di Zootecnia Generale, Università di Milano, Milan, Italy
CHOLNOKY, Eszter, Laboratóriumi Törzsállattenyésztő Intézet, Táncsics M. út, Gödöllő, Hungary
CRIMELLA, C., Istituto di Zootecnia Generale, Università di Milano, Via Celoria 10, Milano 20133, Italy
CSUKA, J., Department of Genetics, Agricultural University, Nitra, Czechoslovakia
DASSAT, P., Osservatorio di Genetica Animale, Via Pastrengo, 28, Torino, Italy
DÉMANT, P., Institute of Experimental Biology and Genetics, Czechoslovak Academy of Sciences, Prague, Budejovicka 1083, Czechoslovakia
DIKOV, V., Institute of Reproduction Biology and Non-Infectious Diseases, Bul. Lenin 73., Sofia 13, Bulgaria
DINKLAGE, H., Institut für Tierzucht und Haustiergenetik der Universität Göttingen, Albrecht-Thaer-Weg 1, 34, Göttingen, GFR
DOLA, Lidia, Wyzsza Szkola Rolnicza, Zaklad Hodowli Bydla, Laboratorium Grup Krwi, Kraków, Al. Mickiewicza 24/28, Poland
DORYNEK, Z., College of Agriculture in Poznan, Chair of Animal Husbandry, Poznan, Wołynska 33, Poland
DOSTAL, J., Department of Animal Science, University of Minnesota, St. Paul, Minnesota, USA
DUNIEC, Maria, Immunogenetics Laboratory, Zootechnical Institute, Balice k/Kraków, Poland
EFREMOV, G. D., Faculty of Agriculture University of Skopje, Skopje, Yugoslavia
ERMENCOVA, Lidia, Research Institute of Animal Breeding, Kostinbrod, Bulgaria
FÉSÜS, L., Bloodgrouping Laboratory, University of Veterinary Science, Rottenbiller u. 50, Budapest VII., Hungary
FIORENTINI, A., Istituto di Fisiologia Veterinaria, Università di Torino, Italy
FLESJÅ, Margrete, Norges Veterinar Høgskole, Postboks 8146., Oslo dep. Oslo I., Norway
FRENTZ, R., Faculté des Sciences de Nancy, Centre de 1er Cycle, 2 bd des Aiguillettes 54-Nancy, France
FRENZL, B., Department of Biology, Medicine Faculty, Charles University, Prague 2, Albertov 4, Czechoslovakia
GAHNE, B., Department of Animal Breeding, Agricultural College, Uppsala 7, Sweden
GASPARSKA, Jolanta, Insitut Genetiki i Hodowli Zwierzat P.A.N. Nowy Swiat 72, Warsawa, Poland
GELDERMANN, H., Tierärztliches Institut der Universität Göttingen, Weserstr. 36, 34. Göttingen, GFR
GIPPERT, Erzsébet, Center of Artificial Insemination, Remény u. 42. Budapest, XIV., Hungary
GLASNÁK, V., C. B. Station, Hradiško, okr. Praha Zapad., Czechoslovakia
GRANCIU, I., Institutul de cercetări zootechnice, Bucureşti-Săftica, România
GÜNTHER, E., Max-Planck Institut für Immunbiologie, Freiburg/Br., Stübeweg 51, GFR

HARDY, Joan, Protein Chemistry Section, Unilever Research Laboratory, Colworth House, Sharnbrook, Bedford, England

HÁLA, K., Institute of Experimental Biology and Genetics, Czechoslovak Academy of Sciences, Prague 6, Flemingovo 2 Czechoslovakia

HOHENBRINK, R., Institut für Tierzucht und Haustiergenetik der Universität Göttingen, 34 Göttingen, Albrecht-Thaer-Weg 1. GFR

HOJNÝ, J., Laboratory of Animal Genetics, Czechoslovak Academy of Sciences, Liběchov, Czechoslovakia

HORVÁTH, I., Center of Artificial Insemination, Remény u. 42. Budapest, XIV., Hungary

HRADECKÝ, J., Laboratory of Animal Genetics, Czechoslovak Academy of Sciences, Liběchov, Czechoslovakia

IVÁNYI, P., Institute of Experimental Biology and Genetics, Czechoslovak Academy of Sciences, Prague 6, Flemingovo 2, Czechoslovakia

JAMIESON, A., Ministry of Agriculture, Fisheries and Food, Fisheries Laboratory, Lowestoft, Suffolk, Great Britain

JOVER, A., Facultad de Veterinaria, Departmento de Zootecnia Cordoba, Spain

KACZMAREK, A., College of Agriculture in Poznan, Chair of Animal Husbandry, Laboratory of Research in Animal Blood Groups, Wołynska 33, Poznan, Poland

KAMINSKI, Marie, Laboratoire d'Enzymologie du CNRS, 91. Gif-sur-Yvette, France

KIDD, K. K., Istituto di Genetica, Università di Pavia, 27100 Pavia, Italy

KLABUKOV, P. G., All-Union Research Institute of Animal Breeding Moskovskaya oblast', Podolskii region, Dubrowitsy, WIZ, USSR

KOCH, J., Tierärztliches Institut der Universität Göttingen, Groner Landstr. 2, 34, Göttingen, GFR

KOVÁCS, G., Department of Animal Husbandry, University of Veterinary Science, Rottenbiller u. 23–25, Budapest, VII., Hungary

KRAAY, G. J., Department of Biomedical Sciences, Ontario Veterinary College, University of Guelph, Guelph, Ontario, Canada

KRAJNOVIĆ-OZRETIĆ, Mirjana, Institute "Rudjer Boškovic" Center for Research, Rovinj, Yugoslavia

KŘEN, V., Department of Biology, Medical Faculty, Charles University, Prague 2, Albertov 4, Czechoslovakia

KŘENOVÁ, Drahomira, Department of Biology, Medical Faculty, Charles University, Prague 2, Albertov 4, Czechoslovakia

KRISTJANSSON, F. K., Animal Research Institute, Canada Department of Agriculture, Ottawa, Ontario, Canada

KŘSIAKOVA, Miloslava, Department of Biology, Faculty of General Medicine, Charles University, Prague, Albertov 4, Czechoslovakia

KÚBEK, A., Laboratory of Animal Genetics, Czechoslovak Academy of Sciences, Liběchov, Czechoslovakia

LARSEN, B., Department of Physiology, Endocrinology and Bloodgrouping, The Royal Veterinary and Agricultural University, 13, Bülowsvej, Copenhagen V, Denmark

LAZAR, P., The Veterinary Institute of Slovenia, Ljubljana, Yugoslavia

LIE, Halldis, Veterinary College of Norway, Norges Veterinaerogskole Institut for Indremedisin, Ullevalsvn 72, Oslo 4, Norway

DE LIGNY, Wilhelmina, Netherlands Institute for Fishery Investigations, Ijmuiden, The Netherlands

LINDSTRÖM, Gunvar, Centr. Assoc. of A. I. Societies, Kuriiritie 11, Tikkurila, Finland

LINHART, J., Animal Breeding Development Institute, Hradistko/Med, okr. Praha-Zapad, Czechoslovakia

LINKLATER, K. A., Blood Group Research Unit, Department of Animal Health, Veterinary Field Station, Easter Bush, Roslin, Midlothian, U.K.

LIPECKA, J., Wyższa Szkoła Rolnicza, Katedra Szczegółowej Hodowli Zwierząt, Lublin, ul. Króla Leszczyńskiego 9, Poland

LOSONCZY Mária, Hungarian Meat Research Institute, Budapest, Hungary

LOSONCZY, S., Bloodgrouping Laboratory, University of Veterinary Sciences, Rottenbiller u. 50., Budapest VII., Hungary

LÖFFELBEIN, H., Institut für Künstliche Besamung 1282 Schönow/Berlin, GDR

LÖHLE, Institut für Künstliche Besammung 1282 Schönow/Berlin, GDR

LUDROVSZKY,F., Department of Animal Husbandry, University of Veterinary Science, Rottenbiller u. 23—25., Budapest VII., Hungary

MÁCHA, J., Department of Genetics, Agricultural College, Zemedelská 3, Brno, Czechoslovakia

MADEYSKA-LEWANDOWSKA, Anna, Institute of Genetics and Animal Breeding, Polish Academy of Sciences, Warsaw, Poland

MAJOR, F., Institute für Tierzucht und Haustiergenetik der Universität Göttingen, Albrecht-Thaer-Weg 1, 34. Göttingen, GFR

MAKAVEEV, Ts., Institute of Animal Breeding, Department of Genetics, Laboratory of Immunogenetics, Kostinbrod-Sofia, Bulgaria

MANSILLA, A. R., Patronato de biologia animal, Avenida de Puerta de Hierro Madrid 3, Spain

MATOUŠEK, J., Laboratory of Animal Genetics, Czechoslovak Academy of Sciences, Liběchov, Czechoslovakia

MÉSZÁROS, I., Center of Artificial Insemination, Remény u. 42., Budapest, XIV., Hungary

MILLOT, P., Station Centrale de Génétique Animale, C.N.R.Z., 78-Jouy-en-Josas, France

MITAT, J., Laboratorio de Immunogenética, Facultad de Ciencias Agropecuarias, Universidad de la Habana, Habana, Cuba

MORVAI, I., Research Institute for Small Animal Breeding, Gödöllő, Hungary

MUNKÁCSY, F., Department of Animal Breeding, University of Agriculture, Debrecen, Hungary

MURAVIEV, I. V., All-Union Poultry Research and Technological Institute, Zagorsk, 11, Moscow-region, USSR

NEMESI, M., Department of Animal Husbandry, University of Veterinary Science, Rottenbiller u. 23—25., Budapest, VII., Hungary

NIKOŁAJCZUK, M., Pracownia Immunopatalogii, Wydz. Wet. WSR, Wrocław, Poland

OOSTERLEE, C. C., Laboratorium voor Bloedgroepen Onderzoek, Landbouwhogeschool, Duivendaal 5, Wageningen, The Netherlands

OSTERHOFF, D. R., Department of Zootechnology, Faculty of Veterinary Science, University of Pretoria, Onderstepoort, Rep. of South Africa

PAPP, M., Bloodgrouping Laboratory, University of Veterinary Science, Rottenbiller u. 50, Budapest VII., Hungary.

PERRAMON, A., Station Centrale de Génétique Animale C.N.R.Z 78.—Jouy-en-Josas, France

PETHES, G., Radioisotope Laboratory, Department of Physiology, University of Veterinary Science, Rottenbiller u. 23—25, Budapest VII., Hungary

PETROVSKÝ, E., Research Institute of Animal Sciences, University of Agriculture, Brno, Czechoslovakia

PILZ, J., Institut für künstliche Besamung 1282, Schönow/Berlin, GDR

PIRCHNER, F., Lehrkanzel und Institut für Tierzucht und Haustiergenetik, Tierärztliche Hochschule Wien, A 1030 Linke Bahngasse 11, Austria

PODLIACHOUK, Luba, Institut Pasteur, 28, Rue du Dr ROUX, Paris 15e, France

RAMOS, J. L. S., Patronato de Biologia Animal, Avenida de Puerta de Hierro, Madrid-3, Spain

RENDEL, J. R., Animal Production and Health Division, FAO, Via delle Terme di Caracalla, Rome, Italy

ROBERTSON, F. W., Department of Genetics, University of Aberdeen, Scotland

RODERO, A., Conseje superios de investigaciones cientificas, Departamento de Zootecnia, Faculdad de Veterinaria, Cordoba, Spain

ROHRBACHER, H., Department of Animal Breeding and Genetics, School of Veterinary Medicine, A 1030 Linke Bahngasse 11, Vienna, Austria

SANDBERG, K. I., Department of Animal Breeding, Agricultural College, Uppsala, Sweden

SARTORE, G., Istituto di Fisiologia Veterinaria, Via Campana 16, Torino, Italy

SCHLEGER, W., Blood Group Laboratory, School of Veterinary Medicine, A 1030 Linke Bahngasse 11, Vienna, Austria

SCHMID, D. O., Institute of Animal Bloodgroup and Resistance Research, Munich, GFR, München-15, Haydnstr. 11.

SCHRÖFFEL, J., Immunologic Laboratory, Animal Breeding Research Institute Hradistko p. Med. Praha-zapad, Czechoslovakia

SCHMIDT, J., College of Agricultural Sciences, Mosonmagyaróvár, Hungary

SCOTT, A. M., Equine Research Station of the Animal Health Trust, Newmarket, Suffolk, England

SEBESTYÉN, G., Research Institute for Animal Husbandry, Kitaibel Pál u. 2—4, Budapest II., Hungary

SHAW, C. R., Department of Biology, The University of Texas M. D., Anderson Hospital and Tumor Institute, Houston, Texas 11025

SIMON, M., Laboratory of Animal Genetics, Czechoslovak Academy of Sciences, Liběchov, Czechoslovakia

SOÓS, Pál, Center of Artificial Insemination, Remény u. 42, Budapest XIV., Hungary

SOOS, Péter, Center of Artificial Insemination, Remény u. 42, Budapest XIV., Hungary
ŠTARK, O., Department of Biology, Medical Faculty, Charles University, Prague 2, Albertov 4, Czechoslovakia
STORMONT, C. J., Department of Veterinary Microbiology, University of California, Davis, California 95616
STUKOVSZKY, J., Center of Artifical Insemination, Remény u. 42, Budapest XIV., Hungary
SUZUKI, Yoshiko, Serology Laboratory, Department of Veterinary Microbiology, University of California, Davis, California 95616
SYKIOTIS, Michèle, Laboratoire d'Enzymologie, CNRS, 91, Gif-Sur-Yvette, France
SZENT IVÁNYI T., Department of Epizootiology, University of Veterinary Science, Hungária krt. 23–25, Budapest XIV., Hungary
TAKÁCS, Erzsébet, Bloodgrouping Laboratory, University of Veterinary Science, Rottenbiller u. 50, Budapest VII., Hungary
THIELE, O. W., Physiologisch-chemisches Institut, Universität, D-34 Göttingen Humboldtallee 7. GFR
TIKHONOV, V. N., Institute of Cytology and Genetics of the USSR Academy of Sciences, Novosibirsk 90, USSR
TJANKOV, S., Chair of Sheep Breeding, High School of Agriculture, Sofia, Bulgaria
TOMASZEWSKA-GUSZKIEWICZ, K., Institute of Genetics and Animal Breeding, Polish Academy of Sciences, Warsaw, Yastrzebiec p-ta Mrokow, Poland
TOSI-LANDUCCI, Simonetta, Osservatorio di Genetica Animale, Via Pastrengo 28, 10128 Torino, Italy
TOSI, R. M., Istituto di Genetica Medica, Università di Torino, Italia
TUCKER, Elizabeth, A. R. C. Institute of Animal Physiology, Babraham, Hall, Cambridge, Great Britain
VERESS, L., High School for Agriculture, Kaposvár, Hungary
VESELSKÝ, L., Laboratory of Animal Genetics, Czechoslovak Academy of Sciences, Liběchov, Czechoslovakia
WIATROSZAK, I., College of Agriculture in Poznań, Department of Animal Husbandry, Laboratory of Research in Animal Blood Groups, Poznań, Wolynska 33, Poland
WILKINS, N. P., Department of Agriculture and Fisheries for Scotland Marine Laboratory Victoria Road Torry Aberdeen, Scotland
ZURKOWSKI, M., Institute of Genetics and Animal Breeding, Polish Academy of Sciences, Warsaw, Jastrebiec p-ta Mrokow, pow. Piaseczno, Poland

# OPENING ADDRESSES

# PROF. I. DIMÉNY

Mr. Chairman, Ladies and Gentlemen,

I was pleased to learn that our proposal to held this conference in Budapest was accepted by the European Society for Animal Blood Group Research two years ago.

Now, it is my privilege and great pleasure to welcome all participants and open the XIIth International Conference on Animal Blood Groups and Biochemical Polymorphism.

In agreement with the peace policy of the Hungarian People's Republic we welcome the guests of any scientific meeting or other international discussion aimed for the benefit of mankind or promoting mutual understanding among the nations.

We regard it as a great privilege to be the hosts of this conference. I assure you that the Organizing Committee, the Rector of University of Veterinary Science and the Executive Committee of E.S.A.B.R., which has acquired international reputation, do their best for the success of this Budapest session. I wish this meeting represented a great step forward in the development of this discipline and were long remembered by all participants.

The proceedings of this rather young branch of science will be discussed in about 130 lectures during the next 4 days. I hope the work of this conference will contribute to the advance of veterinary research and, thus, indirectly also to the welfare of mankind.

By coincidence another conference in a closely related field viz. animal husbandry, organized by the European Association for Animal Production and the Hungarian Ministry of Agriculture and Food, will also be held in this country between the 24th and 28th of August.

We had the intention of shortening the time interval between the two conferences, to facilitate the participation of experts interested in both subjects. However, owing to certain technical difficulties the independent organization of these two meetings had to be decided.

Now, I declare the XIIth International Conference on Animal Blood Groups and Biochemical Polymorphism opened. I hope that despite the tight scientific programme all our guests will find time to get acquainted with our capital and Hungarian hospitality.

Let me wish you every success to the conference and good time in Hungary.

# PROF. DR. A. B. KOVÁCS

Mr. Chairman, Ladies and Gentlemen,

May I also take the opportunity for and on behalf of the University of Veterinary Science and myself to welcome all our guests, participating this conference. Going along with the words of Minister Dimény also I would like to assure you that we regard it a great honour that in cooperation with the European Society for Animal Blood Group Research, this conference has been organized in Budapest.

By way of introduction may I briefly recall some more relevant stages of animal blood group research and that of biochemical polymorphism. As we all know, while discovery of human blood groups by LANDSTEINER immediately resulted in practical utilization, restoring millions of people to life, research on animal blood groups remained a purely academic subject for some decades.

The animal blood group research in Hungary goes back to a past of about 40 years. Basic studies were carried out by GÉZA FETT on cattle, whilst LAJOS BÉRCZY and HENRIK DÖHRMANN worked on pig blood groups in the Animal Husbandry Department of the late Veterinary School in 1926 and 1927, respectively. Natural antibodies were employed for these studies, yielding moderate yet still paramount results, as compared to the contemporary international standards.

The use of isoimmune test-sera, a method elaborated by IRWIN and his colleagues in the 40-es, initiated a considerable progress both in blood group research of various animal species, particularly cattle, as well as in the practical application of the new knowledge so obtained.

The profound knowledge of blood groups enabled us to firecast which characteristics might possibly be transmitted by inheritance and also enabled parentage control tests. Thus, the objectives of the expensive improving and selective procedures in animal breeding could be easier achieved. Hence, investigation of the animal blood group system, generally considered a mere matter of scientific curiosity for some time, turned into a scientific branch of economic significance, amply compensating for the expenditures of the early research period.

In 1954 SZENT-IVÁNYI and SZABÓ, working with hog cholera serum, reported interesting observations on the blood groups of pigs. They have detected four of the so-called isoantigens and the corresponding isoantibodies. Their findings have furnished information both on hemolytic icterus in newborn piglets due to isoimmunization and on the shock-syndrome, occasionally emerging during the process of anti-hog-cholera serum production.

It was professor MÁRKUS in 1958 who with due regard to results obtained abroad, recognized the practical importance of blood group research in cattle and has made many efforts to introduce such methods in Hungary. He set up a small research team at the Animal Husbandry Department of our University, entrusted with blood group studies for mainly practical purposes. By his activities, and the capability of his colleagues trained in the up-to-date trends and methods of animal blood group research abroad, remarkable results have been achieved in the study of cattle-, chicken-, pig-, and goose blood groups. May I express my thanks and appreciation to professor MÁRKUS, who has initiated the blood group research programme at this University.

Further to the University Blood Group Research Laboratory, there was another one founded for cattle blood group work at the Center for Artificial Insemination about 10 years ago. This laboratory, initially engaged exclusively routine work has recently conducted research as well.

I think it is an excellent idea to discuss the problems of animal blood groups along with those of biochemical polymorphisms as the information emerging from these two subjects seems to offer the most reliable basis for a better understanding of the genetic structure of various species, breeds and populations. As a result of such research, changes in various genetic parameters have become measurable which, by their physiological and biochemical actions involved, may become economically important.

The correlation between such data and production characteristics of animals has long been in the focus of interest. Now, we are aware of some pros and cons, encouraging results and failures in this line as well. We all hope that lectures and discussions at the present conference will contribute to a better understanding of many important and rather exciting questions.

Finally may I wish again very useful discussions, good work and good luck to all participants of this conference.

# PROF. DR. M. BRAEND

Your Excellency Professor I. Dimény, Minister of Agriculture and Food, Professor A. B. Kovács, Rector of the University of Veterinary Science, Distinguished Guests, Ladies and Gentlemen,

It is a great pleasure and honour for me, on behalf of the European Society for Animal Blood Group Research, to welcome you to the XIIth European Conference for Animal Blood Groups and Biochemical Polymorphism. Our guestspeakers deserve a particular word of welcome, for their willingness to come and lecture to us. To those who are attending one of our conferences for the first time, I would extend a special welcome. For the rest I would like to say: It is nice seeing you again.

Another two years have passed since we had our Conference at Warsaw. It is natural to ask what has been achieved during this period? Is it justified to have conferences this often? A glance at the program gives the answer for there we see a greater number of papers than ever before. Furthermore, the variety of papers, covering different species as well as specialities, shows a welcome breadth. This is very promising for the future.

It also shows that our Society has the energy, spirit and enthusiasm of a young Society: For this also we should be happy. Certainly our Society is becoming older and therefore we should move into a period of consolidation from that of rapid development. But the scientific activities of our members continue to expand, governed by interest and enthusiasm. This is a proof of the importance of our field and that it has its right of life. It further leads to the strengthening of our Society and its reputation. However, we should not expect the brilliant results of research alone: Science, whether in our field or any other, is not primarily characterized by the few and brilliant findings, but more by the hard and dedicated work of all those who provide the necessary basic information for Science as such to move forward.

One important achievement of our Society since the Warsaw Conference is the launching of our Journal. So far one issue only has appeared but the extensive struggle in laying the foundation is over. The future of the Journal now is primarily dependant upon the members and the contributors; that they will be able to furnish the Editor with important and interesting research results. However seeing the intense scientific activity among our members, I have great faith in its success.

This growth of activity means a heavy burden for those who are directly involved in the planning and organisation. In the matter of Journal our

Society owes its great gratitude to its Secretary Dr. Bouw, and the Editor Dr. Oosterlee. The organization of our conference has also grown to such an extent that it requires much hard work and dedication. We are therefore in great debt to our hosts who have invited us and made the sacrifice going therewith. So far our good hosts have taken excellent care of us, and the program assures us this will continue both for the Scientific part and for the social events. We all know of the Hungarian people, of their hospitality, their music, their dances and their beautiful city of Budapest. So now we are eager to get as much out of our stay here as possible, scientifically as well as experiencing for ourselves the Hungarian people, and their culture.

There are many who deserve our gratitude for making this possible for us. His Excellency the Minister for Food and Agriculture, Professor Dimény, has been so kind as to honour us with his presence in the opening ceremony. We appreciate greatly this interest shown in our Conference, and our particular scientific field.

The organizing Committee deserve our most sincere thanks. I would like to express this to the Chairman of the Committee, Professor A. B. Kovács, Rector of the University of Veterinary Science.

Finally I would like to express our appreciation and gratitude to Professor Márkus: for many years he has shown great interest in our field. His foresight and enthusiasm resulted in the attraction of young co-workers. He further initiated the planning for the Conference here in Budapest and finally was able to invite us two years ago at Warsaw.

# GUEST SPEAKERS' REPORTS

*XIIth Europ. Conf. Anim. Blood Groups Biochem. Polymorph., Bp., 1972 (pp. 29—40)*

# POLYMORPHISMS IN HUMAN SERUM PROTEINS

R. Backhausz

"HUMAN" Institute for Serobacteriological Production and Research, Budapest, Hungary

## INTRODUCTION

The human species has been greatly preferred for the study of polymorphic systems. Well-known examples of human polymorphism are the blood group systems, the hemoglobin variants and the system of leucocyte alloantigens which are responsible for the immunological phenomena following tissue transplantation. There are so many different aspects of human polymorphism that it would be impossible to consider them all in a single lecture. During the past fifteen years several groups of inherited human serum protein traits constituting polymorphic systems have been discovered. The scope of the present report will be limited to the allotypic traits of human serum proteins and I hope that a general survey of the results of newer investigations along this line will be helpful for all participants of this International Conference on animal blood groups and biochemical polymorphism.

The distinction between genetic serum protein types depends partly on differences in electrophoretic mobility, partly on differences in antigenicity. These variations are generally considered to reflect differences in primary protein structure.

The capacity of starch gel electrophoresis to resolve protein mixtures has led to the discovery of genetically determined polymorphisms, such as haptoglobin, transferrin, alpha-1-antitrypsin, beta-1C-globulin etc.

## THE HAPTOGLOBIN VARIANTS

The haptoglobins are a family of serum proteins which bind hemoglobin. As shown by Smithies and Walker in 1955, using starch gel as a supporting medium in electrophoresis, human sera mixed with hemoglobin can be grouped according to haptoglobin type. Three major types, Hp 1–1, 2–1 and 2–2 can be distinguished. Haptoglobin variants differ from each other in electrophoretic mobilities and in the number of demonstrable protein components as well. In type 1–1 only a single haptoglobin component is seen. A component with the same electrophoretic mobility can be detected in type 2–1 but not in type 2–2. The latter types both show a whole series of haptoglobin components. Structural studies on the purified haptoglobins revealed that the multiple components which are characteristic of types 2–1 and 2–2 represent a series of increasing molecular weight (Smithies and Connel, 1959).

When purified serum haptoglobin is subjected to reductive cleavage by mercaptoethanol in the presence of 8 M urea, fast moving (F-) and slow-moving (S-) variants of Hp 1 type can be distinguished. Haptoglobins consist of alpha- and beta-polypeptide chains and Hp-1F, Hp-1S and Hp-2 represent genetically determined allotypes of the alpha-polypeptide chain. The beta-chains are the same in each type. It is interesting that the three Hp 1–1 types cannot be clearly distinguished from one another by routine starch gel electrophoresis of the native haptoglobins, nor can the two different types of Hp 2–1 be distinguished in this manner (SMITHIES et al., 1966; CLEVE et al., 1967).

Rare variants of Hp-2 alpha-chain are designed as JOHNSON (Hp-2J) and modified (Hp-2M) types (SMITHIES and CONNELL, 1959).

Thus, five distinctly dissimilar alpha-haptoglobin chains called Hp-1F, Hp-1S, Hp-2, Hp-2J and Hp-2M have been demonstrated. Investigations of inheritance studies indicate that these five polypeptide chains represent the alleles $Hp^{1F}$, $Hp^{1S}$, $Hp^{2}$, $Hp^{2J}$ and $Hp^{2M}$ (GIBLET and BROOKS, 1963).

In the haptoglobin molecule the chains are cross-linked by disulphide bonds. Studies on the isolated alpha haptoglobin chains revealed remarkable differences between Hp-1F and Hp-1S on the one hand, and Hp-2 on the other hand. It was found that Hp-1F and Hp-1S alpha-chains each had a molecular weight of about 9000, whereas the Hp-2 alpha-chain was almost twice as big as the former types having a molecular weight of about 16 000. The Hp-1F and Hp-1S alpha-chains have identical primary structure, consisting of 83 amino acids and differing only in position 54, where lysin in Hp-1F is replaced by glutamic acid in Hp-1S.

The Hp-2 alpha-chain contains 142 amino acids and the analysis of its primary structure suggests it to represent an end to end fusion of an Hp-1F and an Hp-1S alpha-chain but with the loss at the site of fusion of 12 residues (BLACK and DIXON, 1968). Analysis of the primary structures of immunoglobulin light chains and of haptoglobin alpha-chains showed identical amino acids in certain positions. This parenthood speaks for a common origin of the protein families. Both families are characterized by genetic polymorphism and by the ability to form complexes with macromolecules, with antigens and hemoglobins, respectively (BLACK and DIXON, 1968).

In sera of some individuals of certain populations lack of haptoglobin was noted. There is firm evidence that some forms of ahaptoglobinaemia or hypohaptoglobinaemia are hereditary and this speaks for a silent Hp-gene. On the other hand, in a variety of hemolytic diseases, haptoglobinaemia is due to the hemolysis of red blood cells and loss of haptoglobin (PROKOP and BUNDSCHUCH, 1962; GIBLETT and BROOKS, 1963; WALTER and STEGMÜLLER, 1969).

As shown by PEACOCK (1966), the relative risk of leukemia is approximately fourfold for individuals homozygous for type Hp 1–1 as compared with type Hp 2–2. The observed differences in susceptibility may be related to the chemical action of haptoglobin on some leukemogenic agent.

## THE TRANSFERRIN VARIANTS

The transferrins (siderophilins) are a group of serum proteins which bind iron. Each transferrin molecule is able to form a complex with two atoms of trivalent iron. The molecular weight of transferrin is about 90 000. The genetic control of transferrin types was demonstrated by SMITHIES in 1957. Transferrin types are identified by their mobility in starch gel electrophoresis and by the use of radioactive iron and autoradiography. During the past decade widespread sampling of human populations has demonstrated the existence so far of about 20 transferrins. In succession of decreasing electrophoretic mobility in starch gel electrophoresis, the main types are: $B_0$, $B_{0-1}$, $B_1$, $B_{1-2}$, $B_2$, $B_3$, C, $D_0$, $D_{0-1}$, $D_1$, $D_2$, $D_3$. Further variants are designated as $D_{chi}$, $D_{Montreal}$, $D_{Wigan}$, $B_{Atalanti}$, $B_{Lae}$ etc. (SMITHIES and HILLER, 1959; LAY, 1963; ROBINSON et al., 1963; GLEN-BOTT, et al., 1964; MURRAY et al., 1964; ARENDS and GALLANGO, 1964; SLOPEK et al., 1966; RUOSLAHTI et al., 1968; GIBLETT, 1969).

Differences between transferrin types are consequences of single amino acid substitutions; aspartic residue in transferrin C is e.g. replaced by a glycine residue in transferrin $D_1$ (WANG and SUTTON, 1965).

The position of the slowest haptoglobin variants is in the haptoglobin region, while the fastest variants may be found before the line of coeruloplasmin. The most common phenotype in Europid populations is type C, which corresponds to the genotype $Tf^C/Tf^C$. Transferrin types are controlled by autosomal codominant alleles (SMITHIES and HILLER, 1959). Lack of transferrin in the serum is a rare, hereditary disease characterized by hemosiderosis and severe anaemia (HEILMEYER et al., 1961). In the course of a survey of 212 cord blood sera by starch gel electrophoresis, RAUSEN et al. (1961) obtained genetic evidence that the foetus produced its own transferrin.

## ALLOTYPE OF HUMAN C3

Beta-1C-globulin is the third factor of complement (C3) in human sera. Its molecular weight is about 185 000. The primary site of synthesis is in the liver (ALPER et al., 1969). Its genetically determined polymorphism was demonstrated by high voltage starch gel and agarose gel electrophoresis (AZEN and SMITHIES, 1968; ALPER and PROPP, 1968). Based on family studies, simple autosomal inheritance of five to seven codominant alleles at a single C3 locus is postulated. AZEN and SMITHIES (1968) designate the alleles as 1, 2, 3, 4 and 5, respectively, whereas ALPER and PROPP (1968) use the symbols F, $F_1$, $F_{0.8}$, $F_{0.5}$, $S_{0.6}$ and S, to distinguish the allele products.

## ALLOTYPE OF ALPHA-1-ANTITRYPSIN

The serum alpha-1-antitrypsin (alpha-1 trypsin inhibitor) is a glycoprotein of 45 000 molecular weight. Its genetic variants constitute the Pi-system (Pi is abbreviated for the protease inhibitor). Alpha-1-antitrypsin variants were detected by means of a special starch gel electrophoretic technique (AXELSSON and LAURELL, 1965; FAGERHOLM and BRAEND, 1965). Pi types are products of autosomal codominant alleles of the Pi locus and up to date the following allelic genes have been described: F, I, M, S, V, W, X, Z. The alleles have been given symbols according to the relative electrophoretic mobility of the allele products which decrease in the mentioned order. Phenotype ZZ is associated with a high risk of severe pulmonary disease often resulting in death at about 50 year of age. Lack of alpha-1-antitrypsin in the serum is associated with severe pulmonary emphysema beginning at an early age. Family findings speak for the concept of genetic transmission of the defect as an autosomal recessive trait with emphysema occurring only in the homozygotes (Turino et al., 1969).

## THE COERULOPLASMIN VARIANTS

Allotypic variants of serum enzymes can be demonstrated by combining starch gel electrophoresis with a specific enzyme reaction. In this category of investigations coeruloplasmin pseudo-cholinesterase and alkaline phosphatase groups will be discussed. Coeruloplasmin is a serum alpha-globulin which contains more than 90% of the serum copper. Its molecular weight is about 160 000. Coeruloplasmin oxidizes para-phenylene-diamine to an insoluble brownish dye. Allotypic variants of coeruloplasmin are detected by means of electrophoresis. The most common variants in Europid populations are controlled by three codominant allelic genes designed as $Cp^A$, $Cp^B$ and $Cp^C$. Rare allelic variants are $Cp^{NH}$ (NH stands for New Heaven) and $Cp^{TH}$, a trait found in Thailand (McAllister et al., 1961; Shokeir et al., 1967; Shrefler et al., 1967; Bajatzadeh and Walter, 1969). Deficiency of coeruloplasmin in the serum is a recessive familiar defect characterized by hepatolenticular degeneration (Wilson's disease).

## SERUM CHOLINESTERASE VARIANTS

Serum cholinesterase (pseudocholinesterase) is normally found in quite high activity in human blood serum. Its bulk is probably synthetized in the liver. The new name of this enzyme, proposed by the Enzyme Commission is acylcholine acyl hydrolase. The genetic polymorphism of serum cholinesterase was shown by Kalow and Genest in 1957. The formation of this enzyme is controlled by at least four alleles at one autosomal locus termed $E_1$ (or Ch). The four genes are designated as $E_1^N$ or $E_1^U$ (normal, usual), $E_1^D$ or $E_2^A$ (dibucaine resistant, atypical), $E_1^F$ (fluoride resistant) and $E_1^S$ (silent). In sera of persons homozygous for the silent gene no pseudocholinesterase can be demonstrated. The vast majority of individuals possessing an atypical variant of pseudocholinesterase have been found to be excessively sensitive to succinyldicholinchloride and similar muscle relaxants widely used in surgery. Clinical symptoms of complications are in such cases muscular paralysis and apnea (Kalow and Genest, 1957; Kalow and Staron, 1957; Harris et al., 1960; Liddell et al., 1962; Harris and Whitaker, 1963; Arends et al., 1967).

The four alleles of the $E_1$ locus give rise to ten different genotypes. Their identification may require not only inhibitor tests with dibucaine and fluoride, but also detailed family studies.

When serum cholinesterase is examined electrophoretically, several isozymes (known as $C_1$, $C_2$, $C_3$ and $C_4$) are regularly found. $C_1$, $C_2$ and $C_3$ are minor components and contribute little to the total enzyme activity, most of which is derived from $C_4$. However about 10% of people in European populations carry an extra isozyme, called $C_5$ (HARRIS et al., 1962, 1963; SIMPSON, 1966; ROBSON and HARRIS, 1966). The presence of $C_5$ in the serum appears to be determined by a gene at a locus, called $E_2$. Loci $E_1$ and $E_2$ are not closely linked. Allele $E_2^+$ determines the formation of an extra serum cholinesterase isozyme $C_5$, while the other allele ($E_2^-$) appears to be functionally inactive.

## SERUM ALKALINE PHOSPHATASE TYPES

Polymorphism of serum alkaline phosphatase was demonstrated by BOYER in 1961. Blood serum of normal human adults produces in starch gel electrophoretic zymograms one or two major zones of alkaline phosphatase activity between the transferrin and the haptoglobin 1–1. Human sera can be divided into two groups: type I with a single alkaline phosphatase band and type II with two bands.

All human sera have one alkaline phosphatase band which migrates with a mobility slightly slower than that of transferrin C, and type II sera contain an additional band of slower mobility. It was found that these enzyme types are inheritable. A relationship was found between the alkaline phosphatase type and the Lewis blood group system (AFORS et al., 1963; BAMFORD et al., 1965; SHREFFLER, 1965).

## DETECTION OF SERUM PROTEIN ALLOTYPIC TRAITS BY MEANS OF IMMUNOCHEMICAL METHODS

Further to electrophoretic analysis, another convenient approach to the genetics of serum proteins is based on their antigenic specificity. More than 30 antigens may be distinguished in human serum if immunoelectrophoresis is applied, which combines agar gel electrophoresis with immunodiffusion. Monospecific sera enable the investigation of only one antigenic component. Transferrin variants, too, may be detected by immunoelectrophoresis (SLOPEK et al., 1966).

## GROUP SPECIFIC COMPONENTS

Gc-types (Gc-group specific components) were described by HIRSCHFELD in 1959. Gc is a protein that migrates in the alpha-globulin region on agar gel electrophoresis. By means of immunoelectrophoresis three phenotypes were recognized: a fast moving Gc 1–1 type, a slow moving Gc 2–2 type, and a Gc 2–1 type which has both components distinguishable by the shape of the precipitin line. These three common phenotypes are controlled by two genes, $Gc^1$ and $Gc^2$, segregating at an autosomal locus (HIRSCHFELD et al., 1960). Further rare allelic variants of Gc types are Gc X, Gc Y,

Gc Ab (aborigine) and Gc Chip (Chippewa) (HIRSCHFELD, 1962; CLEVE et al., 1963).

From comparative studies of maternal and cord blood sera it is apparent that the foetus produces its own Gc proteins (USATEGUI-GOMEZ and MORGAN, 1966).

## POLYMORPHISMS OF LOW DENSITY LIPOPROTEINS

Human low density lipoproteins are characterized by a high molecular weight of about 1 300 000 and by a lipid content of 40%. By means of radial double immunodiffusion test or immunoelectrophoresis, polymorphisms of low density lipoproteins can be demonstrated in sera from polytransfused patients or in immune sera from rabbits or horses. Three independent genetic systems of low density lipoproteins, termed as Ag, Ld and Lp have been described till now.

The Ag system was discovered in 1961 by ALLISON and BLUMBERG. Later on it was found that isoantigens constituting the Ag system of low density lipoproteins are controlled by autosomal allelic genes designated as Ag(a), Ag($a_1$), Ag(c), Ag(r), Ag(t), Ag(x), Ag(y) and Ag(z). Genetic studies have shown that some loci of the Ag system are very closely linked (BLUMBERG and RIDDELL, 1963; BUNDSCHUH, 1967; HIRSCHFELD and RITTNER, 1969; WIENER, 1969). VIERUCCI et al. (1968) demonstrated that the Ag genetic factors are not passively transmitted via the placental barrier, but they are produced by the foetus.

The *Lp system* (Lp stands for lipoprotein) is described by BERG (1963). It is characterized by two main antigenic types designated as Lp(a) and Lp(x) and by some minor variants (Lp($a_2$), Lp(2)). For the detection of these antigens rabbit or horse immune sera were used (RITTNER and BUNDSCHUH, 1968).

The *Ld system* (Ld stands for light density) is described by BERG (1965). This system is determined by two alleles, designated as Ld(a) and Ld(x). The corresponding antigens were demonstrated by means of double radial immunodiffusion test in sera from multiple transfused patients.

Congenital deficiency in low-density lipoproteins is a hereditary disorder involving a defect in the absorption and transport of lipids, acanthocyanosis of the red cells and steatorrhoea. It is characterized by dominant autosomal transmission (BACH et al., 1967; ISSELBACHER et al., 1964; LAMY et al., 1960; SALT et al., 1960; RICHET et al., 1969).

## THE Xm-SYSTEM

Alpha-2-macroglobulin is a glycoprotein of human serum, characterized by a molecular weight of about 600 000. BERG and BEARN (1966) found the alpha-2M-globulins of certain individuals to be characterized by Xm(a) specificity. For the demonstration of this antigenic determinant, sera from horses immunized with mixed human serum and adsorbed with Xm(a)-negative human serum can be used, trait-determined by the $Xm^a$ gene localized on chromosome X. As a consequence, the frequency of Xm(a)-positivity is much higher among females than males.

## THE POLYMORPHISMS OF IMMUNOGLOBULINS

Immunoglobulins constitute a family of serum proteins characterized by antibody activity. IgG, IgA, IgM IgD, IgE and IgU classes of human immunoglobulins can be distinguished. The IgG class comprises four subclasses, IgA and IgM comprise two subclasses each. Two symmetrical heavy polypeptide chains and two light chains are connected with each other by interchain disulphide bonds constituting an IgG molecule. Two antigenic types of light chains designed as kappa and lambda may be distinguished. An IgU molecule consists of two identical light chains. Heavy chains of IgG are designated as $\gamma$-, those of IgA, IgD, IgE and IgM as $\alpha$, $\vartheta$, $\varepsilon$, and $\mu$, respectively. In the presence of cystein, papain splits IgG into two fragments, designated as Fab and Fc. The Fab fragment consists of the Fd part of the heavy chain and of a light chain. Primary structural analyses have shown that polypeptide chains of immunoglobulins may be divided into two portions: amino terminal sequences constitute the variable portion, whereas carboxyterminal sequences are constant. It is supposed that V-portions are connected with antibody specificity, whereas allelic forms of immunoglobulins are being attributed to minor variations in the amino acid sequences of the C-portion. Molecular weight of IgU is 40 000, that of IgG, IgA, IgD and IgE is about 150 000, whereas IgM is a macroglobulin consisting of five subunits and possessing a molecular weight of about 1 000 000. Arrangement of the polypeptide chains in IgA, IgD, IgE and in IgM subunits is similar to that of IgG (Awdeh et al., 1970; Backhausz, 1967; Baglioni, 1967; Ceppellini et al., 1964; Fahey, 1962; Fleischman et al., 1963; Fudenberg, 1965; Heide, 1965; Janeway et al., 1967; Kunkel et al., 1967; Lennox and Cohn, 1967; Melli and Mazzei, 1967; Porter, 1959; Smithies, 1967).

## THE Gm GROUPS

The existence of allelic variation in human immunoglobulins was first realized by Grubb in 1956. Gm stands for gamma-globulin. Some Gm-factors are localized on the Fc part of the gamma-chains, whereas others on its Fd part.

The nomenclature of the first fourteen Gm-factors was accepted by the World Health Organization. The number of known Gm factors has risen to 24 till now. The different Gm types are divided between the IgG subclasses. Antigenic differences as expressed by Gm types within a subclass of IgG behave as alleles in genetic crosses. Genetic studies have indicated that the genes of the Gm system are all very closely linked to each other, crossing over being extremely rare and the gene complexes being stable within a given population. There are gene complexes which are characteristic for Europid, for Mongolid, and Negrid populations, respectively (Harboe and Lundevall, 1959; Ceppellini et al., 1965; Mourant, 1965; Oudin, 1966; Van Loghem and Steinberg, 1966; Ropartz, 1969).

## THE Isf-FACTOR

The Isf-factor, where sf stands for San Francisco, corresponds to an allotypic variant of gamma-1-type heavy chains, but it is independent of the Gm loci. In Caucasians, the phenotypic expression of the Isf(1) gene exhibits a variation related to age, on the other hand certain melanodermic population are found to possess the Isf(1) phenotype, the frequency being the same in any age group studied (ROPARTZ et al., 1967, 1968).

## ALLOTYPES OF IMMUNOGLOBULIN LIGHT POLYPEPTIDE CHAINS

Inv-factors are variants of human kappa-type light chains. Three variants, Inv(1), Inv(2) and Inv(3), have been detected until now. Amino acid interchanges are responsible for the antigenic differences (ROPARTZ et al., 1961; OUDIN, 1966; CEPPELLINI et al., 1965; ROPARTZ, 1969).

A similar genetic polymorphism consisting of antigenic subtypes of lambda immunoglobulin chains, has been described recently (EIN and FAHEY, 1967; TISCHENDORF and OSSERMAN, 1969; TISCHENDORF and TISCHENDORF, 1970; HESS and HILSCHMANN, 1970).

Gm-, Inv- and Isf-factors are as a rule demonstrated by hemagglutination inhibition reaction. Human D-positive red cells are coated by incomplete anti D hemagglutinins which carry allotypic specificity. Specific anti-Gm antibodies which may be found in sera of polytransfused patients or patients with chronic rheumatoid arthritis agglutinate D-positive red blood cells coated with Gm-factor carrying anti-D antibodies. This reaction is inhibited by those sera which contain the corresponding Gm-factor.

## DISEASES WITH HEREDITARY IMMUNOLOGICAL DEFICIENCY

Diseases with primary immunological deficiency appear to have a genetic basis. In infantile X-linked recessive agammaglobulinaemia all immunoglobulin classes are extremely deficient. Autosomal recessive alymphocytic agammaglobulinaemia (Swiss type agammaglobulinaemia) is characterized by hypoplastic thymus, absence of lymphocytes and extreme deficiency of all immunoglobulins. Other types of immunological deficiencies are connected with impairment of certain factors of humoral or cellular immune mechanism as selective IgA deficiency, thymoma with agammaglobulinaemia (Good's syndrome), immune deficiency with thrombopenia and eczema (Wiscott–Aldrich syndrome), ataxia teleangiectatica (Barr's syndrome), autosomal recessive lymphopenia with normal plasma cells and immunoglobulins (Nezeloff's syndrome), thymic aplasia (Di-George's syndrome) etc. (BACKHAUSZ, 1967; JANEWAY et al., 1967; MELLI and MAZZEI, 1967; ALLISON et al., 1968).

# REFERENCES

AFORS, K. E., BECKMAN, L. and LUNDIN, L. G. (1963), Genetic variations of human serum phosphatase. *Acta genet.*, **13.** 89.
ALLISON, A. G., ASKONAS, B. A., BENACERRAF, B., CEPPELLINI, R., GOOD, R. A., LENNOCK, E. S., MCDEVITT, H. O., NEZLIN, S. R. and SELIGMANN, M. (1968), Genetics of the immune response. World Health Org., Techn. Rep. Series No. 402. Geneva.
ALLISON, A. C. and BLUMBERG, B. S. (1961), An isoprecipitation reaction distinguishing human serum protein types. *Lancet*, **1.** 176.
ARENDS, T., DAVIES, D. A. and LEHMANN, H. (1967), Absence of variants of usual serum pseudocholinesterase (acylcholine acylhydrolase) in South American Indians. *Acta genet.* **17.** 13–16.
ARENDS, T. and GALLANGO, M. L. (1964), Transferrins in Venezuelan Indians: high frequency of a slow-moving variant. *Sicence*, **143.** 367.
AWDEH, Z. L., WILLIAMSON, A. R. and ASKONAS, B. A. (1970), One cell — one immunoglobulin. *Biochem. J.*, **116.** 241–248.
BACH, C., PLONOVSKI, J., POLONOVSKI, C., LEHUC, R., JOLLY, G. and MOSZER, M. (1967), Absence congénitale de $\beta$-lipoprotéines. *Arch. Fr. Pédiatrie.*, **24.** 1093–1111,
BACKHAUSZ, R. (1967), Immunodiffusion und Immunoelektrophorese. Grundlagen. Methoden und Ergebnisse. VEB G. Fischer Verl. Jena.
BAGLIONI, C. (1967), Homologies in the position of cysteine residues of K and L type chains of human immunoglobulins. *Biochem. Biophys. Res. Commun.*, **26.** 82–89.
BAJATZADEH, M. and WALTER, H. (1969), Studies on the population genetics of the ceruloplasmin polymorphism. *Humangenetik* **8.** 134–136.
BAMFORD, K. F., HARRIS, H., LUFFMAN, J. E., ROBSON, E. B. and CLEGHORN, T. E. (1965), Serum-alkaline-phosphatase and the ABO blood groups. *Lancet*, **1.** 530–531.
BERG, K. (1963), A new serum type in man — the Lp system. *Acta Pathol. Microbiol. Scand.*, **59.** 369–382.
BERG, K. (1965), A new serum type in man. The Ld system. *Vox Sang.*, **10.** 513.
BERG, K. and BEARN, G. (1966), An inherited X-linked serum system in man — the Xm system. *J. Exp. Med.*, **123.** 379.
BLACK, J. A. and DIXON, G. H. (1968), Aminoacid sequence of alpha chains of human haptoglobins. *Nature*, **218.** 736.
BLUMBERG, B. S. and RIDDELL, N. M. (1963), Inherited antigenic differences in human serum beta lipoproteins. A second antiserum. *J. Clin. Invest.*, **42.** 867.
BOYER, S. H. (1961), Alkaline phosphatase in human sera and placentae. *Science*, **134.** 1002.
BUNDSCHUH, G. (1967), Die bisher bekannten Merkmale im Ag-System. *D. Gesundheitsw.*, **22.** 2356–2365.
CEPPELLINI, R., DRAY, S., EDELMAN, G., FAHEY, J., FRANEK, F., FRANKLIN, H., GOODMAN, H. C., GRABAR, P., GURRICK, A. E., HEREMANS, J. F., ISLIKER, H., KARUSH, F., PRESS, E. and TRNKA, Z. (1964), Nomenclature for human immunoglobulins. *Bull. W. H. O.*, **30.** 447.
CEPPELLINI, R., DRAY, S., FAHEY, J., FRANKLIN, E., FUDENBERG, H., GELL, P. G. H., GOODMAN, H. C., GRUBB, R., HARBOE, M., KIRK, R., OUDIN, J., ROPARTZ, C., SMITHIES, O., STEINBERG, A. G. and TRNKA, Z. (1965), Notation for genetic factors of human immunoglobulins. *Bull. W. H. O.*, **33.** 721–724.
CLEVE, H., GORDON, S., BOWMAN, B. H. and BEARN, A. G. (1967), Comparison of the tryptic peptides and amino acid composition of the beta polypeptide chains of the three common haptoglobin phenotypes. *Am. J. Human Genet.*, **19.** 713.
CLEVE, H., KIRK, R. L., PARKER, W. G., BEARN, A. G., SCHACHT, L. E. and KLEINMAN, H. (1963), Two genetic variants of the group-specific component of the human serum: Gc Chippewa and Gc Aborigine. *Am. J. Human Genet.*, **15.** 368–379.
EIN, D. and FAHEY, J. L. (1967), Two types of lambda polypeptide chains in human immunoglobulins. *Science*, **156.** 947.
FAHEY, J. L. (1962), Heterogeneity of gamma-globulin. *Adv. Immunol.*, **2.** 41–109.
FLESICHMAN, J. B., PORTER, R. R. and PRESS, E. M. (1963), The arrangement of the peptide chains in gamma-globulin. *Biochem. J.*, **88.** 220.
FUDENBERG, H. H. (1965), The immune globulins. *Ann. Rev. Microbiol.*, **19.** 301.

GIBLETT, E. R. (1969), Genetic markers in human blood. Blackwell, Oxford.
GIBLETT, E. R. and BROOKS, L. E. (1963), Haptoglobin types in three racial groups. *Nature* **197.** 576–578.
GLEN-BOTT, A. M., HARRIS, H., ROBSON, E. B., BEARN, A. G. and PARKER, W. C. (1964), Transferrin $D_{Wigan}$. *Acta Genet.*, **14.** 52–56.
GRUBB, R. and LAURELL, A. B. (1956), Hereditary serological human serum groups. *Acta Path. Microbiol. Scand.* **39.** 390.
HARBOE, M. and LUNDEVALL, J. (1959), A new type in the Gm-system. *Acta Path. Microbiol. Scand.*, **399.** 357.
HARRIS, H., HOPKINSON, D. A. and ROBSON, E. B. (1962), Two dimensional electrophoresis of pseudo-cholinesterase components in normal human serum. *Nature*, **196.** 1296.
HARRIS, H., ROBSON, E. B., GLEN-BOTT, A. M. and THORTON, J. A. (1963), Evidence for non-allelism between genes affecting human serum cholinesterase. *Nature*, **200.** 1185.
HARRIS, H. and WHITAKER, M. (1963), Differential inhibition of human serum cholinesterase with fluoride: recognition of two phenotypes. *Nature*, **199.** 1115.
HARRIS, H., WHITAKER, M., LEHMANN, H. and SILK, E. (1960), The pseudocholinesterase variants, esterase levels and dibucaine numbers in families selected through suxamethonium sensitive individuals. *Acta Genet.*, **10.** 1.
HEIDE, K. (1965), Die Immunglobuline des menschlichen Serums. *Die gelben Hefte*, **9.** 321–333.
HEILMEYER, L., KELLER, W., VIVELL, O., KEIDERLING, W., BETKE, K., WÖHLER, F. and SCHULTZE, H. E. (1961), Kongenitale Atransferrinämie bei einem sieben Jahre alten Kind. *Dtsche Med. Wschr.*, **86.** 1745.
HESS, M. and HILSCHMANN, N. (1970), Genetischer Polymorphismus im konstanten Teil von humanen Immunglobulin-L-Ketten vom λ-Typ. *Hoppe-Seyler's Z. Physiol. Chem.* **351.** 67–73.
HIRSCHFELD, J. (1959), Immuno-electrophoretic demonstration of qualitative differences in human sera and their relations to the haptoglobins. *Acta Path. Microbiol. Scand.*, **47.** 160–168.
HIRSCHFELD, J. (1962), The Gc-system. *Progr. Allergy* **6.** 155–186.
HIRSCHFELD, H., JOHNSSON, B. and RASMUSON, M. (1960), Inheritance of a new group-specific system demonstrated by means of an immunoelectrophoretic technique. *Nature*, **185.** 931.
ISSELBACKER, K. J., SCHEIG, R., PLOTKIN, G. R. and CAULFIELD, J. B. (1964), Congenital β-lipoprotein deficiency: a hereditary disorder involving a defect in the absorption and transport of lipids. *Medicine*, **43.** 347.
JANEWAY, C. A., ROSEN, F. S., MERLER, E. and ALPER, C. A. (1967), The gamma globulins. Little, Brown and Co. Boston.
KALOW, W. and GENEST, K. (1957), A method for the detection of atypical forms of human serum cholinesterase, as indicated by dibucaine numbers. *Canad. J. Biochem.*, **35.** 229.
KALOW, W. and STARON, N. (1957), On distribution and inheritance of atypical forms of human serum cholinesterase, as indicated by dibucaine numbers. *Canad. J. Biochem.*, **35.** 1305.
KUNKEL, H. G., FAHEY, J. L., FRANKLIN, E. C., OSSERMAN, E. F. and TERRY, W. D. (1967), Système de notation des sous-classes d'immunoglobulines humaines. *Bull. Org. Mond. Santé* **36.** 337–338.
LAMY, M., FRÉZAL, J., POLONOVSKI, J. and REY, J. (1960), L'absence congénitale des β-lipoprotéines. *C. R. Soc. Biol.*, **154.** 1974.
LAY, L. Y. C. (1963), A new transferrin in New Guinea. *Nature*, **198.** 589.
LENNOX, E. S. and COHN, M. (1967), Immunglobulins. *Ann. Rev. Biochemistry*, **36.** 365.
LIDDELL, J., LEHMANN, H. and SILK, E. (1962), A silent pseudocholinesterase gene. *Nature*, **202.** 815.
MCALISTER, R., MARTIN, G. M. and BENDITT, E. P. (1961), Evidence for multiple coeruloplasmin components in human serum. *Nature*, **190.** 927.
MELLI, G. and MAZZEI, D. (1967), Le immunglobuline. Aspetti strutturali e funzionali. Ed. Ist. Sieroterap. Milan.

MOURANT, A. E. (1965), Géographie et anthropologie des hémotypes extraérythrocytaires. *Transfusion*, **8.** 367–376.
MURRAY, R. F., ROBINSON, J. C. and BLUMBERG, B. S. (1964), A new variant of transferrin from Greece. *Nature*, **204.** 382–383.
OUDIN, J. (1966), The genetic control of immunoglobulin synthesis. *Proc. Roy. Soc. B.*, **166.** 207–219.
PEACOCK, A. C. (1966), Serum haptoglobin type and leukemia: An association with possible etiological significance. *J. Nat. Cancer Inst.*, **36.** 631–639.
PORTER, R. R. (1959), The hydrolysis of rabbit gamma globulin and antibodies with crystalline papain. *Biochem. J.*, **73.** 119.
PROKOP, D. and BUNDSCHUH, G. (1962), Die Technik und die Bedeutung der Haptoglobine und Gm-Gruppen in Klinik und Gerichtsmedizin. Walter de Gruyter and Co. Ed. Berlin.
RAUSEN, A. S., GERALD, P. S. and DIAMOND, L. K. (1961), Genetical evidence for synthesis of transferrin in the foetus. *Nature*, **192.** 182–183.
RICHET, G., DUREPAIRE, H., HARTMANN, L., OLLIER, M. P., POLONOVSKI, J. and MAITROT, B. (1969), Hypolipoprotéinémie familiale asymptomatique prédominant sur les bêtalipoprotéines révélée lors de l'étude d'une protèinurie isolée. *Presse Méd.*, **77.** 2045–2048.
RITTNER, C. and BUNDSCHUH, G. (1968), Comparative studies of anti-Lp sera. *Vox Sang.*, **15.** 446–450.
ROBINSON, J. C., BLUMBERG, B. S., PIERCE, J. E., COOPER, A. J. and HAMES, C. G. (1963), Studies on inherited variants of blood proteins. II. Familial segregation of transferrin $B_{1-2}$, $B_2$. *J. Lab. Clin. Med.*, **62.** 762–765.
ROBSON, E. B. and HARRIS, H. (1966), Further data on the incidence and genetics of the serum cholinesterase genotype $C_{5+}$. *Ann. Human Genet.*, **29.** 403.
ROPARTZ, C. (1969), Immunoglobulines humaines et fonctionnement génétique. *Rev. Transfus.*, **12.** 85–96.
ROPARTZ, C., FUDEMBERG, H. H., RIVAT, L., ROUSSEAU, P. Y. and LEBRETON, J. P. (1967), Un troisième lucus participant à la synthèse des yG: Isf. *Rev. Franc. Et. Clin., Biol.*, **12.** 267–269.
ROPARTZ, C., LENOIR, J. and RIVAT, L. (1961), A new inheritable property of human sera: the InV factor. *Nature*, **189.** 586.
ROPARTZ, C., RIVAT, L., RIVAT, C. and ROUSSEAU, P. Y. (1968), Some problems raised by the Isf system of human immunoglobulins. *Vox Sang.*, **14.** 458–459.
RUOSLAHTI, E., SEPPÄLÄ, I. J. T., SIMONS, K. and SEPPÄLÄ, M. (1968), Identity of transferrin $D_{Chi}$ from the Chinese and from the Finns. *Nature*, **220.** 480–481.
SALT, H. B., WOLFF, O. H., LLOYD, J. K. and FOSBROOKE, A. S. (1960), On having no betalipoprotein: a syndrome comprising abetalipoproteinaemia, acanthocyanosis and steatorrhea. *Lancet*, **2.** 325.
SHOKEIR, M. H., SHREFFLER, D. C. and GALL, J. C. (1967), Further electrophoretic variation in human coeruloplasmin. Meeting Amer. Soc. hum. Genet. Toronto.
SHREFFLER, D. C. (1965), Genetic studies of blood group-associated variations in a human serum alkaline phosphatase. *Am. J. Human Genet.*, **17.** 71–86.
SCHREFFLER, D. C., BREWER, G. J., GALL, J. C. and HONEYMAN, M. S. (1967), Electrophoretic variation in human serum coeruloplasmin: A new genetic polymorphism. *Biochem. Genet.* **1.** 101–115.
SIMPSON, E. (1966), Factors influencing cholinesterase activity in a Brasilian population. *Am. J. Human Genet.*, **18.** 243.
SLOPEK, S., LADOSZ, J., HRYNCEWICZ, K. and BRZUCHOWSKA, W. (1966), New methods of identifying serum siderophilin using radioactive iron. *Nature*, **209.** 1036.
SMITHIES, O. (1957), Variations in human serum betaglobulins. *Nature*, **180.** 1482.
SMITHIES, O. (1967), Antibody variability. *Science*, **157.** 267–273.
SMITHIES, O. and HILLER, O. (1959), The genetic control of transferrins in humans. *Biochem., J.*, **72.** 121–126.
SMITHIES, O. and CONNELL, G. E. (1959), Biochemical aspects of the inherited variations in human serum. Ciba Found. Symp. on Biochemistry of Human Genetics. pp. 178–189.
SMITHIES, O., CONNELL, G. E. and DIXON, G. H. (1966), Gene action in the human haptoglobins. I. Dissociation into constituent polypeptide chains. *J. Mol. Biol.*, **21.** 213.

SMITHIES, O. and WALKER, N. F. (1955), Genetic control of some serum proteins in normal humans. *Nature*, **176.** 1255.
TISCHENDORF, F. W. and OSSERMAN, E. F. (1969), Two antigenic subtypes of human lambda immunoglobulin chains. *J. Immunol.*, **102.** 172–178.
TISCHENDORF, F. W. and TISCHENDORF, M. M. (1970), Markers of the variable and constant region of human lambda immunoglobulin chains. *Eur. J. Biochem.*, **13.** 398–402.
TURINO, G. M., SENIOR, R. M., GARG, B. G., KELLER, S., LEVI, M. M. and MANDL, I. (1969), Serum elastase inhibitor deficiency and alpha$_1$-antitrypsin deficiency in patients with obstructive emphysema. *Science*, **165.** 709–711.
USATEGUI-GOMEZ, M. and MORGAN, D. F. (1966), Maternal origin of the group specific (Gc) proteins in amniotic fluid. *Nature*, **212.** 1600–1601.
VAN LOGHEM, E. and STEINBERG, A. G. (1966), A second example of a white family in which Gm(c) occurs, and a further analysis of the first example. *Vox Sang.*, **11.** 38–44.
VIERUCCI, A., DETTORI, M., MORGANTI, G., BEOLCHINI, P. E. and BÜTLER, R. (1968), Synthesis of beta-lipoproteins (Ag groups) in the foetus and the newborn. *Vox Sang.*, **14.** 151–155.
WALTER, H. and STEGMÜLLER, H. (1969), Studies on the geographical and racial distribution of the Hp and Gc polymorphisms. *Human Heredity*, **19.** 209–221.

*XIIth Europ. Conf. Anim. Blood Groups Biochem. Polymorph., Bp., 1972 (pp. 41–54)*

# VALUE AND LIMITATIONS OF RESEARCH IN PROTEIN POLYMORPHISM

F. W. ROBERTSON

Department of Genetics, University of Aberdeen

## INTRODUCTION

The study of genetic differences in the composition of blood has made immense progress as a result of the widespread use of electrophoresis to separate and identify different proteins. So many variants are being reported from so many species that we are in danger of drowning in a flood of facts. I think we are now in a position to take a critical look at this world-wide activity to see whether certain approaches have not already yielded so much essential information that the law of diminishing returns has begun to operate, to consider the ways this information can be put to practical use and to ask whether or not some degree of reorientation and rephrasing of questions may not be desirable. To use an English idiom, there is so much information that it is often difficult to see the wood for the trees; I propose not to pay much attention to the trees, since the rest of the programme will be devoted to this topic, and consider the wood instead.

Let us begin by attempting some kind of evaluation of the great variety of facts which are being gathered on protein variation. Usually one starts from particular instances and tries to work outward to arrive at the significance of the initial observations. But it is also useful to start at the other end and consider some of the possibilities which are inherent in the nucleotide sequences ordered in codons specifying particular amino acids. Such considerations, along with generalisations derived from the comparison of amino-acid sequences in different proteins, including estimates of rates of divergence in evolution, add some kind of perspective to the limited experience of protein variation that one person can have. Also it is often useful to bring together evidence from apparently different fields of enquiry which are so often kept apart by specialisation of techniques and training.

I think we can begin by recalling the immense number of genetic alternatives at a locus. As WRIGHT (1966) has reminded us, if we consider a locus consisting of say 1000 nucleotide pairs, each with four alternatives, and for purposes of illustration choose one arrangement as a standard for reference, then the number of alternatives differing from this standard by a single base substitution will be 3000, while the total number of alternatives will be $4^{1000}$. Clearly, extrapolating to all loci, the number of alternatives in a population will be limited by its size since the number of potential combinations will vastly exceed the number of individuals. Since any change is subject to mutation every site and hence every codon specifying amino acids is subject to challenge by mutation. If, in a given genetic situation, such changes are deleterious, elimination will occur; if they are sufficiently favourable they may spread at rates which will depend on selection pressure, population size, etc. And if they are neutral then, as KIMURA (1968) has shown, the rate of random fixation per species per generation will equal the rate of neutral mutation per gamete per generation. If there are N individuals in a population, there will be 2N copies of a given gene at any given time. The chance of selectively neutral genes being fixed is 1/2N. If $m$ is the mutation rate there will be 2Nm new alleles per generation and hence the rate of fixation of neutral alleles will be proportional to the rate of occurrence of neutral alleles. This raises the question of whether the so-called wild type gene may not include an array of indistinguishable alternative, iso-alleles, some of which we may be able to detect, if they happen to influence electrophoretic mobility, although most will pass unnoticed while many are assumed to be neutral as far as natural selection is concerned.

Now you may think that we do not need to worry much about what we cannot detect but this is not true in this instance. Comparative study of proteins which have diverged during evolution has to take account of the possible non-adaptive fixation of differences. Future improvements in technique will provide more and better determined amino-acid sequences in many proteins. Here I would emphasise the need for quantitative analysis of amino-acid sequences in apparently normal proteins. At present, attention is paid almost exclusively to comparison of sequences in proteins known to be different in one way or another. What we need as well is a study of sequences in proteins which appear identical by existing criteria. Also, it will surely only be a matter of time before we have some information on nucleotide sequences precise enough to compare with amino-acid sequences; already we have some preliminary statistical evidence which I shall consider later.

Now when we come to the topic of fixation of neutral alleles we become controversial, since, right now, there is sharp disagreement between workers about whether or not it is possible. At one extreme there is the view that neutrality is an illusion; natural selection is the supreme arbiter and always has been, so that even the properties of the amino-acid specifying codons are attributed to natural selection. I think this is an extreme view, for several reasons. Naturally, it is almost impossible to disprove the selectionist argument, since one can always postulate subtle selection pressures, or earlier selection, no longer accessible to study, which are responsible for any change we care to mention. On the other hand, while no one would deny the role of natural selection, we can legitimately ask how discriminating it must be and whether or not there is sufficient likelihood of neutral changes to influence our interpretation of rates of amino-acid substitution in proteins during evolution and which also has to be borne in mind when interpreting the variation we presently find in natural populations.

Let us review briefly some of the arguments and counter-arguments.

One of the strongest arguments in favour of neutral changes at the nucleotide level is based on the degeneracy of the genetic code. Since each of the 61 amino-acid specifying codons has 9 alternatives then with respect to the entire set there are 549 possible base substitutions and of these 134 or about a quarter do not involve a change in specification of amino-acid. On this view, natural selection cannot discriminate between alternative synonyms and hence such mutations will occur and be fixed at a rate proportional to their mutation rate, without change of amino-acid sequence. The rival, selectionist view, e.g. CLARKE (1970) will have none of this and holds that if there is heterogeneity between transfer RNAs for a given amino-acid they may differ in degree of attachment to alternative codons for the amino acid in question and hence the balance between frequencies of alternative codons and transfer RNAs could influence rates of protein synthesis sufficiently to become accessible to natural selection. It has also been suggested that particular codon sequences may regulate the secondary structure of tRNA, and that codons may have been favoured which minimise the risk of unfavourable amino-acid substitution by mutation. At present there is no way of knowing how relevant these criticisms are, so it is rather a matter of personal bias as to whether you want to opt for selection at any price, or whether you are ready to allow some other possibilities as well.

If we leave the nucleotide sequences for the while and turn to amino-acid sequences we have the impression that protein chemists are not opposed to the idea that certain amino-acid substitutions may be neutral; indeed it seems fairly clear from numerous comparative studies that there is a wide range of difference between proteins in rates of amino-acid substitution ranging from the extreme conservatism of cytochrome C to the rapid divergence of fibrinogen. Also there are all degrees of sensitivity to amino-acid substitution within a protein, according to which part of the molecule is affected. Naturally some of the best evidence here relates to hemoglobin. For example RIGGS (1959) has made the point that the oxygen dissociation constants of many mammalian hemoglobins are not appreciably different. Also many of the hemoglobin variants in human populations occur as fixed changes in other mammals which suggests that they are functionally equivalent.

PERUTZ and LEHMANN'S (1968) electrophoretic screening of human hemoglobin has demonstrated wide differences between substitutions in their effect on physiology according to their position in the molecule. Changes on the inside of the molecule are generally harmful while changes on the outside appear to be neutral, at least in the heterozygote. Thus, in spite of the electrophoretic charge difference, a good case can be made for selective neutrality of such substitutions and no one can deny, at least, that there is a wide spectrum of differences in effect of substitutions and so it does not seem unreasonable to extend the spectrum into the zone of selective neutrality rather than establish an arbitrary threshold.

Studies on human populations certainly suggest that all the possible alternatives can and do occur. Bearing in mind that of 2217 theoretically possible amino-acids in *a* and *b* chains only 700, i.e. one third, would lead to a charge difference, detectible by electrophoresis, current estimates on 20 000 European subjects (SICK et al., 1967) indicates an observed frequency of 1/1800, which probably corresponds to a real frequency of 1/600 from which it is inferred that there are about 5 million hemoglobin A variants in the human population. PERUTZ and LEHMANN (1968) have brought the total of identified variants to about 100 in less than 0.001% of the human population.

There is no reason to doubt that the same arguments would apply to other serum proteins although the proportion of the changes which can be classified according to degree of effect on the phenotype will vary from protein to protein and this consideration is very relevant to our attitude to searches for protein variation in populations generally. We may have a substantial proportion of electrophoretic differences which are adaptively neutral or, at least have such small differences that they are beyond experimental demonstration.

While on the topic of hemoglobins I think we should note KIMURA's (1969) recent discussion of evolutionary divergence rates. Using the data compiled by ECK and DAYHOFF (1966), KIMURA compared the hemoglobin $\gamma$ chain of man and carp, excluding a few insertions and deletions, with the $\gamma$ chains of mouse, rabbit, horse and ox and came out with remarkably similar estimates of the rate of substitution per amino-acid site, according to probable estimates of times of divergence in the course of evolution. The average figure works out at about $9 \times 10^{-10}$ per amino-acid site per year. Also the rate of divergence of the $\gamma$ and $\beta$ chains from the end of the Ordovician period, some 450 million years ago, works out at a similar rate. KIMURA concludes that for the hemoglobin chains at least, amino-acid substitutions and the primary nucleotide substitutions have proceeded at a constant rate and in a fortuitous manner through different lines of vertebrates and apparently depend mainly on time in years and are independent of generation time, living conditions etc. He concludes that the rate of production of neutral mutations per year is constant, assuming, as we noted earlier, that the rate of neutral gene substitution in a population equals the mutation rate per gamete. These arguments lead to a prediction which is testable. If chance is a major factor in amino-acid substitution, and the degree of change is correlated with time then the rate of change should be as great in conservative living fossils, like coelacanths or *Limulus* as in rapidly evolving lineages. This argument does not imply that selection is unimportant but that there are so many possibilities of neutral mutation that, over a period of time, they make a major contribution to the level of divergence in amino-acid composition.

The selectionist opposition to this kind of argument is that virtually all of the observed changes are the historical evidence of successful selection for alternative amino-acids which react favourably with their biochemical environment and that the multiple effects on biochemical function of substituting one amino-acid for another constitutes a restraining influence on substitutional change generally and has the effect of making rates of substitution, wherever they occur on the molecule, rather similar — an argument which I find unconvincing.

Another line of evidence which should be mentioned here refers to the average amino-acid frequency in a number of vertebrate polypeptides. (KING and JUKES, 1969) have compared the observed frequencies with the expected frequency on the basis of the genetic code and random substitutions of nucleotides. With the exception of the amino-acid arginine, which is much too scarce relative to the number of alternative codons and for which some plausible explanations exist, there is an excellent correlation, with amino-acids like methionine and tryptophan occurring least frequently and serine, leucine and glycine with the greater number of codon alternatives, occurring most commonly. The selectionist criticism, e.g. CLARKE (1970) is that this only holds if polypeptides are averaged and that, if individual proteins are considered, such as bovine cytochrome C, bovine ribonuclease, bovine trypsinogen, horse hemoglobin, papqain, etc., their amino-acid composition deviates significantly from what would be expected from random permutation of bases. Perhaps neither of these approaches is the best that can be devised. It is obviously of dubious statistical validity to hunt through a series of randomly chosen proteins for deviations from the mean. Also it might be better if the correlation with codon alternatives were applied to those regions of the molecules which, on chemical grounds, are either believed to require stability of composition or which have comparative freedom to vary. Perhaps there would be a clear distinction between these categories and the correlation in the latter case might be much improved. Such comparisons of parts of the same protein molecule could provide quite a sensitive test and, when the proportion of the amino-acids falling into either category were taken into account, the analysis would be relevant to KIMURA's argument about rates of neutral substitution.

Before passing on to the consideration of variation in populations I should like to mention briefly current attempts to evaluate nucleotide divergence in the genome directly, since it would be valuable if we could compare amino-acid rates and primary nucleotide rates. The approach here depends on the technique of nucleic acid hybridisation. Thus if a double stranded native DNA is extracted in pure form from a given species and then denatured into single strands by heating or treating at a high pH, then if the single strands are incubated at the right ionic concentration at a suitable temperature they will reform the duplex so that complementary sequences become realigned in their original order, at a rate which is concentration dependent. In actual practice, for the DNA from higher organisms, it is necessary to break the DNA into small pieces by sonication, otherwise with long strands there is such a complex interlocking network that this interferes with duplex formation. This property of reannealing of single strands can be turned to advantage by studying the rate at which duplex is formed between homologous strands, derived by denaturation of a single DNA, compared with the rate of formation of duplex between heterologous strands. This is achieved by immobilising single strands of DNA of a given species, for example by baking on to nitro-cellulose membranes so that it is not free to move but can form duplex stretches of double stranded DNA when incubated in a solution carrying DNA of the same or different origin. It is usual to label the DNA which is held in solution with a suitable isotope and, since we can work out the specific activity i.e. number of disintegrations per $\mu$g DNA, we can estimate how much of the labelled homologous or heterologous DNA is bound to the immobilised material. The greater the level of complementarity between immobilised DNA and the DNA in solution the higher the level of double stranded DNA is formed after incubation. Naturally this is maximal when the DNA in solution is homologous; it will be progressively less the greater the nucleotide divergence between the immobilised sample and that in solution, and therefore, provided we can interpret adequately what differences in rate of duplex formation means in terms of nucleotide substitutions, we have the basis for measurement of DNA base sequence divergence at the primary level. There are various ways in which the discrimination may be enhanced. Thus the stability of duplex at higher temperatures is directly correlated with level of complementarity so it is possible to allow the incubation to run for a period and then progressively raise the temperature to compare how much labelled DNA remains bound to the immobilised DNA in different comparisons. Also it is possible to determine the level of competition between labelled DNA and homologous or heterologous cold DNA for the same complementary sites.

We have not time to pursue this topic in any detail. It is sufficient to note here that estimates by WALKER (1968) and LAIRD, MCCONAUGHY and MCCARTHY (1969) suggest that about 13–15% of the nucleotides differ in mouse and rat for example. By reference to the probable time of divergence of rat and mouse the estimated rate of substitution per codon works out at about an order of magnitude greater than what is estimated from average rate of amino-acid substitution. Of course what we really need here are comparisons with amino-acid sequences on a number of proteins for rat and mouse to compare with the nucleic hybridisation data.

The interpretation of the nucleic acid hybridisation data is open to some confusion at present due to the occurrence in the DNA of higher organisms of stretches of sequences which are reiterated to varying degree and which may collectively account for 10–20% of the DNA per genome or even more in some circumstances. Such sequences, because they are effectively more concentrated, form duplex at a rate which is proportional to their level of reiteration. Current evidence (WALKER, 1968; and ROBERTSON, CHIPCHASE and NGUYEN, 1969) gives the impression that such sequences may diverge more rapidly than the non-reiterated DNA, which presumably includes the DNA which codes for the structural proteins we are familiar with. Also we have no clear idea of what fraction of the DNA codes for proteins; current estimates suggest that it may be only a few per cent of the total, and if this is so, then much depends on how far the rates of divergence of base sequence found in non-reiterated DNA faithfully reflects the rates in DNA which codes for proteins. However, the fact remains, that we have at least an indication from these experiments that the rate of nucleotide substitution may be much higher than the rate of amino-acid substitution and this would favour the idea that neutral base substitutions, especially for synonymous codons, is a major feature of evolutionary divergence at the primary level and this takes us back to the WRIGHT's (1966) suggestion which we used to open this discussion.

The last general point I wish to discuss is the practical evidence of variation in animal populations along with some of the problems posed by the distributions of gene frequencies. Everyone will be familiar with the most recent surveys of polymorphism on a random sample of proteins in man (HARRIS, 1966), *Drosophila* (LEWONTIN and HUBBY, 1966) and the mouse (SELANDER and YANG, 1969) and the finding that the results are remarkably similar in such widely separated species so that about 30–40% of the loci are likely to be polymorphic in the species concerned and 10–15% of the paired loci in an average individual will be heterozygous. Naturally, before considering the wider implications we have to recognise the limitations of such surveys, especially that certain enzyme systems may be inherently more or less likely to be polymorphic and, of course, the general consideration that only soluble enzymes with hydrolytic or oxidative activity or enzymes which can be linked to such in a reaction system (SHAW and KOEN, 1965) are accessible to study by the usual methods. But when all is said and done we are left with the conclusion that there is a high level of genetic variability in natural or domesticated populations of animals and man and hence the origin of this variation calls for comment.

As usual there is a good deal of room for difference of opinion, especially as to the emphasis laid on the precision and role of natural selection. There are sufficient examples of stable polymorphism with high frequencies of alternative alleles over a wide range of environmental conditions and in populations subject to fluctuation in size to justify the conclusion that natural selection is responsible for maintaining the balance while selection of the heterozygotes is the most favoured mechanism. But there is uncertainty at present about the number of loci which can be kept heterozygous in a population by such selection and how far other mechanisms, such as gene frequency dependent selection may be important.

The first reaction to the occurrence of so much variation by some workers such as LEWONTIN and HUBBY (1966) was to pose a dilemma such that if stable equilibria were due to selection of superior heterozygotes then this would involve in intolerable load of lower fitness due to the cumulated, multiplicative effect of the alternative homozygous combinations. This interpretation derived, of course, from HALDANE's (1957) thesis of the cost of natural selection whereby he inferred that there could not be much more than one gene substitution per 300 generations, assuming the death each generation of a constant proportion of individuals carrying a particular gene. However, the high level of polymorphism found in natural populations has shown that loss of fitness due to segregation cannot be excessive and several authors have recently pointed out the error of models which postulate multiplicative effects of homozygous combinations on fitness and performance. In fact there will be pleiotropic effects and interaction between the effects of substitutions at different loci of all degrees of complexity and, by suggesting a particular form of gene interaction SVED, REED and BODMER (1967) have shown how a large number of loci may be maintained in a polymorphic state with selection in favour of the heterozygotes.

Before considering this I think we can draw upon our experience of biometrical genetics to qualify and illustrate some of the theoretical arguments. To anyone, like myself, who has worked extensively in the field of quantitative inheritance, it is no surprise that electrophoretic analysis should have uncovered a wealth of genetic variation. We know from numerous selection experiments in a variety of animals, especially of course in *Drosophila* and the mouse, that, in characters which vary about an intermediate optimum, heritability estimates derived from the regression of progeny mean on parent average are generally high — of the order of 0.4–0.6, while selection in either direction makes an immediate and sustained response. Just in *Drosophila* alone, the traits which have been shown to vary in this way include body size and growth rate, numbers of sternital and sternopleural bristles, the number of egg strings in the ovary, the relative size of body parts, the concentration of red pigment in the eye, the behavioural response to gravity, the cell size and number relations in the wing etc. — the list could be indefinitely extended. Also when we estimate the contribution of genetic segregation to the total phenotypic variance for characters which are components of fitness, whose average level is held near an upper limit and whose genetic behaviour is non-additive, we can show, as in the case of egg production and development time in *Drosophila*, that when we compare the phenotypic variance of genetically uniform individuals with that of individuals from a wild population, exposed to similar environmental conditions, we find a substantially greater variability due to segregation. The very low levels of heritability in such characteristics and other evidence, indicate that the segregating genetic differences are interacting in a non-additive manner as we might expect. There is every reason to suppose that the situation shown in mouse and *Drosophila* are quite general.

Now we do not know at present whether the genetic differences which are responsible for such quantitative variation are qualitatively similar to the electrophoretic variants we encounter in structural proteins. If they were we could combine the evidence from the two fields of enquiry to arrive at a more general synthesis. I will suggest that they are sufficiently alike to obey the same rules and that the small differences in body size or egg production we follow in selection experiments will often be ultimately attributable to the kind of allelic differences we study by different methods when we are following segregation at specific loci. The interrelations between such different manifestations of the same primary changes are very difficult to establish. However, I am particularly inclined to this view from some recently published work by BARTHELMESS and myself (1970) in which we took a typical quantitative character, namely amount of red eye pigment in *Drosophila*, which is due to two or three closely related pteridine compounds, and which behaves as a typical quantitative character in terms of heritability, selection response, reaction to inbreeding etc., but in which we could take the biochemical analysis further than is usually possible by following the changes in precursor compounds in the pteridine pathway leading to the final red pigment by making use of chromatography and the fluorescence of the separated compounds at different excitation wave lengths to obtain estimates of their quantity in different selected strains, inbred lines etc. And this analysis clearly indicated the segregation in the wild population of alleles at several loci which were affecting different steps of the pteridine pathway but which, in terms of red eye pigment, appeared part of a homogeneous distribution of genes affecting pigment content. Also recent work on such characters as sternopleural bristle in *Drosophila*, which is uniquely favourable for critical genetic analysis of allelic differences and detection of linkage relations, has also indicated that a substantial fraction of the normal genetic variation is due to heterozygosity at several loci, which will probably be shown to be different in primary biochemical function (THODAY, GIBSON and SPICKETT, 1964; ROBERTSON, 1967) but which happen to influence the same character. Hence I would strongly recommend that anyone interested in the interpretation of serum variants must take account of the major conclusions from quantitative inheritance since, although the orientation and techniques of the workers concerned seem to be so different, they may in fact be looking at different aspects of the same processes in development. If this view is accepted it has immediate relevance to some of the current interpretations of the polymorphism of serum and other protein variants in man, *Drosophila*, cattle or any other animal you care to mention.

In order to escape from the dilemma of substitutional load SVED, REED and BODMER (1967) proposed that there is a sort of law of diminishing returns with respect to levels of heterozygosity so that provided the level of heterozygosity is high enough, further increase beyond that level has no further advantage. In other words if we consider the change in fitness in relation to level of heterozygosity there will be an asymptotic relation so that most of the individuals will be very similar in fitness and only the minority which falls below the average threshold value will be exposed to natural selection. This model, which postulates epistatic interaction, immediately recalls the evidence from genetic analysis of heterosis in *Drosophila*, in which the genetic properties of hybrid vigour in crosses between inbred lines have been studied by constructing genotypes with different levels of heterozygosity by interchanging chromosomes between lines in a systematic way so that we can test alternative models (ROBERTSON and REEVE, 1955). Without going into the technical details of these experiments, the principal conclusions, which we believe to be quite general for animal populations, were: 1. There is strong evidence of interaction between genes located on different chromosomes which takes the form of a law of diminishing returns with respect to levels of heterozygosity so that, often, the presence of just a single pair of heterozygous chromosomes will increase egg production or growth rate as much as the much more heterozygous $F_1$ of a cross between the pair of inbred lines from which the unlike chromosomes were drawn, and 2. more or less parallel to this increase in performance there is a reduction in sensitivity to environmental conditions, and general evidence that the ability to discriminate between genotypes is very subject to the quality of the environment, so that the better the conditions the less the chance of detecting differences.

Now if we accept the foregoing arguments this has some implications for serum variant studies. There seems to be a widespread feeling that it should be possible to detect the selective bases for particular polymorphisms and, in the case of domesticated livestock, turn this to practical advantage by incorporating the appropriate gene in this or that breed. Since quite low selection differentials in favour of heterozygotes, of the order of 1%, may be compatible with polymorphism at hundreds of loci, it might be expected that the chance of demonstrating conclusively a correlation between genotype and performance would be rather low and the evidence to date supports this view. Natural selection may be ultimately responsible for the occurrence of a given polymorphism in the population even though the likelihood of demonstrating this is low. We are all familiar with the classic case of sickle cell anaemia but you have to look rather hard to find cases as convincing as this in serum variants. For example I think the attempts to associate transferrin differences with performances in cattle etc. have not been particularly convincing and it is highly dubious whether fixing one or other allele in the appropriate breed would lead to a demonstrable difference in economic yield. This is not surprising. Polymorphism at such loci may well have been relatively more important in wild ancestors or at earlier stages of domestication under less favourable conditions of management, but the better and more uniform the environment, the less the likelihood of detecting selective differences and the greater the economic irrelevance of records of serum variation.

The number of cases where gene frequency changes in natural populations of animals have been correlated with environmental changes clearly enough to attribute the changes to natural selection are remarkably few. I might mention a recent report of SEMEONOFF and myself (1968) of seasonal changes at an esterase locus in the short tailed field mouse, *Microtus agrestis* and also changes correlated with the more or less cyclic changes in population density when the drastic decline in population size is likely to be associated with strong selection pressures. And in domesticated animals there is the rather striking case of the A, B and C sheep hemoglobin differences which differ in oxygen affinity and show differences in distribution between hill and lowland breeds (BRAEND, EFREMOV and HELLE, 1964). The general conclusion is that although we may be convinced that natural selection is primarily responsible for establishing a particular polymorphism the prospect of demonstrating this directly and understanding how the selection operates appears to be very low in natural populations and perhaps even lower in domesticated species in which improved environmental conditions may have removed or greatly attenuated the influence of selection. In wild populations the difficulties are aggravated by our ignorance of the ecology of most species and the tremendous element of luck in happening to pick a polymorphism which is amenable to this kind of analysis.

If it is true then that any species we care to work with will be highly heterozygous for loci which determine protein and enzyme differences and if the prospect of demonstrating fitness differences is unencouraging what is the point of recording more and more variants in further species? This activity, unless undertaken to answer specific questions in a precise and quantitative fashion, seems to me to have about the same intellectual content as stamp collecting, without the aesthetic appeal. Hence if we are to escape this criticism, we must decide on the questions and I should like to spend a little time on this aspect.

One of the theoretical aspects which requires attention is to discover whether there are real differences between populations and species in the general incidence of polymorphism. O'BRIEN and MACINTYRE (1969) have suggested that the general level of polymorphism may be higher in *D. melanogaster* than in the closely related sibling species *D. simulans*. Since systematic quantitative study of gene frequencies at a sufficiently large number of loci in several species is quite a laborious undertaking this is certainly a field where it would be sensible for several laboratories to collaborate in a survey of a few species. Naturally, comparisons between species do not mean much unless we define the populations from which the samples are being drawn and as soon as we take up this aspect we encounter the problems of population size, fluctuation in numbers, gene flow between populations etc. Although in man and livestock it should not be difficult to overcome the sampling problems, this is not true in wild species where the information we have about mobility and population size is generally quite inadequate.

Alternatively, the situation may be looked at in reverse and we may decide to use the distribution of gene frequencies as a tool to shed light on breeding structure and mobility. A growing and economically valuable application here is the use of serum and tissue variants to sort out the different geographical races of fish. Given a high enough level of heterozygosity and, preferably, several polymorphic loci there is an encouraging prospect of identifying individuals belonging to different panmictic groups. Evidence of departure from the Hardy–Weinberg distribution can be the starting point for further analysis. Provided we are dealing with populations in equilibrium, which are not undergoing displacement of one allele by another, excess of homozygotes may reflect inbreeding as PETRAS (PETRAS, 1967) has suggested in house mouse populations or may indicate immigration between populations which differ in gene frequency. Systematic study of the special distribution of gene frequency at many loci coupled with detailed analysis of ecology and population size of wild species offers great prospects for analysing population dynamics and of ultimately providing basic information relating to population size and immigration. Here again, the problems are difficult and time consuming and it would be better to concentrate on a comparatively few species in depth rather than collect fragmentary information on many. Perhaps I might mention here, as an example, the practical application of polymorphism in the eider duck to identify two reproductively isolated populations in what at first sight appeared to be a single panmictic unit (MILNE and ROBERTSON, 1965). In this three allele system in one of the egg white proteins it was possible to demonstrate gene frequency differences between a sedentary and migrant groups nesting in the same area. In this case we have every reason to believe that selection is responsible for the polymorphism since it is very widespread and has been encountered in all populations which have been sampled from widely separated localities in Europe, the eastern United States and Northern Canada, but we have not the faintest idea what the selective agent might be or have been, since there are no clues so far. However we can just ignore the origin of the polymorphism and use the allelic differences as handy markers for ecological study.

Alternatively, if we wish to adopt a more analytical approach to the possible role of selection in polymorphic situations then we need to set up experimental populations of known gene frequency and follow changes, if any, in successive generations, preferably sampling from old and young individuals to keep a check on possible selection during life. Obviously the scope for this is rather limited to smaller animals with a short life cycle and in which we can vary the environment, especially by providing a range of competitive conditions.

In this context we need much more experimental information about possible gene frequency dependent selection which has been reported by some workers, e.g. KOJIMA and YARBROUGH (1967) as an additional method of maintaining polymorphism. In this situation the adoptive advantage of an allele is inversely related to its frequency so that when it is rare there is selection for its increase and when it is common the alternative allele is favoured. At present there is considerable uncertainty as to how often this phenomena occurs in nature.

Before finishing this survey I should like to restate a few general conclusions which seem to me, at any rate, to offer some guide to the future orientation of research.

Firstly, the diversity of protein variation which could be demonstrated is inconceivably immense and therefore, if we are not to be utterly overwhelmed in a mass of facts which merely restate what is known or confirm what we might reasonably expect, then we must exercise restraint and only collect information which will provide answers to precisely defined questions.

Secondly, the hope of demonstrating an effect on fitness or economic performance of the vast majority of protein variants is likely to prove illusory.

Thirdly, for those interested in the genetics of natural populations it would seem wisest to concentrate on a few species, whose ecology is amenable to detailed analysis and for which we can obtain accurate estimates of population size. Diffuse and fragmentary observations on many species of unknown ecology is of little value. What we need is a systematic study of gene frequency distributions, related to breeding structure, migration etc. If the species concerned can be bred well enough for the setting up of experimental populations with predetermined gene frequencies, so much the better.

Fourthly, if we are more interested in rates of divergence in proteins, we need comparisons of amino-acid sequence in proteins from related species and this should most certainly include quantitative estimates of possible concealed variation in proteins which appear outwardly identical. Comparison of the incidence of variation in this category with that in proteins known to differ electrophoretically or in other ways, would be most valuable. Here again I think there is much more to be gained by studies in depth, comparing a few proteins in great detail, rather than aiming for incomplete information on many.

Fifthly, such comparisons of amino-acid sequences should be compared with alternative estimates of evolutionary divergence, especially divergence at the primary, nucleotide level, using such techniques as nucleic acid hybridisation which will no doubt become increasingly discriminating in its ability to detect differences.

Sixthly and finally, I would urge that discussion of the circumstances which influence the gene frequency of protein variants should take account of the evidence from quantitative inheritance, which has already accumulated a great deal of information about variation in populations and which has developed a variety of analytical techniques which merit wider application. Indeed, we might hope, before too long, to encounter instances where segregation at a locus, which influences the phenotype of a so-called quantitative character, can be also followed and studied by the methods used by students of protein variation and this would allow us to take the first steps in bringing together two fields of enquiry which have hitherto pursued an independent existence.

## REFERENCES


Barthelmess, L. B. and Robertson, F. W. (1970), *Genet. Res.*, **15.** 65.

Braend, M., Efremov, G. and Helle, O. (1964), *Nature*, **204.** 700.

Clarke, B. (1970), *Science*, **168.** 1009.

Eck, R. V. and Dayhoff, M. C. (1966), *Science*, **152.** 363.

Haldane, J. B. S. (1957), *J. Genet.*, **55.** 511.

Harris, H. (1966), *Proc. Roy. Soc. Lond. B*, **164.** 298.

Kimura, M. (1968), *Nature*, **217.** 624.

Kimura, M. (1968), *Genet. Res.*, **11.** 247.

Kimura, M. (1969), *Genet.*, **63.** 1181.

King, J. L. and Jukes, T. H. (1969), *Science*, **164.** 788.

Kojima, K. I. and Yarbrough, K. M. (1967), *Proc. Nat. Acad. Sci.* USA, **57.** 645.

Laird, C. D., McConaughy, B. I. and McCarthy, B. J. (1969), *Nature*, **224.** 149.

Lewontin, R. C. and Hubby, J. L. (1966), *Genetics*, **54.** 595.

Milne, H. and Robertson, F. W. (1965), *Nature*, **205.** 367.

O'Brien, S. J. and MacIntyre, R. J. (1969), *The Amer. Nat.*, **103.** 97.

Perutz, M. F. and Lehmann, H. (1968), *Nature*, **219.** 902.

Petras, M. L. (1967), *Evolution*, **21.** 259.

Riggs, A. (1959), *Nature*, **183.** 1037.

Robertson, A. (1967), 'Heritage from Mendel', 265 pp. (Brink, R. A. Ed. Univ. Wisconsin. Press), 1967.

Robertson, F. W. and Reeve, E. C. R. (1955), *Zeit. indukt. Abstamn.-u Vererblehre*, **86.** 439.

Robertson, F. W., Chipchase, M. and Nguyen, M. (1969), *Genetics*, **63.** 369.

Selander, R. K. and Yang, S. Y. (1969), *Genetics*, **63.** 653.

Semeonoff, R. and Robertson, F. W. (1967), *Biochem. Genet.*, **1.** 205.

Shaw, C. R. and Koen, A. L. (1965), *J. Histochem. Cytochem.*, **13.** 431.

Sick, K., Beale, D., Irvine, D., Lehmann, H., Goodall, P. T. and MacDougall, S. (1967), *Biochim., Biophys. Acta*, **140.** 231.

Sved, J. A., Reed, T. E. and Bodmer, W. F. (1967), *Genet.*, **55.** 469.

Thoday, J. M., Gibson, J. B. and Spickett, S. G. (1964), *Genet. Res.*, **5.** 1.

Walker, P. M. B. (1968), *Nature*, **219.** 228.

Wright, S. (1966), *Proc. Nat. Acad. Sci.*, **55.** 1074.

## DISCUSSION

J. C. Ashton: I disagree with dr Forbes Robertson in two respects. First, he likens the collection of further data on genetic variants to the collection of postage stamps. I think it would be unfortunate if all activity in this field were to stop. There are still many breeds and populations to be characterized. Second, I think he is unduly pessimistic about finding relationships between genotype and fitness. There is growing evidence inplicating the transferrin locus in various aspects of fitness in such diverse species as tuna, mice, *Microtus*, cattle, pigs and sheep.

W. E. Briles: Our basic objective should be to investigate the role that polymorphisms play in the dynamics of populations. Specifically we should direct our research toward determining specific genotypic effects at the biochemical and physiological level and the genetic interplay between such loci and the remainder of the genome.

In the hopes of encouraging at least a reasonable search for associations between polymorphisms and other seemingly "extraneous" traits would like to report very briefly the essential finding of a paper by Crittenden, Briles and Stone soon to be published. In chickens it has been known for some time that susceptibility to subgroups A and B of lymphoid-sarcoma viruses is controlled by a pair of independent, dominant genes. In testing with a large panel of blood typing reagents the erythrocytes of birds progeny tested for the ability to transmit these two types of susceptibility it was discovered that one reagent designated RI agglutinated the red cells of those birds proven to be transmitting B subgroup susceptibility. Thus, the receptor site which makes possible the penetration of host cells by virus appears to be synonomous with the isoantigen detectibility of this reagent.

F. Pirchner: Instead of searching randomly for polymorphism, an investigation of enzymes which are involved in pathways leading to the front of interest may be rewarding.

D. R. Osterhoff: Would a comparison of biochemical polymorphism in primitive cattle breeds and highly developed and selected breeds give an answer to the possible relationship between polymorphic variants and production traits?

C. Stormont: Attention is called to the direct correlations which can and are being made between specific isozyme variants and the response of animals and man to drugs and their environment in general. Consider for example the two esterase isozymes in rabbit serum, namely atropinesterase and cocainesterase which are all important in determining the response of rabbits to those respective drugs.

J. Rendel: In the selection for productivity in farm animals one has to look for alternative possibilities. For traits with high heritability (e.g. fat content of milk or milk production in cattle), there is no practical need to look for relationships to polymorphic blood variants. Other more efficient methods for progress are available. However for traits with cow heritability there appears to be a considerable non-additive genetic variation and it might therefore be fruitful to look for associations between much (different) traits and heterozygosity for various polymorphic blood characteristics, particularly in cases where the polymorphisms are of enzyme nature.

F. W. Robertson: I should like to comment on the role of Semeonoff and myself on esterase polymorphism in *Microtus*. This has been quoted to illustrate the role of natural selection in genetic polymorphism. But this is a rather special case. I choose *Microtus* because it seemed likely that selection pressures would be especially intense at the period of population decline since this is a species with cyclical fluctuation of density. And also because the ecology of this species is relatively well known — an important point. So the *Microtus* case is not to be regarded as a random sample.

*XIIth Europ. Conf. Anim. Blood Groups Biochem. Polymorph., Bp., 1972 (pp. 55—65)*

# BLOOD GROUPS AND BIOCHEMICAL POLYMORPHISMS IN FISH

Wilhelmina de Ligny

Netherlands Institute for Fishery Investigations, IJmuiden, The Netherlands

It is a pleasure for me to have been invited to present at this meeting a survey of the investigations on blood groups and biochemical polymorphisms in fish, on which I was first asked to give an impromptu report ten years ago, in 1960, at the 7th Animal Blood Group Conference in Edinburgh. In this report, that was based primarily on the information and knowledge collected during a training period in the United States, I mentioned that I know of six laboratories in the world, where studies on fish blood groups had been initiated. In following meetings of the European Society for Animal Blood Group Research evidence of the development of fish blood-typing was provided by the number of contributions on the subject which increased from one progress report in 1962, to the three papers that were presented at the 1966 meeting in Paris. According to the program, this number should again be tripled today.

The increasing number, and the diversification of the subject matter of the papers presented at the meetings of the European Society have however only been an exponent of what has taken place in fish blood-typing since 1960. A review of the literature in the doctoral thesis presented in 1962 at the University of Paris by Lee, contained over 20 references to studies of red cell and serum antigens, and electrophoretic studies of protein polymorphism in fishes. In his review on "Blood groups in marine animals", that was published two years later, in 1964, Cushing discussed about 70 papers in which blood groups and biochemical polymorphisms in fishes, and their application to fishery investigations had been described. Screening the literature four years afterwards, in 1968, in preparing a review article, provided me with an additional 200 references. In a further survey, until April this year, no decline in the publication rate was apparent.

Blood groups and protein polymorphisms have now been investigated in over 200 species of fish. In the literature up to the middle of 1968, that I covered in the review article (de Ligny, 1969), blood groups had been defined in about 20 species. In another 60 species individual differences in the reactivity of the erythrocytes with various agglutinins had been noticed. In about 140 species intraspecific differences in electrophoretic patterns of proteins and enzymes had been searched for, and in 108 such differences, in one or more characters, had been described. In 47 species the frequency distribution of the observed phenotypes indicated the genetic origin of the variation.

Among the characters investigated hemoglobins predominated, and had been studied in 74 species. Transferrins had been looked at in 38 species and serum esterases in 37. Other proteins, in the order in which they were represented in the studies, reviewed in 1969, were muscle myogens, serum albumins and eye-lens proteins. The number of species in which a search for polymorphism of enzymes, other than esterases, had been reported until 1968 was limited. In 7 species polymorphism of lactate dehydrogenase had been described, presumably of genetic origin, and polymorphism of 6-phosphogluconate dehydrogenase and aspartate-amino-transferase had been observed in one species each. More recently polymorphism has also been described for creatine kinase (SCOPES and GOSSELIN-REY, 1969), isocitrate dehydrogenase (QUIROZ-GUTTIERREZ and OHNO, 1970), malate dehydrogenase (BAILEY, COCKS and WILSON, 1969), phosphoglucomutase (LUSH, 1969; ROBERTS, WOHNUS and OHNO, 1969) and sorbitol dehydrogenase (LIN, SCHIPMANN, KITTRELL and OHNO, 1969).

I hope to have shown by this enumeration of figures that, in view of the number of species of fish, and the number of characters investigated, in a survey like this only the essence of the subject can be given. Because of the large number of species studied, it might appear that an attempt could be made at a comparative approach: the more than 200 species studied belong to approximately 50 fish families, which range throughout the entire taxonomic scale, and include the primitive hagfish, as well as the angler fish near the other end of the range. Looking at it from a physiological angle, they range from tropical freshwater fish to marine species from the near-arctic region. In considering such an approach it should be realized however that these 200 species represent a class of the animal kingdom with a total of about 30 000 species, of which they are not even a 1% fraction. The choice of the species in most of the studies moreover has been made on a different basis. And, although there are some interesting recent studies in which evolutionary or functional aspects have been the primary concern of the investigator, an attempt at interpretation of the occurrence and forms of polymorphism observed in these 200 species of fish, from either the phylogenetic or the physiological angle, would be speculative.

Like it was ten years ago, the essence of the majority of the studies is the potential value of blood groups and biochemical polymorphisms for fisheries research, and more in particular for the recognition and differentiation of populations — self-sustaining units, that should be dealt with separately in fisheries assessment and management. In view of the increase in the number of studies and laboratories in which this work has been taken on during the last ten years, an attempt at an evaluation appears worthwhile: have studies of bloodgroups and biochemical polymorphisms been useful to fisheries research, and its final aim — improving and sustaining the yield of fisheries?

Before considering this question we first have to take one step back: what role does differentiation of populations play in world fisheries? This question arises when we compare the species of fish in which population studies by means of blood groups and biochemical polymorphisms have been carried out, with the general picture of the world fish production (Fig. 1). The four species constituting 30% of the total amount of 64 million ton of fish caught yearly, are the Pacific anchovy, a topper with an annual catch of 11 million ton, cod and herring from the North Atlantic, and the Alaskan pollack from the north Pacific waters. The number of studies on blood groups and biochemical polymorphisms in these four species, published or reported until 1968, were respectively 1 for the anchovy, 24 for cod,

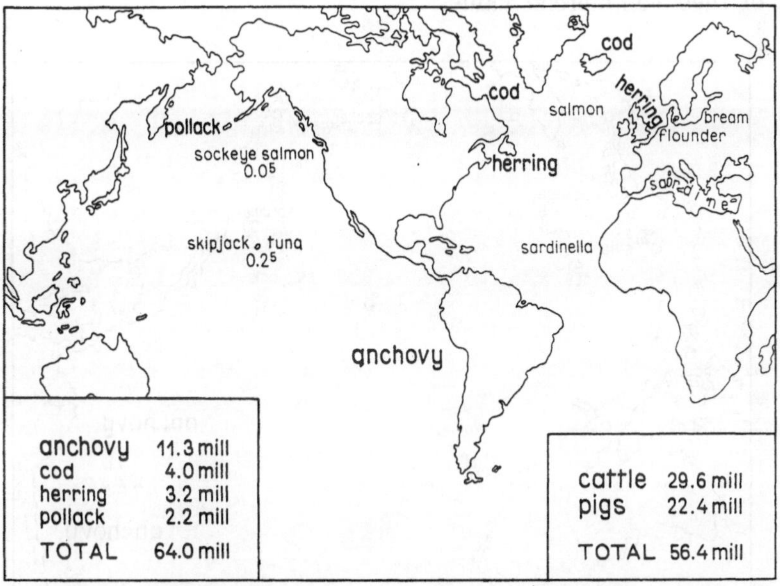
cod
cod
herring
salmon
bream
flounder
sardine
pollack
sockeye salmon
0.05
herring
skipjack tuna
0.25
sardinella
anchovy
anchovy 11.3 mill
cod 4.0 mill
herring 3.2 mill
pollack 2.2 mill
TOTAL 64.0 mill
cattle 29.6 mill
pigs 22.4 mill
TOTAL 56.4 mill

FIG. 1. World fish production. Distribution and 1968 landings of the fish species discussed in the text. (According to FAO, 1969)

16 for herring, while in Alaskan pollack a search for red cell antigens had been made in ten individuals (for references see DE LIGNY, 1969). A contrast between the number of blood-typing studies and the ranking of the species on the FAO list of "Catches and landings" may also be observed in two other species from the Pacific, skipjack tuna and sockeye salmon, of which the annual catches amount to 0.3% and 0.06% of the total world catch. Both these species had, until 1968, been the subject of a similar number of blood-typing studies, approximately 16, as number three on the world list, the Atlantic herring. According to the program for this afternoon's meeting of the Fish section, cod will get the attention that it deserves, ranking second on the world list of fish production, and in agreement with its honorful epithet: "the beef of the sea". Two studies will deal with species, sardine and Sardinella, that belong to the herring family, which ranks third on the world list of catches and landings of fish families. Flounder,

that, according to the program, should be dealt with in two papers, however ranks inbetween the numbers 80 to 100, listed in the FAO catch statistics.

It is clear from this enumeration of facts and figures that a search for blood groups and biochemical polymorphisms in fish species, and their application to population studies, is not correlated with the amount brought ashore. Taking the Atlantic salmon, that also will be dealt with in one of the papers to be presented this afternoon, as an example, in comparison with the sardine, it will be apparent, even if you are only familiar with their price as canned products, that they may be placed on another scale: that of their commercial value.

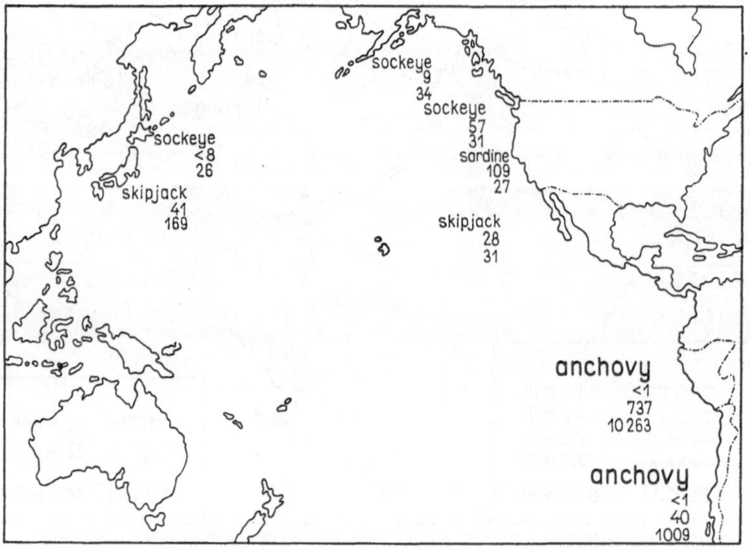
sockeye
9
34
sockeye
57
31
sardine
109
27
skipjack
28
31
sockeye
<8
26
skipjack
41
169
anchovy
<1
737
10 263
anchovy
<1
40
1009

FIG. 2. Landings of Pacific fish species in various countries (sockeye salmon: 1948–'58 skipjack tuna: 1948–'68, sardine: 1958–'68, anchovy: 1948–'58–'68. According to FAO, 1958, 1969)

Returning to Pacific species, on which I will concentrate in the discussion of the subject, because the fishes of the northwest Atlantic will be dealt with sufficiently this afternoon, the contrast between the number of blood group studies, carried out for instance on skipjack tuna and the Pacific anchovy, can also be looked at from a third angle: while tuna, if mostly in its canned form, does appear directly on our dish, the 11 million ton of anchovy, caught yearly, have to go a long way, being turned into fishmeal, and reaching our dish only after having been converted to chicken or pork. I do not know the conversion factor for this pathway, but I may, for your information, mention that, in case the anchovy were fed to the "beef of the sea", the cod, there would be a loss of 90%.

With the thought in mind, that of the total catch of anchovy perhaps only 10% will end up on the human dish, it might occur to you that this is the explanation why in this species so far only one search for biochemical polymorphisms has been reported, and that this might reflect a general lack of interest and activity of fishery biologists with regard to assessment and management of this species. This is not true, however, and we should consider the anchovy fisheries somewhat more in detail. In Fig. 2 the yearly landings of Pacific anchovy in Peru and Chili are given in thousands of tons. The total catch of anchovy, landed in Peru in 1948, was less than 1000 tons.

The increase to 11 million ton took place in less than 20 years. The rapid development of this fishery was due to the improvement of fishing techniques and investments in the fishing fleet. These have been accompanied, however, by biological studies, intended to obtain an estimate of the available stock, and the maximum sustainable yield. On the basis of these studies a yearly quotum has been agreed on between the two countries participating in these fisheries, Peru and Chili. Looking at the annual catch of anchovy landed in both countries in 1968, the 10-fold difference is not a consequence of this agreement however. The 11 million ton landed in Peru reflect the abundance of the anchovy in the 200 miles zone off the Peruvian coast, in which it is fished for, while the yield of a similar zone along the 2700 miles of coastline of Chili is indeed no more than 1 million ton.

In this light, I don't think that it is surprising that the search for a biochemical population marker was initiated in Chili (SIMPSON and SCHLOTFELDT, 1966). If the anchovy caught along the Chilian coast only were an overflow of the abundant Peruvian stock, heavy Peruvian fisheries, or natural causes, might cause a sudden decline in the Chilian catches, and investments in the development of a fishing fleet in Chili would have been in vain. An example of such an event took place before in the eastern Pacific: the disappearance of the sardine stocks from the southern Californian waters, of which the figures, shown in Fig. 2, only reflect the last stage of the process, which caused the close down of the Californian sardine industry.

That in many cases territorial claims, or international competition, have primed the start of genetic population studies, may be illustrated for two more species, which were among the first in which blood-typing was attempted. Sockeye salmon is a so-called anadromous fish, that reproduces, or spawns, and grows up during the first year of its life in rivers. It then migrates into the sea, and comes back again after a variable number of years, to spawn.

Both in American and Canadian rivers, fisheries take place during the spawning run, while another contingent of the salmon are taken on the high seas, in international waters. An increase in the sockeye salmon landings in Japan, between 1948 and 1958, and a simultaneous decline of the yield of the fisheries for sockeye salmon in the United States (Fig. 2), were correlated with an increase of fishing effort of the Japanese fleet on the high seas. The first report by RIDGWAY, CUSHING and DURALL, about a search for a character that might be used to distinguish "American sockeye", from sockeye salmon of Asian origin, dates from 1958.

Japanese fisheries for tunas, including skipjack tuna have, after the second world war, spread out over the entire Pacific, as well as at a later stage, the Atlantic. Landings in Japan tripled from 1948 to 1958, and showed a further increase till 1968 (Fig. 2). Catches landed in the United States hardly increased during the same period. Blood-typing studies of the Pacific skipjack tuna were started in the United States in the nineteen-fifties, and the first blood-group system was described in 1957 by Cushing and Durall. The activities, in particular of the state of Hawaii, to increase its share in the Pacific tuna catches, were accompanied by an extensive blood-typing program of the Tuna Blood Group Centre in Honolulu, one of the most active — if not the most active centre of fish-bloodtyping in the world.

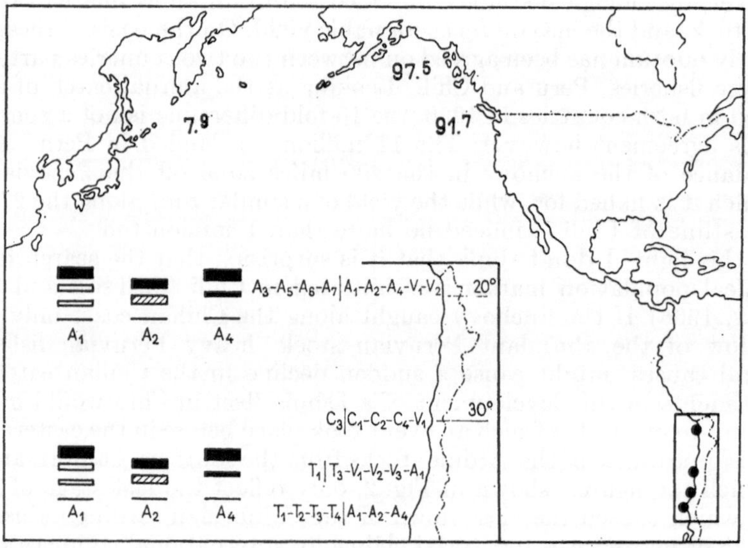
97.5
7.9
91.7
A3-A5-A6-A7|A1-A2-A4-V1-V2
20°
A1
A2
A4
30°
C3|C1-C2-C4-V1
T1|T3-V1-V2-V3-A1
T1-T2-T3-T4|A1-A2-A4
A1
A2
A4

Fig. 3. Frequency of serum antigens in sockeye salmon (Ridgway, Klontz and Matsumoto, 1962); hemoglobin types in Pacific anchovy (after Simpson and Schlotfeldt 1966)

In considering the results of the serological and biochemical studies of fish populations, and trying to assess their value for the solution of the problems of stock differentiation, attention should be given to the variety of methods and characters that have been applied. I will do this in direct connection with the population studies. The numbers shown in the upper part of Fig. 3 represent the percentage of sockeye salmon from Asian waters, and from the coastal range of Alaska to the state of Washington, with either one or both of two serum antigens, detected by Ridgway, Klontz and Matsumoto (1962) by means of rabbit immune sera in immunodiffusion tests. In spite of their succesful results with regard to the differentiation of Asian and North American salmon, the difficulties encountered in the production of sufficient anti-sera with identical specificities, led

RIDGWAY, after considerable effort, to look for another method, typing of red cell antigens by means of iso-immune sera, to which I will come back later.

The first studies on skipjack tuna populations were directed at the detection of red cell antigens by means of plant agglutinins, as mentioned by SCHAEFER (1962). SPRAGUE extended the search for suitable reagents to normal cattle sera and rabbit immune sera. By means of a plant agglutinin and rabbit immune sera he was able to detect three blood-group factors, that, together with a type of cell that was not agglutinated by any of the reagents, appeared to form a blood-group system. Population studies revealed significant differences between three areas in the Central Pacific (SPRAGUE and HOLLOWAY, 1962). In a four-allelic system with a dash-allele, verification of the goodness of fit of the distribution of phenotypes observed, and that expected according to the Hardy–Weinberg law for a random-mating population, is not possible. Subsequent investigations by FUJINO (1967) showed the caution with which data on "blood-group systems" in fish should be considered, unless the mode of inheritance can be verified by breeding studies, or, if this is not possible, the phenotypic distribution can be verified properly. In addition, the possibility of the influence of physiological and/or environmental circumstances on the expression of the blood-group genes should be considered: in the case of the blood-group system described by SPRAGUE and HOLLOWAY, FUJINO found one factor to be length-dependent.

The studies on the anchovy along the coast of Chili consisted of a search for homoglobin variation by means of agar electrophoresis, and were described in 1966 by SIMPSON and SCHLOTFELDT. Samples were collected at four locations, over a total range of 1500 miles, and sent to, and analyzed in a centrally situated laboratory. 19 hemoglobin patterns were described, of which 17 consisted of two or three bands, only slightly differing in migration rate.

A genetic interpretation was not considered possible, because of the number and complexity of the types. Certain types appeared to be specific for one location only, while others were found in different frequencies in more than one location. In my opinion it is remarkable that three of the types, depicted in Fig. 3, predominated in the samples taken at the two outer locations, while they were lacking nearly entirely in samples from the two intermediate locations. It may be noticed that the distance over which the samples had to be transported from the two outer locations to the central laboratory were roughly similar. In spite of the convenience for fish population studies of the method of agar gel electrophoresis of hemoglobins, developed by SICK (1961), the instability of fish hemoglobins led this author to considerable precautions both with regard to the freshness of the samples, and the conditions of electrophoresis (SICK, 1965a).

It appears time now to turn this rather negative account of serological and biochemical studies in Pacific fish species into a more positive one. By the examples given, I have intended to show some of the limitations, and frustrations, that have been encountered by fish blood types and that have accounted for the rather slow progress of the work in the first half of the decade. I should add that the examples by no means are representative either for the methods used, or the investigators.

By means of rabbit immune sera FUJINO and SPRAGUE (1965) have detected another system of red cell antigens in skipjack tuna that, as far as population genetic evidence, and exclusion of physiological influences can proof such, appears to be a genuine genetic system. It consists of three blood-group factors and a number of linear and non-linear subtypes, occurring as six possible "antigenic complexes" (FUJINO and KAZAMA, 1968). The succesful studies by RIDGWAY on red cell antigens of the sockeye sal-

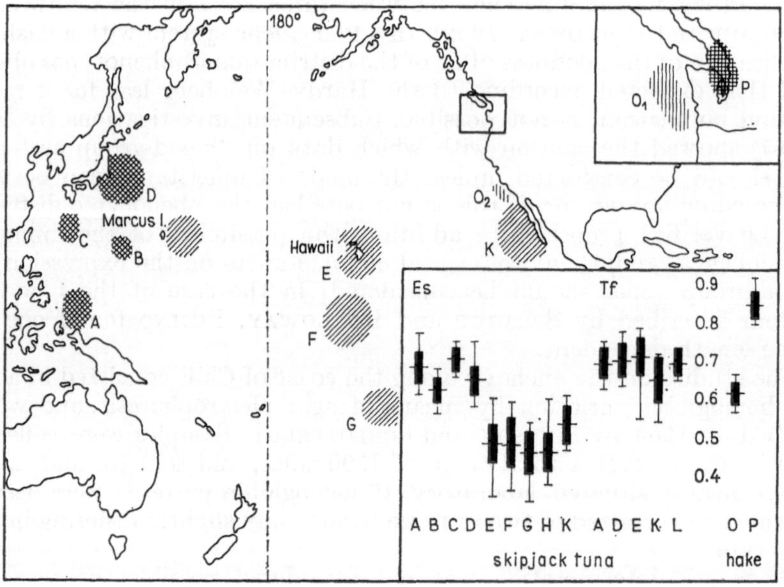
180°
O1
P
D
O2
Marcus I.
C
B
Hawaii
K
E
A
F
G
Es
Tf
0.9
0.8
0.7
0.6
0.5
0.4
A B C D E F G H K A D E K L O P
skipjack tuna
hake

FIG. 4. Sampling locations and frequencies of esterase and transferrin alleles in skipjack tuna (after FUJINO, 1970 and FUJINO and KANG, 1968b); sampling locations and frequencies of an esterase allele in Pacific hake (after UTTER, STORMONT and HODGINS, 1970)

mon, by means of rainbow trout isoimmune sera, have been reported by the author at the Paris meeting in 1966. A complex multiple allelic system, resembling the B system of cattle, was detected and sockeye salmon populations from various tributary areas within one river system could be differentiated.

In contrast with the effective use of hemoglobins in a series of population studies in North Atlantic cod, by SICK and coworkers (SICK, 1965b, FRYDENBERG et al., 1965), MØLLER (1966, 1968) and JAMIESON (JAMIESON and JONSSON, in press), hemoglobins have not been found useful in any of the Pacific species that we have discussed. They were reported by SHARP (1969) to be monomorphic in four species of tuna, including the skipjack tuna. In sockeye salmon RIDGWAY (1964) did not observe hemoglobin polymorphism, while VANSTONE, ROBERTS and TSUYUKI (1964) described ontogenetic variation in the hemoglobin of this species, like it has been observed in other salmonids, as was reported to you at the meeting in Paris in 1966 by Mr. WILKINS.

In the work on skipjack tuna by FUJINO and coworkers the major breakthrough has been brought about by the use of non-specific serum esterases (FUJINO and KANG, 1968a). Figure 4 shows the areas in which skipjack tuna samples have been collected and the frequencies and 95% confidence limits of a serum esterase allele (FUJINO, 1970). The frequencies in samples from four locations in waters ranging from Japan to the Philippines, were found to be significantly higher than in samples collected off the Californian coast, in the Central Pacific and as far west as Marcus Island (FUJINO, in press). The frequency found in the latter area, however, depended on the season in which the samples were taken. A thorough survey led FUJINO to the conclusion that, by means of the serum esterase system, two separate skipjack tuna populations can be distinguished: a population, localized in the western Pacific and a central population, with a similar esterase frequency as skipjack sampled in the eastern Pacific, and migratory in its western range, between Marcus Island and the international date line. I may briefly point to the lack of significant heterogeneity of a transferrin allele, studied by FUJINO and KANG (1968b) in skipjack tuna samples collected throughout the same area (Fig. 4).

The usefulness of esterases for fish population studies is enlarged by the fact that they do not only occur in serum, but also may be detected in extracts of various tissues, like heart, muscle, liver or the vitreous fluid of the eye. In most marine species, organs can easily be obtained, eventually from fish landed frozen, in contrast with blood serum, or red cells, for which blood has to be sampled from live fish. I therefore would like to conclude by referring to the results of the successful application of the esterases of the vitreous fluid of the eye in a recent study, yet to be published, or that just may have appeared, in the second issue of Animal Blood Groups and Biochemical Genetics, by UTTER, STORMONT and HODGINS.

UTTER has studied biochemical polymorphisms in the Pacific hake, a member of the cod family. This species occurs along the entire coast of the United States, from Baja California to the state of Washington and throughout the Puget Sound, the complex of sea arms leading to Seattle (Fig. 4). Together with previous investigations by UTTER (1969) and UTTER and HODGINS (1969), with respect to transferrin and lactate dehydrogenase variants in the hake, the recent studies on the vitreous fluid of the eye showed a similarity between Oceanic samples, taken off south California and off the mouth of the Columbia River (Fig. 4, $O_2$ and $O_1$), which confirmed the fishery biologist's opinion that the hake migrates throughout this range, but in the Puget Sound (Fig. 4, P) significantly deviating frequencies of the alleles involved were found, indicating the existence of separate spawning populations of the hake in the Puget Sound and the Pacific Ocean. The frequency difference of an esterase allele is illustrated in Fig. 4.

The value of bloodgroups and biochemical polymorphisms for the differentiation of fish populations is apparent from the studies described. But these also show the considerable problems that have been encountered by the workers in this field and the perseverance that in many cases was needed before a useful and reliable character was found. Moreover, in a number of population problems serological and biochemical studies have not provided a solution yet, in spite of much effort. I may refer to the problem of the differentiation of the closely related herring groups in the North Sea, although recently the application of the non-specific esterases has provided encouraging results with regard to this species (Simonarson and Watts, 1969; Naevdal, 1970; Ridgway, Sherburne and Lewis, 1970).

The additional attention that fish recently have attracted from workers interested in evolutionary and functional aspects of biochemical polymorphisms, seems well worthwhile, if only because of the variety of species available for study, both in respect to phylogenetic and physiological parameters. Also, because of their relatively short generation time, at least in tropical species, fish may appear suitable experimental animals to those interested in models for genetic studies. The study of biochemical polymorphisms and the differentiation of populations in their natural environment, still appears a major challenge too, however.

## REFERENCES

Bailey, C. S., Cocks, G. T. and Wilson, A. C. (1969), Gene duplication in fishes: Malate dehydrogenases of salmon and trout. *Biochem. biophys. Res. Commun.*, **34.** 605–612.

Cushing, J. E. (1964), The blood groups of marine animals. *Advanc. Mar. Biol.*, **2.** 85–131.

Cushing, J. E. and Durall, G. L. (1957), Isoagglutination in fish. *Am. Nat.*, **91.** 121–126.

FAO, (1958), Production 1957. *Yb. Fish. Statist.*, **7.**

FAO, (1969), Catches and landings 1968. *Yb. Fish. Statist.*, **26.**

Frydenberg, O., Møller, D., Naevdal, G. and Sick, K. (1965), Haemoglobin polymorphism in Norwegian cod populations. *Hereditas*, **53.** 257–271.

Fujino, K. (1967), Review of subpopulation studies on skipjack tuna. Proc. a. Conf. west. Ass. St. Game Fish Commnrs, **47.** 349–371.

Fujino, K. (1970), Immunological and biochemical genetics of tunas. *Trans. Am. Fish. Soc.*, **99.** 152–178.

Fujino, K. (In press), Skipjack subpopulation identified by genetic characteristics in the western Pacific. Proc. CSK Symp., East-West Center, Honolulu, Hawaii, 1968.

Fujino, K. and Kang, T. (1968a), Serum esterase groups of Pacific and Atlantic tunas. *Copeia*, 1968, 56–63.

Fujino, K. and Kang, T. (1968b), Transferrin groups of tunas. *Genetics*, **59.** 79–91.

Fujino, K. and Kazama, T. K. (1968), The Y-system of skipjack tuna blood groups. *Vox Sang.*, **9.** 383–395.

Fujino, K. and Sprague, L. M. (1965), The Y blood groups system of the skipjack tuna (Katsuwonus pelamis) (Abstr.). *Genetics*, **52.** 444.

Jamieson, A. and Jonsson, J. (In press), The Greenland component of spawning cod at Iceland. Rapp. P.-v. Réun. Cons. perm. int. Explor. Mer. **161.**

Lee, J. Y. (1962), Données sur l'application de mèthodes sèrologiques à l'étude des races et des populations chez les tèlèostéens. *Thesis* Fac. Sci. Univ. Paris, 33 p. (Transl. by Dep. of Agric. Fish. Scotl., Mar. Lab., Aberdeen, No. 996, 1965).

Ligny, W. de (1969), Serological and biochemical studies on fish populations. *Oceanogr. Mar. Biol. Ann. Rev.*, **7.** 411–513.

LIN, CHYI-CHYANG, SCHIPMANN, G., KITTRELL, W. A. and OHNO, S. (1969), The predominance of heterozygotes found in wild goldfish of Lake Erie at the gene locus for sorbitol dehydrogenase. *Biochem. Genet.*, **3.** 603–607.

LUSH, I. E. (1969), Polymorphism of a phosphoglucomutase isoenzyme in the herring (Clupea harengus). *Comp. Biochem. Physiol.*, **30.** 391–395.

MØLLER, D. (1966), Genetic differences between cod groups in the Lofoten area. *Nature, Lond.*, **212.** 824 only.

MØLLER, D. (1968), Genetic diversity in spawning cod along the Norwegian coast. *Hereditas*, **60.** 1–32.

NAEVDAL, G. (1970), Distributions of multiple forms of lactate dehydrogenase, aspartate aminotransferase and serum esterase in herring from Norwegian waters. *FiskDir. Skr. Ser. HavUnders.*, **15.** 565–572.

QUIROZ-GUTIERREZ, A. and OHNO, S. (1970), The evidence of gene duplication for S-form NADP-linked isocitrate dehydrogenase in carp and goldfish. *Biochem. Genet.*, **4.** 93–99.

RIDGWAY, G. J. (1964), Salmon serology. *Ann. Rep. int. N. Pacif. Fish. Commn.*, 1962, 107–110.

RIDGWAY, G. J. (1966), A complex blood group system in salmon and trout. Proc. Xth Europ. Conf. Anim. Blood Groups Biochem. Polymorph. Paris, 361–365.

RIDGWAY, G. J., CUSHING, J. E. and DURALL, G. L. (1958), Serological differentiation of populations of sockeye salmon, Oncorhynchus nerka. *Spec. scient. Rep. U.S-Fish Wildl. Serv.* no. 257, 9 pp.

RIDGWAY, G. J., KLONTZ, G. W. and MATSUMOTO, C. (1962), Intraspecific differences in serum antigens of red salmon demonstrated by immunochemical methods. *Bull. int. N. Pacif. Fish. Commn.*, **8.** 1–13.

RIDGWAY, G. J., SHERBURNE, S. W. and LEWIS, R. D. (1970), Polymorphism in the esterase of atlantic herring. *Trans. Am. Fish. Soc.*, **99.** 147–151.

ROBERTS, F. L., WOHNUS, J. F. and OHNO, S. (1969), Phosphoglucomutase polymorphism in the rainbow trout, Salmo gairdneri. *Experientia*, **25.** 1109–1110.

SCHAEFER, M. B. (1962), Report on the investigations of the inter-American tropical tuna commission for the year 1961. *Ann. Rep. inter.-Am Tuna Trop. Commn.*, 1961, **44**–171.

SCOPES, R. K. and GOSSELIN-REY, C. (1968), Polymorphism in carp muscle creatine kinase. *J. Fish. Res. Bd. Can.*, **25.** 2715–2716.

SHARP, G. D. (1969), Electrophoretic study of tuna hemoglobins. *Comp. Biochem. Physiol.*, **31.** 749–755.

SICK, K. (1961), Haemoglobin polymorphism in fishes. *Nature, Lond.*, **192.** 894–896.

SICK, K. (1965a), Haemoglobin polymorphism of cod in the Baltic and the Danish Belt Sea. *Hereditas*, **54.** 19–48.

SICK, K. (1965b), Haemoglobin polymorphism of cod in the North Sea and the North Atlantic Ocean. *Hereditas*, **54.** 49–73.

SIMONARSON, B. and WATTS, D. C. (1969), Some fish muscle esterases and their variation in stocks of the herring (Clupea harengus L.). The nature of esterase variation. *Comp. Biochem. Physiol.*, **31.** 309–318.

SIMPSON, J. G. and SCHLOTFELDT, H. S. (1966), Algunas observaciones sobre las caracteristicas electroforeticas de la hemoglobina de anchoveta, Engraulis ringens, en Chile. *Investnes zool. chil.*, **13.** 21–45.

SPRAGUE, L. M. and HOLLOWAY, J. R. (1962), Studies of the erythrocyte antigens of the skipjack tuna (Katsuwonus pelamis). *Am. Nat.*, **96.** 233–238.

UTTER, F. M. (1969), Transferrin variants in Pacific hake (Merluccius productus). *J. Fish. Res. Bd. Can.*, **26.** 3268–3271.

UTTER, F. M. and HODGINS, H. O. (1969), Lactate dehydrogenase isozymes of Pacific hake (Merluccius productus). *J. exp. Zool.*, **172.** 59–68.

UTTER, F. M., STORMONT, C. J. and HODGINS, H. O. (1970), Esterase polymorphism in vitreous fluid of Pacific hake, Merluccius productus. *Anim. Blood Grps biochem. Genet.*, **1.** 69–82.

VANSTONE, W. E., ROBERTS, E. and TSUYUKI, H. (1964), Changes in the multiple hemoglobin patterns of some Pacific salmon, genus Oncorhynchus, during the parr-smolt transformation. *Can. J. Physiol. Pharmac.*, **42.** 697–703.

WILKINS, N. P. (1966), Immunology, serology and blood group research in fishes. Proc. Xth Europ. Conf. Anim. Blood Groups Biochem. Polymorph. Paris, 355–359.

# I. REPORTS ON GENERAL SUBJECTS

*XIIth Europ. Conf. Anim. Blood Groups Biochem. Polymorph., Bp., 1972 (pp. 69—75)*

# SPECIFICITIES OF REAGENTS FOR BLOODTYPING OF ANIMALS

J. Bouw and C. C. Oosterlee

**Laboratorium voor Bloedgroepen Onderzoek, Landbouwhogeschool Wageningen, The Netherlands**

## SUMMARY

**Bloodtyping reagents have been found to be useful tools for studies on genetic structures in animals. In this report the reliability and value of these reagents is discussed. The first studies have demonstrated that still more attention must be paid to the specificities of the reagents. The authors recommend chicken as a species of animals in which these studies can be made.**

## INTRODUCTION

From recent studies it has become evident that complex blood group systems can reveal interesting information about the composition of genetic structures in animals.

These genetic studies are, however, calling for answers to the question what the reagents used in the bloodgrouping work are detecting and distinguishing.

The reagents are in most cases prepared by injections of red blood cells into animals of the same species. Such serologic reagents have by now been prepared for red blood cell antigens in cattle, pigs, chickens, horses, sheep, fish and various species of experimental animals. In these species of animals a number of different blood group systems — each with its own genetic control — has been established with the produced reagents.

On the basis of data from previous studies in cattle, recent investigations in chickens and some additional data in pigs the authors intend to discuss the specificities of reagents for complex blood group systems in animals. With some examples will be demonstrated that these reagents are to be suspected being polyvalent as a result of which they are not yet reliable as markers for specific parts of a choromosomal area.

## REAGENTS AND THEIR SPECIFICITY

Almost all reagents for bloodgrouping work in animals are produced in iso-immunization. In these immunizations the red blood cells of one animal are repeatedly injected into animals of the same species. In some species these reagents are in vitro causing hemolytic reactions (cattle), in others

TABLE 1

*Number of observed A—E genotypes and percentages in the pure strains*

| | White leghorn strains | | | | | Medium heavy strains | | | | |
|---|---|---|---|---|---|---|---|---|---|---|
| A—E types | $W_1$ | $W_2$ | $W_3$ | $W_4$ | A—E types | A | $R_1$ | $R_2$ | $R_3$ | H |
| $A_1E_2$ | | | 2 ( 3.0) | | $A_2E_2$ | 44 (53.6) | 56 (58.3) | | 21 (22.8) | 25 (33.7) |
| $A_1E_3$ | 16 (20.7) | 2 ( 2.2) | | | $A_2E_3$ | 25 (30.4) | | | | |
| $A_1E_5$ | | | 3 ( 4.5) | 4 ( 6.1) | $A_2E_4$ | | | 11 (12.7) | 26 (28.2) | |
| $A_2E_2$ | 26 (33.7) | 51 (57.3) | 54 (81.8) | 3 ( 4.5) | $A_2E_5$ | | | 2 (2.3). | | |
| $A_2E_3$ | 22 (28.5) | 1 ( 1.1) | | | $A_3E_2$ | 1 ( 1.2) | 7 ( 7.2) | 21 (24.4) | 6 ( 6.5) | |
| $A_2E_5$ | | | 1 ( 1.5) | | $A_3E_4$ | | | 1 ( 1.1) | | |
| $A_4E_1$ | 7 ( 9.0) | 5 ( 5.6) | | | $A_3E_5$ | | 22 (22.9) | 8 ( 9.3) | | 20 (27.0) |
| $A_4E_2$ | 4 ( 5.1) | | 1 ( 1.5) | 11 (16.6) | $A_4E_1$ | | | 14 (16.2) | | 3 ( 4.0) |
| $A_4E_3$ | 2 ( 2.5) | 11 (12.3) | | | $A_4E_2$ | 2 ( 2.4) | 3 ( 3.1) | 15 (17.4) | 1 ( 1.0) | 24 (32.4) |
| $A_4E_6$ | | 19 (21.3) | 5 ( 7.5) | 18 (27.2) | $A_4E_4$ | | 2 (2.0) | 1 ( 1.1) | 26 (28.2) | |
| $A_4E_6$ | | | | 20 (30.3) | $A_5E_1$ | | 1 ( 1.0) | | | 2 ( 2.7) |
| | | | | | $A_5E_2$ | 8 ( 9.7) | | 13 (15.1) | 11 (11.9) | |
| | | | | | $A_5E_4$ | 2 ( 2.4) | 5 ( 5.2) | | 1 ( 1.0) | |
| Total | 77 | 89 | 66 | 66 | | 82 | 96 | 86 | 92 | 74 |

W = White Leghorn
A = Australorp
R = Rhode Island Red
H = New Hampshire

agglutination (chickens) and in some both agglutination and hemolysis (pigs).

The red blood cells of animals are carrying a large variation of substances which can act as antigens when injected into animals of the same species. To avoid the production of a variation of different antibodies in the blood-serum of the recipients the workers on blood groups of chicken usually produce their reagents within inbred lines. From practical experiences it is well-known that within such inbred lines the production of antisera with the same specificity is very well possible for the AE system.

Since there is no question about inbred lines in cattle the workers in this field usually have to work with antisera in which two or more different specificities of antibodies are to be expected. The activity of the antibodies can be eliminated with absorptions. In these absorptions the antibodies to be eliminated are brought together with red blood cells carrying the corresponding antigens and not carrying antigens for which a purified antiserum must be produced. When after the purification by ways of absorptions the sera are reacting only with red blood cells which are capable to absorb all reactivity for all other positively reacting cells the antiserum is considered to be a specific reagent. The methods for the production of reagents for bloodtyping in horses, sheep and pigs are based upon the same principles as those for cattle.

For bloodtyping in cattle 34 hemolytic reagents have been produced for the blood group system B. The substances detected with these reagents are designated: $\underline{\underline{B}}$, $\underline{\underline{G}}$, K, $\underline{I}_1$, $\underline{I}_2$, $\underline{O}_1$, $O_2$, $O_3$, $\underline{P}_1$, $\underline{P}_2$, $\underline{Q}$, $\underline{T}_1$, $\underline{T}_2$, $Y_1$, $\underline{Y}_2$, $\underline{A}'$, $B'$, $D'$, $E'_1$, $E'_2$, $\underline{\underline{E}}'_3$, $\overline{F}'$, $\underline{G}'$, $\overline{\underline{I}}'$, $\underline{J}'_1$, $\underline{J}'_2$, $\underline{K}'$, $\underline{O}'$, $\overline{\underline{D}}'$, $\underline{\underline{Q}}'$, $\overline{\underline{Y}}'$, $\underline{B}''$ and $\underline{G}''$.

For the underlined factors an international agreement has been reached in a series of comparison tests. This means that identical specificities of antibodies for these factors have been produced in a variety of laboratories working with different breeds of cattle. For the non-underlined factors $O_2$, $O_3$, $Y_1$, $E'_1$ and $E'_2$ many futile attempts have been made in order to come to internationally comparable reagents. FIORENTINI and BOUW (1968) demonstrated that a number of these B system reagents — non-underlined as well as underlined ones — are detecting two or more distinctly different antigenic substances.

For the bloodgrouping work in pigs a series of 6 agglutinating reagents for the system E have been accepted internationally in comparison tests. The substances detected with these reagents are designated $E_a$, $E_b$, $E_d$, $E_e$, $E_f$, and $E_g$. HRADECKÝ and HOJNÝ (1970) are now demonstrating on the basis of their studies with mini pigs that the reagent anti-$E_d$ is in fact detecting at least two different antigenic substances designated $E_j$ and $E_k$.

At this moment a series of 11 agglutinating reagents are generally in use at Wageningen for the AE blood group system in chicken. The substances detected with these reagents are designated: $A_1$, $A_2$, $A_3$, $A_4$, $A_5$, $E_1$, $E_2$, $E_3$, $E_4$, $E_5$ and $E_6$. MCDERMID and OOSTERLEE (1970) are reporting about the international comparison tests in which agreements for the specificities of most of these reagents were reached. For the AE system the reagents at Wageningen have been produced in birds of which the composition of the AE blood types is presented in Table 1.

TABLE 2

*Genotypes of donor-recepient combinations*

| Donor/recepient | Genotypes | Relationship | Strain | Reagent |
|---|---|---|---|---|
| 7560 (7553) | $A^1E^3/A^2E^2(A^2E^2/A^4E^3)$ | full sibs | $W_2$ | $A_1$ |
| 7545 (7549) | $A^4E^3/A^2E^2(A^2E^2/A^1E^3)$ | full sibs | $W_2$ | $A_4$ |
| 1951 (1947) | $A^2E^2/A^5E^2(A^2E^2/A^3E^2)$ | half sibs | $R_3$ | $A_5$ |
| 1757 (1745) | $A^2E^2/A^2E^2(A^2E^1/A^2E^3)$ | half sibs | A | $E_2$ |
| 1764 (1744) | $A^2E^3/A^5E^2(A^2E^2/A^5E^2)$ | half sibs | A | $E_3$ |
| 1963 (6115) | $A^2E^2/A^4E^4(A^2E^2/A^2E^2)$ | no relation | $R_1$ | $E_4$ |
| 300 (294) | $A^2E^5/A^2E^2(A^2E^2/A^2E^2)$ | half sibs | $W_3$ | $E_5$ |

In Table 2 a series of donor-recipient combinations are presented which have been found appropriate for the production of some of the AE system reagents.

OOSTERLEE and BOUW (1968) have described the problems which are to be faced when the reagents produced in one laboratory are to be used in tests with birds from other places. In their report is indicated already that immunizations within or between pure bred strains can lead to very well reproducable antisera and that these antisera can react with a large variety of antigens on chicken red blood cells. In the own strains the reagents can react rather specific, their broad specificity becomes apparent in many cases in tests with bloodsamples of other strains of animals.

"Cross-reactivity" is the term used in animal blood group studies when the reagents are showing — usually weak — reactions with antigens other than those for which the antibodies were intended to be produced.

## DETECTION OF ANTIGENS

BOUW and FIORENTINI (1968) demonstrated on the basis of a series of cases of crossing-over that the reagents for the blood group system B of cattle are detecting a number of antigens which are controlled by different parts of the so-called B locus. A tentative map of the location of a series of genes controlling antigenic factors has been made up by BOUW and FIORENTINI (1968).

These authors further demonstrated that some of the internationally well-known reagents are detecting two or more substances with a separate genetic control. FIORENTINI et al. (1970) are showing that the reagent anti-$E_3'$ is in fact reacting with 4 different antigenic factors $E_1'$, F', G" and $H_{12}$, while BOUW and FIORENTINI already demonstrated that at least $E_1'$ and F' are controlled by separate genes.

For the blood group system E in pigs ANDRESEN (1963) already postulated a genetic system with 3 separate parts each controlling a closed system of 2 alleles: $E^a$–$E^d$, $E^b$–$E^c$ and $E^f$–$E^g$. This postulation was confirmed by RASMUSEN (1963) who observed 3 cases of crossing-over between $E^f$–$E^g$ and the rest of the locus.

TABLE 3

*Specificities of reagents for the A—E system in chicken*

| type \ reagent | $A_1$ | $A_2$ GB. | $A_2$ | $A_3$ | $A_4$ GB. | $A_5$ | $A_6$ |
|---|---|---|---|---|---|---|---|
| $A_1$ | + | | + | | | | |
| $A_2$ | | + | + | | | | |
| $A_3$ | | | | + | | | + |
| $A_4$ | | | | | + | | + |
| $A_5$ | | | | | | + | + |

| type \ reagent | $E_2$ | $E_3$ | $E_4$ | $E_1$ | $E_6$ | $E_5$ |
|---|---|---|---|---|---|---|
| $E_2$ | + | | | + | | |
| $E_3$ | | + | | + | | |
| $E_4$ | | + | + | + | | |
| $E_1$ | | | | + | | |
| $E_6$ | | | | | + | + |
| $E_5$ | | | | | | + |

For the AE system in chicken OOSTERLEE and BOUW (1970) are demonstrating that there is a linkage between A and E which is comparable to the linkage on the B locus of cattle and the E locus of pigs. The authors are pointing out that there is not yet evidence that the antigenic factors of A and E are controlled by linked genes within these systems.

In Table 3 is demonstrated how the reagents for the AE system in chickens are reacting with the various blood types.

## DISCUSSION AND CONCLUSIONS

The methods for the production of reagents are indicating that a high rate of specificity is to be expected in the antisera of cattle. The number of antigenic factors and especially the transmission of these factors in combination with each other as genetic units have given support to the idea that the reagents in species like cattle and pigs are detecting single antigenic factors.

The observations of STONE and MILLER (1961) on the S system in cattle, those of FIORENTINI and BOUW on the B system and the findings of HRADECKÝ and HOJNÝ on the E system of pigs have brought us back to LANDSTEINER's (1945) point of view that the antibody key may fit in various similar antigen keyholes.

FIORENTINI and BOUW observed that different types of antigens belonging to the E′ specificity group can induce antibodies which are reacting — or cross-reacting — with all kinds of factors belonging to this specificity group. The specificity of the produced reagents is depending as well upon the structure of the recipient as upon the one of the donor. Since there is no question about pure bred strains of cattle and especially since the immunizations for cattle bloodtyping reagents have been performed in a large variety of breeds of cattle it is understandable that it is extremely difficult to produce reagents with identical specificities for this group of antigens.

Tables 1 and 2 are demonstrating that in chicken bloodgrouping work there are good possibilities to make donor-recipient combinations in which reproducible reagents can be prepared. When such reagents are showing cross-reactivity in tests with other strains of chicken it seems reasonable to assume that these strains are possessing only a part of the antigens present in the donor's blood.

The bloodgrouping work in cattle and pigs has been based mainly upon the control of parentages. The workers in this field have tried to detect as many genetically determined characteristics as possible to improve their possibilities. Each new specificity of reagents was considered as a valuable contribution to the work.

The chicken bloodgrouping work has so far been directed mainly upon the distinction of the genetic variants present in the more or less pure bred lines which had to be investigated. The interest in a further analysis of the antigenic components of the blood groups has been limited.

Especially in cattle it has become evident now that the bloodgrouping work can reveal interesting data for studies on genetic structures. The studies which have been started in this field have strongly emphasized that a profound knowledge of the specificity of the used antisera is indispensable. From the presented data it is clear that even the cattle bloodtyping reagents which were believed to be highly specific are actually to be suspected as polyvalant antibodies.

The reagents for the AE system in chicken as they are in use at Wageningen are presented in Table 3. This table is also showing that some of these reagents like anti-$A_4$ and anti-$E_1$ are reacting with different types of antigens.

On the basis of the presented data it seems likely to assume that further studies on the production of specific reagents for chicken red blood cells can reveal much more data about the antigenic substances of the AE system and about their genetic control.

Oosterlee and Bouw (1970) already demonstrated how chicken can be used in studies on linkage in blood group studies.

In larger farm animals like cattle and pigs the possibilities for studies on linkage of the blood group loci as well as for the production and analysis of reagents are rather limited. On the basis of the reported data the authors strongly recommend chicken for further studies in this field.

## REFERENCES


Andresen, E. (1963), A study of blood groups of the pig. 229 PP Munksgaard — Copenhagen *Thesis.*

Bouw, J. and Fiorentini, A. (1970), Structure of loci controlling complex blood group systems in cattle. Proc. XIth Europ. Conf. Anim. Blood Groups Biochem. Polymorph. Warsaw 1968. 109–114.

FIORENTINI, A. and BOUW, J. (1970), Specificities of antibodies detecting antigenic substances on cattle red blood cells. Proc. XIth Europ. Conf., Anim. Blood Groups Biochem. Polymorph. Warsaw 1968 117–122.

FIORENTINI, A., CARENZI, C., ROGNONI, G. and BOUW, J. (1972), Misleading reagents for cattle bloodtyping. Proc. XIIth Europ. Conf. Anim. Blood Groups Biochem. Polymorph. — Budapest 197.

HRADECKÝ, J. and HOJNÝ, J. (1972), Proc. XIIth Europ. Conf. Anim. Blood Groups Biochem. Polymorph. Budapest 1970.

LANDSTEINER, K. (1945), The specificity of serological reactions Rev. Ed. Harvard University Press — Cambridge.

MCDERMID, E. and OOSTERLEE, C. C. (1972), Developments in comparison of chicken bloodtyping reagents. Proc. XIIth Europ. Conf. Anim. Blood. Groups Biochem. Polymorph. Budapest 1970.

OOSTERLEE, C. C. and BOUW, J. (1970), Detection of alleles in chicken strains and hybrids. Proc. XIth Europ. Conf. Anim. Blood Groups Biochem. Polymorph. Warsaw 1968. 389–396.

OOSTERLEE, C. C. and BOUW, J. (1972), Structure of loci for blood groups in animals. Proc. XIIth Europ. Conf. Anim. Blood Groups Biochem. Polymorph. Budapest 1970.

RASMUSEN, B. A. (1963), Irregularities in transmission of E alleles in pigs. *Immunogenetics Letter* **3.** 31–32.

STONE, W. H. and MILLER, W. J. (1961), Naturally occuring iso-antibodies of the S blood group system in cattle. *J. Immun.* **86.** 165–169.

*XIIth Europ. Conf. Anim. Blood Groups Biochem. Polymorph., Bp., 1972 (pp. 77–82)*

# STRUCTURE OF LOCI FOR BLOOD GROUPS IN ANIMALS

C. C. Oosterlee and J. Bouw

**Laboratorium voor Bloedgroepen Onderzoek, Landbouwhogeschool, Wageningen, The Netherlands**

## SUMMARY

**Irregular transmissions of blood groups of the A and E systems in chickens are described. The irregularities are considered to result from crossing over within the chromosomal region in which the genes controlling the A and E blood groups are closely linked.**

**On the basis of analyses carried out by Bouw and Fiorentini on the structure of the blood group locus B of cattle and the data of the A and E loci of chickens, the authors conclude that a number of complex blood group loci in animals are controlled by series of closely linked genes.**

## INTRODUCTION

At the XIth Conference of Animal Blood Groups at Warsaw 1968, Bouw and Fiorentini (1970) reported analysis of the structure of the blood group locus B in cattle. A tentative map was presented with a linear order for the genes or gene complexes which control the blood groups of this system. The construction of this map was based on irregular transmission of the B locus blood groups. Bouw et al. (1964) proposed that most of these irregularities were explicable by supposing linked genes in which each gene corresponded to a part of the complex blood group.

In chickens twelve autosomal blood group systems were described (Briles, 1962; McDermid, 1964) in which the A and E system (Briles, 1958) and probably the D–H and C–P are linked. Briles estimated the linkage between the A and E systems to have the value of approximately one crossover unit. For the D–H and C–P, the estimates of crossover units are respectively 40.3 and 40.0 (Briles, 1962), indicating that these loci are not closely linked.

Linkage between blood group loci has been reported also in pigs (Andresen and Baker, 1964). Rasmusen (1963) observed three cases of crossing over between parts of the gene complex controlling the E system in pigs. As in our laboratory as yet little information is available on pig blood groups, only data on cattle and chicken blood group loci have been taken into consideration.

Comparing the test results for the genotypes of the A and E systems in 1000 parent-offspring combinations a total of 10 irregular transmissions have been found.

## MATERIAL AND METHODS

In genotype-environment interaction studies in the Department of Poultry Husbandry of the Agricultural University of Wageningen, the blood groups of a White Leghorn (WL) and medium heavy (MH) population of hybrids were determined. The WL and MH populations are both crossbreds of two commercial hybrids. All four commercial hybrids are three way crosses. The construction process of the populations and the bloodtyping procedure were described by OOSTERLEE et al. (1970). The determination of the A and E system alleles were reported by BOUW and OOSTERLEE (1970). The numbers of birds tested are presented in Table 1.

In many cases, no crossovers could be detected due to homozygosity of one parent for one or both systems. The number of exclusive sire/dam combinations indicates the number of birds for which a crossover could be detected for one parent anyhow. On the basis of these combinations the exclusive A–E combinations were calculated.

TABLE 1

*Number of birds tested, number of exclusive sire-dam combinations and number of exclusive A—E gene combinations in White Leghorn and medium heavy hybrids*

| Hybrid | Sires | Dams | Offspring | Exclusive sire/dam combinations | Exclusive A—E combinations |
|---|---|---|---|---|---|
| WL | 53 | 207 | 791 | 507 | 708 |
| MH | 48 | 207 | 794 | 503 | 694 |

The mating groups consisted of one cock with an average of eight hens kept in conventional laying houses. All fertile eggs were hatched in small cages. Doubtful registrations were excluded. Crossovers were accepted exclusively in those cases in which no other sire/dam combination with a normal transmission would fit within the mating group. On the basis of the B system blood groups a second pedigree control was performed.

## RESULTS

Table 2 shows the observed and expected combinations of the A and E alleles of the WL and MH offspring, hatched in 1969, with the calculated chi-square values of the deviations.

The data of Table 2 indicate a linkage between the A and E genes. The combined transmission of A and E genes can be followed up by studying family groups.

An example of a WL and MH family group is presented in Table 3. The underlined AE combinations indicate two recombinants.

In 1010 exclusive sire/dam combinations, with a total of 1402 exclusive A–E gene combinations, 10 irregularities were found. All irregularities can

TABLE 2

*Observed and expected combinations of A and E system alleles in White Leghorn and medium heavy hybrids*

| WL | | | | MH | | | |
|---|---|---|---|---|---|---|---|
| allele | observed | expected | chi-square | allele | observed | expected | chi-square |
| $A^1E^1$ | 13 | 39 | 18.00 | $A^1E^1$ | — | 1 | 1.00 |
| $A^1E^2$ | 4 | 66 | 58.24 | $A^1E^2$ | 9 | 5 | 3.20 |
| $A^1E^3$ | 68 | 15 | 187.25 | $A^1E^3$ | — | 2 | 2.00 |
| $A^1E^5$ | 133 | 48 | 150.52 | $A^1E^4$ | — | 1 | 1.00 |
| $A^1E^6$ | 9 | 60 | 43.35 | $A^1E^5$ | — | 1 | 1.00 |
| $A^2E^1$ | 8 | 66 | 50.95 | $A^2E^1$ | 32 | 59 | 12.35 |
| $A^2E^2$ | 345 | 111 | 493.29 | $A^2E^2$ | 437 | 432 | 0.46 |
| $A^2E^3$ | 26 | 25 | 0.04 | $A^2E^3$ | 257 | 181 | 31.91 |
| $A^2E^5$ | 4 | 81 | 73.19 | $A^2E^4$ | 116 | 109 | 0.45 |
| $A^2E^6$ | 2 | 101 | 97.03 | $A^2E^5$ | 2 | 73 | 69.05 |
| $A^3E^1$ | 19 | 4 | 56.25 | $A^3E^1$ | 20 | 22 | 0.18 |
| $A^3E^2$ | 1 | 6 | 4.16 | $A^3E^2$ | 106 | 156 | 16.02 |
| $A^3E^3$ | — | 1 | 1.00 | $A^3E^3$ | 59 | 67 | 0.95 |
| $A^3E^5$ | — | 4 | 4.00 | $A^3E^4$ | 2 | 40 | 36.10 |
| $A^3E^6$ | — | 5 | 5.00 | $A^3E^5$ | 124 | 27 | 348.48 |
| $A^4E^1$ | 230 | 152 | 40.02 | $A^4E^1$ | 15 | 9 | 4.00 |
| $A^4E^2$ | 106 | 274 | 103.00 | $A^4E^2$ | 97 | 61 | 21.83 |
| $A^4E^3$ | 11 | 63 | 42.92 | $A^4E^3$ | — | 26 | 26.00 |
| $A^4E^5$ | 194 | 200 | 0.18 | $A^4E^4$ | 9 | 16 | 3.06 |
| $A^4E^6$ | 407 | 234 | 127.90 | $A^4E^5$ | — | 11 | 11.00 |
| | | | | $A^5E^1$ | 37 | 15 | 32.26 |
| | | | | $A^5E^2$ | 100 | 104 | 0.15 |
| | | | | $A^5E^3$ | 3 | 44 | 38.20 |
| | | | | $A^5E^4$ | 67 | 27 | 59.25 |
| | | | | $A^5E^5$ | 2 | 18 | 14.22 |
| Total | 1580 | | 1556.29<br>df. 19 | | 1494 | | 734.12<br>df. 24 |

be explained by assumption of a crossover between the linked A and E genes with a crossover percentage of 0.7%.

Comparing the 10 irregular transmissions and the number of the most frequent AE alleles in the two populations (Table 4) it seems likely that in some AE combinations the irregularities are more frequent, WL : $A^1E$, MH : $A^5E$ and MH : $E^5A$ and perhaps WL : $E^3A$.

TABLE 3

*The inheritance of A and E genes in a White Leghorn family and a family of medium heavy birds with two cases of crossing over (The column next to the offspring indicates the observed number of identical genotypes)*

| WL | | | | MH | | | |
|---|---|---|---|---|---|---|---|
| sire | dams | offspring | | sire | dams | offspring | |
| $A^2E^2/A^4E^5$ | $A^1E^5/A^2E^3$ | $A^2E^2/A^1E^5$ | 2× | $A^3E^5/A^5E^1$ | $A^2E^2/A^4E^1$ | $A^3E^5/A^2E^2$ | 2× |
| | | $A^4E^5/A^1E^5$ | 2× | | | $A^5E^1/A^2E^2$ | 2× |
| | | $A^2E^2/\underline{A^1E^3}$ | 1× | | $A^2E^2/A^5E^4$ | $A^3E^5/A^5E^4$ | 3× |
| | $A^4E^5/A^4E^6$ | $A^2E^2/A^4E^5$ | 2× | | | $A^5E^1/A^2E^2$ | 1× |
| | | $A^4E^5/A^4E^6$ | 2× | | | $A^5E^1/A^5E^4$ | 1× |
| | $A^2E^2/A^1E^5$ | $A^2E^2/A^1E^5$ | 2× | | $A^2E^2/A^2E^3$ | $A^3E^5/A^2E^2$ | 2× |
| | | $A^4E^5/A^2E^2$ | 2× | | | $A^5E^1/A^2E^2$ | 1× |
| | | $A^4E^5/A^1E^5$ | 1× | | | $A^5E^1/A^2E^3$ | 1× |
| | $A^2E^2/A^4E^6$ | $A^2E^2/A^2E^2$ | 1× | | | $\underline{A^3E^1}/A^2E^2$ | 1× |
| | | $A^4E^5/A^2E^2$ | 1× | | $A^3E^5/A^5E^4$ | $A^5E^1/A^5E^4$ | 3× |
| | | $A^4E^5/A^4E^6$ | 1× | | | | |
| | $A^2E^2/A^4E^6$ | $A^2E^2/A^2E^2$ | 1× | | | | |
| | | $A^4E^5/A^2E^2$ | 2× | | | | |

TABLE 4

*Observed number of most frequent A—E genotypes in White Leghorn and medium heavy hybrids and the crossovers observed in the parents' A—E genotypes*

| WL | | | MH | | |
|---|---|---|---|---|---|
| most frequent genotypes | observed number | crossing over parents | most frequent genotypes | observed number | crossing over parents |
| $A^1E^3$/[ | 64 | 2×(a, b) | $A^2E^1$/[ | 19 | |
| $A^1E^5$/[ | 82 | 1×(c) | $A^2E^2$/[ | 200 | 4×(d, h, i, j) |
| $A^2E^2$/[ | 208 | | $A^2E^3$/[ | 165 | |
| $A^2E^3$/[ | 26 | 2×(a, b) | $A^2E^4$/[ | 67 | |
| $A^4E^1$/[ | 135 | | $A^3E^2$/[ | 67 | |
| $A^4E^2$/[ | 47 | | $A^3E^3$/[ | 45 | |
| $A^4E^5$/[ | 104 | | $A^3E^5$/[ | 95 | 5×(e, f, g, i, j) |
| $A^4E^6$/[ | 216 | | $A^4E^2$/[ | 56 | |
| | | | $A^5E^1$/[ | 26 | 2×(g, h) |
| | | | $A^5E^2$/[ | 55 | 1×(k*) |
| | | | $A^5E^4$/[ | 54 | 3×(d, e, f) |

A locus ←——×——×——→ E locus

Probable ⟶ $A^5A^1$ $E^3E^5$

* detected in a pure strain

## DISCUSSION AND CONCLUSIONS

According to Briles (1958), the E antigens of chicken are controlled by a locus separated from the A locus but linked to it by about one crossover unit. A crossover unit is defined as the frequency of exchange of 1% between two pairs of linked genes. Our data confirm this crossover percentage of Briles.

No exact proof has as yet been found whether we have to do with one locus or two closely linked loci. Only one possible irregular transmission was observed within the A region of the chromosome part. It must be kept in mind, however, that within the WL and MH hybrids the number of A and E alleles is limited. Also, some alleles do have a rather low gene frequency and certain alleles do give cross-reactivity, perhaps hiding cases of crossovers within loci. Therefore, the number of those sire/dam combinations which can be considered exclusive of detecting crossovers within the A and E regions of the chromosomes is limited to about 2% of the material. The production of a new reagent, detecting a sixth A system allele, will most likely make possible to increase the number of exclusive sire/dam combinations for the A system.

No evidence has been found of a serological relationship between A and E antigens. This does not prove, however, that the genes controlling these antigens do not belong to one locus. A number of antigenic factors of the blood group system B of cattle do not have serological relationship at all and are located on one locus. As has been demonstrated for the end of the X chromosome in *Drosophila melanogaster*, within one crossover unit seven genes with entirely different end results have been located (Gardner, 1968). Some of these genes are separated by 0.1 crossover unit.

Although in our data on chickens show the number of irregular transmissions to be limited, the frequency of irregular transmission of the $A^5$ allele for instance suggests that more than one place of the A region of the chromosome is involved in controlling the A locus alleles. Even the distance between the $A^5$ and other A locus genes can be greater than between A and E genes, although the 10 irregular transmissions indicate a crossing over of the chromosome in the region between the A and E parts. More data have to be accumulated, however, before conclusions on a linear order can be drawn.

On the basis of the data on cattle (Bouw et al., 1970) and those on chickens, the authors conclude that in animals a number of complex blood group loci are controlled by series of closely linked genes.

## REFERENCES

Andresen, E. and Baker, L. N. (1964), The C blood group system in pigs and the detection and estimation of linkage between the C and J systems. *Genetics*, **49**. 379–386.

Bouw, J. and Fiorentini, A. (1970), Structure of loci controlling complex Blood Group Systems in cattle. XIth Europ. Conf. Anim. Blood Groups and Biochem. Polymorph. Warsaw 1968.

Briles, W. E. (1958), A new blood group system E, closely linked with the A system in chickens. *Poult. Sci.* **37.** 1189.

Briles, W. E. (1962), Additional blood group systems in chickens. *Ann. N. Y. Acad. Sci.*, **97.** 173–178.

McDermid, E. M. (1964), Immunogenetics of the chicken. *Vox Sang.*, **9.** 249–267.

Oosterlee, C. C., van Albada, J. and Bouw, J. (1970), Results of chicken blood group research in a genotype-environment interaction study. In press.

Rasmusen, B. A. (1963), Irregularities in transmission of E alleles in pigs. *Immunogenetics Letter*, **3.** (2) 31–32.

*XIIth Europ. Conf. Anim. Blood Groups Biochem. Polymorph., Bp., 1972 (pp. 83—87)*

# SOME RESULTS FROM COMPUTER TREATMENT OF BLOOD GROUP DATA

B. LARSEN

Department of Physiology, Endocrinology and Bloodgrouping
The Royal Veterinary and Agricultural University
Copenhagen, Denmark

## INTRODUCTION

With the aim of analysing the data on bovine blood groups and protein polymorphisms which has been accumulated in our laboratory, the development of computer programs to handle this material has been initiated. So far a series of programs has been developed by means of which pedigrees can be examined for the inheritance of the various alleles controlling blood groups and polymorphic proteins and the results utilized for analyses of genetic linkage between the polymorphic systems (LARSEN and CAWOOD, 1970). Using this set of programs analyses for linkage between ten bovine blood group systems and seven polymorphic protein systems (Hb, Tf, Am, β-Lg, $\alpha_{s_1}$-Cn, β-Cn, and K-Cn) were performed. The lod score method for the detection and estimation of linkage (MORTON, 1955) was applied as described elsewhere together with the details about the data (LARSEN, 1970). In the following the results from these analyses will be mentioned and discussed in relation to previous investigations on linkage of polymorphic systems in cattle (GROSCLAUDE et al., 1964, 1965; LARSEN, 1966, 1969, LARSEN and THYMANN, 1966a, 1966b, HINES et al., 1969).

## RESULTS AND DISCUSSION

In Table 1 a summary of the results are given. The figures indicate the recombination frequency for which linkage of the various systems can be excluded, and for systems where significant lod scores for linkage were obtained, an estimate of the recombination frequency ($\hat{\theta}$) is given. In agreement with previous studies on linkage of polymorphic systems in cattle evidence for a close linkage were observed for the A versus Hb, the J versus β-Lg and between the $\alpha_{s_1}$-Cn, β-Cn and κ-Cn polymorphisms. The families relevant to these comparisons are grouped according to the most probable phase of the sire and summarized in Table 2. For one family in the A-Hb comparison the phase was determined from grandparents.

In the comparison of the A versus Hb comprising families from the present as well as previous investigation (LARSEN, 1966), new possible recombinants were not observed. As apparent from Table 2, three sires not closely related possess the $A^{AZ'}$ allele in combination with the $Hb^A$ gene, thus indicating

TABLE 1

*Summary of linkage analyses between 149 combinations of blood group loci and polymorphic protein loci and sex in cattle. Linkage closer than the recombination frequency indicated for each comparison has been excluded. For the established linkages an estimate of the recombination frequency ($\hat{\theta}$) is given*

| | Sex | β-Lg | κ-Cn | β-Cn | $\alpha_{s_1}$ Cn | Am | Tf | Hb | R'S' | Z | SU | M | L | J | FV | O | B |
|---|---|---|---|---|---|---|---|---|---|---|---|---|---|---|---|---|---|
| A | .43 | .13 | .19 | .14 | .13 | .28 | .34 | $\hat{\theta}\sim.02$ | .39 | .39 | .41 | .33 | .24 | .37 | .33 | .39 | .44 |
| B | .44 | .27 | .28 | .23 | .22 | .35 | .45 | .40 | .40 | .40 | .46 | .44 | .43 | .41 | .43 | .48 | |
| C | .42 | .30 | .27 | .21 | .25 | .34 | .42 | .39 | .38 | .45 | .46 | .41 | .36 | .39 | .38 | | |
| FV | .45 | .28 | .22 | .27 | — | .29 | .30 | .35 | .33 | .39 | .36 | .30 | .37 | .36 | | | |
| J | .29 | $\hat{\theta}\sim.04$ | .15 | .06 | — | .20 | .30 | .23 | .33 | .30 | .29 | .32 | .31 | | | | |
| L | .23 | .12 | .21 | .19 | — | .18 | .24 | .25 | .34 | .25 | .35 | .32 | | | | | |
| M | .35 | .14 | .21 | .08 | .03 | .25 | .30 | .20 | .31 | .37 | .35 | | | | | | |
| SU | .41 | .18 | .22 | .19 | .04 | .33 | .44 | .25 | .38 | .43 | | | | | | | |
| Z | .34 | .10 | .15 | .16 | .02 | .26 | .34 | .18 | .29 | | | | | | | | |
| R'S' | .35 | .22 | .26 | .20 | .19 | .33 | .36 | .29 | | | | | | | | | |
| Hb | .12 | .14 | .19 | .04 | — | .22 | .30 | | | | | | | | | | |
| Tf | .40 | .20 | .23 | .08 | .14 | .25 | | | | | | | | | | | |
| Am | .33 | .23 | .13 | .12 | .13 | | | | | | | | | | | | |
| $\alpha_{s_1}$-Cn | — | — | $\hat{\theta}\sim.00$ | $\hat{\theta}\sim.00$ | | | | | | | | | | | | | |
| β-Cn | — | .08 | $\hat{\theta}\sim.00$ | | | | | | | | | | | | | | |
| κ-Cn | — | .21 | | | | | | | | | | | | | | | |

TABLE 2

*Summary of families showing linkage relations of blood group and polymorphic protein systems in cattle. The families are grouped according to the most probable phase of the sire*

| No. of sires | Most probable phase for sire | Offspring grouped according to the allele combinations received from sire | | Total |
|---|---|---|---|---|
| | | parental | recombinant | |
| | *A system versus hemoglobin types* | | | |
| 15 | $A^A$, $Hb^A$/$A^{AH}$, $Hb^B$ | 92 | 1 | 93 |
| 9 | $A^A$, $Hb^B$/$A^{AH}$, $Hb^A$ | 25 | 0 | 25 |
| 2 | $A^{AZ'}$, $Hb^A$/$A^{AH}$, $Hb^B$ | 49 | 2 | 51 |
| 1 | $A^{AZ'}$, $Hb^A$/$A^A$, $Hb^B$ | 6 | 0 | 6 |
| 27 | total | 172 | 3 | 175 |
| | *$\alpha_{S1}$-Cn versus β-Cn* | | | |
| 4 | $\alpha_{S1}$-$Cn^B$ β-$Cn^B$/$\alpha_{S1}$-$Cn^C$, β-$Cn^A$ | 42 | 0 | 42 |
| | *$\alpha_{S1}$-Cn versus ϰ-Cn* | | | |
| 1 | $\alpha_{S1}$-$Cn^B$, ϰ-$Cn^A$/$\alpha_{S1}$-$Cn^C$, ϰ-$Cn^B$ | 2 | 0 | 2 |
| 3 | $\alpha_{S1}$-$Cn^B$, ϰ-$Cn^B$/$\alpha_{S1}$-$Cn^A$, ϰ-$Cn^A$ | 28 | 0 | 28 |
| 4 | total | 30 | 0 | 30 |
| | *β-Cn versus ϰ-Cn* | | | |
| 20 | β-$Cn^A$, ϰ-$Cn^A$/β-$Cn^B$, ϰ-$Cn^B$ | 102 | 0 | 102 |
| | *J system versus β-lactoglobulin* | | | |
| 6 | $J^J$, β-$Lg^A$/$J^-$, β-$Lg^B$ | 24 | 1 | 25 |
| 3 | $J^J$, β-$Lg^B$/$J^-$, β-$Lg^A$ | 12 | 0 | 12 |
| 3 | $J^s$, β-$Lg^A$/$J^a$, β-$Lg^B$ | 10 | 1 | 11 |
| 12 | total | 46 | 2 | 48 |

that crossing over between the A and the Hb systems may appear, or alternatively that two allele combinations, $A^{AZ'}$, $Hb^A$ and $A^A$ or $A^{AH}$, $Hb^B$, of *Bos indicus* origin remain in the Jersey cattle.

In the comparison of the casein polymorphisms recombinants were not observed and the 20 sires relevant to the β-Cn versus ϰ-Cn comparison were of the same phase. Corresponding to about 4% of recombination only two recombinants appeared in the J versus β-Lg comparison, which differ greatly from the 20% of recombination observed by HINES et al. (1969). In the families relevant to this comparison the segregation of the β-Lg type differs significantly from the expected 1 : 1, while the offspring from all sires determined as β-$Lg^A$/β-$Lg^B$ shows a perfect 1 : 1 segregation. This may suggest that the $\beta$-Lg is not related directly to the J substance or cellular J but to a system being related to or influencing the appearance of J substance. In this respect a parallel may be drawn to the Lutheran-secretor linkage in man, which first appeared as a Lutheran–Lewis linkage due to interaction in phenotypic effect of the Lewis and secretor systems. The correct interpretation of the $\beta$-Lg and J relationship may, therefore, await a better understanding of the genetics underlying the J system and studies on the relation to the Oc substance (SPRAGUE, 1958) and the serum phosphatase A (RENDEL and GAHNE, 1963).

From Table 1 it is seen that less than 25% recombination can be excluded between the Tf and Am loci. A possible loose linkage of these two systems, however, is indicated by a maximum lod score of 1.65 at $\hat{\theta}_1 = .37$, corresponding to a probability of 2.3% for the recombination frequency being .50 rather than .37. Because of the low a priori probability of detecting linkage, the level of significance has been set to .001 or lod score = 3 (MORTON, 1955). Among the 45 families with a total of 362 offspring included in the Tf versus Am comparison, 8 families with known linkage phase showed 36 offspring of parental type to 26 recombinants. Also for families with unknown phase positive lod scores were obtained and the data were homogeneous. Thus a loose linkage of the Tf and Am loci may seem possible.

Apart from the above mentioned the $\beta$-Cn versus Tf comparison (9 families and 57 offspring) yielded a positive lod score of 1.22 at $\hat{\theta}_1 = .26$, while lod score for $\varkappa$-Cn versus Tf were just above zero (23 families and 118 offspring). A similar relation was observed by HINES et al. (1969), indicating that the apparent relation of $\beta$-Cn and Tf is not simply due to chance. Although an explanation cannot be offered at present, the lod score value cannot be regarded as evidence for a possible genetic linkage, since the families relevant to this comparison showed a disturbed segregation of the $\beta$-Cn polymorphisms and a slight excess of $Tf^D$ genes were transmitted from sire to offspring.

For the remaining of the 149 comparisons the lod scores did not approach significance, and the figures of Table 1 indicate that moderate to loose linkage of these systems is not likely.

## REFERENCES

GROSCLAUDE, F., GARNIER, J., RIBADEAU-DUMAS, B. and JEUNET, R. (1964), Etroite dépendance des loci contrôlant le polymorphisme des caséines $\alpha_s$ et $\beta$. *C. R. Acad. Sci. (Paris)*, **259.** 1569–1571.

GROSCLAUDE, F., PUJOLLE, J., GARNIER, J. and RIBADEAU-DUMAS, B. (1965), Déterminisme génétique des caséines $\varkappa$ du lait de vache; étroite liaison du locus $\varkappa$-Cn avec les loci $\alpha_s$-Cn et $\beta$-Cn. *C.R. Acad. Sci. (Paris)*, **261.** 5229–5232.

HINES, H. C., KIDDY, C. A., BRUM E. W. and ARAVE, C. W. (1969), Linkage among cattle blood and milk polymorphisms. *Genetics*, **62.** 401–412.

LARSEN, B. (1966), Test for linkage of the genes controlling haemoglobin, transferrin and blood types in cattle. *Royal Vet. and Agric. University, Yearbook*, 41–48.

LARSEN, B. (1969), On linkage relations of blood groups in cattle. (In Danish, English summary). *Aarsberetn. Inst. Sterilitetsforsk.* 31–38.

LARSEN, B. (1970), Linkage relations of blood group and polymorphic protein systems in cattle. (In Danish, English summary). *Aarsberetn. Inst. Sterilitetsforsk.* (In press)

LARSEN, B. and THYMANN, M. (1966a), Studies on milk protein polymorphisms and the interaction of the controlling genes. Proc. Xth Europ. Conf. Anim. Blood Groups Biochem. Polymorph. Paris, 421–425.

LARSEN, B. and THYMANN, M. (1966b), Studies on milk protein polymorphisms in Danish cattle and the interaction of the controlling genes. *Acta vet. scand.*, **7.** 189–205.

LARSEN, B. and CAWOOD, P. B. (1970), Some computer programs for genetic analyses of bovine blood group data. (In Danish, English summary). *Aarsberetn. Inst. Sterilitetsforsk.* (In press)

MORTON, N. E. (1955), Sequential tests for the detection of linkage. *Amer. J. hum. Genet.*, **7.** 277–318.

RENDEL, J. and GAHNE, B. (1963), Interaction between phosphatases and the J blood groups in cattle. *Immunogenetics Letter*, **2.** 38–43.

SPRAGUE, L. M. (1958), On the recognition and inheritance of the soluble blood group property "Oc" of cattle. *Genetics*, **43.** 906–912.

## DISCUSSION

G. C. ASHTON: With 16 loci and 3 chromosomes marked, is this in keeping with expectation?

B. LARSEN: The number of linkage groups observed seems to agree with the number expected.

*XIIth Europ. Conf. Anim. Blood Groups Biochem. Polymorph., Bp., 1972 (pp. 89—94)*

# DISCRIMINANT FUNCTION AS A METHOD OF COMPARISON BETWEEN ANIMAL POPULATIONS USING BLOOD GROUPS AND PROTEIN VARIATIONS

H. GELDERMANN

Veterinary Institute of University Göttingen, GFR

## SUMMARY

**The differences in the frequencies of numerous phenotypes or genotypes from various gene systems were measured by discriminant function. Different animal populations or inbred lines can be separated and questionable populations or single individuals can be traced. The results can be presented diagrammatically.**

## INTRODUCTION

The discriminant analysis methods project elements of two populations into finite dimensions with the help of transformed criteria. This is done in such a way that both populations appear as point accumulations which are maximally separated (BAUER, 1954). Thus the criteria can be converted into a measurable unit. By means of this unit it is possible to separate different populations and to attach unknown individuals to one of the populations.

By means of a discriminant function, comparisons between animal populations can be carried out by using qualitative data, such as blood groups or protein variants (FISHER, 1950; JOHNSON, 1950). These comparisons can answer the question whether group I is closer to group II than to group III. One way to measure the distance between two groups is the Mahalanobis' Generalized Distance (RAO, 1952; MAJUMDAR and RAO, 1960).

## METHODS

The various criteria of one individual can be integrated by adding the products of $p$ single elements to each weighted factor $b_i$. By the addition of the values the unit $X$ is obtained:

$$X = \sum_{i=1}^{p} b_i x_i .$$

The $x_i$'s are derived from the absolute scores of qualitative data as described by JOHNSON (1950). The distance between means of the two values, $\overline{X}_1$ and $\overline{X}_2$, is defined by:

$$d = \sum_{i=1}^{p} b_i d_i .$$

Hence the mean difference ($d$) of the units is equal to the sum of the mean differences ($d_i$) multiplied by $b_i$ for each criterion. MAHALANOBIS named this expression $pD^2$.

To calculate the $b_i$'s, the sums of the products of the differences in both populations, $q_{ij}$, are derived from the formula

$$q_{ij} = \sum_{k=1}^{2} \sum^{nk} (x_{ik} - \overline{x}_{ik})(x_{jk} - \overline{x}_{jk}) .$$

The number of individuals in the first group are denoted by $n_1$ and those in the second group by $n_2$.

The $q_{ij}$'s are divided by the degrees of freedom, $n_1 + n_2 - 2$:

$$s_{ij} = \frac{q_{ij}}{n_1 + n_2 - 2} .$$

From this we obtain the variances, $s_{ij}$, which are used in the calculations of $b_i$'s:

$$\begin{aligned} b_1 s_{11} + b_2 s_{12} + \dots + b_p s_{1p} &= d_1 \\ b_1 s_{12} + b_2 s_{22} + \dots + b_p s_{2p} &= d_2 \\ \vdots \qquad \vdots \qquad\quad \vdots \quad & \quad \vdots \\ b_1 s_{1p} + b_2 s_{2p} + \dots + b_p s_{pp} &= d_p \end{aligned}$$

The GAUSS–DOOLITTLE process (WEBER, 1967) is used to solve these simultaneous equations. It is

$$\begin{aligned} b_1 &= d_1 c_{11} + d_2 c_{12} + \dots + d_p c_{1p} \\ b_2 &= d_1 c_{12} + d_2 c_{22} + \dots + d_p c_{2p} \\ \vdots \quad & \quad \vdots \qquad\quad \vdots \qquad\quad \vdots \\ b_p &= d_1 c_{1p} + d_2 c_{2p} + \dots + d_p c_{pp} \end{aligned}$$

and the distance of MAHALANOBIS becomes

$$pD^2 = \sum_{i=1}^{p} d_i c_{ii} + 2 \sum_{i=1}^{p} \left( \sum_{j<i}^{p} d_i d_j c_{ij} \right) .$$

To test the significance of the difference between group I and group II the following $F$-test was computed:

$$F = \frac{n_1 n_2}{n_1 + n_2} \cdot \frac{n_1 + n_2 - p - 1}{n_1 + n_2 - 2} \cdot D^2 .$$

For each comparison, the $X$ unit was calculated for each individual. The distributions of the units are presented in diagrams. The computations given were calculated by a special FORTRAN program for the IBM 7040 computer. With this discriminant analysis, up to 50 single criteria can be used per individual.

## RESULTS

The use of the discriminant analysis to compare animal populations on the basis of blood groups and protein variants is demonstrated in the following examples:

*First:* In German registered "Schwarzbunte" (1651 animals), "Rotbunte" (832 animals) and "Fleckvieh" (161 animals) breeds, transferrin and post-albumin types were determined. From the criteria the following values of $D$ were found:

| Comparison between | $D$ | $F$-test |
|---|---|---|
| "Schwarzbunte" and "Rotbunte" | 0.1422 | 11.2 $P < 0.005$ |
| "Schwarzbunte" and "Fleckvieh" | 0.5709 | 46.0 $P < 0.005$ |
| "Rotbunte" and "Fleckvieh" | 0.5855 | 47.7 $P < 0.005$ |

Regarding the $D$ values, the breeds "Schwarzbunte" and "Rotbunte" are closer related to each other than either of them to the "Fleckvieh" breed. The unit distributions are given in Fig. 1. None of the comparisons had distributions which were clearly separable. However, the differences between the investigated groups can be shown graphically.

*Second:* German pigs of the breeds "Landrasse" (157 animals), "Angler Sattelschwein" (128 animals) and "Piétrainschwein" (204 animals) were examined as to the blood group systems A, D, E, F, G, H, I, K, L, and M. Altogether 30 different phenotypes of the 10 systems were used in a discriminant analysis. From this the following values of D were computed:

| Comparison between | $D$ | $F$-test |
|---|---|---|
| "Landrasse" and "Angler Sattelschwein" | 1.5456 | 151.2 $P < 0.005$ |
| "Landrasse" and "Piétrainschwein" | 1.3629 | 151.5 $P < 0.005$ |
| "Angler Sattelschwein" and "Piétrainschwein" | 2.0307 | 295.8 $P < 0.005$ |

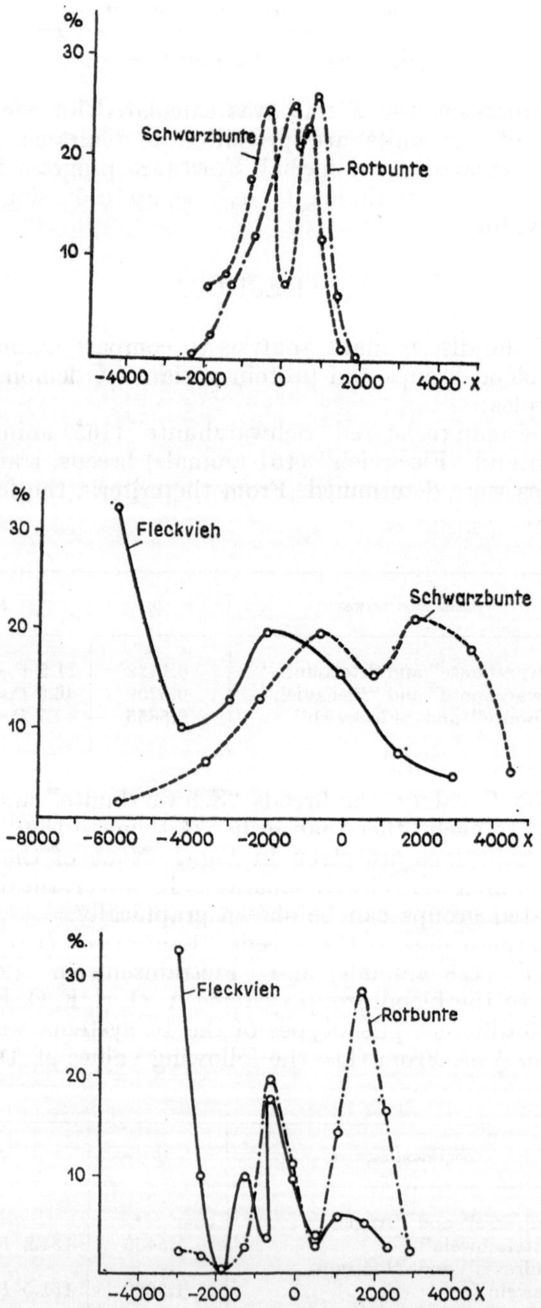
%
30
20
10
Schwarzbunte
Rotbunte
-4000
-2000
0
2000
4000 X
%
30
20
10
Fleckvieh
Schwarzbunte
-8000
-6000
-4000
-2000
0
2000
4000 X
%.
30
20
10
Fleckvieh
Rotbunte
-4000
-2000
0
2000
4000 X

**Fig. 1. Unit frequency distributions for comparisons between cattle breeds**

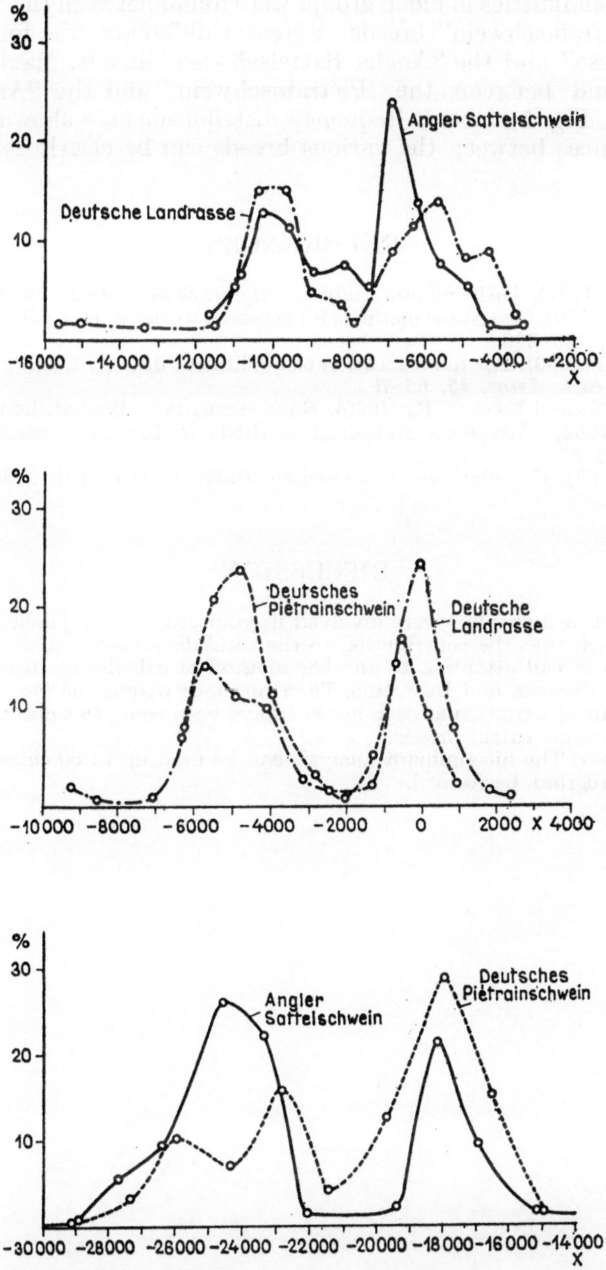
%
30
20
10
Angler Sattelschwein
Deutsche Landrasse
-16000 -14000 -12000 -10000 -8000 -6000 -4000 -2000
X
%
30
20
10
Deutsches Piètrainschwein
Deutsche Landrasse
-10000 -8000 -6000 -4000 -2000 0 2000 X 4000
%
30
20
10
Angler Sattelschwein
Deutsches Piètrainschwein
-30000 -28000 -26000 -24000 -22000 -20000 -18000 -16000 -14000
X

**Fig. 2. Unit frequency distributions for comparisons between pig breeds**

The greatest similarities in blood groups were found between the "Landrasse" and the "Piétrainschwein" breeds. A greater difference was found between the "Landrasse" and the "Angler Sattelschwein" breeds. Maximum difference was found between the "Piétrainschwein" and the "Angler Sattelschwein". In Fig. 2 the unit frequency distributions are shown. The graduated differences between the various breeds can be clearly seen.

## REFERENCES

Bauer, R. K. (1954), Diskriminanzanalyse. *Allgem. Statist. Arch.*, **38.** 205–216.
Fisher, R. A. (1950), Statistical methods for research workers. 11th edition, Edinburgh and London, 285–295.
Johnson, P. O. (1950), The quantification of qualitative data in discriminant analysis. *J. Amer. Statist. Assoc.* **45.** 65–76.
Majumda, D. N. and Rao, C. R. (1960), Race elements in Bengal. London.
Rao, C. R. (1952), Advanced statistical methods in biometric research. London, 236–272; 286–378.
Weber, E. (1967), Grundriß der biologischen Statistik. 6th edition, Jena, 354–355.

## DISCUSSION

K. K. Kidd: How many loci were involved in your studies? Majalanobis distance is calculated such that the contribution to the total distance is equalized for all loci. I would like to call attention to another method of calculating distance proposed by Cavalli—Sforza and Edwards. Their methods weight the loci proportionally to the amount of variation at each locus. I have been using this alternative method in my studies on cattle breeds.

H. Geldermann: The discriminant analysis can be used up to 50 single criteria in a computer program performed.

# II. BLOOD GROUPS AND BIOCHEMICAL POLYMORPHISM IN CATTLE

*XIIth Europ. Conf. Anim. Blood Groups Biochem. Polymorph., Bp., 1972 (pp. 97–101)*

# MISLEADING REAGENTS FOR CATTLE BLOODTYPING

C. Carenzi, A. Fiorentini, G. Rognoni and J. Bouw*

Istitúto di Zootecnia Generale, Universita di Milano, Milan, Italy

* Laboratorium voor Bloodgroepen Onderzoek, Landbouwhogeschool, Vageningen, The Netherlands

## SUMMARY

**On the basis of studies on international tests for the comparison of bloodtyping reagents the authors demonstrate that some of the reagents which were considered to be mono-specific in fact react with 2 or more different antigenic factors. The authors emphasize the importance of further studies on the specificities of the cattle bloodtyping reagents, both for immunologic and genetic studies.**

## INTRODUCTION

Landsteiner et al. (1936) and Landsteiner (1945) already demonstrated that the specificity of reagents in immune sera is not always absolute. Eisen et al. (1964) found that even antibodies produced against single hapten species can be mixtures of antibody molecules with different specificities. Stone and Miller (1961) and Grosclaude (1965) demonstrated that this situation also exists for certain reagents for the blood group system S of cattle. After a complex study on the specificity of reagents for the blood group system B of cattle, Fiorentini and Bouw (1968) were able to demonstrate that some of the reagents for this system react with two or more different antigenic factors.

Starting from this point it will be demonstrated that some of the internationally accepted mono-specific reagents react with more than one single antigenic factor.

Continuing on the basis of the results presented by Fiorentini and Bouw in 1968, we intend to prove the thesis set out previously with some reagents prepared in the laboratories at Milan and Wageningen and used in the 12th International Comparison Test for Cattle Bloodtyping. The results of this comparison test demonstrate that various laboratories have produced reagents with similar specificities.

We further intend to demonstrate how a new reagent can show that a so-called "international reference reagent" in fact detects two or more different antigenic factors.

## MATERIAL AND METHODS

In the present work the reaction patterns of the reagents for the blood group system B as they are found in the 1969 Cattle Comparison Test are used. In Table 1 the patterns of reactions of groups of reagents with the same specificity are compared.

For the internationally acknowledged reagents the reaction pattern is reported following the result of the majority of the participating laboratories. The experimental reagents are labelled according to their identity. This identity is indicated in the table by a number following the symbols adopted by the laboratories in which the reagents are produced.

In Table 1 the reaction patterns of the single reagents have been grouped in conformity with their specificity. We have examined the reagents of E′ and F′ specificities, and also those of O′ specificity group.

Table 2 demonstrates how a newly produced reagent anti-M22 from the Milan laboratory reacts with various phenogroups possessing the internationally adopted factors G and/or $G_1$.

TABLE

| | | 1 | 2 | 3 | 4 | 5 | 6 | 7 | 8 | 9 | 10 | 11 | 12 | 13 |
|---|---|---|---|---|---|---|---|---|---|---|---|---|---|---|
| Part No. 1 | O′ | − | + | − | − | − | − | − | + | − | − | + | + | − |
| | $N_1$ ($Nf_7$/Nl.—$O'_1$/I.Mi) | − | + | − | − | − | − | − | + | − | − | + | − | − |
| | $N_2$ ($H_{10}$/Nl.—$M_{15}$/I.Mi) | − | − | − | − | − | − | − | − | − | − | + | + | − |
| Part No. 2 | $E'_3$ | + | + | + | + | − | − | + | − | − | − | + | + | + |
| | E′ | − | − | − | + | − | − | − | − | − | − | + | − | + |
| | F′ | − | − | − | − | − | − | − | − | − | − | + | + | − |
| | G″ | + | + | + | − | − | − | + | − | − | − | ± | ± | − |
| Part No. 3 | $E'_2$ | − | − | − | + | − | − | − | − | − | − | + | − | + |
| | $E'_1$ | − | − | − | + | − | − | − | − | − | − | − | − | + |
| | $N_3$ ($M_6$/I.Mi—$H_{12}$/Nl.—120/60/W. Gr. Gtt.) | − | − | − | − | − | − | − | − | − | − | + | − | − |
| Part No. 4 | F′ | − | − | − | − | − | − | − | − | − | − | + | + | − |
| | $N_3$ ($M_6$/I.Mi—$H_{12}$/Nl.—120/60/W. Gr. Gtt.) | − | − | − | − | − | − | − | − | − | − | + | − | − |
| | $N_4$ (F′/Nl.—$F'_3$/I.Mi) | − | − | − | − | − | − | − | − | − | − | + | + | − |

## RESULTS AND CONCLUSIONS

In the first part of Table 1 the reagent anti-O′ is clearly found to react only in cases in which also the reagents Nos. 1 and/or 2 react. On this basis it can be concluded that te reactivity of anti−O′ is conditioned by the reactivity of the reagents Nos. 1 and/or 2.

In the second part of Table 1 the reagents belonging to the E′ specificity group are brought together. A comparison of the reaction patterns of these reagents clearly shows that the same applies for anti-$E_3'$ as for anti-O′.

From a more specific study of the reaction patterns of anti-F′ and anti-$E_2'$, which have been shown to condition the reactivity for anti-$E_3'$, we derived the third and fourth part of our table.

In the third part the reaction pattern of anti-$E_2'$ is compared with those of anti-$E_1'$ and reagent No. 3; in the fourth part the same is done with those of anti-F′ with reagents Nos. 3 and 4. In these last two parts both anti-$E_2'$ and anti-F′ are shown to detect more than one antigenic specificity because each of them is conditioned by the reactivity of several reagents.

1

| 14 | 15 | 16 | 17 | 18 | 19 | 20 | 21 | 22 | 23 | 24 | 25 | 26 | 27 | 28 | 29 | 30 | 31 | 32 | 33 | 34 | 35 | 36 | 37 | 38 | 39 | 40 |
|---|---|---|---|---|---|---|---|---|---|---|---|---|---|---|---|---|---|---|---|---|---|---|---|---|---|---|
| | | | | | | | | | | | | | | | | | | | | | | | | | | |
| − | + | − | − | + | + | + | + | + | + | + | + | − | + | − | − | − | − | + | − | − | + | − | − | + | + | + |
| − | − | − | − | − | − | − | − | − | − | + | + | − | + | − | − | − | − | + | − | − | − | − | − | + | − | + |
| − | + | − | − | + | + | + | + | + | + | − | − | − | − | − | − | − | − | − | − | − | + | − | − | − | + | − |
| | | | | | | | | | | | | | | | | | | | | | | | | | | |
| | | | | | | | | | | | | | | | | | | | | | | | | | | |
| + | + | + | + | + | + | − | + | + | + | + | + | + | + | + | + | + | + | + | + | + | + | + | + | + | + | − |
| + | + | + | ± | + | + | − | + | + | − | + | + | + | + | − | + | + | + | + | − | + | − | + | + | − | − | − |
| − | − | − | − | + | + | − | + | + | + | + | + | + | + | + | + | + | + | + | + | − | + | + | − | + | + | − |
| + | + | − | + | + | + | − | − | − | − | − | + | − | + | − | − | − | + | + | − | + | − | + | − | − | − | − |
| | | | | | | | | | | | | | | | | | | | | | | | | | | |
| | | | | | | | | | | | | | | | | | | | | | | | | | | |
| + | + | + | ± | + | + | − | + | + | − | + | + | + | + | − | + | + | + | + | − | + | − | + | + | − | − | − |
| + | + | + | ± | + | + | − | − | − | − | − | − | + | + | − | − | − | − | − | − | + | − | + | + | − | − | − |
| − | − | − | − | + | + | − | + | + | − | + | + | + | + | − | + | + | + | + | − | − | − | − | − | − | − | − |
| | | | | | | | | | | | | | | | | | | | | | | | | | | |
| | | | | | | | | | | | | | | | | | | | | | | | | | | |
| − | − | − | − | + | + | − | + | + | + | + | + | + | + | + | + | + | + | + | + | − | + | + | − | + | + | − |
| − | − | − | − | + | + | − | + | + | − | + | + | + | + | − | + | + | + | + | − | − | − | − | − | − | − | − |
| − | − | − | − | − | − | − | − | − | + | + | − | − | − | + | + | + | − | − | + | − | + | + | − | + | + | − |

TABLE 2

| | Phenogroups of the B System | $G_1$ | M 22 | G |
|---|---|---|---|---|
| 1 | $\mathrm{BGKO_x\ A'\ E'_3\ I'\ O'_1\ G''\ I'_2/GY_2E'_1\ Q'}$ | − | + | + |
| 2 | $\mathrm{I_1\ E'_1\ I'\ G''\ I'_2/GY_2\ E'_1\ Q'}$ | − | + | + |
| 3 | $\mathrm{GY_2\ E'_1\ Q'/.}$ | − | + | + |
| 4 | $\mathrm{I_1\ OQA'\ E'_1\ K'\ Q'\ I'_2/GY_2E'_1\ Q'}$ | − | + | + |
| 5 | $\mathrm{BO_3\ Y_2\ A'\ E'_3\ G'\ P'\ Q'\ I'_2/I'.I'_2}$ | − | − | − |
| 6 | $\mathrm{OJ'_1\ K'\ O'_2/I'_2}$ | − | − | − |
| 7 | $\mathrm{I_2/GY_2\ E'_1\ Q'}$ | − | + | + |
| 8 | $\mathrm{I_2/.}$ | − | − | − |
| 9 | $\mathrm{G_2\ O_1\ T_1\ Y_2\ E'_3\ F'_1\ I'_2/BI'\ P'\ I'_2}$ | − | − | − |
| 10 | $\mathrm{G_1\ O_x\ O'_1\ I'_2/O_x\ D'\ E'_3\ F'_1\ G'\ O'_1}$ | + | − | + |
| 11 | $\mathrm{O_x\ A'\ I'_2/GY_2\ E'_1\ Q'}$ | − | + | + |
| 12 | $\mathrm{OJ'_1\ K'\ O'_2/GY_2E'_1\ Q'}$ | − | + | + |
| 13 | $\mathrm{BI'\ P'\ I'_2/O_x\ A'}$ | − | − | − |
| 14 | $\mathrm{G_1\ I_1/O_x\ Y_2\ A'\ Y'\ I'_2}$ | + | − | + |
| 15 | $\mathrm{BI'\ P'\ I'_2/O_x\ A'}$ | − | − | − |
| 16 | $\mathrm{O_3\ E'_2\ F'_2\ J'_2\ I'_2\ M6\ M13/O_x\ O'_1}$ | − | − | − |
| 17 | $\mathrm{O_3\ E'_2\ F'_2\ J'_2\ I'_2\ M6\ M13/O_x\ A'}$ | − | − | − |
| 18 | $\mathrm{Y_2/.}$ | − | − | − |
| 19 | $\mathrm{BI'\ P'\ I'_2/I_1\ E'_3\ G'\ I'_2}$ | − | − | − |
| 20 | $\mathrm{O_x\ Y_2\ B'\ E'_1\ F'_2\ G'\ O'_1\ I'_2\ M6/.}$ | − | − | − |
| 21 | $\mathrm{O_x\ Y_2\ A'/Q'}$ | − | − | − |
| 22 | $\mathrm{OJ'_1\ K'\ O'_2/GY_2\ E'_1\ Q'}$ | − | + | + |
| 23 | $\mathrm{I_1\ OQA'\ E'_1\ K'\ Q'\ I'_2/GY_2\ E'_1\ Q'}$ | − | + | + |
| 24 | $\mathrm{BO_3\ Y_2\ A'\ E'_3\ G'\ P'\ Q'\ I'_2/BO_1\ Q'}$ | − | − | − |
| 25 | $\mathrm{GO_x\ A'\ D'\ E'_3\ F'_1\ Q'\ I'_2/BO_1\ Q}$ | − | + | + |
| 26 | $\mathrm{BGKO_x\ Y_2\ A'\ O'_1\ I'_2/Y_2\ Y'\ Q'}$ | − | + | + |
| 27 | $\mathrm{M12\ GO_1/E'_3\ G'\ I'_2}$ | − | + | + |
| 28 | $\mathrm{G_2\ O_1\ T_1\ E'_3\ F'_1\ K'/Q'\ I'_2}$ | − | − | − |
| 29 | $\mathrm{BI_1\ Q/GO_x\ A'\ E'_3\ F'_1\ Q'\ I'_2}$ | − | + | + |
| 30 | $\mathrm{Y_2\ D'\ G'\ I'\ Q'\ I'_2/BO_1.}$ | − | − | − |
| 31 | $\mathrm{BGKO_x\ E'_2\ F'_2\ G'\ O'_2/.}$ | − | + | + |
| 32 | $\mathrm{I_2/GY_2\ E'_1\ Q'}$ | − | + | + |
| 33 | $\mathrm{Y_2\ D'\ G'\ I'\ Q'\ I'_2/I_2}$ | − | − | − |
| 34 | $\mathrm{O_x\ Y_2\ A'/I_2}$ | − | − | − |
| 35 | $\mathrm{PI'\ I'_2/Y_2\ E'_3\ G'\ Y'}$ | − | − | − |
| 36 | $\mathrm{G_2\ O_1\ T_1\ E'_3\ F'_1\ K'\ I'_2/OJ'_1\ K'\ O'_2}$ | − | − | − |

Table 2 indicates how the reactivity of the reference reagent anti-G is conditioned by the reactivities of anti-$G_1$ and/or the new experimental reagent from Milan anti-M22.

Therefore, also in this case we can deduce that, on the basis of our present knowledge, at least two antigenic factors condition the final reaction pattern of the international reagent called anti-G.

## DISCUSSION

The presented data show clearly that a further analysis has to be made regarding the specificities of cattle bloodtyping reagents.

Also, the internationally acknowledged reagents do not always react with only one single antigenic factor.

The studies of the results of the comparison tests have given a more clear idea as to the specificities of the reagents which are internationally acknowledged.

The experimental reagents which were presented in the comparison tests have so far not revealed very many results.

The authors have the opinion that a number of these reagents could offer much more information if more details — e.g. donor-recipient combinations etc. — were presented together with these reagents.

The data in this report demonstrate that the studies on specificities of reagents are subject to continuous developments.

The further analyses of the reagents can reveal a good deal of interesting information both for immunologic and genetic studies.

For the field of immunology the analyses of the reagents can offer valuable contributions to the knowledge of antigenic structures on the red blood cells. From the data presented by FIORENTINI and BOUW (1968) it is furthermore evident that the specificities of the produced reagents can offer a good deal of information about the immune responses in various combinations of donors and recipients.

Studies on the genetic control of antigenic factors as performed by BOUW and FIORENTINI (1968) on one hand prove that some reagents detect various antigenic factors which are controlled by different parts of the chromosomal region controlling the blood group system B, on the other hand a thorough knowledge of the specificities of the used reagents is indispensable for studies of the genetic structures of complex blood group loci.

## REFERENCES

BOUW, J. and FIORENTINI, A. (1970), Structure of loci controlling complex blood group systems in cattle, Proc. XIth Europ. Conf. Anim. Blood Groups Biochem. Polymorph. Warsaw 1968, 104–113.

EISEN, H. N., SIMMS, E. S., LITTLE, JR. and STEINER, L. A. (1964), Affinities of 2, 4 dinitro phenyl (DNP) antibodies induced by $\sum$-41-mono-DNP-ribonuclease, *Fed. Proc.*, **23.** 559.

FIORENTINI, A. and BOUW, J. (1970), Specificities of antibodies detecting antigenic substances on cattle red blood cells, Proc. XIth Europ. Conf. Anim. Blood Groups Biochem. Polymorph. Warsaw 1968, 117–122.

GROSCLAUDE, F. (1967), Studies on the blood group system S in French cattle breeds, Proc. 9th Europ. Anim. Blood Group. Conf. Prague 1966, 79–86.

LANDSTEINER, K. (1945), The specificity of serological reactions, Harvard University Press — Cambridge — Mass., **14.** 310–317.

LANDSTEINER, K. and VAN DER SCHEER, J. (1936), On cross-reactions of immune sera to azoproteins, *J. Exp. Med.*, **63.** 325–339.

STONE, W. H. and MILLER, W. J. (1961), Naturally occuring iso-antibodies of the S blood group system in cattle, *J. Immunol.*, **86.** 165–169.

## DISCUSSION

W. SCHLEGER: There are more reagents, which are not monovalent in our nomenclature. We think, that $T_1 + B'' = T_2$, therefore $T_2$ is not a monovalent reagent in our opinion.

*XIIth Europ. Conf. Anim. Blood Group. Biochem. Polymorph., Bp., 1972 (pp. 103–106)*

# OX ERYTHROCYTE AGGLUTINABILITY MODE OF INHERITANCE OF AGGLUTINABILITY

JILL L. COWPERTWAIT and R. L. SPOONER

**Cattle Blood Typing Service, A.R.C. Animal Breeding Research Organisation, Edinburgh**

## SUMMARY

**This paper presents evidence to show that the characteristic of high agglutinability of ox red cells is inherited and recessive to low agglutinability. It also shows its frequency in different breeds of cattle in Britain.**

## INTRODUCTION

Ox erythrocytes are known to be poor agglutinators. When they are added to titrations of isoantibody they have very low agglutination titres, but when added to titrations of rabbit antibovine serum the titres increase and show a gradation of reaction. Differential agglutinability of ox red cells has been shown by GLEESON-WHITE, HEARD, MYNORS and COOMBS (1950) and COOMBS, GLEESON-WHITE and HALL (1951). It has twice been shown that using a particular type of antiglobulin test, in which the cells after being sensitized with a fixed concentration of antibody are added to a titration of antiglobulin serum, there are only two classes of ox red cell, high and low agglutinators (SPOONER, COWPERTWAIT and MADDY, 1970). The one type is highly agglutinable to an antiglobulin dilution of 1 : 4000 and the other type is virtually inagglutinable at 1 : 4.

## MATERIAL AND METHODS

*Bovine Red Cells*. Blood was collected from the jugular vein into Acid Citrate Dextrose and stored at +4°C until used. All the samples were used within one week of collection. The majority were from normal cattle submitted to the Cattle Blood Typing Service.

*Antisera* a) Rabbit antibovine red cell sera R2 (30/12/65), R3 (8/5/66), R4 (19/5/66 and 21/1/69), and R6 (24/5/66). Rabbits were injected intravenously 3 times a week for four weeks with 2 ml of washed bovine red cells. After 2 months the course of injections was repeated. They were bled one week after the last injection in each course. The sera were stored at −30°C and the complement inactivated by heating at 56°C for 30 mins before use.

b) Bovine anti-rabbit globulin — K.21 (17/5/67). Serum was prepared by multiple intramuscular injections of whole rabbit serum in a double emulsion of Freund's complete adjuvant (HERBERT, 1965) and was collected after five monthly immunizations.

*Agglutination test.* A 2% suspension of 3 times washed bovine red cells was added to serial dilutions of the rabbit serum in microtiter plates. The diluent was 0.9% saline. Tests were read after being allowed to settle overnight.

*Antiglobulin test.* A 2% suspension of washed red cells was sensitised for 1 hr at room temperature with a sub-agglutinating concentration of rabbit serum. The cells were washed 3 times in saline and were added to serial dilutions of antiglobulin serum. Tests were read after being allowed to settle overnight.

## RESULTS AND DISCUSSION

A survey of over 2000 cattle was carried out using this antiglobulin test and only the two types of reaction were found, there were no intermediates, the variation being discontinuous rather than a complete gradation from inagglutinable to agglutinable. Whereas the direct agglutination titres, testing the same animals with the rabbit antibovine serum show a complete gradation of agglutinability.

Animals have been sampled as young calves and at different stages in their growth, always giving consistent results showing no variation with season or age. Two seven month foetal samples have also been tested one of these was highly agglutinable and the other inagglutinable.

In a random survey of cattle it was found that there were marked differences between breeds (Table 1). In some, for example the Hereford breed,

TABLE 1

*Antiglobulin survey of British cattle*

| Breed | High | Low | % High |
|---|---|---|---|
| Devon | 26 | 2 | 96 |
| Hereford | 213 | 42 | 80 |
| Aberdeen Angus | 27 | 8 | 77 |
| Lincoln Red | 11 | 5 | 75 |
| Shorthorn | 6 | 2 | 75 |
| Jersey | 103 | 73 | 58 |
| Blue Grey | 48 | 56 | 46 |
| Longhorn | 20 | 24 | 45 |
| Ayrshire | 44 | 61 | 40 |
| South Devon | 24 | 37 | 40 |
| Guernsey | 28 | 88 | 30 |
| Charolais | 10 | 27 | 27 |
| Friesian | 230 | 760 | 23 |
| Sussex | 0 | 26 | 5 |

TABLE 2

*Antiglobulin family studies*

| Parents | | Offspring |
|---|---|---|
| 41 H×H | ⟶ | 41 H |
| 76 H×L | ⟶ | 34 H + 42 L |
| 174 L×L | ⟶ | 17 H + 157 L |

there were found to be a majority of high agglutinators. Of 239 animals tested 80% were high agglutinators. In the Friesian breed the reverse was found of 990 animals there were only 18% high agglutinators (Table 1).

Most of the animals tested were also routinely tested for their blood groups, but the agglutinability of the red cell does not appear to be related to any blood group antigen. On representative panels of animals other polymorphic systems were tested; transferrins, amylases, hemoglobins, carbonic anhydrase and blood Na and K levels. Preliminary studies have shown no clear association between any of these characters and agglutinability. There is however a direct correlation between agglutinability and the sialo protein ratio of the red cell membrane (MADDY and SPOONER, 1970).

From the results of antiglobulin agglutination in the breed survey it appeared probable that the character was genetically controlled. To test this probability the number of families tested within each breed was increased and family data collected whenever possible.

The results of the family studies (Table 2) show that high with high agglutinator matings always produced offspring of high agglutinability never low agglutinability. High with low and low with low matings produced offspring of both types. With heterozygous matings, it made no difference whether the dam of the sire wás the highly agglutinable parent. These results fit the hypothesis that the character of low agglutinability is dominant to that of high.

TABLE 3

*Friesian families*

| Parents | | Offspring |
|---|---|---|
| 2 H×H | ⟶ | 2 H |
| 12 H×L | ⟶ | 4 H + 8 L |
| 44 L×L | ⟶ | 6 H + 38 L |

*Hereford families*

| Parents | | Offspring |
|---|---|---|
| 21 H×H | ⟶ | 21 H |
| 6 H×L | ⟶ | 4 H + 2 L |

*Ayrshire families*

| Parents | | Offspring |
|---|---|---|
| 19 H×L | ⟶ | 8 H + 11 L |
| 25 L×L | ⟶ | 4 H + 21 L |

TABLE 4

| Breed | Parents | Offspring | Expected Nos high | Observed Nos high |
|---|---|---|---|---|
| Friesian | 2 H×H → | 2 H | 2 | 2 |
| | 12 H×L → | 4 H + 8 L | 4 | 4 |
| | 44 L×L → | 6 H + 38 L | 5–6 | 6 |
| Ayrshire | 19 H×L → | 8 H + 11 L | 8 | 8 |
| | 25 L×L → | 4 H + 21 L | 4 | 4 |
| Hereford | 21 H×H → | 21 H | | |
| | 6 H×L → | 4 H + 2 L | | |

As was shown in Table 1 there are big differences in the frequency of high agglutinability in different breeds and so a large number of random families within one breed was necessary to test whether the character was simply inherited (Table 3).

Using the Friesian data the frequency of the genotypes was calculated assuming that there was no selection. From this the probability of homozygous low animals was calculated and the expected number of animals of each type compared with the observed number in the one breed. Similar results are shown for Ayrshire and Hereford families (Table 4).

These results from Friesians indicate that the character for high agglutinability is in fact a recessive trait and this is confirmed in the Ayrshire and Hereford material.

## REFERENCES

COOMBS, R. R. A., GLEESON-WHITE, M. H. and HALL, J. G. (1951), Factors influencing the agglutinability of red cells. II: The agglutination of bovine red cells previously classified as 'Inagglutinable' by the building up of an 'antiglobulin: globulin lattice' on the sensitised cells. *Brit. J. exp. Path.*, **32.** 195–202.

GLEESON-WHITE, M. H., HEARD, D. H., MYNORS, L. S. M. and COOMBS, R. R. A. (1950), Factors influencing the agglutinability of red cells: the demonstration of a variation in the susceptibility to agglutination exhibited by the red cells of individual oxen. *Brit. J. exp. Path.*, **31.** 321–331.

HERBERT, W. J. (1965), Multiple emulsions: a new form of mineral-oil antigen adjuvant. *Lancet*, **2.** 771.

MADDY, A. H. and SPOONER, R. L. (1970), Ox erythrocyte agglutinability. 1. Variation in membrane protein. *Vox Sang.*, **18.** 34–41.

SPOONER, R. L., COWPERTWAIT, J. L. and MADDY, A. H. (1970), Ox erythrocyte agglutinability. 2. Differential agglutinability. *Vox. Sang.*, **18.** 251–256.

*XIIth Europ. Conf. Anim. Blood Groups Biochem. Polymorph., Bp., 1972 (pp. 107–110)*

# ON THE CHEMICAL NATURE OF THE BOVINE J BLOOD-GROUP SUBSTANCE

O. W. THIELE and J. KOCH

Physiologisch-chemisches Institut and Tierärztliches Institut der Universität, Göttingen, GFR

## SUMMARY

**Total lipids prepared from bovine $J^{cs}$ serum exhibit haptenic activity in the homologous J system, while the corresponding lipids prepared from $J^{a}$ serum do not exhibit any haptenic activity. Various fractionation procedures in addition to deacylation led to a purified glycosphingolipid fraction that proved to be highly active as a J hapten. This fraction contained at least five components, two of which seem to be identical with ceramide dihexoside and ceramide trihexoside, respectively.**

**The low-density lipoproteins prepared from $J^{cs}$ serum as well as the total lipids extracted from these lipoproteins inhibit the hemolysis of $J^{cs}$ erythrocytes in the J system. These results suggest that the J blood-group substance dissolved in $J^{cs}$ serum is a lipoprotein and that its determinant group is a glycosphingolipid.**

## INTRODUCTION

In contrast to other blood-group substances the J substance is primarily dissolved (STORMONT, 1949) in the blood plasma and in various body fluids. Bovine serum containing the J substance ($J^{+}$ serum) has therefore been used as the starting material for the investigations described here. Bovine serum lacking the J substance ($J^{-}$ serum) has been used for reasons of comparison.

The J substance corresponds to a naturally occuring antibody (anti-J). There are differences in the reactivity of the anti-J sera of different animals. The presence of J subgroups has therefore been suggested. An international terminology of J subgroups, however, has not been established so far. In the experiments described here $J^{+}$ serum was used throughout which has been tested with anti-J sera that had been checked in international comparison tests.

## MATERIAL AND METHODS

Lipids extracted from $J^{+}$ or $J^{-}$ serum were tested in the J system using a hemolysis-inhibition test. Generally, 1 mg of lipid sample was added to 5 mg of auxiliary lipid (in most cases crude $J^{-}$ glycerophospholipids), 1 ml of isotonic saline was added and the mixture emulsified in a glass homogenizer. A dilution series of this emulsion was prepared so that the first tube contained 200 μg of the sample and each following tube half the amount of sample weight of the preceeding tube.

Five different batches of $J^+$ and $J^-$ serum, respectively, were used. The total lipids were extracted with ethanol-diethyl ether (3 : 1, v/v) or with chloroform-methanol (1 : 1, v/v). The crude total lipids were purified by washing with water (FOLCH et al., 1951) or by passing through Sephadex G-25 (WELLS and DITTMER, 1963). Chromatographic fractionation were carried out on columns of silicic acid (BORGSTRÖM, 1952; HANAHAN et al., 1957; VANCE and SWEELEY, 1967). In order to separate choline-containing lipids from all others an $Al_2O_3$ column was applied (RHODES and LEA, 1957).

Periodate oxidations were performed as follows: 6–10 mg lipid was dissolved in 1 ml chloroform. 0.8 ml 0.1 M phosphate buffer (pH 7.7) and 0.1 ml 0.1 M periodic acid was added. The mixture was stirred for 1 hr. The lipids were extracted with chloroform and washed with water.

In order to prepare purified sphingolipids mild alkaline hydrolyses (SCHMIDT et al., 1946; DEBUCH, 1965) or methanolyses (BROCKERHOFF, 1963; WELLS and DITTMER, 1965) were performed.

Thin-layer chromatography of glycosphingolipids was performed according to SVENNERHOLM (1963). For qualitative assays of sugar and hexosamines the glycosphingolipids were hydrolyzed (SCHWARZ et al., 1961) and subsequently chromatographed on cellulose plates (VON BERG, 1968).

Crude gangliosides were prepared from lyophilized serum (KUHN and WIEGANDT, 1963). Enzymatic hydrolysis by neuraminidase was performed according to KLENK and GIELEN (1963). Low-density lipoproteins were prepared by ultracentrifugation at a density of 1.063 g/ml. Neuraminic acid was assayed according to WARREN (1959), sugar according to RADIN et al. (1955).

## RESULTS AND DISCUSSION

The total lipids of $J^+$ serum inhibit the specific hemolysis in the homologous J system, the total lipids of $J^-$ serum, however, show no inhibitory activity. It is therefore suggested that the total lipids of $J^+$ serum contain the natural J hapten. Mild periodate oxidation destroyed the haptenic activity completely.

After fractionation of the $J^+$ total lipids the individual fractions had to be tested for haptenic activity after admixing suitable quantities of auxiliary lipids (crude $J^-$ glycerophospholipids) in order to get maximum inhibitory effects. After various fractionation procedures (Fig. 1 shows an example) the J hapten was found in a fraction that contained crude sphingolipids. Deacylation by mild alkaline hydrolysis or methanolysis did not succeed in a decrease of haptenic activity. Since acyl ester bonds are cleaved by deacylation while acyl amide bonds (present in sphingolipids) remain stable under the same conditions, the J hapten is suggested to be a sphingolipid. The sphingolipid fraction comprises sphingomyelins and glycosphingolipids. We made use of the fact that chromatography on alumina separates choline-containing lipids (i.e. sphingomyelins) from all other lipids. The J hapten proved to be present in the purified glycosphingolipid fraction only. The amount of glycosphingolipid required for a complete inhibition of specific hemolysis comes near the order of magnitude that is expected for pure haptens.

This glycosphingolipid fraction is still a mixture of at least five components as revealed by thin-layer chromatography, two of them correspond in $R_F$ values to ceramide dihexoside and ceramide trihexoside, respectively, while cerebrosides have not been found in three of four samples. Thin-layer

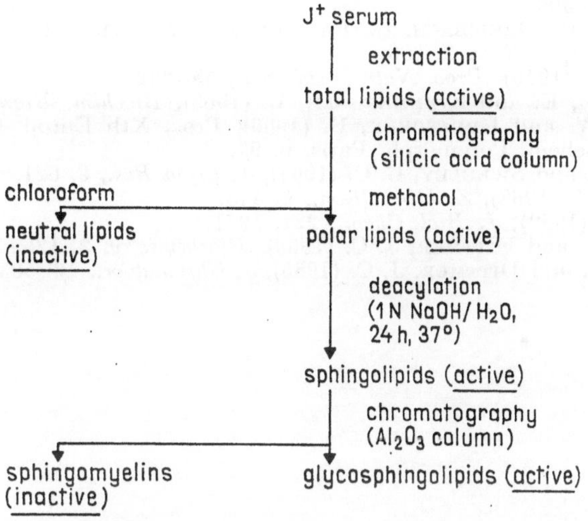
J+ serum
extraction
total lipids (active)
chromatography
(silicic acid column)
chloroform
methanol
neutral lipids
(inactive)
polar lipids (active)
deacylation
(1 N NaOH/H2O,
24 h, 37°)
sphingolipids (active)
chromatography
(Al2O3 column)
sphingomyelins
(inactive)
glycosphingolipids (active)

FIG. 1. Schedule of preparation of glycosphingolipids

chromatography of the sugar moieties revealed one hexosamine and four hexoses in $J^-$ glycosphingolipids (galactose, glucose, mannose, fucose) and an additional unknown sugar in $J^+$ glycosphingolipids.

A crude ganglioside fraction prepared from $J^+$ serum contained only a small percentage ($<1\%$) of neuraminic acid. It showed haptenic activity, but no decrease in activity after the action of neuraminidase.

The low-density lipoproteins prepared from $J^+$ serum by ultracentrifugal flotation as well as the total lipids extracted from these lipoproteins inhibit the specific hemolysis in the J system. These results suggest that the J blood-group substance dissolved in bovine serum is a lipoprotein and that its determinant group is a glycosphingolipid, but probably not a cerebroside or a ganglioside.

In early experiments (THIELE et al., 1966) we studied serum of a single cow, the J subgroup of which was unknown. In this case we found the haptenic activity in a different fraction. This result which differs from the results described here is probably due to a difference in the J subgroup.

## REFERENCES

BORGSTRÖM, B. (1952), *Acta Physiol. Scand.*, **25.** 101.
BROCKERHOFF, H. (1963), *J. Lipid Res.*, **4.** 96.
DEBUCH, H. (1965), *Z. Physiolog. Chemie.*, **343.** 141.
FOLCH, J., ASCOLI, I., LEES, M., MEATH, J. A. and LE BARON, F. N. (1951), **191.** 833.
HANAHAN, D. J., DITTMER, J. D. and WARASHINA, E. (1957), *J. Biol. Chem*, **228.** 685.
KLENK, E. and GIELEN, W. (1963), *Z. Physiolog. Chemie*, **330.** 223.
KUHN, R. and WIEGANDT, H. (1963), *Chem. Ber.*, **96.** 866.
RADIN, N. S., LAVIN, F. B. and BROWN, J. R. (1955), **217.** 790.
RHODES, J. H. and LEA, C. H. (1957), *Biochem. J.*, **65.** 528.
SCHMIDT, G., BENOTTI, J., HERSHMAN, B. and THANNHAUSER, S. J. (1946), *J. Biol. Chem.* **166.** 506.
SCHWARZ, H. P., DREISBACH, L., BARRIONUEVO, M., KLESCHIK, A. and KOSTYK, I. (1961), *J. Lipid Res.*, **2.** 208.
STORMONT, C. (1949), *Proc. Natl. Acad. Sci.*, **35.** 232.
SVENNERHOLM, E. and SVENNERHOLM, L. (1963), *Biochim. Biophys. Acta*, **70.** 432.
THIELE, O. W. and URBASCHEK, B. (1966), Proc. Xth Europ. Conf. Anim. Blood Groups Biochem. Polymorph. Paris, p. 97.
VANCE, D. E. and SWEELEY, C. C. (1967), *J. Lipid Res.*, **8.** 621.
VON BERG, W. (1968), *Z. klin. Chem.*, **6.** 475.
WARREN, L. (1959), *J. Biol. Chem.*, **234.** 1971.
WELLS, M. A. and DITTMER, J. C. (1963), *Biochemistry*, **2.** 1259.
WELLS, M. A. and DITTMER, J. C. (1965), *J. Chromatogr.*, **18.** 503.

*XIIth Europ. Conf. Anim. Blood Groups Biochem. Polymorph., Bp., 1972 (pp. 111–114)*

# ATTACHMENT OF THE BOVINE J BLOOD-GROUP SUBSTANCE AT THE ERYTHROCYTE MEMBRANE

J. SCHRÖFFEL*, O. W. THIELE and J. KOCH

Physiologisch-chemisches and Tierärztliches Institut der Universität, Göttingen, GFR

## SUMMARY

From quantitative studies of the J activity of bovine erythrocytes and of erythrocyte lipids it is concluded that all of the J activity of $J^{cs}$ cells is due to a lipoprotein and that the J substance forms a monolayer on the erythrocyte surface.

In contrast to $J^{cs}$ erythrocytes the J substance dissolved in $J^{cs}$ serum occurs both as a lipoprotein and in a different form, probably as a mucoprotein. It is, therefore, suggested that only the J lipoprotein is absorbed from the plasma onto the erythrocytes during a postnatal period. This is consistent with the fact that in $J^{s}$ animals (with J substance in the serum, but not on the cells) hardly any J activity was found in the serum lipids, almost all the J activity being demonstrable in the residue after lipid extraction.

During the process of hemolysis and subsequent washings bovine erythrocytes release a considerable portion of their membrane constituents in a "soluble" form. To prevent the disruption of ghosts, a small amount of $Mg^{2+}$ is added to the hemolyzing mixture. We observed a loss of about 35% of J activity along with about 15% of various stroma constituents unless $Mg^{2+}$ had been added to the hemolyzing mixture.

On treatment of bovine erythrocyte ghosts with organic solvents of increasing polarity two lipid fractions, "loosely" and "strongly" bound lipids, are obtained. The J activity could be found only in the "strongly" bound fraction along with the majority of glycolipids.

On treatment of bovine $J^{cs}$ ghosts with hypertonic saline, part of the stroma constituents — including the J substance — can be solubilized.

## INTRODUCTION

Previous reports from these laboratories (THIELE and KOCH, 1970) disclosed that the J substance dissolved in bovine $J^{cs}$ serum was a lipoprotein. This does not exclude the occurrence of the J substance in a different form on the erythrocytes and even in the serum in addition to the lipoprotein form. The aim of further studies is to investigate how the J substance is attached to the $J^{cs}$ erythrocytes and why in $J^{s}$ animals the erythrocytes are not coated with the J substance dissolved in $J^{s}$ serum.

* Present address: Animal Breeding Research Institute, Hradištko p. Med. Czechoslovakia

## MATERIAL AND METHODS

J haptenic activity was detected by hemolysis-inhibition tests. Quantitative measurement was performed by photometric assay of the extent of hemolysis in the individual test tubes. The degree of haptenic activity of a certain stroma preparation was expressed in terms of the corresponding volume of packed cells that gave rise to a 50% hemolysis of test erythrocytes.

The methods of preparation of erythrocyte suspension, lipid extraction, assays of phosphorus, cholesterol and protein were essentially the same as described by Burger et al. (1968). Stroma were prepared with hypotonic Veronal buffers with or without added 3 mM $MgCl_2$ (Burger et al., 1968). Sugar was assayed by the method of Radin et al. (1955). Mucoproteins were extracted according to Klenk et al. (1958). "Loosely" and "strongly" bound lipids were extracted from lyophilized stroma by a method similar to that described by Roelofsen (1968). Stroma was treated with hypertonic saline according to Mitchell et al. (1966).

## RESULTS AND DISCUSSION

Burger et al. (1968) observed that during the process of hemolysis and subsequent washings bovine erythrocytes released a considerable portion of their membrane constituents and of their acetylcholine esterase activity. The release is prevented by addition of $MgCl_2$ (in 1–5 mF final concentration) to the hemolyzing mixture. We observed a release of about 15% of stroma dry weight, stroma protein, stroma cholesterol, stroma phospholipid and stroma glycolipid unless $MgCl_2$ had been added to the hemolyzing buffer. This loss of stroma constituents is accompanied by a release of about 35% of the original J activity.

Since the J substance is obviously attached to the outer surface of the erythrocyte membrane, it is clear that during the process of hemolyzing without added $Mg^{2+}$ primarily the superficial layers of the erythrocyte membrane are disrupted. Various membrane components (proteins, phospholipids, cholesterol, glycolipids), if taken as whole classes, do not seem to have special localizations within the membrane. These data show that similar experiments can serve as tools for the elucidation of bovine erythrocyte membrane structure.

For further studies stroma were used that were prepared by hemolysis with added 3 mM $MgCl_2$. By quantitative determinations of the J activities of stroma, total stroma lipids and stroma residues after extraction of total lipids, it was demonstrated that all the J activity of $J^{cs}$ erythrocytes was due to a lipid and that the J activity of total lipids per ml packed cells was virtually identical with the J activity of an equal amount of intact cells. This result suggests that the J substance forms a monolayer on the erythrocyte surface though a mosaic distribution may occur. No J substance seems to be buried in the depth of the membrane.

This result is also consistent with the fact that in mucoproteins extracted from $J^{cs}$ erythrocytes any J activity is missing. In contrast to $J^{cs}$ erythrocytes, bovine $J^{cs}$ serum does not only exhibit J activity of its lipid, but also a considerable activity of its mucoproteins and of its residue after lipid extraction (Table 1). These data evidence the occurrence of the J substance in bovine serum in 2 different forms viz.: J lipoprotein and J mucoprotein. Exclusively the J lipoprotein seems to be able to coat bovine erythrocytes. The reason for the missing absorption of J substance onto erythrocytes in $J^{s}$ animals is that $J^{s}$ serum contains J mucoprotein, but very small amounts of J lipoprotein (Table 1).

TABLE 1

*J activity in various preparations of bovine serum and erythrocytes*

+, active; (+), trace of activity; —, no activity

| | total lipid | residue after lipid extraction | mucoprotein |
|---|---|---|---|
| $J^{cs}$ serum | + | + | + |
| $J^{cs}$ erythrocytes | + | — | — |
| $J^{s}$ serum | (+) | + | + |
| $J^{s}$ erythrocytes | — | — | |

The genetically fixed difference between $J^{cs}$ and $J^{s}$ animals seems to stem from the inability of $J^{s}$ animals to attach that sequence of sugars which is responsible for the determinant to a ceramide moiety.

"Loosely" bound lipids extracted from lyophilized stroma with ether or chloroform contain virtually all of the cholesterol and no J activity. "Strongly" bound lipids subsequently extracted with chloroform-methanol, however, contain the bulk of the glycolipids and virtually all of the J activity. These results point to the mode of binding between the determinant and its protein carrier.

According to MITCHELL and HANAHAN (1966) treatment of human erythrocyte stroma at 4°C with 1.2 M NaCl causes a partial solubilization of stromal lipids and proteins including acetylcholine esterase without complete disruption of the underlying stromal structure. This method was tried on bovine erythrocyte stroma. A solubilization of 22–26% of various stroma constituents including J activity was observed. Though the majority of membrane lipoproteins seem to be held together by hydrophobic bonds, our results suggest that some membrane subunits released on treatment with hypertonic saline are bound in the membrane by electrostatic forces. Moreover, the J lipoprotein, though secondarily absorbed to the erythrocyte surface, seems to be fully integrated to the other surface lipoproteins since no selected release of J substance is noticed on treatment with hypertonic saline.

## REFERENCES

BURGER, S. P., FUJII, T. and HANAHAN, D. J. (1968), *Biochem.*, **7.** 3682.
KLENK, E. and UHLENBRUCK, G. (1958), Z. *Physiol. Chem.*, **311.** 227.
MITCHELL, C. D. and HANAHAN, D. J. (1966), *Biochem.*, **5.** 51.
RADIN, N. S., LAVIN, F. B. and BROWN, J. R. (1955), *J. Biol. Chem.*, **217.** 790.
ROELOFSEN, B. (1968), *Thesis*, Utrecht, p. 12.
THIELE, O. W. and KOCH, J. (1970), *Europ. J. Biochem.* (in press)

## DISCUSSION

**E. TUCKER: Would you expect treatment of J one red cells with phospholipase to remove the J reaction from the cell?**

**J. SCHRÖFFEL: The removal of J reactions by the treatment of the red cells is to be proved.**

**P. MILLOT: In my opinion the J substance in $J^{cs}$ sera and the J substance in the $J^{s}$ sera may not be exactly the same substances but perhaps different antigenic substances having a common reactivity with an antibody of high specificity (I obtained an anti-J cellular serum not inhibited by serum J).**
**What did you think about that?**

**J. SCHRÖFFEL: I do not think that two different substances are responsible for the J-activity of $J^{cs}$ and $J^{s}$ sera. The J haptene (oligosacharide) seems to be the same in both cases but the binding to the membrane components is different and most probably genetically determined.**

*XIIth Europ. Conf. Anim. Blood Groups Biochem. Polymorph., Bp., 1972 (pp. 115–116)*

# PRESENCE OF NATURAL BLOOD GROUP ISOHEMOLYSIN ANTI-$R_1$ IN CATTLE

A. Kaczmarek, D. Golemanov* and Z. Dorynek

**College of Agriculture in Poznan, Chair of Animal Husbandry, Laboratory of Research in Animal Blood Groups, Poznan, Poland**

*** Higher Institue of Veterinary Medicine, Department of Surgery, Sofia Bulgaria**

Many of natural agglutinins anti-A, -G, -X, -J, -D (4) and normal isohemolysins anti-J, -$U_1$, -$U_2$ (2, 3, 4), are known in cattle. According to Osterhoff et al. (1962) normal isohemolysins anti-G, -L and anti-J and -$Y_2$ can be present simultaneously in titres up to 1 : 64 and anti-A even higher titres. We could not find any data on the existence of natural isohemolysin anti-$R_1$ in the available literature. In the course of the search for normal isohemolysins in the sera and the colostrum of pregnant cows and heifers we detected by means of hemolytic test normal blood group isohemolysin anti-$R_1$ in a heifer in the ninth month of pregnancy (No. 135). The titer of the antibody was 1 : 16 and it did not change after calving. After calving the antibody was present in the colostrum and milk for a week in a considerably lower titre 1 : 1 to 1 : 2 while its serum level remained the same.

Anti-$R_1$ was not detected in the serum of the newborn calf until 24 hours after intake of colostrum and subsequently for 35 days, the titer varying from 1 : 1 to 1 : 4.

This evidences that the antibody had passively penetrated the serum of the newborn calf through the colostrum of the dam.

The conclusion that just anti-$R_1$ was concerned was based on a number of absorptions followed by serological reactions. For this purpose tested erythrocytes from 40 adult cows of which only one cow (No. 97) had an antigen $R_1$ which reacted with the serum of the heifer (No. 135) and her calf. Besides, erythrocytes of those calves (Nos. 0.246, 0.286, 0.292 and 0.323) which had in their erythrocyte antigen range an antigen $R_1$ were tested several times at different intervals with the serum, colostrum and the milk of cow No. 137 and the serum of her calf and proved the presence of the natural blood group isohemolysin anti-$R_1$.

## REFERENCES

Osterhoff, D. R. and Van Der Walt, K. (1962), Report VIIIth Europ. Anim. Blood Group Conference, Ljubljana, Yugoslavia, 1962.

Stone, W. H. and Miller, W. J. (1953), *Genetics*, **38.** 693.

Stone, W. H. and Irwin, M. R. (1954), *J. Immunology*, **73.** 397.

Tikhonov, V. N. (1967), Izpolzovanie grupi krovi pri selekcii životnih. Kolos, Moskva, 162–178.

## DISCUSSION

W. SCHLEGER: Have you proved the titer of the natural blood group isohemolysin anti-$R_1$ during the different periods in the year (spring, summer, autumn and winter)? We found that the titer in natural reagents is in autumn higher then in winter or summer.

Z. DORYNEK: This study was carried out in autumn.

*XIIth Europ. Conf. Anim. Blood Groups Biochem. Polymorph., Bp., 1972 (pp. 117–119)*

# CONTRIBUTION TO SEROLOGY OF LYMPHOCYTES IN CATTLE AND PIGS*

D. O. SCHMID and F. OTTO

Institute of Animal Blood Group and Resistance Research, Livestock Breeding Research Organization
Munich, GFR

## SUMMARY

For obtaining specific anti-lymphocyte sera (ALS) it is absolutely necessary to isolate pure lymphocytes.

After discussion of the different methods for obtaining leucocytes and lymphocytes a method is described for the isolation of pure lymphocytes out of the blood of humans, horses, cattle, sheep, pigs and dogs. The method is based on the different specific weights of the blood cells. Jodamid and compounds containing Jodamid were used. For the isolation we recommend especially Uromiro-300 and Uromiro-380, both containing Jodamid.

By means of iso- and heteroimmunizations in cattle and pigs we succeeded in proving different antibodies specific for lymphocytes.

During the last years an intensive work was done concerning the leuco cyte-antigens in humans, subhuman primates and in the small laboratory animals, mouse and rat aiming at the determination of the histocompatibility with regard to donor-selection for the organ-transplantation and concerning autoimmune diseases and allergy.

On the occasion of our last meeting at Warsaw BALNER (1968) gave an excellent survey on this situation. Until recently the leucocyte antigens of farm animals have been nearly unknown.

Our aim was therefore to investigate the lymphocyte-antigens in farm animals with the final destination of marking the phagocytosis. Basis of all research in the field of serology of lymphocytes and especially in the production of specific ALS is the isolation of pure lymphocytes. A considerable experimental work was done in this field. Neither the spontaneous sedimentation nor the differential-centrifugation with the supplement of high molecular substances as dextran, gelatine or PVP, nor the hemolysis results in red cell free lymphocyte-preparations. This is true also for the lymph collected by means of drainage from the ductus thoracicus. Even the most expensive IBM-blood cell separator does not supply us with red cell free fractions. The unique method for isolation of lymphocytes out of the bloodstream has been until recently the Ficoll-Isopaque flotation developed by BÖYUM (1964) and modified by KISSMEYER-NIELSON and KJERBY (1967). This method can be modified according to MAYR in that

* This research work has been carried out with the assistance of the Deutsche Forschungsgemeinschaft, Grant Schm 208/7.

way, that dextran-sedimentation and cotton-procedure is combined with the flotation method.

Our own investigations, terminated only recently, started from the basic consideration of KELLNER and BÖYUM, which tried to find a medium from a substance insoluble in water not resulting in hemolysis and with an account of the different specific weights of the different blood cells, leads to a seperation of the different cells. The compound 3-acetylaminomethyl-5-acetylamino-2,4,6-trijodbenzoicacid, known as Jodamid, fulfils these conditions. In order to isolate lymphocytes we utilize either Jodamid having been neutralized by 5% NaOH with a specific weight of 1.076, or the X-ray-contrast medium Uromiro-Jodamid (Dr. Franz Köhler-Chemie Alsbach-Bergstrasse, Germany), which can be utilized without a preliminary neutralization for an isolation of lymphocytes. Uromiro exists in two trademarks Uromiro-300 and Uromiro-380, with different iodine-content. We recommend Uromiro-380, a 80% sodium salt mixture of N-methylglucamin with an iodine-content of 380 mg/ml. For the production of the necessary solution with a specific weight of 1.076 for the lymphocyte seperation 22.35 ml Uromiro-380 are necessary for 100 ml distilled water. For the isolation of lymphocytes equal parts of a neutral Jodamid solution or Uromiro-380 preparation are overlayed with defibrinated blood and are centrifuged for 30 minutes at 1900 r.p.m. If a sedimentation of the erythrocytes and granulocytes is done at the same time the lymphocytes form a white ring in the separation-layer between serum and Jodamid. If bigger blood quantities are needed it is recommended to overlay the collected lymphocyte-suspension after washing with Tris-BSS and resuspension in saline once more with Jodamid-solution in order to get the same purification as in the case of smaller blood quantities. After one overlayering we get — by means of the Uromiro-Jodamid method with dextran, cotton and Ficoll-flotation — in one step out of 5.0 ml blood of humans, horses, cattle, pigs, sheep and dogs an average amount of 2000–10000 lymphocytes/$\mu$l. There was no contamination with other blood cells. By superlaying twice with Uromiro-380 we were able to produce, for example, out of 20 ml defibrinated pig blood $140 \times 10^6$ lymphocytes, this amount was therefore nearly quantitative. Microscopically we found no damaged cells. The trypanblue-vitality test showed no signs for a cell damage. The lymphocytes remained uncoloured. Lymphocytes treated with Uromiro-Jodamid can still be used as antigens and give rise to antibody production after iso- and heteroimmunizations. The agglutinability of lymphocytes in specific ALS was not disturbed after the isolation with Uromiro-Jodamid. The isolation of granulocyte of peripheral blood shows even more difficulties. The isolation of polynuclear cells has become very important for the research in the field of phagocytosis, of antigens in granulocytes and the production of specific antigranulocyte-sera. Up to now there have been described for the isolation of granulocytes the column-gradient method, the cell-electrophoresis and the Ficoll-flotation. The new method developed in our institute by one of my coworkers, Dr. Otto, is based on a spontaneous sedimentation after adding dextran to the defibrinated blood. Hemolysis is done with ammonium chloride. The separation of the granulocytes and mononuclear cells is effected by means of flotation with Jodamid and centrifugation. We find the granulocytes in the sediment and the lymphocytes in the supernatant.

After the preliminary closing of our work with chicken leucocytes, which showed the possibility to prove specific leucocyte-antigens by means of specific ALS (Schmid and Thein, 1968) we concentrated our work during the last two years on the research of the antigenic structure of lymphocytes in cattle and pigs. By iso- and heteroimmunizations of cattle, pigs, sheep, rabbits and chicken with leuco- or lymphocyte preparations out of the blood stream we produced more than 100 isologous and heterologous ALS. For the immunization we used red cell free lymphocyte-preparations of single animals. For safety purposes we absorbed all raw-sera with lymphocyte-free pig red cells. Only 25% of all immunized pigs formed antibodies. In heteroimmunizations we stated on the contrary a nearly 100% antibody formation. The fractioning of the immune-sera is, from the technical point of view, considerably difficult, as on account of the limited stability of the lymphocytes, the isolation of lymphocytes, the absorption-experiments and the test have to be carried out in a very short time. By means of numerous absorptions of many iso- and heteroimmune sera we succeeded in producing a considerable amount of monovalent and oligovalent lymphocyte-reagents. Actually we dispose of ALS to prove at least three different lymphocyte antigens in cattle (Mü-L[bov] 1–3) and at least eight different ones in pigs (Mü-L[pig] 1–8). But there is a definite indication to believe that on the basis of material we dispose of in further research we will detect some more antigens.

The defined lymphocyte-antigens segregate independently of the red cell antigens and showed no cross-reactions with already known erythrocyte-antigens.

## REFERENCES

Balner, H., (1970), Leukocyte antigens of man and subhuman primates. Proc. XIth Europ. Conf. Anim. Blood Groups Biochem. Polymorph. Warsaw, 1968, 67.

Borowska, M. and Démant, P. (1967), Specificity of cytotoxic antibodies in typing sera against cattle blood group antigens. *Fol. biol.*, **13.** 473.

Böyum, A. (1964), Separation of white blood cells, *Nature*, **21.** 793.

Kissmeyer-Nielsen, F. and Kjerby, K. E. (1967), Lymphocytotoxic-Microtechnique. Purification of lymphocytes by flotation. Report of a conference and workshop Torino and Saint-Vincent 14–24 June 1967.

Lucke, J. N., Immelman, E. J., Symes, M. O. and Hunt, A. C. (1968), Use of horse anti-pig leucocyte serum to suppress the homograft reaction in pigs, *Nature*, **217.** 560.

Otto, F. and Schmid, D. O. (in press), Lymphozytenisolierung aus dem Blut des Menschen und der Tiere. *Blut*

Otto, F. and Schmid, D. O. (1970), Lymphozytenisolierung aus dem Blut des Menschen und der Tiere. Kongress der Deutschen Gesellschaft für Bluttransfusion, Giessen, 1970.

Otto, F. and Schmid, D. O. (1970), Lymphozytenisolierung aus dem Blut des Menschen und der Tiere. XIII. Intern. Kongress für Hämatologie München 3. 8. 1970.

Schmid, D. O. and Thein, P. (1970), Leukocyte antigens in chickens. Proc. XIth Europ. Conf. Anim. Blood Groups Biochem. Polymorph. Warsaw, 1968.

*XIIth Europ. Conf. Anim. Blood Groups Biochem. Polymorph., Bp., 1972 (pp. 121–124)*

# THE INHIBITION GROUPS IN THE SERA OF CATTLE

P. MILLOT

INRA et Institut PASTEUR C.N.R.Z. Jouy-en-Josas, France

## SUMMARY

**Three bovine serum factors, iA, iB and iC are described. These inhibit agglutining power of three iso-immune sera reacting with sensitized red cells. The first two factors are genetically transmitted by two non allelic genes iA[1] and iB[1]. The third one is transmitted by an iC[1] gene, non allelic with iA[1] and representing probably a third locus. A close linkage between these loci is excluded. Contrary to iA and iB, the iC inhibitor is not found in the IgG fraction, and is frequently associated to the J factor in several herds. The iC factor does not seem to be an allotype but it is probably related with the bovine J or bovine R substance. Therefore its specificity is that of a substance of J type fixed on the cells at the moment of the sensitization. The latter changes the J negative cells into J positive ones by fixing on the cells J-like substances which do not fix themselves spontaneously. Anti-J can easily be produced by iso-immunization using serum precipitates. This provides antibodies of various types whose specificities are not identical and which prove a plurality of the J group substances.**

In 1966 at the Paris Conference (MILLOT, 1966) we reported the existence of inhibitor groups in the sera of cattle. Using the technics described in a paper already in press (MILLOT, in press) we have obtained three sera named "antiglobulinic". These agglutinate red cells sensitized by an immune serum and produce specific inhibition when added to increasing dilutions of some ordinary cattle sera.

## I. OBTAINING OF THE INHIBITOR GROUPS' REAGENTS

The three above-mentioned sera were obtained by immunization against: (A) an ethanol-precipitate of the supernatant of a normal cattle serum previously precipitated by rivanol (1 vol. at 1%), (B) a precipitate by 1 to 10 diluted egg-white (3% of the volume) of the supernatant of an anti-egg white cattle serum previously precipitated by 2-mercapto-ethanol (3% of the serum volume), (C) cattle red cells in 20% suspension sensitized by an isoimmune anti-red cell serum. Though antigens A and B were prepared from the same donor's serum the three reagents determined three distinct systems of inhibitor groups. Out of a total of 39 isoimmune sera obtained, there were also three other sera each presenting a special inhibition specificity. These three sera, however, did not permit a clear classification of the cattle into groups because the inhibitor titers were quite different from one another. This phenomenon seems to be fundamental. In all other cases the immune sera agglutinated sensitized red cells but no inhibition was observed.

## II. GENETIC TRANSMISSION OF THREE INHIBITORY FACTORS

The serum groups corresponding to the three inhibitory factors and reacting with the specific immune sera were provisionally named: iA, iB and iC. Their genetics was studied on 178 charolais offspring born from the same number of cows and 8 bulls, each of them being mated to 20 to 25 cows.

A previous study (MILLOT, in press) shows the Mendelian transmission of the three factors, which is dominant on the absence. Two factors, iA and iB are transmitted by two non-allelic genes ($iA^1$ and $iB^1$) each having a frequency of about 0.31. The variance calculated for $iA^1$ is $\pm 0.12$. The iC factor appears to be transmitted in the same way by an $iC^1$ gene, non allelic with $iA^1$. It probably determines a third locus but the genetic analysis shows that about 1/5 of the estimated inhibitors do not occur in the case of anti iC reagent. No irregular genetic transmission has been observed concerning the iA factor. On the contrary the inhibitory character of some heterozygous iB and iC factors may be not expressed as certain offspring deriving from negative parents are sometimes positive. The mating of bifactorial males to double recessive females exclude a close linkage between two inhibitory loci.

## III. RELATIONSHIP BETWEEN THE iC INHIBITORY FACTOR AND THE J SUBSTANCE

Whereas the iA and iB inhibitors are entirely or partially associated to the IgG fraction obtained by chromatography on DEAE cellulose, the iC inhibitor of the same serum is not present in this fraction. Being resistant to 2-mercapto-ethanol, the iC inhibitor cannot be of IgM nature. On the other hand, the distribution of these three inhibitory factors in the charolais breed appears to be different from that of blood group genes and specially from phenogroups on the B locus. But the iC inhibitory factor which appears to be distinct from the J cellular factor in the charolais breed (associated 98 times and did not associate 110 times in 399 cows), appears to be related to this factor in our reference herd composed of cattle from various breeds (ten associated, one isolated iC in 30 cows) and in 50 cows from Aubrac (20 associated, two J and two iC isolated). The relation between the two factors is seemingly not due to chance proven by a serological study on the iC reagent that lead us to the following observation:

1) The reagent which agglutinates the donor's own cells when they are sensitized by an antibody, also agglutinates the non sensitized J cells in 1/3 of the cases and hemolyses them in 4/5 of the cases. Unsensitized J red cells absorb every agglutinine.

2) The inhibition spectrum of anti-iC obtained with our 30 reference sera is identical, whether the reagent agglutinates non sensitized J red cells or the donor's own J negative cells after being sensitized by an immune serum from a J cow. This spectrum nearly agrees with the J cellular factor.

3) When the iC reagent agglutinates sheep red cells, the inhibition by sheep sera reveals what seems to be the soluble R substance.

We have concluded: 1) the iC factor is related to the J globular substance and it must have the same chemical nature; it is probably not an allotype.

2) the anti-iC agglutinin, considered as an antiglobulin because it agglutinates the donor's own cells after sensitization, may have a non-proteinic inhibitory specificity.

3) J negative cells sensitized with an immune serum behave as J positive ones if the sensitizing serum is obtained from a J positive cow. The sensitizing antibody fixes the serum J substance on the cells and the obtained inhibition is of J-anti-J character.

## IV. IMMUNIZATION AGAINST J GROUP FACTOR AND INHIBITION

Recent immunizations of four cows (one globular J and three negative) against ammonium sulphate precipitates or heat coagulates (75°C) of supernatants of J cattle sera previously precipitated by rivanol (1 vol. at 1%) resulted in the following observations: 1) no agglutination of the non sensitized J cells, 2) production of anti-J agglutinin (titer 1/16), 3) increasing titer of anti-J agglutinin (from 1/4 to 1/16), 4) no change in the anti-J titer (1/8) but very strong reaction. The anti-J immunizations proved to be quite difficult in our experimental conditions. (Precipitated antigen was mixed with complete Freund's adjuvant for intramuscular injections.)

One of the obtained sera (III) deserves a special attention. It gives an inhibition spectrum with the non sensitized J cells. This is not in agreement, with the J globular factor detected by the anti-iC reagent but it agrees with the total globular and serum J factor. The inhibition of this serum compared with that of the anti-J hemolysin of our laboratory is in good agreement. One may notice however that the agglutination inhibition is much better than the hemolysin inhibition to determine the J serum factor which we have previously announced (Millot, 1966). (The reaction is more sensitive, the inhibitor titers are about three times higher and the serum J is much more surely revealed.)

The absorption of the studied serum by J cells exhausted the active agglutinin on these cells but kept an active agglutinin on J negative cells sensitized by an immune serum from a cow having soluble J antigen. After absorption the inhibition spectrum remained analogous to serum J but some J globular sera were no longer inhibitors. The same result was obtained without absorption by mixing the studied serum with the J globular cow's serum (1 vol. + 1 vol. of J serum diluted to 1/32). Therefore the studied serum contained two agglutinins of anti-J type: one of them, active on the non-sensitized J cells, was inhibited by the sera containing either cellular J or serum J; the other one, being active only on the sensitized red cells was inhibited by the serum J and by some J globular ones. The latter appeared specific of a serum J substance fixing oneself on the cells only by means of an antibody. It is to be noted that the donor of the anti-iC reagent whose specificity is limited to globular J, ranks among serum J with the studied serum whose specificity is larger.

As for the J globular recipient immunized in the same series (I) it reveals by agglutination of sensitized cells an inhibitory factor different from globular J and serum J but close to serum J (4/5 common inhibitions).

All that shows the existence of several substances of close specificity that we may group under the designation J and among which there is the iC inhibitory factor. Only one of them (globular J) appears to be fitted for spontaneous fixing on red cells in the absence of sensitizing antibody. This confirms some previous studies on the J substance (STONE and IRWIN, 1956; WROBLEWSKI et al., 1958). They concluded that fixing on the cells was generally quantitatively ruled (by plasma concentration) but sometimes qualitatively: we see that the nature of the substance has a basic importance; i.e. some substances of the J group, though they strongly inhibit an anti-total J antibody, appear unable to fix themselves on the cells without the artificial intervening of a sensitizing antibody.

## REFERENCES

MILLOT, P. (1966), Les réactions d'agglutination hétéro- et iso-antiglobuliniques des bovins. *X^e Congrès Européen sur les Groupes sanguins et le Polymorphisme biochiimique des Animaux*, 1 vol., INRA, p. 471–475.

MILLOT, P. (1971), Les groupes d'inhibition de l'agglutination iso-antiglobulinique chez les bovins. I. Etude immunologique. *An. Biol. Anim. biochim. biophys.* (In press)

MILLOT, P. (1971), Les groupes d'inhibition de l'agglutination iso-antiglobulinique chez les bovins. II. Etude de la transmission génétique de 3 facteurs: iA, iB, iC. *An. Biol. Anim. biochim. biophys.* (In press)

STONE, W. H. and IRWIN, M. R. (1956), The J substance of cattle. I. Developmental and immunogenetic studies. *J. Immun.*, **73**. 398–406.

WROBLEWSKI, A., PODLIACHOUK, L. and MILLOT, P. (1958), Recherche de substances de groupes sanguins dans le sérum et la salive de bovins. *An. Inst. Pasteur*, **94** 456–462.

*XIIth Europ. Conf. Anim. Blood Groups Biochem. Polymorph., Bp., 1972 (pp. 125–129)*

# BLOOD GROUPS OF PODOLIAN CATTLE IN ISTRIA AND BOHIJN STRAIN OF CIKA (PINZGAU) CATTLE

P. LAZAR, O. BÖHM, J. SENEGACNIK and ANKA GLIHA

The Veterinary Institute of Slovenia, Ljubljana, Yugoslavia

## SUMMARY

Blood groups of two isolated cattle populations were investigated.

Istrian cattle is a special strain of Podolian breed and Cika cattle. In Bohijn valley there is an isolated subpopulation of the Cika breed.

A trial of their diversity and relation to some other breeds in Yugoslavia was carried out.

## INTRODUCTION

The Bohijn strain of the Cika breed represents a small population in a closed Alpine valley. It represents a branch of brachycerous cattle originating from the prehistoric times of Alpine Illyrians (HANSEN, 1921). The strain has been raised according to the special Alpine system: during the winter the animals are fed and managed in peasant's stables but during summer they are collected in herds of 50–100 animals and grazed on mountain pastures (SPILLER and MUYS, 1962). Geological structure of area (triassic karstic surface) provides only small pastures on very stony and uneven surfaces (MELIK, 1950). Cika is a one purpose breed, for milk production. The average weight of cows is 300 kg, the average milk production is 2000–3000 liters. Today the external appearance of the breed is very similar to that of the Austrian Pinzgau breed. Very much back-crossing with imported bulls had been made in the last two centuries (STEFANČIČ, 1966). The relation of the original population with the native Busha breed of the Balkans has not yet been determined.

The Istrian strain of Podolian cattle (MISON and JARDAS, 1950; OGRIZEK, 1957, 1963) has been raised on the peninsula since the time of the peoples' migration period. Podolian cattle are typical draft animals for farming work on the heavy and deep Istrian soil. The average weight of bulls is 750 kg and of cows 550 kg. In the last two centuries the meat qualities had been improved by Podolian bulls from Italy. Between 1905 and 1931 mostly the bulls from the improved Romagnola strain were imported for back-crossing.

## MATERIALS AND METHODS

Exclusively family material was studied. In both districts the animals have been kept on small peasant farms, 1–3 animals on each. Natural service has been in use in Bohijn valley in Istria; 70–90% of the cows

TABLE 1

*Blood group gene frequencies of the Istrian (Podolian) breed*

| | | | |
|---|---|---|---|
| *B system* | | | |
| $BGI_1O_1A'P'B''$ | .0489 | $I_1E'_1I'$ | .0075 |
| $BGKO_1QB'E'_3G'I'K'O'$ | .0263 | $I_2$ | .0038 |
| $BGKO_1QE'_3G'K'$ | .0940 | $I_2O_1$ | .0075 |
| $BGKO_xB'O'$ | .0188 | $O_1T_1E'_3K'$ | .0113 |
| $BGKQE'_1G'O'$ | .0301 | $O_1E'_1$ | .0133 |
| $BGKQE'_2$ | .0263 | $O_3A'E'_1$ | .0075 |
| $BGKQE'_2O'$ | .0451 | $O_3E'_2J'_2$ | .0338 |
| $BGO_1Y_1$ | .0113 | $O_3J'_1K'O'$ | .0266 |
| $BI_1E'_1$ | .0150 | $O_xQA'$ | .0150 |
| $BO_1QT_1E'_3K'$ | .0865 | $O_xA'$ | .0263 |
| $BO_1Y_2D'$ | .0038 | $O_xO'$ | .0113 |
| $BQT_1G'P'B''$ | .0150 | $PQA'B'E'_1$ | .0113 |
| $BT_1I'P'$ | .0714 | $PE'_3I'$ | .0188 |
| $GO_1$ | .0263 | $Y_2$ | .0038 |
| $GO_xO'$ | .0188 | $Y_2J'_1K'O'Y'$ | .0526 |
| $I_1O_1$ | .1090 | $E'_1G'$ | .0038 |
| $I_1O_1I'$ | .0038 | $E'_2J'_2$ | .0113 |
| $I_1Y_1E'_1G'$ | .0038 | $I'$ | .0038 |
| $I_1Y_1E'_1Y'$ | .0188 | b | .0637 |
| *A system* | | *S system* | |
| A | .4548 | H′ | .5098 |
| H | .0913 | SH′ | .1552 |
| AH | .2215 | U | .0474 |
| AZ′ | .0111 | UH′ | .0560 |
| Z′ | .0107 | UH′H″ | .0749 |
| a | .2266 | UH′H″U″ | .0123 |
| | | SH′U″ | .0238 |
| *FV system* | | SH′H″U″ | .0100 |
| | | H′H″ | .0420 |
| F | .7479 | s | .0686 |
| V | .2521 | | |
| *J system* | | *Hb system* | |
| | | A | .9569 |
| J | .2671 | B | .0431 |
| *L system* | | *Tf system* | |
| L | .2493 | A | .2624 |
| | | D | .6839 |
| *M system* | | E | .0537 |
| M | .0508 | *Am system* | |
| *Z system* | | 1 | .8556 |
| | | 2 | .1444 |
| Z | .5083 | | |
| z | .4917 | | |
| *R′S′ system* | | | |
| R′ | .3440 | | |
| S′ | .6560 | | |

TABLE 2

*Blood group gene frequencies of the Cika breed*

| | | | |
|---|---|---|---|
| $BGKO_xB'O'$ | .0097 | $O_1T_1E'_3K'$ | .0230 |
| $BGKE'_2O'$ | .0230 | $O_1Y_1$ | .0182 |
| $BGO_xA'B'$ | .0098 | $O_1E'_1G'$ | .0133 |
| $BI_1O_1QT_1A'P'$ | .0291 | $O_1E'_2J'_2$ | .0230 |
| $BO_1T_1E'_3K'$ | .0157 | $O_xQD'O'$ | .0145 |
| $BO_1Y_1E'_3P'$ | .0121 | $O_xY_2A'$ | .0048 |
| $BO_3Y_2D'E'_1K'B''$ | .0218 | $O_xY_2A'D'E'_1$ | .0012 |
| $BO_1A'$ | .0073 | $O_xY_2D'E'_1O'$ | .0024 |
| $BO_1E'_2$ | .0182 | $O_xI'O'$ | .0896 |
| $BO_xY_1E'_3O'P'$ | .0339 | $O_xO'$ | .0654 |
| $BO_xA'E'_3P'$ | .0109 | $PY_2G'$ | .0036 |
| $BO_xO'$ | .0024 | $PE'_3$ | .0254 |
| $BQG'O'P'B''$ | .0061 | $PE'_3I'$ | .0048 |
| $BI'$ | .0012 | $T_1E'_3$ | .0412 |
| $GI_1O_xY_2A'G'Y'B''$ | .0073 | $Y_1G'Y'$ | .0085 |
| $GO_xY_2O'$ | .0182 | $Y_1I'Y'$ | .0061 |
| $GI'$ | .0012 | $Y_2$ | .0012 |
| $I_1$ | .0218 | $Y_2G'$ | .0303 |
| $I_1O_3A'G'I'K'$ | .0145 | $E'_1$ | .0097 |
| $I_1O_xY_2A'G'Y'B''$ | .0036 | $E'_2J'_2$ | .0036 |
| $I_1Q$ | .0642 | $E'_3$ | .0073 |
| $I_1E'_1I'$ | .0097 | $G'$ | .0133 |
| $I_2$ | .0291 | $I'$ | .1017 |
| $O_1T_1Y_2E'_3$ | .0206 | b | .0908 |
| $O_1T_1E'_3$ | .0048 | | |
| *A system* | | *S system* | |
| $A_1$ | .1528 | $SH'$ | .0537 |
| $A_2$ | .2312 | $UH'H''$ | .0961 |
| H | .0811 | $UH'H''U''$ | .0589 |
| $A_1H$ | .0811 | $H'$ | .5789 |
| $A_2H$ | .0989 | $H'H''$ | .0939 |
| $A_1Z'$ | .0009 | $H'U''$ | .0499 |
| a | .3640 | $U'$ | .0324 |
| | | s | .0362 |
| *FV system* | | *Am system* | |
| F | .8527 | 1 | .6671 |
| V | .1473 | 2 | .3329 |
| *J system* | | *beta-lactoglobulin system* | |
| J | .0553 | | |
| | | A | .4057 |
| *L system* | | B | .5943 |
| L | .1167 | *alpha-casein system* | |
| *M system* | | A | .0000 |
| M | .0344 | B | 1.0000 |
| *Z system* | | *beta-casein system* | |
| Z | .4265 | A | .8840 |
| z | .5735 | B | .0480 |
| | | C | .0680 |

(TABLE 2 continued)

| | | | |
|---|---|---|---|
| *R′ S′ system* | | *kappa-casein system* | |
| R′ | .1319 | A | .6654 |
| S′ | .8681 | B | .3346 |
| *Hb system* | | | |
| A | .9879 | | |
| B | .0121 | | |
| *Tf system* | | | |
| A | .3568 | | |
| D | .4769 | | |
| E | .1663 | | |

have been inseminated from one center. Great caution was exercised in sampling. Bulls were tested first, cows and their suckling offspring were bled afterwards. For gene frequency calculation exclusively unrelated mother-cows have been taken into consideration.

Samples for Bohijn Cika originated from 9 bulls and 404 cows, those for Istrian cattle from 244 cows.

59 reagents were used for blood-typing ($A_1$ $A_2$ H Z′ B G K $I_1$ $I_2$ $O_1$ $O_3$ $O_x$ P Q $T_1$ $T_2$ $Y_1$ $Y_2$ A′ B′ D′ $E_1'$ $E_2'$ $E_3'$ G′ I′ $J_1'$ $J_2'$ K′ O′ P′ Y′ B″ $C_1$ $C_2$ R W $X_1$ $X_2$ L′ F $\overset{+}{F}$ V J L M S H′ U UU′ U′ H″ U″ R′ Z $\overset{+}{Z}$ YU-L17, 18, 19).

Hemoglobin, transferrin, amylase and beta-lactoglobulin types were determined for all animals, alpha-, beta- and kappa-caseins for 125 Bohijn cows. Gene frequencies were calculated for all systems except C.

## RESULTS AND DISCUSSION

Results are shown in the Tables 1 and 2. Genetic equilibrium was tested on all closed loci. Agreement was found within all loci but for the amylase locus of the Cika breed; there was a lack of heterozygous animals. No explanation can be offered up to date.

Another Slovenian strain of the Cika breed in the valley of Tolmin was studied in 1966 (LAZAR et al., 1966). The two strains are related very closely in respect of gene frequencies and they both do have similarities as much as gene frequencies are concerned with data given by BUSCHMANN (1962) and ERHARD and SCHMID (1964) for the Bavarian Pinzgau breed in 1962. A rather interesting particularity has been found in relation to the $Tf^E$ gene, its frequency being 0.1663 what is an exception for cattle breeds in Yugoslavia and in Central Europe.

The Istrian breed of Podolian cattle differs greatly from all other Yugoslavian cattle breeds studied up to now (LAZAR et al., 1965). In the B system 38 phenogroups have been determined but there must be more. Some of them are typical of other breeds in Central Europe but their frequencies are low; they could be an indication of sporadic immigration. For the S system an allelic richness is characteristic. The allele $S^{SH'H''U''}$ has been found in S system and $A^{Z'}$ in A system which, according to our knowledge, have not yet been described.

Our data for the Istrian breed can be compared with gene frequencies of Podolian cattle in Vojvodina (Danube Plain in Yugoslavia) (LAZAR et al., 1965; SCHMID and MANČIĆ, 1964). The agreement is rather high but there are some greater differences which could be taken as a proof that the Istrian strain must have been considerably influenced by the Italian cattle (FIORENTINI, 1969).

## REFERENCES

BUSCHMANN, H. (1962), Blutgruppengenetische Untersuchungen an süddeutschen Rinderrassen. *Zeitschr. Tierzüchtung Züchtungsbiologie*, **78.** 12–15.

ERHARD, L. and SCHMID, D. O. (1965), Blood group studies in Pinzgau cattle. Proc. 9th Europ. Anim. Blood Group Conf., Prague, 1964, 79–85.

FIORENTINI, A. (1969), U.S.D.A. Comparison test.

HANSEN, J. (1921), Lehrbuch der Rinderzuchtes; Paul Parey, Berlin.

LAZAR, P. et al. (1965), Blood group comparison of some cattle breeds in Yugoslavia; B. Kidrič Foundation report, 1964/65.

LAZAR, P., BÖHM, O. and GLIHA, ANKA (1966), Blood group gene frequencies in three Slovene cattle breeds. Proc. Xth Europ. Conf. Anim. Blood Groups Biochem. Polymorph. Paris, 1966.

MELIK, A. (1950), Planine v Julijskih Alpah; Ljubljana.

MIŠON, I. and JARDAS, F. (1950), Istrian Cattle, *Stočarstvo*, **4.** 345–359.

OGRIZEK, A. (1957, 1963), Contribution to the knowledge of the Istrian indigenous cattle, JAZU, *Acta Biologica* I, III.

SPILLER and MUYS, F. (1962), Planšarstvo in kmetijstvo na naših planinah; Kmetijska tiskovna zadruga v Ljubljani.

STEFANČIĆ, A. (1966), Začetek in razvoj veterinarstva na Slovenskem do prve svetovne vojne; SAZU; Prirodoslovne in medicinske vede: 18, Medicinske vede, 4, Ljubljana.

SCHMID, D. O. and MANČIĆ, D. (1964), Blutgruppen beim Podolischen Steppenrind aus Jugoslavien. *Zeitschr. Tierzüchtung Züchtungsbiologie*, **80.** 216–223.

## DISCUSSION

J. BOUW: You found superiority for the Am homozygote types. Can this not be due to technical failures?

Petrovsky in the session on Biochemical Polymorphism is showing that the quantity of amylase is decreasing strongly during egg production.

P. LAZAR: The technical failure can be excluded on the basis that the Podolian cattle of Istria has been typed partly at the same time and with entirely the same method. In this breed Am data were in genetic equilibrium.

*XIIth Europ. Conf. Anim. Blood Groups Biochem. Polymorph., Bp., 1972 (pp. 131–135)*

# RELATIONSHIPS BETWEEN AUSTRIAN CATTLE BREEDS AS INFERRED FROM BLOOD GROUP AND SERUM PROTEIN FREQUENCIES

K. ZETNER, H. ROHRBACHER, W. SCHLEGER and F. PIRCHNER

Department of Animal Breeding and Genetics, School of Veterinary Medicine, Vienna, Austria

## SUMMARY

Cattle of six Alpine breeds were blood-typed, and transferrins and hemoglobins determined. The frequencies of the respective phenotypes in each breed were correlated with each other and average differences between frequencies in various breeds were computed. From the average distances, a cluster was constructed which depicts the genetic relationships between these breeds and which in all probability is closely related to the genetic relationship between them.

In recent years there has been considerable interest in the application of quantitative methods in taxonomy and in the study of the evolution of breeds and races (SOKAL and SNEATH, 1963; EDWARDS and CAVALLI-SFORZA, 1965; MAIJALA and LINDSTRÖM, 1966). Two different approaches are used. The divisive methods split a set of data into sub-sets and maximize the variance between sub-sets. The second approach, the so-called agglomerative, clusters the observations in such a way that pairs or groups with highest correlations or least differences are combined to one sub-set, and additional observations or clusters connected to these in order of decreasing correlations or increasing distances (GOWER, 1967).

## MATERIAL AND METHODS

Until the early part of this century, many breeds and strains of cattle existed in the Austrian Alps (MÜLLER, 1958; KALTENEGGER, 1889). A trend towards unification operates since then and was intensified in the last decades, so that many of the strains, and some of the breeds, are extinct or near extinction. Information about blood groups, transferrin and hemoglobin frequencies has been collected for the following breeds, among others: Fleckvieh, Braunvieh, Pinzgauer, Tuxer, Pustertaler and Murbodner breed. Table 1 indicates the proportion of Austrian cattle belonging to each of these breeds, and numbers used in the present investigation. The material used to estimate the frequencies in the Murbodner, Fleckvieh, Pinzgauer, Tuxer, Pustertaler breeds was collected by ZETNER (1969) and by ROHRBACHER (1970), that of the Braunvieh comes from routine testing. In the more numerous breeds the frequencies were estimated using results from female animals only, and in the case of Fleckvieh and Murbodner, only results from cows which were dams of female progeny. In the case of very small and nearly extinct breeds Tuxer and Pustertaler, we used all available data and supplemented these with results from crosses between those two breeds with Pinzgauers (the frequency in the two breeds was

TABLE 1

*Breedsize and numbers of animals used in investigation*

| Breed | % of Austrian cattle 1947 | 1969 | Numbers blood typed |
|---|---|---|---|
| Fleckvieh | 36.3 | 62.6 | 175 |
| Braunvieh | 11.8 | 15.7 | 225 |
| Pinzgauer | 16.7 | 10.5 | 133 |
| Murbodner (Gelbvieh) | 24.9 | 6.8 | 112 |
| Tux-Zillertaler | * | | 18 (23)** |
| Pustertaler Sprinzen | *** | | 30 (15)** |

Total cattle population numbers more than 2 million
* Numbers of Tux-Zillertaler in 1947 a few hundred, in 1969 eighteen animals
** In brackets: numbers of crosses used for supplementing information
*** A South-Tyrolean breed, in 1947 several hundred, in 1969 about 100 animals

extrapolated from the difference between the crosses and the Pinzgauer breed). In case of the larger breed, an attempt was made to collect the material from a range of areas. In the case of the Tuxer the samples come from all surviving animals, while the Pustertaler samples derive from nearly all the animals of this breed which are still alive.

Even though in both cases numbers are small, the effective size of these groups is not all that minute since the 18 Tuxer animals are from eight sires, while the Pustertaler are from at least seven sires, the majority of these being unrelated with each other.

Although 55 reagents are used in the laboratory at present, the reactions with only 44 of these were considered in this investigation. In addition to the factor frequencies, the phenotype frequencies for transferrins and hemoglobins are included so that, all in all, frequencies for 52 traits, so to speak, were computed for each of the six breeds. Correlations between the frequencies in the different breeds were estimated as well as average distances which correspond to $\sqrt{(x_1 - x_2)^2}$ which was estimated from the expression

$$\frac{Sx_2^1 + Sx_2^2 - 2Sx_1x_2^{1/2}}{n - 1}$$

where $x_1$ and $x_2$ denote the frequencies of trait $x$ in breeds 1 and 2, respectively.

## RESULTS AND DISCUSSION

Correlations and average distances were used for constructing clusters, whereby the weighted pair-group method as given by SOKAL and SNEATH was employed. Correlations and average distances between the breeds for the 52 phenotype frequencies are given in Table 2. The cluster depicted in Fig. 1 was constructed from the average distances. However, the one constructed from the correlations was essentially similar to this one.

TABLE 2

*Correlations and average distances between breed specific blood group and protein frequencies*

| Breeds | | | Correlations | Differences between breeds |
|---|---|---|---|---|
| Tuxer | — | Pustertaler | 0.797 | 15.4 |
| Tuxer | — | Pinzgauer | 0.833 | 14.7 |
| Tuxer | — | Braunvieh | 0.706 | 18.2 |
| Tuxer | — | Fleckvieh | 0.765 | 16.2 |
| Tuxer | — | Murbodner | 0.804 | 15.1 |
| Pustertaler | — | Pinzgauer | 0.864 | 13.1 |
| Pustertaler | — | Braunvieh | 0.720 | 17.3 |
| Pustertaler | — | Fleckvieh | 0.767 | 15.8 |
| Pustertaler | — | Murbodner | 0.791 | 15.2 |
| Pinzgauer | — | Braunvieh | 0.823 | 14.9 |
| Pinzgauer | — | Fleckvieh | 0.784 | 16.3 |
| Pinzgauer | — | Murbodner | 0.853 | 13.7 |
| Braunvieh | — | Fleckvieh | 0.770 | 15.3 |
| Braunvieh | — | Murbodner | 0.815 | 14.1 |
| Fleckvieh | — | Murbodner | 0.882 | 11.3 |

The cluster does show a very close association between the three spotted breeds of the East Central Alps, Pinzgauer, Pustertaler und Tuxer, which group is distinctly separated from the Fleckvieh and Braunvieh, but also from the Murbodner breed which has its area of origin even further east. The close connection between the Pinzgauer, Pustertaler and Tuxer is to be expected on historical and geographical grounds and agrees with present-day views about breed development. It justifies the conclusion that all three are rather closely interrelated.

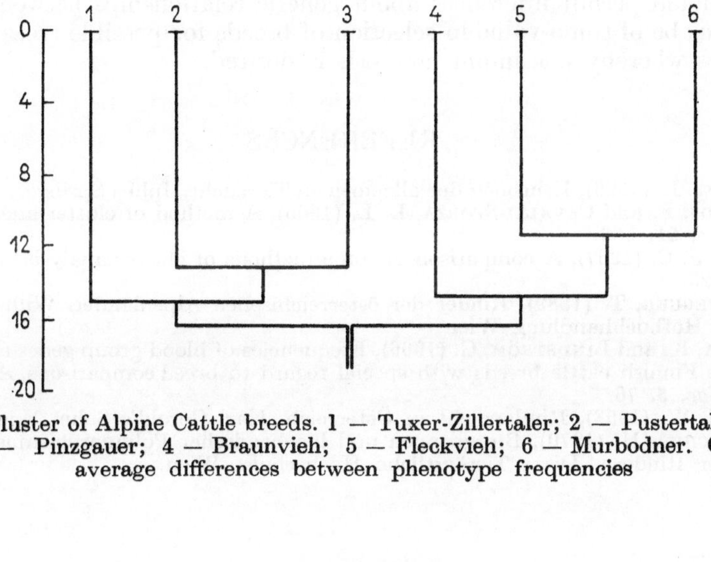
1
2
3
4
5
6
0
4
8
12
16
20

FIG. 1. Cluster of Alpine Cattle breeds. 1 — Tuxer-Zillertaler; 2 — Pustertaler-Sprinzen; 3 — Pinzgauer; 4 — Braunvieh; 5 — Fleckvieh; 6 — Murbodner. Ordinate: average differences between phenotype frequencies

There has been, some time ago, considerable argument over the correct position of the Tux-Zillertaler breed in the system of Alpine cattle breeds. It has been suggested, on the basis of craniological measurements, that this breed is very much distant from the others. In fact, WILCKENS (ADAMETZ, 1926) has given the breed a separate sub-species rank as bos brachycephalus. Those measurements were taken from rather few skulls, and his conclusions certainly do not agree with the results we have got from studying blood groups and protein polymorphism.

However, the cluster puts the Murbodner breed also at a rather large distance from the Pinzgauer, when in fact the areas of distribution of the two breeds adjoin each other and one would expect that the Murbodner would be closer related to the Pinzgauer than to the Fleckvieh which comes from the Western Alps. Although the Murbodner is a composite of Eastern Alpine strains, no evidence of infusion of Fleckvieh genes into the Murbodner is evident.

The cluster describes the phenotypic relationship between the breeds on the basis of blood group, transferrin and hemoglobin frequencies. However, clusters constructed with gene frequencies were essentially alike. The phenotypic relationships shown here should correspond to the similarity between frequencies of blood group, transferrin and hemoglobin genotypes of the various breeds. To the extent that blood group etc. genes are representative for the whole genotype, the phenotypic relationship as found here should reflect the overall genetic relationship between breeds. The genotype frequencies in each breed are the result of the ancestral genotypes as well as of the selection, drift and immigration which occurred since separation from the other breed or breeds. Therefore, distances as found here, reflect genealogical relationships only if the influences mentioned above are of comparable importance in each breed, an assumption hardly fulfilled. However, the approximations may be close enough to yield a rough picture of the real genealogy of Alpine breeds.

Studies such as these appear to be mainly of historical interest. However, they should permit inferences about genetic relationships between breeds, and thus be of some value in selection of breeds for possible cross-breeding schemes whereby maximum heterosis is desired.

## REFERENCES

ADAMETZ, L. (1926), Lehrbuch der allgemeinen Tierzucht. Julius Springer, Wien.

EDWARDS, F. and CAVALLI-SFORZA, L. L. (1965), A method of cluster analysis, *Biometrics*, **21.** 362.

GOWER, J. C. (1967), A comparison of some methods of cluster analysis. *Biometrics*, **23.** 623.

KALTENEGGER, T. (1889), Rinder der österreichischen Alpenländer. Wilhelm Frick, k. k. Hofbuchhandlung Wien.

MAIJALA, K. and LINDSTRÖM, G. (1966), Frequencies of blood group genes and factors in the Finnish cattle breeds with special regard to breed comparisons, *Ann. Agric. Fenniae*, **5.** 76.

MÜLLER, W. (1958), Rinderzucht in Österreich. Karl Gerold's Sohn, Wien.

ROHRBACHER, H. (1970), Blutgruppen und biochemischer Polymorphismus bei Murbodner Rindern. Diss., Tierärztliche Hochschule, Wien.

SOKAL, R. and SNEATH, T. (1963), Principles of numerical taxonomy. W. H. Freeman & Co., San Francisco.
ZETNER, K. (1969), Blutgruppen und biochemischer Polymorphismus bei Tuxer- und Pustertaler Rindern. *Diss.*, Tierärztliche Hochschule, Wien.

## DISCUSSION

K. K. KIDD: With respect to your suggestion that these studies might be used to identify breeds that might be profitably crossed, it is interesting that my results showed Charolais and Hereford as most different and that these two breeds are now commonly used in the United States to produce hybrids for meat production.

*XIIth Europ. Conf. Anim. Blood Groups Biochem. Polymorph., Bp., 1972 (pp. 137—140)*

# DATA ON BLOOD-GROUP PROPERTIES OF THE HUNGARIAN GREY (STEPPE) CATTLE

G. Kovács

**Department of Animal Breeding, University of Veterinary Science, Budapest, Hungary**

## SUMMARY

**This breed of Asiatic origin, being one representative of the oldest cattle breeds o Europe, differs striking from other breeds in both external appearance and blood group characters.**

**A relatively small variety of the B-phenogroups has been detected in the breed, only 21 different B-alleles having been established.**

**This breed can be characterized by the following five predominant B-phenogroups: $BG_2KE'_2O'$, $G_3O_1T_1E'_3K'$, $G_3PQA'D'$, $Y_2A'B'D'E'_3G'$, $BG_3O_1PQB'K'$.**

**Surprisingly the Z′ factor occurs very rarely in the breed, in contrast to the general belief that this blood group factor were of Eastern origin. The frequency of the W factor is much lower than in other European cattle breeds. Also the FV locus shows a noticable difference, the $F^V$ allele exhibiting a relatively high frequency.**

## INTRODUCTION

In order to give a description on the immunogenetical structure of the cattle breeds of Hungary we started blood typing various breeds in the country.

As part of this comprehensive study we have now analyzed the blood types of the Hungarian Grey (Steppe) cattle (HG) and compared them to those of the Hungarian Spotted (HS) breed (Kovács, 1968). These are mostly descriptive data, but they may be of interest to those who are dealing with the origin and history of cattle breeds, as HG is one representative of the ancient primitive breeds of Europe which had originated from Asia at the time of the great invasions as proven by historical records as well as by a transferrin study (Kovács and Soos, 1968), reporting the occurrence and frequencies of $Tf^F$ allele in the HG breed.

Now this breed of cattle consists of less than 1000 individuals altogether, which are kept in three state farms (Hortobágy, Középtisza and Bugac respectively). The HG breed has no practical importance any longer and its maintenance is not economic at all as neither its milk yield nor its meat production are satisfactory when compared with other non-primitive breeds. The survival of the breed has been assured only by its touristic interest as the impressive appearance of HG cattle is an indispensible part of the picture of the Hungarian "puszta".

## MATERIAL AND METHODS

614 individuals belonging to two closed HG populations have been blood-typed in this study with the generally applied hemolytic test, using the following blood typing reagents: $A_2$, H, Z′; B, $G_1$, $G_2$, $G_3$, $I_1$, K, $O_1$, $O_2$, $O_3$, $P_1$, Q, $T_1$, $Y_2$, A′, B′, D′, $E'_2$, $E'_3$, G′, I′, K′, O′; $C_1$, $C_2$, R, W, X, L′, F, V, J, L, $S_1$, $S_2$, U, U′, Z.

Blood typing data were evaluated by gene frequency estimation as described by NEUMANN–SØRENSEN (1958). B phenogroups were established from a considerable amount of family data collected from the herd-books.

Because of the small total number of the whole population examined, cows, breeding bulls and offspring have all been included into the material. The sample can be, nevertheless, regarded as representative of the whole breed, because the total number of the examined animals comprised more than 60% of the individuals belonging to the breed.

Those phenogroups which could not be clearly identified were designated by letter X and treated as a single allele in the course of the gene frequency estimation. Homozygosity was calculated as described by MAIJALA and LINDSTRÖM (1966).

TABLE 1

*Frequencies of B phenogroups in the Hungarian Grey (Steppe) cattle*

| Phenogroups | Frequencies | |
|---|---|---|
| | HG | HS |
| *$BG_2KE'_2O'$ | 0.22698 | 0.01027 |
| $BG_2QTE'_3$ | 0.00432 | |
| $BG_3O_1P_1QB'K'$ | 0.05366 | |
| $BO_3Y_2A'E'_2$ | 0.00649 | |
| $BY_2$ | 0.00435 | |
| *$G_1A'$ | 0.02466 | 0.10768 |
| $G_2Y_2$ | 0.04433 | |
| $G_2Y_2A'E'_2$ | 0.00288 | |
| *$G_3O_1T_1E'_3K'$ | 0.16367 | 0.04525 |
| $G_3PQA'D'$ | 0.09307 | |
| $I_1Q$ | 0.00432 | |
| $I_1T_1Y_2E'_2$ | 0.00288 | |
| $P_1A'$ | 0.02759 | |
| *Q | 0.02274 | 0.02270 |
| *$Y_2$ | 0.04078 | 0.00390 |
| $Y_2A'B'D'E'_3G'$ | 0.08935 | |
| $Y_2E'_2$ | 0.00757 | |
| *A′B′ | 0.00388 | 0.02770 |
| *I′ | 0.01959 | 0.01430 |
| *b | 0.07730 | 0.18788 |
| X | 0.07986 | |

* Alleles occurring in both of the HG and HS breeds

## RESULTS

Altogether 21 B phenogroups have been established which are listed in Table 1. Eight of these groups show relatively low frequencies ($q: 0.01$), which perhaps can be explained by the small total number of the individuals in the breed, the small number of breeding bulls used, and the closed system of breeding in all the three HG populations, and also by the fact, that owing to the small size of the whole population, the animals examined can be considered as one large family material.

26 out of 614 individuals were found to be homozygous for the B locus (4.34%) while the expected percentage of homozygosity was considerably higher (11.63%).

Gene frequency estimations have been carried out on the FV locus as well, as shown in Table 2, parallelly with the corresponding data of the HS breed.

TABLE 2

*Gene frequencies on the FV locus in the HG and HS breeds*

| Breed | $q_F$ | $q_V$ |
|---|---|---|
| HG | 0.7064 | 0.2936 |
| HS | 0.8084 | 0.1916 |

## DISCUSSION

Comparing the phenogroups of HG cattle with those of Hungarian Spotted (HS) cattle, the main breed of Hungary, the most striking difference noticed at the first sight is the relatively small number of B phenogroups in the HG breed. These two breeds have only a few alleles in common, as demonstrated in Table 1, although the HS breed was developed from the HG one, some one hundred years ago when HG cows were mated to Simmental bulls which had been imported from Switzerland. The females of the $F_1$, $F_2$, $F_3$ etc. offspring populations were back-crossed again to Simmental bulls. Despite of these historical facts, the blood group characters which formerly must have been very similar in the two breeds, have by now become so divergent, that similarity cannot be easily recognized.

A further difference in comparison to most of the cattle breeds examined so far in the world is the extremely low frequency of allele $B^b$ (0.07730) which is 0.18788 in the HS breed and even higher in some Western European breeds i.e. 0.3755 in MRY and 0.2985 in the FH (BOUW, 1958) breeds.

At the moment there is no explanation for the low value of the frequency of the *b* allele in the HG population. It can be the consequence of mere chance but it does not seem impossible either that this and other differences have been developed in this breed because of some special circumstances. It should be noted here that the Hungarian Grey cattle have been living under practically the same rather extensive conditions as their ancestors many hundred years ago and that no selection took place in this breed for improvement of either the milking performance or the quality of meat.

On the basis of the striking difference between the observed and expected percentages of homozygous individuals, as far as the B locus is concerned, the question can be raised whether homozygosity and heterozygosity on the B locus have more influence on the survival of the respective individuals in this breed than in others.

The occurence of the W factor in the C system is considerably less frequent in the HG breed than in other European cattle breeds.

The frequency of the $F^V$ allele is also significantly higher than in the HS breed.

Interestingly the Z′ factor which is supposed to be one of the characteristics of cattle originating from Asia, exhibits an extremely low frequency in the HG breed.

## REFERENCES

Bouw, J. (1958), Bloodgroup studies in Dutch cattle breeds. H. Veenman and Zonen Wageningen.

Kovács, G. (1970), Proc. XIth Europ. Conf. Anim. Blood Groups Biochem. Polymorph., Warsaw, 1968, 227–230.

Kovács, G. (1968), *Magy. Ao. L.*, **11.** 574–578.

Maijala, K. and Lindström, G. (1966), *Ann. Agric. Fenniae*, **5.** 76.

Neimann-Sørensen, A. (1958), Blood Groups of Cattle. *Thesis.* København.

## DISCUSSION

F. Pirchner: What is the inbreeding coefficient of this herd?

G. Kovács: I calculated only the degree of homozygosity.

F. Pirchner: How many bulls are used?

G. Kovács: Four in each year, some of them serving for 1–2 years while others for 5–6 years.

J. Bouw: You found a lower rate of homozygosity than expected. How were the bulls used. Each generation of bulls of other origin?

G. Kovács: Bull lines are maintained, and the offspring can be traced back to about twenty bulls or so. The distribution of the offspring among the sires is more or less equal. Of course it may occur that related individuals are mated to each other but they are never closely related.

*XIIth Europ. Conf. Anim. Blood Groups Biochem. Polymorph., Bp., 1972 (pp. 141–143)*

# STUDY ON BLOOD GROUPS IN BULGARIAN CATTLE BREEDS

Ts. Makaveev

Institute of Animal Breeding, Department of Genetics, Laboratory of Immunogenetics, Kostinbrod-Sofia, Bulgaria

## SUMMARY

The erythrocyte antigenic structure was studied in five Bulgarian cattle breeds, Grey Iskar, Rhodope, Bulgarian Brown, Bulgarian Simmental and Bulgarian Red, by means of 38 blood typing reagents.

As a result of our investigations blood group differences were found between the above cattle breeds.

The newly developed Bulgarian cattle breeds which are crosses between the well selected breeds Swiss Brown, Swiss Simmental, Red Danish and Bulgarian Grey Iskar have gene and percent frequencies of some alleles and antigens more close to those of European cattle breeds than to those of the Bulgarian native Grey Iskar breed.

## INTRODUCTION

Evidence that blood groups in cattle exhibit breed differences, was shown by Ferguson (1941). The large number of antigens in cattle allows the study of genetic differences, structure and origin of breeds. Blood group investigations of many authors, among others those by Neimann-Sørensen (1956), Rendel (1958, 1967), Müller (1960) and others were aimed at the study of the origin of, and differences between, various cattle breeds.

Cattle breeds from Bulgaria had not previously been the object of blood group studies.

## MATERIAL AND METHODS

Blood samples of over 3600 cattle were tested with 38 reagents. The animals originated from two native breeds, Grey Iskar and Rhodope and from three newly developed breeds, Bulgarian Red, Bulgarian Simmental and Bulgarian Brown. The last three breeds have been cross-bred from the Grey Iskar breed. The present studies were carried out during the years 1964–1969. All the experimental material was examined by the hemolytic test.

## RESULTS AND DISCUSSION

Percentage frequency of some antigenic factors in investigated populations of Bulgarian cattle breeds is presented in Table 1.

As a result of our investigations blood group differences related to origin were established between the examined Bulgarian cattle breeds.

TABLE 1

*Percentage frequency of antigens in Bulgarian cattle breeds*

| Antigens | Cattle breeds | | | | | | | | | |
|---|---|---|---|---|---|---|---|---|---|---|
| | Iskar | | Rhodope | | Red | | Brown | | Simmental | |
| | *n* | % | *n* | % | *n* | % | *n* | % | *n* | % |
| $A_1$ | 221 | 88.40 | 50 | 57.47 | 743 | 53.15 | 674 | 41.42 | 282 | 67.78 |
| B | 139 | 55.60 | 50 | 57.47 | 840 | 60.08 | 603 | 37.06 | 168 | 40.38 |
| G | 171 | 68.40 | 39 | 44.82 | 154 | 11.01 | 315 | 19.86 | 177 | 42.54 |
| $G_0$ | 201 | 80.40 | 61 | 70.11 | 65 | 11.00 | 373 | 44.88 | 211 | 50.72 |
| $G_2$ | 192 | 76.80 | 44 | 50.57 | 267 | 19.10 | 827 | 50.83 | 167 | 40.14 |
| I | 16 | 6.40 | 3 | 3.44 | 207 | 14.81 | 268 | 16.47 | 45 | 10.81 |
| $O_1$ | 109 | 43.60 | 62 | 71.26 | 278 | 19.88 | 517 | 31.78 | 64 | 15.38 |
| P | 172 | 68.80 | 22 | 25.28 | 236 | 16.88 | 514 | 31.59 | 120 | 28.84 |
| $Q_2$ | 111 | 44.40 | 46 | 52.87 | 121 | 8.65 | 235 | 15.41 | 170 | 40.86 |
| $T_1$ | 102 | 40.80 | 54 | 62.06 | 250 | 17.88 | 549 | 33.95 | 77 | 18.50 |
| $T_2$ | X | X | 46 | 52.87 | X | X | 191 | 29.65 | 131 | 31.49 |
| $Y_1$ | 14 | 5.60 | X | X | 223 | 22.01 | 103 | 8.24 | 4 | 0.96 |
| $Y_2$ | 68 | 27.20 | 50 | 57.47 | 593 | 42.42 | 388 | 23.85 | 139 | 33.41 |
| B′ | 181 | 72.40 | 35 | 40.22 | 161 | 11.52 | 252 | 15.49 | 70 | 16.82 |
| O′ | 67 | 26.80 | 26 | 29.88 | 415 | 29.68 | 138 | 8.48 | 129 | 31.00 |
| G′ | 81 | 32.40 | 78 | 89.65 | 214 | 15.31 | 953 | 58.57 | 215 | 51.68 |
| I′ | 49 | 19.60 | 6 | 6.89 | 73 | 5.70 | 202 | 12.41 | 125 | 30.04 |
| J′ | 22 | 8.80 | 5 | 5.86 | 135 | 9.66 | 193 | 11.86 | 25 | 6.00 |
| K′ | 102 | 40.80 | 23 | 26.43 | 220 | 19.49 | 244 | 14.99 | 62 | 14.90 |
| O′ | 27 | 10.80 | 35 | 40.22 | 158 | 13.28 | 527 | 32.39 | 54 | 12.98 |
| P′ | 55 | 22.00 | 46 | 52.87 | 197 | 16.55 | 437 | 26.86 | 25 | 6,00 |
| Y′ | 52 | 20.80 | 32 | 36.78 | 244 | 17.45 | 186 | 11.43 | 67 | 16.10 |
| $E'_3$ | X | X | 56 | 64.36 | 57 | 35.62 | 246 | 54.91 | 6 | 1.44 |
| $C_1$ | 120 | 48.00 | 32 | 36.78 | 363 | 25.96 | 607 | 37.31 | 196 | 46.39 |
| W | 167 | 66.80 | 61 | 70.11 | 717 | 51.29 | 1237 | 76.02 | 342 | 82.21 |
| $X_1$ | 186 | 74.40 | 49 | 56.31 | 146 | 10.44 | 438 | 26.92 | 138 | 33.17 |
| $X_2$ | 84 | 33.60 | 50 | 57.47 | 75 | 5.36 | 298 | 20.47 | 114 | 27.40 |
| $R_1$ | 2 | 0.80 | — | — | 80 | 5.72 | 74 | 4.97 | 33 | 7.93 |
| F | 117 | 82.97 | 73 | 83.90 | 1026 | 91.71 | 1176 | 87.44 | 114 | 89.76 |
| V | 77 | 54.60 | 53 | 60.91 | 294 | 28.66 | 640 | 47.58 | 57 | 44.88 |
| J | 139 | 55.60 | 48 | 55.17 | 226 | 16.16 | 570 | 35.03 | 148 | 35.57 |
| L | 107 | 42.80 | 44 | 50.57 | 230 | 16.45 | 513 | 31.53 | 176 | 42.30 |
| $S_1$ | 7 | 2.80 | 27 | 31.03 | 238 | 20.89 | 481 | 29.56 | 150 | 36.05 |
| $U_1$ | 47 | 18.80 | 16 | 18.39 | 100 | 7.42 | 138 | 8.04 | 29 | 6.97 |
| $U'_1$ | 93 | 37.20 | 22 | 25.28 | 359 | 25.68 | 502 | 30.85 | 106 | 25.58 |
| U″ | 94 | 37.60 | 45 | 52.87 | 83 | 7.73 | 249 | 15.30 | 40 | 9.61 |
| H″ | 66 | 26.40 | 23 | 26.43 | 70 | 6.14 | 213 | 13.09 | 31 | 7.45 |
| Z | 236 | 94.40 | 76 | 87.35 | 701 | 50.14 | 872 | 53.59 | 287 | 68.99 |

There is evidence (RENDEL, 1967) that some antigenic factors, as $U_1$ Z′ and $V_2$, have low frequencies in the Friesian and Central European cattle breeds, while relatively high frequencies in some African and Asian breeds. The Bulgarian Grey Iskar breed being of Asian origin, shows higher frequencies of the antigens $U_1$ V and Z than the three newly developed breeds Red, Simmental and Brown (Table 2).

After crossing the Grey Iskar breed with European cattle breeds, the frequency of antigens $U_1$ V and Z tended to decrease in the crosses.

TABLE 2

*Percentage frequency of certain antigens in Bulgarian native and newly developed cattle breeds*

| Cattle breeds | n | Blood group antigens, Percentage frequency $U_1$ | V | Z |
|---|---|---|---|---|
| Iskar (native) | 250 | 18.80 | 54.60 | 94.40 |
| Bulgarian Red | 1347 | 7.42 | 28.66 | 50.14 |
| Bulgarian Simmental | 416 | 6.97 | 44.88 | 68.99 |
| Bulgarian Brown | 1627 | 8.04 | 47.58 | 53.59 |

Genetic equilibrium was established on the basis of distribution of blood group factors in the FV-system (Table 3).

The observed and expected values in Grey Iskar, Rhodope, Simmental and Brown populations were similar and the chi-square test showed no statistically significant differences between them. Statistically significant difference of the chi-square test was shown exclusively by the Bulgarian Red population. This was due to predominance of F/V genotypes and diminution of V/V genotypes in comparison to the expected values.

TABLE 3

*Distribution of blood group genotypes of FV system in Bulgarian cattle breeds*

| Cattle breeds | n | Blood group genotypes FF Obs. | FF Exp. | FV Obs. | FV Exp. | VV Obs. | VV Exp. | $\chi^2$ |
|---|---|---|---|---|---|---|---|---|
| Iskar | 141 | 44 | 45.9 | 73 | 69.1 | 24 | 26.0 | 1.19 |
| Rhodope | 87 | 34 | 32.8 | 39 | 41.2 | 14 | 13.0 | 1.32 |
| Simmental | 127 | 70 | 67.3 | 12 | 9.5 | 45 | 50.2 | 1.30 |
| Red | 1072 | 826 | 833.7 | 239 | 223.4 | 7 | 14.9 | 5.34 |
| Brown | 786 | 379 | 372.3 | 342 | 337.3 | 83 | 76.4 | 1.67 |

## REFERENCES

FERGUSON, L. C. (1941), Heritable antigens in the erythrocytes of cattle. *J. Immunol.* **40**.

MAKAVEEV, Ts. (1970), Genetic polymorphism of blood groups and serum protein in native and new developed cattle and buffalo breeds. *Thesis*, 140, Sofia, Bulgaria.

MÜLLER, F. (1960), Contribution à l'étude des groupes sanguins de la race tachetée rouge de Simmental. *Zeitschr. für Tierz. und Züchtungsbiol.*, **74.** 89–105.

NEIMANN-SØRENSEN, A. (1956), Blood groups and breed structure as examplified by three Danish breeds. *Acta Agric. Scandinavica*, VI. 2.

RENDEL, J. (1958), Studies of cattle blood groups. IV. The frequency of blood group genes in Swedish cattle breeds with special reference to breed structure. *Acta Agr. Scandinavica*, VIII. 3.

RENDEL, J. (1967), Studies of blood groups and protein variants as a means of revealing similarities and differences between animal populations. *Anim. Breed. Abstr.*

*XIIth Europ. Conf. Anim. Blood Groups Biochem. Polymorph., Bp., 1972 (pp. 145–149)*

# C ALLELES IN AUSTRIAN CATTLE

W. Schleger

Blood Group Laboratory, Department of Animal Breeding and Genetics, School of Veterinary Medicine, Vienna, Austria

C alleles were studied in 5422 animals of the Austrian breeds: Fleckvieh, Braunvieh, Gelbvieh, Grauvieh and Pinzgauer. Eleven factors, $C_1$ $C_2$ E $R_1$ $R_2$ W $X_1$ $X_2$ C′ L′ and $W_2$, were determined. 56 C alleles occurred with frequencies of more than 1% in at least one of the breeds, and further 13 alleles could be recognized with frequencies of less than 1%. Breed differences are shown in the figures.

Further to gene frequencies in simpler systems frequencies of blood groups in complex systems are suitable to study relationships between and within cattle breeds.

The knowledge of the alleles of the B system has well advanced. C phenogroups were reported first by Stormont and coworkers (1951). In 1964, Nasrat, Kraay and Bouw published a comparison of C-alleles in three Dutch cattle breeds.

In our laboratory eleven reagents of the C system have been used since 1967: $C_1$ $C_2$ E $R_1$ $R_2$ W $X_1$ $X_2$ C′ L′ and $W_2$. Ten of them are internationally acknowledged and comparison-tested. The $W_2$ reagent used by us has not yet been acknowledged. It occurs as an allele $W_2$ for itself and has been demonstrated by us and others. C alleles were determined first in the Fleckvieh and Braunvieh which cover more than 50 and 15%, resp., of the Austrian cattle population, the total number of which is more than two millions. Later on, students have collected material from the less numerous breeds (Zetner, 1969; Rohrbacher, 1970; Erlacher, 1970). It should be pointed out that they found several sires and also dams that did not show reactions with any of the eleven reagents, and thus appear to be $C^c/C^c$ which was advantageous in launching these studies.

Altogether 2569 progeny from 351 different sires and 2466 dams were evaluated (5422 animals). Following the work of Nasrat and coworkers, gene frequencies were computed exclusively from the progeny alleles coming from the dams. In all other breed groups the second allele of the dam was included in the computation of gene frequency, particularly in those cases where several progeny from different sires were available from one dam.

Only directly determined alleles are reported and alleles were not allocated as was done, for example by Neimann-Sørensen (1958). We found 56 C alleles in the Austrian cattle breeds and coded them 1–56 for graphical representation (see Table 1).

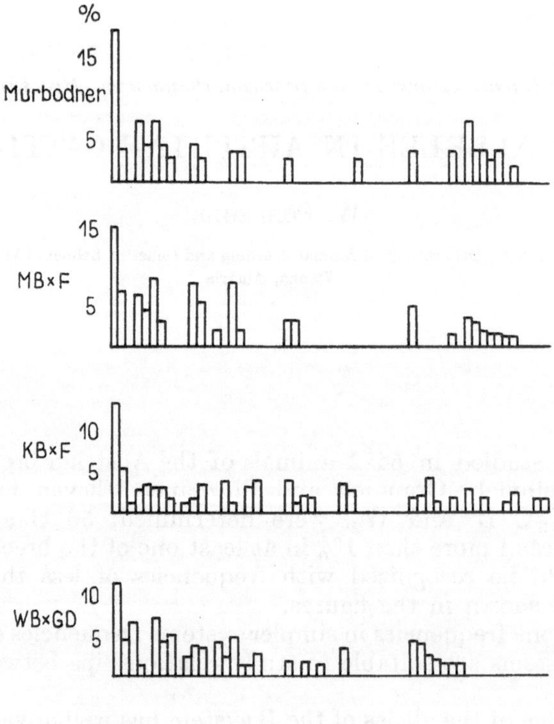
%
15
Murbodner
5
15
MB×F
5
10
KB×F
5
10
WB×GD
5
10
GV
5
12345678910
20
30
40
50
56

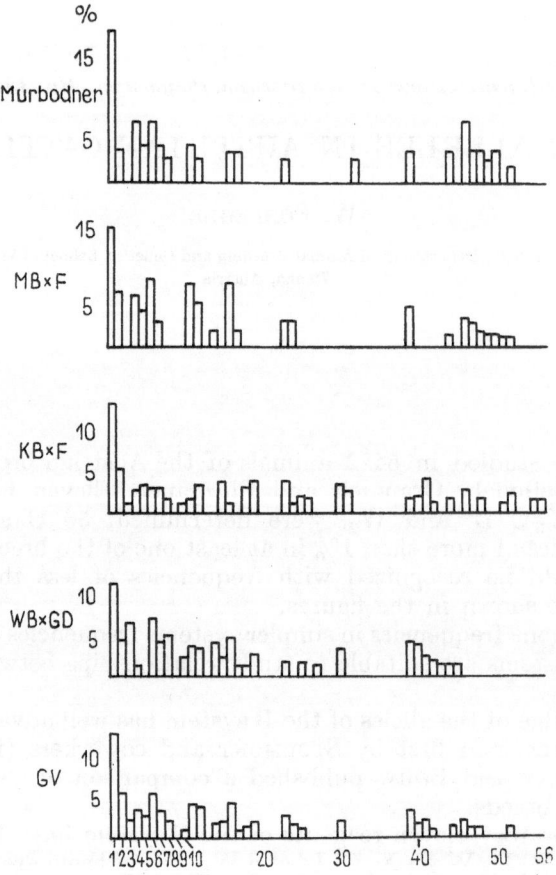

FIG. 1. Frequencies of the most common C alleles in the various Austrian Gelbvieh strains are shown. The values of the Gelbvieh are a composite of the old breeds Waldviertler Blondvieh, Kärntner Blondvieh and Murbodner, as well as the Franken- and Glandonnersberger crosses. The C alleles of Austrian Gelbvieh, if this breed continues to exist, will have roughly these frequencies. The frequencies of the C alleles for the Murbodner were separately computed to indicate the difference of the Murbodner × Franken crosses. Franken- and Glandonnersberger sires were imported from the German Federal Republic and were used for crossing with the various Austrian Gelbvieh strains

Alleles 1–21 occur in the Fleckvieh with frequencies of more than 1%, 22–29 are the Braunvieh alleles which occur in addition to the Fleckvieh alleles, 30—38 are further alleles in Grauvieh, occuring neither in Fleckvieh nor in Braunvieh, 39–45 are additional alleles in the Waldviertler and Waldviertler × Glandonnersberger, 46–52 were those in Murbodner × Frankenvieh, which occur in addition to those in the other breeds, 53–54 in the Pinzgauer, and 55–56 in the Kärtner Blondvieh × Frankenvieh.

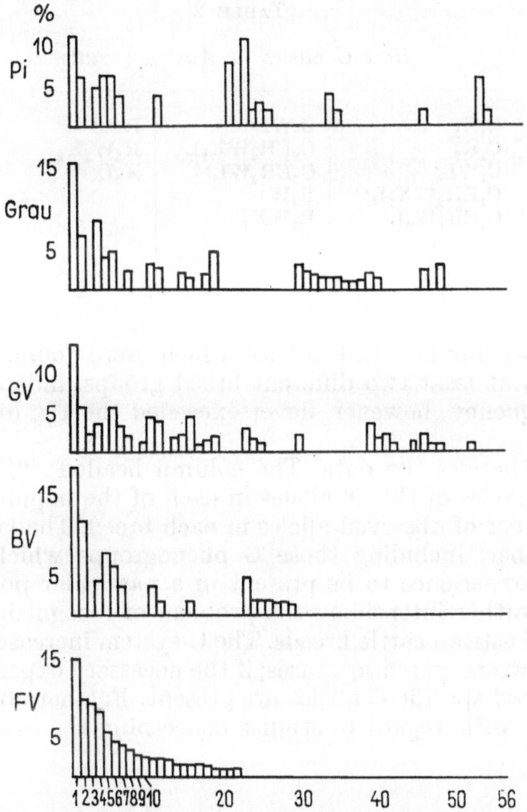
%
10
Pi
5
15
Grau
5
GV
10
5
15
BV
5
15
FV
5
1 2 3 4 5 6 7 8 9 10
20
30
40
50
56

FIG. 2. C-Blood groups in Austrian cattle. Frequencies and distribution of the C alleles in the Austrian breeds Fleckvieh, Braunvieh, Gelbvieh, Grauvieh and Pinzgauer

TABLE 1

*Code numbers of C system alleles in cattle*

| | | | | | | | |
|---|---|---|---|---|---|---|---|
| 1 | c | 15 | $R_2WX_2$ | 29 | $R_1WX_2$ | 43 | $C_2R_2W$ |
| 2 | W | 16 | C′ | 30 | $C_1ER_2W$ | 44 | $C_1ER_1WX_1$ |
| 3 | $R_2WL'$ | 17 | EW | 31 | $C_2ER_2WX_2$ | 45 | $ER_1W$ |
| 4 | $C_2EWX_2$ | 18 | $R_2$ | 32 | $C_1ER_1W$ | 46 | $C_1E$ |
| 5 | $WX_2$ | 19 | $R_2W$ | 33 | $C_2R_2X_2$ | 47 | $C_2W$ |
| 6 | $C_1EW$ | 20 | $R_2WC'$ | 34 | $C_2EWL'$ | 48 | $C_1ER_1WX_2$ |
| 7 | $C_2EW$ | 21 | EWL′ | 35 | E | 49 | $C_2ER_1WX_2$ |
| 8 | L′ | 22 | WC′ | 36 | $EWX_1$ | 50 | $C_2WX_2$ |
| 9 | WL′ | 23 | $C_1$ | 37 | $C_2R_1WX_2$ | 51 | $C_1EW_2C'$ |
| 10 | $R_1WX_1$ | 24 | $C_2$ | 38 | $C_2ER_2X_2$ | 52 | $X_1$ |
| 11 | $X_2$ | 25 | $C_1ER_2WX_2$ | 39 | $C_1W$ | 53 | $C_1EX_2$ |
| 12 | $C_1EWX_2$ | 26 | $W_2$ | 40 | $C_2ER_2W$ | 54 | $C_1EW_2$ |
| 13 | $C_1ER_2WL'$ | 27 | $C_1C'$ | 41 | $C_1ER_2WX_1$ | 55 | $WX_1$ |
| 14 | $C_1EWL'$ | 28 | $C_1R_2W$ | 42 | $EWX_2$ | 56 | $C_2EWX_2L'$ |

TABLE 2

*Rare C alleles in Austrian cattle*

| | | |
|---|---|---|
| $C_1X_2$ | $C_2WL'$ | $R_2L'$ |
| $C_1R_2$ | $C_2ER_2WX_2L'$ | $R_2W_2X_2$ |
| $C_1WL'$ | $C_2ER_2WL'$ | $X_2L'$ |
| $C_1ER_2WX_2L'$ | $R_1W$ | |
| $C_1ER_2W_2L'$ | $R_1WX_2$ | |

Table 2 shows further 13 C alleles which were found in at least three families and in at least two different breed groups, independently of each other. The frequency, however, never exceeded the 1% limit in any of the breeds.

Table 3 synthetizes the data. The column headed "%" shows some of the gene frequencies of the 56 alleles in each of the populations. Column A shows the number of observed alleles in each breed. The last column shows the total number, including those C phenogroups which were assumed, an account of experience to be present in a particular population.

We studied rather intensively the problem of recognizing and certifying of C alleles in Austrian cattle breeds. The C-system increases the probability of detection of wrong parentage cases, if the necessary experience and knowledge of the breed specific C alleles are present. Furthermore, it is of considerable interest with regard to studies concerning the evolution of breeds.

TABLE 3

*C alleles in Austrian cattle breeds*

| Breed | n | ♂ | o | ♀ | d | Alleles | % | Code-A | d. A |
|---|---|---|---|---|---|---|---|---|---|
| FV | 1510 | 132 | 669 | 669 | 669 | 480 | $88^{19}$ | 37 | 62 |
| BV | 1380 | 72 | 656 | 656 | 656 | 455 | $88^{83}$ | 36 | 66 |
| Grau | 854 | 34 | 410 | 410 | 410 | 348 | $92^{80}$ | 41 | 56 |
| Pi | 132 | 9 | 74 | 49 | 74 | 103 | $94^{14}$ | 26 | 32 |
| Gelbvieh | 1384 | 79 | 667 | 638 | 667 | 737 | $94^{31}$ | 47 | 64 |
| MB + F | 517 | 23 | 247 | 247 | 247 | 268 | $95^{79}$ | 30 | 37 |
| WB + GD | 393 | 25 | 184 | 184 | 184 | 209 | $95^{31}$ | 30 | 37 |
| KB + F | 474 | 31 | 236 | 207 | 236 | 260 | $91^{82}$ | 40 | 52 |
| MB | 162 | 25 | 93 | 44 | 93 | 108 | $98^{86}$ | 21 | 22 |
| all | 5422 | 351 | 2569 | 2466 | 2569 | 2231 | — | 56 | 125 (69) |

## REFERENCES

Bouw, J. (1964), Developments of blood group studies in cattle. Blood Groups of Animals, Proc. 9th Europ. Anim. Blood Group Conf., Prague, 25.

Erlacher, J. (in preparation), Blutgruppen und biochemischer Polymorphismus beim Tiroler Grauvieh. *Inaug. Diss.*, Wien.

Kraay, G. J. and Bouw, J. (1964), Frequencies of blood groups of the C-system in Dutch cattle breeds. *Immunogenetics Letter* **3.** 119.

Kraay, G. J. (1967), Verspreiding van Bloedgroepen in het Nederlandse Zwartbonte Rundvee. H. Veenman & Zonen N. V., Wageningen, 1967.

Nasrat, G. E. (1965), The inheritance of blood groups in the blood group system C in cattle. Blood Groups of Animals, Proc. 9th Europ. Anim. Blood Group Conf., Prague, 1964, 69.

Nasrat, G. E., Kraay, G. J. and Bouw, J. (1964), Frequencies of blood groups of the C-system in Dutch cattle breeds. *Immunogenetics Letter*, **3.** 159.

Rohrbacher, H. (1970), Blutgruppen und biochemischer Polymorphismus bei Murbodner Rindern. *Inaug. Diss.*, Wien.

Stormont, C., Owen, R. D. and Irwin, M. R. (1951), The B- and C-systems of bovine blood groups. *Genetics* **36.** 134.

Zetner, K. (1969), Blutgruppen und biochemischer Polymorphismus bei Tuxer und Pustertaler Rindern. *Inaug. Diss.*, Wien.

Zetner, K., Rohrbacher, H., Schleger, W. and Pirchner, F. (1972), Relationships between Austrian cattle breeds, as inferred from blood group and serum protein frequencies. Proc. XIIth Europ. Conf. Anim. Blood Groups. Biochem. Polymorph., Budapest, 1970.

*XIIth Europ. Conf. Anim. Blood Groups Biochem. Polymorph., Bp., 1972 (pp. 151–153)*

# STUDY ON INHERITANCE OF BLOOD CELL ANTIGEN $G_0$ OF THE B SYSTEM IN CATTLE

J. RAPACZ, MARIA DUNIEC, J. TRELA and M. DUNIEC

Immunogenetics Laboratory, Zootechnical Institute, Balice k/Kraków, Poland

## SUMMARY

The reagent anti-$G_2$ obtained after immunization with $G_2$ antigen gave much weaker reactions with red blood cells which possessed the $G_1$ antigen. After absorption with any weaker reacting red cells the immune serum hemolyzed only erythrocytes containing $G_2$ antigen. We called this reagent anti-$G_0$.

Using the reagent anti-$G_0$ over 9000 blood samples of three breeds of cattle (Lowland Black and White, Polish Red and Danish Red) were tested.

It was found that the reagent anti-$G_0$ gave reaction in all cases where the antigen $G_2$ without $G_1$ occurred. On the other hand in a group of 268 cases which were positive for the $G_1$ antigen only 38 samples reacted with the anti-$G_0$ reagent.

When the genotypes were determined for these 38 individuals it appeared that each of them had in one phenogroup the blood factor $G_1$ and in the other the $G_0$ antigen.

The results seem to indicate that the reagent anti-$G_0$ identifies an antigenic product of phenogroups containing the $G_2$ antigen, which does not occur in the product of the phenogroups with the $G_1$ specificity.

## INTRODUCTION

STONE and MILLER (1961) and GROSCLAUDE (1965) described interesting reagents which gave reactions with two and even three antigens from SU system and did not divide during absorption. In 1965 also MILLOT reported a similar case in B system and several years later DUNIEC, RAPACZ and WĘGRZYN (1970) obtained three interesting reagents which they named anti-$I_0$, anti-$G_0$ and anti-$R_0$. In the course of their work DUNIEC et al. presented the results of investigations undertaken in order to establish the relationship between the antigen $I_1$, $I_2$ and $I_0$.

The works mentioned above and, as it seems, also the present paper add some new aspects into the discussion on the genetic control of the blood groups and the specificity of the antibodies.

The work presented here is a continuation of that of DUNIEC, RAPACZ and WĘGRZYN (1970) and contains the results of investigations carried out for establishing the relationship between the antigens $G_1$, $G_2$ and $G_0$.

## RESULTS AND COMMENTS

The polyvalent serum which gave after absorption the reagent anti-$G_2$ was obtained by immunization of the Lowland Black and White cow No. 45 with the red cells of another cow of the same breed. This reagent used in

**Table 1**

*Phenogroups of the B system in which $G_1$ and $G_2$ antigens were encountered*

| No. | Phenogroups | No. of animals | No. | Phenogroups | No. of animals |
|---|---|---|---|---|---|
| 1. | $G_1$ | 27 | 33. | $BG_2KO'E'_2$ | 6 |
| 2. | $G_1O_1$ | 202 | 34. | $BG_2O_1$ | 13 |
| 3. | $G_1O_1Y_2$ | 1 | 35. | $BG_2O_1T_1Y_2Y'$ | 2 |
| 4. | $G_1O_1D'$ | 1 | 36. | $BG_2O_1QT_1K'G''_2$ | 2 |
| 5. | $G_1O_1I'$ | 1 | 37. | $BG_2I_1T_1$ | 1 |
| 6. | $G_1O_1E'_1$ | 2 | 38. | $G_2O_1$ | 18 |
| 7. | $G_1O_xA'$ | 12 | 39. | $G_2O_1QT_1K'G''_2$ | 2 |
| 8. | $G_1I_1$ | 7 | 40. | $G_2O_1T_1Y_2Y'$ | 2 |
| 9. | $G_1Y_2E'_1$ | 1 | 41. | $G_2O_1Y_2$ | 36 |
| 10. | $G_1I'$ | 1 | 42. | $G_2O_1Y_2D'$ | 1 |
| 11. | $BG_1I_1O_1T_1A'P'$ | 13 | 43. | $G_2O_1Y_2Y'$ | 10 |
| 12. | $BG_1KO_1$ | 3 | 44. | $G_2O_1Y_2G''_2$ | 2 |
| 13. | $BG_2KO_xY_1A'G'O'G''_1$ | 5 | 45. | $G_2O_xT_1Y_2Y'A'B'G'$ | 2 |
| 14. | $BG_2KO_xY_1Y'A'G'O'G''_1$ | 10 | 46. | $G_2O_xY_2A'D'G''_2$ | 10 |
| 15. | $BG_2KO_xY_1A'G'G''_1$ | 2 | 47. | $G_2O_xY_2A'P'$ | 2 |
| 16. | $BG_2KO_xY_2A'O'$ | 1393 | 48. | $G_2O_xY_2Y'E'_2$ | 2 |
| 17. | $BG_2KO_xY_2A'J'O'$ | 2 | 49. | $G_2O_xY_1A'G'G''_2$ | 2 |
| 18. | $BG_2KO_xY_2A'O'P'$ | 2 | 50. | $G_2O_xA'D'G''_2$ | 62 |
| 19. | $BG_2KO_xY_2D'O'$ | 2 | 51. | $G_2O_xA'G''_2$ | 14 |
| 20. | $BG_2KO_xY_2O'$ | 7 | 52. | $G_2O_xO'E'_1$ | 3 |
| 21. | $BG_2KO_xPQI'O'E'_1$ | 3 | 53. | $G_2O_xO'E'_1G''_2$ | 112 |
| 22. | $BG_2KO_xQT_1A'O'$ | 2 | 54. | $G_2O_xO'G''_2$ | 4 |
| 23. | $BG_2KO_xQG'O'E'_2G''_1$ | 2 | 55. | $G_2O_xO'E'_2G''_2$ | 10 |
| 24. | $BG_2KO_xA'G'G''_1$ | 11 | 56. | $G_2I_1$ | 2 |
| 25. | $BG_2KO_xG'O'E'_2G''_1$ | 20 | 57. | $G_2Y_1A'$ | 2 |
| 26. | $BG_2KO_xO'$ | 11 | 58. | $G_2Y_1D'$ | 4 |
| 27. | $BG_2KO_xO'E'_1$ | 2 | 59. | $G_2Y_2A'D'G''_2$ | 2 |
| 28. | $BG_2KO_xO'E'_2$ | 118 | 60. | $G_2Y_2I'E'_1$ | 3 |
| 29. | $BG_2KQG'E'_1G''_1$ | 2 | 61. | $G_2Y_2E'_1$ | 1577 |
| 30. | $BG_2KQG'E'_2G''_1$ | 27 | 62. | $G_2A'D'G''_2$ | 16 |
| 31. | $BG_2KQO'E'_2$ | 1 | 63. | $G_2A'G''_2$ | 7 |
| 32. | $BG_2KY_2E'_2$ | 2 | 64. | $G_2E'_1$ | 17 |

higher dilutions gave much weaker reactions with the red cells having $G_1$ antigen. After absorption of this reagent with the weaker reacting red cells there remained antibodies named anti-$G_0$ which still gave reactions with the red cells possessing the $G_2$ antigen.

In order to establish the relation between the antigens detected by the anti-$G_1$, anti-$G_2$ and anti-$G_0$ reagents over 9000 blood samples of Black and White, Polish Red, Danish Red and Simmental cattle were tested using the following reagents: $A_1$, $A_2$, H, Z′, B, $G_1$, $G_2$, $G_0$, K, $I_1$, $I_2$, $O_1$, $O_x$, P, Q, T, $Y_1$, $Y_2$, A′, B′, D′, G′, I′, J′, K′, O′, Y′, $E'_1$, $E'_2$, $G''_1$, $G''_2$, P′, Ba5, $C_1$, $C_2$, $R_1$, $R_2$, W, $X_1$, $X_2$, L′, FV, $J_1$, $J_2$, L, M, S, H′, U, U′, U″, H″, Z.

Within the tested individuals there were 64 different phenogroups in which antigens $G_1$ and $G_2$ occurred (Table 1). It was observed that anti-$G_0$ reagent gave reactions in all cases (3577) where the antigen $G_2$ without $G_1$

**TABLE 2**

*Genotypes in B system of cattle reacted with anti-$G_1$ and anti-$G_0$ reagents*

| No. | Genotype | No. of animals |
|---|---|---|
| 1. | $G_1/BG_2KO_xY_2A'O'$ | 1 |
| 2. | $G_1/BG_2I_1T_1$ | 1 |
| 3. | $G_1/G_2Y_2E'_1$ | 2 |
| 4. | $G_1O_1/BG_2KO_xY_2A'O'$ | 12 |
| 5. | $G_1O_1/BG_2KO_xY_2Y'A'G'O'G''_1$ | 1 |
| 6. | $G_1O_1/G_2O_xO'E'_1G''_2$ | 1 |
| 7. | $G_1O_1/G_2Y_2E'_1$ | 19 |
| 8. | $G_1O_xA'/BG_2KO_xO'E'_2$ | 1 |
| | | 38 |

was present. Only 38 out of 268 cases with $G_1$ antigen gave reactions with anti-$G_0$. When the genotypes were established for these 38 individuals it appeared that each of them had in one phenogroup the blood factor $G_1$ and in the other the $G_2$ antigen. Genotypes for these cows are presented in Table 2. The obtained results seem to indicate that reagent anti-$G_0$ identifies an antigenic product of phenogroups containing the $G_2$ antigen, which does not occur in the product of the phenogroups with the $G_1$ specificity.

The relationship between the antigens described above is the same as that existing between the antigens $I_1$, $I_2$ and $I_0$ already reported by DUNIEC et al. (1970).

The present work has also some practical application for anti-$G_0$ reagent discussed by us that is very useful in determination of genotypes in the B system in cattle.

## REFERENCES

DUNIEC, M., RAPACZ, J. and WĘGRZYN, J. (1970), Investigation on the production of immune antisera in cattle: three new reagents anti-$I_0$, -$G_0$, -$R_0$. Proc. XIth Europ. Conf. Anim. Blood Groups Biochem. Polymorph. Warsaw, 1968, p. 129–133.

GROSCLAUDE, F. (1965), Studies on the S blood-group system in French cattle breeds. Proc. 9th Europ. Anim. Blood Group Conf., Prague, 79–85.

MILLOT, P. (1965), Bovine isohaemolysins seeming to have several specificities. Proc. 9th Europ. Anim. Blood Group Conf., Prague, 75–78.

STONE, W. H. and MILLER, W. J. (1961), Naturally occurring isoantibodies of the S blood group system in cattle. *J. Immun.*, **86.** 165–169.

*XIIth Europ. Conf. Anim. Blood Groups Biochem. Polymorph., Bp., 1971 (pp. 155–158)*

# A STUDY OF PROTEIN AND ENZYME POLYMORPHISM IN BLOOD OF CANADIAN CATTLE

G. J. KRAAY

Department of Biomedical Sciences, University of Guelph, Guelph, Ontario, Canada

## INTRODUCTION

In the past few years, techniques have been adopted in this laboratory to type cattle for the following proteins and enzymes: serum amylase, albumin, transferrin, red cell carbonic anhydrase and hemoglobin. Since June 1969 these five systems have been used in the cattle blood-typing service as an addition to the tests for red cell antigens.

This report deals with the distribution of the genes determining the five polymorphic systems in cattle breeds in Canada.

## MATERIAL AND METHODS

The material consists of animals belonging to nine different breeds. The numbers of animals and their origin are shown in Table 1.

TABLE 1

*Composition and origin of the material*

| Breed | Sex | Number | Origin |
|---|---|---|---|
| Holstein-Friesian | M | 150 | Mostly Canadian; a few from U.S. |
| | F | 250 | All Canadian |
| Jersey | M | 25 | Canadian |
| Hereford | M | 100 | Canadian |
| Aberdeen Angus | M | 100 | Canadian |
| Shorthorn | M | 50 | Canadian |
| Charolais | M | 136 | 118 born in Canada*; 18 imported in 1969 |
| | F | 364 | 80 born in Canada*; 284 imported in 1969 |
| Limousin | M | 14 | Imported in 1969 |
| | F | 24 | Imported in 1969 |
| Simmental | M | 22 | Imported in 1969 |
| | F | 33 | Imported in 1969 |
| Pie Rouge | M | 4 | Imported in 1969 |
| | F | 14 | Imported in 1969 |

*The parents of these animals have been imported in the period 1965–1968.

The animals imported in 1969 were born in the period, December 1968 till April 1969. The Canadian Charolais were selected from the animals born in 1969 out of previously imported French Charolais. The animals of the older Canadian breeds were selected from the material tested in this laboratory for parentage verification, blood-typing of bulls used in A.I. etc.

The methods for determining the polymorphisms in this material have been described in the literature: BRAEND et al. (1962) for Hb; SARTORE et al. (1969) for CA; KRISTJANSSON and HICKMAN (1965) for Tf and GASPARSKI and STEVENS (1968) for Am and Alb.

The gene frequencies were derived by direct gene counting since the genes in all five systems are codominant. The expected numbers of genotypes based on the gene frequencies were compared with the observed numbers.

## RESULTS AND DISCUSSION

The gene frequencies are presented in Table 2.

*The Hb-system.* The data are compared with those compiled by BUSCHMANN and SCHMID (1968) and LUSH (1966). Remarkable is the high frequency

TABLE 2

*Frequencies of genes controlling the Hb-, CA-, Tf-, Am- and Alb-polymorphisms in nine cattle breeds*

| Breed | Holstein Friesian | Jersey | Hereford | Aberdeen Angus | Shorthorn | Charolais | Limousin | Simmental | Pie Rouge |
|---|---|---|---|---|---|---|---|---|---|
| No. of Animals | 400 | 25 | 100 | 100 | 50 | 500 | 38 | 55 | 18 |
| Hb -A | 1.0 | .42 | 1.0 | .99 | 1.0 | .92 | .70 | .86 | .89 |
| -B | .0 | .58 | .0 | .01 | .0 | .08 | .30 | .14 | .11 |
| CA -F | .15 | .28 | .16 | .03 | .30 | .22 | .25 | .21 | .25 |
| -S | .85 | .72 | .84 | .97 | .70 | .78 | .75 | .79 | .75 |
| Tf -A | .52 | .48 | .49 | .36 | .69 | .29 | .27 | .15 | .14 |
| -$D_1$ | .14 | .06 | .10 | .10 | .04 | .19 | .24 | .07 | .06 |
| -$D_2$ | .30 | .46 | .41 | .29 | .27 | .52 | .46 | .75 | .72 |
| -E | .04 | .0 | .0 | .25 | .0 | .0 | .03 | .03 | .08 |
| Am -A | .10 | .06 | .04 | .02 | .04 | .02 | .0 | .02 | .0 |
| -B | .34 | .38 | .44 | .39 | .16 | .57 | .66 | .92 | 1.0 |
| -C | .56 | .56 | .52 | .59 | .80 | .41 | .34 | .06 | .0 |
| Alb -F | 1.0 | 1.0 | 1.0 | 1.0 | .96 | .81 | .91 | .96 | 1.0 |
| -S | | | | | .04 | .19 | .09 | .04 | |

of $Hb^B$ in Jersey's, which is comparable to the frequency found in Jersey's on the Island of Jersey by BANGHAM and BLUMBERG (1958).

*The CA-system.* The results are compared with those published by SARTORE et al. (1969). The frequency of $Ca^F$ is slightly higher in the Canadian than in the U.S. material.

*The Tf-system.* A comparison with the data published by JAMIESON (1966) results in the following conclusions. The frequency of $Tf^E$ in the Aberdeen Angus is very high. The frequencies of Tf-genes of Charolais, British Friesians and Shorthorns in the U.K., are very similar to those in the same breeds in Canada. The Holstein–Friesian data differ from the frequencies in the Dutch Friesian (KRAAY and OOSTERLEE, 1965), but are very similar to those in Italian Friesian (FIORENTINI et al. 1968) and to those reported earlier in Canadian Holsteins by SMITHIES and HICKMAN (1958).

*The Am-system.* The data from this study indicate that the gene $Am^A$ is present in all breeds except the Limousin and Pie Rouge of which, however, a small sample was available. The data compare quite well to those obtained by GASPARSKI and STEVENS (1968).

*The Alb-system.* The data do not support the conclusion of BRAEND and EFREMOV (1965) that the allele $Alb^S$ occurs only in South European and African breeds. In the Shorthorn breed, the allele $Alb^S$ occurs with a low frequency. OSTERHOFF and PRETORIUS (1967) found the allele $Alb^S$ in Herefords and Jerseys in South Africa and SPOONER and OLIVER (1969) in many British breeds.

*Hardy–Weinberg equilibrium.* In general a good agreement was found between the observed and expected numbers. Significant differences were found only in the Am-system in Holsteins and Charolais, where an excess of homozygous and a shortage of heterozygous animals were found.

*Breed comparison.* The gene frequencies in four of the five systems are very similar in Charolais and Limousin. There is a difference only in the Hb-system. The Simmental and Pie Rouge breeds are very similar with respect to the frequencies of the genes in all five systems. This suggests that these breeds have a common origin.

*The usefulness of the five systems in parentage cases.* It is apparent that the usefulness of each of the systems depends on the breeds under study. The Hb-system is of use in the Jersey and the French breeds. The CA-system is comparable to the FV-system of the red cell antigens. The usefulness of the Tf- and Am-systems does not need to be discussed in this respect. The Alb-system has the least variation, but we use it in our work since it is a part of the Amylase gel, on which the albumins can be detected simply by staining with amido black.

## REFERENCES

BANGHAM, A. D. and BLUMBERG, B. S. (1958), Distribution of electrophoretically different haemoglobins among some cattle breeds of Europe and Africa, *Nature,* **181.** 1551.

BRAEND, M. and EFREMOV, G. (1965), Polymorphism of cattle serum albumin, *Nord. Vet. Med.,* **17.** 585.

BRAEND, M., RENDEL, J., GAHNE, B. and ADALSTEINSSON, S. (1962), Genetic studies on blood groups transferrins and hemoglobins in Icelandic cattle. *Hereditas,* **48.** 264.

BUSCHMANN, H. and SCHMID, D. O. (1968), Serumgruppen bei Tieren. Paul Parey, Berlin and Hamburg.

FIORENTINI, A., BARBIERI, V. and CERUTTI, F. (1968), Transferrin types and their frequencies in Brown Alpine and Italian Friesian cattle. *Clinica vet.*. Milano, **91.** 274.

GASPARSKI, J. and STEVENS, R. W. C. (1968), Bovine serum amylase isozymes in several breeds of domestic cattle. *Can. J. Genet. Cytol.*, **10.** 148.

JAMIESON, A. (1966), The distribution of transferrin genes in cattle. *Heredity*, **21.** 191.

KRAAY, G. J. and OOSTERLEE, C. C. (1965), Frequencies of transferrin alleles in Dutch cattle breeds. *Immunogenetics Letter*, **4.** 54.

KRISTJANSSON, F. K. and HICKMAN, C. G. (1965), Subdivision of the allele $Tf^D$ for transferrins in Holstein and Ayrshire cattle. *Genetics*, **52.** 627.

OSTERHOFF, D. R. and PRETORIUS, A. M. G. (1967), Serum albumin variants in cattle. *Proc. S. Afr. Soc. Anim. Prod.*, **6.** 224.

SARTORE, G., STORMONT, C., MORRIS, B. G. and GRUNDER, A. A. (1969), Multiple electrophoretic forms of carbonic anhydrase in red cells of domestic cattle (Bos Taurus) and American buffalo (Bison Bison). *Genetics*, **61.** 823.

SMITHIES, O. and HICKMAN, C. G. (1958), Inherited variations in the serum proteins of cattle. *Genetics*, **43.** 374.

SPOONER, R. L. and OLIVER, R. A. (1969), Albumin polymorphism in British cattle. *Anim. Prod.*, **11.** 59.

*XIIth Europ. Conf. Anim. Blood Groups Biochem. Polymorph., Bp., 1972 (pp. 159–162)*

# BREED STRUCTURES AND SIMILARITIES IN SOME HUNGARIAN CATTLE BREEDS EXAMINED BY MEANS OF THE Tf, Hb AND FV GENE FREQUENCIES

G. KOVÁCS, PÉTER SOOS and M. NEMESI

**Department of Animal Husbandry, University of Veterinary Science and Center of Artificial Insemination, Laboratory of Immunogenetics, Budapest, Hungary**

## SUMMARY

**Having determined the Tf, Hb and FV gene frequencies in seven breeds of cattle the degree of homozygosity and the similarity between breeds was estimated. For this purpose the method developed by MAIJALA and LINDSTRÖM has been adopted. The value of these polymorphic characters in parentage control tests has also been examined.**

Using immunological and biochemical methods, numerous genetic polymorphisms have been detected in the blood and various tissues of the living organisms (RENDEL, 1967).

Blood groups and other inherited polymorphic systems have proved to be valuable markers in characterizing the genetic structure of various breeds or populations.

The present study was undertaken to clarify to which degree Hb, Tf and the FV blood group systems could be used in certain population genetic examinations, as e.g. in estimating the degree of relationships between breeds, determination of homozygosity and calculation of the theoretically possible detectability ratios of erroneous parentage in several Hungarian cattle breeds. We wanted to know, whether the loci with a few alleles, which were responsible for a rather limited genetic variation in most of the cattle breeds were valuable enough for complementary examinations in routine blood-typing.

## MATERIAL AND METHODS

The seven cattle breeds from which the blood samples were collected and the number of individuals examined are listed in Table 1. The samples were subjected to Tf, Hb and FV examinations, using starch-gel electrophoresis and hemolytic tests as described elsewhere (KOVÁCS and SOOS, 1969; KOVÁCS, 1965). On the basis of gene frequency data, breed comparisons and homozygosity estimations were carried out by the method described by MAIJALA and LINDSTRÖM (1966). The theoretically possible ratio of exclusions of erroneous parentage cases were calculated according to the method of WIENER et al. (1930). The actual effectiveness of the three genetic systems

TABLE 1

*Number of animals examined for calculating gene frequencies in the Tf, Hb and FV systems of various breeds*

| Breed | Tf | Hb | FV |
|---|---|---|---|
| Hungarian Grey | 495 | 683 | 407 |
| Hungarian Spotted | 3289 | 1124 | 1127 |
| Simmental | 103 | 98 | 59 |
| Austrian Fleckvieh | 72 | 51 | 51 |
| Brown Swiss | 102 | 82 | 104 |
| Canadian Holstein-Friesian | 56 | 56 | 56 |
| Yugoslavian Holstein-Friesian | 104 | 106 | 106 |

discussed in the above mentioned statistical calculations was tested both separately and also in combination with each other. The combined method is obviously more promising as the gene frequencies in these rather simple systems are poor markers in most of the cattle breeds.

## RESULTS

The Tf, Hb and FV gene frequencies serving as a basis for the different calculations are shown in Table 2, while the results of the $\chi^2$ test in Table 3. Obviously, each of the cattle breeds examined was in genetic equilibrium on the three loci in question. The relationships between nine pairs of breeds were calculated and the results are presented in Table 4. Clearly the Hb and FV loci exhibited a quite insignificant genetic variation and as such contributed little to the differentiation of the individual breeds. The values

TABLE 2

*Tf, Hb and FV gene frequencies in seven cattle breeds of Hungary*

| Breed | Tf | | | | | Hb | | FV | |
|---|---|---|---|---|---|---|---|---|---|
| | A | $D_1$ | $D_2$ | F | E | A | B | F | V |
| Hungarian Grey | 0.4111 | 0.1455 | 0.3808 | 0.0273 | 0.0353 | 0.9348 | 0.0652 | 0.7064 | 0.2936 |
| Hungarian Spotted | 0.1981 | 0.1702 | 0.6085 | 0.0009 | 0.0223 | 0.8644 | 0.1356 | 0.8084 | 0.1916 |
| Simmental | 0.2087 | 0.1019 | 0.6845 | 0.0000 | 0.0049 | 0.8776 | 0.1224 | 0.8814 | 0.1186 |
| Austrian Fleckvieh | 0.1389 | 0.1458 | 0.6875 | 0.0000 | 0.0278 | 0.8627 | 0.1373 | 0.8628 | 0.1372 |
| Brown Swiss | 0.4510 | 0.1225 | 0.4216 | 0.0000 | 0.0049 | 0.8110 | 0.1890 | 0.7933 | 0.2067 |
| Canadian Holstein-Friesian | 0.4821 | 0.1607 | 0.2946 | 0.0000 | 0.0625 | 1.0000 | 0.0000 | 0.8125 | 0.1875 |
| Yugoslavian Holstein-Friesian | 0.4519 | 0.2933 | 0.2163 | 0.0000 | 0.0385 | 0.9858 | 0.0142 | 0.8868 | 0.1132 |

TABLE 3

*$\chi^2$-values calculated on the basis of the Tf, Hb and FV gene frequencies in seven Hungarian cattle breeds*

| Breed | Tf | Hb | FV |
|---|---|---|---|
| Hungarian Grey | 19.51 | 3.62 | 0.04 |
| Hungarian Spotted | 12.24 | 0.14 | 0.26 |
| Simmental | 3.59 | 0.19 | 1.08 |
| Austrian Fleckvieh | 9.65 | 1.25 | 1.25 |
| Brown Swiss | 5.58 | 0.45 | 0.07 |
| Canadian Holstein-Friesian | 3.98 | 0.00 | 0.82 |
| Yugoslavian Holstein-Friesian | 9.11 | 0.00 | 1.68 |

of similarities varied between 70.41 and 99.23 with the above three loci taken into account.

The expected and observed values of homozygosity were in good agreement as can be seen from Table 5. In the case of the Canadian Holstein–Friesian population, homozygosity is 100% on the Hb locus. This can be explained by the relatively small number of animals examined, or by the obvious tendency toward fixation on this locus. The latter explanation seems to be more reliable as the corresponding values of the Yugoslavian Holstein–Friesian population are nearly the same. It is interesting, however, that the summarized value of homozygosity does not exhibit any striking difference between breeds due to the low percentage of homozygosity on the Tf locus in the two breeds. This example stresses the importance of including as many loci as possible in the estimation of homozygosity.

TABLE 4

*Relationship between breeds expressed in percentual terms*

| Breed | Hb | FV | Tf | Hb + FV | FV + Tf | Hb+FV+ Tf |
|---|---|---|---|---|---|---|
| Hungarian Spotted — Hungarian Grey | 99.49 | 98.60 | 88.08 | 98.09 | 86.85 | 86.40 |
| Hungarian Spotted — Austrian Fleckvieh | 99.97 | 99.71 | 98.82 | 99.68 | 98.53 | 98.50 |
| Hungarian Grey — Austrian Fleckvieh | 99.67 | 97.27 | 82.39 | 96.95 | 80.14 | 79.88 |
| Hungarian Grey — Simmental | 99.76 | 96.65 | 73.10 | 96.32 | 70.65 | 70.41 |
| Hungarian Spotted — Simmental | 99.99 | 99.55 | 99.23 | 99.54 | 98.78 | 98.77 |
| Austrian Fleckvieh — Simmental | 99.97 | 99.97 | 99.29 | 99.94 | 99.26 | 99.23 |
| Canadian Holstein Friesian — Yugoslavian Holstein Friesian | 99.99 | 99.50 | 96.33 | 99.49 | 95.85 | 95.84 |
| Simmental — Brown Swiss | 99.59 | 99.28 | 86.86 | 98.87 | 86.23 | 85.88 |
| Canadian Holstein-Friesian — Hungarian Spotted | 98.79 | 99.99 | 77.75 | 98.78 | 77.74 | 76.80 |

TABLE 5

*Percentages of homozygosity in seven cattle breeds*

| Breed | | Tf | Hb | FV | Summarized percentage |
|---|---|---|---|---|---|
| Hungarian Grey | exp. | 33.72 | 87.80 | 58.52 | 17.33 |
| | obs. | 30.30 | 86.97 | 58.97 | 15.54 |
| Hungarian Spotted | exp. | 43.90 | 76.53 | 69.00 | 23.18 |
| | obs. | 44.80 | 76.24 | 69.48 | 23.73 |
| Simmental | exp. | 52.25 | 78.52 | 79.09 | 32.45 |
| | obs. | 52.43 | 77.55 | 76.27 | 31.01 |
| Austrian Fleckvieh | exp. | 51.40 | 76.43 | 76.32 | 29.98 |
| | obs. | 56.94 | 72.55 | 72.55 | 29.97 |
| Brown Swiss | exp. | 39.62 | 69.34 | 67.20 | 18.76 |
| | obs. | 43.14 | 67.07 | 66.35 | 19.20 |
| Canadian Holstein-Friesian | exp. | 34.89 | 100.00 | 69.53 | 24.26 |
| | obs. | 35.71 | 100.00 | 73.21 | 26.14 |
| Yugoslavian Holstein-Friesian | exp. | 33.85 | 97.20 | 79.92 | 26.29 |
| | obs. | 31.73 | 97.17 | 77.36 | 23.85 |

As far as the theoretically possible highest ratio of exclusions of erroneous parentage cases is concerned, the data of Table 6 indicate the suitability of using different loci in parentage control tests.

The data show that Tf and Hb examination, as supplements to blood-typing, significantly increase the effectiveness of parentage control tests.

TABLE 6

*Theoretically possible percentage of exclusions on the basis of the Tf, Hb and FV systems*

| Systems used | Exclusion ratios in the different breeds | | | | | | |
|---|---|---|---|---|---|---|---|
| | Hungarian Grey | Hungarian Spotted | Simmental | Austrian Fleckvieh | Brown Swiss | Canadian Holstein-Friesian | Yugoslavian Holstein-Friesian |
| Hb | 5.72 | 10.35 | 9.59 | 10.44 | 12.98 | 0.00 | 1.38 |
| FV | 16.44 | 13.09 | 9.36 | 10.44 | 13.71 | 12.91 | 9.03 |
| Tf | 38.37 | 30.65 | 21.34 | 22.05 | 26.02 | 29.94 | 30.19 |
| Hb + FV | 21.22 | 22.09 | 18.05 | 19.80 | 24.91 | 12.91 | 10.29 |
| Hb + Tf | 41.90 | 37.83 | 28.88 | 30.54 | 35.62 | 29.94 | 31.15 |
| Tf + FV | 48.50 | 39.73 | 28.70 | 30.54 | 36.16 | 38.98 | 36.49 |
| Hb + FV + Tf | 51.45 | 45.97 | 35.54 | 44.29 | 44.45 | 39.98 | 37.37 |

## REFERENCES

KOVÁCS, G. (1965), *Magy. Áo. L.*, **8.** 343–347.
KOVÁCS, G. and SOOS, P. (1969), *Magy. Áo. L.*, **12.** 662–669.
MAIJALA, K. and LINDSTRÖM, G. (1966), *Ann. Agric. Fenniae*, **5.** 76.
RENDEL, J. (1967), *Anim. Breed. Abstr.*, **35.** 371–383.
WIENER, A. S., LEDERER, M. and POLAYES, S. H. (1930), *J. Immun.*, **19.** 259–282.

*XIIth Europ. Conf. Anim. Blood Groups Biochem. Polymorph., Bp., 1972 (pp. 163–171)*

# POLYMORPHISM OF TRANSFERRINS IN GERMAN CATTLE

H. GELDERMANN

Tierärztliches Institut der Universität Göttingen, GFR

## SUMMARY

**German registered "Schwarzbunte" and "Rotbunte" cattle populations were used for a genetical study of transferrins. Calculations of gene and genotype frequencies, homozygosity and inbreeding coefficients and their relationship to age, sex, breed and regional breed associations were made. Transferrin polymorphism was explained using the values obtained from the Hardy–Weinberg equilibrium.**

## INTRODUCTION

In registered German "Schwarzbunte" and "Rotbunte" cattle populations the horizontal starch-gel electrophoresis was used for the study of different transferrin types. The types were named according to the groups

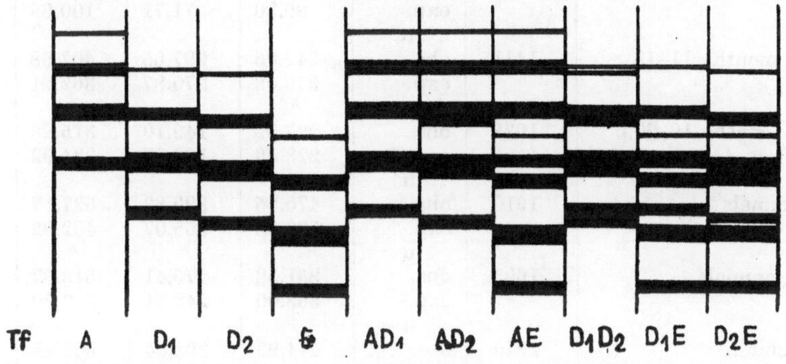
Tf
A
D1
D2
AD1
AD2
AE
D1D2
D1E
D2E

FIG. 1. The transferrin types in the "Deutsche Schwarzbunte" and "Deutsche Rotbunte" breeds

of electrophoretic fractions as A, $AD_1$, $AD_2$, AE, $D_1$, $D_1D_2$, $D_1E$, $D_2$, $D_2E$ or E (Fig. 1). These ten phenotypes are controlled by four codominant and autosomal alleles $Tf^A$, $Tf^{D_1}$, $Tf^{D_2}$ and $Tf^E$ (ASHTON, 1957, 1958, 1965; HICKMAN and SMITHIES, 1957; SMITHIES and HICKMAN, 1958; KRISTJANSSON and HICKMAN, 1965).

## MATERIAL AND METHODS

The investigations of the factors influencing the transferrin polymorphism were made on 3818 German registered cattle. For each animal, age, sex, breed and regional breed associations were recorded.

Animals with distinct criteria were collected and analyzed in equal groups. On such groups the following calculations were made:

1. Computation of the genotype and gene frequencies. The observed genotype frequencies were compared with the expected Hardy–Weinberg frequencies which were calculated from the gene frequencies.
2. Estimation of the proportional homozygous transferrin genotypes in the so-called degree of homozygosity for the observed and for the expected genotype frequencies.
3. Comparison between the observed and the expected frequencies of homo- or heterozygotic genotypes by means of the inbreeding coefficient (WRIGHT, 1921, 1922).

The differences between values from two groups or between observed and expected frequencies were assessed by Chi-square, *t*-test and *u*-values (FISHER, 1948; WEBER, 1967). The equations used were given by GELDERMANN (1969).

TABLE

*Observed and expected genotype*

| Group | Number of animals | | Frequencies of A | $AD_1$ | $AD_2$ | AE |
|---|---|---|---|---|---|---|
| Total population | 3818 | obs. | 858.60 | 720.80 | 1103.13 | 182.43 |
| | | exp. | 907.87 | 649.02 | 1061.81 | 197.00 |
| | | u | | ** | | |
| Age of months 1–10 | 375 | obs. | 82.96 | 75.69 | 113.15 | 11.27 |
| | | exp. | 89.30 | 71.71 | 100.68 | 15.02 |
| | | u | | | | |
| Age of months 11–18 | 1411 | obs. | 342.06 | 297.65 | 407.58 | 68.51 |
| | | exp. | 376.58 | 270.87 | 362.21 | 71.63 |
| | | u | * | | ** | |
| Age of months 19–203 | 1038 | obs. | 227.82 | 149.10 | 315.88 | 41.75 |
| | | exp. | 223.06 | 136.07 | 334.92 | 45.25 |
| | | u | | | | |
| Male animals | 1910 | obs. | 470.25 | 399.83 | 527.17 | 99.75 |
| | | exp. | 506.55 | 365.07 | 482.92 | 106.15 |
| | | u | | * | * | |
| Female animals | 1695 | obs. | 351.38 | 275.41 | 515.87 | 74.76 |
| | | exp. | 363.00 | 248.51 | 512.90 | 81.40 |
| | | u | | | | |
| Schwarzbunte | 1975 | obs. | 474.93 | 392.04 | 566.21 | 87.25 |
| | | exp. | 503.99 | 360.53 | 526.50 | 100.36 |
| | | u | | | * | |
| Rotbunte | 1014 | obs. | 240.72 | 228.89 | 269.47 | 60.87 |
| | | exp. | 267.01 | 198.52 | 242.63 | 65.50 |
| | | u | | * | * | |

* = $P < 0.05$ ** = $P < 0.01$ *** = $P < 0.001$

## RESULTS

In the investigated cattle populations the frequencies of the transferrin types showed deviations from the Hardy–Weinberg equilibrium frequencies (Table 1). In the population investigated a significant difference was found between the genotypes $Tf^A/Tf^{D_1}$ ($P < 0.005$) and between $Tf^{D_1}/Tf^{D_2}$ ($P < < 0.001$). These differences were evident in each breed as well.

The distribution of the transferrin type frequencies in the different ages showed that in the age group from 11 to 18 months, the genotypes $Tf^A/Tf^A$ and $Tf^{D_1}/Tf^{D_2}$ were less proportional ($P < 0.05$ and $P < 0.001$, respectively) whereas the genotype $Tf^A/Tf^{D_2}$ was more frequent than expected ($P < 0.001$). The animals in the age group from 19 to 203 months had fewer $Tf^{D_1}/Tf^{D_2}$ types ($P < 0.005$) and more of the type $Tf^{D_2}/Tf^{D_2}$ ($P < 0.025$). The age-related changes in the transferrin type frequencies act on the gene frequencies (Fig. 2). The frequencies of $Tf^A$ and $Tf^{D_1}$ decreased ($P < 0.001$) in the age group of 11 to 18 months whereas the frequency of $Tf^{D_2}$ increased ($P < 0.001$).

1

*frequencies in transferrins*

| transferrin types | | | | | | $\chi^2$ |
|---|---|---|---|---|---|---|
| $D_1$ | $D_1D_2$ | $D_1E$ | $D_2$ | $D_2E$ | E | |
| 122.55 | 278.95 | 86.11 | 340.97 | 113.46 | 11.00 | *** |
| 115.99 | 379.53 | 70.42 | 310.46 | 115.20 | 10.69 | |
| | *** | | | | | |
| 13.61 | 34.79 | 9.25 | 24.06 | 10.27 | 0.00 | |
| 14.39 | 40.42 | 6.03 | 28.38 | 8.47 | 0.63 | |
| 48.27 | 97.90 | 32.24 | 82.87 | 29.91 | 4.00 | ** |
| 48.71 | 130.27 | 25.76 | 87.10 | 34.45 | 3.41 | |
| | *** | | | | | |
| 27.94 | 72.81 | 15.76 | 150.88 | 32.04 | 4.03 | * |
| 20.75 | 102.15 | 13.80 | 125.72 | 33.97 | 2.29 | |
| | ** | | * | | | |
| 65.97 | 127.70 | 49.43 | 118.98 | 44.92 | 6.02 | ** |
| 65.77 | 174.02 | 38.25 | 115.10 | 50.60 | 5.56 | |
| | *** | | | | | |
| 52.66 | 126.65 | 29.61 | 202.14 | 61.51 | 5.00 | ** |
| 42.53 | 175.56 | 27.86 | 181.17 | 57.51 | 4.56 | |
| | *** | | | | | |
| 67.06 | 146.42 | 41.12 | 136.70 | 56.22 | 7.04 | * |
| 64.48 | 188.32 | 35.90 | 137.51 | 52.42 | 5.00 | |
| | ** | | | | | |
| 28.97 | 67.07 | 32.98 | 54.24 | 27.79 | 3.00 | * |
| 36.90 | 90.20 | 24.35 | 55.12 | 29.76 | 4.02 | |
| | * | | | | | |

Sbt = Deutsche Schwarzbunte
Rbt = Deutsche Rotbunte
VRS = Verband Rheinischer Schwarzbuntzüchter
WH = Westfälisches Herdbuch
VOSt = Verband Ostfriesischer Stammviehzüchter
SSH = Schwarzbunte Schleswig-Holsteins
VRR = Verband Rheinischer Rotbuntzüchter
RSH = Rotbunte Schleswig-Holsteins

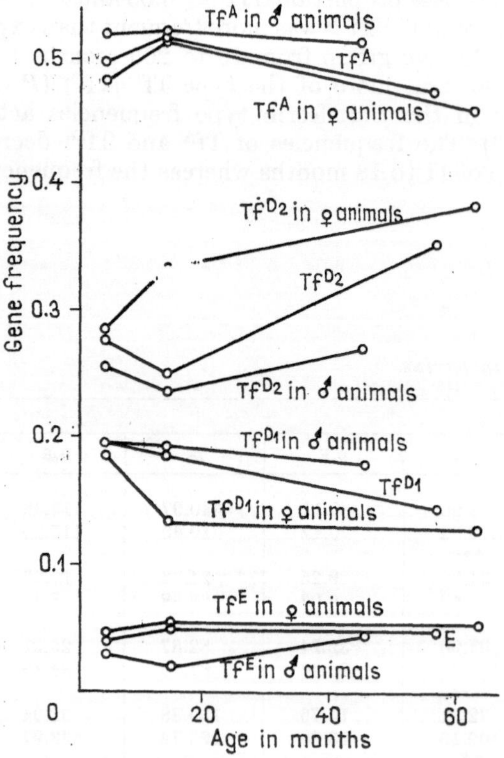

TfA in ♂ animals
TfA
TfA in ♀ animals
TfD2 in ♀ animals
TfD2
TfD2 in ♂ animals
TfD1 in ♂ animals
TfD1
TfD1 in ♀ animals
TfE in ♀ animals
E
TfE in ♂ animals
Gene frequency
0.5
0.4
0.3
0.2
0.1
0
20
40
60
Age in months


FIG. 2. The gene frequencies of transferrins in the three age groups

The frequencies of the transferrin types seem to be sex related. Bulls have the genotypes $Tf^A/Tf^{D_1}$ and $Tf^A/Tf^{D_2}$ more often than expected ($P < 0.05$ and $P < 0.025$, respectively) and have fewer $Tf^{D_1}/Tf^{D_2}$ types ($P < 0.001$). A smaller than expected observed frequency for the $Tf^{D_1}/Tf^{D_2}$ type could be found also in cows ($P < 0.001$). The gene frequencies of $Tf^A$ and $Tf^{D_1}$ were higher in bulls ($P < 0.001$ and $P < 0.05$, respectively), whereas the females had a higher frequency of the transferrin allele $Tf^{D_2}$ ($P < 0.001$) (Figs 2 and 3).

The relation between the portion of homozygous transferrin types and the age of the cattle was very interesting. Figure 4 shows that there are less homozygous transferrin types than expected according to the gene frequencies at equilibrium conditions at the age of 1 to 18 months ($P <$

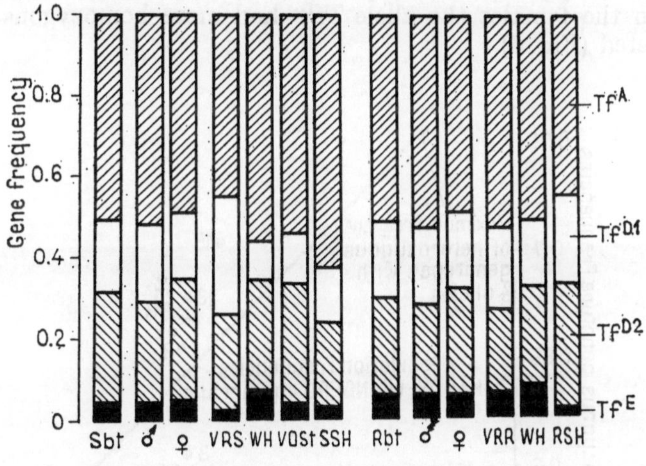
Gene frequency
1.0
0.8
0.6
0.4
0.2
0
Sbt ♂ ♀ VRS WH VOSf SSH Rbt ♂ ♀ VRR WH RSH
Tf^A
Tf^D1
Tf^D2
Tf^E

FIG. 3. The frequencies of transferrin alleles in the investigated cattle breeds and regional breed organizations

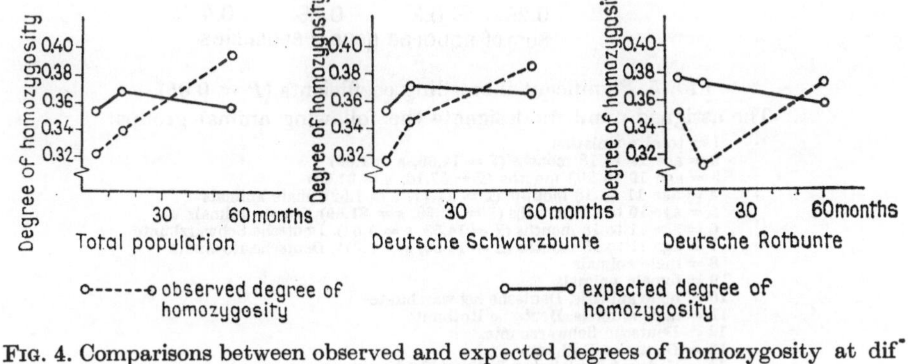
Degree of homozygosity
0.40
0.38
0.36
0.34
0.32
30
60 months
Total population
Deutsche Schwarzbunte
Deutsche Rotbunte
observed degree of homozygosity
expected degree of homozygosity

FIG. 4. Comparisons between observed and expected degrees of homozygosity at different ages

$< 0.001$). However the degree of homozygosity at the age of 19 to 203 months showed an increase ($P < 0.025$).

These findings were confirmed by calculating the inbreeding coefficients. Figure 5 shows, that irrespective of the group 19 to 203 months old, the significant inbreeding coefficients were always negative. This can be interpreted in such a way that more heterozygous genotypes are present than expected with the Hardy–Weinberg equilibrium. Considering the transferrin alleles separately (Fig. 6), allele $Tf^A$ showed negative inbreeding coefficients whereas allele $Tf^{D_2}$ had positive inbreeding coefficients. No significant inbreeding coefficients were found for the alleles $Tf^{D_1}$ and $Tf^{E}$.

Also for a part of homozygous and heterozygous genotypes it was possible to detect sex related differences (Fig. 4). In the males there was a higher heterozygosity than expected. This was related to the allele $Tf^{A}$ (Fig. 6). In the females the allele $Tf^{D_2}$ had more homozygous genotypes than expected (Fig. 6).

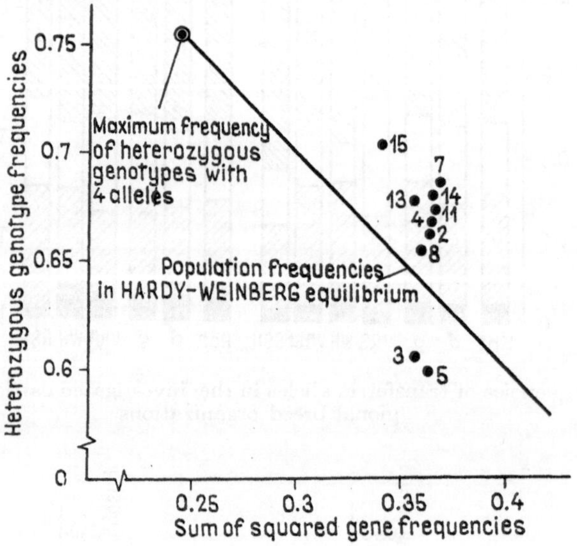
Heterozygous genotype frequencies
0.75
0.7
0.65
0.6
0
Maximum frequency of heterozygous genotypes with 4 alleles
Population frequencies in HARDY-WEINBERG equilibrium
15
7
13
14
4
11
2
8
3
5
0.25
0.3
0.35
0.4
Sum of squared gene frequencies

FIG. 5. Significant inbreeding coefficients ($P < 0.05$)
The assigned numbers designate the following animal groups:

1 = total population
2 = age 11 to 18 months ($\bar{x} = 14.69$, $s = 1.59$)
3 = age 19 to 203 months ($\bar{x} = 57.18$, $s = 31.58$)
4 = age 11 to 18 months ($\bar{x} = 14.71$, $s = 1.52$), male animals
5 = age 19 to 203 months ($\bar{x} = 63.36$, $s = 31.69$), female animals
6 = age 11 to 18 months ($\bar{x} = 14.53$, $s = 1.64$), Deutsche Schwarzbunte
7 = age 11 to 18 months ($\bar{x} = 14.90$, $s = 1.31$), Deutsche Rotbunte
8 = male animals
9 = female animals
10 = male animals, Deutsche Schwarzbunte
11 = male animals, Deutsche Rotbunte
12 = Deutsche Schwarzbunte
13 = Deutsche Rotbunte
14 = Verband Rheinischer Rotbuntzüchter
15 = Verband Rheinischer Schwarzbuntzüchter
16 = Schwarzbunte Schleswig-Holsteins
17 = Rotbunte Schleswig-Holsteins

The deviations of the transferrin genotype frequencies from the Hardy-Weinberg equilibrium and the observed inbreeding coefficients suggest that the fitness of the animals had an influence on the transferrin genotypes. The minimized expect value for a genotype can be compared with the favoured genotypes. Taking these differences as a relative reduction in the gamete development, one can calculate the coefficient of selection, $s$

(FALCONER, 1964). Using the following terms for each transferrin allele an example for the coefficients of selection of the allele $Tf^A$ is shown:

$s_1$ = coefficient of selection of genotype –/–
$s_2$ = coefficient of selection of genotype $Tf^A/Tf^A$
$q$ = frequency of $Tf^A$
$p$ = common frequency of $Tf^{D_1}$, $Tf^{D_2}$ and $Tf^E$

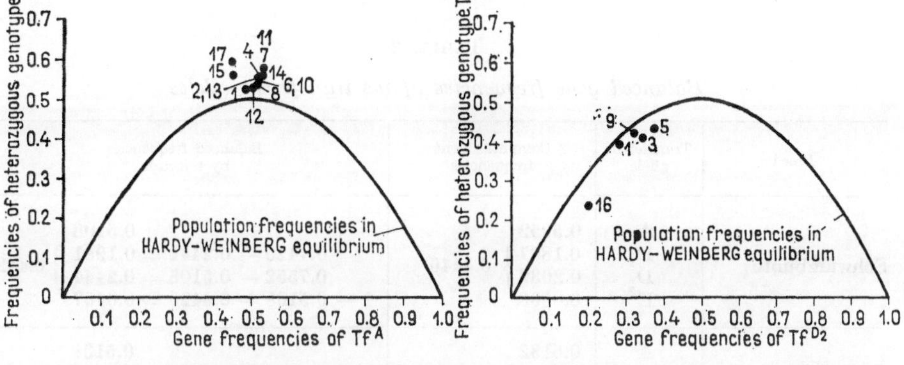
Frequencies of heterozygous genotype TfA/-
Population frequencies in HARDY-WEINBERG equilibrium
Gene frequencies of TfA
Frequencies of heterozygous genotype TfD2/-
Population frequencies in HARDY-WEINBERG equilibrium
Gene frequencies of TfD2

FIG. 6. Significant inbreeding coefficients ($P < 0.05$) with separated consideration of $Tf^A$ or $Tf^{D_2}$. The numbers are explained in Fig. 5

Assuming that there is a constant frequency in both breeds of cattle we can derive the following equation:

$$s_1 p = s_2 q$$

$$q = \frac{s_1}{s_1 + s_2}.$$

The calculation of the selection coefficients for allele $Tf^A$ in the "Deutsche Schwarzbunte" breed was carried out by using the differences between observed and expected genotype frequencies:

| Transferrin genotypes | Gametic contribution | Coefficient of selection |
|---|---|---|
| $Tf^A/Tf^A$ | $1 - \frac{29.06}{503.99}$ | $s_2 = \frac{29.06}{503.99} = 0.0577$ |
| $Tf^A/-$ | 1 | |
| $-/-$ | $1 - \frac{29.07}{483.63}$ | $s_1 = \frac{29.07}{483.63} = 0.0601$ |

From the coefficient of selection calculations $Tf^A$ frequency is

$$q = \frac{0.0601}{0.0601 + 0.0577} = 0.5108$$

Considering only two different allele situations at one time, the selected balanced frequencies for the 4 transferrin alleles can be evaluated. The balanced frequencies are given in Table 2 which shows that they are in agreement with the observed frequencies.

TABLE 2

*Balanced gene frequencies of the transferrin alleles*

| Breed | Transferrin allele | Computed gene frequencies | | Balanced frequencies by selection | |
|---|---|---|---|---|---|
| Schwarzbunte | A | 0.5052 | | 0.5108 | |
| | $D_1$ | 0.1807 | 0.4446 | 0.4425 — 0.2444 = 0.1981 | 0.4425 |
| | $D_2$ | 0.2639 | | 0.7552 — 0.5108 = 0.2444 | |
| | E | 0.0503 | | 1 — 0.5108 — 0.4425 = 0.0467 | |
| Rotbunte | A | 0.5132 | | 0.5138 | |
| | $D_1$ | 0.1908 | 0.4239 | | 0.4194 |
| | $D_2$ | 0.2331 | | | |
| | E | 0.0629 | | 1 — 0.5138 — 0.4194 = 0.0668 | |

From that the frequencies of allele $Tf^A$ in one case and the alleles $Tf^{D_1}$ and $Tf^{D_2}$ in another case were maintained by their superior heterozygous genotypes. The combined frequencies of $Tf^A$ and $Tf^{D_2}$ were maintained in many of the genotypes with both alleles or without both alleles. With the above selection mechanisms, the frequencies of $Tf^{D_1}$ and $Tf^E$ may be fixed and their genotypes are in agreement with the Hardy–Weinberg equilibrium. In the populations investigated the heterozygous genotypes seemed to be more prevalent in animals under one year of age, whereas the homozygous genotypes appeared more often in older animals.

## REFERENCES


ASHTON, G. C. (1957), Serum protein differences in cattle by starch gel electrophoresis. *Nature* (Lond.) **180.** 197–199.

ASHTON, G. C. (1958), Genetics of beta-globulin polymorphism in British cattle. *Nature* (Lond.) **182.** 370–372.

ASHTON, G. C. (1965), Serum transferrin d alleles in Australian cattle. *Austral. J. biol. Sci.*, **18.** 665–670.

FALCONER, D. S. (1964), Introduction to quantitative genetics. Oliver and Boyd, Edinburgh and London.

FISHER, R. A. (1948), Statistical Methods for Research Workers. London, 10th ed.

GELDERMANN, H. (1969), Darstellung und Beschreibung des Transferrin- und Postalbuminpolymorphismus bei einigen Deutschen Rinderrassen. *Diss.* Göttingen.

HICKMAN, C. G. and SMITHIES, O. (1957), Evidence for inherited differences in the serum proteins of cattle (Abstr.). *Proc. Genetics Soc. Canada,* **2.** 39.

KRISTJANSSON, F. K. and HICKMAN, C. G. (1965), Subdivision of the allele $Tf^D$ for transferrins in Holstein and Ayrshire cattle. *Genetics,* **52.** 627–630.

SMITHIES, O. and HICKMAN, C. G. (1958), Inherited variations in the serum proteins of cattle. *Genetics,* **43.** 374–375.

WEBER, E. (1967), Grundriss der biologischen Statistik. VEB Gustav Fischer Verlag, Jena, 6th ed.

WRIGHT, S. (1921), Systems of mating. *Genetics,* **6.** 111–178.

WRIGHT, S. (1922), Coefficients of inbreeding and relationship. *Amer. Nat.*, **56.** 330–338.

## DISCUSSION

D. R. OSTERHOFF: It was of interest to notice the increase of $Tf^{D_2}$ with age. The same increase had been found a long time ago in our studies and was later on confirmed by Ashton. What would your explanation be for this increase?

H. GELDERMANN: The changes of the transferrin alleles $Tf^{D_2}$ and $Tf^E$ in animals of different age may be influenced by the selection of the alleles $Tf^A$ and $Tf^{D_1}$. There was no statistical significance for any selection acting on the alleles $Tf^{D_2}$ and $Tf^E$ found in the material investigated.

*XIIth Europ. Conf. Anim. Blood Groups Biochem. Polymorph., Bp., 1972 (pp. 173—179)*

# ELECTROPHORETIC INVESTIGATIONS ON POLYMORPHISM OF AMYLASE IN ITALIAN CATTLE BREEDS

C. CRIMELLA, F. CERUTTI and G. ROGNONI

Istituto di Zootecnia Generale, Università di Milano, Milan, Italy

## SUMMARY

Some aspects of the electrophoretic behaviour of amylase are discussed.

The genetic manifestations and gene frequencies of the amylase system have been studied in the Italian cattle breeds.

This genetic system has been studied also with regard to its associations with the other protein-enzyme polymorphisms of cattle blood.

## INTRODUCTION

Studying enzymic polymorphisms we have wished to obtain more information on the electrophoretic manifestations of amylase, as earlier investigations in this laboratory brought to light electrophoretic pattern situations which led to the anticipation of interesting results (CRIMELLA, 1969).

## MATERIAL AND METHODS

Amylase was studied by the method suggested by ASHTON (1958, 1965, 1966a, 1966b), adapted to our requirements in certain details favouring a greater differentiation of the electrophoretic bands.

Here we describe only those parameters which underwent variation (see Table 1). At a later date, as all the subjects considered for the investigation

TABLE 1

*a)* Thickness of gel 6 mm (13% Connaught Starch Gel 262/1).

*b)* Paper insert MN 866 mm $10 \times 6 \times 1.7$.

*c)* Insertion time 30′ = voltage should be sufficient to produce a current of 2.7 mA per cm.

*d)* Running time 2h 30′ = voltage 13/14 V per cm width (5 mA); final current should be about 6.5/7 mA per cm.

*e)* 1% NN-dimethyl-p-phenylenediamin (p-Amino-NN-dimethyl anilin) in aqueous solution.

*f)* Incubation at 40°C—41°C for 18 hours.

*g)* Washing solution: 5 : 5 : 1; methanol : water: acetic acid, for two hours, after this period gel transferred to water.

TABLE 2

*Serum amylase phenotypes and estimated gene frequencies in Italian Cattle**

| Breeds | Number of animals | Phenotype frequencies | | | $\chi^2$ | P** | Gene frequencies | |
|---|---|---|---|---|---|---|---|---|
| | | BB | BC | CC | | | Am B | Am C |
| Italian Friesian | 544 | 175 | 260 | 109 Obs. | | | | |
| | | 171.000416 | 267.996160 | 105.002336 exp. | 0.484326 | 0.98 P 0.95 | 0.56066 | 0.43934 |
| Italian Brown | 290 | 176 | 96 | 18 obs. | | | | |
| | | 173.020670 | 101.958490 | 15.020840 exp. | 0.990390 | 0.7 P 0.50 | 0.77241 | 0.22759 |
| Chianina | 256 | 166 | 78 | 12 obs. | | | | |
| | | 164.160100 | 81.679600 | 10.160200 exp. | 0.519517 | 0.8 P 0.7 | 0.80078 | 0.19922 |
| Piemontese | 88 | 36 | 42 | 10 obs. | | | | |
| | | 36.920400 | 40.159152 | 10.920448 exp. | 0.18497 | 0.95 P 0.90 | 0.64773 | 0.35227 |
| Valdostana | 64 | 47 | 14 | 3 obs. | | | | |
| | | 45.562494 | 16.874944 | 1.562496 exp. | 1.857660 | 0.5 P 0.3 | 0.84375 | 0.15625 |

* only two alleles were found in this study
** two degrees of freedom

TABLE 3

*Considered loci*

| No. of Sire | Tf×Ca | | Tf×Am | | Am×Ca | |
|---|---|---|---|---|---|---|
| | conc. | disc. | conc. | disc. | conc. | disc. |
| 1 | 4 | 1 | 3 | 1 | 6 | — |
| 2 | — | — | 8 | 1 | — | — |
| 3 | — | — | 4 | 1 | — | — |
| 4 | 3 | — | 1 | 1 | 1 | 1 |
| 5 | — | — | 12 | 5 | — | — |
| 6 | — | — | — | — | 3 | 1 |
| 7 | — | — | 1 | 1 | — | — |
| 8 | — | — | 7 | 5 | — | — |
| 9 | — | — | 2 | 2 | — | — |
| 10 | 2 | — | 3 | — | 3 | — |
| 11 | 1 | 1 | 2 | 2 | 1 | 1 |
| 12 | 2 | 1 | 2 | — | 1 | 1 |
| 13 | 3 | 1 | — | — | — | — |
| 14 | 6 | 5 | 7 | 2 | 6 | 4 |
| 15 | 4 | 4 | 6 | 4 | 4 | 4 |
| 16 | 8 | 7 | 10 | 6 | 9 | 4 |
| 17 | — | — | 8 | 6 | — | — |
| 18 | 3 | 3 | 1 | — | 4 | — |
| 19 | — | — | 3 | 2 | — | — |
| 20 | 2 | — | — | — | — | — |
| 21 | — | — | 3 | 2 | — | — |
| 22 | 4 | — | 4 | 1 | 2 | — |
| 23 | — | — | 4 | 2 | — | — |
| 24 | — | — | 9 | 1 | — | — |
| 25 | — | — | 4 | 3 | — | — |
| 26 | — | — | 6 | 4 | — | — |
| 27 | — | — | 5 | 5 | — | — |

TABLE 4

*Considered loci Tf; Ca*

| No. of Sire | Ratio | Theoretic system | | |
|---|---|---|---|---|
| | | Number of discordant subjects | Theoretic frequencies computed from data experimental | Theoretic frequencies according to the binomial $\left[\frac{1}{2}+\frac{1}{2}\right]^5$ |
| 1 | 4 : 1 | 0 | 0.058182 | 0.03125 |
| 14 | 6 : 2 | 1 | 0.230303 | 0.15625 |
| 15 | 4 : 4 | 2 | 0.353333 | 0.31250 |
| 16 | 8 : 7 | 3 | 0.259091 | 0.31250 |
| 18 | 3 : 3 | 4 | 0.088485 | 0.15625 |
| | | 5 | 0.010606 | 0.3125 |

$Z = 1.022090 \quad 0.31 > P > 0.30$

TABLE 5

*Considered loci Ca; Am*

| No. of Sire | Ratio | Theoretic system | | |
|---|---|---|---|---|
| | | Number of discordant subjects | Theoretic frequencies computed from data experimental | Theoretic frequencies according to the binomial $\left[\frac{1}{2}+\frac{1}{2}\right]^4$ |
| 1 | 6 : 0 | 0 | 0.207692 | 0.0625 |
| 14 | 6 : 4 | 1 | 0.438461 | 0.2500 |
| 15 | 4 : 4 | 2 | 0.292307 | 0.3750 |
| 16 | 9 : 4 | 3 | 0.061538 | 0.2500 |
| | | 4 | 0.000000 | 0.0625 |

$Z = 2.409709 \quad 0.02 > P > 0.01$

of amylases had been tested also for other polymorphisms, and according to the available literature no systematic study has yet been made on the conjoined genetic behaviour of this polymorphism, it seemed worthwile to study the possibilities of some linkage between amylase and other polymorphisms of the blood (BRAUNER–NIELSEN, 1966; RENDEL, 1967).

TABLE 6

*Considered loci Tf; Am*

| No. of Sire | Ratio | Theoretic system | | |
|---|---|---|---|---|
| | | Number of discordant subjects | Theoretic frequencies computed from data experimental | Theoretic frequencies according to the binomial $\left[\frac{1}{2}+\frac{1}{2}\right]^{16}$ |
| 2 | 8 : 1 | — | 0.001445 | 0.0000152 |
| 3 | 4 : 1 | 1 | 0.012171 | 0.0002441 |
| 5 | 12 : 5 | 2 | 0.46302 | 0.0018312 |
| 8 | 7 : 5 | 3 | 0.110555 | 0.0085456 |
| 14 | 7 : 2 | 4 | 0.182359 | 0.0277732 |
| 15 | 6 : 4 | 5 | 0.214604 | 0.0666556 |
| 16 | 10 : 6 | 6 | 0.190681 | 0.1222020 |
| 17 | 8 : 6 | 7 | 0.128666 | 0.1745744 |
| 19 | 3 : 2 | 8 | 0.068391 | 0.1785420 |
| 21 | 3 : 2 | 9 | 0.029156 | 0.1745744 |
| 22 | 4 : 1 | 10 | 0.010664 | 0.1222020 |
| 23 | 4 : 2 | 11 | 0.003470 | 0.0666556 |
| 24 | 9 : 1 | 12 | 0.000938 | 0.0277732 |
| 25 | 4 : 3 | 13 | 0.000179 | 0.0085456 |
| 26 | 6 : 4 | 14 | 0.000019 | 0.0018312 |
| 27 | 5 : 5 | 15 | 0.00000091 | 0.0002441 |
| | | 16 | 0.0000000039 | 0.0000152 |

$Z = 4.207251 \quad P < 0.001$

For this second investigation, three polymorphisms were used the amylase, the transferrins and the carbonic anhydrases, showing in the populations studied alleles of almost identical frequency and such providing for a high number of heterozygous individuals which was essential for the investigation in question.

To demonstrate statistically this possible linkage, it was necessary to use a sampling of definitely related subjects, in which the relationship declared by the herd book was confirmed by bloodtyping. In the last analysis, 457 subjects subdivided into 215 family groups descending from 27 bulls were used; these subjects were selected from a sample of approximately 3400 individuals.

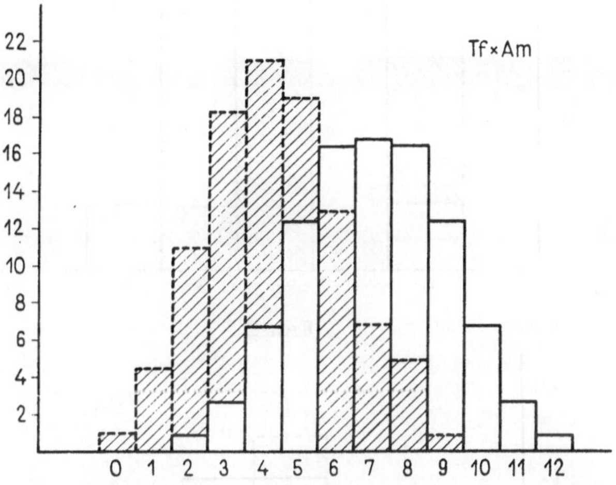
Tf×Am
22
20
18
16
14
12
10
8
6
4
2
0 1 2 3 4 5 6 7 8 9 10 11 12

FIG. 1

To establish the means of transmission of the alleles of the pairs of loci considered, examination was made only of the heterozygous bulls for the polymorphisms under observation.

The sons of these bulls were divided into classes with regard to the pair of loci studied, and with regard to the type of alleles received from the sire.

Thus, considering two pairs of loci at a time, one founder can produce only two classes of gametes; each of which having a frequency of 0.5, if the two loci are independent.

The ratio between the frequencies of the two classes of gametes will thus be the estimate of the probability of the existence of a linkage state. The probabilities that the pairs of polymorphisms considered were linked were therefore given by the ratio between the two classes of male offspring considered, concordant or discordant with the hypothesis (H o), which was pre-established for no-linkage and therefore for an equal frequency between the classes of gametes transmitted from the father to the single sons.

In order to arrive at an overall estimate of this probability the compound probability was calculated for this method (see CERUTTI et al., 1970 in press; CERUTTI and CRIMELLA, 1970).

The descents, in which it was not possible to determine from which parent the alleles originated and which showed fewer than five family groups, were eliminated from the calculation of the compound probability.

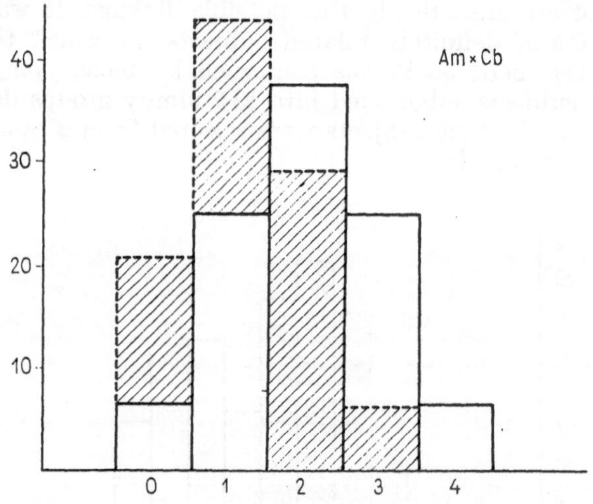

40
30
20
10
Am × Cb
0
1
2
3
4


FIG. 2

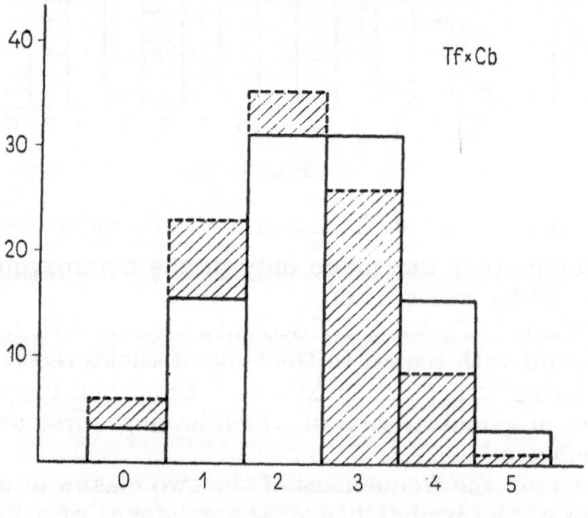

40
30
20
10
Tf × Cb
0
1
2
3
4
5


FIG. 3

## RESULTS

We studied the electrophoretic behaviour of amylase in certain Italian cattle populations, for calculation of the allelic frequency.

If being necessary to give to the frequencies found in the breed an estimated value of the real parameter of the population we used only those subjects entered in the herd book, for which it was not possible to show the degree of relationship (Table 2). Table 3 shows the results from which the data used subsequently for the calculation of the probabilities were collected. Tables 4, 5 and 6, on the other hand, show the calculation of the probabilities for the possible combinations of the polymorphisms considered, and the related values of the distribution of "Z" and their levels of significance. Figures 1, 2 and 3 plot the histograms representing the theoretical distributions of concordance frequencies in the transmission of the pair of factors considered.

## DISCUSSION

The results obtained might appear conflicting, as they may lead to suppose an association between the loci Tf and Am, Ca and Am, while an association between Tf and Ca is not significant.

Such result may be explained by the fact that the difference between the probability 0.5 and that found for the distribution considered, more exactly 0.424242, were too small with regard to the smallness of the sample studied, thus, although an association can exist, it might be so limited as not to be disclosed by the few data available.

Efforts are being made in this laboratory to extend the material of research, in order to prove or disprove the results which for the time being are regarded hypothetical.

## REFERENCES

ASHTON, G. C. (1958), A genetic mechanism for protein polymorphism in cattle, *Nature*, **182.** 65–66.

ASHTON, G. C. (1965), Serum amylase (thread protein) polymorphism in cattle, *Genetics*, **51.** 431–437.

ASHTON, G. C. (1966a), Cattle serum amylase polymorphism, Proc. Xth Europ. Conf. Anim. Blood Groups Biochem. Polymorph. Paris. 1966, 289–292.

ASHTON, G. C. (1966b), Distribution of transferrin, albumin, postalbumin, amylase and hemoglobin genotypes in Droughtmaster cattle, *Anim. Breed. Abstract*, **35.** 1234.

BRAUNER-NIELSEN, P. (1966), Studies on the genetic relationship between the serum amylase system and other blood group systems in pigs, Proc. Xth Europ. Conf. Anim. Blood Groups. Biochem. Polymorph. Paris, 1966. 449–462.

RENDEL, J. (1967), Studies of blood groups and protein variants as a means of revealing similarities and differences between animal population, *Anim. Breed. Abstracts*, **35.** 3.

CRIMELLA, C. (1969), Previous considerations on amylase system in Frisona Italian race, *Atti Soc. Ital. Sci. Vet.* XXIII (in press).

CERUTTI, F. and CRIMELLA, C. (1970), Studio del comportamento genetico di alcuni polimorfismi proteici ed enzimatici del sangue bovino. *Genetica Agraria*, (in press).

*XIIth Europ. Conf. Anim. BloodGroups Biochem. Polymorph., Bp., 1972 (pp. 181–182)*

# SERUM AMYLASE ISOZYMES IN WISENTS AND CATTLE-WISENT HYBRIDS

J. M. Gasparski

Department of Biomedical Sciences, University of Guelph, Canada

## SUMMARY

**Serum amylase isoenzyme phenotypes have been studied in wisents (European bison), cattle-wisent $F_1$ and $F_2$ hybrids and their bovine parents. In wisents only one amylase type was found. Bison amylase band migrated in the same region as cattle amylase isozyme type B but slightly slower. In all hybrids both wisent and cattle parental, bands were observed.**

## INTRODUCTION

European bisons or wisents (*Bison bonasus — Linneus*, 1758) reared in Poland form today a relatively small strongly inbred relict population. Research on the wisent blood properties was first suggested in Poland by Gasparski. It was hoped that results of this study might form a biological basis for constructive breeding of this species of animals.

Heretofore, for several years investigations on the blood properties of wisents have been concerned with: (a) the polymorphism of the red-cell antigenic factors (Czaja and Gasparski, 1960; Gasparski and Dubiski, 1962; Gasparski and Gerner-Nowakowa, 1963; Gasparski et al., 1963; Gasparski, 1964a, 1964b, 1965). (b) The serologic specificity of wisent normal sera with reference to red-cell antigenic factors in human and in several species of animals (Gasparski et al., 1967). (c) The serologic properties of the serum globulins (Gasparski and Dubiski, 1963). (d) The polymorphism of hemoglobins and transferrins (Braend and Gasparski, 1967; Braend, Gasparski and Krasinika, 1969).

The studies on the wisent blood properties were conducted in the direct form as well as in comparison with blood properties of domestic cattle and their hybrids.

The study which is reported here was undertaken to determine the amylase isozymes in wisents.

## MATERIAL AND METHODS

Serum samples from 18 wisents, 16 cattle-wisent $F_1$ and $F_2$ hybrids and their 7 bovine parents were tested for serum amylase isozymes.

Serum samples were taken few years ago (1961–1966) and stored at −20°C until used.

Starch gel electrophoretograms were produced in a discontinuous buffer system. The technique used was the same as described by GASPARSKI (GASPARSKI and STEVENS, 1968; GASPARSKI, 1970). Gel buffer was prepared with 0.013 M tris solution adjusted with 0.05 M citric acid to pH 6.85. The vessel buffer was prepared with 0.3 M boric acid and 0.05 M sodium hydroxide, pH 8.08. The starch concentration was 13%. An initial voltage of 145 volts was increased to 350 volts when the borate boundary had migrated 4–5 cm from the line of insertion. The run was completed when the borate boundary had migrated 9 cm from the line of insertion. The stain was prepared just before use by dissolving 0.3 g p-phenylenediamine dihydrochlorid in 300 ml 0.1 M sodium acetate solution adjusted with glacial acetic acid to pH 5.7.

## OBSERVATIONS

Only one type of serum amylase was observed in each sample of wisent. This type was designated B because bison amylase band migrated in the same region as cattle amylase type B but slightly slower. The same types were observed in buffalo (American bison). About 100 samples of bison sera were tested and was observed only one type of amylase similar to cattle type B. (GASPARSKI and STORMONT, unpublished).

In all hybrids both wisent and cattle parental bands were observed.

## REFERENCES

CZAJA, M. and GASPARSKI, J. M. (1960), *Nature*, **185.** 185–186.
GASPARSKI, J. M. and DUBINSKI, S. (1962), *Ann. N. Y. Acad. Sci.* **97.** 285–295.
GASPARSKI, J. M. and GERNER-NOWAKOWA, A. (1963), *Immunogenetics Letter*, **3.** 65–67.
GASPARSKI, J. M. et al. (1963), *Acta thesiol.* **7.** 311–316.
GASPARSKI, J. M. and DUBISKI, S. (1963), *Acta thesiol.* **7.** 16, 314–320.
GASPARSKI, J. M. (1964a), *Biulet. ZHDZ.* **4.** 47–62.
GASPARSKI, J. M. (1964b), *Genetica pol.*, **5.** 141–142.
GASPARSKI, J. M. (1965), Proc. 9th Europ. Anim. Blood Group Conf. Prague, 1964. 93–97.
GASPARSKI, J. M. et al. (1966) Proc. Xth Europ. Conf. Anim. Blood Groups Biochem. Polymorph. Paris, 1966.
BRAEND, M. and GASPARSKI, J. M. (1967), *Nature*, **214.** 98–99.
BRAEND, M., GASPARSKI, J. M. and KRASINIKA, M. (1969), *Hereditas*, **62.** 185–191.
GASPARSKI, J. M. and STEVENS, R. W. C. (1968), *Can. J. of Genetics a. Cyt.* **10.** 1, 148–151.
GASPARSKI, J. M. (1970), Proc. XIth Europ. Conf. Anim. Blood Groups, Biochem. Polymorph. Warsaw, 1968.
GASPARSKI, J. M. and STORMONT, C. (unpublished).

*XIIth Europ. Conf. Anim. Blood Groups Biochem. Polymorph., Bp., 1972 (pp. 183—184)*

# QUANTITATIVE DETERMINATION OF ACID ERYTHROCYTE PHOSPHATASE (AEP) IN CATTLE

H. ROHRBACHER and E. BAMBERG

Department of Animal Breeding and Genetics and Department of Biochemistry, School of Veterinary Medicine, Vienna, Austria

## SUMMARY

**Considerable polymorphism of the AEP was revealed in cattle. A slight modification of the method used by DOGRUL showed an additional type (types). The quantitative determination of the amount of enzymes disclosed a remarkable quantitative variation which, however, did not appear to be related to the AEP type.**

DOGRUL (1968, 1969) found evidence for genetic polymorphism of AEP in cattle. He could easily differentiate two types, I and II, but a subdivision of the latter into types IIa and IIb frequently yielded ambiguous results and, furthermore, this variation did not appear to show simple Mendelian inheritance. HOPKINSON, SPENCER and HARRIS (1964) applied quantitative methods for enzyme determination, and succeeded in showing definite quantitative differences between types of AEP. Therefore it was decided to try a similar approach on erythrocytes from cattle.

## METHODS

For the quantitative investigation we used the method described by DOGRUL, with the modification that we used for the quantitative determination of AEP: after centrifugation of blood, the serum was withdrawn and the erythrocytes were washed three times with physiological saline. The erythrocytes were hemolized with distilled water (erythrocytes: distilled water = 1 : 6), three times frozen and thawed. One-tenth of one ml of this hemolysate was mixed with 0.4 ml distilled water and 0.5 ml substrated solution (0.01 M tri-sodium phenyl phosphate in 0.05 ml citrate buffer) in three tubes. The first tube was deproteinized immediately, the second after 15 minutes at 37°C, and the third tube after 45 minutes' incubation (2 ml 10% tri-chlor-acetate). After five minutes' centrifugation at 1000 r.p.m. we added 4 ml 0.5 M NaOH-solution which induced appearance of paranitrophenolate ion, whic h in turn was determined at 405 nm. One unit AEP as used here is equivalent to the quantity of enzyme which sets free one micro-Mol paranitrophenol at 37°C.

## RESULTS

The quantitative investigation revealed a considerable variability of the AEP. In contrast to DOGRUL we found four rather distinct types consisting of three zones (Table 1). As an explanation we should suggest that

TABLE 1

*Zones of activity of acid erythrocyte phosphatase (AEP) in cattle*

| AEP type | Activity | | |
|---|---|---|---|
| 1) A | | + | |
| B | + | + | + |
| C | + | + | |
| D | (+) | + | + |
| 2) I | | + | |
| IIa | (+) | + | |
| IIb | + | + | |

1) present investigation
2) investigation by DOGRUL

the point of DOGRUL's type I has been split into two points, probably on account of our use of phosphate starch. Assuming this to be so, we found agreement with DOGRUL's results in eight of ten cases, but not in two.

The quantitative investigation revealed pH 5.7, and substrate concentration of 0.02 M to be optimal. Table 2 shows the estimated AEP activity given as deviation from the mean of the run in erythrocytes of various types. Survey of the results immediately revealed a rather wide variation within each type, which was confirmed by an analysis of variance. The conclusions, therefore is that no relationship exists between the quantity and type of AEP, in contrast to the results of HOPKINSON and coworkers.

TABLE 2

*Quantity of AEP in various types*

| Type (present investigation) | Number of animals | Units AEP[1] |
|---|---|---|
| A | 2 | —.22 |
| B | 9 | —.05 |
| C | 4 | +.24 |
| D | 11 | +.13 |
| I[2] | 5 | +.21 |
| IIa | 4 | +.45 |
| IIb | 1 | —.24 |

[1] deviation from mean of run
[2] types as characterized by Dogrul


## REFERENCES

DOGRUL, F. (1970), Proc. XIth Europ. Conf. Anim. Blood Groups Biochem. Polymorph. Warsaw, 1968, 223.

DOGRUL, F. (1969), *Wiener Tierärztl. Msch.*, **56.** 329.

HOPKINSON, D., SPENCER, N. and HARRIS, H. (1964), *Amer. J. Human Genetics*, **16.** 141.

*XIIth Europ. Conf. Anim. Blood Groups Biochem. Polymorph., Bp., 1972 (p. 185)*

# CARBONIC ANHYDRASE IN CATTLE TISSUES

G. SARTORE and R. BRUNO

**Istituto di Fisiologia e Chimica, Biologica Facoltà di Medicina Veterinaria dell'Università di Torino, Torino, Italy**

## SUMMARY

Blood and tissue samples from a variety of cattle breeds have been collected at a local slaughterhouse. The red cell lysates and gastric mucosa, pancreas and kidney homogenates were subjected to alcohol-chloroform treatment and assayed for carbonic anhydrase activity by the electrometric method of WILBUR and ANDERSON (1948, *J. Biol. Chem. 176*, 147). The activity units were calculated according to the formula given by RICKLI et al. (1964, *J. Biol. Chem.*, *239*, 1065) slightly modified.

Grouping the samples according to the CA type of their red cells, as determined by starch gel electrophoresis, it has been found that tissues follow a rather definite trend. The activity in samples belonging to CA-S type was higher than in those of type CA-F, while in type CA-SF was intermediate. The differences, between types, were significant for red cells, gastric mucosa and pancreas, but not for kidney.

## DISCUSSION

C. STORMONT: What tissue was it which showed no statistically significant variation in activity of the three carbonic anhydrase phenotypes?

G. SARTORE: No significant differences were found in kidney where we must note the high content of proteins in the tissue extracts so that it is marking at least in fact the CA activity.

J. RENDEL: Is it not possible to use quantitative determination for classifying CA types?

G. SARTORE: It is rather difficult to classify CA types by quantitative determination because there is an extensive overlapping of the activities of CA-SF types with CA-S and CA-F.

G. C. ASHTON: It might be worth using acrylamide gel in slabs to overcome your problems with destruction of starch gel by amylase.

*XIIth Europ. Conf. Anim. Blood Groups Biochem. Polymorph., Bp., 1972 (pp. 187–189)*

# A NEW PHENOTYPE IN THE CARBONIC ANHYDRASE SYSTEM OF CATTLE

C. STORMONT, B. G. MORRIS and YOSHIKO SUZUKI

**Serology Laboratory, Department of Veterinary Microbiology, University of California, Davis, California**

A polymorphic system of carbonic anhydrase has been described for cattle in which three phenotypes: F, FS and S were recognized. These phenotypes appeared to be under the control of a pair of co-dominant autosomal alleles $CA^F$ and $CA^S$ (SARTORE, 1966; SARTORE et al., 1969). In 1968 SARTORE discovered a new variant zone, which migrates just below zone S, and identified it as $S_{Piedmont}$. As a consequence of this finding, a third allele $CA^F_{Piedmont}$ was postulated.

Following the procedures for starch-gel electrophoresis described in the previous studies and using only the protein stain, we have routinely determined the carbonic anhydrase phenotypes of all cattle blood samples submitted to us in connection with our service blood typing program. This has enabled us to extend the data on the distribution on the three phenotypes in eight breeds of cattle and to collect data of a ninth breed, the Charolais (Table 1).

The frequencies of $CA^F$ remained unchanged for the Ayrshire, Brown Swiss and Polled Hereford breeds. For the remaining breeds in which changes

TABLE 1

*Distribution of carbonic anhydrase phenotypes in 3062 males representing nine breeds of cattle*

| Breeds | Number sampled | Phenotypes | | | Frequency of $CA^F$ | *Prior Study | |
|---|---|---|---|---|---|---|---|
| | | F | FS | S | | Number sampled | Frequency of $CA^F$ |
| Aberdeen Angus | 78 | 0 | 4 | 74 | .02 | 114 | .01 |
| Ayrshire | 37 | 0 | 10 | 27 | .14 | 86 | .14 |
| Brown Swiss | 70 | 0 | 10 | 60 | .07 | 95 | .07 |
| Charolais | 76 | 11 | 35 | 30 | .38 | — | — |
| Guernsey | 202 | 2 | 25 | 175 | .07 | 352 | .04 |
| Hereford | 257 | 19 | 90 | 148 | .25 | 408 | .24 |
| Holstein-Friesian | 1538 | 43 | 427 | 1068 | .17 | 1102 | .20 |
| Jersey | 242 | 33 | 124 | 85 | .39 | 395 | .41 |
| Polled Hereford | 281 | 5 | 51 | 225 | .11 | 365 | .11 |

*SARTORE, et al. 1969

were noted the figures did not vary greatly from those of the previous studies.

No new variants were observed in any of the registered animals which were tested. However, a new phenotype was observed in samples from three commercial grade animals of the Angus breed.

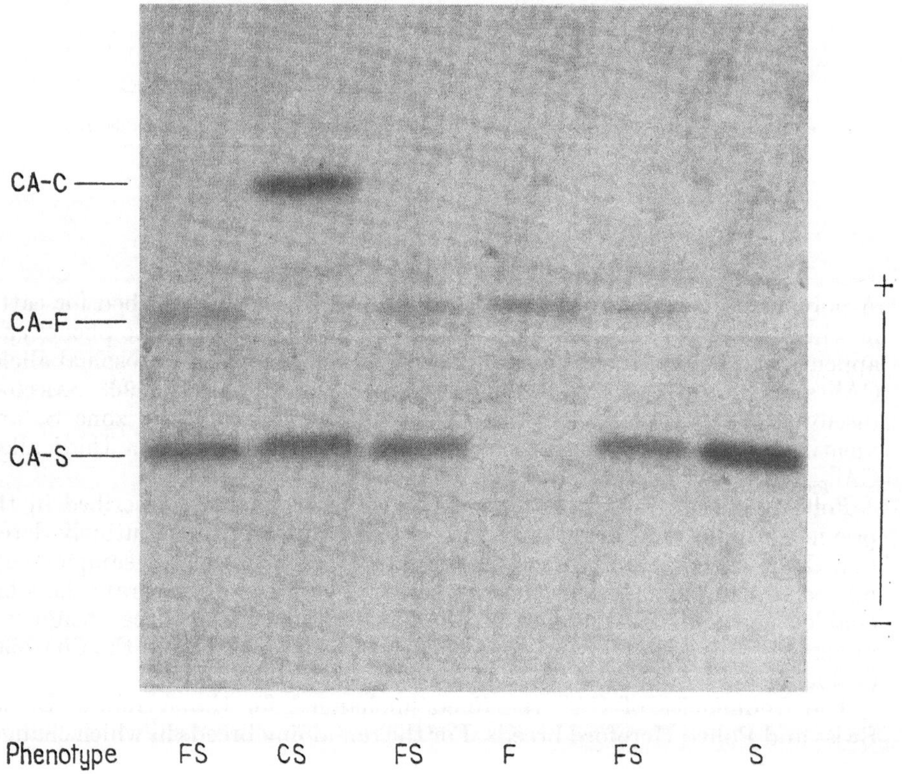
CA-C
CA-F
CA-S
+
−
Phenotype FS CS FS F FS S

FIG. 1. A starch gel slice stained with Buffalo Black, showing the carbonic anhydrase system of cattle. Phenotypes F, FS, S and CS are represented in the six samples

The new phenotype was designated CS, the C indicating a zone which migrates well ahead of zone F. Zone C conformed to the same criteria found for zones F and S: esterase activity with the substrate alpha naphthyl acetate, specific inhibition of the esterase activity by acetazolamide ($10^{-4}$) and stainability with protein dyes, e.g., buffalo black or nigrosin.

Phenotypes CS, F, FS and S are shown in Figure 1. The samples run on the gel were alcohol-chloroform extracts prepared from red cells according to a procedure described by ROUGHTON and BOOTH (1946). The staining of zones C and S is of equal intensity and the staining of zone F is much reduced in comparison to the other two. SARTORE (1968) has found that the weaker staining of zone F is reflected in the lower average levels of enzyme activity for phenotypes F and FS as compared to that for phenotype S. On the basis of the staining intensity alone, one might, therefore, predict that the enzyme activity of phenotype CS would fall within the range found for phenotype S.

Although no family data were available on the individuals of phenotype CS, we are confident that the new zone, C, will eventually be shown to be under the control of a new allele, $CA^C$, in the carbonic anhydrase system.

## REFERENCES

ROUGHTON, F. J. W. and BOOTH, V. H. (1946), The manometric determination of the activity of carbonic anhydrase under varied conditions. *Biochem. J.* **40.** 319–330.

SARTORE, G. (1966), Ricerche su un nuovo polimorfismo genetico riguardante una esterasi degli eritrociti bovini, *Atti Ass. Genet. It.* XI. 217–222.

SARTORE, G. (1970), Carbonic anhydrase types of cattle red cells. Proc. XIth Europ. Anim. Conf. Blood Groups Biochem. Polymorph. Warsaw, 1968, 211–216

SARTORE, G., STORMONT, C., MORRIS, B. G. and GRUNDER, A. A. (1969), Multiple electrophoretic forms of carbonic anhydrase in red cells of domestic cattle *(Bos Taurus)* and American buffalo *(Bison Bison)*. *Genetics.* **61.** 823–831.

## DISCUSSION

C. R. SHAW: The quantitative differences in the F + S types are interestingly parallel to the carbonic anhydrase variant in man, in which the rare variant is produced in much less amount than the normal.

*XIIth Europ. Conf. Anim. Blood Groups Biochem. Polymorph., Bp., 1972 (pp. 191–195)*

# CARBONIC ANHYDRASE POLYMORPHISM IN SOME HUNGARIAN CATTLE BREEDS

PÉTER SOOS

Center for Artificial Insemination, Laboratory of Immunogenetics, Budapest, Hungary

## SUMMARY

The variants of red cell carbonic anhydrase (CA) were examined by starch-gel electrophoresis in two native cattle breeds of Hungary, Hungarian Grey and Hungarian Spotted, and in imported breeds. Gene frequencies for $CA^F$ and $CA^S$ have been estimated in each breed and were compared with the respective data of other breeds, examined previously. The value of this polymorphic system in parentage control tests has also been scrutinized.

In the hemolysates from eight buffaloes *(Bos bubalus L.)*, tested for comparison, two CA bands were demonstrable which showed the same mobility as the corresponding bands in cattle.

## INTRODUCTION

Carbonic anhydrase (CA) one of the important enzymes of the red blood cells after separation by starch-gel electrophoresis is demonstrated by its esterase activity. Owing to its relatively high concentration in the red blood cells, bands of CA can be identified in the gel also by staining with common protein stains. This provides a simple method for routine serial examinations.

In cattle, the CA types are determined by a co-dominant autosomal pair of genes designated $CA^F$ and $CA^S$ (SARTORE, 1966; SARTORE et al., 1969). These two alleles make possible the occurrence of three genotypes ($CA^F/CA^F$, $CA^S/CA^S$ and $CA^F/CA^S$) which have been found in all the cattle breeds examined up to now. Some authors reported also gene frequency data. In the Italian Piedmont breed, a third CA allele with lower mobility than common $CA^S$ ($CA^{S}_{Piedmont}$) has been found by SARTORE (1968). Recently STORMONT et al. (1970) found a new CA allele in three individuals of the Aberdeen–Angus breed which appeared in heterozygous form together with the $CA^S$ allele. The new allele is located behind the band of $CA^F$ at a distance similar to that between $CA^F$ and $CA^S$.

The aim of the present study was to determine the frequency of the CA alleles in two indigenous Hungarian and in some imported cattle breeds as well as to investigate their population genetic importance and suitability in parentage control tests.

## MATERIAL AND METHODS

The blood samples of the Hungarian Grey cattle were collected from individuals of a herd in the Hortobágy State Farm. Blood samples from the Hungarian Spotted breed were materials sent to our laboratory for parentage control tests from various parts of the country. Samples from Swiss–Brown, Simmental, Yugoslavian Holstein–Friesian and the Austrian Fleckvieh collected for comparison were taken at the Budapest slaughterhouse.

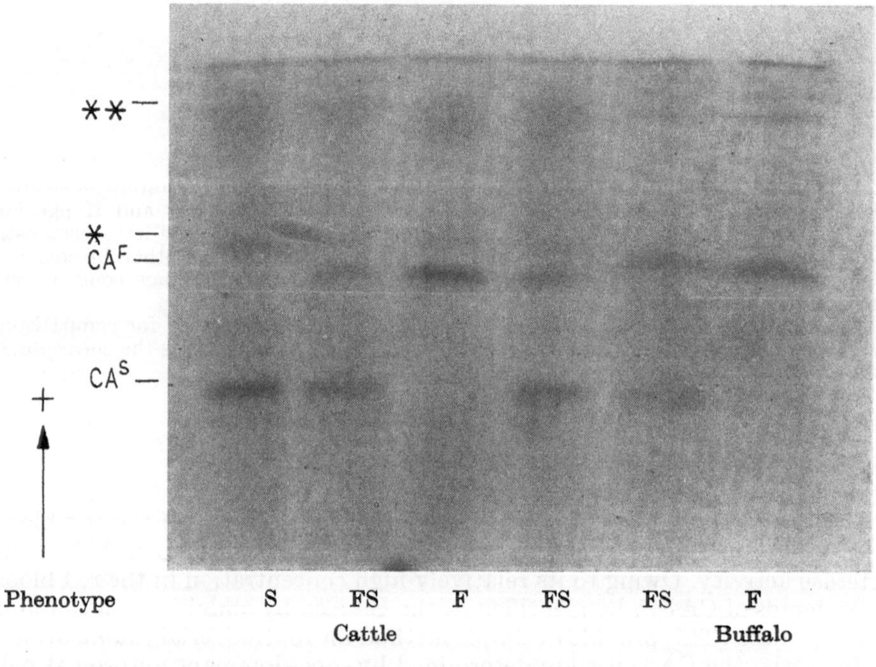
**
*
$CA^F$
$CA^S$
+
Phenotype S FS F FS FS F
Cattle
Buffalo

FIG. 1. CA phenotypes of cattle and buffalo, stained with protein stain. * = "additional band", ** = A esterase region

Starch-gel electrophoresis was carried out according to a modified method of SARTORE et al. (1969), with the pH of the gel reduced from 7.3 to 6.9 and using hydrolized starch prepared in our laboratory. To prevent staining of the enzyme in the routine examinations, the thickness of the gel was reduced to 3 mm and the electrophoresis was allowed to take place at room temperature. Amido Black 10-B—Nigrosin solution was used for protein staining.

The results were evaluated by conventional statistical methods.

## RESULTS

Each of the three common CA phenotypes were found in all breeds examined (Fig. 1). The distribution of phenotypes as well as the values of gene frequencies are shown in Table 1. According to the results of the $\chi^2$ tests, a significant deviation ($P < 0.01$) from the Hardy–Weinberg equilibrium occurred exclusively in the Hungarian Spotted population.

TABLE 1

*Expected and observed distribution of the CA phenotypes and the frequencies of the two CA alleles in the breeds examined*

| Breed | Number of animals | Distribution of CA phenotypes | FF | FS | SS | $\chi^2$ (d.f.: 1) | $q_{CA^F}$ | $q_{CA^S}$ |
|---|---|---|---|---|---|---|---|---|
| Hungarian Grey | 158 | obs. | 5 | 39 | 114 | 0.53 | 0.1550 | 0.8450 |
| | | exp. | 3.80 | 41.38 | 112.82 | | | |
| Hungarian Spotted | 890 | obs. | 56 | 270 | 564 | 8.91* | 0.2146 | 0.7854 |
| | | exp. | 40.99 | 300.00 | 549.01 | | | |
| Simmental | 92 | obs. | 14 | 32 | 46 | 4.01 | 0.3261 | 0.6739 |
| | | exp. | 9.78 | 40.44 | 41.78 | | | |
| Austrian Spotted | 41 | obs. | 1 | 6 | 34 | 1.17 | 0.0975 | 0.9025 |
| | | exp. | 0.39 | 7.22 | 33.39 | | | |
| Brown Swiss | 92 | obs. | 1 | 6 | 85 | 4.42 | 0.0435 | 0.9565 |
| | | exp. | 0.17 | 7.66 | 84.17 | | | |
| Holstein-Friesian | 105 | obs. | 2 | 26 | 77 | | | |
| | | exp. | 2.14 | 25.72 | 77.14 | 0.01 | 0.1429 | 0.8571 |

* Significance at 0.01 probability level.

A faster and a slower allele was found also in the samples from the 8 buffalos *(Bos bubalus L.)* (Fig. 1). A considerable difference was that the so called "additional band" which in cattle occurs before the band of the faster allele could not be detected in the electrophoretogram of buffalo blood samples. The phenotypic distribution of the 8 buffalos was 7 FF and 1 FS. Decisive conclusions remain to be drawn by further examination on a larger population.

## DISCUSSION

The results of the examination carried out in two Hungarian and in some imported cattle breeds show that the structure of the CA locus is similar to that in other breeds examined earlier, i.e. the frequency of the $CA^S$ allele is higher than that of the $CA^F$ allele. Gene frequencies established in the cattle breeds examined so far are shown in Table 2.

TABLE 2

*The frequencies of CA alleles in cattle breeds examined up to now*

| Breed | Country | Number of animals | $q_{CA^F}$ | $q_{CA^S}$ | References |
|---|---|---|---|---|---|
| Jersey | USA | 395 | 0.41 | 0.59 | SARTORE et al. (1969) |
| Simmental | Switzerland | 92 | 0.33 | 0.67 | this paper |
| Longhorn | USA | 94 | 0.28 | 0.72 | SARTORE et. al. (1969) |
| Hereford | USA | 408 | 0.24 | 0.76 | SARTORE et al. (1969) |
| Hungarian Spotted | Hungary | 890 | 0.21 | 0.79 | this paper |
| Holstein-Friesian | USA | 1102 | 0.20 | 0.80 | SARTORE et al. (1969) |
| Aosta Red Pied | Italy | 116 | 0.19 | 0.81 | SARTORE and BERNOCO (1966) |
| Hungarian Grey | Hungary | 158 | 0.16 | 0.84 | this paper |
| Holstein—Friesian | Yugoslavia | 105 | 0.14 | 0.86 | this paper |
| Ayrshire | USA | 86 | 0.14 | 0.86 | SARTORE et al. (1969) |
| Polled Hereford | USA | 365 | 0.11 | 0.89 | SARTORE et al. (1969) |
| Austrian Spotted | Austria | 41 | 0.10 | 0.90 | this paper |
| Brown Swiss | USA | 95 | 0.07 | 0.93 | SARTORE et al. (1969) |
| Guernsey | USA | 352 | 0.04 | 0.96 | SARTORE et al. (1969) |
| Brown Swiss | Switzerland | 92 | 0.04 | 0.96 | this paper |
| Aberdeen Angus | USA | 114 | 0.01 | 0.99 | SARTORE et al. (1969) |

The gene frequency values of the Hungarian Spotted breed are intermediate between those of the Hungarian Grey and Simmental breeds which played a role in the development of the Hungarian Spotted breed. It can be noticed, however, that the frequency of the $CA^F$ allele is relatively high in the Simmental breed, while it is rather low in the Austrian Fleckvieh in the development of which the Simmental breed has also been involved. This can be probably explained by the fact that the Austrian Fleckvieh samples used in this study derived from a rather closed population of the breed and that the number of individuals examined was too small. A more representative gene frequency value could be obtained only on the basis of a random material including more individuals. By using the $2 \times j$ statistical method (MATHER, 1951) the difference between the Holstein–Friesian and Brown Swiss stocks, completely dissimilar in character, was not significant, although each stock of animals derived from completely separated lines of the respective breeds.

The CA system can be more effectively used than the hemoglobin (Hb) system in the parentage control of the Hungarian Spotted breed, which is the most important cattle breed in Hungary. The theoretically possible highest value of the exclusion calculated on the basis of hemoglobin-allele frequencies (KOVÁCS et al., 1970) is as much as 10.35% while that for the CA system is 14.01%. Combining the two systems with the transferrin (Tf) polymorphism (KOVÁCS and SOOS, 1969), this value can be increased to 53.46%. As the proposed modifications used in the CA examination are relatively simple, the conclusion may be drawn that in addition to the hemoglobin and transferrin systems, the CA results can also yield useful information on the Hungarian Spotted breed in parentage control tests.

## REFERENCES

Kovács, G. and Soos, P. (1969), A szarvasmarha transferrin vizsgálatok újabb eredményei. II. Populációgenetikai vizsgálatok hazai szarvasmarha fajtákban serumtransferrin-típusok felhasználásával. *Magy. Áo. L.*, **24.** 662.

Kovács, G., Soos, P. and Nemesi, M. (1972), Breed structures and similarities of some Hungarian cattle breeds examined by means of the Tf, Hb and FV gene frequencies. Proc. XIIth Europ. Conf. Anim. Blood Groups Biochem. Polymorph. Budapest 1970.

Mather, K. (1951), Statistical analysis in biology. Methuen and Co. Ltd. London.

Sartore, G. (1966), Richerche su un nuovo polimorfismo genetico riguardante una esterasi degli eritrociti bovini. *Atti Ass. Genet. It*,. **11.** 217.

Sartore, G. (1970), Carbonic anhydrase types of cattle red cells. Proc. XIth Europ. Conf. Anim. Blood Groups Biochem. Polymorph. Warsaw 1968, 211.

Sartore, G. and Bernoco, D. (1966), Research on biochemical polymorphisms in the indigenous cattle of Piedmont. *La Ricerca Scientifica*, **36.** 1368.

Sartore, G., Stormont, C., Morris, B. G. and Grunder, A. A. (1969), Multiple electrophoretic forms of carbonic anhydrase in red cells of domestic cattle (Bos-taurus) and American buffalo (Bison bison). *Genetics*, **61.** 823.

Stormont, C., Morris, B. G. and Suzuki, Y. (1972), A new phenotype in the carbonic anhydrase system of cattle. Proc. XIIth Europ. Conf. Anim. Blood Groups Biochem. Polymorph., Budapest, 1970.

*XIIth Europ. Conf. Anim. Blood Groups Biochem. Polymorph., Bp., 1972 (pp. 197–200)*

# ALBUMIN POLYMORPHISM IN BELGIAN CATTLE BREEDS

Y. Bouquet and A. Van De Weghe

**Department of Animal Genetics and Breeding, Faculty Of Veterinary Medicine of Ghent, Belgium**

## INTRODUCTION

The number of proteins with inherited structural variation is increasing very fast, but many systems are nearly monomorphic in a given species, the occurrence of genetic variants being so rare that they are of little use as genetic markers.

A system becomes polymorphic and interesting as a genetic marker when the less frequent variant is present at a rather moderate frequency in a particular ethnical group. The variability and usefullness of each system must first be tested in each population. One system can be polymorphic and useful in one population but not in another. This is the case with albumin in cattle. 43% of the serum proteins of cattle are constituted by albumin, one of the proteins with fast electrophoretic mobility. Although incompletely known, transport is the main physiological function attributed to this fraction, as it links easily with vitamins, hormones, metal ions and fatty acids. Polymorphism in albumin of cattle was demonstrated in different European Lowland, South European, African and Australian breeds by Braend and Efremov (1965), Ashton (1964), Ashton and Lampkin (1965), Sartore and Bernoco (1966), Carr (1966), Osterhoff (1968), Spooner and Oliver (1969). Mainly three phenotypes, controlled by two alleles, A and B, with very unequal frequencies, can be found. In addition, at least two other alleles can be detected in heterozygous state, as very rare mutants in some individuals belonging to the same family, approximating in this way the situation found occasionally in human isolates.

## MATERIAL AND METHODS

The usefulness of the system in connection with other marker system for the genetic characterization of individuals in the five Belgian cattle breeds was investigated. Samples for bloodtyping purposes were collected with a citrate anticoagulant. From these samples the plasma was isolated. All the young elite bulls allowed for reproduction in 1969–1970 on a total 5774 individuals were checked together with their mothers for albumin variants. The sample may be considered as highly representative for the elite population and primarily responsible for all kinds of genetic improvement in production abilities of the various breeds.

Zone electrophoresis was performed with discontinuous buffers. The vessel buffer was composed of 20 g NaOH and 92.8 boric acid for 5 l. distilled water, pH 8.6 while the gel buffer consisted of 4.2 g citric acid and 5.0 g. TRIS for 5 l. distilled water, pH 5.6. Hydrolized starch of Connaught was used at a concentration of 13.5%. The electrophoresis was initiated at 250 V and 25 mA for each plate of $12 \times 18 \times 0.5$ cm for 5 minutes. After removal of the insert papers, electrophoresis was continued for 2 hours at 300 V and 40 mA. The boundary buffer-line migrated over at least 8 cm. The gels were treated with nigrosine and de-stained with the classical washing solution 40/40/20 water, methanol, glacial acetic acid.

## RESULTS

Mainly two phenotypes, AA and AB, could be demonstrated in the Belgian cattle. Also an AD individual was detected. The AA type appears on the gels as a weakly staining faster migrating and an intensely dark staining slower zone. The AB type is constituted by two intensively stained zones, one with mobility of the AA type and another with a slower mobility and two faint coloured zones migrating just in front of the mentioned heavily stained zones. No animal of BB-phenotype was available in our material. The effect of storage of plasma samples on the duplication of the dark

TABLE 1

*Frequency of albumin alleles in Belgian cattle breeds*

| Breed | | Number of animals | Phenotypes | | | Frequency | | |
|---|---|---|---|---|---|---|---|---|
| | | | AA | AB | AD | $A/^A$ | $A/^B$ | $A/^D$ |
| Middle and High Belgian | Bulls | 1824 | 1683 | 141 | — | 0.961 ±0.003 | 0.039 ±0.003 | — |
| | Mothers | 1033 | 974 | 59 | — | 0.971 ±0.004 | 0.029 ±0.004 | — |
| Red Pied of East-Flandern | Bulls | 722 | 720 | 2 | | 0.999 | 0.001 | — |
| | Mothers | 503 | 500 | 2 | 1 | 0.997 | 0.002 | 0.001 |
| Red of West-Flandern | Bulls | 266 | 265 | 1 | — | 0.998 | 0.002 | — |
| | Mothers | 155 | 154 | 1 | — | 0.997 | 0.003 | — |
| Black Pied | Bulls | 352 | 352 | — | — | 1.000 | — | — |
| | Mothers | 174 | 174 | — | — | 1.000 | — | — |
| Red Pied of Campine | Bulls | 480 | 480 | — | — | 1.000 | — | — |
| | Mothers | 265 | 265 | — | — | 1.000 | — | — |
| | | 5774 | | | | | | |

staining bands, perhaps related with the molarity of the buffer, was noticed several times. The true differentiation of the phenotypes was not disturbed, especially when a long stored BB-sample was running next to a fresh AB sample, which may cause some difficulty. The bulls and cows belonging to the breeds of Black pied, and Red pied of Campine were all fixed and homozygous for type AA. Only one bull and one cow of the red breed of West-Flanders and two bulls with two cows, both daughters of the same sire, of the red pied breed of East Flandern, were of the type AB. One cow of the red pied breed of East Flandern appeared to possess the AD type. However, in the Middle and High Belgian blue-pied breed the phenotype variation was more pronounced. The gene frequency of allele B reached 4% and 3% for the bulls and the cows, respectively (see Table 1). Since at the end of the past century many shorthorn animals were imported to the continent and especially to Belgium, for improvement of the beef quality of the local breeds, special attention was paid to the eventual occurence of the C- and G-alleles, as mentioned by SPOONER and OLIVER (1969). No other phenotypes than AA, AB, AD could, however, be demonstrated.

## DISCUSSION

The albumin allele B could not be demonstrated in the majority of the Belgian cattle breeds, except for the breed of Middle and High Belgium. The Belgian cattle belong to the lowland cattle group of Europe which differs from the cattle populations of central Europe and those of the mediterranean area in respect of many marker-systems. This was confirmed for the British cattle breeds and the European breeds exported to South Africa (SPOONER and OLIVER, 1969; OSTERHOFF, 1968).

In other parts of Europe, where the breeds originate from another stock, a notable frequency of the gene B appears to be not an occasional, but a deep-rooted genetic character. This seems to be the case for the Charolais. The breed of Middle and High Belgium, nearly half of the total cattle population in the country, is completely surrounded by lowland breeds in which no albumin B is present. The question may arise from what kind of population the B allele would originate. During the past half century, selection for a dual purpose ability, with predominance for meat quality, was conducted in the cattle of Middle and High Belgium without noticable external influence. Before this period, however, Shorthorns and Lowland Black Pied animals were crossed with the local cattle populations. One of the results of these crosses was the appearance of the blue pied coat colour due to the introduction of the silver and black factors.

Because no B-albumin allele could be demonstrated in the above mentioned breeds, the probability seems rather low that gene immigration could have been the main reason for the appearance of AB-phenotypes in the breed of Middle and High-Belgium. Nevertheless, in populations apparently isolated geographically or by other genetic barriers, a local immigration of foreign genes still remains possible. This is apparently the case for the Sussex in Great Britain. Also the breeding policy can be responsible for the extension of a genetic character. The diffusion of a genetic character can be favoured by genetic drift, when popular bulls from a definite line are propagated in the population. Pedigree analyses of the herd-book material in the cattle of Middle and High Belgium proved that this could be apparently the reason for the extension of the albumin allele B. The recent intensive use of bulls originating from the line of Trill, which was of the AB type, was responsible for the 1% difference in frequency between the older cows, mothers of the bulls, and their sons. Probably, in the near future the frequency of the B allele will increase rapidly. One cow of the Red pied breed of East-Flanders was detected to have AD type. This type occurs occasionally in Ayrshire cattle (SPOONER and OLIVER, 1969) and may support the view of some authors about an eventual common origin of both breeds, as suggested by the typical pigment localization on the necks (BUTAYE, 1969).

The absence of any homozygous-BB animal gives a deviation from the genotype repartition as expected by chance. No relationship could be confirmed between Hb B and Al B as pointed out by OSTERHOFF (1969) since Hb A is predominant in the Belgian cattle breeds in which only a few individuals can be detected with hemoglobin type AB.

## REFERENCES

ASHTON, G. C. (1964), Serum albumin polymorphism in cattle. *Genetics*, **50.** 1421–1426.

ASHTON, G. C. and LAMPKIN, G. H. (1965), Serum albumin and transferrin polymorphism in East African cattle. *Nature*, **205.** 209–210.

BRAEND, M. and EFREMOV, G. (1964), Polymorphism of cattle serum albumin. *Nord. Vet. Med.* **11.** 585–588.

BUTAYE, R. (1969), Personal communication.

CARR, W. R. (1966), Serum albumin polymorphism of some breeds of cattle in Zambia. Proc. Xth Europ. Conf. Anim. Blood Groups. Biochem. Polymorph. Paris, 1966, 293–297.

OSTERHOFF, D. R. (1968), Immunogenetic studies in South African Cattle breeds *Thesis*, Univ. Pretoria, 240.

SARTORE, G. and BERNOCO, D. (1966), Research on biochemical polymorphisms in the indigenous cattle of Piedmont. Proc. Xth Europ. Conf. Anim. Blood Groups Biochem. Polymorph. Paris, 1966, 283–288.

SPOONER, R. and OLIVER, R. (1969), Albumin polymorphism in British cattle. *Anim. Prod.* **11.** 59–63.

*XIIth Europ. Conf. Anim. Blood Groups Biochem. Polymorph., Bp., 1972 (pp. 201—205)*

# THE SEPARATION OF BOVINE HEMOGLOBIN BY ISOELECTRIC FOCUSING IN POLYACRYLAMIDE GEL

W. R. CARR

**Animal Productivity Research Laboratory, National Council for Scientific Research**
**Agricultural Research Unit, Chilanga, Zambia**

## SUMMARY

**A method of separating bovine hemoglobin phenotypes by isoelectric focusing in polyacrylamide gel is described. Differences with the separation obtained with zone electrophoresis in starch gel are discussed. A major and minor components were detected in each of the four hemoglobin types examined.**

The method of choice for the detection of hemoglobin polymorphism is usually a modification of zone electrophoresis on starch gel as originally developed by SMITHIES (1955). Simple rapid, techniques have been devised by a number of different workers and have resulted in the detection of at least five hemoglobin alleles in cattle blood (EFREMOV and BRAEND, 1965; CARR, 1964, 1965; CARR and PARSHOTAM, 1967; OISHI, ABE, SUZUKI and NAMIKAWA, 1968). In starch gel electrophoresis, hemoglobins are applied on filter paper placed in a slit in the gel, and move continuously through the gel as a result of the applied electrical potential. In this process they are fractionated according to the charge on the protein molecule and the gel pore size. The separation thus achieved depends mostly on the charge on the hemoglobin and its molecular size. At the same time, diffusion causes spreading of the zones. It is possible that alleles could be present which have similar characteristics and are not separated by this technique.

The technique of isoelectric fractionation of macromolecules to allow migration of protein molecules to their isoelectric points in a suitable pH gradient has been studied in details by SVENSSON (1961). This can be achieved by low molecular weight ampholytes stabilized by sucrose density gradients. The development in recent years have been reviewed by HAGLUND (1967). Isoelectric focusing in a liquid medium has advantages in the preparative fractionation of protein mixtures, but is rather expensive and slow for analytical applications. However, it is possible to form pH gradients in polyacrylamide gels, using carrier ampholytes in apparatus designed for disc electrophoresis (WRIGLEY, 1968; DALE and LATNER, 1968) and in thin layers (AWDEH, WILLIAMSON and ASKONAS, 1968). This technique overcomes the difficulties of convection and rapid diffusion which occur in the liquid media and also requires much smaller quantities of protein and carriers ampholytes.

Thin layer plates of polyacrylamide gel are more suitable than discs for comparison of different hemoglobin types, and a modification of the method of Awdeh et al. (1968) was developed to examine blood samples from cattle containing four different hemoglobin alleles.

## MATERIAL AND METHODS

Blood was obtained from experimental herds of cattle which had all been typed for hemoglobin polymorphism by zone electrophoresis on starch gel (Carr and Parshotam, 1967). The lysed erythrocytes were adjusted with glycerol to give an approximate hemoglobin content of 20% in 50% glycerol. This solution could be further diluted with water before separation.

Polyacrylamide gel was prepared from a 30% acrylamide stock solution and a 1% catalyst solution containing riboflavin as described by Wrigley (1968). 9.0 ml of the acrylamide solution, 2.5 ml of the catalyst solution and 0.9 ml carrier ampholytes (40%) pH 5.8 (LKB – Produkter AB, Bromma, Sweden) were made up to 37 ml with distilled water and the resulting mixture poured on to a glass plate 15×20 cm. The glass plate was fitted with an edging of flexible plastic material 1 mm thick, sealed to the plate with stopcock grease. A further glass plate of the same dimensions was lowered carefully onto the plate to exclude air bubbles, and the whole irradiated by fluorescent light for about an hour until the gel had set (Awdeh et al., 1968).

After removing one glass plate, the other plate containing the gel was placed on graph paper and the hemoglobin solutions were soaked into squares of filter paper 1.0×0.5 cm which were transferred to the surface of the gel 3 cm from one end, using the graph paper to aid positioning. Each filter paper square was separated by 2 mm. Allowing 1 cm clearance at either side these dimensions permitted the examination of nine samples. Two carbon electrodes were prepared by placing them parallel 18 cm apart on a sheet of plastic wrap ("Gladwrap" Union Carbide, U.S.A.) in a refrigerator or cold room. Before each run the anode was painted with a 5% v/v phosphoric acid and the cathode with 5% v/v ethanolamine solution. The gel was inverted and placed on the electrodes so that the samples were nearer the anode. The wrap was folded around the plate to prevent evaporation and a DC potential of 300 V applied for 16 hours. The pH gradient was determined at the end of the run by cutting disks as described by Awdeh et al. (1968). No staining was necessary at hemoglobin concentrations of 4% and above and the gel could be examined immediately. When staining was required it was necessary to wash the gel exhaustively with 10% v/v trichloracetic acid to remove the carrier ampholytes before staining with a protein stain.

## RESULTS AND DISCUSSION

A typical example of the separation achieved with approximately 5% and 20% hemoglobin solutions is shown in Figs 1 and 2 respectively. Unlike starch gel separations, higher concentrations of the hemoglobin solutions

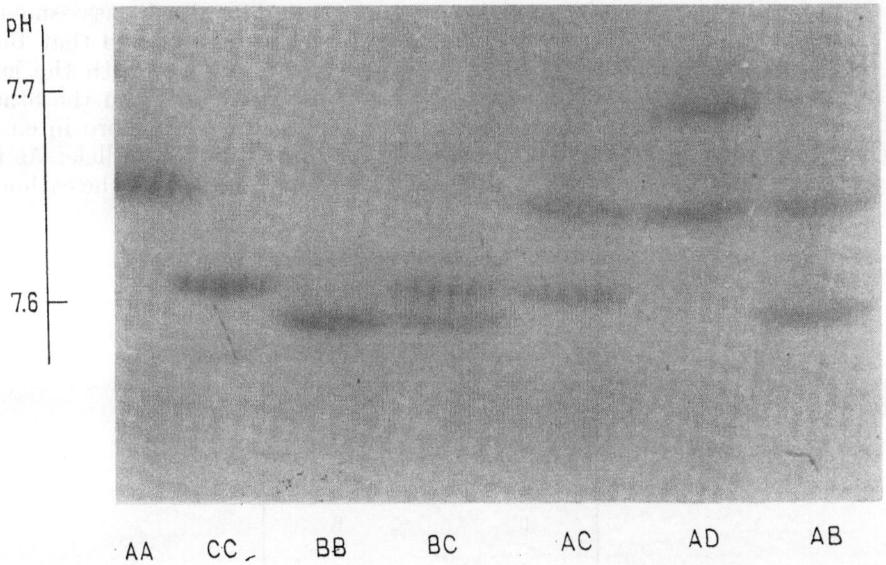
pH
7.7
7.6
AA
CC
BB
BC
AC
AD
AB

FIG. 1. Separation of 5% hemoglobin solutions by isoelectric focusing (unstained)

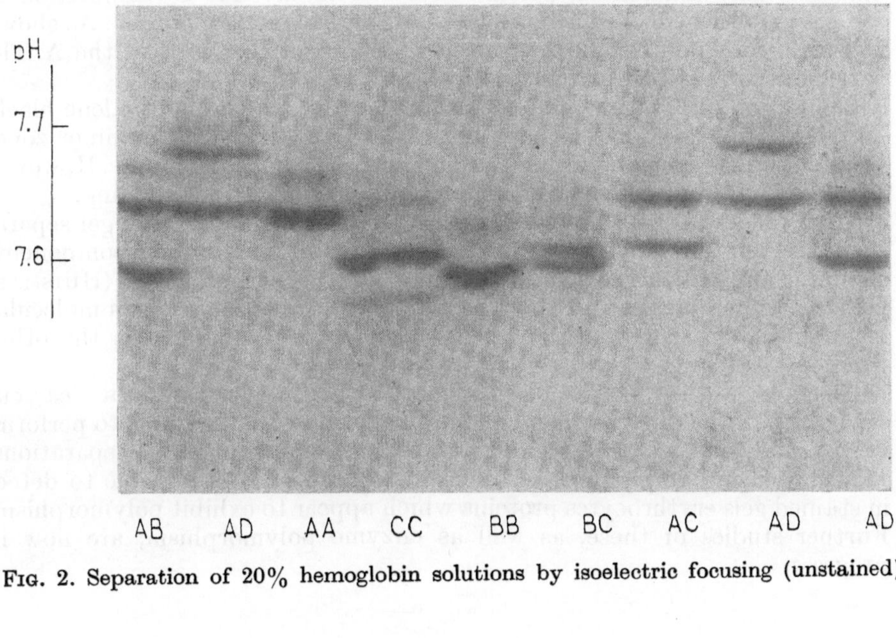
pH
7.7
7.6
AB
AD
AA
CC
BB
BC
AC
AD
AD

FIG. 2. Separation of 20% hemoglobin solutions by isoelectric focusing (unstained) do not diffuse appreciably and thus cause great difficulty in assessment. The approximate pH gradient is shown, and it will be observed that Bov $Hb^{A}$, Bov $Hb^{B}$ and Bov $Hb^{D\ Zambia}$ each show two zones with the less intense giving the higher pH in each case. Bov $Hb^{C\ Rhodesia}$ on the other hand, also has a less intense zone at a lower pH than the more intense. The separation is shown diagrammatically in Fig. 3 for each allele. As in starch gel electrophoresis, the separation of the main zones from the cathode

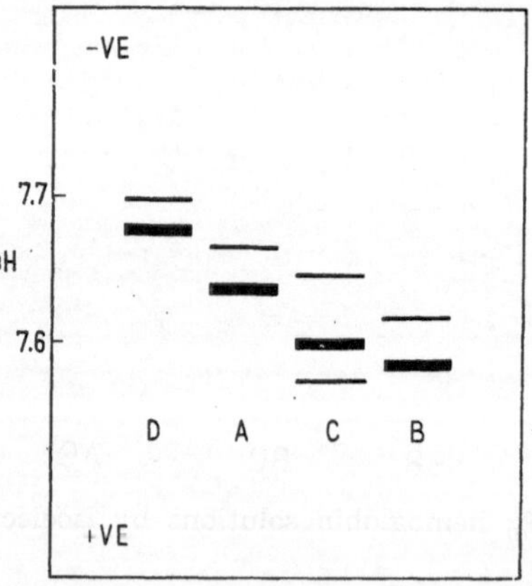
-VE
7.7
pH
7.6
D A C B
+VE

FIG. 3. Diagrammatic representation of bovine hemoglobin alleles, as separated by isoelectric focusing in polyacrylamide gel

end of the gel follows the order D, A, C and B, although the spacing is different. Any differences, presumably, are mainly due to elimination of the gel pore size effect which occurs in starch gel electrophoresis. As shown in Fig. 3, the main C zone was closer to the main B zone than the A. The reverse occurred with starch gel electrophoresis (CARR, 1964).

Separation of 5% hemoglobin solutions, stained with naphthalene black, are shown in Fig. 4. It will be noted that the intensity of the minor zones at higher pH is much increased compared with the unstained gels. However, the zone at lower pH in type C is relatively unchanged by staining.

With the exception of type C, isoelectric focusing and starch gel separations are very similar, except for the presence of these minor components similar components are also apparent in gel filtration separations (HUISMAN, personal communication). Type C merits further investigation at a molecular level, as it does not appear to behave in the same manner as the other types.

Although the isoelectric focusing method as described gives a clear cut separation of the different hemoglobin phenotypes and is simple to perform, it has no great advantages over starch gel electrophoresis separations. However, using ampholytes of different pH ranges it is possible to detect in stained gels erythrocytes proteins which appear to exhibit polymorphism. Further studies of these, as well as enzyme polymorphism, are now in progress.

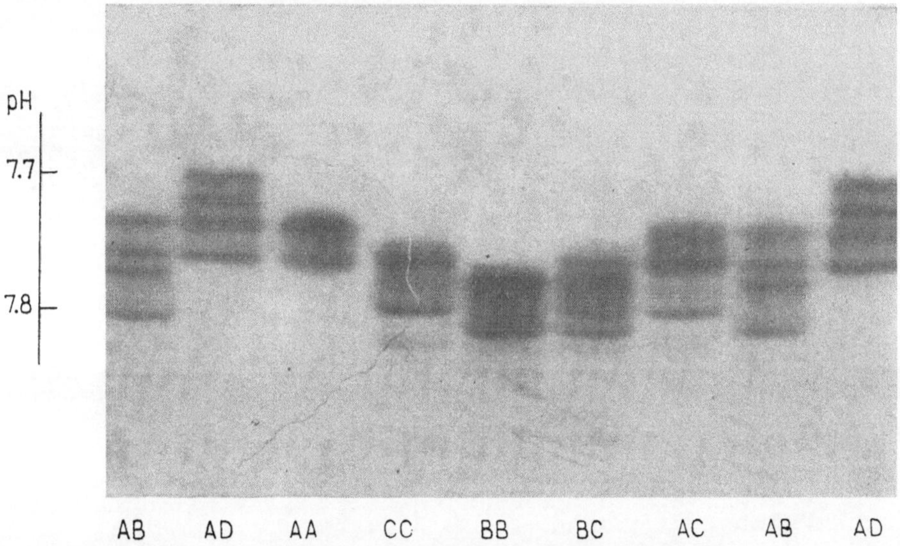
pH
7.7
7.8
AB
AD
AA
CC
BB
BC
AC
AB
AD

FIG. 4. Separation of 5% hemoglobin solutions by isoelectric focusing (stained)

## ACKNOWLEDGEMENTS

The assistance of Mssrs. J. PARSHOTAM and A. NKHOMA is gratefully acknowledged.

## REFERENCES

AWDEH, Z. L., WILLIAMSON, A. R. and ASKONAS, B. A. (1968), Isoelectric focusing in polyacrylamide gel and its application to immunoglobulins. *Nature*, **219**. 66–67.

CARR, W. R. (1964), The haemoglobins of indigenous breeds of cattle in Central Africa. *Rhod. J. Agric. Res.*, **2**. 93–94.

CARR, W. R. (1965), A new bovine haemoglobin variant. *Rhod. J. Agric. Res.*, **3**. 62.

CARR, W. R. and PARSHOTAM, J. (1967), Improvements in the separation of cattle haemoglobins and transferrins by zone electrophoresis on starch gel. *Proc. S. Africa Soc. Anim. Prod.*, **6**, 230–233.

DALE, G. and LATNER, A. L. (1968), Isoelectric focusing in polyacrylamide gels. *Lancet*, **1**. 847–848.

EFREMOV, G. and BRAEND, M. (1965), A new haemoglobin in cattle. *Acta vet. Scand.*, **6**. 109–111.

HAGLUND, H. (1967), Isoelectric focusing in natural pH gradients — a technique of growing importance for fractionation and characterization of proteins. *Science Tools.*, **14**. 17–23.

OISHI, T., ABE, T., SUZUKI, S. and NAMIKAWA, T. (1968), Haemoglobin polymorphism in East Asian cattle and Formosan water buffalo. *Immunogenetics Letter*, **5**. 170–173

SMITHIES, O. (1955), Zone electrophoresis in starch gel: Group variations in the serum proteins of normal human adults. *Biochem. J.*, **61**. 629–641.

SVENSSON, H. (1961), Isoelectric fractionation, analysis and characterization of ampholytes in natural pH gradients. I. The differential equation of state of solute concentrations at a steady state and its solution for simple cases. *Acta Chem. Scand.*, **15**. 325.

WRIGLEY, C. W. (1968), Analytical fractionation of plant and animal protein by gel electrofocusing. *J. Chromatog.*, **36**. 362—365.

*XIIth Europ. Conf. Anim. Blood Groups Biochem. Polymorph., Bp., 1972 (pp. 207–210)*

# IRON SATURATION OF DIFFERENT CATTLE TRANSFERRIN PHENOTYPES

R. C. Buis

Laboratorium voor Bloedgroepen Onderzoek, Landbouwhogeschool, Wageningen, The Netherlands

## SUMMARY

**Iron binding capacity of different cattle transferrin phenotypes was investigated on preparations, gained by a combined rivanol-ethanol-fractionation. The colorimetric estimation of this capacity revealed no significant differences between the groups.**

## INTRODUCTION

However widespread research has been done on genetically determined polymorphism in cattle transferrins, little attention has been paid to detect quantitative differences in iron binding capacity between different phenotypes. Neethling and Osterhoff (1966) demonstrated such differences after starch gel electrophoresis of $Fe^{59}$-saturated serum.

In our study, we made an attempt to investigate these differences by means of a colorimetric method.

## MATERIAL AND METHODS

From 33 cows at several ages and with different transferrin phenotypes, 100 ml blood plasma was taken. Phenotypes were: 7×AA, 10×AD, 10×DD, 4×AE and 2×DE; $D_1$ and $D_2$ were not separated in this experiment.

Transferrin was isolated according to the method of Kistler et al. (1960), as used for human transferrin isolation and iron binding assay. The method depends on plasma fractionation by means of rivanol (2-aethoxy-6,9-diamin-acridin-lactate), followed by two ethanol fractionation steps. The procedure was carried out twice for each preparation.

Starch gel electrophoresis after Ashton (1965) was used to confirm the purity of the preparations.

The colorimetric determination of iron binding capacity depended upon the formation of a brownish transferrin-iron-complex, when $Fe^{++}$ was added to the preparations at pH $> 6.5$. Iron binding capacity was estimated by means of a modified method after Kistler et al.: at 12-minute intervals, 0.025 ml of a solution of 40 $\mu$g $Fe^{++}$ (as $FeSO_4(NH_4)_2SO_4$) per ml Veronal buffer (pH 7.5) was added to test tubes, containing 4 ml 1% transferrin preparation. 10 minutes after each addition, the extinction at $\lambda = 465$ m$\mu$ was measured in a 1 cm cuvette. As blancs, preparations were used from which iron had been removed by 4-hour incubation at 37°C with 0.0125 M EDTA and acetic acid up to pH = 5.5. Test tubes and blancs were kept at 37°C during the test. 12 additions of iron were enough to reach full saturation in each preparation. Corrections were made for increase in volume in the test tubes.

The wavelength of 465 m$\mu$ was used after a control experiment, in which both human and cattle transferrin showed an increase at $\lambda = 465$ m$\mu$ after iron saturation.

A statistical evaluation of differences in extinction increase and iron saturation amounts between the groups of phenotypes was carried out with a test after WILCOXON (1945). This distribution-free ranking method was chosen for in our results no normal frequency distribution could be expected, due to the little number in the groups.

## RESULTS

Protein recovery was rather constant: Starting from 7.20 ± 0.10 g (mean ± standard error), the isolation procedure yielded 0.21 ± 0.01 g. The electrophoretical picture showed an increase in density in the transferrin region, and a total disappearance of albumin. However, a faint new band was found, migrating somewhat slower than albumin.

The saturation curves of all preparations gave a picture like represented in Fig. 1. Each preparation showed an initial, partial saturation (A). Additional $Fe^{++}$ caused an increase of $E_{465}$ until a bend in the curve was reach-

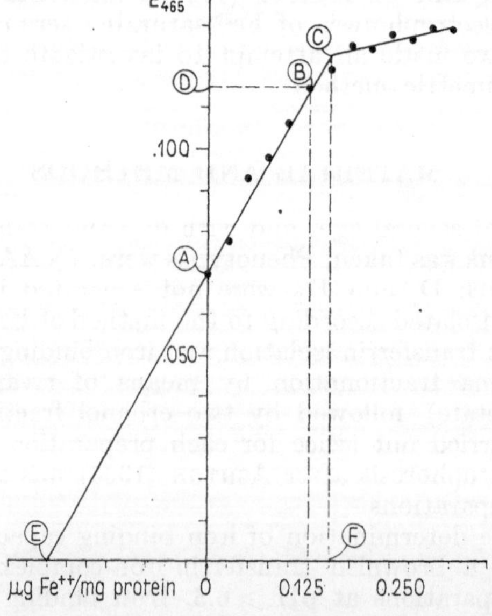

E465
.100
.050
A
B
C
D
E
F
µg Fe++/mg protein
0
0.125
0.250


**FIG. 1. Characteristic saturation curve after addition of 12 × 0.025 µg $Fe^{++}$/mg protein to a 1% protein concentration cattle transferrin preparation (prep. No. 1, Tf AA)**

TABLE 1

*Increase of $E_{465}$ after addition of 0.125 μg $Fe^{++}$/mg protein to 33 transferrin preparations ("AD" in Fig. 1)*

| Increase of $E_{465}$ (class centres) | Frequency in transferrin preparations of different phenotypes | | | | |
|---|---|---|---|---|---|
| | AA | AD | DD | AE | DE |
| .030 | 1 | 3 | 2 | — | — |
| .040 | 1 | 2 | 2 | 2 | 1 |
| .050 | 3 | 1 | 5 | 1 | 1 |
| .060 | 1 | 1 | 1 | 1 | — |
| .070 | 1 | 2 | — | — | — |
| .080 | — | 1 | — | — | — |
| Mean and standard error* | .049 ±.004 | .052 ±.006 | .047 ±.003 | .044 ±.004 | .048 ±.004 |

* calculated from individual values

ed (C). After further addition, only a slight increase was measured, presumably due to $Fe(OH)_3$ formation. Measurements concerned two facts of the curves:

(1) Each curve was straight until at least 0.125 μg $Fe^{++}$/mg protein had been added (B). The increase of $E_{465}$ was calculated for this section (AD). From Table 1 it will be clear that only slight differences occur in $E_{465}$ increase, between the five groups of phenotypes. The Wilcoxon test revealed no significant differences ($p > 0.1$).

Now this increase seemed to be equal for each phenotype, a second group of measurements was carried out:

(2) The total amount of iron, required for full saturation, was estimated after extrapolation of AC to the abcissa. EF represented the amount concerned. The result of estimation saturation amounts has been given in Table 2. The slight differences between the groups of phenotypes seemed

TABLE 2

*Amounts of iron, required to reach full saturation of transferrin preparations ("EF" in Fig. 1)*

| μg $Fe^{++}$/mg protein (class centres) | Frequency in transferrin preparations of different phenotypes | | | | |
|---|---|---|---|---|---|
| | AA | AD | DD | AE | DE |
| 0.20 | — | 2 | 2 | — | 1 |
| 0.25 | 1 | 2 | 3 | 2 | — |
| 0.30 | 4 | 4 | 2 | 1 | — |
| 0.35 | 1 | 2 | 3 | — | — |
| 0.40 | — | — | — | — | — |
| 0.45 | — | — | — | 1 | 1 |
| 0.50 | 1 | — | — | — | — |
| Mean and standard error* | 0.34 ±0.03 | 0.29 ±0.02 | 0.30 ±0.02 | 0.31 ±0.05 | 0.34 ±0.11 |

* calculated from individual values

to be due to chance ($p > 0.1$). The average saturation amount was 0.32 μg $Fe^{++}$/mg protein.

By means of three control preparations, the whole preparation and estimation procedure appeared to yield a reproducibility of about 80% between sample pairs from the same cow.

## DISCUSSION

Our results (no differences between the groups of phenotypes) can be due to loss of transferrin "activity" during fractionation. This is supported by a comparison between our average saturation amount with that from alcohol-fractionated human transferrin preparations, obtained by SCHADE and CAROLINE (1946) and SURGENOR et al. (1949): resp. 0.44 and 1.25 μg $Fe^{++}$/mg protein.

It was not possible to check iron binding capacity during fractionation, for only after one repeatment of the procedure, the saturation amount could be estimated colorimetrically. Only a radioactive assay seems to be able to yield a continuous purification control, and presumably as well a more sensitive and reliable procedure to estimate iron binding capacity of transferrin.

## REFERENCES

ASHTON, G. C. (1965), Serum transferrin D alleles in Australian cattle, *Nature*, **198.** 117.

KISTLER, P., NITSCHMANN, H., WYTTENBACH, A., STUDER, M., NIEDEROST, CH. and MAUERHOFER, M. (1960), Humanes Siderophilin: Isolierung mittels Rivanol aus Blutplasma und Plasmafractionen, analytische Bestimmung und Kristallisation, *Vox Sang.*, **5.** 403–415.

NEETHLING, L. P. and OSTERHOFF, D. R. (1966), Radio-isotope studies on Transferrins in Cattle. Proc. Xth Europ. Conf. Anim. Blood Groups Biochem. Polymorph. Paris 1966, 261–266.

SCHADE, A. L. and CAROLINE, L. (1946), An Iron-binding Component in Human Blood Plasma. *Science*, **104.** 340–341.

SURGENOR, D. M., KOECHLIN, B. A. and STRONG, L. E. (1949), Chemical, clinical and immunological studies on the products of human plasma fractionation: XXXVII, The metalcombining globulin of human plasma. *J. Clin. Invest.*, **28.** 73–78.

WILCOXON, F. (1945), Individual comparisons by ranking methods. *Biometrics Bull.*, **1.** 80–83.

## DISCUSSION

B. GAHNE: What is the age of the animals in your experiment?

R. C. BUIS: All were older than one year. The AA, AD and DD types derived from animals in a lactation period.

*XIIth Europ. Conf. Anim. Blood Groups Biochem. Polymorph., Bp., 1972 (pp. 211–215)*

# BLOOD POLYMORPHIC SYSTEMS AND STRESS IN CATTLE

D. R. OSTERHOFF and I. S. WARD-COX

Department of Zootechnology, Faculty of Veterinary Science, University of Pretoria, Onderstepoort, Rep. of South Africa

## SUMMARY

In view of the climatic conditions in tropical and subtropical areas where a great number of cattle live throughout dry winters on a very low level of nutrition and exist on the borderline between life and death during long periods of drought, studies of the possible relationship between genes being responsible for certain serum types and those responsible for fitness factors are indicated.

Two groups of cows and two groups of bulls, altogether 55 animals, were selected according to their transferrin genotypes and subjected to hunger stress. Series of hematological observations, including ATP-, radioactive Iodine T3- and cholesterol values were made at regular intervals.

No statistically clear superiority of the one genotype over the other could be found, but the information gained on the practical side, methods of rationing the limited available feed and treating of physically weak animals, is of great importance for farmers in drought stricken ranching areas.

## INTRODUCTION

Following the work on transferrin types and adaptability published earlier (OSTERHOFF, 1966) another experiment was performed in order to obtain more evidence of the possible superiority of the animals possessing the transferrin allele $Tf^E$. It was thought that the easiest way to test the relationship between the presence of $Tf^E$ with better adaptability including higher heat tolerance, lower minimum nitrogen requirements and higher rumen fermentation rates was a comparison of animals being similar with regard to breed, type and condition but with different transferrin types under the heaviest possible stress conditions — starvation. The hemoglobin, albumin and amylase types were also taken into consideration and correlated with the body weight loss and to changes in physiological blood values.

## MATERIAL AND METHODS

Experimental animals were selected from a group of Afrikaner cattle according to weight, stage of pregnancy and transferrin types. Although great care was taken to ensure reasonably good similarity within and between the experimental groups certain differences existed (see Table 1).

TABLE 1

*Description of the different groups of animals used*

| Groups Transferrin genotypes | AA $Tf^A/Tf^A$ | AD $Tf^A/Tf^D$ | AE $Tf^A/Tf^E$ | DD $Tf^D/Tf^D$ | DE $Tf^D/Tf^E$ | EE $Tf^E/Tf^E$ |
|---|---|---|---|---|---|---|
| No. of animals at start of the experiment | | | | | | |
| cows | 7 | 8 | 5 | 6 | 8 | 5 |
| bulls | 3 | 4 | 0 | 4 | 2 | 2 |
| Average age (years) | | | | | | |
| cows | 7.4 | 8.6 | 7.0 | 8.6 | 7.5 | 7.6 |
| bulls | 8.0 | 6.5 | — | 5.0 | 8.0 | 6.0 |
| Average weight (lbs)* | | | | | | |
| cows | 942 | 965 | 956 | 982 | 949 | 874 |
| bulls | 1660 | 1339 | — | 1212 | 1620 | 1341 |
| No. of cows in calf | 1 | 2 | 2 | 3 | 2 | 1 |
| No. of animals at end of the experiment | | | | | | |
| cows | 7 | 8 | 4 | 5 | 6 | 5 |
| bulls | 3 | 4 | 0 | 4 | 2 | 2 |

* 1 lb. = 453.6 g

Unfortunately only eleven cows were in calf at the beginning of the experiment, but in each group there was at least one pregnant cow represented.

The total number of 39 cows and 15 bulls was divided into four different groups and fed as follows:

Group I: 20 cows (10 in each kraal), 20 lbs of hay per head per day.
Group II: 19 cows (10 in one, 9 in the other kraal), 70 lbs of hay per head on Tuesdays and 70 lbs on Fridays.
Group III: 9 bulls (in one kraal) 20 lbs of hay per head per day.
Group IV: 6 bulls (in one kraal) normal feed: 5–6 lbs of green lucerne, 10–11 lbs of teff hay, ±7 lbs of maize silage and 4–5 lbs of concentrate mixture.

Only 20 lbs of hay of extremely low nutritional value with a crude protein content of only 2.28% was fed per head per day. It was obvious that the animals had to loose weight very rapidly on this ration and the actual losses during the period of four months ranged between 20 and 25%.

The usual techniques (OSTERHOFF, 1968) were applied in the determination of the transferrin, albumin and hemoglobin types. Hematological values including hemoglobin, hematocrit, albumin, globulin, blood urea nitrogen (B.U.N.), serum alcaline phosphatase (S.A.P.), total cholesterol, blood sugar and adenosine triphosphate (A.T.P.) were collected at monthly intervals, while the thyroxin binding globulins, total lipids and red cell fragility were only measured on the last specimens, i.e. during the last week of the starvation experiment. The experiment was concluded when the first animal was at the point of falling down from malnutrition.

## RESULTS

Since a more even distribution of different transferrin types was possible in the larger female group, most attention is directed to the description of the results in these animals. At the conclusion of the experiment the clinical condition of animals could be expressed in terms of the percentage change from the normal values at the beginning of the experiment. The results for the cows falling into different transferrin groups are fully given in Table 2.

TABLE 2

*Percentage change of blood values during the stress period from August to November in different transferrin groups*

| Transferrin types | AA | AD | AE | DD | DE | EE |
|---|---|---|---|---|---|---|
| Body weight | −19.5 | −24.5 | −22.6 | −23.6 | −23.6 | −24.5 |
| Albumin | − 7.9 | −18.7 | −23.5 | −16.9 | −17.5 | −22.1 |
| Globulin | −20.7 | −16.2 | −21.6 | −12.6 | −24.2 | −23.3 |
| Hemoglobin | −15.9 | −15.9 | −16.0 | −17.8 | −12.6 | −20.6 |
| Hematocrit | −17.6 | −16.9 | −18.5 | −15.4 | −19.0 | −28,5 |
| B.U.N. | + 1.2 | + 3.1 | − 2.0 | −19.1 | −29.4 | − 4.7 |
| S.A.P. | + 7.6 | − 3.1 | −19.6 | − 7.1 | − 4.8 | −19.3 |
| T. cholesterol | − 6.0 | −19.0 | − 1.5 | − 5.2 | −13.8 | −22.7 |
| Blood sugar | − 5.9 | +13.0 | −13.0 | − 1.8 | − 0.4 | −16.8 |
| A.T.P. | −37.3 | −41.4 | −36.4 | −40.7 | −41.6 | −55.3 |

The mode of feeding had also very little influence on the blood values and no further statistical evaluation was necessary.

From Fig. 1 it is evident that there was only a minor difference between the transferrin and also between the different feeding groups with regard

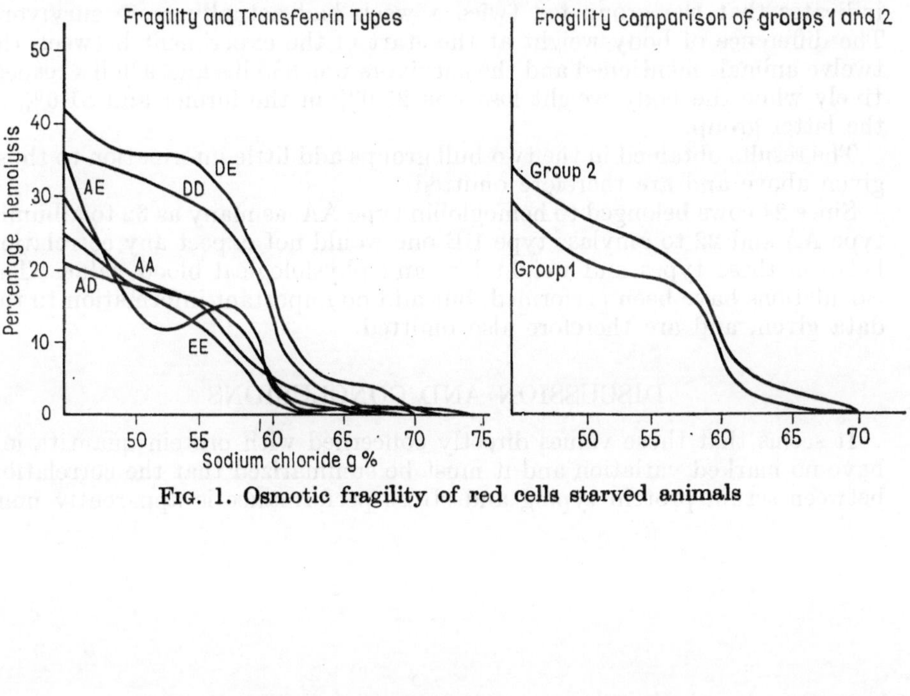
Fragility and Transferrin Types
Fragility comparison of groups 1 and 2
Percentage hemolysis
50
40
30
20
10
0
DE
AE
DD
AA
AD
EE
Group 2
Group 1
50
55
60
65
70
75
Sodium chloride g %

**FIG. 1. Osmotic fragility of red cells starved animals**

to the resistance of the erythrocytes to hemolysis in decreasing strength of hypotonic saline solutions.

At the end of the experiment indirect measurements of the thyroid activity were taken by the T3-uptake test. In the Thyopac T3 test liothyronine labelled with radioactive iodine is mixed with a serum sample. The capacity of the serum to bind the labelled T3 is a reflection of the degree of prior saturation of the serum proteins and thus an indirect measure of the activity of the thyroid gland.

Plasma lipids have recently been studied in different breeds including Afrikaner (O'KELLY, 1967) with the main aim to find relationships of the lipid mechanism and adaptation to tropical environments. Therefore it was decided to investigate the lipid levels of starving animals also in this study.

TABLE 3

*T3-values and plasma lipids in starving cattle*

| Transferrin types | AA | AD | AE | DD | DE | EE |
|---|---|---|---|---|---|---|
| T3-values (counts) | 104 | 100 | 103 | 98 | 101 | 102 |
| Plasma lipids (mg/100ml) | 351 | 347 | 304 | 346 | 433 | 311 |

Also these blood values did not show any significant differences between animals belonging to different transferrin types.

The most remarkable observation was made directly after the conclusion of the experiment. At the end of November the raining season started with cold weather, which many of the animals could not stand. Before more losses were encountered the weakest animals were slaughtered. Altogether nine animals had to be taken out of group I while only three had to leave group II, a difference which is significant at the $P < 0.05$-level and which indicates that the group fed twice a week had actually more survivors. The difference of body weight at the start of the experiment between the twelve animals mentioned and the survivors was 885 lbs and 975 lbs respectively while the body weight loss was 25.0% in the former and 21.6% in the latter group.

The results obtained in the two bull groups add little information to those given above and are therefore omitted.

Since 24 cows belonged to hemoglobin type AA, as many as 32 to albumin type AA and 22 to amylase type BB one would not expect any correlation between these types and weight loss and physiological blood values. The calculations have been performed, but add no important information to the data given, and are therefore also omitted.

## DISCUSSION AND CONCLUSIONS

It seems that those values directly concerned with protein quantitation have no marked variation and it must be summarized that the correlation between serum protein typing and stress performance is apparently nonexistent. One point of criticism, however can be brought forward to the experiment performed, i.e. Afrikaner cows and also oxen do not compare favourably with other animals in feedlots. The experiment if performed on veld should give clearer results and then possibly the animals with the transferrin allele $Tf^E$ would show their superiority. One could say, all animals on the diet given here, never had a chance to show their genetic ability. The experiment should, therefore, be repeated with a great number of animals on the veld, or better, gene frequency studies should be performed on large groups of animals before and after a severe drought.

The values showing the greatest degree of variation, like the B.U.N., S.A.P. and blood sugar values are directly concerned with enzyme action, especially transaminases, dehydrogenases and phosphatases. Therefore it would appear that enzyme typing would possibly give a more decisive answer to animal performance under these conditions.

The most practical result from the whole experiment is, however, that animals should in a period of drought not be fed before they have lost 20% of their body weight. One should keep the reserves until the raining season starts, because then the greatest losses are encountered. One should only feed twice a week to keep the losses as low as possible. It is amazing, how constant most of the physiological blood values and also the weights of all internal organs remained. The latter can after periods of weight losses easily be activated to make up for the previous decreases in body weight.

## ACKNOWLEDGEMENTS

We express our thanks to Mr. H. J. Walzl and his colleagues from the Department of Medicine, Faculty of Veterinary Science for many of the blood analyses.

## REFERENCES

O'Kelly, J. C. (1968), *Aust. J. biol. Sci.* **21.** 1013.

Osterhoff, D. R. (1966), Proc. Xth Europ. Conf. Anim. Blood Groups Biochem. Polymorph. Paris, 1966, 273.

Osterhoff, D. R. (1968), Immunogenetical studies in South African cattle breeds. *Thesis*, 340 pp. Univ. Pretoria.

## DISCUSSION

B. Gahne: Is there any indication of genetic differences in the ability to resist your stress conditions?

D. R. Osterhoff: No heritability estimates have been performed on performance of animals and stress conditions and adaptability. We certainly are starting further for the correlations of polymorphic systems, to adaptability and it would be of great value to find a relationship between any of our polymorphic systems and the performance of animals under the mentioned conditions.

*XIIth Europ. Conf. Anim. Blood Groups Biochem. Polymorph., Bp., 1972 (pp. 217–223)*

# CASEIN TYPES IN HUNGARIAN SPOTTED CATTLE AND OTHER BREEDS PLAYING A ROLE IN ITS DEVELOPMENT

I. Horváth and I. Mészáros

**Center of Artificial Insemination, Laboratory of Immunogenetics, Budapest, Hungary**

## INTRODUCTION

We examined the milk casein polymorphism in Hungarian Spotted cattle breed and in those cattle breeds which played a significant role in its evolution. The ancestor of the Hungarian Spotted breed had been the Hungarian Grey cattle breed brought along by the ancient Hungarians immigrating from the East in 896.

Individuals of Hungarian Grey breed show low or medium milk producing capacity. This stock had been crossed with the "Busa" and "Riska" cattle characterized by higher milk yields than common for the cattle population of the Carpathian Basin at that time. Later they were crossed also with other breeds of Western origin i.e. Pinzgau, Brown Swiss, Simmental, Holstein-Friesian and Austrian Fleckvieh breeds, having higher milk productivity.

The greater part of the stock was pure-bred to maintain the valuable properties of the Hungarian Grey cattle which was still popular in the 19th century.

However, the newer tendencies of breeding for higher milk yield with good meat production maintained, have gradually superseded the tendencies for pure breeding of the stock which has been overcrossed and at present only a protected population of 600–800 cows is preserving the genetic features of the original breed.

## METHODS AND OBSERVATIONS

Milk samples were obtained from Hungarian Grey cattle from two state farms, and from Hungarian Spotted cows of two herds, in which also Simmental bulls were used for service. The Simmental breed consisted of imported cows which have been selected for various aspects.

Milk samples of Brown Swiss cows were obtained also from the stock of two state farms; in one, considerable inbreeding has been practised for many years, in the other the cattle were imported from Switzerland a few years ago. The Friesian cows were imported from the Netherlands. Also Austrian Spotted cattle were imported during the past ten years. Milk samples from Pinzgauer individuals were supplied from the Slovakian Livestock Breeding Institute at Nitra.

The milk samples were delivered to our Institute without preserving material in the winter while in the summer with sublimate added as preservation solution. The samples were examined by starch-gel-urea electrophoresis, using the method of ASCHAFFENBURG and THYMANN with a slight modification. Electrophoresis was continued for 6 hours at +4°C and 6 V/cm. Then the gel blocks were sliced longitudinally into two parts and stained with Amido Black 10 B.

The primary aim of our studies was the determination of $\alpha$-$_{S1}$ and $\beta$-casein types.

TABLE

*Relationship between $\alpha_{S1}$-*

| Breeds | No. | $\alpha_{S1}$-casein / $\beta$-casein | 1. BB/AA | 2. BB/BB | 3. BB/CC | 4. BB/AB |
|---|---|---|---|---|---|---|
| Hungarian Grey | 96 | observed | 50 | — | — | 3 |
| | | expected | (49.9) | (2.4) | (1.0) | (13.6) |
| | | difference | +0.1 | −2.4 | −1.0 | −10.6 |
| | | % | 52 | — | — | 3 |
| Pinzgau | 137 | observed | 70 | — | — | 10 |
| | | expected | (65.6) | (3.4) | (1.4) | (19.2) |
| | | difference | +4.4 | −3.4 | −1.4 | −9.2 |
| | | % | 52 | | | 7 |
| Holstein-Friesian | 99 | observed | 91 | — | — | 6 |
| | | expected | (47.4) | (2.4) | (1.0) | (13.9) |
| | | difference | +43.6 | −2.4 | −1.0 | −7.9 |
| | | % | 92 | — | — | 6 |
| Austrian-Fleckvieh | 89 | observed | 54 | — | — | 20 |
| | | expected | (42.4) | (2.2) | (0.9) | (12.5) |
| | | difference | +11.6 | −2.2 | −0.9 | +7.5 |
| | | % | 60 | — | — | 23 |
| Brown-Swiss | 235 | observed | 59 | 35 | — | 92 |
| | | expected | (112.5) | (6.0) | (2.5) | (33.0) |
| | | difference | −53.5 | +28.7 | −2.5 | +59.0 |
| | | % | 25 | 15 | — | 40 |
| Simmental | 242 | observed | 98 | 1 | 11 | 17 |
| | | expected | (115.9) | (6.0) | (2.6) | (17.6) |
| | | difference | −17.9 | −5.0 | +8.4 | −0.6 |
| | | % | 40.5 | 0.5 | 5 | 7 |
| Hungarian-Spotted I + II | 574 | observed | 280 | 4 | 5 | 65 |
| | | expected | (274.9) | (14.4) | (6.2) | (80.7) |
| | | difference | +5.1 | −10.4 | −1.2 | −15.7 |
| | | % | 50 | 1 | 1 | 11 |
| Total: | 1472 | — | 702 | 40 | 16 | 213 |

## RESULTS

The distribution of various casein types in the Hungarian Grey cattle breed indicated a homogenous stock. In addition to the homogeneity proved by the qualitative similarity of alleles in the Hungarian Grey and the Friesian breeds, an ancestral background may be supposed.

Percentual distribution of $\alpha_{S1}$ casein alleles was found characteristic, the high percentage of casein B allele suggests the inbreeding within Friesian

1

*and β-casein allele pairs*

| 5. | 6. | 7. | 8. | 9. | 10. | FG | $\chi^2$ | P=% |
|---|---|---|---|---|---|---|---|---|
| BB/AC | BC/BC | BC/AA | BC/AB | BC/AC | CC/AA | | | |
| —<br>(13.7)<br>−13.7<br>— | —<br>(1.7)<br>−1.7<br>— | 37<br>(13.7)<br>+23.3<br>39 | 3<br>(1.8)<br>+1.2<br>3 | —<br>(1.0)<br>−1.0<br>— | 3<br>(0.9)<br>+2.1<br>3 | 54 | 55.2 | 60 |
| 16<br>(19.5)<br>−3.5<br>12 | —<br>(2.4)<br>−2.4<br>— | 25<br>(19.6)<br>+6.4<br>18 | 7<br>(2.6)<br>+4.4<br>5 | 8<br>(1.4)<br>+6.6<br>5.3 | 1<br>(1.4)<br>−0.4<br>0.7 | 54 | 38.8 | 95 |
| —<br>(14.1)<br>−14.1<br>— | —<br>(1.7)<br>−1.7<br>— | 2<br>(14.1)<br>−12.1<br>2 | —<br>(1.9)<br>−1.9<br>— | —<br>(1.2)<br>−1.2<br>— | —<br>(0.9)<br>−0.9<br>— | 54 | 78.1 | 8 |
| 5<br>(12.6)<br>−7.6<br>6 | —<br>(1.5)<br>−1.5<br>— | 8<br>(12.7)<br>−4.7<br>10 | 2<br>(1.7)<br>+0.3<br>1 | —<br>(1.0)<br>−1.0<br>— | —<br>(0.9)<br>−0.9<br>— | 54 | 24 | 99 |
| 1<br>(33.7)<br>−32.7<br>0.4 | 4<br>(3.1)<br>+0.9<br>1.6 | 20<br>(33.7)<br>−13.7<br>10 | 17<br>(2.7)<br>+14.3<br>1 | —<br>(2.5)<br>−2.5 | 7<br>(2.3)<br>+4.7 | 54 | 388.8 | — |
| 79<br>(34.5)<br>+44.5<br>31 | 10<br>(4.3)<br>+5.7<br>5 | 20<br>(34.7)<br>−14.7<br>9 | —<br>(4.7)<br>−4.7<br>— | 3<br>(2.6)<br>+0.4<br>1 | 3<br>(2.5)<br>+0.5<br>1 | 54 | 109.6 | 0.1 |
| 110<br>(81.9)<br>+28.1<br>20.5 | 6<br>(10.1)<br>−4.1<br>1 | 96<br>(82.3)<br>+13.7<br>16 | —<br>(11.3)<br>−11.3<br>— | 5<br>(6.2)<br>−1.2<br>1 | 3<br>(5.8)<br>−2.8<br>0.5 | 54 | 36.9 | 96 |
| 211 | 20 | 208 | 29 | 16 | 17 | | | |

breed and qualitative differences between breeds. The same is indicated also by the linkage studies of alleles related to matings in Friesian breeds, showing great differences between the observed and expected results. From it may be concluded that inbreeding as well as the high value of $\chi^2$ shows a significant deviation against other genetic balance featuring stocks (Table 1).

This appears to explain the significant differences between the data of gene frequencies (Table 2). Qualitative differences of Hungarian Grey

TABLE 2

*The gene frequencies in cows from the examined cattle breeds*

| Breeds | No. | $\alpha_{S1}$-casein | | | $\beta$-casein | | |
|---|---|---|---|---|---|---|---|
| | | $-Cn^A$ | $-Cn^B$ | $-Cn^C$ | $-Cn^A$ | $-Cn^B$ | $-Cn^C$ |
| Hungarian Grey | 96 | — | 0.7605 | 0.2395 | 0.9687 | 0.0313 | — |
| Pinzgau | 137 | — | 0.8467 | 0.1533 | 0.8503 | 0.0620 | 0.0877 |
| Austrian Fleckvieh | 89 | — | 0.9438 | 0.0562 | 0.8315 | 0.1404 | 0.0281 |
| Brown-Swiss | 235 | — | 0.8915 | 0.1085 | 0.6000 | 0.3895 | 0.0105 |
| Simmental | 242 | — | 0.9400 | 0.0600 | 0.7046 | 0.0599 | 0.2355 |
| Holstein Friesian | 99 | — | 0.9899 | 0.0101 | 0.9697 | 0.0303 | — |
| Hungarian Spotted I. + II. | 574 | — | 0.9068 | 0.0320 | 0.8188 | 0.0670 | 0.1142 |

cattle as compared to the Pinzgauer and Austrian Spotted breeds are indicated by the $\beta$-AC alleles which allow to conclude for a Simmental background. The gene frequencies in the three breeds as well as the results of the linkage studies of $\alpha_{S1}$ and $\beta$-casein matings show great differences owing partly to the different origin and partly to the qualitative differences between these breeds. The Simmental cattle exerted the greatest influence on milk casein distribution in Hungarian Spotted cattle. The appearance of $\beta$-C allele in different genotypes can be ascribed to the Simmental influence.

The $\beta$-C is considered to be characteristic of the Simmental breed. From its frequency conclusions can be drawn on the degree of Simmental influence (Figs 1 and 2).

Study on correlation indicates that the casein phenotypes of the Hungarian Grey breed show significant differences both qualitatively and quantitatively when compared to other breeds (Table 3).

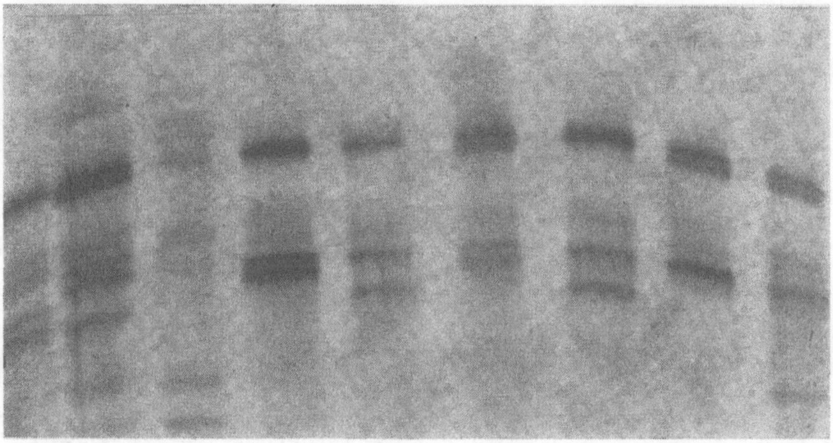

$\alpha$-s$_1$-Cn BB BC BB BB BB BC BB BC BC
$\beta$-Cn BC AC AA AB AC AB AC AA AA

FIG. 1. $\alpha$-s$_1$-Cn and $\beta$-Cn phenotypes from the herd in which Simmental bulls were preferred

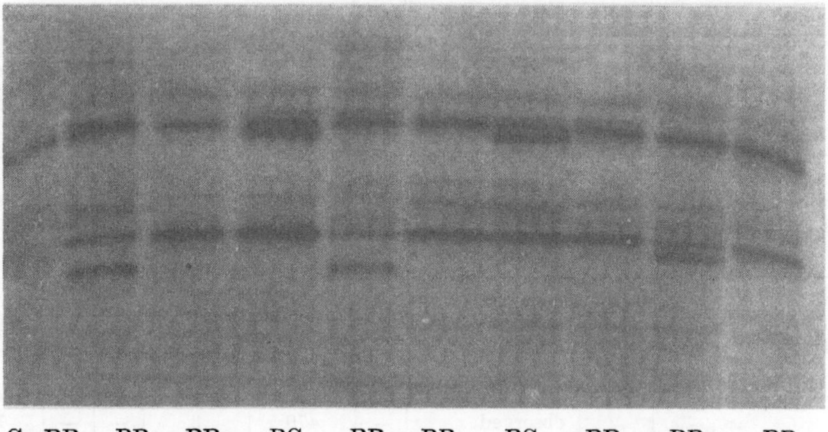

$\alpha$-s$_1$-Cn BB BB BB BC BB BB BC BB BB BB
$\beta$-Cn AB AC AA AA AC AA AA AA AB AA

FIG. 2. $\alpha$-s$_1$-Cn and $\beta$-Cn phenotypes from the herd where Simmental bulls were used less frequently

TABLE

*The distribution of $\alpha_{S1}$*

| Breeds | No. | | $\alpha_{S1}$-casein | | | | | |
|---|---|---|---|---|---|---|---|---|
| | | | AA | BB | CC | AD | AC | BC |
| Hungarian Grey | | observed | — | 53 | 3 | — | — | 40 |
| | 96 | expected | — | (55.51) | (5.5) | — | — | (34.9) |
| | | difference | | —2.5 | —2.5 | — | — | +5.1 |
| | | % | — | 55 | 3 | — | — | 42 |
| | | observed | — | 96 | 1 | — | — | 40 |
| Pinzgau | 137 | expected | — | (98.2) | (3.1) | — | — | (35.5) |
| | | difference | — | —2.2 | —2.1 | — | — | +4.5 |
| | | % | — | 70 | 0.7 | — | — | 29.3 |
| | | observed | — | 97 | — | — | — | 2 |
| Holstein-Friesian | 99 | expected | — | (97.01) | (0.009) | — | — | (1.9) |
| | | difference | — | 0.01 | —0.009 | — | — | +0.1 |
| | | % | — | 98 | — | — | — | 2 |
| | | observed | — | 79 | — | — | — | 10 |
| Austrian | 89 | expected | — | (79.2) | (0.2) | — | — | (9.4) |
| Fleckvieh | | difference | — | —0.2 | —0.2 | — | — | +0.6 |
| | | % | — | 89 | — | — | — | 11 |
| | | observed | — | 191 | 7 | — | — | 37 |
| Brown Swiss | 235 | expected | — | (186.7) | (2.7) | — | — | (45.4) |
| | | difference | — | +4.3 | +4.3 | — | — | —8.4 |
| | | % | — | 81 | 3 | — | — | 16 |
| | | observed | — | 216 | 3 | — | — | 23 |
| Simmental | 242 | expected | — | (212.9) | (0.8) | — | — | (27.1) |
| | | difference | — | —3.1 | +2.2 | — | — | —4.1 |
| | | % | — | 89 | 1 | — | — | 10 |
| | | observed | — | 244 | 1 | — | — | 49 |
| Hungarian Spotted | 294 | expected | — | (241.1) | (2.1) | — | — | (46.8) |
| I. | | difference | — | +2.9 | —1.1 | — | — | +2.2 |
| | | % | — | 83 | 0.5 | — | — | 16.5 |
| | | observed | — | 226 | 2 | — | — | 52 |
| Hungarian Spotted | 280 | expected | — | (224.8) | (2.8) | — | — | (50.4) |
| II. | | difference | — | +1.2 | —0.8 | — | — | +1.6 |
| | | % | — | 80 | 1 | — | — | 19 |
| | | observed | — | 470 | 3 | — | — | 101 |
| Hungarian Spotted | 574 | expected | — | (471.9) | (4.9) | — | — | (97.1) |
| I. + II. | | difference | — | —1.9 | —1.9 | — | — | +3.9 |
| | | % | — | 82 | 0.5 | — | — | 17.5 |
| Total: | 1472 | | | | | | | |

3

*and β-casein phenotypes*

| β-casein | | | | | | | | |
|---|---|---|---|---|---|---|---|---|
| AA | BB | CC | AB | AC | BC | FG | $\chi^2$ | P = % |
| 90<br>(90.0)<br>—<br>94 | —<br>(0.08)<br>−0.08<br>— | —<br>—<br>—<br>— | 6<br>(5.8)<br>+0.2<br>6 | —<br>—<br>—<br>— | —<br>—<br>—<br>— | 4 | 1.28 | 90 |
| 96<br>(97.3)<br>−1.3<br>70 | —<br>(0.5)<br>−0.5<br>— | —<br>(0.1)<br>−0.1<br>— | 17<br>(14.3)<br>+2.7<br>12.5 | 24<br>(21.6)<br>+2.4<br>17.5 | —<br>(1.5)<br>−1.5<br>— | 8 | 4.7 | 80 |
| 93<br>(93.1)<br>−0.1<br>94 | —<br>(0.08)<br>−0.08<br>— | —<br>—<br>—<br>— | 6<br>(5.8)<br>+0.2<br>6 | —<br>—<br>—<br>— | —<br>—<br>—<br>— | 5 | 0.20 | 99 |
| 62<br>(61.5)<br>+0.5<br>70 | —<br>(1.6)<br>−1.6<br>— | —<br>(0.06)<br>−0.06<br>— | 22<br>(20.7)<br>+1.3<br>24 | 5<br>(4.1)<br>+0.9<br>6 | —<br>(0.6)<br>−0.6<br>— | 8 | 2.6 | 90 |
| 86<br>(84.6)<br>+1.4<br>36 | 35<br>(35.9)<br>−0.9<br>15 | —<br>(0.2)<br>−0.2<br>— | 109<br>(109.8)<br>−0.8<br>47 | 1<br>(2.9)<br>−1.9<br>0.5 | 4<br>(1.8)<br>+2.2<br>1.5 | 8 | 12.4 | 15 |
| 121<br>(116.5)<br>+4.5<br>50 | 1<br>(0.4)<br>+0.6<br>0.4 | 11<br>(13.13)<br>−2.3<br>5.0 | 17<br>(19.4)<br>−2.4<br>7.0 | 82<br>(74.4)<br>+7.6<br>33 | 10<br>(5.3)<br>+4.7<br>0.6 | 8 | 7.1 | 50 |
| 201<br>(199.5)<br>+1.5<br>68 | —<br>(2.1)<br>−2.1<br>— | —<br>(1.7)<br>−1.7<br>— | 45<br>(41.7)<br>+3.3<br>15 | 44<br>(39.4)<br>+4.6<br>14.5 | 4<br>(3.5)<br>+0.5<br>2.5 | 8 | 7.1 | 50 |
| 178<br>(179.2)<br>−1.2<br>64 | 2<br>(0.5)<br>+1.5<br>1 | 5<br>(5.3)<br>−0.3<br>2 | 22<br>(22.4)<br>−0.4<br>7 | 71<br>(61.6)<br>+9.4<br>25 | 2<br>(3.9)<br>+1.9<br>1 | 8 | 7.0 | 50 |
| 379<br>(383.1)<br>−4.1<br>66 | 2<br>(3.7)<br>−1.7<br>0.3 | 5<br>(7.4)<br>−2.4<br>1 | 67<br>(62.9)<br>+5.9<br>11.7 | 115<br>(106.9)<br>+8.1<br>20 | 6<br>(8.4)<br>−2.4<br>1 | 8 | 3.3 | 90 |

*XIIth Europ. Conf. Anim. Blood Groups Biochem. Polymorph., Bp., 1972 (pp. 225—228)*

# BLOOD PROTEIN POLYMORPHISM OF THE NATIVE CATTLE, HORSES AND PIGS IN EASTERN ASIA

T. ABE, K. MOGI, T. OISHI, K. TANAKA* and S. SUZUKI*

National Institute of Animal Industry, Department of Animal Genetics, Laboratory of Immunogenetics, Chiba, Japan

Blood protein polymorphism of the native cattle, horses and pigs in Eastern Asia was examined, and on the basis of gene frequencies, the relationship among breeds was discussed.

## (A) CATTLE

The blood samples were taken from Taiwan Yellow cattle *(B. indicus)*, Korean cattle *(B. taurus)* and three major improved Japanese breeds and a Japanese native cattle, "Mishima" cattle. Starch gel electrophoresis was carried out by the method of JAMIESON (1965); ASHTON (1964).

Fourteen phenotypes of transferrin were found in the Asian cattle sera surveyed. Ten of them corresponded to those observed in European cattle sera and were genetically controlled by four alleles; $Tf^{A}$, $Tf^{D_1}$, $Tf^{D_2}$ and $Tf^{E}$. Other four different patterns were observed in eight animals. All of them had slower bands than $Tf^{E}$ bands and appeared to be controlled by a different allele, which was named temporarily $Tf^{X}$, whose bands were similar to $Tf^{G}$ bands (JAMIESON, 1965).

Of cattle hemoglobin, two variants other than, HbA, B and C were found in Korean cattle. One of them had a slightly slower band than HbA, and the other was still slower one, which we named temporarily HbX and HbY. It is supposed that HbX accords with HbD reported by EFREMOV (1965), and HbY with Hb Khillali reported by NAIK (1965).

Four phenotypes of albumin were found, three of which were controlled by $Alb^{A}$ and $Alb^{B}$ alleles, which had already been reported by ASHTON (1965). Another new variant tentatively called Alb BX was observed in two samples of Yellow cattle (ABE, 1968).

Absolute gene frequency differences for each of the two-breed comparisons indicated that the Japanese breeds, especially Japanese Brown cattle, are more like Korean cattle than Yellow cattle. Clines of gene frequencies of Tf, Hb and Alb genes among breeds were also recognized as shown in Table 1.

* Tokyo University of Agriculture, Tokyo, Japan.

TABLE 1

*Gene frequencies of serum transferrin, hemoglobin and albumin types in East Asian cattle*

| Alleles | Frequencies in | | | | | | | |
|---|---|---|---|---|---|---|---|---|
| | Yellow cattle | Korean cattle | Japanese Brown | Japanese Black | Japanese Shorthorn | Mishima cattle | Holstein* | Charolais* |
| $Tf^A$ | 0.163 | 0.304 | 0.202 | 0.373 | 0.535 | 0.480 | 0.356 | 0.346 |
| $Tf^{D_1}$ | 0.057 | 0.242 | 0.281 | 0.286 | 0.160 | 0.510 | 0.219 | 0.106 |
| $Tf^{D_2}$ | 0.564 | 0.253 | 0.447 | 0.327 | 0.244 | | 0.407 | 0.548 |
| $Tf^E$ | 0.167 | 0.196 | 0.067 | 0.014 | 0.061 | 0.010 | 0.019 | 0 |
| $Tf^X$ | 0.049 | 0.005 | 0.003 | 0 | 0 | 0 | 0 | 0 |
| $Hb^A$ | 0.586 | 0.905 | 0.910 | 0.975 | 0.990 | — | 1.000 | 0.894 |
| $Hb^B$ | 0.319 | 0.082 | 0.087 | 0.025 | 0.010 | — | 0 | 0.106 |
| $Hb^C$ | 0.095 | 0.013 | 0.003 | 0 | 0 | — | 0 | 0 |
| $Alb^A$ | 0.183 | 0.980 | 1.000 | 1.000 | 0.997 | — | 1.000 | 0.894 |
| $Alb^B$ | 0.810 | 0.020 | 0 | 0 | 0.003 | — | 0 | 0.106 |
| $Alb^X$ | 0.007 | 0 | 0 | 0 | 0 | — | 0 | 0 |
| No. of animals | 127 | 159 | 178 | 122 | 156 | 171 | 1066 | 52 |

* The European breeds in Japan.

From these results, it is supposed that Japanese native cattle originated from Korean cattle, which might have been somewhat influenced by Yellow cattle. However, the drastic decrease of $Alb^B$, $Hb^B$ and $Hb^C$ gene frequency from Yellow cattle to Korean cattle suggests that Korean cattle might have originated from other breed than Yellow cattle. Therefore it is difficult to support the Southern Route theory on the origin of Japanese cattle that, with the spread of rice paddy farming, Yellow cattle migrated into Japan directly from China via Korea.

## (B) HORSE

Using a modified starch gel method for cattle transferrin, we determined horse transferrin and albumin types in the same running gel.

Seventeen of the possible 21 phenotypes of transferrin were encountered in our Laboratory and the genetic data confirmed that they were controlled by six co-dominant alleles, $Tf^D$, $Tf^F$, $Tf^H$, $Tf^M$, $Tf^O$ and $Tf^R$ as reported by BRAEND and STORMONT (1964).

In Kiso-uma, one of the Japanese native ponies, gene $Tf^M$ was found like in Shetland ponies, but the gene frequency was not so high.

Three phenotypes of albumin; A, AB and B, were observed and the gene frequency of $Alb^A$ in Kiso-uma was the highest among the breeds examined.

TABLE 2

*Gene frequencies of serum transferrin and albumin types in three breeds of horses in Japan*

| Alleles | Frequencies in | | |
|---|---|---|---|
| | Kiso-uma* | Thoroughbreds | Anglo-Arabs |
| $Tf^D$ | 0.311 | 0.317 | 0.298 |
| $Tf^F$ | 0.417 | 0.483 | 0.443 |
| $Tf^H$ | 0.065 | 0.018 | 0.129 |
| $Tf^M$ | 0.007 | 0 | 0 |
| $Tf^O$ | 0.076 | 0.115 | 0.086 |
| $Tf^R$ | 0.119 | 0.067 | 0.044 |
| $Alb^A$ | 0.482 | 0.251 | 0.379 |
| $Alb^B$ | 0.518 | 0.749 | 0.621 |
| No. of animals | 139 | 358 | 228 |

* Japanese native ponies.

## (C) PIG

Serum transferrin, ceruloplasmin, amylase and hemopexin types in the four breeds of Taiwan native swine were examined by starch gel electrophoresis technique (BAKER, 1968; KRISTJANSSON, 1961). The observed gene frequencies were compared with those in the improved European breeds in Japan.

As shown in Table 3, $Tf^A$ gene was found very rarely in Taiwan native swine, while $Tf^C$ gene, not present in three European breeds examined,

TABLE 3

*Gene frequencies of serum transferrin, ceruloplasmin and amylase types in the four breeds of Taiwan native swine*

| Alleles | Frequencies in | | | | | | | |
|---|---|---|---|---|---|---|---|---|
| | Taoyuan | Meinung | Ting-shuanghsi | Short-ear | Total | Yorkshire* | Landrace* | Berk-shire* |
| $Tf^A$ | 0 | 0 | 0 | 0.046 | 0.004 | 0.114 | 0 | 0.257 |
| $Tf^B$ | 0.831 | 0.896 | 0.853 | 0.818 | 0.846 | 0.886 | 1.000 | 0.725 |
| $Tf^C$ | 0.169 | 0.104 | 0.147 | 0.136 | 0.150 | 0 | 0 | 0 |
| $Cp^a$ | 0.069 | 0.037 | 0.088 | 0 | 0.062 | 0 | 0.037 | 0 |
| $Cp^b$ | 0.931 | 0.963 | 0.912 | 1.000 | 0.938 | 1.000 | 0.963 | 1.000 |
| $Am^A$ | 0.444 | 0.500 | 0.125 | 0.458 | 0.385 | 0.093 | 0.156 | 0 |
| $Am^B$ | 0.356 | 0.259 | 0.680 | 0.417 | 0.415 | 0.907 | 0.830 | 0.990 |
| $Am^C$ | 0.050 | 0.204 | 0 | 0.125 | 0.070 | 0 | 0.007 | 0.010 |
| $Am^Y$ | 0.150 | 0.037 | 0.194 | 0 | 0.130 | 0 | 0 | 0 |
| No. of animals | 76 | 27 | 34 | 12 | 149 | 206 | 150 | 100 |

* The improved European breeds in Japan.

was found commonly in Taiwan native breeds. In ceruloplasmin, gene frequency of $Cp^a$ in Taiwan breeds was higher than those of the European breeds. In amylase, extensive polymorphism was observed in Taiwan native breeds. $Am^A$ and $Am^C$ gene are comparatively common in Taiwan native breeds except Tingshuanghsi breed. A new variant tentatively called $Am^Y$ was discovered in the three breeds of Taiwan native ones except Short-ear breed. This AmY band has a little slower mobility than that of AmB, and was found in the form of YY, AY or BY respectively.

In hemopexin, eight new bands other than Hp-O, Hp-1, Hp-1F, Hp-2 and Hp-3 were detected in Taiwan breeds, whose gene frequencies could not be calculated. However, their phenotypic variety suggested that Taiwan breeds were more polymorphic in hemopexin than the European breeds.

From these results, Taiwan native swine is considered to belong to a taxonomic group, *Sus vittatus*, differing from the improved European breeds, *Sus scrofa ferus*.

## REFERENCES

ABE, T., OISHI, T., SUZUKI, S., AMANO, T., KONDO, K., NOZAWA, K., NAMIKAWA, T., KUMAZAKI, K., KOGA, K., HAYASIDA, S. and OTSUKA, J. (1968), *Jap. J. Zootech. Sci.*, **39.** 523–535.

ASHTON, G. C. (1964), *Genetics*, **50.** 1421–1426.

ASHTON, G. C. (1965), *Aust. J. Biol. Sci.*, **18.** 665–670.

BAKER, L. N. (1968), *Vox Sang.*, **15.** 154–158.

BRAEND, M. and STORMONT, C. (1964), *Nord. Vet. Med.*, **16.** 31–37.

EFREMOV, G. and BRAEND, M. (1965), *Acta Vet. Scand.*, **6.** 109–112.

JAMIESON, A. (1965), *Heredity*, **20.** 419–441.

KRISTJANSSON, F. K. (1963), *Genetics*, **16.** 907–910.

NAIK, S. N. and SANGHVI, L. D. (1965), Proc. 9th Europ. Conf. Anim. Blood Groups, Prague, 1964, 295–299.

*XIIth Europ. Conf. Anim. Blood Groups Biochem. Polymorph., Bp., 1972 (pp. 229—231*

# TRANSFERRIN AND HEMOGLOBIN POLYMORPHISM IN THE RUMANIAN SPOTTED CATTLE BREED

I. GRANCIU

Cattle Breeding Research Institute, Corbeanca-Ilfov, Rumania

## SUMMARY

**Transferrins and hemoglobins were studied in 1028 individuals including sires, dams and offspring of Rumanian Spotted breed in two farms.**

**Using starch gel electrophoresis three Tf alleles and six Tf genotypes and two Hb alleles and three Hb genotypes were identified. Figures are presented on phenotypic distribution.**

**The estimated gene frequencies of Tf and Hb alleles were as follows: $Tf^A$ 0.09, $Tf^D$ 0.90, $Tf^E$ 0.01 and $Hb^A$ 0.85, $Hb^B$ 0.15, respectively.**

**Comments are made on age differences in gene frequencies and genotype distributions.**

## INTRODUCTION

Many scientific reports describe the existence of genetic polymorphism in proteins such as those found in animal blood serum, milk etc.

Protein polymorphism was found to vary in populations and breeds. This variation is thought to be related in a certain respect to those structures which determine genetic and phenotypic differences between individuals for traits such as adaptability, resistance to adverse environment etc.

Aspects like those mentioned above are studied in many laboratories in order to characterize populations and to make use of genetic particularities in the biochemical components of the animal body for both scientific and practical purposes (OSTERHOFF et al., 1965).

In this report results of transferrin (Tf) and hemoglobin (Hb) investigations on the Rumanian Spotted cattle breed (RS) are presented.

## MATERIAL AND METHODS

The Tf and Hb types in RS breed were studied on 1028 individuals at the Bontida and Braşov experimental stations.

Starch gel electrophoresis was used to determine Tf and Hb types (ASHTON, 1957; SMITHIES, 1955).

## RESULTS AND DISCUSSION

At Tf locus the presence of three allelomorphic co-dominant genes was demonstrated, $Tf^A$, $Tf^D$, $Tf^E$ and six phenotypes were identified whose distribution is given in Table 1.

TABLE 1

*Distribution of transferrin phenotypes in the Rumanian Spotted breed*

| Farm. | Specification | A/A | D/D | E/E | A/D | A/E | D/E | Total | $\chi^2$ test 4. d.f. |
|---|---|---|---|---|---|---|---|---|---|
| **Bonțida** | observed | 4 | 595 | 2 | 112 | 3 | 12 | 726 | 1.5226 |
| | expected | 5.35 | 594.48 | 0.40 | 110.29 | 1.27 | 13.53 | 725.92 | $P > 0.50$ |
| **Brașov** | observed | 4 | 241 | — | 53 | 1 | 3 | 302 | 1.2659 |
| | expected | 3.18 | 236.60 | — | 55.0 | 0.41 | 3.55 | 301.94 | $P > 0.50$ |
| **Total** | observed | 8 | 836 | 2 | 165 | 4 | 15 | 1028 | 3.3144 |
| | expected | 8.33 | 834.6 | 0.40 | 164.18 | 1.75 | 17.04 | 1027.96 | $P > 0.30$ |

For transferrin types there was a good agreement between the observed and expected figures under assumption of genetic equilibrium ($P > 0.50$).

The frequencies of both Tf genotypes and genes were estimated in parents, progeny and in the entire breed. Results of these estimations are given in Tables 2 and 3.

TABLE 2

*Transferrin genotype frequencies in the Rumanian Spotted breed*

| Specification | A/A | D/D | E/E | A/E | A/D | D/E |
|---|---|---|---|---|---|---|
| Progeny | 0.01 | 0.83 | 0.00 | <0.01 | 0.15 | <0.01 |
| Adults | 0.01 | 0.77 | <0.01 | <0.01 | 0.18 | 0.02 |
| Total | 0.01 | 0.80 | <0.01 | <0.01 | 0.16 | 0.02 |

TABLE 3

*Tf gene frequency in the Rumanian Spotted breed*

| Specification | TfA | TfD | TfE |
|---|---|---|---|
| Progeny | 0.08 | 0.91 | 0.01 |
| Adults | 0.10 | 0.88 | 0.02 |
| Total | 0.09 | 0.90 | 0·01 |

The homozygous $Tf^D/Tf^D$ and heterozygous $Tf^A/Tf^D$ genotypes were found to be most common, the $Tf^D$ allele showed the highest gene frequency. It should be mentioned that like in many other breeds, the frequency of $Tf^E$ allele was very low (0.01) in the RS breed as well.

Due to influence of sires (all were homozygous $Tf^D/Tf^D$) the frequency of Tf D type tended to increase in the offspring.

The hemoglobin types were determined on an identical number of animals.

Two co-dominant alleles ($Hb^A$ and $Hb^B$) and three phenotypes were identified at the Hb locus.

In Table 4 the phenotype distributions are shown separately for each farm and for the entire breed.

TABLE 4

*Hemoglobin phenotype distribution in the Rumanian Spotted breed*

| Farm | Specification | A/A | A/B | B/B | Total | $\chi^2$ test 4. d.f. |
|---|---|---|---|---|---|---|
| Bonţida | observed | 516 | 196 | 15 | 727 | 2.0431 |
| | expected | 518.60 | 190.34 | 17.56 | 726.50 | $P > 0.30$ |
| Braşov | observed | 232 | 64 | 5 | 301 | 0.0675 |
| | expected | 231.56 | 64.89 | 4.35 | 301 | $P > 0.95$ |
| Total | observed | 748 | 260 | 20 | 1028 | 0.2187 |
| | expected | 749.21 | 256.20 | 21.88 | 1027.29 | $P > 0.50$ |

There is a rather close agreement between the observed and expected numbers of individuals. Existing differences are not significant statistically.

The parameters for both genotypes and gene frequencies of Hb locus are given in Table 5.

TABLE 5

*Hemoglobin types in the Rumanian Spotted breed*

| Frequency of: | Specification | B/B | A/B | A/A |
|---|---|---|---|---|
| Genotypes | offspring | 0.02 | 0.24 | 0.74 |
| | adults | 0.03 | 0.27 | 0.70 |
| | total | 0.02 | 0.25 | 0.73 |
| Genes | offspring | 0.14 | — | 0.86 |
| | adults | 0.16 | — | 0.84 |
| | total | 0.15 | — | 0.85 |

The $Hb^A$ allele is the most common one ($q = 0.85$) in the RS population and its gene frequency increases in the offspring generation ($q = 0.86$) due mostly to sires' influence.

These observations contribute to the knowledge of the RS breed in respect of polymorphic characters. They offer new possibilities for studies of genetic variability under selection pressure, on origin, structure and relationship to other breeds and also for practical purposes.

## REFERENCES

ASHTON, G. C. (1957), *Nature*, **180.** 917

OSTERHOFF, D. R., VON HERRDEN. J. R. (1965), Proc. 9th Eur. Conf. Anim. Blood Groups, Prague. 1964, 301.

SMITHIES, O. J. (1955), *Biochem.*, **61.** 629.

*XIIth Europ. Conf. Anim. Blood Groups Biochem. Polymorph., Bp., 1972 (pp. 233–236)*

# A NON-SPECIFIC INHIBITING FACTOR IN THE SEMINAL PLASMA OF BULLS

RUTH SAISON and G. J. KRAAY

Department of Biomedical Sciences, University of Guelph, Guelph, Ontario, Canada

Various possible causes of infertility have been under investigation for many years, and more recently the possible involvement of immune mechanisms has been studied. There are many aspects of immunity which could interfere with fertilization in specific cases, or under specific conditions.

In humans the present trend seems to follow a course of preventing conception in the cause of economic and population stability. However, in the interests of economics where domestic animals are concerned, fertility and increased production are the goal.

In order to implicate an immune mechanism in cases of sterility it must be possible to demonstrate the presence of antigen and antibody capable of forming antigen/antibody complexes. Seminal plasma and sperm provide sources of antigens. Seminal plasma may also provide the antibody to form such complexes as the presence of auto-antibody in seminal plasma has been demonstrated in many cases. Foreign antigens introduced into the female genital track are capable of stimulating and producing an immune response. Here, in the simplest terms are definite possibilities for lowered fertility or sterility as a result of incompatible matings. The presence of antibody, whether in seminal plasma or female vaginal or cervical secretions, is capable of immobilizing sperm, and so preventing its progress to the ova.

The most obvious conditions for incompatibility could be matings between animals with unlike blood group patterns, should blood group antigens be present on sperm. If this were the case, blood group antigens on sperm which were not present on the cells of the female could be sensitized by naturally occurring antibodies or immune antibodies in the serum and secretions of the female and immobilize the sperm.

Tests were devised to demonstrate the presence or absence of blood group antigens and substances on sperm and in seminal plasma. Cattle antibodies to blood group antigens act as hemolysins, so hemolytic tests were set up with cattle blood group reagents, using sperm cells, sperm extracts and seminal plasma as inhibitors and cattle red blood cells as antigens. The reciprocal of the first dilution of inhibitor giving a 2 reading after 3 hours was taken as the inhibition titre.

Seminal plasma is known to contain spermatozoa coating antigens and it is reasonable to assume that if blood group antigens are present in seminal plasma, they could be adsorbed onto sperm cells. The presence of human blood group antigens and substances on sperm and in seminal plasma of secretors has been reported. Human ABO blood group antigens are believed to be present in the seminal plasma of secretors and are adsorbed onto the sperm. As far back as 1952 DOCTON and coworkers demonstrated blood group antigens on sperm cells of bulls by inhibition and absorption tests.

Table 1 shows the results of the tests of nine blood group reagents with the seminal plasma of three bulls used as inhibiting substances. The reactions of the red cells of each bull to the antisera used are given for each factor.

TABLE 1

*Inhibition titres of bull seminal plasma against cattle blood group antigens*

| Factor | Dilution | Bull No. 1 | Red cell specificity | Bull No. 2 | Red cell specificity | Bull No. 3 | Red cell specificity |
|---|---|---|---|---|---|---|---|
| J | 1/40 | 128 | + | 128 | + | 32 | + |
| J′ | 1/25 | 64 | — | 256 | + | 64 | + |
| $T_1$ | 1/25 | 64 | — | 256 | — | 64 | — |
| I′ | 1/30 | 128 | — | 256 | — | 256 | — |
| $X_2$ | 1/80 | 64 | + | 256 | — | 128 | + |
| $A_1$ | 1/40 | 32 | + | 128 | + | 16 | + |
| O′ | 1/80 | 64 | — | 128 | + | 64 | — |
| K′ | 1/160 | 128 | — | 256 | + | 64 | + |
| $Y_2$ | 1/60 | 64 | + | 256 | — | 256 | — |

It is clear that inhibition of the reaction of antisera and red cells was non-specific. The seminal plasma of all three bulls inhibited the reactions with all factors tested. That of bull 2 had a greater inhibiting ability than those of bulls 1 and 3. This ability is not related to the specificity of the inhibitor, as the seminal plasma of bull 2 inhibits to high titres regardless of his red cell specificity for the factors tested. Bull 1, although positive to anti-$Y_2$, inhibits its reaction to a titre of 64, and bull 3, which is negative for this factor inhibits its reaction to a titre of $>256$.

The non-specific nature of this inhibiting factor in seminal plasma does not rule out the presence of blood group specific inhibition. Any specific inhibition that takes place in these tests may be masked by the non-specific inhibiting factor.

The ability of this substance to inhibit could be the result of several reactions. It could be affecting the cell surface, preventing the union of antibody with antigen; it could be acting on the antigen/antibody complex, preventing the reaction with complement, or it could be acting upon the complement.

Inactivation of the seminal plasma for 1/2 hour at 56°C lowered the inhibition titre but did not completely remove it, proving the substance to have some resistance to heat.

The reaction of cattle red blood cells to seminal plasma was investigated. Seminal plasma from 72 bulls was titrated and tested against a panel of cattle cells chosen at random from samples sent in to the cattle blood typing laboratory. Duplicate sets of tests were done using unheated seminal plasma, with and without complement, and seminal plasma boiled for five minutes. All lysing ability was removed from the seminal plasma after boiling. Table 2 shows the results of these tests with and without complement using the seminal plasmas diluted 1/10. Approximately 40% of the seminal

TABLE 2

*Tests of the seminal plasma of 72 bulls with cattle red blood cells with and without complement**

| No. of animals tested | No. of cells per sample | No. of Tests | Results | | | |
|---|---|---|---|---|---|---|
| | | | Without C′ | % | With C′ | % |
| 72 | 10 | 720 | +287 | 39.9 | +72 | 10.0 |
| | | | −433 | 60.1 | −648 | 90.0 |

* Samples of seminal plasma and red blood cells represent random samples from several breeds.

plasma/red cell combinations resulted in lyses without complement at this dilution. With the addition of complement this number dropped by 30%. The possibility that an enzyme is implicated must be considered. Some enzymes are capable of inactivating complement. In the lower dilutions of seminal plasma the active component may be utilized in inactivating the complement, thus inhibiting the reactions with seminal plasma or with blood group antisera. In the higher dilutions free complement is present in increasing amounts and is available for the haemolytic reaction when antisera is present.

The seminal plasma from one bull was fractionated by column chromatography, using G-200 Sephadex. There were four peaks and nine fractions in all were collected, the last two being column buffer.

These fractions were tested by three methods; immunoelectrophoresis against rabbit-anti-seminal plasma serum, starch-gel electrophoresis and hemolytic tests with cattle red blood cells. Lines were obtained by immunoelectrophoresis with fractions number 1, 2, 3, 4, 5 and 7. By starch-gel electrophoresis stained for proteins, bands were present in fractions 1, 3, 4, 5 and 7. Five lines were present in the whole seminal plasma; fraction 4 had four of these bands, fractions 1 and 5 had the two fastest bands in varying amounts, fraction 3 had the 3rd and 5th and fraction 7 the 2nd. The serological activity was in fractions 2, 3, 4 and 5, 3 and 4 giving the strongest reactions. The titre of these two fractions was lowered with the addition of complement to the test, but not removed. These components, were not strong, and the dilution of the fractions may not have provided a sufficient amount of the active component to completely inactivate the complement.

The presence of an antigen/antibody complex in these tests cannot be ruled out. Further investigation is necessary to clarify and characterize the reaction and substances involved. Its significance must be pursued and its possible implication, if any, in reproduction considered.

## DISCUSSION

J. MATOUŠEK: Did you test only complement in inhibiting the lysing ability of seminal plasma?

J. KRAAY: Yes. We tested only complement.

J. MATOUŠEK: We have studied other proteins as regards inhibition of the lysing ability of seminal plasma, egg yolk, egg white and milk, they showed the same or more intensive inhibition.

*XIIth Europ. Conf. Anim. Blood Groups Biochem. Polymorph., Bp., 1972 (pp. 237—240)*

# THE EFFECT OF SOME INHIBITORS ON ESTERASES IN THE GENITAL TRACT FLUID OF BULLS

A. KÚBEK and L. VESELSKÝ

Laboratory of Animal Genetics, Czechoslovak Academy of Sciences, Libečhov, Czechoslovakia

## INTRODUCTION

Esterases in body fluids form a complex. AUGUSTINSSON (1958, 1959, 1961) divides esterases, according to separating activity of substrates, sensitivity towards inhibitors and mobility in the electrophoretic field, into arylesterases, aliesterases and cholinesterases.

The most frequently used inhibitors of esterases are organophosphatic compounds, eserines and its derivates, EDTA, bivalent ions of heavy metals, fluorides, diethylparanitrophenyl phosphate, arzanilic acid, and iodoacetate (ALDRIDGE, 1951; URIEL, 1963). Some substances inhibit only one group of enzymes, whereas others are not inhibited by them and there are also some cases where these substances become activators for certain enzymes (KOMA, 1963).

The aim of the present work was to differentiate, by the use of some inhibitors, individual kinds of esterases from the complex which is present in the seminal vesicle and ampullar fluids of bulls.

## MATERIAL AND METHODS

Spermatozoa and the fluids from the genital organs of bulls were obtained from slaughtered animals as described by VESELSKÝ and KÚBEK (1970). Homologous antisera were prepared by heteroimmunizations of rabbits according to VESELSKÝ (1969).

The immunoelectrophoretic method of SCHEIDEGGER (1955) in Veronal-lactate buffer, pH 8.6, was used for the inhibition test. Inhibitions were performed by 50-minute incubation of precipitations on agar gels in 0.15 M phosphate buffer, pH 7.4, which contained the inhibitor. A $10^{-3}$ M physostigmine salicylate solution was used to inhibit cholinesterase, EDTA as inhibitor for arylesterase and L-serine to inhibit aliesterase. The staining method according to KÚBEK (1968) was used for the detection of esterases with the difference that the buffer contained the inhibitor at $10^{-3}$ concentration.

Spectrophotometric measuring on a spectrophotometer Spekol, C. Zeiss, Jena, was carried out according to SELIGMAN and NACHLAS (1950). Inhibition tests for quantitative determination of esterases were performed in physostigmine salicylate solution at a concentration of $10^{-5}$ M, in EDTA at $10^{-5}$ M, in L-serine at $10^{-5}$ M, in sodium taurochlorate at $3.5 \times 10^{-4}$ M and in cupreous sulphate at a concentration of $10^{-3}$ M.

## RESULTS

In the seminal vesicle fluid esterase activity was considerably reduced after inhibition with $10^{-3}$ M solution of physostigmine salicylate in the fraction migrating from the antigen reservoir to the $\alpha_1$ globulin region.

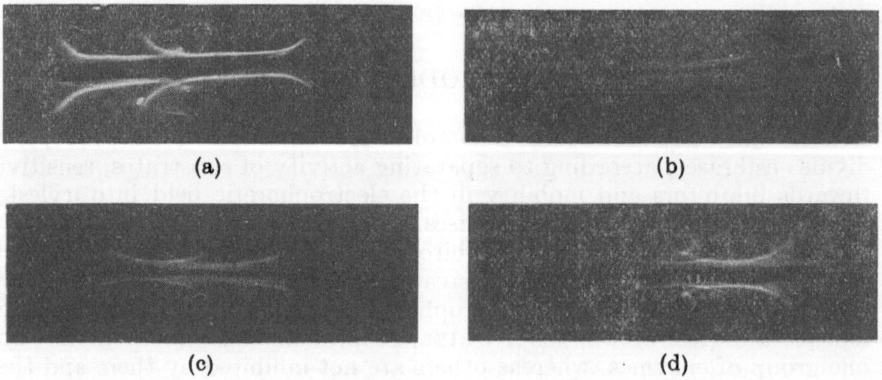
(a)
(b)
(c)
(d)

FIG. 1. Total esterase activity of seminal vesicle fluid of bulls (a) inhibited by $10^{-3}$ M solution of physostigmine salicylate (b), EDTA (c) and L-serine (d)

Esterase activity bound to the fraction localized in the $\beta_2$ region was inhibited almost completely. The use of $10^{-3}$ M EDTA solution effects only partial reduction of esterase activity both in the fraction migrating from the $\beta_1$ into the $\alpha_1$ globulin region and in fractions localized in the $\beta_2$ globulin region. The $10^{-3}$ L-serine solution causes reduction of activity in the seminal vesicle fluid in the fraction located in the $\beta_1$—$\alpha_1$ region and an almost complete disappearance of the fraction localized in the $\beta_2$ globulin region (Fig. 1).

After inhibition by $10^{-3}$ M solution of physostigmine salicylate in ampullar fluid of bulls, activity was reduced in the fraction migrating from the antigen reservoir up to the albumin region and in the fraction in the $\beta_2$ globulin region. The fraction in the $\beta_1$—$\alpha_1$ globulin region was almost completely inhibited by this inhibitor. Inhibition with $10^{-3}$ M EDTA solution caused no reduction of esterase activity in the ampullar fluid. With $10^{-3}$ M L-serine solution it was possible to inhibit the activity of the fraction passing from the antigen reservoir into the $\alpha_1$ globulin region and also the fraction in the $\beta_2$ region. The activity of the fraction in the $\beta_1$ globulin region up to the albumins was completely inhibited (Fig. 2).

Quantitative determination of esterase activity confirmed the results obtained by immunoelectrophoresis. By the use of $10^{-5}$ M physostigmine salicylate esterase activity in seminal vesicle fluid decreased by 19% from a total activity of 710 units, and in the ampullar fluid by 12% from a total activity of 990 units. When using $10^{-5}$ M L-serine, activity decreased by 21% on average in the seminal vesicle fluid and by 15% in the ampullar fluid. Inhibition with $10^{-5}$ M EDTA solution resulted in a decrease of activity in the seminal vesicle fluid by 9% on average. In the ampullar fluid, however, EDTA did not cause any decrease of activity.

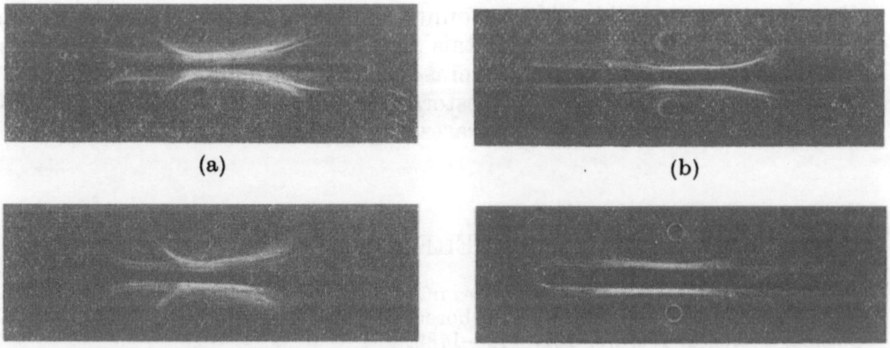

(a) (b) (c) (d)

**FIG. 2.** Total esterase activity of ampullar fluid of bulls (a) inhibited by $10^{-3}$ M solution of physostigmine salicylate (b), EDTA (c) and L-serine (d)

We have achieved no inhibition with $3.5 \times 10^{-4}$ M sodium taurochlorate as inhibitor of lipases, on the contrary, activity in the seminal vesicle fluid increased by 12% and in the ampullar fluid by 15%. $10^{-3}$ M solution of cupreous sulphate caused a decrease of esterase activity in the seminal vesicle fluid by 12% and in the ampullar fluid by 17%.

## DISCUSSION

It is apparent from the mentioned results that reduction of esterase activity in the seminal vesicle fluid is caused mainly by the effect of physostigmine salicylate and L-serine. EDTA inhibited considerably less. Inhibition of esterase activity also occurs due to the effect of cupreous sulphate. We can therefore assume that the seminal vesicle fluid contains mostly cholinesterase and aliesterase and has a considerably lower contents of arylesterase and esterases inhibited by cations of bivalent metals.

The ampullar fluid with the highest contents of esterases among the fluids of the genital tract, contains a high percentage of aliesterases and about the same amount of cholinesterases and esterases inhibited by bivalent metals. Arylesterases were not detected.

The seminal vesicle and ampullar fluids lack esterases inhibited by sodium taurocholate which, according to SELIGMAN and NACHLAS (1950) is the inhibitor of lipases. AUGUSTINSSON (1959) found the similar composition of esterases complex in the blood serum of cattle. We have, however, not found lipases in the genital tract fluid, although they are present in the blood serum.

## CONCLUSION

By means of inhibition tests we have found that seminal vesicle fluid of bulls contains a considerable amount of cholinesterase, aliesterase and esterases inhibited by bivalent metals. The contents of arylesterase is low, lipases are not present at all. Esterases in the ampullar fluid are mostly represented by aliesterases and esterases inhibited by bivalent metals. The ampullar fluid lacks arylesterases and lipases.

## REFERENCES

ALDRIDGE, W. N. (1954), Some esterases of the rat. *Bioch. J.* **57.** 692–702.

AUGUSTINSSON, K. B. (1958), Electrophoretic separation and classification of blood plasma esterases. *Nature,* **181.** 1786–1789.

AUGUSTINSSON, K. B. (1959), Electrophoresis studies on blood plasma esterases. I. Mammalian plasmata. *Acta Chem. Scand.,* **13.** 571–592.

AUGUSTINSSON, K. B. (1961), Multiple forms of esterase in vertebrate blood plasma. *Ann. N. Y. Acad. Sci.* **94.** 844–860.

KOMA, D. J. (1963), Characteristics of the esterases of human cells grown in vitro. *J. Histochem. Cytochem.,* **11.** 619–623.

KÚBEK, A. (1970), Electrophoretical study of the esterases in pig serum. Proc. XIth Europ. Conf. Anim. Blood Groups Biochem. Polymorph. Warsaw 1968, 355.

SCHEIDEGGER, J. (1955), Une micro-méthode de l'immunoélectrophorèse. *Int. Arch. Allergy Appl. Immun.,* **7.** 103–109.

SELIGMAN, A. M. and NACHLAS, M. M. (1950), The calorimetric determination of lipase and esterase in human serum. *J. Clin. Invest.,* **29.** 31–35.

URIEL, J. (1963), Characterisation of enzymes in specific immunoprecipitates. *Ann. N. Y. Acad. Sci.,* **103.** 956–959.

VESELSKÝ, L. (1969), Studium antigenních vlastností esterázových a fosfatázových aktivit krevního séra, spermií a tekutin pohlavního aparátu býk a kanců a folikulární tekutiny krav a prasnic. *Kandidátská práce.,* LCŽ ČSAV Liběchov 1969.

VESELSKÝ, L. and KÚBEK, A. (1970), Immunoelectrophoretical study of esterases in blood serum, spermatozoa and genital tract fluids of bulls. *Anim. Blood Grps biochem. Genet.* (In press)

## DISCUSSION

MRS. M. KAMINSKI: When do you perform your inhibition tests: 1. Do you add the inhibitor before or after the addition of antiserum? 2. Did you observe the protective effect by antibody?

L. VESELSKÝ: 1. We perform inhibition test after antigen-antibody reaction. 2. We did not find any fraction of antibodies when we tested an antiserum in inhibition tests.

*XIIth Europ. Conf. Anim. Blood Groups, Biochem. Polymorph., Bp. 1972 (pp. 241–244)*

# GENETIC RELATIONSHIPS AMONG CATTLE BREEDS

K. K. KIDD* and LAURA A. SGARAMELLA-ZONTA

Istituto di Genetica, Università di Pavia, Pavia, Italy
* Laboratory of Genetics, University of Wisconsin, Madison, Wisconsin, USA

Methods are now available for using gene frequency data to determine the probable evolutionary histories of populations. The methods are based on the work of CAVALLI-SFORZA and EDWARDS (1967) and assume that gene frequencies vary randomly through time and that no hybridization has occurred in the populations studied. However, the methods are rather robust and meaningful results occur using data that do not completely fit those original assumptions (KIDD, 1971). Immunogenetic data (including biochemical polymorphisms) on cattle breeds are ideal for such studies, because this material more closely conforms to the assumptions than do data on man.

Each population to be studied is characterized by its gene frequencies. All populations must have comparable data on all loci. This restriction limits the data available for comparisons among breeds typed by different laboratories, especially for the newer systems. This is a major problem at present since the accuracy of these analyses is largely a function of the number of loci used. The formulae for calculating genetic distances from gene frequency data and the two different methods we use to evaluate evolutionary trees on the basis of genetic distances are given elsewhere (KIDD, 1971).

One study (KIDD, 1969) has used data on 6 loci to study 16 cattle breeds. Here we present results involving 8 cattle breeds using 14 loci (9 blood group, hemoglobin transferrin, and three allotype loci). These data were all collected in the Immunogenetics Laboratory at the University of Wisconsin specifically for this study. Some of the gene frequencies used here have already been published (KIDD, 1969). Figure 1 shows the two best trees obtained using two different distance transformations. Figure 2 shows the two best trees obtained according to two different methods of analysis.

In each case, our methods do not allow us to state absolutely which of the two trees is correct, by our statistics they are not significantly different. Therefore, we base our conclusions on comparison. In Fig. 1 the relationships among the Icelandic, Brown Swiss, Longhorn, and Charolais are consistent and indicate that Icelandic (and hence Norwegian) cattle probably share a common origin distinct from that of the other two breeds. However,

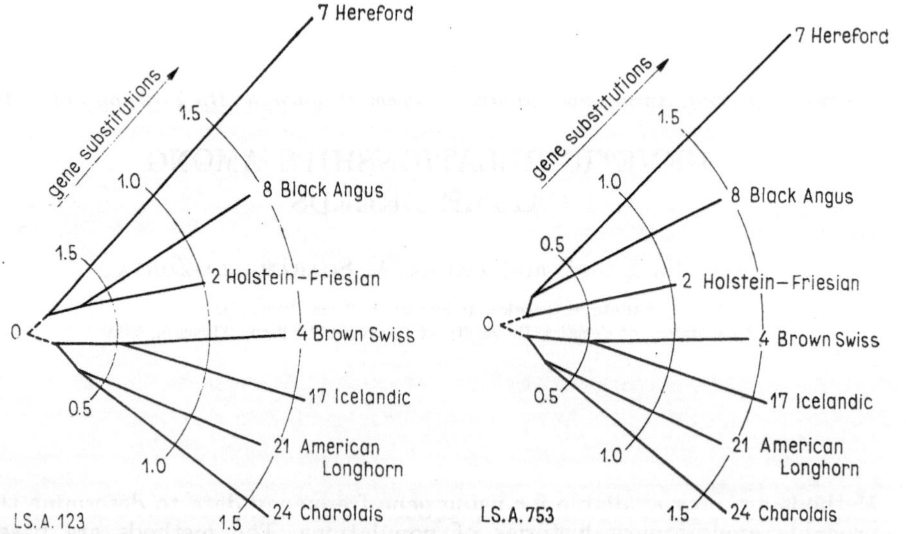
7 Hereford
gene substitutions
1.5
1.0
8 Black Angus
1.5
2 Holstein-Friesian
0
4 Brown Swiss
0.5
17 Icelandic
21 American Longhorn
1.0
LS. A. 123
1.5
24 Charolais
7 Hereford
gene substitutions
1.5
1.0
8 Black Angus
0.5
2 Holstein-Friesian
0
4 Brown Swiss
0.5
17 Icelandic
21 American Longhorn
1.0
LS.A.753
1.5
24 Charolais

**Fig. 1.** The two best trees for seven populations. In this polar plot the specific arrangement of the trees and position of the origin are arbitrary. The actual length of a segment is its radial length. These two trees were the best by least squares analysis on both $A$ and $G$ distances. The differences between the trees indicate the relationships most indecisive. Each breed is also identified by its computer code number. The data used here can be found elsewhere (Kidd, 1969)

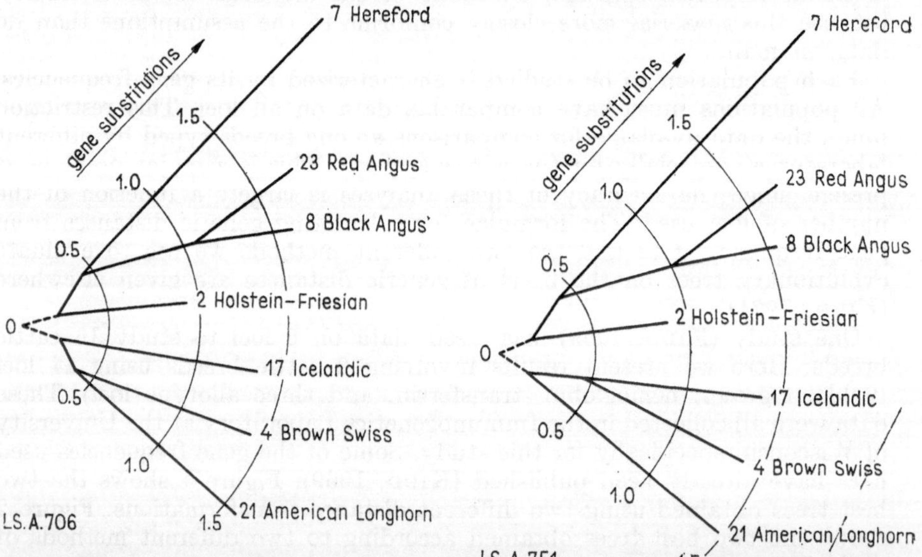
7 Hereford
gene substitutions
1.5
23 Red Angus
1.0
8 Black Angus'
0.5
2 Holstein-Friesian
0
17 Icelandic
0.5
4 Brown Swiss
1.0
LS.A.706
1.5
21 American Longhorn
7 Hereford
gene substitutions
1.5
23 Red Angus
1.0
8 Black Angus
0.5
2 Holstein-Friesian
0
17 Icelandic
0.5
4 Brown Swiss
1.0
21 American Longhorn
LS.A.751
1.5

**Fig. 2.** The two best trees for seven populations. Unpublished data on Red Angus was added to the data used in Fig. 1. These trees are the best two by both least squares and minimum path analyses of $A$ distances. See Fig. 1 for a general description of these trees. The statistical distributions for this data are described elsewhere (Kidd, 1969) because of the short lengths of the internal segments between the two groups, the relationships must be considered tentative. The relationships of Hereford, Black Angus, and Holstein are definitely ambiguous. The most definitive relationship is the clear separation of these latter three breeds from the four former ones.

Figure 2 shows a consistent close relationship between Red and Black Angus. Indeed, had it not been so we could not consider any of the relationships valid. The addition of Red Angus cattle also clarifies the relationships with Hereford and Holstein: now it seems more likely that the breeds from Great Britain are more closely related. The relationship of Icelandic cattle to Brown Swiss and Longhorn is however now ambiguous. The main division shown in Fig. 1 remains clear in this analysis as well. In both figures the extreme length of the Hereford segment probably results from the greater inbreeding and hence higher genetic drift in this breed. Conversely, the short length of the Holstein segment may reflect a smaller amount of drift in this population.

## REFERENCES

CAVALLI-SFORZA, L. L. and EDWARDS, A. W. F. (1967), Phylogenetic analysis: Models and estimation procedures. *Evolution*, **21.** 550–570; also *Amer. J. Human Genet.*, **9.** 234–257.

KIDD, K. K. and SGARAMELLA-ZONTA, L. A. (1971), Phylogenetic Analysis: Concepts and Methods. *American J. Human Genetics*, **23.** 235—252.

KIDD, K. K. (1969), Phylogenetic analysis of cattle breeds. *Ph. D. thesis*, University of Wisconsin.

## DISCUSSION

G. C. ASHTON: Is it true that this approach is completely stochastic, and that no provision is made for differential selection at any of these loci?

K. K. KIDD: The methods are based on the assumption that the differences among populations are stochastic; however, differential selective forces may be stochastic. It is true that any differential selective forces that are not stochastic cannot yet be considered and, if present, would distort the results.

J. BOUW: Breeds cannot be compared on the basis of gene frequencies when the sampling methods are not comparable.

K. K. KIDD: The variance of the gene frequency estimates is a function of the method of sampling. However, even small samples containing animals of various relationships give an unbiassed estimate of the gene frequencies of the population. Another related problem is the degree to which the population sampled represents the breed as a whole. Since large breeds are not generally panmictic, an estimate based on one subsection may not seem likely to represent the whole breed. However, here too such a gene frequency estimate is an unbiassed estimate for the whole breed. Any unbiassed gene frequency estimates can be used for these analyses. Of course, the variance of the comparisons increases as the variances of the gene frequency estimates increases.

D. R. OSTERHOFF: You are using 6 loci only, what would happen if you add another 6 loci in these calculations?

K. K. KIDD: The results to be published in these proceedings are based on 14 loci. Comparison of these relationships with those based on 6 loci shows some inconsistencies. These studies should be based on as many loci as possible and I prefer the 14 loci results to the 6 loci results. I feel that 6 is too few for great reliability. That is why I am pleasantly surprised that those results were in such close agreement with most of the expected relationships.

N. P. Wilkins: How many loci are required in a species of fish in order to process data obtained at sea in a research vessel computer using this method?

K. K. Kidd: As far as I know, these methods can be applied to any species to determine the relationships among populations. The exact number of polymorphic loci required is still an unanswered question, but possibly fewer than 10 will give a high probability of reconstructing the correct tree. I doubt if a small computer could handle this type of analysis if more than 6 populations were being considered.

*XIIth Europ. Conf. Anim. Blood Groups Biochem. Polymorph., Bp., 1972 (p. 245)*

# ACTIVITY REPORT OF THE IMMUNOGENETICS LABORATORY, THE UNIVERSITY OF HAVANA, CUBA

J. MITAT

Immunogenetics Laboratory, The University of Havana, Cuba

Our laboratory was founded in March 1968 after my visit in the Laboratory of Physiology and Genetics of Animals in Liběchov, Czechoslovakia. Dr. MATOUŠEK kindly permitted me to study blood groups in his laboratory and supplied us a reagents stock for blood typing of cattle (47) and pigs (45). I am much obliged to him for his ready help.

Cattle isoimmunizations were started in April 1968. So far, the following reagents have been produced: $A_1$ $A_2$ $Z'$ B $G_2$ $I_1$ $O_1$ $O_3$ $O_x$ Q T $Y_2$ $A'$ $B'$ $D'$ $E'_2$ $E'_3$ $G'$ $I'$ $K'$ $O'$ $Y'$ $C_2$ $R_1$ $R_2$ $X_1$ $X_2$ W $L'$ F V J L $S_1$ $U_1$ $U_2$ $U'_1$ $U''$ Z/– and some Cu (Cuba) experimental preparations.

These were compared with international reagents by colleagues Drs SCHRÖFFEL and PILZ.

We obtained also some other cattle reagents last year by the courtesy of Drs. BOUW, SCHRÖFFEL and PILZ.

Three cattle breeds, Holstein, Criollo and Zebu were used for isoimmunization. The most outstanding breed in the production of antibodies was the Zebu. We still have nine more breeds with which to work for the production of reagents being in progress, and I hope they will contribute to the extension of the spectrum of factors, especially in the B system.

The production of reagents for the investigation of cattle blood groups was begun only two years ago. Electrophoretic studies are still more recent. Aid of personnel from Liběchov has been available for only six months. Since then we have studied cattle hemoglobin, transferrin and amylase by means of starch gel electrophoresis on more than 5000 animals from several breeds and cross-breeds.

At present our principal service activity comprises the proof of identity and determination of paternity in cattle, based on blood groups and electrophoretic investigations.

*XIIth Europ. Conf. Anim. Blood Groups Biochem. Polymorph., Bp., 1972 (pp. 247 – 250)*

# RELATIONSHIPS BETWEEN BLOOD GROUPS, TRANSFERRIN AND HEMOGLOBIN TYPES AND DAIRY PERFORMANCE IN AUSTRIAN CATTLE

F. PIRCHNER, H. ROHRBACHER, W. SCHLEGER and G. MAYRHOFER

**Department of Animal Breeding and Genetics, School of Veterinary Medicine, Vienna, Austria**

## SUMMARY

The associations between genes at the blood group loci FV, Z, R′S′ and the transferrin and hemoglobin locus, with performance traits (milk yield, fat percentage) were investigated in the Austrian Fleckvieh, Braunvieh and Gelbvieh. The blood group, transferrin and hemoglobin genotypes of the bulls were determined from their daughters. The influence of the genes at the respective loci on milk yield and fat percentage in the progeny group of the bulls was estimated by means of Least Squares analysis. No significant associations were found. It is concluded that genes at the loci which were investigated do not have effects upon performance traits as large as, or larger than, half of the genetic standard deviation of the respective trait.

Almost since the discovery of blood groups in domestic animals, it has been suggested that blood groups and biochemical traits could be associated with performance, and should be valuable for animal improvement. NEIMANN-SØRENSEN and ROBERTSON (1961), among others, found that variation in performance characters which was associated with biochemical polymorphism, accounted for but little variation in the performance traits, and that biochemical polymorphism was of rather small help in selection work. Meanwhile a number of studies have been published. Notably ASHTON (1960) and ASHTON and HEWETSON (1968) reported significant associations between transferrin type and milk yield in cattle. JAMIESON and ROBERTSON [1967] confirmed these, but their investigation showed a connection rather smaller than that reported by ASHTON and his colleagues. In the United States, recent work (RAUSCH, BRUM, and LUDWICK, 1969, BRUM and coworkers, 1969) indicated that only few blood groups were associated with statistically significant yield and percentage increments and that the variation altogether accounted for by blood groups was rather small.

## MATERIAL

In the course of routine work and investigations of biochemical polymorphism in various Austrian cattle breeds, we collected information about the biochemical and blood group genotypes of bulls. Usually this was inferred from a sample of ten to twenty daughters and/or some eight to twelve daughter-dam pairs which were blood-typed. Only in isolated instances could we determine the blood type of the bull directly. At the same time we collected progeny tests of these bulls. Those of the Fleckvieh and Gelbvieh bulls were routinely computed by the Austrian Federation of Cattle Breeders, using a modified contemporary comparison method within classes of farms of comparable performance, while the progeny tests of the Braunvieh bulls were computed by us, using a modified Cornell methods (PIRCHNER, 1970). We used only bulls with twenty or more

TABLE 1

*Distribution of genotypes*

| System | Genotype | Breed: Fleckvieh | Braunvieh | Murbodner | Total | |
|---|---|---|---|---|---|---|
| Hemoglobin (Hb) | AA | 29 | 9 | 11 | 49 | |
| | AB | 2 | 2 | 6 | 10 | |
| | BB | 1 | | | 1 | 60 |
| Transferrin (Tf) | AA | 1 | | | 1 | |
| | $AD_1$ | 4 | | | 4 | |
| | $AD_2$ | 4 | 1 | | 5 | |
| | $D_1D_2$ | 11 | 5 | | 16 | |
| | $D_2E$ | 2 | 2 | | 4 | |
| | $D_2F$ | | 1 | | 1 | 31 |
| FV | FF | 22 | 3 | 9 | 34 | |
| | FV | 9 | 7 | 8 | 24 | |
| | VV | 1 | | 2 | 3 | 61 |
| R′S′ | R′R′ | 1 | | | 1 | |
| | R′S′ | 8 | 3 | 6 | 17 | |
| | S′S′ | 23 | 8 | 11 | 42 | 60 |
| Z | ZZ | 2 | 1 | 4 | 7 | |
| | Z/– | 8 | 5 | 7 | 20 | |
| | –/– | 9 | 2 | 1 | 12 | 39 |
| Average group size 3 | | 290 | 176 | 88 | | |

daughters. Some of the bulls had, however, more than a thousand daughters. Table 1 shows the distribution of the total material. The varying numbers of bulls studied for effects in different systems are due to the inability to determine particular biochemical and/or blood group genotypes of some bulls from the available information.

On account of the numerically small material we did not attempt to analyse associations between genes at complex loci and performance traits. Further, we restricted ourselves to loci where there was significant genetic polymorphism.

The data were analysed by means of a statistical model which took into account the effects of breeds and of the genes at that particular locus. The analysis was done separately for each locus. It was similar to the analysis employed by JAMIESON and ROBERTSON. The error sum of squares was computed as the reduction due to fitting the mean, the breed effects, and the gene effects, from the total weighted sum of squares of sire progeny means. The weights were the inverse of the variance of the sire means

$$\left(\frac{n}{1 + (n-1)\,h^2/4}\right).$$

## RESULTS AND DISCUSSION

The standard deviations between progeny groups were 205 kg milk and 0.20% fat. Table 2 shows the Least Square estimates of the genes at various loci, together with their errors. The estimated gene effects describe the change in progeny mean caused by introducing a particular gene from the sire. A bull homozygous for such a gene would cause an effect twice the one indicated in the table. This, incidentally, would also be the effect of a single gene upon the performance of the individual itself, in other words, the gene effects at the individual level are twice as large as those given in the table.

The results are statistically insignificant. Therefore a detailed discussion of size and direction of effects is not indicated. Nevertheless, it should be pointed out that the direction of the effects of genes Z and F agrees with reports of TOLLE (1960) and BRUM et al. (1969). As for transferrin genes, there appears to be rough agreement with the results published by ASHTON and HEWETSON, and JAMIESON and ROBERTSON, while MEYER (1967) failed to find effects due to genes at this locus. However, differences between breed-specific diversions of effects have been suggested (ASHTON and HEWETSON).

The lack of statistical significance, however, need not be interpreted to mean that effects are necessarily absent, but merely reflects the numerical limitation of our data.

In the comparisons where about 60 bulls are available, the errors of estimates are between 20 and 30 kg milk, and .03 and .04% fat. Since multiple comparisons are performed the *t*-value necessary for statistical

TABLE 2

*Gene effects on milk yield and fat%*

| System | Gene | Milk (kg) | | Fat % | |
|---|---|---|---|---|---|
| | | $b_i$ | $s_{bi}$ | $b_i$ | $s_{bi}$ |
| Hb | A | +7 | 30 | −.01 | .04 |
| | B | −7 | 30 | +.01 | .04 |
| Tf | A | −17 | 68 | −.01 | .10 |
| | $D_1$ | −37 | 64 | −.01 | .09 |
| | $D_2$ | +82 | 102 | +.12 | .15 |
| | E | −96 | 94 | +.03 | .14 |
| | F | +68 | 109 | −.12 | .16 |
| R′S′ | R′ | −33 | 27 | +.04 | .03 |
| | S′ | +33 | 27 | −.04 | .03 |
| Z | Z | −22 | 26 | +.02 | .03 |
| | (−) | +22 | 26 | −.02 | .03 |
| FV | F | −50 | 21 | | |
| | V | +50 | 21 | | |

significance ($p < .05$) must be about 2.7 rather than 2.0 as for single comparisons and 60 d.f. Therefore, effects to be recognized as significant when using the standard errors of Table 2 with between 50 and 60 d.f., would have to be of the order of 70 kg and .09% which is nearly half of the standard deviation between larger progeny groups in case of milk yield, and more than that in case of fat percentage. To recognize effects of 20–30 kg, as appear in Table 2, as statistically significant, would require a volume of data roughly seven times larger.

In conclusion, therefore, we can state that in our material the genes of the blood group systems R'S', FV, Z, and those of the transferrin and hemoglobin systems do not show effects on performance traits which are as large as, or larger than, half of the genetic standard deviation of the latter.

## REFERENCES

Ashton, J. C. and Hewetson, R. W. (1970), Transferrin type and milk yield in dairy cattle. Proc. XIth Europ. Conf. Anim. Blood Groups Biochem. Polymorph. Warsaw, 1968., 591.

Ashton, J. C. (1960), β-globuline polymorphism and economic factors in dairy cattle. *J. Agric. Sci.*, **54.** 321.

Brum, E. W., Rausch, W. H., Heins, H. C. and Ludwick, T. M. (1969), Associations between milk and blood polymorphism types and lactation traits of Holstein cattle *J. Dairy Sci.*, **51.** 1031.

Jamieson, A. and Robertson, A. (1967), Cattle transferrin and milk production. *Animal Production*, **9.** 491.

Meyer, H. (1967), Untersuchungen zum Transferrin-Polymorphismus beim Rind. *Zbl. Vet. Med.*, A, **14.** 335.

Neimann-Sørensen, A. and Robertson, A. (1961), The association between blood groups and several production characters in Danish cattle breeds. *Acta Agric. Scand.*, **11.** 163.

Pirchner, F. (1970), Eignung verschiedener Vergleichsdurchschnitte zur Nachkommenprüfung beim Rind. *Z. Tierz. Zb.* In press.

Rausch, W. H., Brum, E. W. and Ludwick, T. M. (1969), Relationship between blood type and predicted differences in production of Holstein sires in artificial insemination. *J. Dairy Sci.*, **1951.** 445.

Tolle, A. (1960), Grundlagen und Untersuchungsergebnisse von Beziehungen zwischen Blutgruppenfaktoren und Färsenlaktation. *Züchtungskunde*, **32.** 324.

## DISCUSSION

J. C. Ashton: I wish to congratulate *Dr. Pirchner* for clearly stating that insignificant results do not mean that there is no effect.

Recent results suggest that there is a breed difference in relation to the effect of $D_1$ and $D_2$. It is important that breeds are considered separately.

*XIIth Europ. Conf. Anim. Blood Groups Biochem. Polymorph., Bp., 1972 (pp. 251–254)*

# THE MAINTENANCE OF POLYMORPHISM IN CATTLE

PH. LHERMINIER

I.T.E.B. Paris, France

## INTRODUCTION

The study of correlations between the polymorphism of blood constituents and economic characteristics has been the object of extensive research for ten years. The principal purpose of this work has been to predict from a study of blood the selective value of animals at birth that is, to find a "martingale" for a rapid and simple genetic test of economic value.

Correlations observed in three French breeds are reported. These results are interesting in themselves, but still more for their contradictions among themselves and in relation to findings in other breeds.

## MATERIAL

The analyses of blood groups, transferrins, hemoglobins, caseins and lactoglobulins were carried out in the "Laboratoire des Groupes Sanguins de Jouy en Josas", using also protocols from this laboratory. 280 FFPN bulls (French Frisian black and white), 240 Normand bulls and 90 Montbéliard bulls, progenytested by 50 000 daughters, plus 1000 Normand cows, 630 Montbéliard cows and a family of 15 Montbéliard bulls with 570 daughters were studied.

## RESULTS

Fat percentage was the same in all examined cases. The differences shown in Table 1 are related to milk quantity and fat quantity which varied almost similarly.

Dissimilar results that require confirmation have been obtained with antigen Z, whose presence seemed advantageous in only one Montbéliard family, and with antigen F, whose presence was advantageous in the High-Normand sub-breed and disadvantageous in the Low-Normand sub-breed.

Results are presented in full detail in "Elevage-Insémination No 112, Paris, 1969".

TABLE 1

*Effects of presence of allele systems listed below on quantity of milk and fat production*

| Systems \ Breeds | Normande | F.F.P.N. | Montbéliarde |
|---|---|---|---|
| A | 0 | 0 | 0 |
| B | 0 | 0 | 0 |
| C | . | . | 0 |
| F (Presence of F) | + and − | +3.5% | −2% |
| J (Presence of J) | 0 | 0 | +3 to +6% |
| L (Presence of L) | 0 | −3.5% | 0 |
| M | 0 | . | . |
| S | 0 | 0 | 0 |
| Z (Presence of Z) | 0 | 0 | +3.5% |
| R′ (Presence of R′) | −6% | . | 0 |
| T′ | 0 | . | 0 |
| Hemoglobins (A and B) | . | . | NS |
| Transferrins (A, D and E) | 0 | . | 0 |
| Caseins (BAB and CAA) | . | . | NS |
| Lactoglobulins (A and B) | . | . | NS |

. not tested
0 very slight difference
N.S. number of observations too limited

Significance
2%: ——  1%: ═══
0.1%: ═══  0.01%: ════

## DISCUSSION

Of the fifteen loci examined, five showed measurable physiological differences. Scrutiny of statistical calculation permits the conclusion that these results are inexplicable by sampling errors or familiar pseudo-correlations. Using serious experimental and statistical precautions, many investigators demonstrated correlations which nullify or reverse themselves from one breed to another. A complete review of these findings would fill several pages. It would show the variation of these results which are obviously unrelated to artefacts.

Why are the alleles of blood groups or of biochemical polymorphisms often associated with physiological differences?

This question poses another one: the presence of blood groups and biochemical polymorphisms is a problem in itself. It definitely has a cause and a role; it cannot be senseless.

Almost everywhere one correctly looked, one found a polymorphism (ROBERTSON, 1966) and very often there was at least a correlation with a physiological criterion (milk production, length of gestation, age at first calving, etc.). These results are not only surprising, but also help to explain the phenomenon of polymorphism. One must admit that these substances have different physiological properties, for example diverse enzymatic powers which is the case for red cell acid phosphatase variants (HOPKINS, SPENCER and HARRIS, 1964). Thus the presence or absence of a certain allele is translated by a modification of fitness; that is to say, under the effect of selection the frequency of the gene corresponding to this protein will develop. The correlation between the presence of a certain allele and a gain in milk production is not surprising: the presence itself of the allele is the consequence of its physiological activity. If the selection constantly favours one allele, why are the others conserved and why does not the system go toward a total homozygosity? The answer to this question rests precisely within the heterogeneity of the measured correlations.

The double problem posed by the existence of polymorphisms of the blood constants and their capricious correlations with economic parameters is not paradoxical. An allele that is not associated with any selective advantage has practically no chance of being maintained. If this allele is submitted to a constant or very strong pressure, it is rapidly fixed (WRIGHT, 1968). Therefore all the authors prove that the observed correlations always correspond to the selective, weak but not negligible advantage of 1 to 6%. It is reasonable that the weakest escape statistical analysis while the strongest give way to an indirect selection over the blood groups, accompanying the fixation of the allele.

The theory of polymorphism is now well known. One can resume it by saying that it can only exist if the alleles are submitted to heterogenous selective forces, or equivalently if they possess a variable selective value; that is to say they are alternatively advantageous or disadvantageous according to the conditions within which they function. The polymorphism disappears if the alleles are neutral or constant in their effects. These conditions certainly apply to the blood groups and to the biochemical types here studied (ASHTON, 1965; BOESIGER, 1967; DOBZHANSKY, 1957).

If the observed correlations are a measure of the selective values of the alleles, it is necessary that they be versatiles. The variety of alleles and the exuberance of the polymorphisms are the answer to the heterogeneity of the selective forces, which determines their frequency and which is classed in this manner:

1. Diversity of the topographical relations among the genes: chromosomic alterations can make the frequencies of certain alleles fluctuate, but the linkage desequilibrium does not explain a long term polymorphism (MATHER, 1963; DOBZHANSKY, 1957).

2. Diversity of the physiological relations among the genes: the hybrid vigor in establishing a relationship of choice between alleles is the clearest example of the realization of a stable equilibrium. We have shown (LHERMINIER, 1969 and unpublished) the maintenance of an elevated frequency in the F.F.P.N. breed of the lethal defect "amputation of limbs". This gene has a negative, lethal effect on the homozygote, and a positive one on the heterozygote (+15% fat). The same gene placed in different genetic backgrounds can have opposite effects on the complex physiological equilibriums.

3. Diversity of selective factors: the heterogeneity of the environment carries ipso facto that of the selective forces, and therefore to the genes

sensitive to the selection (FORD, 1955). Though advantageous in one environment an allele can be disadvantageous in another. The production of 4000 kg milk does not require the same sets of functions according to the nature of the soil, the climate and raising methods.

## CONCLUSION

The discovery that several alleles possess selective values, judged dubious because contradictory, actually constitutes the most logical justification of the existence of these multiple alleles. Some null or constant selective values contribute to the disappearance of the polymorphism. This does not mean that the action of blood groups is limited to slight modifications in milk production. It should be evident in many other aspects of fitness. Taken first practical purpose, these first experiments point to the use of polymorphisms to better determine the selective values of animals; it is, on the contrary, the diversity of these selective values which can serve, from a theoretical point of view, to better understand the maintenance of the polymorphism.

## ACKNOWLEDGEMENTS

The author express his thanks to the staff of the "Laboratoire des Groupes Sanguins du C.N.R.Z., Jouy en Josas" for the aid they have given him in the understanding and collection of the blood group material.

## REFERENCES

ASHTON, G. C. (1965), Cattle serum polymorphism, a balanced polymorphism? *Genetics*, **52.** 983.

BOESIGER, E. (1967), La signification évolutive du polygénotypisme des populations naturelles. *Ann. Biol.*, VI. 9–10.

DOBZHANSKY, T. (1957), Mendelian populations as genetic systems. Cold Spring Harbor Symp. Quant. Biol., XXII, 385–393.

HOPKINSON, SPENCER and HARRIS (1964), Red cell acid phosphatase variants. *Amer. J. Hum. Genet.*, **16.** 141–154.

LHERMINIER, P. (1969), Constantes sanguines et production laitière. *Elevage-Insémination*, 112.

MATHER, K. (1963), The measurement of linkage in heredity. Methuen & Co. London.

MORTON, N. E. and CHUNG, C. S. (1959), Are the MN blood groups maintained by selection? *Amer. J. Hum. Genet.*, **11.** 237.

ROBERTSON, A. (1966), Biochemical polymorphism in animal improvement. Proc. Xth Europ. Conf. Anim. Blood Groups Biochem. Polimorph., Paris 1966, 35.

WRIGHT, S. (1968), Evolution and the genetics of populations. Vol. I. The University of Chicago Press.

FORD, E. B. (1955), Rapid evolution and the conditions which make it possible. Cold Spring Harbor Symp. Quant. Biol. XX, 230.

## DISCUSSION

J. C. ASHTON: I think the best argument against "neutral" genes left over from an ancestral animal comes from the distribution of gene frequencies in the same breed located in various parts of the world. If these stocks all come from a parental stock, and the numbers were small, and the genes are neutral, why has genetic drift not occured? The frequencies of the transferrin genes for example, for any breed are remarkedly consistent. This implies a maintenance mechanism, otherwise we would found populations with all $Tf^A$ or all $Tf^E$ and this does not happen.

*XIIth Europ. Conf. Anim. Blood Groups Biochem. Polymorph., Bp.. 1972 (pp. 255 – 259)*

# IMMUNOGENETIC STUDIES ON COWS OF HIGH AND LOW BUTTERFAT PRODUCTION

A. Kaczmarek, H. Balbierz, Z. Dorynek, M. Nikołajczuk, M. Switek and T. Szałajko

**College of Agriculture in Poznan, Department of Animal Husbandry, Poznan, Poland**
**College of Agriculture in Wrocław, Department of Obstetrics and Gynecology, Laboratory of Immunopathology, Wrocław, Poland**

## SUMMARY

Results of four-year observations on a total of 3904 heifers suggested the frequencies of some B alleles to be characteristic of groups of heifers with low or high butterfat yield (in kg and %). However, no difference was found between the tested groups of animals in the frequencies of individual transferrin types.

Investigations were undertaken to clarify whether the selection of breeding stock can be based on the results of immunogenetic study. Since our findings did not agree with those of other investigators, especially concerning B allele $BO_1Y_2D'$ and frequency of transferrin type DD, the immunogenetic indices cannot be used in selection. However further studies are recommended to find out whether there is a relationship between blood group structure and productive characteristics in cows.

## INTRODUCTION

As yet there has been no information of breeding work based on the study of correlations between blood groups or transferrin types and productive traits in cattle. But research along this line has been in progress and it is impossible to enumerate the papers published on this subject. Most investigations deal with the correlation of blood groups and milk yield or butterfat per cent (Conneally and Stone, 1965; Rausch et al., 1968; Bunsch, 1969; Sapiro, 1969; Gurjanova, 1969; Karmanova, 1969; McClure, 1952; Ashton, 1968; Heidler, 1968). Only a few of them showed the absence of such correlation. Thus according to Lherminier (1968), in France a scheme has been established for checking, and in future maybe also for selecting, bulls by their blood groups.

Similar correlations between transferrins and fattening ability in cattle have been pursued (Makarechian, 1966, 1967) without positive result up to date. Bogdanow and Obuchowskij (1967) confirmed Ashton's assumption as to existing influence of transferrin locus on reproduction in cattle. Slightly different were the results of Boyd's study on this subject (1969). Mitscherlich et al. (1966) found a correlation between blood groups in cows and their susceptibility to mastitis.

At the previous European Conference on Animal Blood Groups and Biochemical Polymorphism in Warsaw, the authors presented a report (1968) on the first part of investigations into differences in blood antigenic structure in cows in relation to butterfat content in milk. This is a report on the second part of our study.

TABLE 1

*Differences in frequencies of B alleles in heifers with low and high butterfat yield (kg and %)*

| B allele | Heifers with low butterfat yield kg: Investigation I | Investigation II | Total number | Total frequency | Heifers with high butterfat yield kg: Investigation I | Investigation II | Total number | Total frequency | |
|---|---|---|---|---|---|---|---|---|---|
| $BO_3Y_2A'E_3'G'G''P'Q'$ | 0.0833 | 0.1211 | 42 | 0.1111 | 0.1229 | 0.1546 | 59 | 0.1505 | 2.3115 |
| $BGKO_xY_2A'O'$ | 0.1499 | 0.1453 | 57 | 0.1485 | 0.1077 | 0.1352 | 53 | 0.1340 | 0.5046 |
| I′ | 0.0595 | 0.1211 | 35 | 0.0912 | 0.1028 | 0.0834 | 32 | 0.0886 | 0.3721 |
| $GO_1$ | 0.0224 | — | 5 | 0.0125 | 0.0044 | — | 1 | 0.0024 | 1.6774 |
| $GY_2E_1'Q'$ | 0.0988 | 0.1271 | 39 | 0.1020 | 0.1328 | 0.1996 | 70 | 0.1816 | 8.3495 |
| $BO_1Y_2D'$ | 0.0642 | 0.0052 | 14 | 0.0355 | 0.0044 | 0.0178 | 5 | 0.0119 | 4.8715 |
| | | | $n = 201$ | | | | $n = 212$ | | |
| | Heifers with low butterfat % in milk | | | | Heifers with high butterfat % in milk | | | | |
| $BO_3Y_2A'E_3'G'G''P'Q'$ | 0.0450 | 0.0892 | 23 | 0.0762 | 0.1554 | 0.0837 | 41 | 0.1415 | 5.7679 |
| $BGKO_xY_2A'O'$ | 0.1213 | 0.1902 | 47 | 0.1630 | 0.1743 | 0.1235 | 48 | 0.1680 | 0.0211 |
| I′ | 0.1213 | 0.0728 | 34 | 0.1149 | 0.0850 | 0.1069 | 31 | 0.1049 | 0.1338 |
| $GO_1$ | — | 0.0157 | 2 | 0.0064 | 0.0206 | — | 4 | 0.0129 | 0.1748 |
| $GY_2E_1'Q'$ | 0.1851 | 0.1339 | 43 | 0.1479 | 0.1138 | 0.1152 | 35 | 0.2124 | 0.8798 |
| $BO_1Y_2D'$ | 0.0399 | 0.0316 | 12 | 0.0388 | 0.0103 | 0.0072 | 3 | 0.0097 | 6.7173 (5.4309) |
| | | | $n = 157$ | | | | $n = 156$ | | |

## MATERIAL AND METHODS

During the years 1968 and 1969, in the same farms as specified in the first part of this work, we collected information as to low or high butterfat yield (kg) and content in milk (%) during the first lactation period of heifers. Our calculations cover productive data from 1799 heifers sired by 98 bulls of Lowland Black and White breed. Animals were typed according to upper or lower limits of tolerance, i.e. blood antigenic structure was determined in heifers of extremely high or low butterfat yield. Another segregation of heifers was based on calculating the value of butterfat per cent in milk and standard deviation. Two groups were selected, one for mean yield plus standard deviation, the other for mean yield minus standard deviation: The selected heifers originated from 57 bulls. There were 92 heifers with very low and 104 heifers with very high butterfat yield; 62 heifers with low butterfat per cent in milk and 63 heifers with high butterfat per cent were chosen from tested herds.

The blood samples from these animals were examined by means of 50 test sera. The transferrin types were detected by starch gel electrophoresis, without differentiating D type into $D_1$ and $D_2$. The gene frequency in blood-group system B was calculated according to the method described by BRAEND. A total of 40 B alleles were found but only the more frequent alleles were used for comparison.

TABLE 2

*Differences in gene frequencies in blood group systems A, J, L, M, Z, FV in heifers with low and high butterfat yield (kg and %). Gene frequency in %*

| Antigen | Heifers with low butterfat yield kg | | | Heifers with high butterfat yield kg | | |
|---|---|---|---|---|---|---|
| | investigation | | Total frequency | investigation | | Total frequency |
| | I | II | | I | II | |
| A | 13.79 | 14.67 | 14.20 | 18.92 | 15.18 | 17.30 |
| J | 15.40 | 7.34 | 11.62 | 14.48 | 12.30 | 13.40 |
| L | 21.60 | 24.82 | 23.06 | 22.42 | 21.36 | 22.20 |
| M | 15.40 | 10.32 | 13.62 | 9.23 | 14.50 | 12.32 |
| F | 86.70 | 89.67 | 88.06 | 85.65 | 89.90 | 87.74 |
| V | 13.30 | 10.33 | 11.94 | 14.35 | 10.10 | 12.26 |
| Z | 27.98 | 29.29 | 28.42 | 24.85 | 32.07 | 27.65 |
| | | | $n = 201$ | | | $n = 212$ |
| | Heifers with low butterfat % in milk | | | Heifers with high butterfat % in milk | | |
| A | 9.09 | 11.10 | 9,71 | 13.10 | 11.81 | 12.66 |
| J | 14.28 | 8.42 | 11.85 | 12.49 | 10.92 | 11.96 |
| L | 23.07 | 21.72 | 22.63 | 27.07 | 27.63 | 27.50 |
| M | 14.29 | 9.31 | 12.21 | 14.33 | 10.03 | 12.66 |
| F | 89.80 | 83.06 | 86.43 | 89.89 | 82.54 | 84.28 |
| V | 10.20 | 16.94 | 13.57 | 10.11 | 17.46 | 15.72 |
| Z | 6.33 | 19.68 | 11.49 | 8.33 | 22.34 | 13.78 |
| | | | $n = 157$ | | | $n = 156$ |

## RESULTS

Table 1 shows the differences in frequencies of B alleles of heifers with low and high butterfat yields. The occurence of certain alleles of high frequency was identical or similar in parts I and II of the study. They were: $BO_3Y_2A'E'_3G'G''P'Q'$; $BGKO_xY_2A'O'$ and $GY_2E'_1Q'$. Slightly higher differences between parts I and II were found in the frequencies of the remaining 3 alleles: L'; $GO_1$ and $BO_1Y_2D'$. Comparing the results of both parts together one can see clearly the higher frequency of allele $GY_2E'_1Q'$ in the group of more butterfat yielding cows, while the allele $BO_1Y_2D'$ was more frequent in the group of cows giving less kg of butterfat. A similar relation was observed already in the first part of the study. The differences in the remaining B alleles were insignificant.

The second part of Table 1 shows the percentual proportions of butterfat in milk. The frequencies of the studied B alleles were the same or similar in both parts of the study. There was, however, a fundamental difference in that the B allele $BO_3Y_2A'E'_3G'G''P'Q'$ was more frequent in the group of heifers with high butterfat per cent in milk, while allele $BO_1Y_2D'$ was more often found in the group of heifers with lower butterfat per cent in milk. The occurence of the remaining B alleles as well as of antigens in other systems (Table 2) did not show any difference between the studied groups of heifers.

TABLE 3

*Differences in frequencies of transferrin types*

| Transferrin type | Heifers with low butterfat yield (kg) | | | | Heifers with high butterfat yield (kg) | | | |
|---|---|---|---|---|---|---|---|---|
| | Investigation | | Number | Total frequency | Investigation | | Number | Total frequency |
| | I | II | | | I | II | | |
| AA | | | 26 | | | | 39 | |
| AD | $q^A$=0.378 | $q^A$=0.365 | 85 | $q^A$=0.372 | $q^A$=0.479 | $q^A$=0.410 | 82 | $q^A$=0.430 |
| AE | $q^D$=0.568 | $q^D$=0.594 | 6 | $q^D$=0.575 | $q^D$=0.510 | $q^D$=0.552 | 6 | $q^D$=0.531 |
| DD | $q^E$=0.053 | $q^E$=0.050 | 62 | $q^E$=0.052 | $q^E$=0.040 | $q^E$=0.036 | 57 | $q^E$=0.038 |
| DE | | | 12 | | | | 9 | |
| EE | | | 1 | | | | — | |
| | | $n$ = | 192 | | | $n$ = | 193 | |
| | Heifers with low butterfat % in milk | | | | Heifers with high butterfat % in milk | | | |
| AA | | | 24 | | | | 26 | |
| AD | $q^A$=0.406 | $q^A$=0.390 | 73 | $q^A$=0.403 | $q^A$=0.426 | $q^A$=0.385 | 67 | $q^A$=0.409 |
| AE | $q^D$=0.532 | $q^D$=0.546 | 8 | $q^D$=0.537 | $q^D$=0.540 | $q^D$=0.567 | 7 | $q^D$=0.551 |
| DD | $q^E$=0.062 | $q^E$=0.054 | 44 | $q^E$=0.059 | $q^E$=0.034 | $q^E$=0.045 | 49 | $q^E$=0.038 |
| DE | | | 11 | | | | 5 | |
| EE | | | — | | | | — | |
| | | $n$ = | 160 | | | $n$ = | 154 | |

Analyzing the two parts of Table 1, it is noticed that the B allele $BO_3 Y_2 A'E_3'G'G''P'Q'$ was more frequent in the group of heifers with higher yield and higher per cent of butterfat in milk. $GY_2E_1'Q'$ was more frequent in the heifers giving more butterfat (kg), while $BO_1Y_2D'$ in the heifers with low per cent and low yield of butterfat.

The frequencies of transferrin types in two tested group are shown in Table 3. The overall results disclose no difference in their frequencies in the groups compared.

## REFERENCES


ASHTON, G. C. and HEWETSON, R.W. (1970), Transferrin Type and Milk Yield in Dairy Cattle. Proc. XIth Europ. Conf. Anim. Blood Groups Biochem. Polymorph. Warsaw, 1968.

BALBIERZ, H., KACZMAREK, A., NIKOŁAJCZUK, M., SWITEK, M., DORYNEK, Z. (1970), Immunogenetical Studies on Cows of High and Low Butterfat Production. Proc. XIth Europ. Conf. Anim. Blood Groups Biochem. Polymorph., Warsaw, 1968.

BOGDANOW, L. W. and OBUCHOWSKIJ, W. M. (1967), Izuczenje tipow transferrinow i tipow giemogłobinow u krupnogo rogatogo skota. *Z. Obszcz. Biol.* **28.** 76–81.

BOYD, H., JAMIESON, A. and HALL, J. G. (1969), Fertilization rate and embryonic survival in dairy cows in relation to transferrin and J antigen groupings of sires and dams. *J. Reprod. Fert.* **18.** 317–324.

BUNSCH, B. (1969), Untersuchungen über die Serumtransferrintypen beim Deutschen Schwarzbunten Rind. *Arch. Tierz.*, **12.** 115–125.

BUNSCH, B. (1969), Untersuchungen über die Serumtransferrintypen und Merkmalen der Milchleistung. *Arch. Tierz.* **12.** 199–209.

CONNEALLY, P. M. and STONE, W. H. (1965), Association between a blood group and butterfat production in dairy cattle. *Nature*, **206.** 115.

GURYANOVA, A.S. and PILKO, V. V., ŠAPIRO, JU. O. (1969), Ob ispolzovanii tipov gemoglobina i transferrina dlya rannei otsenki mlodnyaka. *Zivotnovodstvo*, **31.** 64–65.

HEIDLER, W. (1968), Die Beziehungen zwischen den Serumproteinfraktionen sowie dem Serumproteingehalt und der Milchleistung. *Arch. Tierz.* **11.** 449–457.

KARMANOV, E. P. and NIKOLAJEVA, G. N. (1969), O nasledovanii belkovych frakcij krovi i ich svjazi s žirnomolaćnostju u krupnogo rogatogo skota. *Genetika*, **5.** 57–64.

LHERMINIER, P. (1968), Connaissances des groupes sanguins et choix des taurillons. *Elevage Insém.* **106.** 9–15.

MAKARECHIAN, M. and HOWELL, W. E. (1967), Relationship between transferrin type and productive traits in beef steers. *J. Anim. Sci.*, **26.** 27–30.

MAKARECHIAN, M. and HOWELL, W. W. (1966), Relationship between transferrin type and feedlot performance in beef heifers. *Canad. J. Anim. Sci.*, **46.** 177–180.

MCCLURE, T. J. (1952), Correlation Study of Bovine Erythrocyte Antigen A and Butterfat Test. *Nature*, No. 4321.

MITSCHERLICH, E., HONGREVE, F., KOCH, J. and SCUPIN, E. (1966), Studies on the correlation between the resistance to mastitis and the bloodgrouping factors in Holstein-Frisian cattle. *Deutsche Tierärztliche Wochenschrift*, **5.** 73.

RAUSCH, W. H., BRUM, E. W. and LUDWICK, T. (1968), Relationship between blood type and predicted differences in production of Holstein sires in artificial insemination. *J. Dairy Sci.* **51.** 445–451.

ŠAPIRO, JU. O. (1969), O polimorfizme transferrinov u krupnogo rogatogo skota kostromskoj i svickoj porod i ego svjazi s nekotorymi pokazateljami krovi produktivnost ju. *Genetika*, **5.** 43–52.

*XIIth Europ. Conf. Anim. Blood Groups Biochem. Polymorph., Bp., 1972 (pp. 261–266)*

# THE ROLE OF POLYMORPHIC PROTEINS IN PRODUCTION OF CATTLE

J. DOSTÁL* and A. G. HUNTER

Department of Animal Science, University of Minnesota, St. Paul, Minnesota, U. S. A.

## SUMMARY

1. Neuraminidase affects the electrophoretic mobility of transferrin $D_1$ and $D_2$ variants such that the distance between the $D_1$ and $D_2$ phenotypes is increased.

2. The transferrin molecule has more than one polypeptide sub-unit.

3. The iron binding capacity of the transferrins is controlled by some factors other than the phenotype of the transferrin gene.

4. The transferrin binds in vitro Estrone, $\beta$-Estradiol and Estriol. The binding capacity of the transferrins for the $\beta$-Estradiol is affected also by some factors other than the phenotypes of the transferrin genes.

## INTRODUCTION

Relationships between protein polymorphisms and milk production or reproduction have been reported by many laboratories. This study investigated the role of transferrin polymorphism on iron and hormone transport, two physiological processes that affect lactation and reproduction.

## MATERIAL AND METHODS

The AA, $AD_1$, $AD_2$, AE, $D_1D_1$, $D_1D_2$, $D_1E$, $D_2D_2$, $D_2E$ and EE cattle serum transferrins were isolated and purified as described by DOSTÁL and HUNTER (1970).

Each transferrin variant was incubated with the enzyme neuraminidase (GAHNE, 1962) and the electrophoretic mobility of each transferrin variant was compared by starch gel electrophoresis (KRISTJANSSON and HICKMAN, 1965).

Using starch gel electrophoresis with urea (ASCHAFFENBURG, 1966), the polypeptide chain sub-units of the transferrin molecule were studied.

Iron content of the transferrin variants before and after addition of ferric trichloride was determined with a Perkin Elmer Emission Spectrophotometer. The iron transferrin complex after the addition of ferric trichloride was separated by column chromatography on Sephadex G-25 (FLODIN, 1961).

* Present address: Laboratory of Animal Genetics, Czechoslovak Academy of Sciences, Liběchov, Czechoslovakia.

The hormone Estrone, β-Estradiol and Estriol (SPOFA) were added to transferrin solutions in vitro and the hormone transferrin complex was separated from free hormone by column chromatography on Sephadex G-25 (DOSTÁL, 1970). The amount of hormone bound to transferrin was detected colorimetrically using the Kober colour reaction (BROWN et al., 1968).

## RESULTS AND DISCUSSION

Treatment with neuraminidase produced some electrophoretic mobility changes in the transferrin $D_1$ and $D_2$ alleles. Comparison of Figs 1 and 2 revealed the difference in electrophoretic mobility of $D_1$ and $D_2$ transferrins

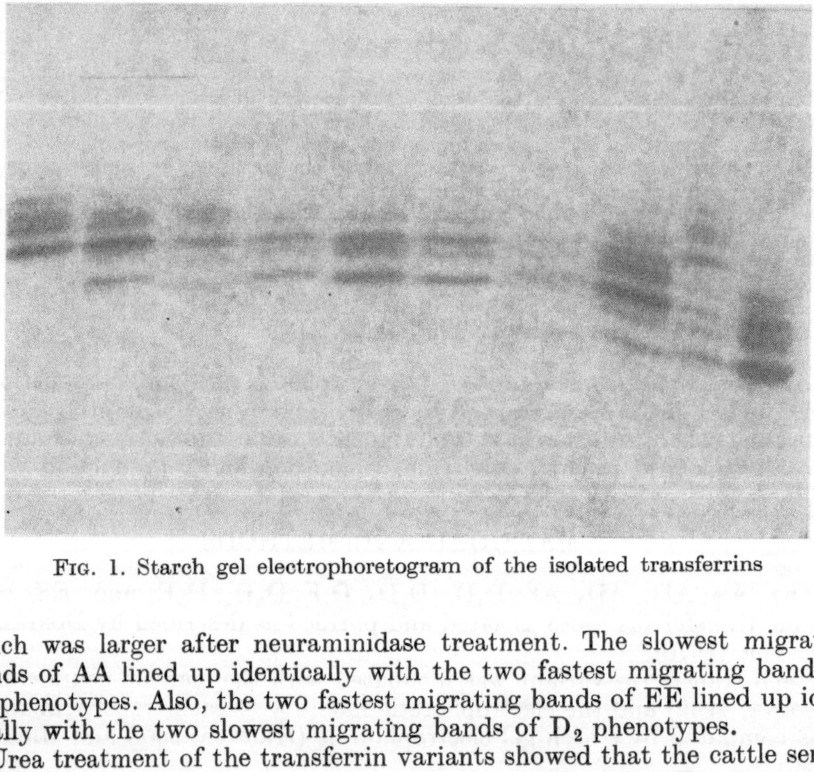

FIG. 1. Starch gel electrophoretogram of the isolated transferrins

which was larger after neuraminidase treatment. The slowest migrating bands of AA lined up identically with the two fastest migrating bands of $D_1$ phenotypes. Also, the two fastest migrating bands of EE lined up identically with the two slowest migrating bands of $D_2$ phenotypes.

Urea treatment of the transferrin variants showed that the cattle serum transferrins consisted of more than one polypeptide chain (Fig. 3).

Iron binding capacity was highly variable between different phenotypes and also varied within the same phenotype Table 1. Therefore, binding capacity of the transferrins for iron was affected by some phenomenon other than the transferrin genotype.

The affinity of the transferrins for estrogenic hormones is presented in Fig. 4. Practically all of the added saturated Estrone to solution was

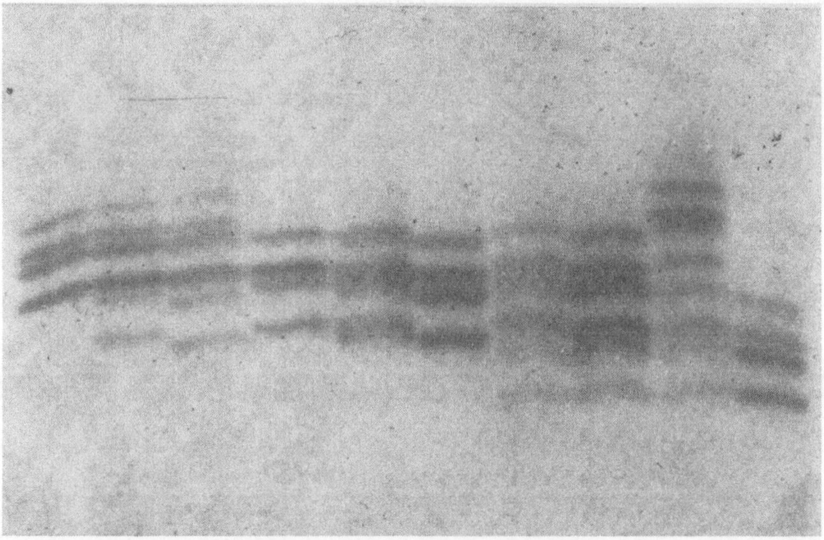

FIG. 2. Starch gel electrophoretic comparison of transferrin variants after 65 hours' incubation at 37°C with neuraminidase

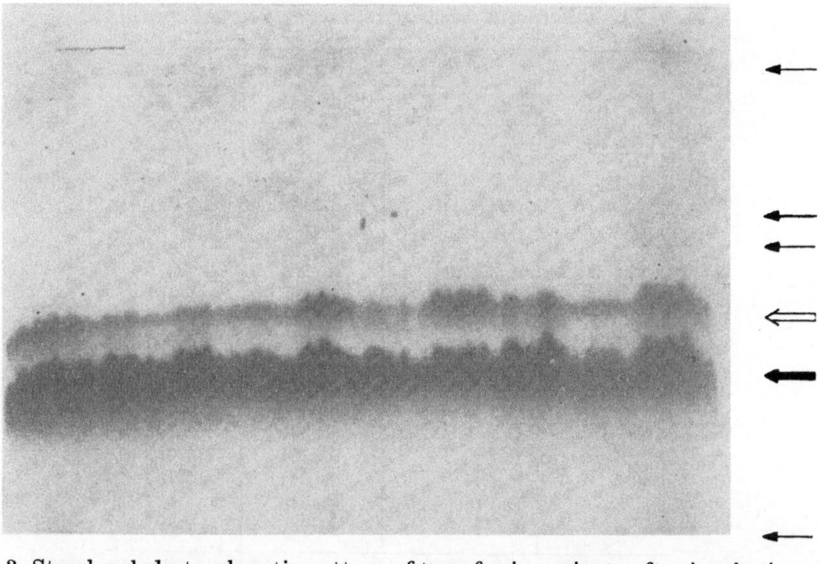

FIG. 3. Starch gel electrophoretic pattern of transferrin variants after incubation with six molar urea and 1% 2-mercaptoethanol

TABLE 1

*The iron binding capacity of transferrins*

| Cow No. | Transferrin | The μg Fe bound per 1 mg transferrin | |
|---|---|---|---|
| | | before saturation | after saturation |
| M 574 | AA | 0.7896 | 2.8133 |
| E 513 | AA | 0.6115 | 2.5211 |
| E 390 | $AD_1$ | 0.7308 | 2.4997 |
| AM 512 | $AD_1$ | 0.7137 | 3.3206 |
| E 397 | $AD_2$ | 0.6722 | 2.4440 |
| AM 541 | AE | 0.7575 | 4.3583 |
| E 599 | $D_1D_1$ | 0.5433 | 7.0032 |
| E 602 | $D_1D_2$ | 0.6150 | 3.5869 |
| 510 | $D_1E$ | 1.4361 | 1.4574 |
| E 405 | $D_2D_2$ | 0.7249 | 3.0034 |
| E 598 | $D_2E$ | 0.7343 | 2.4992 |
| 808 | EE | 0.9528 | 3.4203 |
| Standard deviation between duplicates | | | ±0.2540 |

TABLE 2

*The β-Estradiol binding capacity of transferrins*

| Cow No. | Transferrin | μg β-Estradiol per 1 mg transferrin | μg β-Estradiol per 10 mole transferrin |
|---|---|---|---|
| M 570 | AA | 0.1621 | 5.003 |
| E 513 | AA | 0.1598 | 4.962 |
| M 569 | AA | 0.2446 | 7.592 |
| M 564 | AA | 0.0954 | 2.962 |
| M 574 | AA | 0.0598 | 1.856 |
| M 567 | AA | 0.1504 | 4.670 |
| E 390 | $AD_1$ | 0.0793 | 2.461 |
| AM 512 | $AD_1$ | 0.1455 | 4.516 |
| E 397 | $AD_2$ | 0.0855 | 2.654 |
| AM 541 | AE | 0.0989 | 3.072 |
| E 599 | $D_1D_1$ | 0.1347 | 4.181 |
| E 602 | $D_1D_2$ | 0.0510 | 1.583 |
| 510 | $D_1E$ | 0.2093 | 6.496 |
| M 501 | $D_2D_2$ | 0.2301 | 7.043 |
| E 405 | $D_2D_2$ | 0.1847 | 5.732 |
| E 598 | $D_2E$ | 0.0546 | 1.697 |
| 808 | EE | 0.3026 | 9.392 |
| Standard deviation between duplicates | | ±0.0139 | ±0.432 |

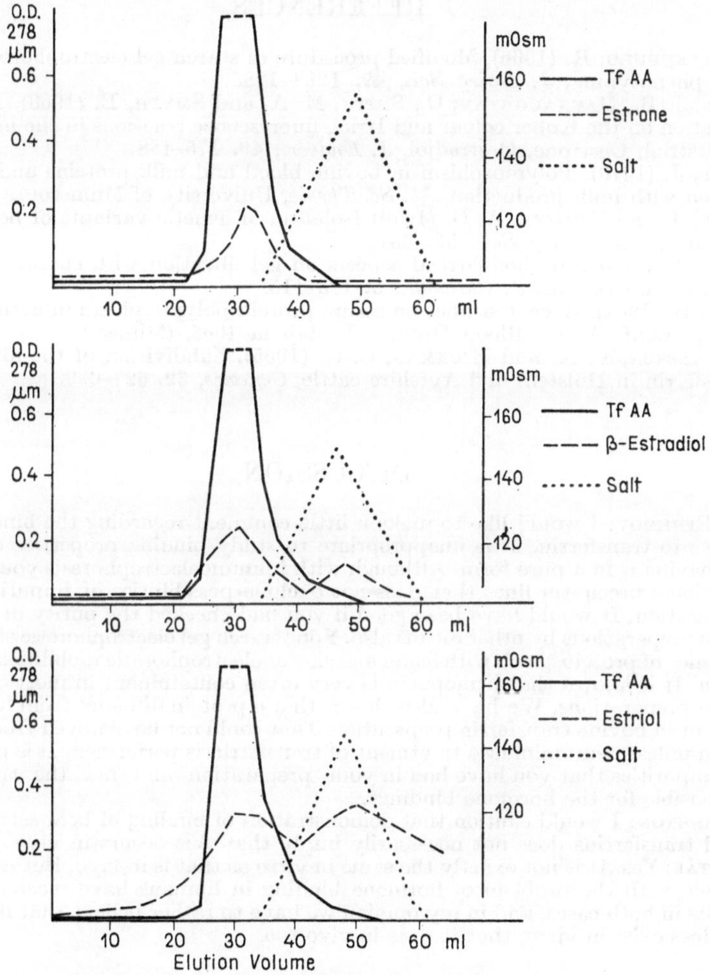
O.D. 278 μm
mOsm
Tf AA
Estrone
Salt
β-Estradiol
Estriol
10 20 30 40 50 60 ml
Elution Volume

FIG. 4. The separation of transferrin-estrogen complex by column chromatography on Sephadex G-25

bound by transferrin. The β-Estradiol was bound to transferrin (saturated) and the excess or free Estradiol appeared as the second peak. A similar situation occurred with Estriol but the transferrin-Estriol complex was not completely separated from the free hormone. Therefore, only transferrin binding capacity for β-Estradiol was compared using the isolated phenotypes. These results are presented in Table 2. The variation among phenotypes and within the same phenotype showed that the genetically controlled phenotype did not control the degree of binding capacity of transferrins for β-Estradiol.

## REFERENCES

ASCHAFFENBURG, R. (1966), Modified procedure of starch gel electrophoresis for β-casein phenotyping, *J. Dairy Sci.*, **49.** 1284–1285.

BROWN, J. B., MACNAUGHTAN, C., SMITH, M. A. and SMYTH, B. (1968), Further observation on the Kober colour and Itrich fluorescence reactions in the measurement of Oestriol, Oestrone, Oestradiol. *J. Endocr.*, **40.** 175–188.

DOSTÁL, J. (1970). Polymorphism in bovine blood and milk proteins and their association with milk production, *M. Sc. Thesis*, University of Minnesota.

DOSTÁL, J. and HUNTER, A. G. (1970) Isolation of genetic variants of bovine serum transferrin, *J. Dairy Sci.*, **53.** 651.,

FLODIN, P. (1961), Methodological aspects on gel filtration with special reference to desalting operations, *J. Chromat.*, **5.** 103–115.

GAHNE, B. (1962), Recent studies on serum protein polymorphism in cattle, Rep. 8th Europ. Conf. Anim. Blood Groups, Ljubljana 1962, (Mimeo.).

KRISTJANSSON, F. K. and HICKMAN, C. G. (1965), Subdivision of the allele $Tf^D$ for transferrin in Holstein and Ayrshire cattle, *Genetics*, **52.** 627–630.

## DISCUSSION

G. D. EFREMOV: I would like to make a little comment regarding the binding of hormones to transferrin. It is unappropriate to study binding properties of a protein not having it in a pure form. Although with immunoelectrophoresis you did not see additional precipitin lines that does not exclude possibilities of impurities in your preparation. It would have been good if you had checked the purity of your transferrin preparations by other means also. Your starch gel electrophoretic slides showed presence of proteins both with faster and slower electrophoretic mobilities than transferrin. It is known that hemopexin is very often contaminant in most of the transferrin preparations. We have also shown that a protein different from hemopexin is present in bovine transferrin preparation. That could not be removed from the transferrin unless neuraminidase treatment of transferrin is performed. It is possible that the impurities that you have had in your preparation and not the transferrin is responsible for the hormone bindings.

G. C. ASHTON: I would caution that demonstration of binding of beta-estradiol to isolated transferrins does not necessarily imply that this occurs in vivo.

J. DOSTÁL: Yes. It is not exactly the same in vitro as that is in vivo. But many reports dealing with the problem of hormone binding in humans have received the same results in both cases, and in my opinion we have to rather believe that if some binding does exist in vitro, that it does in vivo too.

*XIIth Europ. Conf. Anim. Blood Groups Biochem. Polymorph., Bp., 1972 (pp. 267–270)*

# TRANSFERRIN TYPES IN HUNGARIAN A. I. BULLS AND THEIR RELATIONSHIP TO FERTILITY DATA

ERZSÉBET GIPPERT, PÁL SOÓS, PÉTER SOOS and J. STUKOVSZKY

Center of Artificial Insemination, Laboratory of Immunogenetics, Budapest, Hungary

## SUMMARY

**The transferrin gene frequencies of the A. I. bulls used in Hungary are compared to those of a randomly chosen Hungarian Spotted cattle population that had been established earlier.**

**The $Tf^{D_1}$ and the $Tf^{E}$ alleles showed higher frequencies in the bulls than in the cattle examined.**

**To clarify the possible relationship between Tf phenotypes and the fertility of bulls, one-year fertility data and the Tf phenotypes of 194 bulls were analyzed.**

**The fertility of the bulls was apparently not influenced by the Tf locus.**

## INTRODUCTION

After describing the transferrin (Tf) types of cattle and the investigation of the mode of inheritance, their possible relationship with various production and physiological characteristics was examined. This seemed reasonable considering the fundamental role of transferrin in metabolism (ASHTON, 1958, 1960; LARSEN, 1961; HICKMAN and DUNN, 1961). Although on the basis of the statements of these first reports a positive effect seemed to exist between Tf types and milk production, Tf types and climatic resistance, Tf types and fertility of A.I. bulls, many contradictory findings were published during the recent years in this field (see JAMIESON and ROBERTSON, 1967; ASHTON and HEWETSON, 1969).

JAMIESON (1966) detected a significant difference between the distribution of the individual Tf alleles in the A.I. bulls and the random population of a certain breed. If there is actual relationship among Tf genes and other production characters, then such a difference may be caused by the selection of the bulls.

The aim of the present study was to investigate whether there was an actual difference between the frequencies of Tf genes in the A.I. bulls and those of a random Hungarian Spotted (HS) cattle population, and whether the Tf types of the sires had any influence on the conception rate or fertility independently of the Tf types of dams following a single insemination.

## MATERIAL AND METHODS

Serum samples collected from 486 A.I. bulls were examined with the method of MAKARECHIAN and HOWELL (1966) as modified by KOVÁCS and SOOS (1969). The Tf gene frequency data of the random HS population used in this study were derived from the report of the above authors.

The fertility data of 194 A.I. bulls were collected from the 1967 records of the Artificial Insemination Stations in the country. The two populations were compared by the 2×j test suggested by MATHER (1951), while the Tf types and the fertility data by Student's *t*-test.

## RESULTS AND DISCUSSION

The Tf gene frequencies of the A.I. bulls and their standard deviations are presented in Table 1. The expected and observed values of the phenotypical distribution of the bulls can be seen from Table 2. The rare $Tf^F$ allele was not encountered in the examined bull population. The $\chi^2$ value showing the deviation from the Hardy–Weinberg law was not significant thus the bull population examined was considered to have been in the state of genetic equilibrium. Comparison by the 2×j test of the random HS and the A.I. bull populations revealed a highly significant difference in the distribution of Tf genes ($P < 0.001$; d.f. 4). This can be explained chiefly by the higher frequency of $Tf^{D_1}$ and $Tf^E$ alleles in the bull population as compared to those in the HS cattle (Table 3). This is probably the result of the preference of certain bull lines.

TABLE

*Expected and observed values of the transferrin phenotypes and*

| No. | | AA | $AD_1$ | $AD_2$ | AE | $D_1D_1$ |
|---|---|---|---|---|---|---|
| 486 | obs. | 27 | 26 | 119 | 14 | 9 |
| | exp. | 23.33 | 27.59 | 128.19 | 10.52 | 8.16 |

* $H$ = degree of homozygosity (%).

TABLE

*Phenotypes and respective fertility percentages, total*

| | AA | $AD_1$ | $AD_2$ | AE | $D_1D_1$ |
|---|---|---|---|---|---|
| Fertility % | 51.29 | 50.67 | 52.01 | 51.40 | 58.40 |
| No. of services | 10473 | 7759 | 24814 | 2411 | 2845 |
| No. of bulls | 23 | 8 | 35 | 6 | 3 |

TABLE 1

*Transferrin gene frequencies in a random Hungarian Spotted cattle population and in the A.I. bulls*

| | Hungarian Spotted | A.I. bull |
|---|---|---|
| | population | |
| $Tf^A$ | 0.1981 ±0.0049 | 0.2191 ±0.0133 |
| $Tf^{D_1}$ | 0.1702 ±0.0046 | 0.1296 ±0.0108 |
| $Tf^{D_2}$ | 0.6085 ±0.0060 | 0.6019 ±0.0157 |
| $Tf^F$ | 0.0009 ±0.0004 | — |
| $Tf^E$ | 0.0223 ±0.0018 | 0.0494 ±0.0069 |

The differences between the average fertility percentage of the bulls with different phenotypes were not significant in any group (Table 4). Similarly, there was no significant difference when comparing individuals being homo-, or heterozygous on the Tf locus. The above results suggest the fertility of bulls to be unrelated to the Tf locus.

2

*those of the degree of homozygosity in the A.I. bulls*

| $D_1D_2$ | $D_1E$ | $D_2D_2$ | $D_2E$ | EE | $\chi^2$ | H* |
|---|---|---|---|---|---|---|
| 76 | 6 | 183 | 24 | 2 | 4.11 | 45.47 |
| 75.82 | 6.22 | 176.07 | 28.91 | 1.19 | | 42.95 |

4

*numbers of first inseminations and of bulls examined*

| $D_1D_2$ | $D_2D_2$ | $D_1E$ | $D_2E$ | EE | Homozygotes | Heterozygotes |
|---|---|---|---|---|---|---|
| 54.34 | 54.28 | 54.20 | 52.56 | — | 54.06 | 52.66 |
| 19655 | 76055 | 1453 | 4212 | — | 89373 | 60304 |
| 27 | 83 | 2 | 7 | — | 109 | 85 |

TABLE 3

*The distribution of transferrin types in the Hungarian Spotted cattle breed and in the A.I. bulls as assessed by the 2 × j test*

| | No. of animals | $Tf^A$ | $Tf^{D_1}$ | $Tf^{D_2}$ | $Ff^F$ | $Tf^E$ | $\chi^2_{2\times j}$ | d.f. | *P* |
|---|---|---|---|---|---|---|---|---|---|
| | | genes | | | | | | | |
| Hungarian Spotted population | 3289 | 1303 | 1119 | 4003 | 6 | 147 | 37.37 | 4 | $<0.001$ |
| A.I. bulls | 486 | 213 | 126 | 585 | — | 48 | | | |

# REFERENCES


ASHTON, G. C. (1958), Genetics of beta-globulin polymorphism in British cattle. *Nature*, **182.** 370.

ASHTON, G. C. (1960), Beta-globulin polymorphism and economic factors in dairy cattle. *J. Agric. Sci.*, **54.** 321.

ASHTON, G. C. and HEWETSON, R. W. (1969), Transferrins and milk production in dairy cattle. *Anim. Prod.*, **11.** 533.

HICKMAN, C. G. and DUNN, H. O. (1961), Differences in percentage of non-returns to service between transferrin types of bulls. *Can. J. Genet. Cytol.*, **3.** 391.

JAMIESON, A. (1965), The distribution of transferrin genes in cattle. *Heredity* (Lond.), **21.** 191.

JAMIESON, A. and ROBERTSON, A. (1967), Cattle transferrins and milk production. *Anim. Prod.*, **9.** 491.

KOVÁCS, GY. and SOÓS, P. (1969), A szarvasmarha transferrin vizsgálatok újabb eredményei. II. Populációgenetikai vizsgálatok hazai szarvasmarhafajtákban serumtranszferrin típusok felhasználásával. *Magy. Áo. Lapja*, **24.** 662.

LARSEN, B. (1961), Serum-, hemoglobin- og maelketypers mulige indflydelse pa den Kvantitative og Kvalitative maelkeproduktion hos Kvaeg. *Aarsberetning Inst. Sterilitetsforskning*, (Copenhagen).

MAKARECHIAN, M. and HOWELL, W. E. (1966), Relationship between transferrin type and feedlot performance in beef heifers. *Can. J. Anim. Sci.*, **46.** 177.

MATHER, K. (1951), Statistical analysis in biology. Methounen and Co. Ltd. London.


# DISCUSSION

F. PIRCHNER: It would appear justified to look for transferrin effects upon bull fertility as for effects upon milk yield.

J. C. ASHTON: The difference in fertility between transferrin types is due to an interaction between cow type and bull type. It is virtually impossible to get a significant difference between bull types alone.

However it is noteworthy that there was an excess of homozygous types, and that homozygous bulls had higher average conception rates. This implies a superiority of homozygous bulls, which is what previous reports would suggest.

*XIIth Europ. Conf. Anim. Blood Groups Biochem. Polymorph., Bp., 1972 (pp. 271–272)*

# DATA ON THE DYNAMICS OF THE SYNTHESIS OF HbF DURING ONTOGENESIS IN CALVES

V. Dikov

Institute of Reproduction Biology and Non-Infectious Diseaes, Sofia, Bulgaria

Information is lacking whether the so-called primitive hemoglobin (HbP) established in human embryo in its early phases of development would exist in animals. The evidence of foetal hemoglobin (HbF) in animals, however, has not been questioned (Mikle et al., 1963).

The character of the two basic hemoglobin types in animals — foetal and adult — is determined by the physiological conditions under which the combination of the oxygen is taking place during the different stages of ontogenetic development of the organism (Tringer, 1961).

Investigations on the genetics of hemoglobin polymorphism in cattle are being carried out along different lines. The data on the dynamics of the synthesis of foetal hemoglobin have not yet been properly evaluated.

## MATERIAL AND METHODS

The present study was carried out on 337 calves, 1–120 days old, and on 26 cows. All animals were of the Bulgarian Brown cattle breed. The method of Gahne, Rendel and Venge was used (after Schmid, D. O.) to receive pure hemoglobin. The other methods for experimental work and some slight modifications undertaken in them were reported earlier (Dikov, 1965; Dikov and Georgieva, 1969; Dikov, 1970).

## RESULTS AND DISCUSSION

During the postnatal development of calves in the period from 1 to 120 days of age, 7 types of hemoglobin were demonstrated in their erythrocytes:

| | | |
|---|---|---|
| HbF | HbFA | HbA |
| | HbFB | HbB |
| | HbFAB | HbAB |

At birth and during the first and second day after birth only foetal hemoglobin was demonstrable in the blood of the calves. HbFA appeared first between the 10th–15th day of life. HbFA was rarely found in the period from the 50th to the 75th day of age, and HbFAB was found in only two calves older than 75 days.

No isolated existence of HbF was established in calves older than 15 days.

A decisive event in the development of adult hemoglobin is the day of its first appearance and the dynamics of this process. HbA is the first one which appears. Rare cases of HbA were established after 60 days of age. Its regular appearance, however, is only after 90 days of age of the calves.

The analysis of the results reported above shows a regularity in the gradual transition of foetal to adult hemoglobin.

The mechanisms controlling the synthesis of the adult hemoglobin types comes into effect during the second month after birth and reaches a maximum in the 4th month. Data on hemoglobin in cows immediately before and after delivery are of interest, too.

The fact that we failed to demonstrate HbF in the blood of calves older than 120 days and in cows pregnant and newly calved as well, is no sufficient reason to assume that the synthesis of this type of hemoglobin has been fully suppressed in these periods. In support of this we refer to the work of Manitasi and Ingram (1970) and to some earlier studies by Charnoff (1961), Boglioni (1963) and Hosoi (1965).

This partly answers the question whether more than one type of hemoglobin can be established on erythrocytes.

The data obtained in this study may serve as indicator in determining the degree of the physiological maturity of calves and also in establishing certain forms of hemoglobinopathy.

*XIIth Europ. Conf. Anim. Blood Groups Biochem. Polymorph., Bp., 1972 (pp. 273–276)*

# REPORT ON THE 1970 CATTLE COMPARISON TEST, PERFORMED IN HUNGARY AT THE CENTER OF ARTIFICIAL INSEMINATION, LABORATORY OF IMMUNOGENETICS

PÁL SOÓS and J. STUKOVSZKY

**Center of Artificial Insemination, Laboratory of Immunogenetics, Budapest, Hungary**

After performing a Comparison Test in Hungary in 1965 and having seen the 1966 U.S.D.A. Comparison Test, this year we decided to carry out the processing of the results by using computer technique. Having discussed the organization of this work with our Secretary Dr. BOUW, he agreed with this idea.

In this Comparison Test 40 blood samples were sent out for testing to 25 participating laboratories, between the 5th and 10th of April, 1970. We used a cyanide containing anticoagulant:

| | |
|---|---|
| sodium citrate | 20 g |
| sodium chloride | 5 g |
| sodium cyanide | 0.4 g |
| distilled water ad | 1000 ml |

16 laboratories informed us, that the parcels had arrived in good condition. No remark was made about hemolyzed or broken samples. Only Dr. SCHLEGER (Austria) complained in his letter that because of the cyanide solution, results of the electrophoretic examinations were imperfect. According to our own experience, electrophoretic examinations to detect different serum and red cell polymorphisms can be carried out without difficulty regardless of the cyanide solution.

The blood samples were collected from the following breeds:

1–21. Hungarian Spotted.
22–29. Hungarian Grey.
30–35. Brown Swiss.
36–38. Simmental.
39. Austrian Spotted.
40. Jersey.

Many participants had problems with our allele list. Our $O_1$ and G″ reagents had indeed given rise to some errors. Secondly we put stress in first line to the selection of animals with a wide variety of factors and biochemical polymorphisms. Our intention was that as much factors as possible be included.

In some cases laboratories sent in their reports without using the factor designation accepted at the Paris Conference and therefore the items were included in the wrong factor column. For example $T_2$ was designated $T_1$ or T.

## REAGENTS

We do not want to comment much on the results and this can be considered simply as personal impressions.

Most of the reagents were in good agreement with each other (Table 1), only few showing serious discrepancies. With 56 reagents most of the laboratories produced reliable results which we believe can satisfy us.

TABLE 1

*Reagents in good agreement*

| System | Factors |
|---|---|
| A | $A_1$ $A_2$ H Z′ |
| B | $I_1$ $I_2$ K $O_1$ $O_3$ $P_1$ $P_2$ Q $T_1$ $T_2$ Y A′ B′ D′ $E_2'$ F′ G′ I′ J′ K′ O′ P′ Q′ Y′ B″ G″ |
| C | $C_1$ $C_2$ E $R_1$ $R_2$ W $X_1$ $X_2$ L′ |
| F/V | F V |
| J | J |
| L | L |
| M | M |
| S | S H′ U U′ U″ H″ S″ |
| Z | Z |
| R′/S′ | R′ S′ |
| T′ | T′ |

Problems were encountered with the following factors:

B — in sample 24 — 10 laboratories had positive reactions and 18 ones negative reactions. On this basis it should be considered whether $B_1$ and $B_2$ reagents with a shorter and longer reaction pattern exist or not. To this problem more attention should be paid in the next Reference Test.

According to the results of the present Comparison Test, G and $G_2$ reagents were found identical in 13 laboratories. As far as $G_1$ is concerned, 12 laboratories found the same reactions and therefore now $G_1$ could be accepted as a subtype of the former reagents.

Most of the laboratories have disagreed in respect of the following reagents:

$$O_2,\ O_x,\ E_1',\ E_3',\ I_2'$$

It should be noted that $U_2$ of Göttingen and $A_7$ of Austria showed the same reactions.

It was found that S and S″ — as far as the 40 samples are considered — had the same reaction patterns and, therefore, they could not be differentiated.

We analyzed to what extent experimental reagents are in agreement with other known factors (Table 2) or with each other (Table 3).

TABLE 2

*Possible relationship between experimental reagents and known factors*

| Known reagents | Reagents with preliminary designation | Laboratory symbol |
|---|---|---|
| $G_1$ | F-11 | F/J |
| | K-45 | SU/K |
| K | GÖ-12 | D/GÖ |
| $K_2$ | F-3 | F/J |
| $O_1$ | 70/2 | SA |
| $P_1$ | 70/5 | SA |
| $Y_2$ | Mü-7 | D/Mü |
| $A'$ | H-7B | NL |
| $G''$ | Sch-15 | D/Sch |
| $G'_1$ | SU-16 | SU/D |
| $X_1$ | GÖ-11 | D/GÖ |
| | X-18 | I/M |
| | LJ-17 | YU/Lj |
| $J_1$ | 70/6 | SA |
| | 70/10 | SA |
| $S'$ | $S'_2$ | I/M |

TABLE 3

*Experimental reagents with identical reaction patterns*

| Reagent | Laboratory symbol | Reagent | Laboratory symbol | Reagent | Laboratory symbol |
|---|---|---|---|---|---|
| DK 23 | DK | D/Gö 10 | D/Gö | A′ | NL |
| D/Gö 9 | D/Gö | F′ | NL | H 7 B | NL |
| PL/B 3 | PL/B | | | | |
| | | D/Gö 12 | D/Gö | D/Gö 11 | D/Gö |
| F 8 | F/J | M 6 | I/M | X 1 B | I/M |
| I″ | NL | H 12 | NL | LJ 17 | Yu/Lj |
| F 11 | F/J | M 12 | I/M | T 1 | I/T |
| G 1 | I/M | M 12 | NL | 70/15 | SA |
| | | CS 22 | CS | | |
| F 16 | F/J | | | 70/2 | SA |
| M 13 | I/M | G″ | NL | LJ 19 | Yu/Lj |
| | | SU 16 | SU/D | | |
| F 18 | F/J | | | | |
| D/Sch 17 | D/Sch | | | | |

## CONCLUSIONS AND SUGGESTIONS

The classification of the different reagents' reaction patterns by manual work, as done in the 1965 Comparison Test, is a tremendous job. The computer technique, used in this year was not perfect either. Owing to some special difficulties we could not compare the experimental reagents between themselves and with other well-known reagents.

For well trained workers and beginners alike, it is essential to see clearly how their reagents work. This is most important for producing internationally comparable results.

Furthermore I would like to propose to publish again the nomenclature in our new Journal and to inform people how preliminary designations should be used.

## DISCUSSION

G. Kovács: In Paris in 1966 a nomenclature "sub-committee" was established, which presented suggestions. These suggestions have been accepted but have not been used as yet. I think this will present still more problems in the future.

Pál Soós: The $E'_3$ should not be regarded an independent reagent as it is a mixture of reagents. Dr. Bouw says he has seen the same thing by $O_x$ where reactions appear only where there is $O_3/O'$ or A′.

F. Pirchner: The German speaking laboratories have devised a code for B-alleles and would invite other laboratories to join.

J. Rendel: With regard to nomenclature we cannot solve the problems to-day. It is necessary to have a committee. At least one "old-timer" in the field of cattle blood grouping should be on the committee. May I suggest Bent Larsen.

D. R. Osterhoff: All problems regarding any types of changes in nomenclature should be discussed now, while we are all together. If new reagents are included in tests, the donor-recipient-combinations should be given, otherwise these test-results should not be recorded in the final lists. Should the reagent anti-$E'_3$ be taken out of our list and replaced by anti-F′, anti-G″, anti-$E'_1$ and anti-$H_{12}$ ($E'_2$) respectively?

B. Larsen: I do not think that we should take any reagents out of the comparison test. Doing that we cannot get any further information on new reagents belonging to the same group. With regard to comparison tests we have now for many years had one every year. As the number of laboratories participating in these tests are increasing it has become our heavy and expensive duty for the laboratory being in charge of sending out blood samples and collecting the results. In the Committee we have, therefore, discussed this problem and reached at the conclusion that a comparison test in cattle every second year would be sufficient. We have, furthermore, contacted Dr. E. Kiddy, Beltsville, about this, and he is willing, every second year, to take care of the compilation of the results by his data processing equipment, while a European laboratory on the other hand should be in charge of sending out the blood samples. If there is serious objection about having a comparison test every second year only, the Committee would be glad to know.

# III. BLOOD GROUPS AND BIOCHEMICAL POLYMORPHISM IN PIGS

*XIIth Europ. Conf. Anim. Groups Blood Biochem. Polymorph., Bp., 1972 (pp. 279–283)*

# NATURAL BLOOD-GROUP ANTIBODIES IN PIGS

JOAN HARDY

**Protein Chemistry Section, Unilever Research Laboratory, Colworth House, Sharnbrook, Bedford, England**

"Natural" blood-group antibodies are usually considered to be those which may be found in the serum of an animal who does not possess and who, has never been exposed to the corresponding red cell antigen. Antibodies derived from the mother either transplacentally or from colostrum and which usually disappear rapidly from the serum of the newborn are not included in this definition.

Red cell antibodies are common in pigs who have been vaccinated with crystal violet swine fever vaccine and when investigating such antibodies GOODWIN, SAISON and COOMBS (1955) reported low titre antibodies in unvaccinated sows. Naturally occurring antibodies and what they believed to be naturally occurring hemolytic disease of piglets were described by SZENT-IVÁNYI and SZABÓ (1954) but their extensive investigation has not been repeated nor have the antibodies found by any of these earlier workers been identified in terms of the present pig blood group nomenclature. We have recently examined sera from the following:

1. 50 unvaccinated and unmated hogs and gilts aged between 24 weeks to 10 months.
2. 14 unvaccinated, mated gilts aged 9–10 months.
3. 2 unvaccinated boars.
4. 20 sows all of whom had had at least one litter but who were unvaccinated.

Apart from not having had crystal violet swine fever vaccine none of these animals had had injections of any kind except for one gilt in our immunisation herd who had penicillin injections for pneumonia.

All the A negative animals examined, except two, had natural anti A in their serum. This was always active in direct, Coombs and hemolytic tests but in six animals some A positive cells failed to react with their sera except in the antiglobulin test. When the six sera were tested in parallel they always reacted and failed to react with the same cells suggesting that the anti A possessed by them differed in some respect from that possessed by the remaining 39 animals with anti A. The animals concerned were not of any particular breed.

Of the two A negative animals who did not possess anti A one had no antibodies and the second had a very interesting hemolytic antibody which reacts with the majority of A negative cells but has not so far been identified further.

Apart from anti A the commonest antibodies found were anti Ea and Eb. In category (1)n 7/50 of the animals possessed anti Ea and 3/50 anti Eb. 2/14 of the mated gilts possessed anti Eb and 1/13 of the sows possessed either anti Ea or Eb making a total of 8/86 animals with anti Ea and 6/86 with anti Eb. Anti La was found in two animals and Anti Lg in one. Ten unidentified antibodies were found, nine in category (1) one in category (2) and five in category (3). Two of these antibodies have the same specificity although one reacts only in the direct and the other only in the indirect agglutination test. All the unidentified antibodies have proved to be univalent once the anti A or other antibody present has been removed by absorption.

The first question that arises is whether we are justified in calling these "natural" antibodies. We accept that many immunologists dislike the use of this term, with some justification, but in the present context we simply mean can we positively exclude the possibility that these antibodies were present because of:

a) The absorption of antibody from the maternal colostrum.

b) Immunization with foreign pig red cells either by injection or as the result of an incompatible mating?

Only one pig in the first group received any injections of any kind. All the pigs in this group were either slaughtered at the end of a feeding trial and their sera obtained at slaughter or were animals in our immunization herd whose sera we tested before starting immunisation. The pigs on feeding trial were aged 24–36 weeks. Antibodies from colostrum usually disappear rapidly, although LINKLATER (1969) has found anti Ea apparently derived from the dam in a pig as late as 20 weeks. In these feeding trial pigs we had no information on the blood type of sire or dam so cannot state definitely that they were not products of an incompatible mating. It only seems unlikely that antibodies originating from such a mating would still be present as late as 36 weeks. One hog from our immunization herd still possesses an anti Eb first found at 10 months, before we started immunization, at a titre of 1 : 4. He is now over a year old has been immunized with Eb negative cells but the anti Eb persists at the same titre. A Lincolnshire Curly Coat × Welsh crossbred gilt also in our herd was not immunized at all because of a severe attack of pneumonia for which she was given penicillin injections with disposable plastic syringes i.e. they had not been used on any other pig. When we eventually tested her serum she was at least ten months old and we found to our surprise that she had a strong direct acting anti Lg in her serum. She was a La homozygous animal and came from an interesting family who were all either L adhj homozygous or else L adhjlbdfi. We unfortunately have not the boar's or sow's red cell type but it does not seem likely that this antibody could have been maternally derived.

In the second group of gilts that had been mated but who had not farrowed we found two animals with anti Eb in their serum. One of them also possessed a weak unidentified indirect antibody. We knew the blood type of the dam and sire in these two instances. Both had E def/def sires and the dams were:

a) E aeg deg — gilt E aeg def
b) E deg deg — gilt E deg def.

Here there was no question at all of an Eb incompatible mating. In both instances the gilts have been mated with an E def/def boar so even if the passage of foetal cells across the placenta were likely during pregnancy, these foetuses could not possess the Eb red cell antigen.

LINKLATER (1968) has demonstrated convincingly that isoimmunization of the parturient sow by foetal cells is possible. It seems most likely that this will occur late in pregnancy, or during farrowing. This is now believed to be the most likely time for foetal cells to enter the maternal blood stream in humans and this fact has indeed made it possible to avoid the formation of anti rhesus in rhesus negative women by removing rhesus positive cells with the injection of anti rhesus serum into the maternal circulation.

In our final group of 20 sows, most of whom had farrowed several times antibodies could well have been formed in this way. The actual percentage of animals with antibodies other than anti A was slightly above that in the first group, 40 as opposed to 36%, but the numbers are too small to be of significance. In no case was the titre of the antibody high but the sows were all tested after their piglets were weaned as they were old sows who were sent for slaughter and whose blood we then tested before using it for absorptions. All were animals from our farm and the fact that they were sent for slaughter was in some instances due to a poor breeding record. This was so in the case of the Pietrain sow who had had six litters by the same Pietrain boar. Her record sounds exactly like that of a typical case of progressive hemolytic disease of the newborn:

1st litter no details but all alive
2nd litter 10 piglets, 1 neonatal death
3rd litter 10 piglets, 2 neonatal deaths
4th litter 11 piglets, 2 neonatal deaths
5th litter 16 piglets, 14 neonatal deaths or stillborn
6th litter aborted.

Unfortunately we had no opportunity of testing earlier litters but were able to test the sows' sera with the boar's red cells. These were agglutinated in the antiglobulin test but we have so far not identified the antibody although it reacts with other cells from other breeds. The blood types of this boar and sow were:

Boar Aneg Bb neg E defdeg Fa neg Da neg Gab H–/–Iaa. Kb. Lbcgi.
Sow Aneg Bb neg E dbgdef Fa neg Da neg Gab H–/–Iaa. Ka. Lbcgi.

We are certain the antibody is not anti Kb, the only obvious incompatibility in the above typing. Both sow and boar had anti A in their serum.

It would on this evidence appear very likely that anti Eb, anti Ea and other antibodies can occur "naturally" in pigs. The titre of such antibodies could well be increased specifically by an incompatible mating or immunization or non-specifically during immunization for other antibodies. We have in our herd a pig who was immunized with Ea positive red cells two years ago and who has repeatedly produced anti Ea after six subsequent immunizations with incompatible but Ea negative red cells. Anti Ea has not been detectable in her serum between immunizations.

TABLE 1

*Antibodies found in sera of unvaccinated pigs*

| Category | No. | Age | Breed | Mated | Litters | Antibodies found: A | Ea | Eb | La | Lg | X | No. with antibodies other than A |
|---|---|---|---|---|---|---|---|---|---|---|---|---|
| Hogs | 8 | 9 months | Wessex | No | | 1 | | 1 | | | | 1 |
| Hogs | 4 | 24.36 w | L. W. | No | | 2 | | 1 | | | 1 | 2 |
| Hogs | 4 | 24.36 w | Cotswold | No | | 3 | 1 | | | | 1 | 2 |
| Hogs | 4 | 36 weeks | L. W. | No | | 1 | 2 | | | | 1 | 2 |
| Gilts | 6 | 24.36 w | L. W. | No | | | 1 | | | | 1 | 1 |
| Gilts | 6 | 24.36 w | Cotswold | No | | 2 | 1 | | | | | 1 |
| Gilts | 4 | 9 months | L. W. | No | | 2 | 2 | | | | 2 | 3 |
| Gilts | 4 | 6 months | LCC/CY | No | | 2 | | | | | | 0 |
| Gilts | 3 | 10 months | LCC/CY | No | | 3 | | | | 1 | | 1 |
| Gilts | 3 | 6 months | Pietrain | No | | 3 | | 1 | | | | 1 |
| Gilts | 4 | 36 weeks | L/W/LW | No | | 3 | | | 1 | | 3 | 4 |
| | 50 | | | 0 | | 22 | 7 | 3 | 1 | 1 | 9 | 18 (36%) |
| Gilts | 7 | 9 months | L. W. | Yes | No | 5 | | 1 | | | 1 | 2 |
| Gilts | 7 | 9 months | L/W/LW | Yes | No | 3 | | 1 | | | | 1 |
| Boar | 1 | 2–3 years | Pietrain | Yes | | 1 | | | | | | 0 |
| Boar | 1 | 9 months | L. W. | Yes | | 1 | | | | | | 0 |
| | 16 | | | | | 10 | | 2 | | | 1 | 3 |
| Sow | 1 | 2–3 years | Pietrain | Yes | 6 | 1 | | | | | 1 | 1 |
| Sows | 7 | 2 years | L. W. | Yes | Yes | 6 | | | 1 | | 1 | 2 |
| Sows | 12 | 2 years | LWX | Yes | Yes | 6 | 1 | 1 | | | 3 | 5 |
| | 20 | | | | | 13 | 1 | 1 | 1 | | 5 | 8 (40%) |

Key LW Large White
Costwold Hybrid Large White Cross
L.C.C. Lincolnshire curly coat
CY Welsh
L/W/LW Landrace, Large White, Wessex Cross

It is also possible that these antibodies may in some circumstances protect the sow against isoimmunization by foetal red cells in the same way as natural anti A may be removing A positive rhesus positive foetal cells rapidly from her circulation protects an A negative, rhesus negative woman from rhesus isoimmunization.

## REFERENCES

Goodwin, R. F. W., Saison, R. and Coombs, R. R. A. (1955), J Blood Groups of the Pig II Red Cell antibodies in the sera of pigs injected with Crystal violet swine fever vaccine. *J. Comp. Path.*, **65.**

Linklater, K. (1968), Iso-immunization in the Parturient Sow by Foetal Red Cells. *Vet. Rec.*, 203.

Szent-Iványi, T., Szabó, S. (1954), Blood Groups in Pigs. *Acta Vet. Hung.*, **4.** 429.

## DISCUSSION

H. Dinklage: Did you test the extracts of soya beans against red cells of pigs? Does it parallel to any of the already known factors?

J. Hardy: The crude soya bean extract and also various stages in the processing of the soya meal were tested. All showed the same type of reaction.

R. Hohenbrink: Did you found parallel reactions of the soya bean extract and anti-Eb with a crude extract or an absorbed extract?

J. Hardy: Soya bean extract has been shown to react with most $Eb^+$ and O negative cells but only 100 cells have as far been tested. So far only A-positive cells which are also Eb positive have reacted with the bean extract.

*XIIth Europ. Conf. Anim. Blood Groups Biochem. Polymorph., Bp., 1972 (pp. 285—287)*

# PRELIMINARY INVESTIGATIONS ON TWO BLOOD SERUM ANTIGENS IN PIGS

MARIA DUNIEC

Immunogenetics Laboratory, Zootechnical Institute Balice k/Kraków, Poland

## SUMMARY

The isoimmune sera which were used for detecting the two antigens discussed here were obtained by immunizing the pig donors with the whole serum of pig recipients, with the addition of Freund's adjuvant.

The sera were used for testing over 1000 pigs (Polish Landrace and Polish Large White) by double diffusion (Ouchterlony) test on microscope slides.

It was found that the frequency distribution of the L (preliminary designation) antigen detected by one serum was 90% in Polish Large White and 99.3% in Polish Landrace. For the second antigen (preliminarily denoted as G) the frequency distribution was 9.3% and 40% respectively.

On the basis of the results obtained so far it may be concluded that the antigens tested by us are genetically controlled.

## INTRODUCTION

The application of immunodiffusion and immunoelectrophoresis techniques in immunology offered ample opportunity for studying the antigenic differentiation of the blood serum and other tissue proteins.

The antigenic differences of the blood serum can be detected not only by using the heteroprecipitins but also by isoprecipitins. Using the latter, the allotypic differences were detected in some species, among others in rabbits (OUDIN, 1956; DRAY, 1958; DUBISKI et al., 1958), in hens (SKOŁBA, 1964, 1966), in cattle (RAPACZ et al., 1968; SPOONER, 1970) and in pigs (RAPACZ et al., 1968, 1970a, b; LANG, 1970; TIKHONOV et al., 1970).

This paper includes also preliminary studies on two blood serum antigens detected by using the sera which contained isoprecipitins.

## MATERIAL AND METHODS

Ten adult Polish Large White pigs were treated with intramuscular injections of whole serum from other pigs for the period of 2–5 months at one-week intervals.

Five ml of serum with the addition of 0.5 ml of Freund's adjuvant were used for one immunization. Before each successive injection the serum of the pig recipients was examined for isoprecipitins by means of Ouchterlony's double diffusion technique in agar gel. The pigs in the sera of which the presence of isoprecipitins was established were bled 7 days after the last injection and the serum was separated and stored at −18°C.

The sera obtained after immunization were used for testing the normal sera of other Polish Large White and Polish Landrace pigs with double diffusion test in agar gel, using the modified technique described by RAPACZ et al. (1968).

The precipitates were stained with Amido Black for proteins and with Sudan Black for lipids. Immunoelectrophoresis was performed according to the method described by RAPACZ et al. (1968).

## RESULTS

In the serum of pig No. CH-8/0, the presence of antibodies which reacted with the donor serum and normal serum of some other pigs was found after the 5th injection. The antigen detected by the use of this serum was named

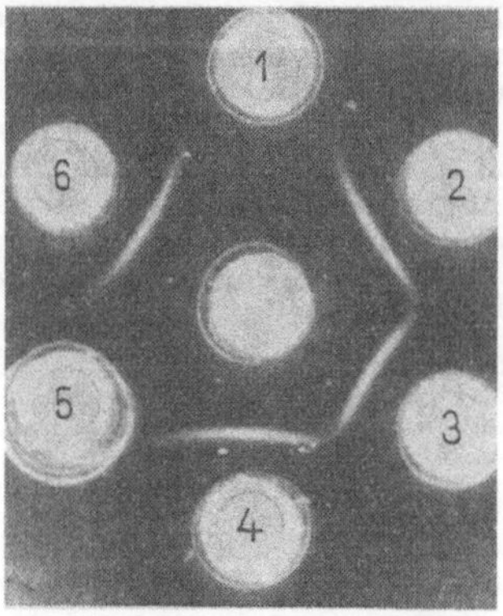
1
2
3
4
5
6

FIG. 1. Precipitation reactions observed between immune serum anti-L (central reservoir) and positive sera of normal pigs (2, 3, 4, 6)

L. Precipitation, reactions of this serum are shown in Fig. 1. This antigen was also identified in the serum of pig No. CH-2/6 after immunization. The third pig, the serum of which contained isoprecipitins was pig No. CH-2/3. In the remaining 7 pigs no isoprecipitins were detected. The serum of pig CH-2/3 contained the antigen named G.

The precipitation line of antigen G was, nevertheless, much weaker than that of antigen L. It was anticipated that reimmunization of the pig CH-2/3 would cause an intensification of the reaction but, reimmunization performed after 3 months failed to confirm this assumption. The precipitation lines of antigen L stained very distinctly with Sudan Black but weakly with Amido Black, while those of antigen G stained weakly with Amido Black and did not stain with Sudan Black.

The results of immunoelectrophoresis and staining indicate that the L antigen is a lipoprotein. The immunoelectrophoresis of the G antigen proved that after electrophoretic separation this antigen located in the area occupied by $\alpha$-globulines.

More than 1000Polish Landrace and Polish Large White pigs were tested with anti-L and anti-G reagents. The frequency of occurence of the L antigen was 90% in Polish Large White and 99.3% in Polish Landrace pigs, while the frequency of G antigen was 9.3% and 40% respectively.

Both of the allotypes described were of hereditary character. Each of them could be detected in the offspring only in that case when it was present in at least one of the parents.

## REFERENCES


Dray, S. and Johns, G. O. (1958), Differences in the antigenic components of sera of individual rabbits as shown by induced isoprecipitin, *J. Immunol.*, **81.** 142–149.

Dubiski, S., Dudziak, Z., Skałba, D. and Dubiska, A. (1959), Serum groups in rabbits. *Immunology*, **2.** 84–92.

Oudin, J. (1956), Specific precipitation reaction between sera from animals of the same species. (Réaction de précipitation spécifique entre des sérums d'animaux de même espèce). *C. R. Acad. Sci. Paris*, **242.** 2489–2490.

Lang, B. G. (1970), Globulin allotyping in pigs using iso-precipitins., Proc. XIth Europ. Conf. Anim. Blood Groups Biochem. Polymorph. Warsaw, 1968, 301–306.

Rapacz, J. and Hasler, J. (1970a), Allotypes (serum antigens) in farm animals. Proc. XIth Europ. Conf. Anim. Blood Groups and Biochem. Polymorph. Warsaw, 1968, 101–105.

Rapacz, J., Korda, N. and Stone, W. H. (1968), Serum antigens of cattle I. Immunogenetics of A macroglobulin allotype. *Genetics*, **58.** 387–398.

Rapacz, J., Grummer, R. H., Hasler, J. and Shackelford, (1970b), Allotype polymorphism of low density B-lipoproteins in pig serum. (LDLpp 1, LDLpp 2). *Nature*, **225.** 941.

Skałba, D. (1964), Allotypes of hen serum proteins, *Nature*, **204.** 894.

Skałba, D. (1966), Antigenic differences of hen serum proteins detected by anti-allotypic immune sera. Proc. Xth Europ. Conf. Anim. Blood Groups Biochem. Polymorph. Paris, 1966, 477–480.

Spooner, R. L. (1970), Studies on $I_{gG}$ allotypes in cattle, Proc. XIth Europ. Conf. Anim. Blood Groups Biochem. Polymorph. Warsaw, 1968, 195–199.

Tikhonov, V. N., Valdman, S. M. and Savina, M. A. (1970), Immunogenetic study of allotypes of pig serum protein using isoprecipitins, Proc. XIth Europ. Conf. Anim. Blood Groups Biochem. Polymorph., Warsaw, 1968, 307–309.

*XIIth Europ. Conf. Anim. Blood Groups Biochem. Polymorph., Bp., 1972 (pp. 289–292)*

# MATERNAL ANTI-Eb IN THE SERUM OF NEWBORN PI GLETS WITH OR WITHOUT OVERT HEMOLYTIC DISEASE AS A POSSIBLE SOURCE OF RED CELL TYPING ERRORS

**JOAN HARDY and SANDRA SHAW**

Protein Chemistry Section, Unilever Research Laboratory, Colworth House, Sharnbrook, Bedford, England

## INTRODUCTION

A series of potentially incompatible matings for Eb have been examined. They showed an initial screening:

1. An increase in total mean litter size at birth from 1st–3rd litters.
2. A gradually increasing piglet death rate with subsequent incompatible matings.
3. A tendency for the final litter size to drop because of deaths.
4. An increase in rejection rate. This last could have been due to increased selectiveness in the Breeding Station but should still affect all types of piglets equally.

All these sows had been given an annual injection of crystal violet swine fever vaccine but as they were also mated with an Eb positive boar on each occasion. LINKLATER's (1968) evidence that the titre of a maternal antibody can be increased after mating with a boar possessing the corresponding antigen is not irrelevant.

These findings were suggestive but not statistically conclusive and are not of immediate concern here. If 2–4 were due to hemolytic disease of the newborn one would naturally expect piglets with the paternal factor, foreign to the sow to be affected rather than piglets with the same blood type as the sow. In fact it was found that this was the case in some individual matings but the reverse was often also present and overall the piglets were affected about equally.

The next query was "Do piglet's red cells type correctly when they have been in a serum containing an antibody to one or more factors on their red cells?" It is known that maternal anti-rhesus can affect rhesus typing of an infant's red cells. This seemed an important point to establish and the experiment described below was performed.

a) We had available, in bulk, serum from a gilt which contained anti Eb only. We believe this to be a "natural" antibody i.e. not due to immunization, or to passive transfer from the mother sow because:

i) The animal was over five months old.

ii) It had never been pregnant or immunised in any way. It was a slaughter house specimen, the gilt having been on a standard feeding trial.

This antibody gave a clear anti-Eb pattern in direct agglutination test whenever tested against a panel of cells of known blood type. The titre was 1 : 2–1 : 3.

b) Red cells of the following types were also available:
E dbg dbg
E dbg deg and E dbg aeg
E deg deg
c) Routine high titre anti-Eb.
d) Routine anti-Ee.

## METHOD

a) The antiserum with natural anti-Eb was diluted as follows:

1) 1 : 1 (neat), (2) 1 : 2, (3) 1 : 4

b) Suspensions of the four red cell samples in anticoagulant (citrate) were washed three times with normal (0.92%) saline. After the final spin the supernatant liquid was removed and "packed" red cells left.

c) A volume of .5 ml packed red cells +5 ml of antiserum was incubated for 1 hour at 32°C. A second incubation was set up for 2 hours.

d) The incubated cells were removed washed three times and resuspended in saline to form a 2% suspension.

e) This suspension was tested with

i) anti-Eb and anti-Ee routine at normal working dilutions
ii) anti-Eb (routine) neat 1 : 1

Untreated red cell samples were typed as controls at the same time.

It was hoped that these conditions would simulate those to which piglets red cells might be subjected when mixed with a maternally derived antibody in vivo and then typed routinely in the laboratory.

## RESULTS

These are given in detail in Table 1 and can be summarised as follows:

a) No visible agglutination was observed when washing the incubated samples.

b) The treated Eee cells agglutinated with anti-Ee and did not agglutinate with anti-Eb, i.e. their reactions were like those of the control. This was the case after incubation with all dilutions of the anti-Eb (natural).

c) The treated Eeb cells agglutinated fairly normally with anti-Ee but did not agglutinate strongly or in some case at all with anti-Eb at the normal working dilution. When agglutination occurred, in some instances it was so weak that it could be missed in normal routine typing conditions. Using the anti-Eb 1 : 1 (neat) (its working dilution is 1 : 4) a stronger agglutination could be produced but even then the treated cells reacted much less strongly than did the control cells.

d) A similar but less marked effect was noticed with Ebb cells.

e) The diminution in agglutination was more marked after 2 hours incubation than after 1 hour.

f) The most marked inhibiting effect was obtained with natural anti-Eb diluted 1 : 4 and two hours incubation.

TABLE 1

*Effect of incubation with natural anti Eb on typing of Eee, Eeb and Ebb red cells*

| Routine anti: | Incubated cells | | | | | | | | | | | | | | | | | | Normal cells | | |
|---|---|---|---|---|---|---|---|---|---|---|---|---|---|---|---|---|---|---|---|---|---|
| | 1 hour's incubation with anti-Eb at: | | | | | | | | | 2 hour's incubation with anti-Eb at: | | | | | | | | | | | |
| | 1:1 | | | 1:2 | | | 1:4 | | | 1:1 | | | 1:2 | | | 1:4 | | | | | |
| | Ee | Eb | | Ee | Eb | | Ee | Eb | | Ee | Eb | | Ee | Eb | | Ee | Eb | | Ee | Eb | |
| | 1:3 | 1:1 | 1:4 | 1:3 | 1:1 | 1:4 | 1:3 | 1:1 | 1:4 | 1:3 | 1:1 | 1:4 | 1:3 | 1:1 | 1:4 | 1:3 | 1:1 | 1:4 | 1:3 | 1:1 | 1:4 |
| Cells: | | | | | | | | | | | | | | | | | | | | | |
| Edegdeg | 4 | . | . | 3 | . | . | 3 | . | . | 3 | . | . | 3 | . | . | 4 | . | . (Eee) | 4 i.e. | . (Eee) | . |
| Edbgdeg | 2 | 3 | 2 | 2 | 2 | 2 | 2 | 1 | 1 | 3 | 2 | 1 | 3 | 2 | 1 | 4 | 2 | + Ee?b | 2 | 4 (Eeb) | 4 |
| Eaegdbg | 3 | 3 | 1 | 3 | 3 | 1 | 3 | 2 | 1 | 3 | 2 | 2 | 2 | 2 | + | 3 | 1 | . (Eee) | 3 | 4 (Eeb) | 3 |
| Edbgdbg | . | 2 | 2 | . | 3 | 2 | . | 3 | 1 | . | 3 | 2 | . | 2 | 2 | . | 1 | 1 (Ebb?) | . | 4 (Ebb) | 3 |
| *With Bromelin:* | | | | | | | | | | | | | | | | | | | | | |
| Edegdeg | 4 | . | . | 3 | . | . | 3 | . | . | 4 | . | . | 2 | . | . | 3 | . | . (Eee) | 4 | . | . |
| Edbgdeg | 2 | 3 | 2 | 2 | 3 | 1 | 2 | 3 | + | 3 | 2 | 1 | 2 | 2 | 1 | 2 | 1 | + (Ee?b) | 2 | 4 | 4 |
| Eaegdbg | 3 | 3 | 1 | 2 | 2 | 1 | 3 | 2 | + | 2 | 1 | 1 | 2 | 1 | + | 2 | 1 | . (Eee) | 3 | 4 | 3 |
| Edbgdbg | . | 3 | 1 | . | 3 | 3 | . | 3 | 2 | . | 4 | 2 | . | 2 | 1 | . | 2 | + (Ebb?) | . | 3 | 3 |

Probable routine typing result given in last column

*Normal working dilution of anti-Eb (routine typing) is 1 : 4*

4 = Complete agglutination. No free cells.
3 = Moderate clumps of agglutinated cells. Some free cells.
2 = Finer aggregations of cells. Many free cells.
1 = Very fine agglutination. Mostly free cells.
\+ = Mostly free cells. Very doubtful.
? = Fine agglutination. Probably negative.
. = No agglutination. Negative.

## DISCUSSION AND CONCLUSIONS

Mistyping of red cells from piglets of the genotype Eeb could occur if these piglets' blood contained maternal anti-Eb. Such piglets would be recorded as Eee.

It is worth noting that the antibody titre would not have to be very high to produce this effect as we obtained it at a dilution of 1 : 4 — a dilution at which the natural anti-Eb used would not produce visible agglutination of Eb+ red cells. We assumed that the maximal inhibiting effect which was obtained at this dilution was due to the fact that at this strength the antibody could not agglutinate cells (visibly or otherwise) and could only 'coat' the cells by becoming attached to the Eb factor sites on the erythrocyte.

This "blocking" action is well recorded in human red cell typing but usually in relation to indirect acting ("incomplete") antibodies which normally do not produce agglutination in vitro unless antiglobulin is added (Coombs test).

Maternal antibodies have been shown to persist in piglet's blood up to 20 weeks after birth (Linklater, 1968). They could therefore be a source of typing errors in routine red cell typing of piglets at three weeks or even later and could certainly affect typing of piglets at birth.

Finally if it is accepted that red cell antibodies occur in a proportion of sows and this proportion is increased by crystal violet immunization for swine fever and by pregnancy the passage of these antibodies into the suckling piglets serum must be a fairly frequent occurrence. We have so far only demonstrated that natural anti-Eb can act as a blocking antibody and prevent the correct typing of Eeb cells but it would seem probable that other antibodies may act in the same way and we propose to test this theory further with other red cell factors and antibodies.

## REFERENCES

Bruner, D. W., Brown, R. G., Hull, F. E. and Kinkaid, Alice (1949), Blood Factors and Baby Pig Anaemia. *J. Amer. Vet. med. Ass.*, **115.** 94–96.

Goodwin, R. F. W., Coombs, R. R. A. and Saison, Ruth (1955–56), The Blood Group of the Pig I–IV all in *J. Comp. Path.* In particular II Red cell iso-antibodies in the sera of pigs injected with crystal violet swine fever vaccine. *J. Comp. Path.*, **65.** 72–92, and IV. The A antigen-antibody system and hemolytic disease in newborn piglets. *J. Comp. Path.*, **66.** 317-331.

Linklater, K. A. (1968), Isoimmunisation of the parturient sow by foetal red cells. *Vet. Rec.*, **83.** 2034.

## DISCUSSION

K. A. Linklater: In routine typing this phenomenon would presumably show up on the saline controls for the indirect sensitisation test.

In my work I detected anti-Ea in Ea negative piglets up to twenty weeks of age but in Ea positive piglets the anti-Ea disappeared within the first three days after birth and was therefore not a source of error in typing the piglets at an older age.

Joan Hardy: The direct acting antiglobulin test would in fact indicate that the piglet red cells were coated with maternal antibody.

It is agreed that normally maternal antibody would disappear rapidly in a piglet with the corresponding antigen so that typing errors at three weeks could occur. However the error would be possible in piglets typed at birth, and in dead piglets.

*XIIth Europ. Conf. Anim. Blood Groups Biochem. Polymorph., Bp., 1972 (pp. 293–297)*

# E BLOOD GROUP SYSTEM IN MINIATURE PIGS

J. Hradecký and J. Hojný

Laboratory of Animal Genetics, Czechoslovak Academy of Sciences, Liběchov, Czechoslovakia

## SUMMARY

New antibodies of the E blood group system of pigs were prepared by immunization of miniature pigs. Blood factor $CZ_{20}$ (Ek) is in *contrast* to Ea, but also to Ej. Genetically it is determined by alleles $E^{bdg(k)}$, $E^{edgh(k)}$, $E^{edfh(k)}$ and probably also by allele $E^{bdf(k)}$.

Blood factor $CZ_{21}$ (El) is a subgroup of factor Ea. Its presence on erythrocytes is determined by allele $E^{aeg(l)}$ and apparently also by allele $E^{aegi(l)}$. In pigs of the Large White, Landrance and Cornwall breeds, it reacts identically with antibody Ea.

The case of reagent Ed proves that standard antibodies prepared from sera of other pig breeds may be non-specific for miniature pigs.

## INTRODUCTION

Up to now nine blood factors have been disclosed in the E system as follows: Ea, Eb, Ed, Ee, Ef, Eg, Eh, Ei, Ej determined by alleles $E^{bdg}$, $E^{edgh}$, $E^{aeg}$, $E^{edfh}$, $E^{bdf}$, $E^{aef}$, $E^{eg}$, $E^{aegi}$ and $E^{edghj}$. The first five alleles were described by Andresen (1962). He found that blood factors Ea-Ed, Eb-Ee and Ef-Eg are contrasts and form closed subsystems. The sixth allele was detected in Yorkshire pigs by Rasmusen (1965). Andresen (1965) described allele $E^{eg}$ in pigs of the Hampshire breed. Blood factor Eh determined by alleles $E^{edgh}$ and $E^{edfh}$ was disclosed by Hojný et al. (1966). The next two alleles $E^{aegi}$ and $E^{edghj}$ were detected by Dinklage and Major (1968) and Dinklage et al. (1969).

Schahmirzadi (1967) studied blood groups in miniature pigs and published the results in his thesis.

## MATERIAL AND METHODS

The new antibodies described in this report were obtained from miniature pigs immunized with erythrocytes (in one case lymphocytes) of donors, selected according to blood groups within and outside this breed. Serological verification of antibodies and genetic studies were carried out on mini pigs and also on the pigs of Landrace, Large White and Cornwall breeds, and their crosses.

Immunizations, preparations of specific reagents and classification of new factors into genetic systems were performed according to Hojný et al. (1966). The immunizing dose of lymphocytes obtained from lymph nodes of a killed animal, consisted of 2 ml leucocyte suspension at a concentration of $75 \times 10^{-6}$ (maximal RBC contamination 2%) and 2 ml of complete Freund's adjuvant. Injections were administrated at 7-day intervals. Repeated blood collections were carried out on the 14th day after the 3rd injection.

## RESULTS

Antibody $CZ_{20}$ (Ek) was prepared from the serum of boar, Z 84, reimmunized with lymphocytes from the same donor, S 194, of the Large White breed. The raw serum contained, besides antilymphocytic cytotoxic antibodies, complete agglutinins, reacting at a titre of 1/16–1/32 with erythrocytes (further on RBC) of a large number of pigs from all breeds. After its verification the new reagent was added to the standard antibodies and used for current testing. So far 520 pigs of different breeds were tested from which 494 were positive and 26 were negative. Negative were only Ea and Ej homozygous animals and heterozygotes $E^{aeg}/E^{edghj}$. Heredity of this factor was studied on families by means of double backcross mating (Table 1). The results confirm the belonging of $CZ_{20}$ (Ek) factor to the E-system and prove that it is determined by the complex alleles $E^{bdg(k)}$, $E^{edfh(k)}$ and $E^{edgh(k)}$. The presumed allele $E^{bdf(k)}$ could not yet be verified because it was not found in the set of tested animals.

TABLE 1

*Segregation of alleles $E^{bdg(k)}$, $E^{edgh(k)}$, $E^{edfh(k)}$ in selected types of double backcross matings*

| Mating type | Number of offspring | Reactivity with $CZ_{20}$ (Ek) + | | − | |
|---|---|---|---|---|---|
| | | $E^{bdg}$, $E^{edgh}$, $E^{edfh}$ | | | |
| | | + | − | + | − |
| $CZ_{20}$/—, $E^{bdg}$/— × —/—, —/— | 36 | 14 | 0 | 0 | 22 |
| $CZ_{20}$/—, $E^{edgh}$/— × —/—, —/— | 15 | 9 | 0 | 0 | 6 |
| $CZ_{20}$/—, $E^{edfh}$/— × —/—, —/— | 24 | 14 | 0 | 0 | 10 |

TABLE 2

*Segregation of alleles $E^{aeg(l)}$ in selected types of double backcross matings*

| Mating type | Number of offspring | Reactivity with $CZ_{21}$ (El) + | | − | |
|---|---|---|---|---|---|
| | | $E^{aeg}$ | | | |
| | | + | − | + | − |
| $CZ_{21}$/— , $E^{aeg}$/— × —/—, —/— | 63 | 34 | 0 | 0 | 29 |

Another reagent was obtained from the serum of sow Z 6, immunized with RBC from sow 138/19/2 (Landrace). The formation of antibodies anti-A, anti-$CZ_{10}$ (Jb) was merely expected. The analysis, however, showed three antibodies: anti-A, incomplete anti-$CZ_{10}$ and a new, unknown antibody designated as $CZ_{21}$. This antibody reacted in the direct agglutination test at a titre of 1/4–1/8. At parallel comparisons with the other reagents in 525 pigs of all breeds but miniature pigs, it appeared to be identical with Ea, although it was produced by a Ea-positive animal of $E^{aeg}/E^{edghj}$ genotype. In miniature pigs the result was different. From 167 tested Ea-positive RBC samples, 120 were positive also with $CZ_{21}$. The belonging of $CZ_{21}$ to the E system was reliably confirmed on 10 litters with 63 offspring, by means of double backcross matings (Table 2). These results prove that blood factor $CZ_{21}$ (El) is determined by allele $E^{aeg(l)}$. Orientative comparison with antibody $E_l$ which was kindly supplied by Dr. Dinklage, suggested that in miniature pigs there are three alleles ($E^{aeg}$, $E^{aeg(l)}$, $E^{aegi(l)}$).

When boar Z2 was immunized, we expected the formation of anti-Lh, Lf, Lj and Lk antibodies. The genotypes of donor and recipient were supposedly identical – $E^{aeg}/E^{edgh(k)}$. The serum, however, contained a complete agglutinin at a titre of 1/32 which had the same reaction with RBCs from other pig breeds as Ed. In miniature pigs, however, it reacted with a smaller number of RBCs than standard Ed reagents. For working purposes this antibody was designated $Ed_1$. Together with anti-$Ed_1$ of miniature pig RBC, we have also tested several Ed reagents from pigs of the Large White and Landrace breeds. The RBC of Ed-positive, $Ed_1$-negative pigs considerably weakened Ed reagents (e.g. from a titre of 1/32 to 1/2), but anti-Ed was not fully absorbed even by five consecutive absorptions (Table 4). From matings Ed + $Ed_1$ × Ed + $Ed_1$ all animals were always Ea-positive, which gave further evidence for the fact that anti-Ed is non-specific for RBC of miniature pigs without additional absorption.

TABLE 3

*An example of occurrence of new alleles $E^{aegl}$, $E^{bdgk}$ in offsprings from family Z12 (miniature pigs)*

| Pig No | Phenotypes | | | | | | | | | | | Genotypes |
|---|---|---|---|---|---|---|---|---|---|---|---|---|
| | Ea | Eb | Ed | Ee | Ef | Eg | Eh | Ei | Ej | Ek | El | |
| ♂ Z45 | + | + | + | + | . | + | . | . | . | + | . | aeg/bdgk |
| ♀ Z12 | + | . | . | + | . | + | . | . | . | . | + | aeg/aegl |
| 201 | + | . | . | + | . | + | . | . | . | . | . | aeg/aeg |
| 202 | + | + | + | + | . | + | . | . | . | + | . | aeg/bdgk |
| 203 | + | . | . | + | . | + | . | . | . | . | . | aeg/aeg |
| 204 | + | + | + | + | . | + | . | . | . | + | + | aegl/bdgk |
| 205 | + | + | + | + | . | + | . | . | . | + | . | aeg/bdgk |
| 206 | + | . | . | + | . | + | . | . | . | . | + | aegl/aeg |
| 207 | + | . | . | + | . | + | . | . | . | . | + | aegl/aeg |

TABLE 4

*Results of 1 to 5 times repeated absorption of antibodies Ed and $Ed_1$ with RBC from miniature pigs of types Ed+, $Ed_1$−*

| Reagent | Titre before absorption with RBC | | Titre after absorption with RBC Ed+, $Ed_1$− | | | | |
|---|---|---|---|---|---|---|---|
| | (Ed+ $Ed_1$+) | (Ed+ $Ed_1$−) | 1× | 2× | 3× | 4× | 5× |
| | | | RBC (Ed+ $Ed_1$+) | | | | |
| Ed (81/6) | 1/32 | 1/16 | 1/4 | 1/4 | 1/2 | 1/2 | 1/2 |
| $Ed_1$ ($Z_2$) | 1/64 | — | 1/64 | 1/64 | 1/64 | 1/32 | 1/32 |

To complete our results, we state the allele frequencies of the E-system in 378 miniature pigs from our herd: $E^{aeg} = 0.335$; $E^{aegl} = 0.171$; $E^{bdgk} = 0.148$; $E^{edghk} = 0.189$; $E^{edghj} = 0.157$. We should like to emphasize, however, that these frequencies characterize exclusively our herd of miniature pigs originating from several imported individuals.

## DISCUSSION AND CONCLUSIONS

The aim of immunizing pigs with lymphocytes was to prepare antilymphocyte antibodies which are discussed in another paper. The formation of antibodies reacting in the direct agglutination tests with RBC surprised us at first, particularly when the first analyzed serum contained an antibody, disclosing a new blood factor. The question, however, is, what impulse caused the formation of Ek, *viz.* whether the erythrocytes present at insignificant concentration in the immunizing dose or the lymphocyte antigens were responsible. RBC contamination in the lymphocyte suspension was negligible in comparison with the currently used immunizing dose. We are rather inclined to accept the second alternative which supports the hypothesis of HÁLA (1967) and also the results of SIMON and HOJNÝ (1970) who consider the E system factors as common to both erythrocytes and lymphocytes.

The conclusion arrived at by means of Ed antibody formed in the miniature pig and by means of verifying the standard Ed reagent with miniature pig erythrocytes, is an objective phenomenon often encountered in blood group serology of pigs. Monospecific reagents need not to be monovalent for the determination of corresponding blood factors in other breeds, especially if they are phylogenetically distant. Thus it is possible to explain the differences which occurred in the determination of the Ed factor at the Comparison Test (Copenhagen 1966) and the Reference Test (Liběchov 1970). In all cases at issue the blood samples came from miniature pigs.

The detection of blood factors Ei and Ej (DINKLAGE and MAJOR, 1968; DINKLAGE et al., 1969) and factors Ek, El described in the present report means that the E system becomes again the most complex system of blood groups in pigs as regards the number of factors and alleles.

## REFERENCES

ANDRESEN, E. (1962), Blood groups in pigs. *Ann. N. Y. Acad. Sci.*, **97.** 205–225.

ANDRESEN E. (1963), A Study of Blood Groups of the Pig. Munksgaard, Copenhagen.

ANDRESEN, E. (1965), Minus-minus (—/—) Ea Ed phenotypes and a new allele $E^{eg}$ (= $E^7$) in pigs of the Hampshire breed. *Vox Sang.* **10.** 738–741.

DINKLAGE, H. and MAJOR, F. (1968), $E^{aegi}$ (= $E^8$), a new allele in the E blood-group system of the pig. *Vox Sang.* **14.** 315–317.

DINKLAGE, H., SCHAHMIRZADI, H., HRADECKÝ, J. and HOJNÝ, J. (1969), $E^{edghj}$ (= $E^9$), a new allele in the E blood-group system of the pig. *Vox Sang.*, **17.** 129–133.

HÁLA, K. (1967), The occurrence of erythrocyte antigens on cells from various tissues of the pig. *Folia Biol.* (Praha) **13.** 190–192.

HOJNÝ, J. and HRADECKÝ, J. (1970), Dextran test in the study of blood groups in pigs. XIth Europ. Conf. Anim. Blood Groups and Biochem. Polymorph., Warsaw, 1968, 259–264.

HOJNÝ, J., GAVALIER, M., HRADECKÝ, J. and LINHART, J. (1966), New blood factors in pigs. In: Proc. Xth Europ. Conf. Anim. Blood Groups Biochem. Polymorph., Paris, 1966, 151–158.

RASMUSEN, B. A. (1965), $E^{aef}$ ($E^6$), a sixth allele at the E blood group locus in Yorkshire pigs. *Vox Sang.*, **10.** 242–245.

SCHAHMIRZADI, H. (1967), Blutgruppenfaktoren beim Göttinger Zwergschwein. *Diss.* Göttingen.

SIMON, M. and HOJNÝ, J. (1972), A study on lymphocyte antigens in pigs by means of anti-erythrocyte reagents. XIIth Europ. Conf. Anim. Blood Groups Biochem. Polymorph., Budapest, 1970.

## DISCUSSION

H. DINKLAGE: In which breeds did you find the new factors?

J. HRADECZKÝ: The new factors Ek and El occurred also in other pigs (Large White, Landrace and Cornwall).

*XIIth Europ. Conf. Anim. Blood Groups Biochem. Polymorph., Bp., 1972 (pp. 299—303)*

# A CONTRIBUTION TO THE STUDY ON H, J AND M BLOOD GROUP SYSTEMS IN PIGS

J. HOJNÝ and J. HRADECKÝ

Laboratory of Animal Genetics, Czechoslovak Academy of Sciences, Liběchov, Czechoslovakia

## SUMMARY

**Three new blood factors of pigs, detected by isoimmune antibodies, are described. Genetic investigation of these factors has shown that they belong to the H, J and M system, respectively.**

**From the serum of a miniature sow died on hemorrhagic diathesis during immunizations, a hemolysin was obtained which was related to reagent Hc. After serological verification it was designated as anti-Hd. Factor Hd is genetically determined by alleles $H^{cd}$ and $H^{bd}$. Allele $H^{bd}$ occurs in miniature pigs.**

**By immunization within the inbred Landrace line an antibody was gained which discloses a new factor $CZ_{10}$(Jb). It forms, within this line, a closed system with factor Ja. This finding, however, does not apply in general and the J-system with alleles $J^a$, $J^b$ and $J^-$ remains open.**

**The M-system was extended by blood factor Mg which is contrasting to Ma. Its presence on erythrocytes is determined by allele $M^{dg}$. This allele occurs in miniature pigs of Göttingen origin. A so far non-described allele $M^{ad}$ was proved in miniature pigs, too.**

## INTRODUCTION

During the last years a considerable progress has been made in the study of pig blood groups. More than 60 blood factors determined by allelic genes from 16 blood group loci are known.

The present report offers some new information on systems H, J and M.

The factors Ha, Hb, Hc, and alleles $H^a$, $H^b$, $H^{ab}$, $H^c$ and $H^-$ in the H-system have been described (ANDRESEN, 1957; ANDRESEN and WROBLEWSKI, 1961; NIELSEN, 1962; ANDRESEN, 1964). Antigen $CZ_9$, classified in the H-system under the designation Hd (HOJNÝ et al., 1966) proved to be identical with Hc (HOJNÝ et al., 1967).

The J-system, discovered already in 1957, is still known as an open system with factor Ja and alleles $J^a$ and $J^-$ (ANDRESEN, 1957).

The M-system comprises the blood factors Ma, Mb, Mc, Md, Me and Mf (NIELSEN, 1961; BRUMMERSTEDT-HANSEN et al., 1962; HOJNÝ and HRADECKÝ, 1968; DINKLAGE and MAJOR, 1969). According to the authors these factors are genetically determined by at least 8 different alleles.

## MATERIAL AND METHODS

Immunizations, testing of antibodies and genetic investigations were carried out on a highly inbred line of the Landrace breed (imported from Canada) bred through more or less closely related matings since 1962, and on miniature pigs (imported from GFR, Göttingen and the United States, Minnesota). Blood samples from other breeds and samples obtained from the slaughter-houses were also tested.

The immunization procedure was the same as described in our previous paper (Hojný et al., 1966), new antibodies and standard reagents were used in one of the four current serological tests (Hojný and Hradecký, 1968).

Classification of blood factors into systems was carried out on the basis of family studies of selected matings.

## RESULTS

*H system.* The sow Z 24 (miniature, pig) was reimmunized after a short interval of only one month with erythrocytes (further only RBC) from the same donor 314/16 (Landrace). Six days after the second injection she died. In the serum which we managed to obtain, three from the five expected antibodies were proved, namely E(l), J(b) and Hc. The Hc antibody lysed every Hc-positive RBC of all tested breeds of pigs. In miniature pigs it reacted with and was completely absorbed by some other samples. Under the designation Hd it was used together with Ha, Hb and Hc for testing parents and offspring. A total of 62 litters with 388 offspring were tested. Table 1 shows only 21 selected types of backcross matings proving the existence of alleles $H^{cd}$ and $H^{bd}$. The first row of the Table giving proof for allele $H^{cd}$ comprises families of the Large White, Landrace and Cornwall breeds while the other rows miniature pigs only.

From a total of 234 miniature pigs tested 77 animals, (32.9%) had phenogroup Hbd.

Table 1

*Inheritance of phenogroups Hcd and Hbd in selected types of mating in 21 families and 173 offspring*

| Mating type | Number of families | Offspring | |
|---|---|---|---|
| Hcd/— × H—/— | 9 | 44 Hcd/— | 50 H—/— |
| Ha/bd × H—/— | 1 | 4 Hbd/— | 3 Ha/— |
| Ha/bd × Hb/— | 5 | 15 Hbd/(b)(c—) | 6 Ha/b<br>6 Ha/— |
| Hab/bd × Hb/— | 5 | 19 Hbd/(b)(c—) | 14 Hab/(b)(—) |
| Hab/bd × Ha/— | 1 | 4 Hbd/a<br>1 Hbd/— | 7 Hab/(b)(—) |

*J system.* A weak, incomplete antibody designated as $CZ_{10}$ was first formed by the boar 83/6 immunized with RBC 81/9 (HOJNÝ, 1965). A sound reagent was obtained much later from the sow 134/13/8 by planned reimmunization with RBC 134/13/3 (all Landrace). After absorption of anti-A antibody the reagent should have been monospecific. Nevertheless, just like the original anti-$CZ_{10}$, it reacted weakly also with Ca-positive RBC, even when both donors proved to be Ca-negative. This reagent was really specific after a single absorption with Ca-positive RBC. Genetic classification of the factor $CZ_{10}$ was hindered by its rare occurrence, outside of our Landrace herd, (e.g. from 519 blood samples collected at the slaughter-house, only 20 (3.8%) were $CZ_{10}$ positive), and by the fact that the reaction of RBC from piglets with $CZ_{10}$ was unreliable. Our attention was directed to our herd, and young pigs were tested and retested at the age of over two months. Gradually collected data have shown the $CZ_{10}$ factor to belong to the J-system. Study of 39 families with 319 offspring, divided according to mating types into 6 groups, proved the supposition that within our semi-inbred population the factors Ja and $CZ_{10}$(Jb) formed a closed system (Table 2). This does not apply to Large White, Cornwall and miniature pig breeds in which the Jb factor has not yet been found. According to a preliminary investigation of a total of 93 Landrace pigs from 3 farms factor Jb was found in 6 animals (6.4%), 15 pigs were both Ja and Jb negative, i.e. of genotype $J^-/J^-$.

*M system.* The serum from the boar Z 69 reimmunized with RBC from the boar Z 2 (both miniature pigs) yielded after complicated absorptions an antibody against an unknown antigen. The new factor found inminiature pigs only was present together with factor Md. All available M reagents (Ma, Mc, Md), including the new antibody, were used to test 33 families with 191 offspring. Nine selected families of backcross mating are presented as evidence that the new blood factor, designated Mg, is exclusive of Ma and is inherited together with factor Md through the newly detected allele $M^{dg}$ (Table 3). Besides proving the existence of allele $M^{dg}$ these selected litters also provide evidence for another so far undescribed allele $M^{ad}$. (Before designating the new factor with the symbol Mg, we have first demonstrated the absence

TABLE 2

*Inheritance of blood factors Ja and Jb($CZ_{10}$) in the inbred Landrace line of pigs in 39 families and 319 offspring*

| Mating type | Number of families | Offspring phenotypes | | | $\chi^2$ | d.f. |
|---|---|---|---|---|---|---|
| | | Ja | JaJb | Jb | | |
| Ja×Ja | 6 | 54(54) | — | — | — | — |
| Ja×JaJb | 7 | 23(25) | 27(25) | — | 0.3 | 1 |
| Ja×Jb | 5 | — | 42(42) | — | — | — |
| JaJb×JaJb | 7 | 10(12) | 25(24) | 13(12) | 0.4 | 2 |
| JaJb×Jb | 11 | — | 53(50.5) | 48(50.5) | 0.2 | 1 |
| Jb×Jb | 3 | — | — | 24(24) | — | — |

TABLE 3

*Inheritance of phenogroups Mdg and Mad in selected types of mating in 9 families and 49 offspring*

| Mating type | Number of families | Offspring | |
|---|---|---|---|
| Mad/dg×M–/– | 5 | 15 Mad/– | 12 Mdg/– |
| Mad/dg×Md/– | 1 | 2 Mad/(d)(–) | 2 Mdg/(d)(–) |
| Mad/dg×Md/d | 1 | 3 Mad/d | 2 Mdg/d |
| Mdg/d×M–/– | 2 | 7 Mdg/– | 6 Md/– |

of the possible agreement between the factors Me-Md and Mf-Mg. The anti-Me and Mf reagents were kindly supplied by Dr. Dinklage.)

Out of a total of 234 miniature pigs 61 (i.e. 26%) were Mg-positive.

## DISCUSSION AND CONCLUSIONS

We have never seen a pig killed as a result of intramuscular RBC immunization and first we refused to admit this reason for the pig's death. Such a possibility had to be considered, however, when under the same circumstances 3 more pigs died of hemorrhagic diathesis. The sera always contained strong antibodies against RBC donors. The considerable decrease of RBC was apparent but the antibodies did not react with the homologous erythrocytes in vitro. Immunization procedures differed from those used for other pig breeds, only in relatively higher doses of antigens. During the next immunizations the dose was reduced from 6 ml of 50% RBC suspension to a half of the amount and during reimmunizations the incomplete adjuvant was omitted. These measures have been effective so far.

The sensitivity of miniature pigs towards the formation of antibodies may be regarded as one of the reasons for the pigs' death. In general antibodies reach higher titres and as compared to the other breeds they are more frequently of hemolytic character. For this reason miniature pigs are suitable for the production of blood group reagents. Using them as donors and recipients several new factors have been discovered (SCHAHMIRZADI, 1967; DINKLAGE and MAJOR, 1968; HRADECKÝ and HOJNÝ, 1970). Some of these factors including also Hd and Mg seem to be of limited value for other pig breeds.

Gene frequencies of the new alleles were not calculated because the herd of miniature pigs resulting from matings of a limited number of individuals cannot be regarded as representative. The same goes for the tested pigs of the Landrace breed.

The reactions anti-$CZ_{10}$(Jb) with Ca-positive RBC at first suggested that the new factor would belong to the C system. This serological curiosity drew our attention indirectly also to the J system, which is known to be in close linkage with the C system (ANDRESEN and BAKER, 1964). The occurrence of Jb factor in our Landrace breed facilitated the classification of the rare factor $CZ_{10}$ into the J system. The shifting of frequency in favour of this factor was caused by the boar 81/4 who was Jb-homozygous. It was also found that allele $J^-$ was eliminated from this population after two generations of inbreeding.

The interpretation of the results obtained in the study of M system in pigs presents certain difficulties, because all M-antibodies, have never been tested together and those which had been compared often differed from each other. This is given by the character of the M-antibodies which, perhaps with the exception of Ma, form agglutinates difficult to judge. Some of them, e.g. Md and Mc, like Jb, caused difficulties when piglets are tested. The solution of the problems concerning the number of alleles in the M system requires international cooperation.

## REFERENCES

ANDRESEN, E. (1957), Investigations on blood groups of the pig. *Nord. Vet. Med.* **9.** 274–284.

ANDRESEN, E. (1964), Further studies on the H blood group system in pigs with special reference to a new red cell antigen Hc. *Acta Genet.* (Basel) **14.** 319–326.

ANDRESEN, E. and WROBLEWSKI, A. (1961), The G and H blood group systems of the pig. *Acta Vet. Scand.* **2.** 267–280.

ANDRESEN, E. and BAKER, L. N. (1964), The C blood group system in pigs and the detection and estimation of linkage between the C and J systems. *Genetics* **49.** 379–386.

BRUMMERSTEDT-HANSEN, E., HESSELHOLT, M., LARSEN, B., MOUSTGAARD, J., MØLLER, I., BRÄUNER-NIELSEN, P. and PALLUDAN, B. (1962), Recent progress in immunogenetic research. Report VIIIth Europ. Conf. Anim. Blood Groups (Mimeo.) Ljubljana, 1962.

DINKLAGE, H. and MAJOR, F. (1968), $E^{aegi}$ ($= E^8$), a new allele in the E blood-group system of the pig. *Vox Sang.* **14.** 315–317

DINKLAGE, H. and MAJOR, F. (1969), New factors and alleles in the M blood-group system of the pig. *Vox Sang.* **17.** 316–319.

HOJNÝ, J. (1965), Erytrocytární antigeny prasat chovaných v ČSSR. (Erythrocytic antigens of pigs kept in Czechoslovakia). *Thesis.* Lab. Anim. Genet., Libechov, p. 101.

HOJNÝ, J. and HRADECKÝ, J. (1970), Dextran test in the study of blood groups in pigs. Proc. XIth Europ. Conf. Anim. Blood Groups Biochem. Polymorph., Warsaw, 1968. 259–264.

HOJNÝ, J., GAVALIER, M., HRADECKÝ, J. and LINHART, J. (1966), New blood factors in pigs. Proc. Xth Europ. Conf. Anim. Blood Groups Biochem. Polymorph., Paris, 1966. 151–158.

HOJNÝ, J., MATOUŠEK, J., SCHRÖFFEL, J. and BAKER, L. N. (1967), Comparative pig serum and blood typing test Liběchov and Ames. *Immunogenetics Letter* **5.** 96.

HRADECKÝ, J. and HOJNÝ, J. (1972), E blood group system in miniature pigs. Proc. XIIth Europ. Conf. Anim. Blood Groups Biochem. Polymorph., Budapest, 1970.

NIELSEN, P. B. (1961), The M blood group system of the pig. *Acta Vet. Scand.* **2.** 246–256.

NIELSEN, P. B. (1962), A new allele in the H blood group system in pigs. *Annual Report,* The Royal Vet. and Agr. Coll., Ster. Res. Institute, pp. 201–204 (Mortensen, Copenhagen).

SCHAHMIRZADI, H. (1967), Blutgruppenfaktoren beim Göttinger Zwergschwein. *Diss.* Göttingen.

*XIIth Europ. Conf. Anim. Blood Groups Biochem. Polymorph., Bp., 1972 (pp. 305–309)*

# COLLOIDAL TEST IN DETERMINING BLOOD GROUPS IN PIGS

I. Wiatroszak

College of Agriculture in Poznań, Department of Animal Husbandry,
Laboratory of Research in Animal Blood Groups, Poznań, Poland

## SUMMARY

An attempt at application of a colloidal test in determining pig blood groups mainly with incomplete antibodies is described. The method consists in evoke agglutination reaction in high colloidal medium with simultaneous use of papain activated with cysteine hydrochloride. Normal pig blood serum without antibodies against erythrocytes was used to dilute the reagents and to make a suspension of red blood cells.

The effectivity of enzyme test of one stage using papain activated with cysteine hydrochloride was also investigated in determining pig blood groups. In this test the reagents should be diluted not in saline but in normal pig sera.

The comparison of five methods proved that in many cases both investigated methods could be successfully used for determining pig blood groups instead of other more time-consuming tests. The effectivity, however, depended also on quality of substances used, mainly on activity of papain and cysteine hydrochloride.

## INTRODUCTION

Determination of blood cell antigens in pigs using incomplete antibodies presents some difficulties. The methods used up to date, mainly the indirect anti-globulin test of Coombs and enzyme test of two stages are effective but very time-consuming, as stated by Hardy (1970), Hojný (1970) and others. In modern animal breeding there is a continually increasing demand for determination of blood groups in animals. Thus the necessity arose of developing a method which would be superior, by its simplicity and rapidity, to those used at present. We wanted to make use of the fact that the methods applied in determination of human blood groups could also be used in pigs. We considered the possibility of application of agglutination reaction using proteolytic enzymes (Coombs and Gell, 1963; Hekker et al., 1957; Löw, 1955; Morton and Pickles, 1963) and agglutination reaction in colloidal medium (Dausset and Vidal, 1950; Diamond and Abelson, 1945; Hirszfeldowa, 1958; Unger, 1951). In Poland Hirszfeld introduced the application of human serum of blood group AB containing no antibodies against erythrocytes for diluting reagents and for suspending blood cells, mainly O when determining factor Rh. (cit. after Bratkowska–Seniów, 1966.)

The aim of this study was to investigate the efficacy of using, in enzyme and colloidal tests, the normal pig serum without antibodies against erythrocytes, and papain activated with cysteine hydrochloride.

## MATERIAL AND METHODS

In the investigations we used blood cells of known antigenic structure, taken into conventional anticoagulant. They were tested with 27 of our test sera with incomplete antibodies. We also used six reagents obtained from Göttigen (GFR).

The following tests were used for comparison:

1. Coombs' test,
2. Dextran test according to Hojný and Hála,
3. Enzyme test of two stages using papain,

TABLE 1

*Serological characteristics and titers of incomplete antibodies in some blood systems*

| Reagents | | Antibody titers | | | | | |
|---|---|---|---|---|---|---|---|
| Designation | No. | Agglutination test | Dextran test | Coombs test | Papain test of two stages | Papain test of one stage | Colloidal test |
| 1 | 2 | 3 | 4 | 5 | 6 | 7 | 8 |
| Ca | 5283 | 0 | 0 | 2 | 2 | 2 | 2 |
| | 4366 | 0 | 0 | N | N | N | N |
| Ha | W/2 | 0 | 0 | 8 | 4 | 4 | 4 |
| | 3371 | 0 | 0 | 4 | 2 | 2 | 2 |
| Hb | W/1 | 0 | 2 | 8 | 4 | 4 | 4 |
| Hc | 7 | 0 | 0 | N | 0 | 0 | 0 |
| Ia | 5888 | 0 | 2 | 4 | 2 | 2 | 2 |
| Ib | 5244 | 0 | 0 | 4 | 2 | 2 | 2 |
| | 5739 | 0 | 0 | 16 | 8 | 8 | 8 |
| Ja | 7559 | 0 | 8 | 4 | 2 | N | N |
| La | 5/7 | 0 | 32 | 32 | 8 | 16 | 32 |
| Lb | 6783 | 0 | 32 | 16 | 16 | 16 | 16 |
| Lc | 9099 | 0 | 32 | 16 | 16 | 16 | 16 |
| | D/Gö | 0 | 4 | 4 | 4 | 4 | 4 |
| Ld | 3371 | 0 | 4 | 4 | 2 | 2 | 2 |
| | D/Gö | 0 | 4 | 2 | N | N | N |
| Lf | 5133 | 0 | 4 | 2 | N | N | N |
| | D/Gö | — | 4 | 2 | N | N | N |
| Lg | 7045 | 0 | 128 | 64 | 64 | 64 | 64 |
| Lh | W/6 | 0 | 32 | 16 | 16 | 16 | 16 |
| | D/Gö | 0 | N | 2 | N | N | N |
| Li | 6728 | 0 | 32 | 16 | 8 | 8 | 8 |
| Lj | W/5 | 0 | 0 | 4 | 2 | 2 | 2 |
| Lk | W/7 | 0 | 32 | 16 | 16 | 16 | 16 |
| | D/Gö | 0 | 8 | 8 | 8 | 8 | 8 |
| Ll | W/0 | 0 | 4 | 2 | 2 | 2 | 2 |
| | D/Gö | 0 | 4 | N | N | N | N |
| Ma | 4317 | 0 | 0 | 32 | 16 | 16 | 16 |
| | 2741 | 0 | 0 | 32 | 32 | 32 | 32 |
| | 4 | 0 | 0 | 4 | 0 | 0 | N |
| Mc | 7752 | 0 | 0 | 8 | 8 | 8 | 8 |
| | W/1 | 0 | 0 | 16 | 8 | 8 | 8 |
| Na | W/1/7 | 0 | 0 | N | 0 | 0 | 0 |

4. Enzyme test of one stage using papain activated by addition of cysteine hydrochloride,
5. Colloidal test, simultaneously using papain activated with cysteine hydrochloride.

In the test we used 1% solution of papain activated by addition of cysteine hydrochloride in phosphate buffer according to Michaelis 1/15 M with pH 6.2.

Buffer was prepared as follows:
$KH_2PO_4$ – 9.078 g/1 litre distilled water 8 parts
$Na_2HPO_4$ – 11.188 g/1 litre distilled water 2 parts
Papain was prepared as follows:

10 g of papain was dissolved in 250 ml buffer. Then 4.85 g cysteine hydrochloride was dissolved in 25 ml buffer. These solutions were mixed and buffer was added to 1000 ml volume. Then it was incubated in a water bath of 37°C for an hour, after which it was filtrated by asbestos filter K 5 or Seitz Filter EK, or a simple filter paper, but it could also be centrifuged. The papain when checked for activity, was poured into small ampoules or vials, corked and stored at −20°C. It can be defrosted only once and used in a few hours. The exact descriptions of preparing, checking and storing papain are to be found in medical publications concerning human immunology (Rudowski and Pawelski, 1968).

In the enzyme test of one stage the reagents were diluted in normal pig sera which did not contain antibodies against erythrocytes. Here we used 2.5% suspension of blood cells in physiological solution (0.92 NaCl). While in the colloidal test the medium of high viscosity was created by using normal pig serum for suspending blood cells as well as for diluting the reagents. For this test a 3% suspension was prepared immediately before testing, from blood cells washed three times in saline.

The procedure was as follows: To one drop of reagent we added one drop of activated papain and one drop of blood cell suspension. One reading was made after 30–45 minute incubation in water bath at 37°C.

## RESULTS

When the positive results of the studied methods were obtained in a large number (thousands) of tests, the comparative analyses were carried out.

As seen in the table, in this comparative test we used thirty-three reagents which did not give positive reaction in complete agglutination. The results clearly indicate the possibility of using both methods. The comparison of five various tests proved correctness of reactions obtained. It appeared that in all cases the two studied tests could be used instead of the enzyme test of two stages. If test sera of low titer are used, in the enzyme test of two stages the agglutination reaction obtained was slightly more distinct than in the enzyme test of one stage.

Basing on the results obtained, we may suppose that colloidal test, uniting two methods, enlarges the range of sensitivity. In this test the reactions were more distinct. Some reagents had to be more absorbed because they gave additional reactions. Coombs' test could not be substituted only in the cases of test sera anti-Hc and anti-Na. These sera had low titers and gave positive reactions only in the anti-globulin test.

It appeared that the colloidal test required much care in preparing blood cell suspension, which had to be done very exactly just before testing. Our observations showed that a normal serum suitable for using in these tests could most often be obtained from pigs possessing A and Mc blood groups.

As in all other serological methods, a control test should be carried out simultaneously with these tests.

In both methods one cannot use reagents diluted in saline. In the case when the reagents became over diluted with NaCl solution, due to numerous absorptions, they gave positive results only when a suitable amount of normal serum was added to them.

The investigations showed that reliability of the tests depended also on the substance used, mainly on the degree of activity of papain and of cysteine hydrochloride. In our work we often encountered too low activity of papain as well as unsuitable cysteine. According to our observation, the most suitable was Hungarian DL cysteine and West German papain.

In the course of our work we noticed that some antibodies could be detected in immune sera exclusively by Coombs' test, while others by colloidal test or dextran test. Thus a general conclusion can be drawn that in investigations on production of test sera all possible tests should be used. It is specially important in detecting reagents determining still unknown blood cell antigens.

The methods discussed are superior to the remaining methods in simplicity and rapidity of procedure. The most practical seems to be the first method in which blood cell suspension in saline is used. Both methods are specially efficient in the case of using test sera of high titer of incomplete antibodies. We would like to recommend these methods believing that they will be helpful in producing test sera and in practical utilization of blood groups in modern animal breeding.

## CONCLUSIONS

The normal pig serum containing no antibodies against erythrocytes, as well as papain activated with cysteine hydrochloride can be successfully used in test for determination of blood cell antigens in pigs. In this test, however, special attention should be paid to the quality of substances used, as the results obtained by these methods depend to a large degree on activity of papain and of cysteine hydrochloride.

Another practical conclusion is that in the research on pig blood groups all available methods should be used as this would widen the possibility of detecting antibodies.

## ACKNOWLEDGEMENTS

These investigations were carried out with financial support of a grant from the Department of Agriculture of the United States. The author expresses his gratitude for this support.

The author would like to thank Dr. Dinklage for six reagents which ge have for this investigation.

## REFERENCES

Bratkowska-Seniów, B. (ed.), (1966), Immunologia kliniczna, Warszawa, PZWL.

Coombs, R. R. A. and Roberts, F. (1959), *Brit. Med. Bull.*

Dausset, J. and Vidal, G. (1950), *C. r. Soc. Biol.*, **144.** 679.

Diamond, L. K. and Abelson, N. M. (1945), The detection of Rh sensitization evaluation test for Rh antibodies. *Journ. Lab. Clin. Med.*, **3.** 668.

Hardy, J. (1970), Enzymes in pig red cell typing. Proc. XIth Europ. Conf. Anim. Blood Groups Biochem. Polymorph. Warsaw, 1968, 317–320.

Hekker, A. C., Klomp-Magne, W., Krijen, H. W. and Van Loghem, J. J. (1957), *Vox Sang.*, **2.** 128.

Hirszfeldowa, H. (1958), (ed.) Grupy krwi. Warszawa 1958. PZWL.

Hojný, J. and Hradecký, J. (1970), Dextran test in the study of blood groups in pigs. XIth Europ. Conf. Anim. Blood Groups and Biochem. Polymorph. Warsaw, 1968, 257–264.

Löw, B. (1955), *Vox Sang.*, **5.** 94.

Morton, J. A. and Pickles, M. M. (1951), The proteolytic enzyme test for detecting incomplete antibodies. *Journ. Clin. Path.*, **4.** 189.

Rudowski, W. and Pawelski, S., (ed.) (1968), Transfuzjologia kliniczna. Warszawa PZWL.

Unger, L. J. (1951), A method for detecting Rh antibodies in extremely low titre. *J. Lab. Clin. Med.*, **39.** 246.

*XIIth Europ. Conf. Anim. Blood Groups Biochem. Polymorph., Bp., 1972 (pp. 311–314)*

# IMMUNOGENETIC RECONSTRUCTION OF THE PHYLOGENESIS OF PIG BREEDS

V. N. TIKHONOV

Institute of Cytology and Genetics of the USSR Academy of Sciences, Novosibirsk, USSR

Most of pig breeds reproduced at present in all countries were created by initial use of wild European and Asiatic boars (*Sus Scrofa Ferus* and *Sus Vittatus*). Our immunogenetical study of over twenty breeds reproduced in the USSR and other countries has shown a different degree of influence of Asiatic pigs on the investigated breeds in respect of blood groups (TIKHONOV, 1966, 1967). The pig breeds created by the use of the Berkshire breed, as well as Berkshire itself (Berkshire is originated, as is known, by use of *Sus Vittatus)* always demonstrated a very high frequency of antigen Fa (Table 1). Accordingly, the frequency of antigen Fb was the lowest. As it may be seen from the above data, the same tendency was found while studying black Poland-Chinese breed in Cuba during recent joint studies

TABLE 1

*Immunogenetical evidence of the genelogical relation of some pig breeds to Sus Scrofa Ferus and Sus Vittatus in system F*

| Breeds | No. | Gene frequencies | | Genotype frequencies, % | | |
|---|---|---|---|---|---|---|
| | | *$F^a$ | $F^b$ | $F^a/F^a$ | $F^a/F^b$ | $F^b/F^b$ |
| Vietnamese Black** | 41 | 0.7317 | 0.2683 | 48.78 | 48.78 | 2,44 |
| Vietnamese Spot** | 8 | 0.5625 | 0.4375 | 25.00 | 62.50 | 12.50 |
| Berkshire | 280 | 0.7926 | 0.2074 | 62.86 | 32.85 | 4.29 |
| Poland—China Black** | 200 | 0.6625 | 0.3375 | 43.50 | 45.50 | 11.00 |
| Kemerovskaya | 288 | 0.3599 | 0.6401 | 12.95 | 46.08 | 40.97 |
| Ukrainian Stepnaya Spot | 426 | 0.3269 | 0.6731 | 10.69 | 44.01 | 45.30 |
| Landrace | 723 | 0.0677 | 0.9323 | 0.47 | 12.62 | 86.91 |
| Large White | 609 | 0 | 1.0000 | 0 | 0 | 100.00 |
| Mangalica (Georgia) | 154 | 0.1298 | 0.8702 | 0 | 24.88 | 75.12 |
| Mangalica Lasasta*** | 202 | 0.0460 | 0.9540 | 0 | 8.78 | 91.22 |
| Mangalica White*** | 105 | 0.0050 | 0.9950 | 0 | 1.00 | 99.00 |
| Kakhetian | 153 | 0 | 1.0000 | 0 | 0 | 100.00 |
| Sus Scrofa Ferus**** | 726 | 0 | 1.0000 | 0 | 0 | 100.00 |

* Antigen Fa is not found among Large White breed and other breeds originated from it (Lithuanian, Latvian, Siberian Severnaya, Muromskaya, Urzumskaya and some others),

** V. Tikhonov and A. Camacho (Cuba, 1970),

*** V. Jovanovič and Z. Stojanovič (1966),

**** H. Buschmann (238 animals, 1964) and I. Wiatroszak (488 animals, 1969).

of the Cuba and USSR Academies of Sciences (TIKHONOV and CAMACHO, 1970).

For the direct evidence of the participation of the pigs of the Asiatic genealogical descent in the phylogenesis of up-to-date breeds it was decided to study blood groups of Vietnamese pigs that are now the closest relatives to *Sus Vittatus*. This investigation was carried out with the black Vietnamese pigs of Zoos in Habana, Santiago de Cuba and Tallin.

Two ancient primitive European breeds: Mangalica and Kakhetian from Georgia were taken for study as representatives of pig breeds which are genealogically most closely related to *Sus Scrofa Ferus*. In both of these breeds the newborn piglets often have dark and light longitudinal stripes and this demonstrates the direct origin of these breeds from *Sus Scrofa Ferus*. These breeds were developed from European wild pigs without considerable crossing with other races.

Immunogenetic studies of Mangalica breed, imported in 1948 have confirmed the data of JOVANOVIČ and STOJANOVIČ (1964–1966) and FÉSÜS (1968) of the exclusively high frequency of antigen Ga, and almost complete absence of factors Gb and Fa. The data sharply contrast with the corresponding character of the pig breeds studied so far. On the other hand, the data are in good agreement with the frequency of blood group genes in the European wild boar (BUSCHMANN, 1965; WIATROSZAK, 1969). This tendency is also confirmed by the F system of blood groups in which completely reverse patterns were observed. In Mangalica breed from Georgia the frequency of Fa antigen is very low and in Mangalica breed from Jugoslavia it is zero. According to our data antigen Fa is widely distributed among all pig breeds developed from crosses with pigs from South-East Asia (for example, with the Vietnamese and Black Poland-Chinese pigs we have studied).

The data permit to suppose that the above-mentioned immunogenetic characters of the Mangalica breed are the consequence of the direct origin of this breed from the European wild pigs. In this case it can be assumed that the origin of genetic polymorphism at least in some blood group systems in up-to-date breeds, was determined by the use of wild European and Asiatic pigs for the initial development of the mentioned breeds. To test this hypothesis we have studied the immunogenetics of pigs of the native Kakhetian breed in the Eastern regions of Georgia. This breed was developed by the domestication of the European wild swine and had not been crossed during developmental process with any other breeds. The data on very high (almost 1) frequency of $G^a$ and $F^b$ genes show that among all the pig breeds that have been studied up to now, Kakhetian pigs according to the mentioned blood group systems are genealogically most closely related to the European wild boar. This corresponds to the historical, zootechnical, morphological data on the Kakhetian breed.

The correspondence of immunogenetic data on blood groups of the two most ancient pig breeds directly originating from the wild European boar and retaining its immunogenetical specificity, is important in understanding the phylogenesis of up-to-date pig breeds. This coincidence is of special interest because both Vietnamese breeds have just opposite gene frequency of alleles of systems F and G as compared to the wild European boar

TABLE 2

*Immunogenetic evidence of the genealogical relation of some pig breeds to Sus Scrofa Ferus and Sus Vittatus in system G*

| Breeds | No. | Gene frequencies | | Genotype frequencies, % | | |
|---|---|---|---|---|---|---|
| | | $G^a$ | $G^b$ | $G^a/G^a$ | $G^a/G^b$ | $G^b/G^b$ |
| Sus Scrofa Ferus*** | 726 | 1.0000 | 0 | 100.00 | 0 | 0 |
| Mangalica Lasasta** | 202 | 0.9875 | 0.0125 | 97.52 | 2.48 | 0 |
| Mangalica White** | 105 | 0.9605 | 0.0395 | 92.32 | 7.68 | 0 |
| Mangalica (Georgia) | 154 | 0.9513 | 0.0487 | 91.56 | 8.44 | 0 |
| Kakhetian | 153 | 0.9531 | 0.0469 | 91.94 | 8.06 | 0 |
| Poland—China Black* | 200 | 0.8725 | 0.1275 | 75.00 | 24.50 | 0.50 |
| Kemerovskaya | 288 | 0.7031 | 0.2969 | 50.00 | 40.62 | 9.38 |
| Ukrainian Stepnaya Spot | 426 | 0.6092 | 0.3908 | 36.39 | 49.06 | 14.55 |
| Berkshire | 279 | 0.4122 | 0.5878 | 12.90 | 56.60 | 30.50 |
| Large White | 609 | 0.3517 | 0.6483 | 16.50 | 37.00 | 46.50 |
| Landrace | 723 | 0.3396 | 0.6604 | 14.25 | 39.42 | 46.33 |
| Vietnamese Black* | 41 | 0.2073 | 0.7927 | 4.88 | 31.71 | 63.41 |
| Vietnamese Spot* | 8 | 0.0625 | 0.9375 | 0 | 12.50 | 87.50 |

* See Table 1
** See Table 1
*** See Table 1

(Tables 1 and 2). It should be added that Poland-Chinese, Berkshire and other breeds which are exactly known to originate by use both *Sus Scrofa Ferus* and *Sus Vittatus*, are occupying the intermediate position.

Some deviations from this tendency are easily explained by the influence of crossing (for instance, Mangalica breed with Berkshire one, Poland-China Spot and Vietnamese Spot breeds with Large White breed and so on). Even single crossing can have large influence due to genetic-automatical processes (genetic drift) in isolation conditions of these populations.

The correspondence of the immunogenetic data on the blood groups in primitive pig breeds deriving directly from the European wild pigs and maintaining the latters' immunogenetic features is obvious. The comparative study of different breeds showed that the frequency of Fa- and Gb-antigens (and of Fb and Ga, respectively) can probably serve as a criterion of the relative contribution of the Asiatic and, correspondingly, of the European wild pigs *(Sus Scrofa Ferus* and *Sus Vittatus)* to the breedforming process of all up-to-date breeds.

## REFERENCES

BUSCHMANN, H. (1965), Blood group studies in pigs. Proc. 9th Europ. Anim. Blood Group Conf., Prague. 1964.

FÉSÜS, L. (1970), Blood groups and serum proteins in the Hungarian Mangalica pig. Proc XIth Europ. Conf. Anim. Blood Groups Biochem. Polymorph., Warsaw, 1968.

JOVANOVIČ, V. and STOJANOVIČ, Z. (1966), Proc. Xth Europ. Conf. Anim. Blood Groups Biochem. Polymorph., Paris, 1966.

TIKHONOV, V. N. (1966), Studies on blood group factors for genetical analysis of selection processes. Proc. Xth Europ. Conf. Anim. Blood Groups Biochem. Polymorph., Paris. 1966.
TIKHONOV, V. N. (1967), The blood groups in animals and their application for animal breeding. Moskow.
TIKHONOV, V. N. (1968), Immunological incompatibility as a cause of fertility decrease in animal crossbreeding. VI Congrès de reproduction et insémination artificielle. Paris.
WIATROSZAK, I. (1969), Immunogeneticzna charakterystyka świń hodowanych w Polsce, Poznan.

*XIIth Europ. Conf. Anim. Blood Groups Biochem. Polymorph., Bp., 1972 (pp. 315–322)*

# SERUM PROTEIN POLYMORPHISM IN PIG BREEDS IN SOUTH AFRICA

E. H. H. MEYER

Animal and Dairy Science Research Institute, Blood Group Laboratory, Irene, South Africa

## SUMMARY

Seven pig breeds in South Africa (Landrace, Large White, Large Black, Minnesota No. 1, Wessex Saddleback, Bantu pig and Kolbroek) have been compared with respect to gene and phenotype frequencies of the following polymorphic serum markers: amylase ($Am^A$, $Am^{BF}$, $Am^B$, $Am^C$), ceruloplasmin ($Cp^A$, $Cp^B$), hemopexin ($Hp^0$, $Hp^{1F}$ $Hp^1$, $Hp^2$, $Hp^3$) and transferrin ($Tf^A$, $Tf^B$). The results compare well with those of populations of the same breed in other countries.

## INTRODUCTION

Breed comparisons in pigs are greatly facilitated by the large number of genetically controlled polymorphic serum proteins that have been described. In the present study seven commercial and "indigenous" pig breeds occurring in South Africa are compared with respect to four serum markers: amylase (Am), ceruloplasmin (Cp), hemopexin (Hp) and transferrin (Tf). The comparison is also extended to populations in other countries.

## MATERIAL AND METHODS

Serum samples of 1132 pigs were obtained from the following sources:

| | | |
|---|---|---|
| Landrace: | 27 | breeders at a boar performance testing station |
| | 9 | breeders at a pig progeny testing station |
| | 2 | commercial herds |
| | 2 | experimental herds |
| Large White: | 7 | breeders at a boar performance testing station |
| | 3 | breeders at a pig progeny testing station |
| | 1 | commercial herd |
| | 1 | experimental herd |
| Large Black: | 2 | stud herds |
| | 1 | commercial herd |
| Wessex Saddleback: | 2 | commercial herds |
| Minnesota No. 1: | 2 | experimental herds |

Bantu pig: 10 different samplings at the abbatoir, Pretoria
Kolbroek: 1 commercial herd (Western Cape province)

The Bantu pig and the Kolbroek are the only breeds that can to some extent be regarded as "indigenous" in South Africa.

The samples were examined by horizontal starch gel electrophoresis according to the method of KRISTJANSSON (1963), with modifications by BUSCHMANN and SCHMID (1968). For staining, the gel was sliced horizontally into four parts. The bottom part was stained for transferrin, the second part from the bottom for ceruloplasmin and amylase, and the second part from the top for hemopexin.

## RESULTS AND DISCUSSION

### 1. AMYLASE

All the four known amylase alleles were observed in the Landrace, Minnesota No. 1 and Bantu pig (Table 1). In the Large Black and Large White only $Am^A$ and $Am^B$ alleles are present. The $Am^B$ allele has the highest

TABLE 1

*Gene frequencies of amylase alleles in pig breeds in South Africa*

| Breed | Number of animals tested | Amylase allele | | | |
|---|---|---|---|---|---|
| | | $Am^A$ | $Am^{BF}$ | $Am^B$ | $Am^C$ |
| Landrace | 578 | 0.178 | 0.010 | 0.809 | 0.003 |
| Large Whire | 131 | 0.023 | — | 0.977 | — |
| Large Black | 83 | 0.072 | — | 0.928 | — |
| Minnesota No. 1 | 78 | 0.006 | 0.006 | 0.974 | 0.013 |
| Wessex Saddleback | 19 | 0.553 | — | 0.421 | 0.026 |
| Bantu pig | 230 | 0.035 | 0.102 | 0.715 | 0.148 |
| Kolbroek | 13 | — | 0.154 | 0.692 | 0.154 |

frequency in all the breeds except for the Wessex Saddleback, where the $Am^A$ allele is most frequent. Of the ten possible phenotypes, all except AmABF were observed (Table 2). As shown in Table 3, the gene frequencies in the Landrace and Large White compare very well with those in other countries.

### 2. CERULOPLASMIN

As indicated in Table 4, both the $Cp^A$ and $Cp^B$ alleles were found in all the breeds except the Large White, which are homogeneous for $Cp^B$. Compared to the CpBB phenotype, the CpAA and CpAB phenotypes are rare in all the breeds (Table 5). In the Landrace a far higher proportion of homozygotes are observed than is expected. A comparison of the Landrace, Large White and Saddleback pigs with those in Denmark and Britain reveals only minor differences (Table 6).

TABLE 2

*Observed and expected distribution of amylase phenotypes*

| Breed | | Amylase phenotype | | | | | | | | | |
|---|---|---|---|---|---|---|---|---|---|---|---|
| | | AA | ABF | AB | AC | BFBF | BFB | BFC | BB | BC | CC |
| Landrace | obs. | 24 | 0 | 158 | 0 | 3 | 5 | 0 | 384 | 4 | 0 |
| | exp. | 18.3 | 2.1 | 166.5 | 0.6 | 0.1 | 9.4 | 0.0 | 378.3 | 2.8 | 0.0 |
| Large White | obs. | 2 | — | 2 | — | — | — | — | 127 | — | — |
| | exp. | 0.1 | — | 5.9 | — | — | — | — | 125.0 | — | — |
| Large Black | obs. | 0 | — | 12 | — | — | — | — | 71 | — | — |
| | exp. | 0.4 | — | 11.1 | — | — | — | — | 71.5 | — | — |
| Minnesota No. 1 | obs. | 0 | 0 | 1 | 0 | 0 | 1 | 0 | 75 | 0 | 1 |
| | exp. | 0.0 | 0.0 | 0.9 | 0.0 | 0.0 | 0.9 | 0.0 | 74.0 | 2.0 | 0.0 |
| Wessex Saddleback | obs. | 9 | — | 2 | 1 | — | — | — | 7 | 0 | 0 |
| | exp. | 5.8 | — | 8.9 | 0.6 | — | — | — | 3.4 | 0.4 | 0.0 |
| Bantu pig | obs. | 2 | 0 | 9 | 3 | 10 | 18 | 9 | 127 | 48 | 4 |
| | exp. | 0.3 | 1.6 | 11.5 | 2.4 | 2.4 | 33.6 | 6.9 | 117.6 | 48.7 | 5.0 |
| Kolbroek | obs. | — | — | — | — | 2 | 0 | 0 | 9 | 0 | 2 |
| | exp. | — | — | — | — | 0.3 | 2.8 | 0.6 | 6.2 | 2.8 | 0.3 |

TABLE 3

*Comparison of amylase allele frequencies in pig breeds in various countries*

| Breed | Country | Amylase allele | | | | Number of animals tested |
|---|---|---|---|---|---|---|
| | | $Am^A$ | $Am^{BF}$ | $Am^B$ | $Am^C$ | |
| Landrace | South Africa[1] | 0.178 | 0.010 | 0.809 | 0.003 | 578 |
| | Denmark[2] | 0.4 | — | 0.83 | 0.03 | 200 |
| | Britain[3] | 0.20 | — | 0.80 | — | 95 |
| | Czechoslovakia[4] | 0.158 | — | 0.833 | 0.009 | 174 |
| | Germany[5] | 0.113 | 0.024 | 0.845 | 0.018 | 225 |
| | Germany[6] | 0.040 | 0.025 | 0.928 | 0.007 | 986 |
| Large White | South Africa[1] | 0.023 | — | 0.977 | — | 131 |
| | Britain[3] | 0.05 | — | 0.95 | — | 161 |
| | Czechoslovakia[4] | 0.250 | — | 0.737 | 0.013 | 236 |

[1] Present study
[2] Hesselholt et al., 1966
[3] Imlah, 1965
[4] Gavalier et al., 1966
[5] Dinklage, 1968
[6] Schmid, 1968

## 3. HEMOPEXIN

The gene frequencies for five hemopexin alleles are shown in Table 7. The $Hp^{3F}$ allele (HESSELHOLT and HRISTIC, 1966) and $Hp^4$ allele (BAKER, 1967) are absent in the breeds in South Africa. The frequency of the $Hp^1$ allele is the highest in all the breeds. Phenotypes with the $Hp^{1F}$ allele are very infrequent (Table 8). An excess of homozygotic phenotypes is observed in the Bantu pig. Comparing the Landrace and Large White populations in various countries (Table 9), quite a variation is apparent in the $Hp^1$ and $Hp^3$ frequencies. The $Hp^0$, $Hp^{1F}$ and $Hp^2$ alleles are uniformly low in frequency.

TABLE 4

*Gene frequencies of ceruloplasmin alleles in pig breeds in South Africa*

| Breed | Number of animals tested | Ceruloplasmin allele | |
|---|---|---|---|
| | | $Cp^A$ | $Cp^B$ |
| Landrace | 578 | 0.056 | 0.944 |
| Large White | 131 | — | 1.000 |
| Large Black | 83 | 0.042 | 0.958 |
| Minnesota No. 1 | 78 | 0.051 | 0.949 |
| Wessex Saddleback | 19 | 0.132 | 0.868 |
| Bantu pig | 230 | 0.011 | 0.989 |
| Kolbroek | 13 | 0.269 | 0.731 |

TABLE 5

*Observed and expected distribution of ceruloplasmin phenotypes*

| Breed | | Ceruloplasmin phenotype | | |
|---|---|---|---|---|
| | | AA | AB | BB |
| Landrace | obs. | 17 | 31 | 530 |
| | exp. | 1.8 | 61.1 | 515.1 |
| Large White | obs. | — | — | 131 |
| | exp. | — | — | 131.0 |
| Large Black | obs. | 1 | 5 | 77 |
| | exp. | 0.1 | 6.7 | 76.2 |
| Minnesota No. 1 | obs. | 1 | 6 | 71 |
| | exp. | 0.2 | 7.6 | 70.2 |
| Wessex Saddleback | obs. | 1 | 3 | 15 |
| | exp. | 0.3 | 4.0 | 14.3 |
| Bantu pig | obs. | 1 | 3 | 226 |
| | exp. | 0.0 | 5.0 | 225.0 |
| Kolbroek | obs. | 2 | 3 | 8 |
| | exp. | 0.9 | 5.1 | 7.0 |

TABLE

*Observed and expected*

| Breed | | Hemopexin | | | | | |
|---|---|---|---|---|---|---|---|
| | | 0–0 | 0–1F | 0–1 | 0–2 | 0–3 | 1F–1F |
| Landrace | obs. | 7 | 0 | 28 | 2 | 20 | 0 |
| | exp. | 1.8 | 0.5 | 31.7 | 4.1 | 23.8 | 0.0 |
| Large White | obs. | 0 | 0 | 1 | 0 | 0 | 0 |
| | exp. | 0.0 | 0.0 | 0.9 | 0.0 | 0.1 | 0.1 |
| Large Black | obs. | 1 | 0 | 0 | 0 | 0 | 0 |
| | exp. | 0.0 | 0.0 | 1.5 | 0.1 | 0.4 | 0.0 |
| Minnesota No. 1 | obs. | 1 | — | 3 | 1 | 1 | — |
| | exp. | 0.2 | — | 4.2 | 1.8 | 0.6 | — |
| Wessex Saddleback | obs. | — | — | — | — | — | — |
| | exp. | — | — | — | — | — | — |
| Bantu pig | obs. | 20 | 1 | 22 | 6 | 2 | 0 |
| | exp. | 5.5 | 0.1 | 43.0 | 10.5 | 6.3 | 0.0 |
| Kolbroek | obs. | — | — | — | — | — | — |
| | exp. | — | — | — | — | — | — |

TABLE 6

*Comparison of ceruloplasmin allele frequencies in pig breeds in various countries*

| Breed | Country | Ceruloplasmin allele | | Number of animals tested |
|---|---|---|---|---|
| | | $Cp^A$ | $Cp^B$ | |
| Landrace | South Africa[1] | 0.056 | 0.944 | 578 |
| | Denmark[2] | 0.0154 | 0.9846 | 130 |
| | Britain[3] | 0.06 | 0.94 | 95 |
| Large White | South Africa[1] | — | 1.000 | 131 |
| | Britain[3] | — | 1.00 | 161 |
| Saddleback | South Africa[1] | 0.132 | 0.868 | 19 |
| | Britain[3] | — | 1.00 | 28 |

[1] Present study
[2] Graetzer et al., 1965
[3] Imlah, 1965

TABLE 7

*Gene frequencies of hemopexin alleles in pig breeds in South Africa*

| Breed | Number of animals tested | Hemopexin allele | | | | |
|---|---|---|---|---|---|---|
| | | $Hp^0$ | $Hp^{1F}$ | $Hp^1$ | $Hp^2$ | $Hp^3$ |
| Landrace | 578 | 0.055 | 0.007 | 0.499 | 0.064 | 0.375 |
| Large White | 131 | 0.004 | 0.031 | 0.889 | 0.023 | 0.053 |
| Large Black | 83 | 0.012 | 0.012 | 0.735 | 0.066 | 0.175 |
| Minnesota No. 1 | 78 | 0.045 | — | 0.603 | 0.263 | 0.090 |
| Wessex Saddleback | 19 | — | — | 0.553 | 0.184 | 0.263 |
| Bantu pig | 230 | 0.154 | 0.002 | 0.607 | 0.148 | 0.089 |
| Kolbroek | 13 | — | — | 0.500 | 0.038 | 0.462 |

8

*distribution of hemopexin phenotypes*

| Hemopexin phenotype | | | | | | | | | |
|---|---|---|---|---|---|---|---|---|---|
| | 1F–1 | 1F–2 | 1F–3 | 1–1 | 1–2 | 1–3 | 2–2 | 2–3 | 3–3 |
| | 8 | 0 | 0 | 143 | 30 | 225 | 1 | 40 | 74 |
| | 4.0 | 0.5 | 3.0 | 143.9 | 36.9 | 216.3 | 2.4 | 27.7 | 81.3 |
| | 7 | 1 | 0 | 108 | 2 | 7 | 1 | 1 | 3 |
| | 7.2 | 0.2 | 0.4 | 103.5 | 5.4 | 12.3 | 0.1 | 0.3 | 0.4 |
| | 2 | 0 | 0 | 51 | 1 | 17 | 5 | 0 | 6 |
| | 1.5 | 0.1 | 0.4 | 44.8 | 8.1 | 21.4 | 0.4 | 1.9 | 2.5 |
| | — | — | — | 30 | 27 | 4 | 6 | 1 | 4 |
| | — | — | — | 28.4 | 24.7 | 8.5 | 5.4 | 3.7 | 0.6 |
| | — | — | — | 9 | 2 | 1 | 2 | 1 | 4 |
| | — | — | — | 5.8 | 3.9 | 5.5 | 0.6 | 1.8 | 1.3 |
| | 0 | 0 | 0 | 99 | 42 | 17 | 7 | 6 | 8 |
| | 0.6 | 0.1 | 0.1 | 84.7 | 41.3 | 24.9 | 5.0 | 6.1 | 1.8 |
| | — | — | — | 3 | 1 | 6 | 0 | 0 | 3 |
| | — | — | — | 3.3 | 0.5 | 6.0 | 0.0 | 0.5 | 2.8 |

TABLE 9

*Comparison of hemopexin allele frequencies in pig breeds in various countries*

| Breed | Country | Hemopexin allele | | | | | Number of animals tested |
|---|---|---|---|---|---|---|---|
| | | $Hp^0$ | $Hp^{1F}$ | $Hp^1$ | $Hp^2$ | $Hp^3$ | |
| Landrace | South Africa[1] | 0.055 | 0.007 | 0.499 | 0.064 | 0.375 | 578 |
| | Denmark[2] | 0.041 | — | 0.342 | 0.115 | 0.503 | 900 |
| | Britain[3] | 0.06 | — | 0.51 | 0.05 | 0.38 | 55 |
| | Czechoslovakia[4] | 0.040 | 0.075 | 0.276 | 0.014 | 0.595 | 174 |
| | U.S.A.[5] | — | — | 0.67 | 0.05 | 0.28 | |
| | Germany[6] | 0.018 | — | 0.696 | 0.076 | 0.210 | 255 |
| Large White | South Africa[1] | 0.004 | 0.031 | 0.889 | 0.023 | 0.053 | 131 |
| | Britain[3] | 0.05 | — | 0.84 | 0.04 | 0.06 | 85 |
| | Czechoslovakia[4] | 0.007 | 0.042 | 0.722 | 0.000 | 0.229 | 236 |

[1] Present study
[2] Graetzer et al., 1965
[3] Imlah, 1965
[4] Gavalier et al., 1966
[5] Kristjansson, 1961
[6] Buschmann and Schmidt, 1968

TABLE 10

*Gene frequencies of transferrin alleles in pig breeds in South Africa*

| Breed | Number of animals tested | Transferrin allele | |
|---|---|---|---|
| | | $Tf^A$ | $Tf^B$ |
| Landrace | 578 | 0.003 | 0.997 |
| Large White | 131 | 0.099 | 0.901 |
| Large Black | 83 | 0.012 | 0.988 |
| Minnesota No. 1 | 78 | 0.013 | 0.987 |
| Wessex Saddleback | 19 | 0.105 | 0.895 |
| Bantu pig | 230 | 0.139 | 0.861 |
| Kolbroek | 13 | 0.423 | 0.577 |

## 4. TRANSFERRIN

Only the $Tf^A$ and $Tf^B$ alleles are present in the breeds studied (Table 10). The $Tf^C$ allele (KING, 1962), $Tf^D$ allele (SCHRÖFFEL, 1966) and the $Tf^E$ allele (BAKER, 1968) have been reported to occur with low frequencies. Except for the Kolbroek breed, the $Tf^B$ allele frequency is much higher than the $Tf^A$ one. The observed and expected distribution of phenotypes are well in agreement (Table 11). As demonstrated in Table 12, the population of Landrace and Saddleback are comparable in various countries. However, the $Tf^B$ frequency in the Large White pigs in South Africa is exceedingly high.

TABLE 11

*Observed and expected distribution of transferrin phenotypes*

| Breed | | Transferrin phenotype | | |
|---|---|---|---|---|
| | | AA | AB | BB |
| Landrace | obs. | 0 | 4 | 574 |
| | exp. | 0.0 | 3.5 | 574.5 |
| Large White | obs. | 2 | 22 | 107 |
| | exp. | 1.3 | 23.4 | 106.3 |
| Large Black | obs. | 1 | 0 | 82 |
| | exp. | 0.0 | 2.0 | 81.0 |
| Minnesota No. 1 | obs. | 1 | 0 | 77 |
| | exp. | 0.0 | 2.0 | 76.0 |
| Wessex Saddleback | obs. | 0 | 4 | 15 |
| | exp. | 0.2 | 3.6 | 15.2 |
| Bantu pig | obs. | 5 | 54 | 171 |
| | exp. | 4.4 | 55.1 | 170.5 |
| Kolbroek | obs. | 3 | 5 | 5 |
| | exp. | 2.3 | 6.4 | 4.3 |

TABLE 12

*Comparison of transferrin allele frequencies in pig breeds in various countries*

| Breed | Country | Transferrin alleles | | | | Number of animals tested |
|---|---|---|---|---|---|---|
| | | $Tf^A$ | $Tf^B$ | $Tf^C$ | $Tf^D$ | |
| Landrace | South Africa[1] | 0.003 | 0.997 | — | — | 578 |
| | Denmark[2] | — | 1.000 | — | — | 500 |
| | Britain[3] | — | 0.97 | 0.03 | — | 187 |
| | Czechoslovakia[4] | — | 0.991 | 0.009 | — | 174 |
| Large White | South Africa[1] | 0.099 | 0.901 | — | — | 131 |
| | Britain[3] | 0.33 | 0.67 | — | — | 394 |
| | Czechoslovakia[4] | 0.295 | 0.684 | — | 0.021 | 236 |
| Saddleback | South Africa[1] | 0.105 | 0.895 | — | — | 19 |
| | Britain[3] | — | 0.91 | 0.09 | — | 46 |

[1] Present study
[2] Moustgaard and Hesselholt, 1965
[3] Imlah, 1970
[4] Gavalier et al., 1966

## REFERENCES

BAKER, L. N. (1967), *Vox Sang.*, **12.** 397.
BAKER, L. N. (1968), *Vox Sang.*, **14.** 446.
BUSCHMANN, H. and SCHMID, D. O. (1968), Serumgruppen bei Tieren. Berlin: Paul Parey.
DINKLAGE, H. (1968), *Züchtungskunde*, **40.** 228.
GAVALIER, M., HOJNÝ, J., HRADECKÝ, J., LINHART, J. and SCHRÖFFEL, J. (1966), Proc. Xth Europ. Conf. Anim. Blood Groups Biochem. Polymorph. Paris, 1966.
GRAETZER, M. A., HESSELHOLT, M., MOUSTGAARD, J. and THYMANN, M. (1965), Proc. 9th Europ. Conf. Anim. Blood Groups, Prague, 1964, 279.
HESSELHOLT, M. and HRISTIC, V. (1966), *Acta Vet. Scand.* **7.** 187.
HESSELHOLT, M., LARSEN, B. and NIELSEN, P. B. (1966), *Ann. Report*, Royal Vet. Agric. College, Copenhagen, 78.
IMLAH, P. (1965), Proc. 9th Europ. Conf. Anim. Blood Groups, Prague, 1964, 109.
IMLAH, P. (1970), *Anim. Blood Grps Biochem. Genet.* **1.** 5.
KING, J. W. B. (1962) Cited by IMLAH, P. (1965), Proc. 9th Europ. Conf. Anim. Blood Groups, Prague, 1964, 109.
KRISTJANSSON, F. K. (1961), *Genetics*, **46.** 907.
KRISTJANSSON, F. K. (1963), *Genetics*, **48.** 1059.
MOUSTGAARD, J. and HESSELHOLT, M. (1965), *Proc. Int. Symp.* "Swine in Biomedical Research". Washington, 25.
SCHMID, D. O. (1968), *Zentbl. Vet. Med.*, B **15.** 990.
SCHRÖFFEL, J. (1966), *Nature*, **210.** 1274.

*XIIth Europ. Conf. Anim. Blood Groups Biochem. Polymorph., Bp., 1972 (pp. 323 – 325)*

# ISOPRECIPITINS AGAINST SERUM ANTIGENS IN NORMAL SERA OF SWINE

J. Rapacz, Judith Hasler, Anna Kazana and M. Duniec

Immunogenetics Laboratory, Zootechnical Institute, Balice k/Kraków, Poland

Isoprecipitins against several serum lipoprotein and protein antigens of unknown origin were discovered in the sera of adult pigs. Studies have been made in an attempt to explain the presence of these antibodies in the sera. It has been demonstrated that crystal violet swine fever vaccine stimulates the production of isoprecipitins.

The original discovery was made when normal sera of 156 adult and 416 young pigs were tested by the immunodiffusion technique, to study the genetics of low density beta-lipoprotein antigens (LDLpp-3) (Rapacz et al., 1970). An unexpected, strong precipitate was observed in agar gel between peripheral wells in which normal sera of two sows had been placed. Immunoelectrophoretic analysis showed that the isoprecipitins were present in the serum of sow G-6918 and were directed against the antigen of the beta-globulin fraction of serum 522. Serum number G-6918 gave similar precipitation with 18% of the normal sera tested.

Sow G-6918 of the Yorkshire breed was one of a litter produced by a pregnant sow imported from Great Britain. The serum was collected after her second farrowing, when she was two years of age, and then stored for six months before the discovery was made.

The antigenic stimulus was unknown and the antibodies detected in serum G-6918 may be either naturally occurring or immune in origin. The latter could arise through injection with blood or serum antigens, or administration of crystal violet swine fever vaccine.

Normal serum of pigs with blood group O in most cases contains naturally occurring antibodies against blood group antigen A and other blood group antigens (Goodwin and Coombs, 1956; Kaczmarek and Wiatroszak, 1967; Saison et al., 1955; Szent-Iványi and Szabó, 1954; Szymanowski et al., 1926). 20 immune sera from the Immunogenetics Laboratory in Poznan, which were produced against red blood cell antigens were tested in our laboratory and six of them had isoprecipitins against serum lipoprotein antigens (paper in preparation). In humans, Allison and Blumberg (1961) were first to report isoprecipitins against lipoproteins in the serum of a multitransfused patient.

It has been well established by Goodwin et al. (1955), Goodwin and Saison (1956), and Joysey et al. (1959), that crystal violet swine fever vaccine can stimulate the production of many red cell iso-antibodies when injected into pigs. DOLL et al. (1952) found that red cell agglutinins appeared in the serum of mares receiving an equine virus abortion vaccine during pregnancy. In mink, RAPACZ and SHACKELFORD (1963a), RAPACZ et al. (1968) and SAISON (1962) showed that isoagglutinins were produced as a side effect of virus enteritis vaccination. In these three species the respective species tissue is a component of the vaccine.

An extensive survey has been made to determine the factors responsible for isoprecipitin production. It was suspected at that time that the most probable agent responsible for isoprecipitins was crystal violet swine fever

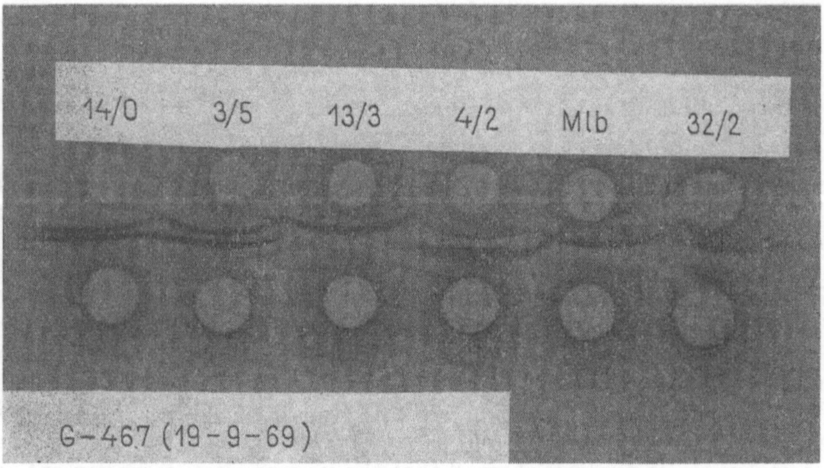
14/0
3/5
13/3
4/2
Mlb
32/2
G-467 (19-9-69)

FIG. 1. The immune sera G-467 obtained from a pig vaccinated with crystal violet swine fever vaccine, tested by gel precipitation with six normal pig sera

vaccine. Pooled blood from 300 pigs from two breeds (Polish Large White and Polish Landrace) with a very wide range of antigens was used for production of the vaccine in the Veterinary Institute, Puławy, Poland (personal communication).

All normal sera were retested against each other in search of antibodies, and in two of them precipitins were found. Search for isoprecipitins were extended to 437 adult pigs from 14 herds, and all data concerning vaccination and inoculation were recorded. Pigs from four herds where vaccination was practiced, and from three control herds, were repeatedly bled and their sera tested. The results present evidence that isoprecipitins are stimulated by crystal violet swine fever vaccine. The sera were collected before and after (7–21, 30–45 and 150–180 days) vaccination. Isoprecipitins were found in only 28 of the sera collected at first bleeding after vaccination. Seven sera had isoprecipitins at the 45th day, and only two on the 180th day. Most of them gave the same reaction patterns as immune sera anti-Lpp-3 (RAPACZ et al., 1970). A comparison study showed that these sera had antibodies against at least five different serum antigens, which are carried on eight different molecules. The results demonstrated that with exeption of two sera (G-467 and Mł-11) all have isoprecipitins against only serum lipoprotein antigens. Two pigs (G-467 and Mł-11) with strong isoprecipitins were reinjected three months after the original preventive vaccination in an attempt to increase the level of antibodies. Both responded in production of strong precipitins. The reactions of one of these sera (G-467) with six randomly chosen normal sera are shown in Fig. 1.

All precipitating sera were tested against human normal sera in the Central Laboratory of the Blood Transfusion Service of the Swiss Red Cross, Bern, Switzerland. Serum G-467 also gave precipitation with human serum lipoprotein components.

## REFERENCES


Allison, A. C. and Blumberg, B. S. (1961), An iso-precipitin reaction distinguishing human serum protein types, *Lancet*, 634–637.

Doll, E. R., Richards, M. G., Wallace, M. E. and Bryans, J. T. (1952), The influence of an equine fetal tissue vaccine on heamagglutination activity of mare serums: its relation to hemolytic icterus of newborn foals, *Cornell Vet.* **42.** 495.

Goodwin, R. F. W., Saison, R. and Coombs, R. R. A. (1955), The blood groups of the pig. (II). *J. Comp. Path.*, **65.**

Goodwin, R. F. W. and Saison, R. (1956), The blood groups of the pig, (III). *J. Comp. Path.*, **66.**

Goodwin, R. F. W. and Coombs, R. R. A. (1956), The blood groups of the pigs. (IV). *J. Comp. Path.*, **66.**

Joysey, V. C., Goodwin, R. F. W. and Coombs, R. R. A. (1959), The blood groups of the pig. (VII). *J. Comp. Path.*, **69.**

Kaczmarek, A. and Wiatroszak, I. (1967), Investigations on blood groups in new racial group of the Złotnicka pig. College of Agriculture, Poznan.

Rapacz, J. and Shackelford, R. M. (1963a), Immunogenetic studies in the domestic mink, *Immunogenetics Letter*, **3.** 55–62.

Rapacz, J., Shackelford, R. M. and Hasler, J. (1970), Naturally occurring and immune antibodies as a possible cause of hemolytic disease in the domestic mink. Proc. XIth Europ. Conf. Anim. Blood Groups Biochem. Polymorph., Warsaw, 1968, 585.

Rapacz, J., Hasler, J., Duniec, M. and Kazana, J. (1972), Serum antigens of beta-lipoproteins in pigs (LDLpp-3). Proc. XIIth Europ. Conf. Anim. Blood Groups Biochem. Polymorph., Budapest, 1970.

Saison, Ruth, Goodwin, R. F. W. and Coombs, R. R. A. (1955), The blood groups of the pig. (I). *J. Comp. Path.*, **65.**

Saison, Ruth (1962), The blood groups in mink. I. Six blood group systems in mink. *J. Immunol.*, **89.** 881–885.

Szent-Iványi, T. and Szabó, I. (1954), Blood groups in pigs, *Acta vet. Hung.*, **4.** 429–446.

Szymanowski, Z., Stetkiewicz, S. and Wachler, B. (1926), Les groupes sérologiques dans le sang du porc et leur relation avec les groupes du sang humain, *C.r. Sc. Biol.* **94.** 204.

*XIIth Europ. Conf. Anim. Blood Groups Biochem. Polymorph., Bp., 1972 (pp. 327–330)*

# BLOOD GROUP SPECIFIC PHYTHEMAGGLUTININS IN THE PIG

H. DINKLAGE and R. HOHENBRINK

Institute of Animal Husbandry and Animal Genetics, University of Göttingen, GFR

## SUMMARY

**Extracts from 329 well-ripened plant seeds were used to test 2% pig erythrocyte suspensions. Twenty-seven strong specific agglutinins, which can be divided into thirteen additional groups according to their reaction, were identified. One extract was treated by systematic absorption and finally converted into a monospecific antiserum. Swine family studies showed that this factor occurs only in the offspring where at least one parent possesses the factor. A gene frequency estimation was calculated for two swine breeds. No similarity with the blood group factors identified so far was evidenced.**

## INTRODUCTION

The exact mechanism of the specific reactions of natural phythemagglutinins is still unidentified. There are common features between antibodies and plant "lectins" (BOYD, 1954). The reaction of specific blood groups and so-called "unspecific" phythemagglutinins depends on the structural-chemical configurations, just as in the antibodies (KRÜPE, 1952; 1956). Hemagglutinins from plants are intracellular plasma proteins with antigenic structures for carbohydrate determinants. They are thus capable of agglutinating blood cells (KRÜPE and ENSGRABER, 1967). For practical blood typing in man, several group specific phythemagglutinins have been identified (KRÜPE, 1955; KOULUMIES, 1949; KRÜPE, 1963; KRÜPE et al., 1963). Erythrocytes of variuos animal species were tested with unspecific phythemagglutinins and it appeared, that in the pig differences could be found (BHATIA and ALLEN, 1962; JAFFÉ et al., 1965; OTTENSOOSER, 1955).

## MATERIAL AND METHODS

*Plant extracts*. Through an international exchange of plant seeds among a number of botanical gardens, it was possible to obtain 329 well-ripened seeds of differing species. The origin of each species was treated separately. Extracts were prepared as follows: Seeds were ground in a coffee-mill or reduced to small pieces in an Ultra-Turrax in a 0.9% sodium chloride solution. The active powder was extracted using a physiological sodium chloride solution at the ratio of 1 : 20 and by shaking at ambient temperature for two hours. The suspension was centrifuged and the supernatant liquid filtered. The extracts were conserved at −30°C. The frozen plant extracts did not loose activity or change titre after two years of storage.

*Erythrocyte suspension.* The 329 extracts were used to test 2% erythrocyte suspensions from 29 Landrace and 5 Göttingen Miniature pigs. Since some plant substances showed agglutination only in the presence of pig serum, papain or bromelin, all extracts were tested against a 2% erythrocyte suspension to which the same quantity of serum was added from an A positive pig.

## RESULTS

When the 329 extracts were tested with the 2% erythrocyte saline suspension, they were classified into the following categories:

I. Strong specific agglutinins (27 extracts).
II. Weak specific agglutinins (30 extracts).
III. Unspecific agglutinins (93 extracts).
IV. Inactive types (179 extracts).

The addition of serum from an A positive pig to the 2% erythrocyte suspension showed the following results:

(a) formerly inactive reactions remained inactive.
(b) formerly inactive reactions became active.
(c) formerly active reactions became inactive.

Among the strong specific agglutinins there were some, which showed identical specificity when tested against the sodium chloride erythrocyte suspension. As shown in Table 1, the 27 strong specific agglutinins can be divided into 13 groups. The reactions were repeatable. The extracts were derived from different varieties and origins of vicia, pisum, caragana, swainsonia galegifolia, galega orientalis, erythrina, lathyrus, robinia pseudoacacia, psoralea bituminosa, virgilia oroboides.

One extract (vicia villosa, Winterwicke Afra) was tested by systematic cross absorption. Using titration it appeared that the crude extract contained three antibody fractions. Further absorption resulted in monospecific antiserum. Erythrocytes from animals of the following breeds Landrace, German Yorkshire, Piétrain, Göttingen Miniature pig, German Pasture pig, American Yorkshire, Mangalica, wild pig and various crossbreds were tested with the crude extract and both absorbed fractions.

The pattern of inheritance of the fraction with two antibodies was studied on 47 families involving 144 offspring and that of the monospecific fraction on 79 families involving 167 offspring.

These studies showed that the factor identified by the phythemagglutinin occurs only in offspring which had at least one parent with this factor. There is no evidence of any similarity with the blood group factors identified so far.

For the breeds Landrace and American Yorkshire, gene frequencies were estimated using the monospecific antiserum (Table 2).

TABLE 1

*Blood group specific phythemagglutinins in the pig*

| Phythemagglutinin | Erythrocytes | | | | | | | | | |
|---|---|---|---|---|---|---|---|---|---|---|
| | 1 | 2 | 3 | 4 | 5 | 6 | 7 | 8 | 9 | 10 |
| Vicia villosa | 4 | . | 4 | 4 | 4 | 4 | 4 | 4 | 4 | 4 |
| Pisum elatius | 4 | . | 4 | 4 | 4 | 4 | 4 | 4 | 4 | 4 |
| Vicia villosa | 4 | 4 | 4 | 4 | 4 | 4 | 4 | . | 4 | 4 |
| Caragana boisii | 4 | 4 | 4 | 4 | 4 | 4 | 4 | . | 4 | 4 |
| Caragana brevispina | 4 | 4 | 4 | 4 | 4 | 4 | 4 | . | 4 | 4 |
| Caragana decorticans | 4 | 4 | 4 | 4 | 4 | 4 | 4 | . | 4 | 4 |
| Caragana arborescens (2 bases) | . | . | 4 | 4 | 4 | 4 | 4 | . | 4 | 4 |
| Swainsonia galegifolia | . | . | 4 | 4 | 4 | 4 | 4 | . | 4 | 4 |
| Caragana decorticans | . | . | 4 | 4 | 4 | 4 | 4 | . | 4 | 4 |
| Galega orientalis | . | . | . | . | . | 4 | 4 | . | . | . |
| Vicia hirsuta (2 bases) | . | . | . | . | . | 4 | 4 | . | . | . |
| Vicia hirsuta | . | . | . | . | . | . | 4 | . | . | . |
| Erythrina crista galli | . | . | 4 | . | . | . | 4 | . | . | . |
| Lathyrus pratensis | . | . | . | . | . | 4 | . | . | . | . |
| Robinia pseudoacacia | . | 4 | . | . | . | . | . | . | . | . |
| Vicia articulata | 4 | . | . | 4 | . | . | 4 | . | . | 4 |
| Psoralea bituminosa (2 bases) | . | . | . | . | . | . | . | . | 4 | . |
| Virgilia oroboides | 4 | 4 | 4 | 4 | 4 | 4 | . | 4 | 4 | . |
| Erythrina crista galli | . | . | . | . | 4 | . | . | . | . | . |
| Erythrina batissima | . | . | . | . | 4 | . | . | . | . | . |
| Vicia tetrasperma | 4 | . | . | . | . | . | . | . | . | . |
| $GO_{20}$ | . | . | 4 | . | . | . | 4 | . | . | 4 |

TABLE 2

*Frequencies of positive or negative reactions for the German Landrace and American Yorkshire swine*

| Breed | + | — | Total |
|---|---|---|---|
| Landrace | 83 | 213 | 296 |
| American Yorkshire | 52 | 70 | 122 |

## REFERENCES


BHATIA, H. M. and ALLEN, F. H. (1962), *Vox Sang.*, **7.** 83.
BOYD, W. C. and SHAPLEY, E. (1954), *J. Immunol.*, **73.** 216.
JAFFÉ, W., MONTBRUN, M., CALLEJAS, A. and JAFFÉ, M. (1965), *Z. Immun. Forsch.*, **129.** 196.
KOULUMIES, R. (1949), *Ann. Med. exp. Biol. Fenn.*, **27.** 185.
KRÜPE, M. (1952), *Z. Immun. Forsch.*, **123.** 355–366.
KRÜPE, M. (1955), *Habil. Schrift* (Marburg/Lahn).
KRÜPE, M. (1956), Blutgruppenspezifische Eiweißkörper (Phytagglutinine), *Ferd. Enke Verl.*, Stuttgart.
KRÜPE, M. (1963), *Zbl. Bakt.* I. O. **160.** 289.
KRÜPE, M., MÜLLER, G. J. and FENNER, G. (1963), *Z. Immun. Forsch.*, **125.** 64.
KRÜPE, M. and ENSGRABER, A. (1967), *Bibl. haemat.*, **27.** 34–49.
OTTENSOOSER, F. (1955), *An. Acad. Bras. Cien.*, **27.** 519.


## DISCUSSION

P. MILLOT: Have you the same reactions on the different red cells when you make dilutions of your reagents?

R. HOHENBRINK: Titration of the extracts is possible but gives shorter reactions and leads us to the opinion that more than one antibody fraction is included. Special absorptions give a monospecific reagent ($GO_{20}$).

J. HARDY: Have you tested Ulex?

R. HOHENBRINK: No.

J. HARDY: Do you find variation in the same species of seed collected from different areas?

R. HOHENBRINK: Yes. Extracts of seeds from different origin may have different reaction patterns as shown in Table 1.

E. ANDRESEN: Have you observed reactions also with withe cell antigens?

R. HOHENBRINK: The reactions with white cells antigens we haven't studied till now.

*XIIth Europ. Conf. Anim. Blood Groups Biochem. Polymorph., Bp., 1972 (pp. 331–335)*

# EVIDENCE FOR THE ISOIMMUNIZATION OF SOWS BY INCOMPATIBLE FOETAL RED CELLS

K. A. LINKLATER

Blood Group Research Unit, Department of Animal Health, Veterinary Field Station, Easter Bush, Roslin, Midlothian, Scotland

## SUMMARY

Studies have been carried out in four Ea-negative sows mated to an Ea-positive boar and anti-Ea titres have been boosted after parturition in three of them. In the fourth, there was a marked rise in titre of anti-Ea four weeks before parturition. In addition, in one, anti-Kb, and in another, anti-La, were also produced. Subsequent mating of these two to an Ea-negative, Kb-positive, La-positive boar resulted in boosting of the anti-Kb and anti-La titres respectively but not the anti-Ea titre after parturition. Two Ea-negative gilts were mated to an Ea-positive boar and anti-Ea was produced in one after parturition.

On examining the sera from 135 sows and boars in the routine typing service carried out by the Blood Group Research Unit Laboratory, no antibodies other than anti-A have been found in the boars' sera, while, in the sows' sera red cell antibodies other than anti-A have been found in 20% of them. As crystal violet vaccination for swine fever was discontinued in 1964 in this country, it does not now confuse the picture.

All these results support further the hypothesis that isoimmunization of the sow by foetal red cells can occur. It does seem also that the factors involved are those to which antibodies are most readily produced by the injection of red cells from a donor into a recipient. Of these antibodies anti-Ea and anti-Eb are by far the most common and are also found at the highest titres.

## INTRODUCTION

In 1954, SZENT-IVÁNYI and SZABÓ, when studying the pathogenesis of hemolytic disease of newborn piglets, reported finding four red cell antibodies in pigs, one of which was anti-A. They called the three others anti-B, -C and -D and postulated that these could be produced by immunization of the sows by incompatible red cell antigens of foetuses in utero. Unfortunately these antibodies were never classified into present day nomenclature for blood group systems in pigs. Similarly, GOODWIN, SAISON and COOMBS (1955), while concentrating on iso-antibodies produced in pigs injected with crystal violet swine fever vaccine, did not preclude the possibility of naturally occurring hemolytic disease in newborn piglets by transplacental immunization of the sow and found red cell antibodies other than anti-A at low titres in a small number of unvaccinated pigs.

It was not until after 1964, however, when crystal violet vaccination for swine fever was discontinued in the United Kingdom, that the opportunity to study isoimmunization of sows by foetal red cells arose and LINKLATER (1968) reported evidence on this. Using a sow in a Minimal Disease Pig Unit it was demonstrated that antibodies were produced to the blood group factor Ea after giving birth to Ea-positive piglets but not after giving birth to Ea-negative piglets alone. A complete life history of the sow was available and the possibility of previous injection of pig red cells was precluded. The results presented here are a continuation of this work and lend further evidence to it.

## MATERIAL AND METHODS

1. *Minimal Disease piggery*. Two sows (88 and 405) were found to have low levels of anti-Ea in their serum (LINKLATER, 1968). These were both inseminated by an Ea-negative boar and then by an Ea-positive one.

Two Ea-negative gilts (1347 and 1350) were also inseminated by an Ea-positive boar.

2. *Conventional piggery*. Two sows (301 and 302) which were found to have red cell antibodies (anti-A, anti-Ea and anti-Kb, and anti-A, anti-Ea and anti-La respectively) were bought in from a commercial herd and kept in the Blood Group Research Unit piggery. Both were inseminated by a boar which was Ea-positive, Kb-positive, La-positive once and then by a boar which was Ea-negative, Kb-positive and La-positive.

### *Porcine sera sent in for routine parentage checking*

135 samples of adult pig serum sent into the Blood Group Research Unit for routine parentage checking were tested for the presence of red cell antibodies. The same techniques as mentioned above using red cells from five animals of known red cell type, were used in the analysis of these samples. This was in order to cover all possible combinations of known red cell types in our blood group panel.

### *Blood sampling*

Blood samples were taken from the sows as described by LINKLATER (1968). These were taken at fortnightly intervals for the first three months of pregnancy and weekly for the last month. For the first three weeks after parturition, samples were taken twice weekly, thence weekly until weaning and service.

Piglets were sampled at birth before sucking and typed according to the red cell reagents of the Blood Group Research Unit.

### *Serology*

Serum was removed and tested for red cell antibodies as described by LINKLATER (1968).

## RESULTS

Sow 88. The serum from this animal showed a marked rise in anti-Ea titres after each parturition which produced Ea-positive piglets but not after the one which produced Ea-negative ones alone. This titre reached a peak at 10–17 days post partum which was maintained for about a month after which time there was a gradual fall until it was almost down to the preparturition level. This picture was repeated over three parturitions.

Sow 405. This animal showed a similar response to 88. However, the antibody rise after parturition was not as great. It gradually fell off as well and was at a very low level by the time the next parturition came around.

Gilt 1347. In samples taken throughout one pregnancy and after parturition, no antibodies to Ea were detected.

Gilt 1350. No antibodies were detected during the first pregnancy, but ten days after parturition anti-Ea was detected at a low titre (1/2).

Sow 301. When purchased, this animal had a low titre of anti-Ea (1/2) as well as anti-A. For the first time another antibody, anti-Kb, was found to be present in the neat serum but not at any higher dilution. This animal was different from the others in that a marked rise in titre of anti-Ea to 1/64 and a slight rise in titre of anti-Kb to 1/2 was apparent after 3 months of pregnancy, indicative of some immunological stimulus at this stage. In the subsequent mating to the Ea-negative, Kb-positive boar, the picture was repeated for the anti-Kb (titre to 1/8) but there was no boost in the anti-Ea titre during pregnancy or after parturition.

Sow 302. Anti-La was present in this sow's serum as well as anti-A and anti Ea. As with sows 88 and 405, the rise in titres occurred post partum (1/128) and had fallen to a low level by the succeeding parturition (1/8). When mated to an Ea-negative La-positive boar, anti-La was boosted after parturition, but not the anti-Ea.

*Routine samples*. 135 serum samples were examined. Of the 61 boars none had evidence of antibodies other than anti-A. On the other hand, of the 74 from sows, 60 (81%) either contained anti-A alone or were completely negative while 14 (19%) had other antibodies as shown in Table 1. Reciprocal titres of the latter were all fairly low and lay between 2 and 16. The most common antibodies present were anti-Ea and anti-Eb.

TABLE 1

| Iso-antibody | No. of times isolated | Titre range |
|---|---|---|
| Anti-Ea | 4 | 2–16 |
| Anti-Eb | 7 | 2–8 |
| Anti-Ea and anti-Ka | 1 | 8 |
| Anti-Ea and anti-La | 2 | 4–8 |
| Total | 14 | 2–16 |

## DISCUSSION

In the present investigation, the possibility of vaccination with porcine erythrocytes in any form can be completely excluded. Anti-Ea is the antibody most studied in the experimental sows and, in three cases, this reached its highest level around ten days after parturition. In one of the gilts, anti-Ea was also present for the first time after parturition. These findings are consistent with immunization taking place at the time of parturition, presumably by leakage of foetal red cell antigens into the maternal circulation. That this rise in titre is not simply physiological is shown by the fact that when mated to Ea-negative boars, no rise in titre occurred in the two sows in the Minimal Disease Pig Unit. Similarly when sows 301 and 302 were mated to the Ea-negative, Kb-positive, La-positive boar, no rise in titre of anti-Ea occurred but anti-Kb and anti-La respectively were both boosted.

That almost 20% of the sera from sows in the routine typing service showed red cell antibodies other than anti-A while the sera from the boars did not contain any other than anti-A is further evidence that iso-immunization by foetal red cells is occurring. This figure is surprisingly high especially as all the sows are bred with boars of the same breed, and all are young animals being blood sampled at 4–6 weeks after their first or second parturitions. A higher incidence might be expected where crossbreeding is practiced and LINKLATER (1970) has shown that the sera of 50% of sows which had produced litters suffering from thrombocytopenic purpura, a condition of piglets thought to result from isoimmunization of the sow by foetal thrombocytes, contained red-cell antibodies other than anti-A. Most of the latter sows had been mated to boars of a different breed.

All the results support further the hypothesis that iso-immunization of the sow by foetal red cells can occur. It does seem also that the factors involved are those to which antibodies are most readily produced by the injection of red cells from a donor into a recipient. Of these, anti-Ea and anti-Eb are by far the most common. In the present investigation anti-Ka, anti-Kb and anti-La were also noted. In a previous report anti-Ed and anti-Ee have also been found (LINKLATER, 1970).

## REFERENCES

GOODWIN, R. F. W., SAISON, RUTH and COOMBS, R. R. A. (1955), *J. comp. Path.* **65.** 79–92.

LINKLATER, K. A. (1968), *Vet. Rec.*, **83.** 203–204.

LINKLATER, K. A. (1970), *Vet. Rec.*, **86.** 630–631.

SZENT-IVÁNYI, T. and SZABÓ, S. (1954), *Acta vet. Hung.*, **4.** 429–446.

## DISCUSSION

T. SZENT-IVÁNYI: In Hungary early observations (1952) on hemolytic disease in pigs were made before crystal violet vaccine was used in this country. This is a proof that iso-immunization of the sow can occur without previous sensibilization by pig blood containing vaccines. Iso-immunization of the sow, evidently by antigens o the fetuses inherited from the boar, led to clinical cases of the new-born especially in crosses: Mangalica sows × Berkshire boars.

K. A. LINKLATER: I have not seen hemolytic disease of newborn piglets on a clinical level but only subclinically.

JOAN HARDY: The titre of natural antibodies found by us in most cases did not exceed 1 : 2, or 1 : 4. In only two cases were the titres as high as 1:8. In the cases described by Dr. Linklater the titre following mating with incompatible boar rose to much higher levels.

K. A. LINKLATER: I have tried to exclude the possibility of injections as far as possible. Certainly the animals studied in the Minimal Disease Unit have not been injected with anything but I cannot be sure of those in the routine parentage checking. However, I point out again the fact that in the latter no iso-antibodies were seen in the boars while almost 20% of the sows had iso-antibodies.

JOAN HARDY: The recent discoveries in the mechanism of human hemolytic disease i.e. the immunization of the mother around parturition by a transplacental hemorrhage removes most of the philosophical objections of veterinarians to the possibility of maternal immunization in animals such as pigs.

K. A. LINKLATER: I think the antigens get into the maternal circulation at or soon after parturition perhaps not as whole erythrocytes but as broken down fragments.

*XIIth Europ. Conf. Anim. Blood Groups Biochem. Polymorph., Bp., 1972 (pp. 337—341)*

# BIOCHEMICAL POLYMORPHISM IN MINNESOTA AND MINNESOTA × VIETNAMESE PIGS

W. Schleger and E. Dworak

Blood Group Laboratory, Department of Animal Breeding and Genetics, School of Veterinary Medicine, Vienna, Austria

Blood sera of 119 miniature pigs were investigated, using horizontal starch gel electrophoresis. Phenotypes were determined with regard to ceruloplasmines, amylases, transferrins, hemopexins, pre-albumins, post-albumins, and $S\alpha_2$-globulins. Gene frequencies were computed. The combined probability of exclusion of wrong parentage in populations with genetic equilibrium would amount to 79.4%.

Biochemical systems showing simple Mendelian inheritance are well suited for control of identity, for pedigree control, and clarification of disputed parentage.

The present study was initiated to investigate the biochemical systems in the miniature pig herd of the Veterinary School, to study their inheritance, and to compute the probability of excluding wrong parentage. The herd was created in 1966, and derives from miniature pigs and pure Vietnamese pigs. The population is rather heterogenous and is not yet in the genetic equilibrium. Blood sera from 15 boars and 23 sows as well as from 81 progeny from various matings, were investigated for eight different biochemical systems, using the available methods.

We did not succeed in isolating the albumine fraction as described by Kristjansson (1965). However, the phenotypes of the systems Cp, Am, Tf,

Table 1

*Gene frequencies of the Cp in various pig breeds*

| Breed | N | Author | $Cp^1$ | $Cp^2$ | Probability of excluding wrong parentage |
|---|---|---|---|---|---|
| Danish Landrace | 250 | Graetzer and coworkers, 1964 | .015 | .985 | |
| Large White | 161 | Imlah, 1964 | . | 1.000 | |
| Landrace | 95 | Imlah, 1964 | .06 | .94 | |
| Hung. Mangalica | 360 | Fésüs, 1968 | . | 1.000 | |
| Minnesota + Vietnamese | 119 | Schleger and Dworak, 1970 | .42 | .58 | 18.4% |

Hp, pre-alb, post-alb, and $S\alpha_2$ could be identified. The determination of the ceruloplasmins, and the amylases is carried out according to the method given by GRAETZER and co-workers (1964), and HESSELHOLT (1968).

The high frequency of $Cp^1$ in the miniature-pig herd is due to the homozygosity of the Vietnamese pigs for that gene. In pure Minnesotas $Cp^1$-frequency should be around 0.1–0.2.

Homozygous $Am^3/Am^3$ types predominate in the Vietnamese pigs. However, the pure Minnesotas also show a fairly high $Am^3$ frequency. The $Am^{2F}$ allele as described by HESSELHOLT was not found in the miniature pigs; we did, however, find that type in some animals of the Austrian Yorkshire.

The investigations of hemopexin (formerly haptoglobin) were accomplished according to the method of KRISTJANSSON, as modified by HESSELHOLT. Gene frequencies of the animals in the herd were:

$Hp^0$ 0.114, $Hp^1$ 0.639, $Hp^2$ 0.136, and $Hp^3$ 0.109. The distribution of the gene frequencies is very favourable from the point of view of parentage control. These frequencies are similar to frequencies in other pig breeds (BUSCHMANN and SCHMID, 1968). Probability of excluding wrong parentage is 25.4%.

TABLE 2

*Gene frequencies of amylases in various pig breeds*

| Breed | N | Author | $Am^1$ | $Am^2$ | $Am^3$ | Probability of excluding wrong parentage |
|---|---|---|---|---|---|---|
| Danish Landrace | 466 | Graetzer, 1964 | .13 | .84 | .03 | |
| Large White | 161 | Imlah, 1964 | .05 | .95 | | |
| Landrace | 95 | Imlah, 1964 | .20 | .80 | | |
| Hung. Mangalica | 360 | Fésüs, 1968 | 1.00 | 1.00 | | |
| Minnesota × Vietnamese | 119 | Schleger and Dworak, 1970 | .003 | .445 | .552 | 18.8% |

TABLE 3

*Gene frequencies of pre-albumins in various pig breeds*

| Breed | N | Author | $Pa^A$ | $Pa^B$ | Probability of excluding wrong parentage |
|---|---|---|---|---|---|
| Bohemian Yorkshire | 392 | Schröffel, 1965 | .584 | .416 | |
| Prestice | 164 | Schröffel, 1965 | .680 | .320 | |
| Cornwall | 94 | Schröffel, 1965 | .878 | .122 | |
| Minnesota × Vietnamese | 100 | Schleger and Dworak, 1970 | .370 | .630 | 17.9% |

The transferrin system showed the following frequencies: $Tf^A$ 0.226, $Tf^B$ 0.747, $Tf^C$ 0.027. The probability of excluding wrong parentage is 16.4%. The distribution of the gene frequencies is similar to most other breeds (BUSCHMANN and SCHMID, 1968).

Pre-albumins were investigated first by KRISTJANSSON (1963) whose methods and interpretation we followed.

It should be pointed out that the pre-albumins are difficult to handle, and that interpretation has been difficult in some cases. We are working on a modification of the methods.

SCHRÖFFEL (personal communication 1964), has the opinion that the post-albumins show a polymorphism. IMLAH (1964) showed the position of post-albumins in some of his graphs. BUSCHMANN (1968) did recognise individual variation in the post-albumins in some electrophoretograms. KÚBEK (1968) studied this system in 64 progeny of the matings of various breeds, using the methods of KRISTJANSSON (1963), and differentiated two post-albumin

TABLE 4

*Distribution of Pst A types in progeny from various matings*

| ♂ | Mating | ♀ | Number of matings | Number of progeny | AA obs. | AA exp. | AB obs. | AB exp. | BB obs. | BB exp. |
|---|---|---|---|---|---|---|---|---|---|---|
| AA | × | AB | 1 | 8 | 4 | 4 | 4 | 4 | — | — |
| BB | × | AA | 1 | 3 | 1 | 1.5 | 2 | 1.5 | — | — |
| AB | × | AB | 7 | 35 | 8 | 8.75 | 19 | 17.5 | 8 | 8.75 |
| AB | × | BB | 7 | 38 | — | — | 16 | 19 | 22 | 19 |
| Total | | | 16 | 84 | 13 | 19.44 | 41 | 41.94 | 30 | 22.63 |

Expected equilibrium frequencies $\chi^2 = 4.71$, 1 d.f., $0.05 < P < 0.02$

zones. In zone 1, he differentiated an A- and a B-fraction. We succeeded in showing, in some families of the miniature pigs, a polymorphism in zone 1, as was first indicated by KÚBEK. One of the bands was moving faster (band A) and the other band slower (band B). With regard to zone 2 we would prefer to wait for further results.

In 111 investigated animals, 28 AA, 51 AB, and 32 BB types were found. The corresponding gene frequencies are: Pst $A^A$ 0.481, Pst $A^B$ 0.519. The probability of excluding wrong parentage is therefore 18.7%.

SCHRÖFFEL (1964) discovered the $S\alpha_2$ system. In our investigation we followed his method. The identification of the various types is rather difficult when the gels are not perfect.

The progeny types of the mating sire BC × dam AB deviate significantly from the frequencies expected from Mendelian segregation, assuming heterozygosity in both parents. Progeny of the other nine mating combinations show frequencies corresponding to the expected values. SCHRÖFFEL found in none of the 13 mating combinations in 278 progeny from 34 litters a significant deviation.

TABLE

*Distribution of $S\alpha_2$ globulin types in*

| ♂Mating♀ | Number of litters | Number of progeny | AA obs. | AA exp. | AB obs. | AB exp. | BB obs. | BB exp. |
|---|---|---|---|---|---|---|---|---|
| AA×AB | 3 | 29 | 13 | 14.5 | 16 | 14.5 | | |
| AB×AA | 1 | 4 | 2 | 2 | 2 | 2 | | |
| AB×AB | 1 | 2 | 1 | 0.5 | 1 | 1 | | 0.5 |
| AB×AC | 1 | 8 | 2 | 2 | 5 | 2 | | |
| AB×CC | 1 | 2 | | | 1 | 1 | | |
| BC×AA | 2 | 10 | | | 8 | 5 | | |
| BC×AB | 5 | 17 | | | 11 | 4.25 | 5 | 4.25 |
| BC×BB | 1 | 3 | | | | | 3 | 1.5 |
| BC×BC | 2 | 8 | | | | | 5 | 2 |
| CC×AC | 1 | 8 | | | | | | |
| Total | 18 | 91 | 18 | 26.93 | 44 | 32.38 | 13 | 9.73 |

Expected equilibrium frequencies
* = significant
++ = highly significant

$\chi^2 = 18.917$, 3 d.f., $.005 < P < .002^{++}$
The probability of excluding wrong parentage = 25.2%

TABLE 6

*Gene frequencies of $S\alpha_2$ globulins*

| Breed | $N$ | Author | $S\alpha_2{}^A$ | $S\alpha_2{}^B$ | $S\alpha_2{}^C$ | Probability of excluding wrong parentage |
|---|---|---|---|---|---|---|
| Bohemian Yorkshire | 209 | Schröffel, 1964 | .344 | .357 | .299 | |
| Prestice | 115 | Schröffel, 1964 | .604 | .278 | .118 | |
| Cornwall | 86 | Schröffel, 1964 | .401 | .442 | .157 | |
| Minnesota×Vietn. | 113 | Schleger and Dworak, 1970 | .544 | .327 | .128 | 25.2% |

The Hardy—Weinberg equilibrium values were computed for all systems. The amylase system is the only one where a genetic equilibrium is present. All other systems show statistically significant deviations of the observed values from the equilibrium. This is obviously due to the heterogeneity of the herd.

The study of biochemical systems via electrophoresis was found to be an excellent tool in checking the accuracy of pedigrees, even without blood grouping. The combined probability of excluding wrong parentage, using seven systems investigated in this study, amounts to 79.4%

## REFERENCES

ASHTON, G. C. (1957), Zone electrophoresis of mammalian sera in starch gels. *Nature* **179**, 823.

BUSCHMANN, H. and SCHMID, D. O. (1968), Serumgruppen bei Tieren. Paul Parey Verlag, Berlin und Hamburg.

5

*91 offspring from various matings*

| AC | | BC | | CC | | | | |
|---|---|---|---|---|---|---|---|---|
| obs. | exp. | obs. | exp. | obs. | exp. | $\chi^2$ | | d.f. |
| 1 | 2 | | 2 | | | 5.500 | .25<P<.10 | 3 |
| 1 | 1 | | | | | | | |
| 2 | 5 | | | | | 3.600 | .10<P<.05 | 1 |
| 2 | 4.25 | 1 | 4.25 | | | 11.009 | .02<P<.01 | 3* |
| | 1.5 | | | | | | | |
| | | 3 | 4 | | 2 | 5.250 | .10<P<.05 | 2 |
| 4 | 4 | | | 4 | 4 | | | |
| 10 | 12.68 | 4 | 7.62 | 4 | 1.49 | | | |


FÉSÜS, L. (1970), Blood Groups and Serum Proteins in the Hungarian Mangalica Pig. Proc. XIth Europ. Conf. Anim. Blood Groups Biochem. Polymorph., Warsaw 1968, 275.

GRAETZER, M. A., HESSELHOLT, M., MOUSTGAARD, J. and THYMANN, M. (1965), Studies on protein polymorphism in pigs, horses and cattle. Proc. 9 th Europ. Anim. Blood Group Conf., Prague, 1964, 279.

HESSELHOLT, M. (1963), Haptoglobin polymorphism in pigs. *Acta vet. Scand.*, **4.** 238.

HESSELHOLT, M. (1970), Additional studies into serum amylase in Swine, Proc. XIth Europ. Conf. Anim. Blood Groups Biochem. Polymorph. Warsaw 1968, 347.

IMLAH, P. (1965), A study of blood groups in pigs. Proc. 9 th Europ. Anim. Blood Group Conf., Prague 1964, 109.

KRISTJANSSON, F. K. (1961), Genetic control of three haptoglobins in swine. *Genetics*, **46.** 907.

KRISTJANSSON, F. K. (1963), Genetic control of two pre-albumins in pigs. *Genetics*, **48.** 1059.

KÚBEK, A. (1970), A study of Post-albumines in Pigs. Proc. XIth Europ. Conf. Anim. Blood Groups Biochem. Polymorph., Warsaw, 1968 315.

KÚBEK, A. (1970), Electrophoretical Study of the Esterases in Pig Serum. Proc. XIth Europ. Conf. Anim. Blood Groups Biochem. Polymorph., Warsaw 1968, 355.

SCHRÖFFEL, J. (1965), Genetic determination of the serum "thread proteins" and the slow-$\alpha_2$-Globulin polymorphism in pigs. Proc. 9 th Europ. Anim. Blood Group Conf., Prague 1964, 321.


## DISCUSSION

H. DINKLAGE: Why did you not participate in the Comparison Test 1970?

W. SCHLEGER: Till now we had been working only in cattle. This is the first study in pigs. In the future we will participate in the comparison tests for biochemical polymorphism in pigs.

*XIIth Europ. Conf. Anim. Blood Groups Biochem. Polymorph., Bp., 1972 (pp. 343—346)*

# DEGREE OF HETEROZYGOSITY IN PURE-BRED AND CROSS-BRED PIGS AS RELATED TO BREEDING PERFORMANCE CHARACTERISTICS

R. HOHENBRINK, H. DINKLAGE and RUTH GRUHN

Institute of Animal Husbandry and Animal Genetics, University of Göttingen, GFR

## SUMMARY

The degree of heterozygosity in purebred and crossbred pigs was estimated for ten blood group gene loci. Relationship between the degree of heterozygosity and breeding performance characteristics was investigated.

## INTRODUCTION

The estimation of the degree of heterozygosity, the ratio between heterozygous and all marked gene loci, is possible because the distinction can be made between loci having homo- and those having heterozygous blood group and protein variants.

Indications that heterozygosity of blood group genes is linked with various performance characteristics have been suggested for farm animals by GILMOUR (1960), MITSCHERLICH (1965), TIKHONOV (1966), KOCH et al. (1968), SCHYIA (1969). However, most workers considered only one gene locus which does not provide any information on the degree of heterozygosity. The consideration of several gene loci will give information on the degree of heterozygosity of the genome of an individual animal or a population. The present investigations were conducted to determine if there were relationships of economic importance between the degree of heterozygosity, measured by means of ten blood group systems, and some breeding performance characteristics.

## MATERIAL AND METHODS

The degree of heterozygosity was estimated on 616 German Landrace sows taken from the different breed associations in Northern Germany (DINKLAGE and HOHENBRINK, 1970). The following characteristics for the first four litters were considered: number of piglets born, number of piglets weaned and average litter weight at four weeks of age.

The degree of heterozygosity was estimated on a further 107 German Landrace (GL) sows and 54 $F_1$ sows mated to German Landrace, German Yorkshire (GY) and Piétrain (Pi) boars, resulting in 404 Landrace-, 411 Piétrain × Landrace-, 408 Yorkshire × Landrace-, 401 GL × (GY × GL)-,

401 Pi×(GY×GL)- and 249 GY×(GY×GL)-offspring. The following data on breeding performance were taken into consideration: number of piglets born (alive and stillborn), and number of piglets weaned.

Blood group factors were determined for the ten blood group systems A, D, E, F, G, H, I, K, L and M.

## RESULTS

### *1. Relationship between degree of heterozygosity of sows and breeding performance characteristics*

It appears from Table 1 that the number of heterozygous gene loci of a sow, at least as far it is measurable by means of blood group factors, has no effect on breeding performance.

TABLE 1

*Breeding performance in relation to the number of heterozygous blood group gene loci in 616 sows*

| No. of heteroz. loci | No. of sows | 1st litter No. of piglets born | 1st litter No. of piglets weaned | Average of 2nd–4th litter No. of piglets born | Average of 2nd–4th litter No. of piglets weaned |
|---|---|---|---|---|---|
| 1 | 13 | 10.1±0.9 | 9.3±1.4 | 11.7±1.8 | 10.5±1.9 |
| 2 | 66 | 10.6±1.9 | 9.6±1.8 | 11.9±2.3 | 10.3±2.2 |
| 3 | 126 | 10.2±1.5 | 9.1±1.6 | 11.7±2.0 | 10.3±2.0 |
| 4 | 167 | 10.2±1.7 | 9.3±1.7 | 11.5±2.0 | 10.2±1.9 |
| 5 | 137 | 10.1±1.6 | 9.3±1.6 | 11.6±1.9 | 10.4±1.8 |
| 6 | 73 | 10.4±1.5 | 9.4±1.6 | 11.7±1.6 | 10.4±1.7 |
| 7 | 32 | 10.3±1.7 | 9.4±2.0 | 11.1±1.6 | 10.1±1.6 |
| 8 | 2 | 10.5±1.4 | 9.1±0.4 | 11.7±1.6 | 10.1±2.1 |

There were only a few sows in the marginal classes and no indication of a trend was found in the intermediate classes.

### *2. Influence of parental genotype on breeding performance characteristics*

A comparison of all mating combinations, from which 100% homo- or 100% heterozygous offspring were expected because of the parental genotypes, was made for the E, G, I, L and M blood group systems. It appeared that matings resulting in 100% heterozygotic offspring were superior in the number of piglets born and weaned to those resulting in 100% homozygotic offspring. This relationship was found in the system E, G, I and M. In the L system, the $L^{bcgi}$ allele showed a very high frequency, and it was assumed that this results from particular selection advantages and that in this case homozygosity was superior to heterozygosity (Table 2).

TABLE 2

*Matings of Landrace boars to Landrace sows resulting in 100% homo- or 100% heterozygous offspring in the E, G, I, L and M systems*

| Blood group system | Genotype | No. of matings | 1st litter No. of piglets | | Average of 2nd–4th litter No. of piglets | |
|---|---|---|---|---|---|---|
| | | | born | weaned | born | weaned |
| E | heteroz. | 87 | 10.3±1.4 | 9.3±1.6 | 11.6±1.8 | 10.3±1.8 |
| | homoz. | 8 | 9.8±0.9 | 8.9±0.8 | 10.9±2.0 | 9.7±1.7 |
| G | heteroz. | 31 | 10.2±1.5 | 9.5±1.4 | 11.9±1.9 | 10.7±1.9 |
| | homoz. | 63 | 10.1±1.4 | 9.0±1.6 | 11.3±1.6 | 10.1±1.7 |
| I | heteroz. | 20 | 11.1±1.7 | 10.1±1.7 | 11.9±1.9 | 10.4±1.9 |
| | homoz. | 36 | 10.5±1.3 | 9.6±1.3 | 11.8±2.0 | 10.5±2.1 |
| L | heteroz. | 16 | 10.5±1.4 | 9.8±1.4 | 11.2±1.4 | 10.2±1.6 |
| | homoz. | 139 | 10.7±1.5 | 9.7±1.6 | 11.8±1.8 | 10.5±1.9 |
| M | heteroz. | 18 | 10.6±1.2 | 9.5±1.1 | 12.2±1.7 | 10.7±1.7 |
| | homoz. | 81 | 10.6±1.4 | 9.5±1.6 | 11.6±2.0 | 10.2±2.0 |

### *3. Relationship between degree of heterozygosity of offspring and breeding performance of their mothers*

In the various cross-bred and pure-bred control progeny, piglet losses declined with increasing degree of heterozygosity. Offspring from the five cross-bred combinations showed on the average higher degrees of heterozygosity, better litter performance and lower piglet losses (HOHENBRINK, 1970).

TABLE 3

*Degree of heterozygosity, litter performance and piglet losses of pure-bred and cross-bred offspring*

| | GL×(GY×GL) | GY×GL | Pi×(GY×GL) | GY×(GY×GL) | Pi×GL | GL×GL |
|---|---|---|---|---|---|---|
| No. of progeny | 401 | 408 | 401 | 249 | 411 | 404 |
| Degree of heterozygosity | 51.0±26.3 | 50.6±29.0 | 50.5±30.2 | 49.4±32.7 | 47.7±26.0 | 45.7±24.8 |
| Piglets born | 10.2± 2.4 | 10.9± 2.5 | 10.5± 1.9 | 10.1± 1.6 | 10.8± 2.1 | 9.9± 2.1 |
| Piglets born alive | 9.6± 2.2 | 10.1± 2.2 | 9.6± 2.0 | 9.6± 1.5 | 10.0± 1.7 | 9.1± 2.1 |
| Piglets weaned | 8.6± 2.2 | 9.4± 2.1 | 8.5± 1.9 | 8.9± 1.4 | 8.9± 1.8 | 7.9± 1.9 |
| Piglet losses (%) | 15.7 | 13.7 | 19.1 | 11.9 | 17.6 | 20.2 |

## REFERENCES

DINKLAGE, H., and HOHENBRINK, R. (1970), Untersuchung über den Einfluß heterozygoter Blutgruppengenorte auf Merkmale der Zuchtleistung bei der Deutschen Landrasse (*in press Züchtungskunde).*

GILMOUR, D. G. (1960), Bloodgroups in chickens. *Brit. Poultry Sci.,* **1.** 75–100.

HOHENBRINK, R. (1970), Der Heterozygotiegrad von Schweinen aus Reinzucht und Kreuzung. *Diss.* Göttingen.

KOCH, J., SCUPIN, E., STRANZINGER, G. and MITSCHERLICH, E. (1968), Untersuchungen über Beziehungen zwischen Blutgruppenfaktoren und Mastitisresistenz beim schwarzbunten Niederungsrind. *Zeitschr. f. Tierzücht. u. Züchtbiol.*, **85.** 36–46.

MITSCHERLICH, E. (1965), Genetische Beziehungen zwischen Eigenschaften des Blutes und Leistungsmerkmalen bei verschiedenen Haustierarten. *Züchtungskunde,* **37.** 375–387.

SCHYIA, D. (1969), Untersuchungen über Beziehungen zwischen Blutgruppenfaktoren und Nachkommenverlusten beim Deutschen Schwarz- und Rotbunten Niederungsrind. *Diss.* Gießen.

TIKHONOV, V. N. (1966), Studies on blood group factors for genetical analysis of selection processes. Proc. Xth Europ. Conf. Anim. Blood Groups Biochem. Polymorph. Paris, 1966.

## DISCUSSION

P. MILLOT: Are the differences between the numbers statistically significant?

H. DINKLAGE: No they are not!

D. R. OSTERHOFF: Would the picture of the results obtained be clearer by including the serum polymorphic systems in these studies?

R. HOHENBRINK: We have separated the material into the offspring of each boar and there was the same tendency in results for the purebred control group as over the whole material.

*XIIth Europ. Conf. Anim. Blood Groups Biochem. Polymorph., Bp., 1972 (pp. 347–357)*

# STUDY ON CORRELATION BETWEEN BLOOD GROUPS AND SOME PRODUCTIVE CHARACTERISTICS IN THE PIG

I. WIATROSZAK

**College of Agriculture in Poznań, Department of Animal Husbandry, Laboratory of Research in Animal Blood Groups Poznań, Poland**

## SUMMARY

Investigations were carried out on 4415 bacon pigs evaluated as to their productive characteristics at the Pig Progeny Testing Stations. The following features were taken into consideration: 1. Weight of lean in primary cuts; 2. Weight of lean in ham; 3. backfat thickness, mean from 5 measurements; 4. mean daily gain in weight; 5. length of carcass; 6. slaughter age; 7. area of loin "eye".

Blood samples from all animals and from their parents were tested with at least 36 test sera belonging to 12 blood group systems. The statistical calculations were carried out for the systems: A, E, F, G, K. Special attention was paid to system E.

It was found that the homozygotic animals $E^{bdg}/E^{bdg}$ had less lean in primary cuts and in ham and thicker backfat than homozygotic animals $E^{edh}/E^{edh}$. The heterozygotic animals $E^{bdg}/E^{edh}$ did not significantly differ from either of the homozygotic groups mentioned above, showing medium values of these productive traits.

A statistically significant interaction was found only in the case of allele $G^b$.

The homozygotes $E^{bdg}/E^{bdg}$ had also shorter carcasses and lower mean daily gains in weight than homozygotes $E^{edh}/E^{edh}$. These differences, however, were slight and in most cases statistically insignificant. The tested animals did not differ in the area of loin "eye" and in slaughter age.

## INTRODUCTION

Studies on correlation between blood groups and productive features in pigs were carried out by many investigators (BALTZER, 1963; SCHRAPE, 1966; TIKHONOV, 1966; GAVALIER, 1969; WIATROSZAK and ALEXANDROWICZ, 1970), but their results are not in agreement. That is why we undertook this study on as large animal material as available.

## MATERIAL AND METHODS

Animal material included 4415 bacon pigs (2216 barrows and 2194 gilts) being progeny of 430 boars and 1123 sows. They were purebred Polish Large White (2672 animals) and Polish Landrace (1743 animals) bacon type pigs from Pig Progeny Testing Stations at Pawłowice and Mełno, where they were evaluated as to 16 productive characteristics.

The methods used were described in the work by WIATROSZAK and ALEXANDROWICZ (1970). The blood antigenic structure in animals was confronted with their carcass evaluation concerning seven traits of relatively high heritability (DUNIEC, 1960). They were as follows:

1. Weight of lean in primary cuts
2. Weight of lean in ham
3. Backfat thickness, mean from five measurements
4. Mean daily gain in weight
5. Length of carcass
6. Slaughter age
7. Area of loin "eye".

In all tested baconers and their parents the blood antigens were determined using at least 36 test sera belonging to 12 blood group systems (A, B, C, E, F, G, H, I, J, K, L, M) and their genotypes were also determined.

For a preliminary orientation a number of productive traits were confronted with blood groups. On the basis of the mean values, blood group system E appeared to be the most interesting. In this system the animals were divided in five heterozygotic and homozygotic groups.

We considered the possibility of interactions between these groups and alleles $A^A$, $F^a$, $G^b$, $K^{ac}$ and $K^{ace}$, as seen in Table 1.

The statistical calculations were carried out according to the method of analysis of variance of designing with two non-orthogonal classifications when interactions are present (FEDERER and ZELEN, 1966). Significance of more interesting comparisons of means was checked by Student $t$ test.

TABLE 1

*Blood groups considered in statistical calculations*

| Blood group system | Allele | Genotypes Group I | Genotypes Group II | Notes |
|---|---|---|---|---|
| E | $E^{aeg}$<br>$E^{bdg}$<br>$E^{edh}$ | aeg/bdg<br>aeg/edh<br>bdg/bdg<br>bdg/edh<br>edh/edh | | $E^{edh}$ includes individuals $E^{efdh}$ and $E^{edgh}$ |
| A | $A^A$ | –/– | A/– | $a^0$ allele was not considered |
| F | $F^a$ | –/– | a/– | |
| G | $G^b$ | a/a | a/b<br>b/b | |
| K | $K^{ac}$ | b/b<br>b/–<br>–/– | ac/b<br>ac/– | $K^{ac}$ includes individuals $K^{ac}$ and $K^{ace}$ |

## RESULTS

The results of investigations are presented in Tables 2 to 9. As they show, statistically significant differentiation in productive traits was found in E blood group system between homozygotic animals bdg/bdg and homozy-

TABLE 2

*Mean weight lean in primary cuts in barrows*

| No. | Genotypes in E system | Animals Gb negative | | Animals Gb positive | | Comparison negative vs. positive |
|---|---|---|---|---|---|---|
| | | number | mean (kg) | number | mean (kg) | (kg) |
| (1) | aeg/bdg | 7 | 16.14 | 86 | 15.88 | 0.26 |
| (2) | aeg/edh | 11 | 16.56 | 163 | 16.06 | 0.50 |
| (3) | bdg/bdg | 34 | 15.63 | 260 | 15.81 | −0.18 |
| (4) | bdg/edh | 66 | 16.03 | 715 | 16.02 | 0.01 |
| (5) | edh/edh | 69 | 16.27 | 805 | 16.09 | 0.18 |
| | Total | 187 | 16.08 | 2029 | 16.02 | 0.06 |
| Comparison | | | | | | |
| (1) vs (2) | | | −0.42 | | −0.18 | |
| (3) vs (5) | | | −0.64* | | −0.28** | |
| (4) vs (3) and (5) | | | 0.08 | | 0.07 | |

F = test for the significance of interaction E×G: 3.85**
* Significant at 5% level
** Significant at 1% level

TABLE 2a

*Mean weight of lean in primary cuts in gilts*

| No. | Genotypes in E system | Animals Gb negative | | Animals Gb positive | | Comparison negative vs. positive |
|---|---|---|---|---|---|---|
| | | number | mean (kg) | number | mean (kg) | (kg) |
| (1) | aeg/bdg | 10 | 17.41 | 91 | 16.87 | 0.54 |
| (2) | aeg/edh | 11 | 17.20 | 141 | 17.04 | 0.16 |
| (3) | bdg/bdg | 25 | 16.71 | 233 | 16.75 | −0.04 |
| (4) | bdg/edh | 52 | 17.08 | 758 | 16.94 | 0.14 |
| (5) | edh/edh | 69 | 17.55 | 809 | 17.10 | 0.45** |
| | Total | 167 | 17.25 | 2032 | 16.98 | 0.27 |
| Comparison | | | | | | |
| (1) vs (2) | | | 0.21 | | −0.17 | |
| (3) vs (5) | | | −0.84** | | −0.35** | |
| (4) vs (3) and (5) | | | −0.05 | | 0.02 | |

F = test for the significance of interaction E×G: non significant

TABLE 3

*Mean weight of lean in ham in barrows*

| No. | Genotypes in E system | Animals Gb negative | | Animals Gb positive | | Comparison negative vs. positive |
|---|---|---|---|---|---|---|
| | | number | mean (kg) | number | mean (kg) | (kg) |
| (1) | aeg/bdg | 7 | 4.58 | 86 | 4.46 | 0.12 |
| (2) | aeg/edh | 11 | 4.86 | 163 | 4.58 | 0.28* |
| (3) | bdg/bdg | 34 | 4.48 | 260 | 4.44 | 0.04 |
| (4) | bdg/edh | 66 | 4.60 | 715 | 4.49 | 0.11 |
| (5) | edh/edh | 69 | 4.90 | 805 | 4.50 | 0.40** |
| | Total | 187 | 4.71 | 2029 | 4.49 | 0.22** |
| Comparison | | | | | | |
| (1) vs (2) | | | −0.28 | | −0.12* | |
| (3) vs (5) | | | −0.42** | | −0.06 | |
| (4) vs (3) and (5) | | | −0.09 | | 0.02 | |

F = test for the significance of interaction E×G: 4.96**

TABLE 3a

*Mean weight of lean in ham in gilts*

| No. | Genotypes in E system | Animals Gb negative | | Animals Gb positive | | Comparison negative vs. positive |
|---|---|---|---|---|---|---|
| | | number | mean (kg) | number | mean (kg) | (kg) |
| (1) | aeg/bdg | 10 | 4.94 | 91 | 4.75 | 0.19 |
| (2) | aeg/edh | 11 | 4.87 | 141 | 4.81 | 0.06 |
| (3) | bdg/bdg | 25 | 4.96 | 233 | 4.67 | 0.29** |
| (4) | bdg/edh | 52 | 5.04 | 758 | 4.74 | 0.30** |
| (5) | edh/edh | 69 | 5.16 | 809 | 4.80 | 0.36** |
| | Total | 167 | 5.06 | 2032 | 4.76 | 0.30 |
| Comparison | | | | | | |
| (1) vs (2) | | | 0.07 | | 0.06 | |
| (3) vs (5) | | | −0.20* | | −0.13** | |
| (4) vs (3) and (5) | | | −0.02 | | 0 | |

F = test for the significance of interaction E×G: non significant

TABLE 4

*Backfat thickness mean from five measurements in barrows*

| No. | Genotypes in E system | Animals Gb negative | | Animals Gb positive | | Comparison negative vs. positive |
|---|---|---|---|---|---|---|
| | | number | mean (cm) | number | mean (cm) | (cm) |
| (1) | aeg/bdg | 7 | 2.40 | 86 | 2.70 | −0.30 |
| (2) | aeg/edh | 11 | 2.84 | 163 | 2.67 | 0.17 |
| (3) | bdg/bdg | 34 | 2.80 | 260 | 2.69 | 0.11 |
| (4) | bdg/edh | 66 | 2.77 | 715 | 2.62 | 0.15** |
| (5) | edh/edh | 69 | 2.80 | 805 | 2.60 | 0.20** |
| | Total | 187 | 2.78 | 2029 | 2.63 | 0.15** |
| Comparison | | | | | | |
| (1) vs (2) | | | −0.44* | | 0.03 | |
| (3) vs (5) | | | 0 | | 0.09** | |
| (4) vs (3) and (5) | | | −0.03 | | −0.02 | |

F = test for the significance of interaction E×G: 8.03**

TABLE 4a

*Backfat thickness mean from five measurements in gilts*

| No. | Genotypes in E system | Animals Gb negative | | Animals Gb positive | | Comparison negative vs. positive |
|---|---|---|---|---|---|---|
| | | number | mean (cm) | number | mean (cm) | (cm) |
| (1) | aeg/bdg | 10 | 2.12 | 91 | 2.33 | −0.21 |
| (2) | aeg/edh | 11 | 2.43 | 141 | 2.36 | 0.07 |
| (3) | bdg/bdg | 25 | 2.48 | 233 | 2.40 | 0.08 |
| (4) | bdg/edh | 52 | 2.56 | 758 | 2.38 | 0.18 |
| (5) | edh/edh | 69 | 2.44 | 809 | 2.33 | 0.11* |
| | Total | 167 | 2.47 | 2032 | 2.36 | 0.11** |
| Comparison | | | | | | |
| (1) vs (2) | | | −0.31 | | −0.03 | |
| (3) vs (5) | | | 0.04 | | 0.07* | |
| (4) vs (3) and (5) | | | −0.10 | | −0.02 | |

F = test fo r the significance of interaction E×G: non-significant

TABLE 5

*Mean weight of lean in primary cuts in barrows*

| No. | Genotypes in E system | Animals Fa negative | | Animals Fa positive | | Comparison negative vs. positive |
|---|---|---|---|---|---|---|
| | | number | mean (kg) | number | mean (kg) | (kg) |
| (1) | aeg/bdg | 87 | 15.94 | 6 | 15.35 | 0.59 |
| (2) | aeg/edh | 161 | 16.10 | 13 | 15.94 | 0.16 |
| (3) | bdg/bdg | 270 | 15.84 | 24 | 15.27 | 0.57* |
| (4) | bdg/edh | 727 | 16.04 | 54 | 15.84 | 0.20 |
| (5) | edh/edh | 835 | 16.12 | 39 | 15.84 | 0.28 |
| Total | | 2080 | 16.04 | 136 | 15.73 | 0.31** |
| Comparison | | | | | | |
| (1) vs (2) | | −0.16 | | −0.59 | | |
| (3) vs (5) | | −0.28** | | −0.57 | | |
| (4) vs (3) and (5) | | 0.06 | | 0.28 | | |

F = test for the significance of interaction E×F: non significant

TABLE 5a

*Mean weight of lean in primary cuts in gilts*

| No. | Genotypes in E system | Animals Fa negative | | Animals Fa positive | | Comparison negative vs. positive |
|---|---|---|---|---|---|---|
| | | number | mean (kg) | number | mean (kg) | (kg) |
| (1) | aeg/bdg | 94 | 16.95 | 7 | 16.57 | 0.38 |
| (2) | aeg/edh | 143 | 17.08 | 9 | 16.61 | 0.47 |
| (3) | bdg/bdg | 237 | 16.81 | 21 | 15.99 | 0.82** |
| (4) | bdg/edh | 748 | 16.96 | 62 | 16.77 | 0.19 |
| (5) | edh/edh | 838 | 17.15 | 40 | 16.65 | 0.50** |
| Total | | 2060 | 17.03 | 139 | 16.60 | 0.43** |
| Comparison | | | | | | |
| (1) vs (2) | | −0.13 | | −0.04 | | |
| (3) vs (5) | | −0.34** | | −0.66* | | |
| (4) vs (3) and (5) | | −0.02 | | 0.45 | | |

F = test for the significance of interaction E×F: non significant

TABLE 6

*Backfat thickness, mean from five measurements in barrows*

| No. | Genotypes in E system | Animals Fa negative | | Animals Fa positive | | Compariosn negative vs. positive |
|---|---|---|---|---|---|---|
| | | number | mean (cm) | number | mean (cm) | (cm) |
| (1) | aeg/bdg | 87 | 2.67 | 6 | 2.89 | —0.22 |
| (2) | aeg/edh | 161 | 2.67 | 13 | 2.79 | —0.12 |
| (3) | bdg/bdg | 270 | 2.69 | 24 | 2.84 | —0.15 |
| (4) | bdg/edh | 727 | 2.63 | 54 | 2.74 | —0.11 |
| (5) | edh/edh | 835 | 2.60 | 39 | 2.90 | —0.30** |
| | Total | 2080 | 2.63 | 136 | 2.81 | —0.18** |
| Comparison | | | | | | |
| | (1) vs (2) | 0 | | 0.10 | | |
| | (3) vs (5) | 0.09** | | —0.06 | | |
| | (4) vs (3) and (5) | —0.01 | | —0.13 | | |

F = test for the significance of interaction E×F: non significant

TABLE 6a

*Backfat thickness, mean from five measurements in gilts*

| No. | Genotypes in E system | Animals Fa negative | | Animals Fa positive | | Comparison negative vs. positive |
|---|---|---|---|---|---|---|
| | | number | mean (cm) | number | mean (cm) | (cm) |
| (1) | aeg/bdg | 94 | 2.30 | 7 | 2.39 | —0.09 |
| (2) | aeg/edh | 143 | 2.36 | 9 | 2.38 | —0.02 |
| (3) | bdg/bdg | 237 | 2.40 | 21 | 2.47 | —0.07 |
| (4) | bdg/edh | 748 | 2.38 | 62 | 2.45 | —0.07 |
| (5) | edh/edh | 838 | 2.33 | 40 | 2.47 | —0.14* |
| | Total | 2060 | 2.36 | 139 | 2.45 | —0.09* |
| Comparison | | | | | | |
| | (1) vs (2) | —0.06 | | 0.01 | | |
| | (3) vs (5) | 0.07* | | 0 | | |
| | (4) vs (3) and (5) | 0.02 | | —0.02 | | |

F = test for the significance of interaction E×F: non significant

TABLE 7

*Mean weight of lean in primary cuts in barrows*

| No. | Genotypes in E system | Number of animals | Mean (kg) |
|---|---|---|---|
| (1) | aeg/bdg | 93 | 15.90 |
| (2) | aeg/edh | 174 | 16.09 |
| (3) | bdg/bdg | 294 | 15.79 |
| (4) | bdg/edh | 781 | 16.03 |
| (5) | edh/edh | 874 | 16.10 |
| | Total | 2216 | |
| Comparison | | | |
| (1) vs (2) | | | −0.19 |
| (3) vs (5) | | | −0.31** |
| (4) vs (3) and (5) | | | 0.09 |

TABLE 7a

*Mean weight of lean in primary cuts in gilts*

| No. | Genotypes in E system | Number of animals | Mean (kg) |
|---|---|---|---|
| (1) | aeg/bdg | 101 | 16.93 |
| (2) | aeg/edh | 152 | 17.05 |
| (3) | bdg/bdg | 258 | 16.75 |
| (4) | bdg/edh | 810 | 16.95 |
| (5) | edh/edh | 878 | 17.13 |
| | Total | 2199 | |
| Comparison | | | |
| (1) vs (2) | | | −0.12 |
| (3) vs (5) | | | −0.38** |
| (4) vs (3) and (5) | | | 0.01 |

gotes edh/edh. The homozygotes $E^{bdg}/E^{bdg}$ had less lean in primary cuts and in ham, and thicker backfat than homozygotes $E^{edh}/E^{edh}$. The group of heterozygotic animals $E^{bdg}/E^{edh}$ did not significantly differ from homozygotic animals in groups mentioned above, showing medium values in these productive features. On the basis of these findings we may suppose that none of these alleles is dominant. The tables show similar results in the groups of gilts and barrows as to studied features.

Statistical analysis of interactions between blood group system E and alleles of 4 other systems showed a statistically significant interaction only

TABLE 8

*Mean weight of lean in ham in barrows*

| No. | Genotypes in E system | Number of animals | Mean (kg) |
|---|---|---|---|
| (1) | aeg/bdg | 93 | 4.47 |
| (2) | aeg/edh | 174 | 4.60 |
| (3) | bdg/bdg | 294 | 4.44 |
| (4) | bdg/edh | 781 | 4.50 |
| (5) | edh/edh | 874 | 4.53 |
| | Total | 2216 | |
| Comparison | | | |
| (1) vs (2) | | | −0.13* |
| (3) vs (5) | | | −0.09** |
| (4) vs (3) and (5) | | | 0.02 |

TABLE 8a

*Mean weight of lean in ham in gilts*

| No. | Genotypes in E system | Number of animals | Mean (kg) |
|---|---|---|---|
| (1) | aeg/bdg | 101 | 4.77 |
| (2) | aeg/edh | 152 | 4.82 |
| (3) | bdg/bdg | 258 | 4.70 |
| (4) | bdg/edh | 810 | 4.76 |
| (5) | edh/edh | 878 | 4.83 |
| | Total | 2199 | |
| Comparison | | | |
| (1) vs (2) | | | −0.05 |
| (3) vs (5) | | | −0.13** |
| (4) vs (3) and (5) | | | 0 |

in the case of allele $G^b$ in barrows. It would mean that the differences found in barrows of blood group system E depended on the presence of factor $G^b$ in them. In gilts this interaction was insignificant.

As regards other productive characteristics studied, the homozygotic animals $E^{bdg}/E^{bdg}$ had also lower mean daily gains in weight and shorter carcasses than homozygotes $E^{edh}/E^{edh}$. These differences, however, were slight and in most cases statistically insignificant.

No significant differences were found concerning the two remaining characteristics studied, namely area of loin "eye" and slaughter age.

TABLE 9

*Backfat thickness, mean from five measurements in barrows*

| No. | Genotypes in E system | Number of animals | Mean (cm) |
|---|---|---|---|
| (1) | aeg/bdg | 93 | 2.68 |
| (2) | aeg/edh | 174 | 2.68 |
| (3) | bdg/bdg | 294 | 2.70 |
| (4) | bdg/edh | 781 | 2.63 |
| (5) | edh/edh | 874 | 2.61 |
| Total | | 2216 | |
| Comparison | | | |
| (1) vs (2) | | 0 | |
| (3) vs (5) | | 0.09** | |
| (4) vs (3) and (5) | | −0.03 | |

TABLE 9a

*Backfat thickness, mean from five measurements in gilts*

| No. | Genotypes in E system | Number of animals | Mean (cm) |
|---|---|---|---|
| (1) | aeg/bdg | 101 | 2.31 |
| (2) | aeg/edh | 152 | 2.36 |
| (3) | bdg/bdg | 258 | 2.40 |
| (4) | bdg/edh | 810 | 2.39 |
| (5) | edh/edh | 878 | 2.34 |
| Total | | 2199 | |
| Comparison | | | |
| (1) vs (2) | | −0.05 | |
| (3) vs (5) | | 0.06* | |
| (4) vs (3) and (5) | | 0.02 | |

The results of these investigations permit the supposition that the correlations found are not merely accidental. This supposition, however, needs further study to be confirmed.

The differences found are not large from a practical point of view as mostly they do not exceed 5%, but even such correlations of blood groups and highly inheritable productive features may appear to be helpful when searching for indices for early selection of pigs.

## ACKNOWLEDGEMENT

These investigations were carried out with the financial support of a grant from the Department of Agriculture of the United States. The author expresses his gratitude for this support.

## REFERENCES

BALTZER, J. (1963), Untersuchungen über das Bestehen von Beziehungen zwischen Blutgruppenfaktoren und Daten des Schlachtkörperwertes und Mastleistung des Schweines. *Dissertation.* Göttingen.

DUNIEC, M. (1960), Korelacje fenotypowe i genetyczne między niektórymi cechami użytkowymi oraz ich wskaźniki odziedziczalności u świń typu mięsnego. Kraków.

FEDERER, W. T., ZELEN, M. (1966), Analysis of multifactor classifications with unequal numbers of observations. *Biometrics,* **22.** 3.

GAVALIER, M. (1969), Prispevek ke studiu korelaci mezi erytrocytarnimi antigeny a nekterymi ukazateli vykrmnosti a jatecne hodnoty prasat. *Živocisna vyroba,* **14.** 2.

SCHRAPE, H. (1966), Untersuchungen über Beziehungen zwischen Blutgruppenfaktoren und Leistungseigenschaften beim Schwein. *Dissertation.* Göttingen.

TIKHONOV, V. N. (1966), Immunogenetic analysis of polymorphism in blood groups in connection with some problems of selection. *Thesis.* Novosibirsk.

WIATROSZAK, I. and ALEXANDROWICZ, S. (1970), Study on correlation between blood groups and some productive characteristics in pig. Proc. XIth Europ. Conf. Anim. Blood Groups Biochem. Polymorph. Warsaw, 1968.

*XIIth Europ. Conf. Anim. Blood Groups Biochem. Polymorph., Bp., 1972 (pp. 359–362)*

# INDUCTION OF GENETIC CORRELATION BETWEEN BLOOD GROUPS AND PRODUCTIVE CHARACTERS BY LINE-BREEDING

V. N. Tikhonov and Zoja K. Burlak

Institute of Cytology and Genetics of the USSR Academy of Sciences,
Siberian Research Institute of Animal Breeding, Novosibirsk, USSR

## SUMMARY

Selection for blood groups permits to design models of transmission of hereditary characters from the parents to the offspring. It provides a basis for the prediction of the best combinations of parents. Thus, the immunogenetic method may considerably modernize line-breeding and promote the solution of the main problem, namely to retain and increase the genetic similarity to the line ancestor, using minimum inbreeding.

There is, as a rule, no true direct genetic correlation between productive performances and blood groups, but there are possibilities of using false genetic correlations induced by line-breeding for improving the productivity of a population.

The relation between antigens of blood groups and economic characters can be due to three genetic mechanisms: true and false genetic correlations and heterozygotic effect (partially related to the mentioned correlations).

True correlation determined by pleiotropy and sometimes by complete linkage is based on the physical and biological nature of gene effect. Significant correlations between blood groups and productive characters have been very rarely established so far; the value of these correlations is rather low and the relative contribution of their effect among the other factors determining productivity is small as well (Tikhonov and Burlak, 1969).

False genetic correlations arise as the consequence of the loss of independent distributions between genetic formulas of blood groups and productive characters in populations (Le Roy, 1968).

We consider this connection can be artificially induced directly by line-breeding, using animal mating with different degrees of inbreeding (Lush, 1949). Immunogenetic control of transmission of inherited factors from generation to generation by means of blood groups, permits to retain genetical similarity of the animals of line with line ancestor using minimum inbreeding (Figs 1 and 2).

Most alleles of the animal blood groups studied up to the present do not appear to be connected with productive characters .However if the greatly productive line ancestors are marked by the blood group alleles, the latter will always pass to the progeny together with its hereditary basis of high productivity. Consequently, the false genetic correlations of the blood groups with the productive characters may be induced by the line-breeding.

The immunogenetic control of induction of above-mentioned correlations is realized by the following methods:

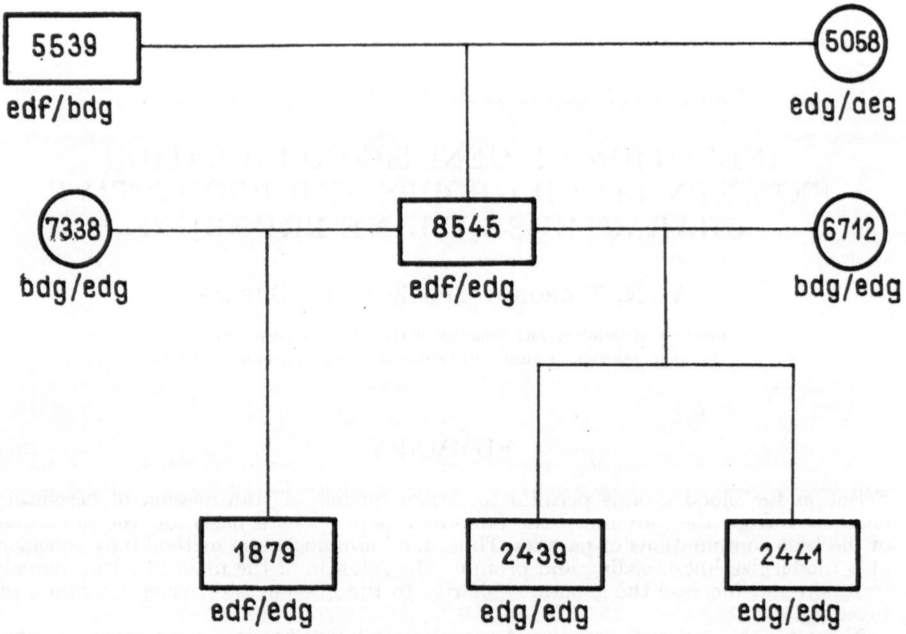
5539
edf/bdg
5058
edg/aeg
7338
bdg/edg
8545
edf/edg
6712
bdg/edg
1879
edf/edg
2439
edg/edg
2441
edg/edg

FIG. 1

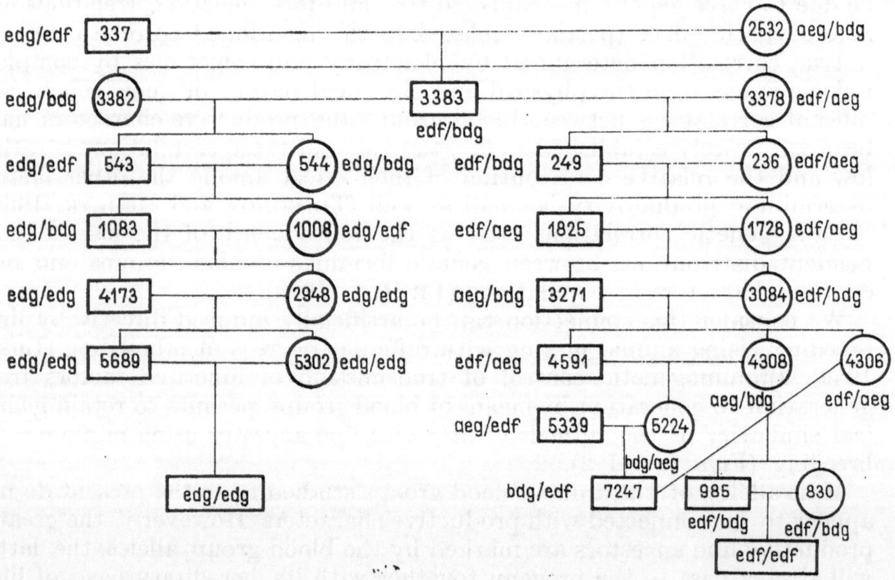
edg/edf 337
2532 aeg/bdg
edg/bdg 3382
3383
edf/bdg
3378 edf/aeg
edg/edf 543
544 edg/bdg
edf/bdg 249
236 edf/aeg
edg/bdg 1083
1008 edg/edf
edf/aeg 1825
1728 edf/aeg
edg/edg 4173
2948 edg/edg
aeg/bdg 3271
3084 edf/bdg
edg/edg 5689
5302 edg/edg
edf/aeg 4591
4308
aeg/bdg
4306
edf/aeg
aeg/edf 5339
5224
bdg/aeg
bdg/edf 7247
985
edf/bdg
830
edf/bdg
edg/edg
edf/edf

FIG. 2

1. The definition of specific immunogenetic characters according to blood group antigens of a sire-ancestor of mark. For this purpose both well-known genetic systems and separate antigens, especially those being controlled by the alleles with low gene frequency, may be used (TIKHONOV, 1962).

2. The performance progeny testing of the main sires in the line including the definition of blood group alleles, which are transmitted to the progeny simultaneously with higher productivity (Table 1).

3. The definition of the specific immunogenetic characters of the line, including the total correlation according to animal blood groups of this line, with the line ancestor. For this purpose, the correlation formula-index of similarity $r = \frac{\Sigma X_i \cdot Y_i}{\sqrt{\Sigma X_i^2 \cdot \Sigma Y_i^2}}$ (MAIJALA and LINDSTRÖM, 1966) can be used, which demonstrates the difference of the line from the rest line of the breed as well (Table 2).

4. The control of marker allele transmission from generation to generation, to avoid the loss of specific line marker antigenes (compare animals boars N 1879 with NN 2439 and 2441, Fig. 1, the two last animals cannot be considered as the members of this line, as, according to immunogenetical test, they have not got from the father the marker antigen Eedf).

5. The reconstruction of the line ancestor genotype by the selection of corresponding animals for mating. Similarly, one can perform the improvement of the inbred line by the introduction of boars and female swines, remotely related but similar in genotypes, into the line (Fig. 2).

TABLE 1

*Differentiation of the offspring productive performance according to sire transmitting marker alleles of E-, G- and F-loci*

| Sire | Marker allele | No. of offspring | Age of achieving the weigth of 95 kg | Age of achieving the weigth of 100 kg | Average daily gain (g) | Food conversion (kgs) | Length of sternum (cm) |
|---|---|---|---|---|---|---|---|
| | | | Locus E | | | | |
| Viking 6615 | $E^{aeg}$ | 3 | 173.3 | 178.6 | 730 | 3.84 | 97.0 |
| | $E^{edg}$ | 9 | 184.7 | 187.9 | 627 | 4.18 | 91.7 |
| Erk 7727 | $E^{aeg}$ | 3 | 192.0 | 198.7 | 627 | 4.02 | 93.0 |
| | $E^{edg}$ | 8 | 180.6 | 186.8 | 657 | 3.87 | 93.1 |
| | | | Locus G | | | | |
| Viking 6615 | $G^{a}$ | 3 | 179.0 | 184.7 | 700 | 3.82 | 95.7 |
| | $G^{b}$ | 9 | 182.8 | 189.9 | 637 | 4.19 | 92.1 |
| Pjardik 9485 | $G^{a}$ | 8 | 182.1 | 188.9 | 662 | 3.94 | 89.8 |
| | $G^{b}$ | 5 | 196.4 | 204.0 | 619 | 3.98 | 92.4 |
| | | | Locus F | | | | |
| Viking 6615 | $F^{a}$ | 2 | 168.5 | 175.0 | 757 | 3.79 | 97.5 |
| | $F^{b}$ | 10 | 183.2 | 191.8 | 632 | 4.16 | 92.1 |
| Pjardik 9485 | $F^{a}$ | 14 | 189.3 | 196.0 | 632 | 4.03 | 91.9 |
| | $F^{b}$ | 6 | 184.3 | 191.1 | 661 | 3.82 | 93.7 |

TABLE 2

*Index of immunogenetic likeness between lines of Siberian Severnaya breed according to frequency of crosses*

| Line | No. of genea-logic groups | Sibiryak | Kedr | Taeznyi | Dobryi | Enisei | Tiger | Dikyi | Nalim |
|---|---|---|---|---|---|---|---|---|---|
| Sibiryak | 4 | — | 0.84 | 0.86 | 0.86 | 0.88 | 0.89 | 0.90 | 0.91 |
| Kedr | 3 | 0.84 | — | 0.92 | 0.89 | 0.92 | 0.94 | 0.93 | 0.91 |
| Taeznyi | 2 | 0.86 | 0.92 | — | 0.98 | 0.94 | 0.99 | 0.96 | 0.96 |
| Dobryi | 2 | 0.86 | 0.89 | 0.98 | — | 0.98 | 0.94 | 0.99 | 0.96 |
| Enisei | 2 | 0.88 | 0.92 | 0.94 | 0.98 | — | 0.97 | 0.98 | 0.99 |
| Tiger | 1 | 0.89 | 0.94 | 0.99 | 0.94 | 0.97 | — | 0.99 | 0.98 |
| Dikyi | 1 | 0.90 | 0.93 | 0.96 | 0.99 | 0.98 | 0.99 | — | 0.99 |
| Nalim | 1 | 0.91 | 0.91 | 0.96 | 0.96 | 0.99 | 0.98 | 0.99 | — |

## REFERENCES

LE ROY, H. L. (1968), Elemente der Tierzucht. Genetik, Mathematik, Populationsgenetik. München.

LUSH, J. L. (1949), Animal Breeding Plans. Iowa State University Press, Ames, Iowa.

MAIJALA, K. and LINDSTRÖM, G. (1966), Frequencies of blood group genes and factors in the Finnish cattle breeds with special regard to breed comparison. *Annales Agric. Fenniae*, **5.** 76.

TIKHONOV, V. N. (1962), The study of blood groups of the pig and the possibilities of their applying for immunogenetic control in the USSR. Report. 8th Europ. Conf. Anim. Blood Groups, Ljubljana.

TIKHONOV, V. N. (1967), The blood groups in animals and their application for animal breeding. Moscow, Kolos.

TIKHONOV, V. N. and BURLAK, Z. K. (1969), The immunogenetic analysis of the crossing of animals and for studying of heterosis. *Zivotnovodstvo*, **12.** 48.

*XIIth Europ. Conf. Anim. Blood Groups Biochem. Polymorph., Bp., 1972 (pp. 363 – 367)*

# DATA ON THE INFLUENCE OF PARENTAL IMMUNOGENETIC BLOOD INDICES ON SOME ECONOMICALLY VALUABLE CHARACTERISTICS OF OFFSPRING IN PIG BREEDING

NINA O. SUKHOVA, M. O. SIMON, N. I. SEMENOV,
I. M. MASLUKOV and IRINA S. LISINA

**Laboratory of Immunogenetics and Department of Pig-Breeding of Sibirian Research and Designing-Technological Institute of Animal Breeding, Novosibirsk, USSR**

## SUMMARY

**Analysis of experimental studies on fattening control and observations made in pig breeding have shown definite influence of differences in parents' blood types on litter size and growth rate of offspring. In pure breeding, the litter size and growth rate of piglets increased with the deviation in the parents' blood types.**

## INTRODUCTION

The problem of increasing the production of farm animals and poultry by control of blood group factors has attracted the attention of specialists of various branches.

Investigations have shown a definite connection of blood group genotypes with economic traits (RENDEL, 1959, 1961; BALTZER, 1964; STORMONT, 1966; TIKHONOV, 1967; etc.). Some authors, nevertheless, failed to confirm this relationship (BRUCKS, 1964; SMITH, JENSEN, BAUER and COX, 1968; etc.).

The contradictory data have necessitated additional investigations in this field.

The most feasible resolution seems to be the establishment of connection between the immunogenetic indices of animals and their productivity by means of the analysis of data obtained by comparison of offspring from parents displaying different degree of similarity in the antigenic blood properties.

Such an approach is based on Darwin's and his followers' well-known idea on biological superiority of individuals representing a result of distant (non-consanguineous) interbreedings.

Based on Landsteiner's data (LANDSTEINER, 1945), who established the changes of cross reactions by the degree of distant phylogenetic relationship between animals, the authors have chosen the index of antigenic properties of blood as the index of genealogical closeness.

## MATERIAL AND METHODS

The extent of the influence of antigenic properties of parental blood on vitality and growth of the offspring has been studied in pedigree pig breed, and the development of the animals has been observed during inbreeding.

In the experiment a group of sows of the White-Black breed have been inseminated with the mixed sperms of two boars, one of Large White breed, the other of White-Black breed. The boars of each breed were produced by inbred and outbred animals. The origin of the breed was determined by immunogenetic methods (SUKHOVA, DMITRIEVA, et al., 1968). A total of 60 offspring from one breed have been bred for the control fattening to the age of six months while another 72 pigs from 6 boars and 18 sows of the Large White breed were also under control for fattening. Each boar was mated to three sows from which one considerably differed in blood type from the boar, the second sow differed in a minor degree and the third one differed hardly. The sows and the boars of both breeds were selected according to age, body conformation and productivity. All animals were kept under control for fattening and fed on the standard ration formulated by the All Union Institute of animal Breeding (mixed feed 55–5, and skim milk).

In addition to the experimental animals and the animals under control for fattening, a total of 2632 pigs of the Large White breed derived from the breeding farm "Bolshevik" in the Novosibirsk area were under observation for three years. All animals were managed uniformly according to the standards.

Results were analyzed according to the parents' blood type on the basis of cross-reactions using a set of blood-typing reagents (SUKHOVA, 1968; SIMON, SUKHOVA et al., 1970). The differences between parents were considered to be more pronounced if the erythrocytes of one of the parents gave strong reaction with the reagent that has previously been absorbed with the red blood cells of the other partner. On the contrary, the blood types were considered more similar if the erythrocytes of the parents did not show any agglutination in cross-reactions. The agglutination reactions were made by Coomb's usual method in saline and with dextran according to ANDRESEN (1963) and TIKHONOV (1967).

## RESULTS

The results of comparative studies on the growth of pure-bred and cross-bred offspring from White-Black sows mated with inbred and outbred sires are shown in Table 1.

The Table shows a certain increase of growth rate in cross-bred pigs, being particularly pronounced in the offspring of inbred boars. Irrespective whether the sire was inbred or outbred, the group of White-Black breed had similar indices of development.

Grouping the data on Sibirian Black and White breed according to the blood types of the parents, the differences between such groups were considerably greater (Table 2).

The data clearly show the higher growth rates of pure-bred pigs deriving from parents with marked differences in blood type. In the groups studied the differences in live weight persisted until the 6th month of age. Differences between cross-bred pigs in groups of the same blood type were insignificant.

TABLE 1

*Results of fattening control in pure-bred and cross-bred pigs deriving from the Sibirian White Black sows*

| Indices / Origin | Average live weight (kg) Age (month) No. | two | four | six | Difference in live weight at 6 months of age td 1–4 2–4 3–4 | td 1–3 2–3 | td 1–2 | σ |
|---|---|---|---|---|---|---|---|---|
| Cross-bred offspring from Large White inbred boar | 14 | 19.17 | 48.06 | 90.29 | 5.70*** | 5.60*** | 2.54 | 7.29 |
| Cross-bred offspring from Large White outbred boar | 19 | 19.34 | 45.58 | 83.46 | 3.51** | 3.3** | 0 | 7.52 |
| Black and White inbred boar | 16 | 16.63 | 38.13 | 75.30 | 0.37 | 0 | — | 6.61 |
| Black and White outbred boar | 11 | 15.16 | 40.90 | 74.35 | 0 | — | — | 6.00 |

TABLE 2

*Growth of the Black and White pigs with regard to similarities in parents' blood types*

| Degree of difference in parents' blood types | | New-born weight (kg) M+ —m | One-month weight increase (kg) M+ —m | Two-month weight increase (kg) M+ —m |
|---|---|---|---|---|
| Essential | 8 | 1.42 +–0.10 | 7.72 +–0.55 | 19.99 +–2.08 |
| Average | 14 | 1.55 +–0.07 | 6.46 +–0.69 | 15.40 +–1.42 |
| Absent | 60 | 1.24 +–0.03 | 6.20 +–0.18 | 14.63 +–0.30 |
| | 82 | 1.32 | 6.33 | 15.09 |

Control fattening results in the Large White breed, divided into groups with regard to similarities in the parents' blood types, are given in Table 3.

Data of Table 3, as well as of the previous experiment, show the higher growth rate of pigs with marked differences in parents' blood type.

Analogous results on growth rate were obtained in a larger group of pigs studied under farm conditions over a three-year period. Weight data of pigs recorded in 1970 are shown in Table 4.

Table 4 shows the increased litter size and growth rate of pigs under farm conditions in the case of marked differences in the parents' blood type.

TABLE 3

*Weight increase of pigs under fattening control*

| Group | Degree of difference of parents' blood | Quantity | | Daily increase of pigs' live weight (g) | td 1–3 | P |
|---|---|---|---|---|---|---|
| | | litter | pigs | M+ —m | | |
| 1 | Strong | 6 | 24 | 651 + –9.2 | 5.6 | 0.001 |
| 2 | Average | 6 | 24 | 603 + –4.3 | | |
| 3 | Absent | 6 | 24 | 549 + –14.9 | | |

TABLE 4

*Average performances of the Large White pigs in the State Farm. "Bolshevik" (1970) in connection with the parents' blood type*

| Group | Degree of difference in parents' blood types | Number of | | Litter size (M) | Body weight | | | |
|---|---|---|---|---|---|---|---|---|
| | | Litter | Pigs | | newborn (M) | one-month old (M) | two-month old (M) | four-month old (M) |
| 1 | 7 | 11 | 118 | 10.7 | 1.25 | 8.5 | 19.0 | 38.84 |
| 2 | 6 | 9 | 99 | 11.0 | 1.21 | 8.5 | 19.1 | 37.2 |
| 3 | 5 | 22 | 223 | 10.2 | 1.18 | 8.1 | 19.3 | 36.1 |
| 4 | 4 | 22 | 224 | 10.1 | 1.17 | 7.6 | 16.4 | 34.2 |
| 5 | 3 | 17 | 176 | 10.3 | 1.03 | 7.5 | 17.9 | 34.7 |
| 6 | 2 | 20 | 169 | 8.4 | 1.01 | 7.5 | 17.8 | 33.8 |
| 7 | 1 | 11 | 105 | 9.5 | 1.00 | 7.8 | 17.8 | 32.7 |
| 8 | 0 | 8 | 67 | 8.3 | 1.03 | 7.3 | 17.5 | 32.0 |

## DISCUSSION

The relationship between similarities of parental blood types and litter size and growth rates of offspring in swine breeds has been disclosed by a series of observations. Considerably greater differences of weight were found in pig groups having some degree of difference in parental blood type as compared to those having identical parental blood types. The results of the fattening control also show higher growth rate of the offspring of those sows which differed considerably in blood type from the sires.

The relationship of productive traits with the stronger reaction with absorbed sera, which is explicable by the larger number of blood factors specific for the residual antibodies, agrees with the literary data on summation of favourable effects of allele heterozygosis (PLUM, 1959; SCHULTZ and BRILES, 1953, SOKOL, 1970).

More pronounced complementary or additive effect may be assumed under definite difference in genetic information.

## REFERENCES


ANDRESEN, E. (1963), A study of blood groups of the pig. Munksgaard, Copenhagen.

BALTZER, I. (1964), Relationships between blood group factors and carcass quality and fattening performance in pigs. *Züchtungskunde*, **36.**

BRUCKS, R. (1964), Die Blutgruppensystemed es Schweines unter besonderer Berücksichtigung des L-System.

LANDSTEINER, K. (1945), The Specificity of Serological Reactions, Harvard Univ. Press, Cambridge.

PLUM, M. (1959), Hetero blood types and breeding performance. *Science*, **129.** 3351.

RENDEL, J. (1959), A study on relationship between blood groups and production characters in cattle. Report VIth Europ. Conf. Anim. Blood Groups, Munich, 1959. 8–23.

RENDEL, J. (1961), Recent studies on relationships between blood groups and production characters in farm animals. *Z. Tierzucht und Züchtungsbiol.* **75.** 2.

SCHULTZ, F. T. and BRILES, W. E. (1953), The adaptive value of blood group gene in chickens. *Genetics*, **38.** 34.

SIMON, M. O., SUKHOVA, NINA O., LISINA, IRINA S., SEMENOV, N. I., GORELOVA, GALINA K. and DMITRIEVA, GALINA, L. (1970), Usage of antigenic properties of blood in selective breeding work. *Animal Breeding*, **1.** 57.

SMITH, C., JENSEN, E. L., BAKER, L. W. and COX, D. F. (1968), Quantitative studies on blood group and serum protein systems in pigs. *J. Anim. Sci.* **27.** 4.

SOKOL, V. I. (1970), Genetic polymorphous systems and some possibilities of their use in genetics and selection of cattle. *Thesis.*

STORMONT, C. (1966), Contribution of blood typing to dairy science progress. *J. Dairy Sci.*, **50.** 2.

SUKHOVA, NINA O. (1970), An immunogenetic comparison of blood antigens and their value in selection of breeding stock. Proc. XIth Europ. Conf. Anim. Blood Groups and Biochem. Polymorph. Warsaw, 1968.

SUKHOVA, NINA O., DMITRIEVA, GALINA L., GORELOVA, GALINA K. and KOLOMNIKOV, V. A. (1968), Immunogenetic analysis of origin of pigs obtained by heterospermous insemination of sows. "Animal Breeding in Siberia during 50 Years", Novosibirsk, 221–231.

TIKHONOV, V. N. (1967), Usage of blood group in selection of animals. Publishing House "Kolos", Moscow.

*XIIth Europ. Conf. Anim. Blood Groups Biochem. Polymorph., Bp., 1972 (pp. 369–374)*

# A STUDY ON LYMPHOCYTE ANTIGENS IN PIGS BY MEANS OF ANTI-ERYTHROCYTE REAGENTS

M. SIMON and J. HOJNÝ

Laboratory of Animal Genetics, Czechoslovak Academy of Sciences, Liběchov, Czechoslovakia

## SUMMARY

Anti-erythrocyte reagents were absorbed by lymphocytes obtained from lymph nodes. 41 antibodies currently used for the determination of erythrocyte antigens in 15 blood group systems in pigs were used in our experiments. The lymphocytes from animals whose erythrocytes had the corresponding antigen, absorbed or considerably reduced the titre of antisera in the A, E and N blood group systems. There was no marked decrease of titre in the other antisera not even after repeated absorption by lymphocytes. The lymphocytes of animals whose erythrocytes did not carry the corresponding antigen did not absorb the antibody.

It can therefore be assumed that the antigens of the A, E and N systems are common to both erythrocytes and lymphocytes.

## INTRODUCTION

The possible presence of erythrocyte antigens on lymphocytes was detected both by direct (agglutination) and indirect (absorption) methods. Convincing evidence for the presence of ABO factors on human leucocytes by means of the direct method was given by GURNER and COOMBS (1958) and by BERROCHE et al. (1955) by means of the indirect method. DAUSSET and TANGÜN (1965) summarized reports dealing with the detection of erythrocyte isoantigens on leucocytes and platelets. The presence of AB antigens on leucocytes and platelets is already beyond doubt. As to the other erythrocyte antigens results were often contradictory. Special attention has been paid to antigen D, but in spite of various methods used the conclusions are not unequivocal. In animals very little has been done on possible relations between erythrocyte and leucocyte antigens. HÁLA (1967) absorbed antierythrocyte antibodies of pigs with lymphocytes from lymph nodes and some tissue homogenates. Positive results with lymphocytes were obtained in anti Ea, Eb and Na, no decrease of titre following absorption was found with anti Gb, Ia, Ib, Ha, Hb, and Ma.

## MATERIAL AND METHODS

Anti-erythrocyte antibodies of pigs (HOJNÝ, HRADECKÝ, 1968) were absorbed with lymphocytes obtained from lymph nodes of pigs killed at the slaughterhouse. Blood samples were collected from the same animals and tested for the presence of erythrocyte antigens.

The nodes were cut into little cubes, the lymphocytes released and washed three to four times in saline and at the same time the unwanted erythrocytes were discarded. After 10 min centrifugation at 3800 r.p.m. sedimented lymphocyte mass was obtained the upper part of which was only slightly contaminated by erythrocytes. This layer was transferred to another test-

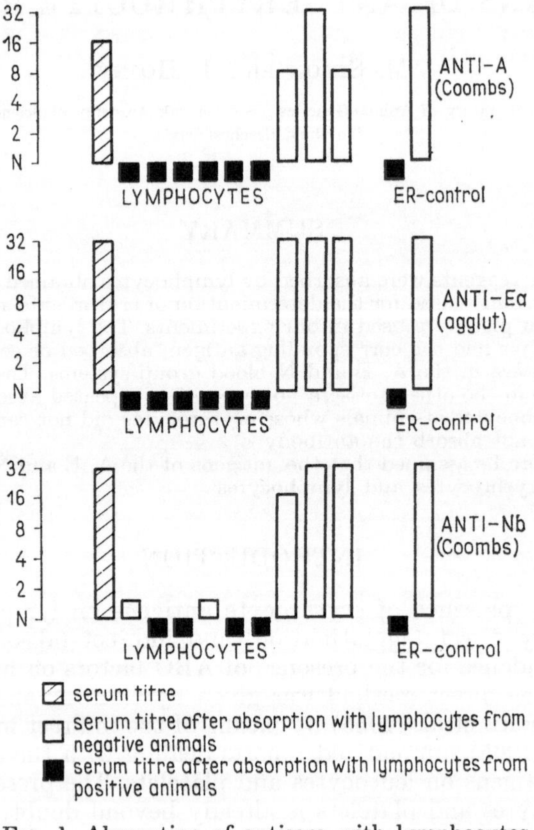
32
16
8
4
2
N
ANTI-A
(Coombs)
LYMPHOCYTES
ER-control
ANTI-Ea
(agglut.)
ANTI-Nb
(Coombs)
serum titre
serum titre after absorption with lymphocytes from negative animals
serum titre after absorption with lymphocytes from positive animals

FIG. 1. Absorption of antisera with lymphocytes

tube, re-suspended and centrifuged again. After the fourth washing the contamination of lymphocytes by red blood cells (RBC) was not higher than 0.3%. During the whole procedure the suspensions were kept in an ice bath.

Antisera were absorbed once or twice (1 part of lymphocyte sediment and 1 part of serum) 30 min at 37°C and after absorption they were kept another 30 min in the refrigerator at 4°C. During the whole period of absorption the test-tubes were shaken several times. Besides positive animals, antisera were also absorbed with lymphocytes of individuals whose erythrocytes were lacking the corresponding antigen. Absorptions by erythrocytes from positive and negative animals served as further controls.

After absorption the titre of eventually remaining antibodies in the reagents was determined together with the determination of the control titre of the corresponding non-absorbed reagent.

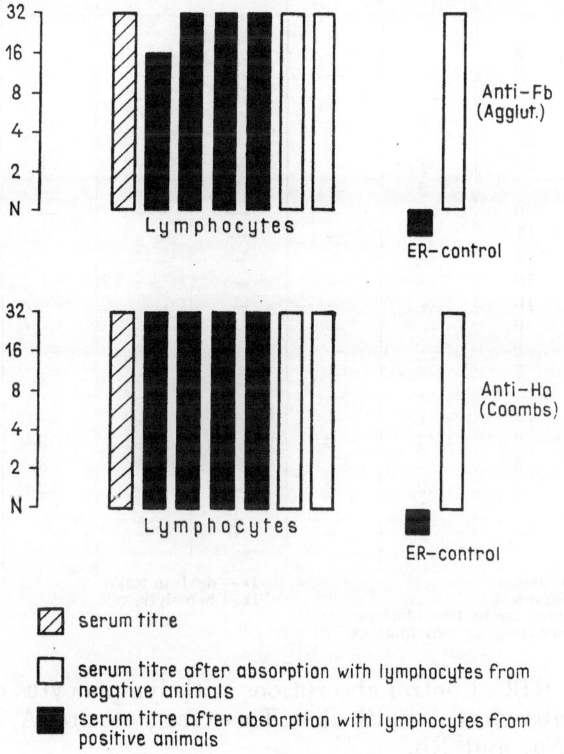
32
16
8
4
2
N
Lymphocytes
ER-control
Anti-Fb
(Agglut.)
32
16
8
4
2
N
Lymphocytes
ER-control
Anti-Ha
(Coombs)
serum titre
serum titre after absorption with lymphocytes from negative animals
serum titre after absorption with lymphocytes from positive animals

FIG. 2. Absorption of antisera with lymphocytes

We have tested a total of 41 antisera. Those with an unsatisfactory titre were excluded from investigation. The titre of strong antisera was adjusted to a level of 1/8–1/32. Antibodies were tested by the four current serological procedures i.e. direct agglutination, indirect Coombs, hemolytic and dextran tests.

## RESULTS

On the basis of the results obtained by absorbing blood group reagents with lymphocytes (see Table 1a, b), antisera could be divided into 3 groups:

a) The first group comprises the reagents the antibodies of which were fully absorbed by lymphocytes of animals carrying the corresponding anti-

TABLE 1a

*Absorption of antisera with lymphocytes*

| Reagent | Test | Positive animals | | | | Negative animals | | | |
|---|---|---|---|---|---|---|---|---|---|
| | | Abs. | Titre reduced | | Not abs. | Abs. | Titre reduced | | Not abs. |
| | | | slightly | considerably | | | slightly | considerably | |
| A | A | 2 | | | | | | | 2 |
| | C | 7 | — | — | — | — | — | — | 7 |
| Ba | A | — | — | — | 8 | — | — | — | — |
| Bb | A | — | 1 | — | 6 | — | — | — | 4 |
| Ca | C | — | — | — | 6 | — | — | — | 4 |
| Da | A | — | 1 | — | 6 | — | 1 | — | 4 |
| Db | A | — | 2 | — | 8 | — | — | — | — |
| Ea | A | 9 | — | — | — | — | — | — | 6 |
| Eb | A | 11 | — | — | — | — | — | — | 7 |
| Ed | A | 11 | — | — | — | — | — | — | 2 |
| Ee | A | 10 | — | — | — | — | — | — | 4 |
| Eg | D | 10 | — | — | — | — | — | — | 2 |
| Ej | A | 6 | — | — | — | — | — | — | 6 |
| Ek | A | 8 | — | — | — | — | — | — | — |
| El | A | 7 | — | — | — | — | — | — | 5 |
| Fa | A | — | 1 | — | 7 | — | — | — | 4 |
| Fb | A | — | 1 | — | 6 | — | — | — | 1 |
| Ga | C | 2 | 3 | 4 | — | — | 3 | — | 2 |
| Gb | A | 6 | | | | | 3 | | |
| | C | 2 | 9 | 4 | — | — | 2 | 3 | 2 |
| Ha | C | — | 1 | — | 8 | — | — | — | 3 |
| Hb | H | — | 1 | — | 6 | — | — | — | 5 |

A — direct agglutination
C — indirect Coombs test
slightly — not more than by two dilutions
considerably — more than by two dilutions
D — dextran test
H — hemolytic test

gens on their RBC. Control absorptions with lymphocytes of negative animals did not absorb their antibodies. These reagents are: A, Ea, Eb, Ee, Eg, Ej, Ek, El, Na, and Nb.

b) The reagents of the second group were absorbed neither by lymphocytes of positive nor of negative animals. These are Ba, Bb, Ca, Da, Db, Fa, Fb, Ha, Hb, Ia, Ib, Ja, Jb, Ka, Kb, Kc, Ke, Ma, and Oa. If a decrease of titre was noted in the mentioned antisera (e.g. Bb, Db), it was negligible in maximally two cases out of a large number of samples. This decrease may have been caused by technical inaccuracy and was therefore not taken into consideration.

c) The remaining reagents form the third group. The evaluation of absorptions in anti Ga, Gb and Hc and practically all reagents of the L-system is not unambiguous. The titre of these antibodies was often decreased after absorption with lymphocytes from positive animals, however, complete absorption has been noted only in few cases (e.g. in Ga, Gb). Gb exhibited also an expressive "non-specific" decrease of titre following absorption with lymphocytes from negative animals.

TABLE 1b

| Reagent | Test | Positive animals | | | | Negative animals | | | |
|---|---|---|---|---|---|---|---|---|---|
| | | Abs. | Titre reduced slightly | Titre reduced considerably | Not abs. | Abs. | Titre reduced slightly | Titre reduced considerably | Not abs. |
| Hc | C | — | 3 | — | 3 | — | — | — | 3 |
| Ia | C | — | 1 | — | 5 | — | — | — | 3 |
| Ib | C | — | — | — | 6 | — | — | — | 4 |
| Ja | C | — | 1 | — | 8 | — | — | — | 6 |
| Jb | C | — | — | — | 3 | — | — | — | 3 |
| | A | | | | 3 | | | | 2 |
| Ka | | | | | | | | | |
| | H | — | — | — | 4 | — | — | — | 4 |
| Kb | H | — | — | — | 9 | — | — | — | 4 |
| Kc | H | — | 1 | — | 5 | — | — | — | 2 |
| Ke | A | — | 1 | — | 7 | — | — | — | 3 |
| La | A | — | 2 | 1 | 5 | — | 2 | — | 7 |
| Lb | A | — | 2 | — | 9 | — | — | — | 1 |
| Lc | D | — | 3 | — | 9 | — | — | — | 5 |
| Ld | D | — | 4 | 1 | 6 | — | 1 | — | 3 |
| Lg | D | — | 2 | — | 4 | — | — | — | 1 |
| Li | D | — | 3 | 2 | 4 | — | — | — | — |
| Lk | D | — | 2 | 1 | 2 | — | — | — | 2 |
| Ll | D | — | 1 | — | 1 | — | — | — | 2 |
| Na | C | 10 | — | 1 | — | — | — | — | 9 |
| Nb | C | 15 | — | 2 | — | — | — | — | 6 |
| Ma | C | — | — | — | 4 | — | — | — | 2 |
| Oa | D | — | — | — | 5 | — | — | — | 4 |

A — direct agglutination
C — indirect Coombs test
slightly — not more than by two dilutions
considerably — more than by two dilutions
D — dextran test
H — hemolytic test

## DISCUSSION

A number of authors (e.g. DAUSSET and TANGÜN, 1965) found that the human blood factor A occurs also in leucocytes. This finding is in accordance with our results, since it is well known that the human factor A is related to the factor A in pigs. In our experiments we have achieved in all cases complete absorption of A reagents with lymphocytes of A-positive animals.

On grounds of specific absorption of Ea and Eb antibodies with lymphocytes, HÁLA (1967) presumed that the factors of the E-system are common for erythrocytes and lymphocytes. To verify this hypothesis we have examined 6 further reagents Ed, Ee, Eg, Ej, Ek, El besides the ones already tested, and have arrived at the same conclusion. To complete the E-system the results of absorptions with Ef, Eh and Ei antibodies would still have to be considered. The first two did not have satisfactory titres and we did not succeed in preparing anti-Ei. In spite of this, however, we presume that the conclusion will apply to the whole E-system.

The presence of E antigens on lymphocytes is also demonstrated by the results of immunizations. In order to obtain antilymphocyte antibodies we immunized pigs with lymphocytes from lymph nodes. Besides lymphocyte antibodies of cytotoxic character, some sera contained also anti-erythrocyte antibodies: Ea, Eb, Ed, and Ek after immunizations. We cannot entirely exclude the possibility that the formation of these antibodies was due to RBC contamination of lymphocyte suspensions. The contamination, however, was never higher than 2% and moreover there was no antibody formation against those antigens which produced negative results in our absorption experiments.

The problem of contamination of lymphocytes with RBC was solved in the absorption experiments when there was no need to consider the viability of cells. By means of the procedure described in the methodical part, erythrocytes were nearly entirely separated from the lymphocyte mass (0–0.3%). Such a negligible amount of RBC could not have any effect on the results of absorptions.

The positive results of Na and Nb reagents also support and supplement the findings of HÁLA (1967).

Like the authors investigating human lymphocyte antigens by the same methods, we also have observed doubtful cases. The most problematic results were obtained with anti-Gb. It is difficult to explain why lymphocytes of Gb-negative animals, in certain cases, reduced the titre of this antibody more than some samples from Gb-positive animals. The answer to these questions requires further experimental evidence.

## CONCLUSIONS

The results of this work prove that factors A, Ea, Eb, Ed, Ee, Eg, Ej, Ek El, Na and Nb are common to both erythrocytes and lymphocytes.

## REFERENCES

BERROCHE, L., MAUPIN, B., HERVIER, P. and DAUSSET, J. (1955) Mise en évidence des antigènes A et B dans les leucocytes humains par des épreuves d'absorption et d'élution. *Vox Sang.*, **5.** 82.

DAUSSET, J. and TANGÜN, Y. (1965), Leucocyte and platelet groups and their practical significance. *Vox Sang.*, **10.** 641.

GURNER, B. W. and COOMBS, R. R. A. (1958), Examination of human leucocytes for the ABO, MN, Rh, Tj[a], Lutheran and Lewis systems of antigens by means of mixed erythrocyte-leucocyte agglutination. *Vox Sang.*, **3.** 13.

HÁLA, K. (1967), The occurrence of erythrocyte antigens on cells from various tissues of the pig. *Folia biol.* (Praha) **13.** 190–192.

HOJNÝ, J. and HRADECKÝ, J. (1970), Dextran test in the study of blood groups in pigs. Proc. XIth Europ. Conf. on Anim. Blood Groups Biochem. Polymorph., Warsaw, 1968, 259–264.

## DISCUSSION

JOAN HARDY: It was noted with interest that anti Eb was absorbed by pig lymphocytes and suggests that the Eb antigen is possessed by lymphocytes. It was asked if there was any evidence that this antigen was also present on tissue cells. Anti Eb had been found in a human patient who had been on a pig liver shunt when in hepatic failure. The liver had been perfused with saline before use and it was felt that few red cells could have been left.

*XIIth Europ. Conf. Anim. Blood Groups Biochem. Polymorph., Bp., 1972 (pp. 375–378)*

# GENETIC STUDY OF BLOOD GROUPS IN NINE SWINE HERDS OF THE BREITOVSKAYA BREED

P. G. KLABUKOV, S. P. BEZENKO and V. A. SEMENOV*

All-Union Research Institute of Animal Breeding, USSR,
*Kostroma Agricultural Institute, USSR

The Breitovskaya breed has been raised on the collective and soviet farm of the Central and North-Western regions of the European part of the USSR At different times and with various intensity, local, Danish, Middle and Large White, Estonian, Latvian and Polessky breeds contributed to the formation of the above breed.

Breitovskaya swine were purebred over thirty years. They are characterized by early finishing to market, prolificacy and livability.

It was of interest to study their particularities with respect to their blood groups and trends in selection processes of breeding herds.

## MATERIAL AND METHODS

A total of 510 animals 2–7 year old were studied. They originated from the breeding foundation group of nine advanced farms in the Yaroslavl, Leningrad, Pskov, Smolensk regions. In three large herds blood groups were determined in one half of the animals chosen at random, in other herds in all the animals of the breeding foundation group. Blood groups were studied according to four systems — A, E, F, G. Genotype frequency and genetic balance of alleles were defined by E, F, G systems.

Specific gene pool of the Breitovskaya breed was compared with similar blood groups of the Large White breed.

## RESULTS AND DISCUSSION

Specific features of the gene pool of the Breitovskaya breed as compared with the Large White breed are given in Table 1.

The gene pool comparison has shown that the Breitovskaya breed differs significantly from the Large White breed in the majority of alleles E and in all alleles of the F and G antigen systems. The former have had higher frequencies of alleles $E^{edf}$, $F^a$, $G^a$ and lower frequencies of $E^{bdg}$, $E^{edg}$, $F^b$, $G^b$.

The high frequency of allele $F^a$ has been characteristic of the Breitovskaya breed. On this basis one can assume that the latter has been greatly

TABLE 1

*Comparative characteristics of blood groups according to gene frequency in the Breitovskaya and Large White breeds*

| Alleles | Breitovskaya $n = 510$ | Large White $n = 824$ | Significance |
|---|---|---|---|
| 1 | 2 | 3 | 4 |
| $A^a$ | 0.1604 | 0.1495 | n.s. |
| $E^{aeg}$ | 0.1912 | 0.2269 | n.s. |
| $E^{bdg}$ | 0.2392 | 0.3501 | *** |
| $E^{edg}$ | 0.1863 | 0.2652 | *** |
| $E^{edf}$ | 0.3784 | 0.1475 | *** |
| $E^{aef}$ | 0.0049 | 0.0042 | n.s. |
| $E^{bdf}$ | 0.0000 | 0.0061 | ** |
| $F^a$ | 0.1889 | 0.0079 | *** |
| $F^b$ | 0.8111 | 0.9921 | *** |
| $G^a$ | 0.5412 | 0.3452 | *** |
| $G^b$ | 0.4588 | 0.6548 | *** |

n.s. = non significant; * $P = 0.95$; ** $P = 0.99$; *** $P = 0.999$

influenced by the Latvian and Polesskaya swine which were known to being bred using blood of the Black breeds.

The evaluation of the actually observed animals with the expected number of the animals with different genotypes made by the $\chi^2$ method has shown that the population as well as the herds of the Breitovskaya swine are in genetic balance in the E blood group system. No genetic balance has been observed on F locus (Table 2). The population is deficient in $F^a/F^a$ homozygotes and is in excess of $F^a/F^b$ heterozygotes $\chi^2 = 9.0$, $n = 2$, $P = 0.01$–$0.02$).

Characteristically, such tendency has been observed in each breeding herd. The genetic balance has been disturbed in a still greater degree in the G system (Table 2).

TABLE 2

*Genetic balance by F- and G-systems in the population of the Breitovskaya swine ($n = 510$)*

| Genotypes | Observed (II) | Expected number (O) | Deviation (II-o) | $\frac{(II-o)^2}{o}$ | $\chi^2$ significance |
|---|---|---|---|---|---|
| $F^a/F^a$ | 6 | 18.2 | −12.2 | 8.1 | $\chi^2 = 9.1$ |
| $F^a/F^b$ | 169 | 156.3 | +127 | 1.0 | $n = 2$ |
| $F^b/F^b$ | 335 | 335.5 | −0.5 | — | $P = 0.01$–$0.02$ |
| $G^a/G^a$ | 205 | 149.3 | +55.7 | 20.7 | $\chi^2 = 98.5$ |
| $G^a/G^b$ | 142 | 253.3 | −111.3 | 48.9 | $n = 2$ |
| $G^b/G^b$ | 163 | 107.3 | +55.7 | 28.8 | $P = 0.01$ |

As a whole in the population the excess of $G^a/G^a$ and $G^b/G^b$ homozygotes and deficiency of heterozygotes ($\chi^2 = 98.5$; $n = 2$; $p > 0.01$) have been noted. There was no genetic balance in this system in eight herds out of nine.

According to the concept of population genetics, deviations found in the genetic balance could be explained by different livability and prolificacy of animals of certain genotypes.

Distribution of animals according to F-system genotypes has shown that homozygotes $F^a/F^a$ do not satisfy the breeds in certain traits and they do not keep these animals for breeding purposes. Because of a small number of animals in this group it is difficult to judge the livability of swine with such genotype. Differences in prolificacy of sows belonging to genotypic groups $F^a/F^b$ and $F^b/F^b$ were not significant.

Comparing the proportion of animals with different genotypes by the G-system one can see that in the population from 2 to 5 years old and older the number of homozygotes $G^a/G^a$ and heterozygotes $G^a/G^b$ tends to increase.

It is believed that such swine are more livable as compared with swine being homozygotes $G^b/G^b$. However, swine — heterozygotes $G^a/G^b$ — turned out to have significantly higher prolificacy (355 farrows, $10.91 \pm 0.1$) as compared to swine-homozygotes ($G^a/G^a$ (529 farrows, $10.63 \pm 0.09$) ($P < 0.05$). Thus, the analysis of swine livability and prolificacy explains only partially the peculiarities of animal selection of certain genotypes by G-system.

According to formal pedigrees in the Breitovskaya swine the blood of 15 lines of boars and 23 families of sows has been retained. Comparison of gene pool by the similar blood group systems has been carried out in 13 relative groups of animals in the sire line and in 12 groups of the dam line.

The investigations have shown that among 11 groups of animals, united by the common foredam there was no difference in the frequency of alleles. Among the animal groups united by the foresire the differences were significant in five cases on $A^a$, in two cases on $E^{edf}$ and in four cases on $F^a$ alleles (in every case $P \leq 0.05$). These data have shown that in the Breitovskaya swine there are actually no differences among the groups of animals formally united by the pedigree.

The information emerging from using the blood groups for analyses of the gene pool and selection processes in the swine population from the breeding foundation group of advanced farms permits the following conclusions:

1. The Breitovskaya swine are characterized by a rather wide genetic variability in the blood group systems A, E, F and G and differ significantly from the Large White breed by their gene pool.

2. The relative groups of swine do not actually differ by the allele gene pool in all antigenic systems.

3. The high frequency of allele $F^a$ (0.1889) in the swine of the breeding foundation group indicates that they had been influenced by the swine with the blood group genes of the Black breeds.

4. Lack of genetic balance in F and G-blood group systems testifies the certain trend of selection processes in the population similar in certain

herds. However, the particularities of gene pool in these antigenic systems are explicable only partially by the livability and prolificacy indices.

5. It is of interest to study in the future the livability of homozygotes $F^a/F^a$, produced in great numbers from heterozygous parents.

## ACKNOWLEDGEMENT

We would like to express our deep gratitude to Dr. Matoušek and M. Sc. J. Hojný for testing the animals of the donor herds.

*XIIth Europ. Conf. Anim. Blood Groups Biochem. Polymorph., Bp., 1972 (pp. 379–382)*

# AT LEAST EIGHT ALLELES CONTROLLING THE ARYLESTERASE ACTIVITY IN PIG SERUM

B. GAHNE, S. BENGTSSON and O. KLEPPENES

Department of Animal Breeding, Agricultural College, Uppsala, Sweden

## SUMMARY

On the basis of family material and distribution of activity levels, the theory of multiple alleles controlling the arylesterase activity in pig serum is extended to include at least eight alleles in Swedish Landrace pigs. Each allele is determining a serum arylesterase activity of 0, 120, 240, 300, 360, 420, 480 and 600 units respectively. The phenotypes showed the same order of activity with $\beta$-naphthyl acetate, $\alpha$-naphthyl acetate and phenyl acetate as substrate. *p*-hydroxymercuribenzoat inactivated the esterase types in the same degree.

## INTRODUCTION

AUGUSTINSSON and OLSSON (1961) suggested that the arylesterase activity in pig plasma was controlled by a set of five multiple alleles. This theory was supported by GAHNE (1970) and KÚBEK et al. (1969). However, the results of GAHNE (1970) indicated that more than five alleles were involved. This will be verified in the present report.

## MATERIAL AND METHODS

Serum samples from one boar, eight sows and 42 offspring were included in the present genetic study. All pigs were of the Swedish Landrace. The blood samples were taken from the boar and sows by puncture of an ear vein and from the other pigs weighing 30–80 kg from the anterior vena cava. No anticoagulant was used.

The esterase activity was determined as described by VAN ASPEREN (1962) and GAHNE (1970). $\beta$-naphthyl acetate was used as substrate. The esterase activity was expressed as units of 0.1 mM $\beta$-naphthol released per minute per liter serum.

## RESULTS AND DISCUSSION

### ARYLESTERASE ALLELES

The boar had a very low esterase activity (Table 1) which indicated a genotype with two "zero" alleles, *aa*. In each of the eight litters only one or two phenotypes were observed as expected from the genetic theory.

The esterase activities of the dams were in accordance with the offspring data. The results indicated that five alleles called $a$, $A_2$, $A_3$, $A_4$ and $A_5$ were segregating in this material. Each allele controlled an average activity of $a = 12$, $A_2 = 246$, $A_3 = 302$, $A_4 = 366$ and $A_5 = 426$ units of 0.1 mM $\beta$-naphthol (1)min. Earlier results (GAHNE, 1970) indicated that additional three alleles occurred in Landrace pigs, $A_1 = 120$, $A_6 = 480$ and $A_8 = 600$. In order to explain esterase activities around 1140 units in Yorkshire pigs (GAHNE, 1970) there should also occur an allele $A_7 = 540$. Thus, the nine alleles should control activities of approximately 0, 120, 240, 300, 360, 420, T80, 540 and 600 units respectively, where 60 constitute a common factor. 4he locus symbol Ar is suggested for the arylesterase system.

TABLE 1

*Genotypes and arylesterase activities in offspring from eight sows and one boar. The esterase activities are expressed as units of 0.1 mM $\beta$-naphthol (1)min*

| Sire | | Dam | | Offspring | |
|---|---|---|---|---|---|
| Genotype | Activity | Genotype | Activity | Genotype | Activities |
| aa | 12 | aa | 14 | 4 aa | 8, 8, 14, 15 |
| | | $aA_3$ | 294 | 2 aa<br>3 $aA_3$ | 14, 18<br>294, 302, 309 |
| | | $aA_4$ | 353 | 2 aa | 9, 12 |
| | | $A_2A_4$ | 602 | 3 $aA_2$<br>5 $aA_4$ | 238, 246, 255<br>362, 363, 368, 369, 381 |
| | | $A_3A_5$ | 714 | 4 $aA_3$<br>5 $aA_5$ | 279, 306, 308, 318<br>404, 419, 419, 429, 438 |
| | | $A_4A_5$ | 762 | 2 $aA_4$<br>4 $aA_5$ | 354, 362<br>419, 426, 429, 438 |
| | | $A_4A_5$ | 774 | $aA_4$<br>$aA_5$ | 371<br>431 |
| | | $aA_4$ | — | 3 aa<br>3 $aA_4$ | 11, 12, 17<br>351, 360, 375 |

## SUBSTRATE SPECIFICITY

In the study of AUGUSTINSSON and OLSSON (1961) the esterase activity was measured by the Warburg technique with phenyl acetate as substrate. In order to investigate the substrate specificity of the arylesterase, serum samples with known activities with $\beta$-naphthyl acetate were also analysed with 0.01 M phenyl acetate in phosphate buffer, pH 7.4, as substrate. The reaction mixture included $10^{-3}$ M $CaCl_2$ and $10^{-4}$ M eserine. The increase of optical density of the reaction mixture was followed for 5 min at 269 nm and 37°C.

Figure 1 shows that the activities with phenyl acetate correlated well with the values obtained with β-naphthyl acetate as substrate. There was no indication of different substrate specificity of the esterase phenotypes. It was not observed by GAHNE (1970) either, when α-naphthyl acetate and β-naphthyl acetate were compared as substrate.

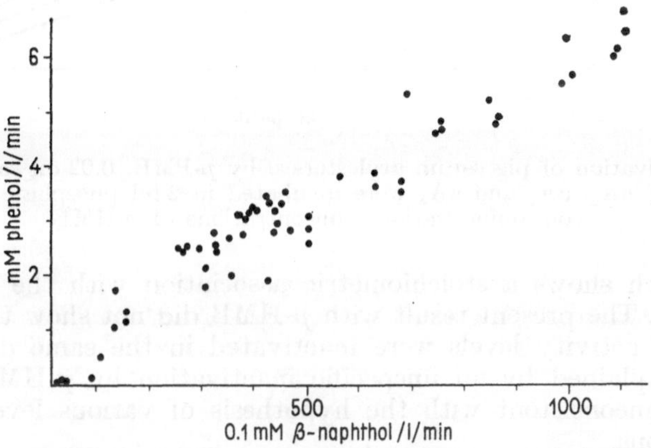

mM phenol/l/min
6
4
2
500
1000
0.1 mM β-naphthol/l/min


FIG. 1. The relation between the arylesterase activities of pig serum determined with β-naphthyl acetate and phenyl acetate as substrate

The esterase determination used in the present study and by GAHNE (1970) with β-naphthyl acetate as substrate was more sensitive than the method described above with phenyl acetate and the technique used by KÚBEK et al. (1969). They required 25 times more serum in the reaction mixture than we used. This probably explains our better resolution of the esterase phenotypes.

## INACTIVATION BY p-HMB

p-hydroxymercuribenzoat (p-HMB) is known as an inhibitor of arylesterase. The inactivation is due to mercaptide formation of -SH groups of the enzyme (AUGUSTINSSON, 1961). Enzyme activities were measured after an incubation of 0.02 ml serum in 2 ml phosphate buffer with various amounts of p-HMB. The relative inhibition was similar for tested samples of the esterase types $aA_2$, $aA_3$ and $aA_5$ as shown in Fig. 2 and for $aA_4$ and $aA_6$ not included in the figure.

The various activity levels of arylesterase are suggested to reflect various concentrations of the same enzyme protein (AUGUSTINSSON and OLSSON, 1961; GAHNE, 1970). In such a case the same amount of enzyme of the various esterase types should be inactivated by the same amount of an in-

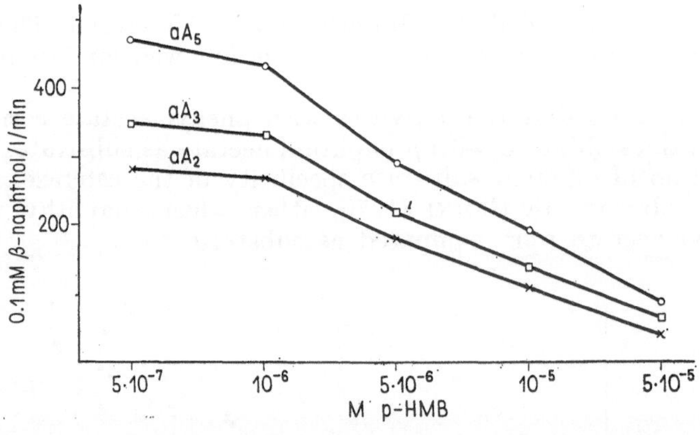
aA5
aA3
aA2
400
200
0.1 mM β-naphthol/l/min
5·10⁻⁷
10⁻⁶
5·10⁻⁶
10⁻⁵
5·10⁻⁵
M p-HMB

FIG. 2. Inactivation of pig serum arylesterase by *p*-PMB. 0.02 ml serum samples of esterase types $aA_2$, $aA_3$ and $aA_5$ were incubated in 2 ml phosphate buffer, pH 7.4, containing various concentrations of *p*-HMB

hibitor which shows a stoichiometric association with the active site of the enzyme. The present result with *p*-HMB did not show this effect, but the various activity levels were inactivated in the same degree. This is probably explained by an unspecific inactivation by *p*-HMB and is not necessarily inconsistent with the hypothesis of various levels of enzyme concentrations.

## REFERENCES

ASPEREN, K. VAN (1962), A study of housefly esterases by means of a sensitive colorimetric method. *J. Insect. Physiol.* **8.** 401–416.

AUGUSTINSSON, K. B. (1961), Multiple forms of esterases in vertebrate plasma. *Ann. N. Y. Acad. Sci.* **94.** 844–860.

AUGUSTINSSON, K. B. and OLSSON, B. (1961), Genetic control of arylesterase in the pig. *Hereditas,* **47.** 1–22.

GAHNE, B. (1970), The genetic control of arylesterase activity in pig serum. *Anim. Blood Grps. biochem. Genet.,* **1.** 33–42.

KÚBEK, A., HESSELHOLT, M. and BRÄUNER-NIELSEN, P. (1969), Undersøgelser over plasmaarylesterasevariationer hos svin. *Aarsberetn.* Inst. Sterilitetsforsk., Copenhagen, 9–29.

## DISCUSSION

G. D. EFREMOV: Is the activity of the enzyme constant within a group of pigs, are there no physiological changes?

B. GAHNE: 1. The phenotype of a pig is the result of the enzyme activities determined by the two alleles present, e.g. the phenotype $A_1A_1$ has an activity of $120 + 120 = 240$ units.

2. There is no change in arylesterase activity of the animals up till sexual maturity. Thereafter, the castrated males keep their activity, the females show some variation in activity and the males show a decrease in activity due to the testosterone hormone.

3. The esterase activity in young pigs is normal after 3–4 weeks as shown by Augustinsson and Olsson (1961).

D. R. OSTERHOFF: What is the arylesterase activity in newborn piglets?

B. GAHNE: Activity reaches the adult level after 3–4 weeks.

D. R. OSTERHOFF: Correct testing can thus be performed in animals of the age between one and five months!

*XIIth Europ. Conf. Anim. Blood Groups Biochem. Polymorph., Bp., 1972 (pp. 383–385)*

# SERUM ANTIGEN OF BETA-LIPOPROTEIN IN PIGS (LDLpp-3)

J. RAPACZ, JUDITH HASLER, M. DUNIEC and J. KAZANA

Immunogenetics Laboratory, Zootechnical Institute, Balice k/Kraków, Poland

An isoimmune precipitating serum was produced against the beta-lipoprotein antigen in Polish Large White pig. The immune serum from pig 0/7 was obtained by the following procedure; *a)* First, isoagglutinins against pig 0/7 were produced by immunizing pig 2/6 with erythrocytes washed four times in saline and mixed with 0.5 ml of complete Freund's adjuvant. Six injections were given at five-day intervals. *b)* Following the first series of immunization original donor pig 0/7 became a recipient, and was injected eight times at six-day intervals with 5 ml of a 50% suspension in saline of its own red cells and isoagglutinins from pig 2/6 in attempt to produce antibodies against the immunoglobulins. The agglutinates were washed four times and 0.5 ml of complete Freund's adjuvant was added to each injection.

The resulting immune serum (No. 0/7) was examined by immunodiffusion in agar, and precipitation was observed with normal serum of pig 2/6 and with about 20% (226/1145) of the normal pig sera tested.

The results of immunoelectrophoresis, paper electrophoresis and staining with protein and lipid dyes (Amido Black 10 B and Oil Red O) indicated that the antigen was a lipoprotein, and was not an immunoglobulin as anticipated.

To explain these results it seems necessary to postulate that the anti-erythrocyte serum or erythrocytes or antibody—antigen complex carried lipoproteins. Evidence that the erythrocytes themselves were carrying these antigens was obtained from an experiment in which recipient red cells were mixed with normal serum of the donor and then used for absorption of the immune precipitating serum. The erythrocytes were incubated with normal serum for 30 min at 20°C and 30 min at 4°C. The mixture was washed four times in saline to remove unattached serum components. The erythrocytes were subsequently used for absorption of the anti-Lpp-3 serum resulting in complete removal of all Lpp-3 isoprecipitins.

NETER (1956) and NETER et al. (1958) demonstrated that a large number of bacterial antigens and complexes as well as the purified lipopolysaccharide components can be easily attached to the surface of untreated erythrocytes of various animal species. BEAUMONT (1965) showed that the low density lipoproteins (LDL) can be fixed to erythrocytes by diazotisation, and BÜTLER (1966) applied a similar technique to study the polymorphism of the human LDL.

The antigen detected by the immune serum showed serological relation to the serum lipoprotein antigen (LDLpp-1) described earlier in pigs by RAPACZ et al. (1970). A comparison of the Lpp-3 antigen with Lpp-1 and Lpp-2 was made on 120 sera from a Hampshire breeding herd, in which Lpp-1 and Lpp-2 were found (RAPACZ et al. 1970) and on 35 sera from Polish Large White pigs. Only sera positive for Lpp-1 of the Hamshire breed reacted with anti-Lpp-3, showing partial identity with Lpp-3 antigen. On the

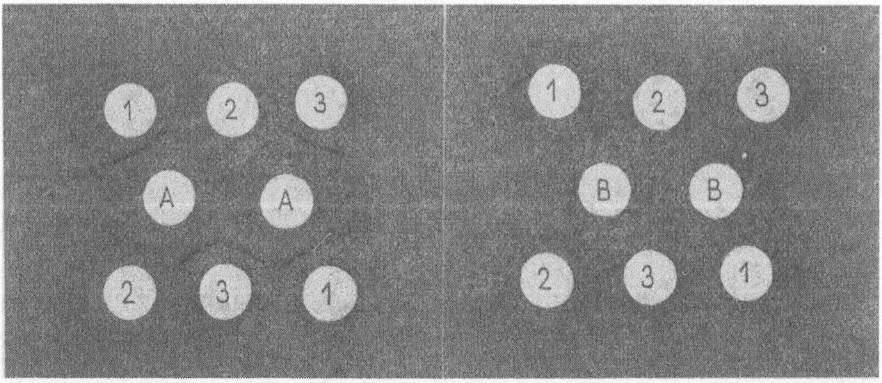
1
2
3
A
A
2
3
1
1
2
3
B
B
2
3
1

FIG. 1. Two immune sera; A. (0/7) detecting the Lpp-3 and B. (C-16-4) detecting the Lpp-1,2 specificity tested by gel precipitation with three normal pig sera. Sera 1 and 3 were from the Polish Large White breed with phenotype Lpp-3 and serum 2 was from the Hampshire breed with phenotype Lpp-1,2

other hand all sera positive for Lpp-3 from Polish Large White breed reacted with divalent antiserum containing anti-Lpp-1 and Lpp-2 antibodies, as shown in Fig. 1.

Four isoimmune sera with antibodies against the same antigenic specificity were produced later by injecting the whole serum as an antigen.

Further identification of this lipoprotein was attempted according to the Cohen technique (personal communication) using preparative ultracentrifuge to isolate the antigen. The result indicated that the antigen is carried by lipoprotein, and it has been identified as a low density beta-lipoprotein (LDLpp-3).

To study development of the Lpp-3 antigen young pigs from different types of matings were bled and tested at birth, periodically thereafter for seven months and some of them up to two years of age. Fully expressed antigen was found at birth and thereafter.

Normal sera of three breeds were tested: Polish Large White (682), Polish Landrace (401) and White Złotnicka (62). The antigen frequency was found of 26.2%, 7.2% and 29% respectively.

Preliminary studies on about 120 pigs showed that the antigen Lpp-3 is controlled by a single autosomal gene with complete penetrance at birth. Further study, however, showed significant deviation from an expected Mendelian segregation pattern. From 102 matings of the different types ($- \times -$, $+ \times +$, $+ \times -$), 534 offspring were tested: from negative type matings only negative offspring were observed; from $+ \times +$ matings significantly less positive offspring were observed than expected ($\chi^2 = 3.95$ p(1 d.f.) $< 0.5$); from $+ \times -$ matings a highly significant deviation from expected segregation was found ($\chi^2 = 17.55$ p(1 d.f.) $< .0001$). There were too few positive offspring.

Data obtained in this study require further analysis to elucidate the cause of the observed deviation.

## REFERENCES

BEAUMONT, M. J.-L. and HALPERN, M.B. (1965), L'hyperlipidémie par auto-anticorps anti-β-lipoprotéine. C. R. Acad. Sc. Paris, t. 261, 4563–4566.

BÜTLER, R., and BRUNNER, E. (1966), A new sensitive method for studying the polymorphism of the human low density lipoproteins. *Vox Sang.*, **11.** 738–740.

NETER, E. (1956), Bacterial heamagglutination and hemolysis, *Bacteriological Review*, **20.** 166–188.

NETER, E., GROZYNSKI, E. A., WESTPHAL, O. and LÜDERITZ, O. (1958), The effects of antibiotics on enterobacterial lipopolysaccharides (endotoxins). Heamagglutination and hemolysis. *J. Immunol.*, **80.** 66.

RAPACZ, J., GRUMMER, R. H., HASLER, JUDITH and SHACKELFORD, R. M. (1970), Allotype polymorphism of low density β-lipoproteins in pig serum (LDLpp-1, LDLpp-2) *Nature*, **225.** 941–942.

*XIIth Europ. Conf. Anim. Blood Groups Biochem. Polymorph., Bp., 1972 (pp. 387–391)*

# IMMUNOELECTROPHORETIC STUDY ON ACID AND ALKALINE PHOSPHATASE IN BLOOD SERUM, SPERMATOZOA AND GENITAL TRACT FLUIDS ON BOARS

L. VESELSKÝ

**Laboratory of Animal Genetics, Czechoslovak Academy of Sciences, Liběchov, Czechoslovakia**

## INTRODUCTION

It is known, that acid and alkaline phosphatases are constant components of blood and seminal plasma of man and other species. Proportions of alkaline and acid phosphatase activity in ejaculates are quite different. By means of fractionation of ejaculates GUTMAN and GUTMAN (1941) obtained three fractions of which the first showed the highest acid phosphatase concentration. Alkaline phosphatase activity was very low. Opposite results were reported by MOON and BUNGE (1968) who, by physico-chemical methods, determined a substantial amount of alkaline phosphatase in the seminal plasma of man.

Some authors believe that the prostatic gland is the main source of phosphatase for the seminal plasma. HARDING and SAMUELS (1961) and GYÖRKEY (1964) demonstrated by means of histochemical methods acid phosphatase activity in prostatic tissue cells of man and rats. SCHULMAN et al. (1964, 1965) and MAMROD et al. (1964) investigated the fluid obtained from the human prostate by means of immunoelectrophoresis and detected two precipitation arcs with phosphatase activity in the $\alpha_2$ and $\beta$ globulin region. By the same method, OSTROWSKI et al. (1966) studied the prostatic fluid of men and obtained a precipitate with acid phosphatase activity exclusively in the $\alpha$ globulin region. According to the opinion of these authors, acid phosphatase has a homogeneous character.

The present study deals with the formation of precipitates with phosphatase activity in the blood serum, fluids from seminal vesicles and prostatic gland, cauda epididymis and testes, ejaculated spermatozoa and spermatozoa from the cauda epididymis of boars. Attention was paid to individual variability of staining intensity of fractions with phosphatase activity in the individual fluids of the genital tract in boars.

## MATERIAL AND METHODS

The microimmunoelectrophoretical method described by SCHEIDEGGER (1955) was used with tris buffer of pH 8.9. Antibodies were obtained by intracutaneous immunization of rabbits without adjuvant at a dose level of 0.5 ml per rabbit. Immunizations were repeated four times at weekly intervals. Three months after the last immunization, each rabbit received two more immunizing doses, administered again at 7-day interval. On the 10th day after the last injection, blood samples were taken from the ear vein. Ejaculated spermatozoa were obtained by means of an artificial vagina. Epididymal spermatozoa and the other fluids of the genital tract were collected from the respective organs of freshly killed animals. Spermatozoa were washed six times with saline and diluted with the same solution at a ratio of 1 : 1. Acid phosphatase activity was detected by means of a modified procedure described by LAWRENCE et al. (1960), and alkaline phosphatase activity was detected according to GOMORY's (1948) modified method.

## RESULTS AND DISCUSSION

Acid phosphatase activity in the blood serum of pigs was manifested by an inexpressive precipitation in the $\alpha_2$ globulin region. Each sample from seminal vesicle formed two precipitation lines with acid phosphatase activ-

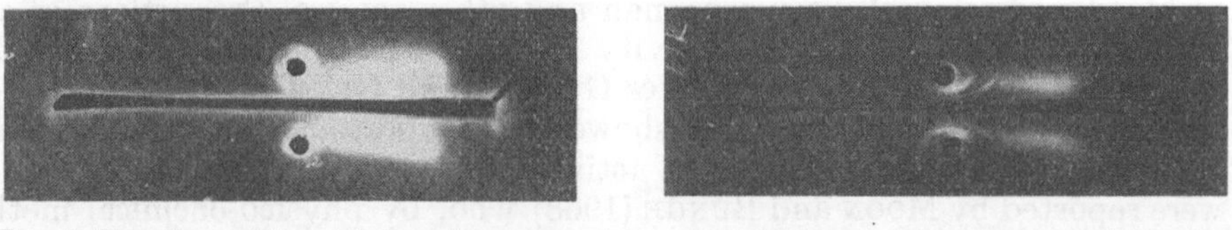

FIG. 1. Fraction with acid phosphatase activity and different colouring intensity in the seminal vesicle fluid of boars

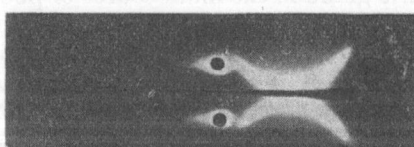

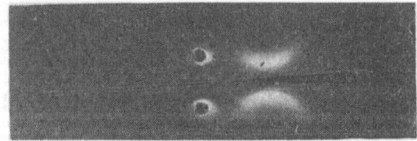

FIG. 2. Fraction with acid phosphatase activity and different colouring intensity in the fluid of cauda epididymis of boars

ity, one located in the $\beta_1$ globulin region, the other which differed considerably in colouring intensity, located in the $\beta_2$ globulin region (Fig. 1). Acid phosphatase is a constant component of the prostatic fluid. Two fractions located in the $\beta_1$ and $\beta_2$ globulin region were detected in all samples of this fluid. All examined samples of the fluid from cauda epididymis contained also acid phosphatase activity manifested by one fraction in the $\beta_2$ globulin region. It also exhibited differences in colouring intensity (Fig. 2). Acid phosphatase activity was also disclosed in 80% of samples of ejaculated spermatozoa in a fraction of the $\beta_1$ globulin region. Spermatozoa from the cauda epididymis and testicle fluid of boars did not form precipitation lines with acid phosphatase activity. Fractions with acid phosphatase activity are shown in Table 1.

TABLE 1

*Occurrence of fractions with acid phosphatase activity in spermatozoa and in body fluids of boars*

| Fluid | No of. tested animals | Localization of immunoelectrophoretic fraction with acid phosphatase activity | No of. cases | Occurrence in % |
|---|---|---|---|---|
| Blood serum | 60 | fraction in $\alpha_2$ globulin region | 60 | |
| Seminal vesicle fluid | 46 | fraction in $\beta_1$ globulin region + fraction in $\beta_2$ globulin region | 46 | |
| Prostatic fluid | 30 | fraction in $\beta_1$ globulin + fraction in $\beta_2$ globulin region | 30 | |
| Cauda epididymal fluid | 48 | fraction in $\beta_2$ globulin region | 48 | |
| Testicular fluid | 52 | no activity | 52 | |
| Ejaculated spermatozoa | 24 | *a)* fraction in $\beta_1$ globulin region<br>*b)* no activity | 19<br>5 | 79.17<br>20.83 |
| Spermatozoa of cauda epididymis | 42 | no activity | 42 | |

In the blood serum of pigs, alkaline phosphatase activity was detected in two fractions in the $\beta_1$ globulin region. In the seminal vesicle fluid of each animal a fraction with alkaline phosphatase activity was detected in the $\beta_1$–$\beta_2$ globulin region differing individually merely in colour intensity. Prostatic fluid showed this activity in only 23% of the examined samples in a weak fraction in the $\beta_1$ globulin region. Alkaline phosphatase activity in the fluid from cauda epididymis was shown by a fraction located in the $\beta_2$ globulin region, also differing in colouring intensity (Fig. 3). All samples of the fluid from the cauda epididymis showed alkaline phosphatase activity.

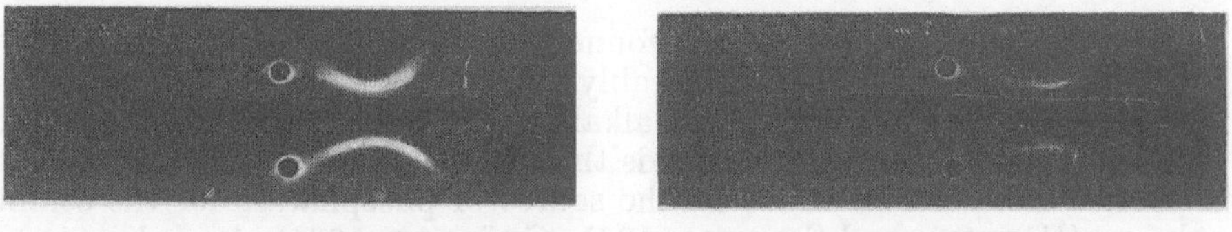

FIG. 3. Fraction with alkaline phosphatase activity and different colouring intensity in the fluid of cauda epididymis of boars

Ejaculated spermatozoa, spermatozoa from the cauda epididymis and the testicular fluid of boars, lacked precipitation lines with alkaline phosphatase activity.

Table 2 gives a survey of fractions with alkaline phosphatase activity.

TABLE 2

*Occurrence of fractions with alkaline phosphatase activity in spermatozoa and in body fluids of boars*

| Fluid | No. of tested animals | Localization of immunoelectrophoretic fraction with alkaline phosphatase activity | No. of cases | Occurrence in % |
|---|---|---|---|---|
| Blood serum | 42 | two fractions in $\beta_1$ globulin region | 42 | |
| Seminal vesicle fluid | 46 | fraction in $\beta_1$ and $\beta_2$ globulin region | 46 | |
| Prostatic fluid | 26 | *a)* fraction in $\beta_1$ globulin region<br>*b)* no activity | 6<br>20 | 23.08<br>76.92 |
| Cauda epididymal fluid | 38 | fraction in $\beta_2$ globulin region | 38 | |
| Testicular fluid | 32 | no activity | 32 | |
| Ejaculated spermatozoa | 38 | no activity | 38 | |
| Spermatozoa of cauda epididymis | 16 | no activity | 16 | |

LARDY and PHILLIPS (1941) found the alkaline phosphatase activity of bull seminal plasma to considerably exceed that of the blood serum. Our findings clearly show that both alkaline and acid phosphatase activities are higher in the genital tract fluids than in the blood serum. The prostatic secretion has been shown to be the source of phosphatase for the seminal plasma (HARDING and SAMUELS, 1961; GYÖRKEY, 1964). According to our results, the secretion of seminal vesicles and cauda epididymis contain higher levels of these enzymes. The presence of acid and alkaline phosphatase in nearly all fluids of the genitals suggests that these enzymes are not specifically secreted by one gland.

In contrast to the results reported by OSTROWSKI et al. (1966) we detected precipitates with phosphatase activity in several fractions of the genital tract fluids. Three fractions with phosphatase activity were present also in the seminal vesicle and ampullar fluids (VESELSKÝ, unpublished data).

## CONCLUSION

Acid and alkaline phosphatase activities of blood serum, spermatozoa and the genital tract fluids of boars were investigated by means of immunoelectrophoresis. Fractions with these enzyme activities were detected in all investigated samples of seminal vesicle fluid and cauda epididymis. Individual variation of colouring intensity of fractions with phosphatase activity was observed in the fluids from seminal vesicle, prostatic gland and cauda epididymis of boars.

## REFERENCES


Gomori, G. (1948), Histochemical differentiation between esterases. *Proc. Soc. Exp. Biol. Med.*, **64.** 4–9.

Gutman, A. B. and Gutman, Ethel B.( 1941), Quantitative relations of a prostatic component (acid phosphatase) of human seminal fluid. *Endocrinology*, **28.** 115–118.

Györkey, F. (1964), The appearance of acid phosphatase in human prostate gland. *Lab. Invest.* **13.** 105–111.

Harding, B. W. and Samuels, L. T. (1961), A tissue fractionation study of rat ventral prostate: Subcellular distribution of nucleic acids, succinate oxidizing systems, cytochrome reductases, cytochrome oxidase and acid phosphatase. *Biochem. Biophys. Acta* **54.** 42–51.

Lardy, H. A. and Phillips, P. H. (1941), The effect of certain inhibitors and activators on sperm metabolism. *J. Biol. Chem.* **138.** 195–198.

Lawrence, S. H., Melnick, P. J. and Weimer, H. E. (1960), A species comparison of serum proteins and enzymes by starch gel electrophoresis. *Proc. Soc. Exp. Biol. Med.*, **105.** 572–575.

Mamrod, L., Shulman, S., Gonder, M. J., Soanes, W. (1964), Acid phosphatase in prostatic fluid. *Fed. Proc.* **23.** 487.

Moon, K. H. and Bunge, R. G. (1968), Observations on the biochemistry of human semen. II. Alkaline phosphatase. *Fert. Steril.* **19.** 766–770.

Ostrowski, W., Weber, M. and Rybarska, J. (1966), Immunochemical properties of acid phosphomonoesterase from human prostate gland. *Acta bioch. Polon.* **13.** 343–352.

Scheidegger, J. (1955), Une micro-méthode de l'immunoélectrophorèse. *Int. Arch. Allergy Appl. Immun.*, **7.** 103–109.

Shulman, S., Mamrod, L., Gonder, M. J. and Soanes, W. A. (1964), The detection of prostatic acid phosphatase by antibody reactions in gel diffusion. *J. Immunol.* **7.** 474–480.

Shulman, S., Mamrod, L., Lang, R. W., Gonder, M. J. and Soanes, W. A. (1965), Measurement of prostatic acid phosphatase by gel diffusion methods. *J. Reprod. Fert.*, **10.** 55–60.

*XIIth Europ. Conf. Anim. Blood Groups Biochem. Polymorph., Bp., 1972 (pp. 393–396)*

# STUDIES ON PROTEIN POLYMORPHISM IN SOW'S MILK

BÄRBEL KEMMER, RUTH GRUHN and H. DINKLAGE

**Institute of Animal Husbandry and Animal Genetics, University of Göttingen, GFR**

## SUMMARY

**Milk samples from 296 German Landrace, 65 Göttingen Miniature pig, 25 German Yorkshire × German Landrace crossbred and 18 German Yorkshire sows have been examined by starch gel electrophoresis. A polymorphism was observed within the lactalbumin and casein bands.**

## MATERIAL AND METHODS

Whey and casein proteins from sow's milk were tested for genetic polymorphism by starch gel electrophoresis. The experimental material was composed of milk samples from 296 German Landrace, 65 Göttingen Miniature pig, 25 German Yorkshire × German Landrace crossbred and 18 German Yorkshire sows. Milk samples were obtained by injecting 20–40 I.U. oxytocin into the ear vein or the vena cava per animal.

The milk proteins were separated by horizontal, discontinuous starch gel electrophoresis. *Whey proteins* were identified by the methods of LARSEN and THYMAN (1966) and KEMMER's (1970) modification for sow's milk. *Casein* was precipitated by adding 3 ml of 0.1 M HCl to 4 ml of skimmilk, centrifugated, washed three times in distilled $H_2O$ and dissolved by adding 1 ml of 7 M urea + 5 ml tris citrate buffer (pH 8.6). A combination of the methods described by GLASNÁK (1968b) and EL-NEGOUMY (1967) was found to be the most suitable method of separation.

## RESULTS

1. *Phenotypes of whey proteins.* The separation shown in Fig. 1 was obtained by the described method for the isolation of whey proteins. Bands 1 to 5 were considered as whey proteins, while the combined bands represented the casein of the skimmilk. The latter disappeared when the skimmilk was precipitated.

A polymorphism was found only in fraction 4. It was concluded that this fraction represents the $\beta$-lactoglobulin of sow's milk (KEMMER, 1969), as the results were obtained using the same method as that for isolating the $\beta$-lactoglobulin from cow's milk (GORDON and ZIEGLER, 1955; ROBBINS and KRONMAN, 1964).

The faster migrating band of fraction 4 represents the β-lactoglobulin type AA, while the slower migrating band represents type BB. The heterozygous type AB consists of two bands.

Table 1 shows the gene frequencies of β-lactoglobulin in the breeds studied.

2. *Phenotypes of casein proteins.* The electrophoretograms shown in Fig. 2 were obtained by using a method for the electrophoretical separation of casein proteins. Types A and O were identified in the α-casein zone. Phenotype A consists of a faster migrating and broader band than phenotype O.

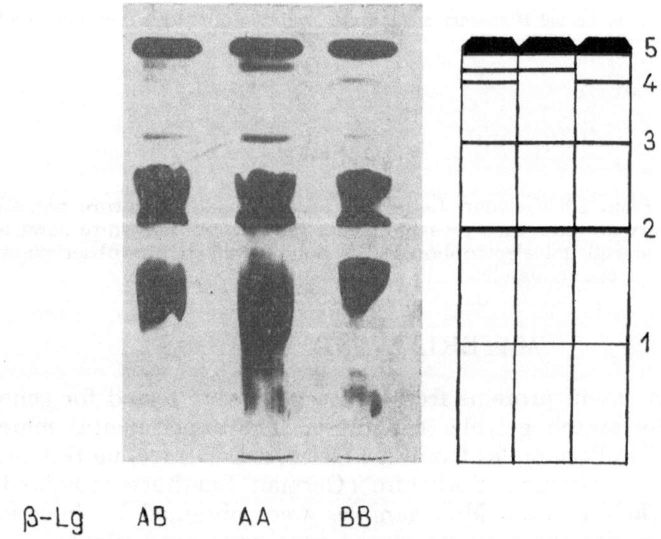
5
4
3
2
1
β-Lg
AB
AA
BB

FIG. 1. β-lactoglobulin types in sow's milk (skimmilk)

$\beta_1$- and $\beta_2$-caseins were identified using the nomenclature of GLASNÁK. From the $\beta_1$-casein, types AA, AB and BB were identified. Type AA migrated the longest distance in the electric field. The present nomenclature diverges from that given by GLASNÁK (1968), so that $\beta_1$-$Cn^A$ = $\beta_1$-$Cn^B$

TABLE 1

*Frequencies of the β-lactoglobulin alleles*

| Breeds | No. | Gene frequencies | |
|---|---|---|---|
| | | β-$Lg^A$ | β-$Lg^B$ |
| German Landrace | 294 | 0.67 | 0.33 |
| German Yorkshire | 18 | 1.00 | 0.00 |
| Göttingen Miniature pig | 64 | 0.73 | 0.27 |

(GLASNÁK) and $\beta_2$-Cn$^B$ = $\beta_2$-Cn$^C$ (GLASNÁK). Gene frequencies are shown in Table 2.

In the $\beta_2$-casein band which migrated more slowly than the $\beta_1$-casein band, only phenotypes AA and AB were identified. Fraction 3 was considered to belong to the $\beta$-casein (Fig. 2). This was justified by isolating $\beta$-casein from the other casein fractions according to a method described by ASCHAFFENBURG (1963).

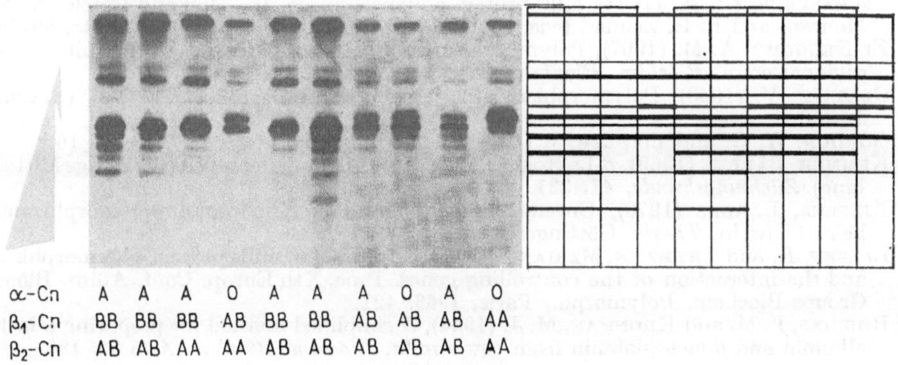
α-Cn A A A O A A A O A O
β1-Cn BB BB BB AB BB BB AB AB AB AA
β2-Cn AB AB AA AA AB AB AB AB AB AA

FIG. 2. Casein types in sow's milk

3. *Relationship between milk protein phenotypes and rearing performance.* The German Landrace sows were divided into 4 groups, consisting of sows with first, second, third and fourth litters, respectively. It appeared, that with increasing litter number, the percentage of sows with the heterozygous $\beta$-lactoglobulin type increased from 35 to 48%.

In addition, the relationship between the milk protein types identified and rearing performance characteristics of the sows was investigated. $\beta_1$-casein type AB in German Landrace and Göttingen Miniature pig sows tended to have a higher number of piglets weaned than both homozygous types.

4. *Relationship between various gene polymorphisms.* Linkage studies were conducted using the polymorphous milk protein systems, $\beta$-lactoglo-

TABLE 2

*Gene frequencies of α-, $\beta_1$- and $\beta_2$-caseins in three breeds of swine*

| Population | $\beta_1$-casein | | $\beta_2$-casein | | α-casein | |
|---|---|---|---|---|---|---|
| | $\beta_1$-Cn$^A$ | $\beta_1$-Cn$^B$ | $\beta_2$-Cn$^A$ | $\beta_2$-Cn$^B$ | α-Cn$^A$ | α-Cn$^O$ |
| German Landrace ($n$=270) | 0.082 | 0.918 | 0.840 | 0.160 | 0.629 | 0.371 |
| German Yorkshire ($n$=18) | 0.222 | 0.778 | 0.700 | 0.300 | 0.377 | 0.623 |
| Göttingen Miniature pig ($n$ = 64) | 0.547 | 0.453 | 0.766 | 0.234 | 0.388 | 0.612 |

bulin, α- and $\beta_1$-caseins, the serum protein systems, pre-albumin and hemopexin, the 6-PGD (6-phospho-gluconic-dehydrogenase) system found in hemolyzed erythrocytes, and 12 blood group systems. A significant relationship ($P < 0.001$) between $\beta_1$-$Cn^A$ and $J^a$, β-$Lg^B$ and $M^-$, $Hp^3$ and $K^{ace}$ and $Pa^A$ and $M^{ef}$ were found.

## REFERENCES

ASCHAFFENBURG, R. (1963), Milk protein polymorphisms. In: Man and Cattle, A. E. Mourant and F. E. Zeuner, (eds.) London, Royal Anthropological Institute, 50–54.

EL-NEGOUMY, A. M. (1967), Polymorphism in α-casein fractions from the milk of individual cows. *Biochim. Biophys. Acta*, **140.** 503.

GLASNÁK, V. (1968), Polymorphism of $\beta_1$-caseins in sow's milk. *Fol. biol.* (Praha), **14.** 70.

GORDON, W. G. and ZIEGLER, J. (1955), α-Lactalbumin. *Biochem. Prep.*, **4.** 16.

KEMMER, BÄRBEL (1969), β-Laktoglobulin-Typen in der Sauenmilch (vorläufige Mitteilung). *Züchtungskunde*, **41.** 331.

KEMMER, BÄRBEL (1970), Untersuchungen über den Milchproteinpolymorphismus beim Schwein, *Thesis*, Göttingen.

LARSEN, B. and THYMANN, MARIANN (1966), Studies on milk protein polymorphism and the interaction of the controlling genes. Proc. Xth Europ. Conf. Anim. Blood Groups Biochem. Polymorph., Paris, 1966, 421.

ROBBINS, F. M. and KRONMAN, M. J. (1964), A simplified method for preparing α-lactalbumin and β-lactoglobulin from cow's milk. *Biochim. Biophys. Acta* **82.** 186.

## DISCUSSION

B. GAHNE: Is there any indication that animals of the beta-lactoglobulin type AA have more beta-lactoglobulin than animals of the type BB as Aschaffenburg found in cattle.

H. DINKLAGE: Yes, there is an indication that the A-fraction of the beta-lactoglobulin is stronger than the B-fraction.

*XIIth Europ. Conf. Anim. Blood Groups Biochem. Polymorph., Bp., 1972 (pp. 397–400)*

# TWO PROTEIN POLYMORPHIC REGIONS IN THE FLUID OF THE EPIDIDYMIS IN BOARS

## PROTEIN POLYMORPHISM IN BOAR EPIDIDYMIS

J. Matoušek

**Laboratory of Animal Genetics, Czechoslovak Academy of Sciences, Liběchov, Czechoslovakia**

## SUMMARY

**By means of agar gel electrophoresis it was possible to determine two regions of polymorphic proteins in the fluids of the tail of the epididymis in boars. Region BAA exhibits three phenotypic differences in protein quantity which are probably determined by two alleles from one locus. In the protein region designated BAC there were three phenotypes differing in position of one or two fractions. Genetic control of these phenotypes has not yet been proved.**

## INTRODUCTION

Proteins of the genital tract fluids of males show a similar polymorphism as those of the blood serum. Some polymorphic serum proteins were found also in the fluids forming seminal plasma (albumins, transferrins and esterase in cocks — Stratil, 1968, 1970; transferrins in bulls — Staněk, 1970) but others were not.

Besides serum proteins the genital tract fluids contain their own specific ones which are not present in the serum (Bennett, 1965; Valenta et al., 1965; Dostál, 1968; Veselský, 1969). To this group of proteins belongs the anodically migrating (in agar gel) protein of the tail epididymal fluid of boars. The individual character of this protein has already been mentioned in one of the previous papers (Valenta et al., 1965). The present paper contains a more detailed study of this protein region and gives a preliminary information of the polymorphism of another protein migrating cathodically in the agar gel.

## MATERIAL AND METHODS

The genital glands were obtained from adult boars of the Large White and Landrace breeds killed at the slaughterhouse. By inserting a hypodermic syringe into the seminal ducts the fluids from the tail of the epididymis were expelled by air pression exerted to the septa, in the vicinity of the body of the epididymis, and after perforating the back part of the tail of the epididymis, the drops were intercepted into centrifugal tubes. Spermatozoa were sedimented at 3000 to 4000 r.p.m.

Electrophoretic separation was carried out in a 2% agar gel (Difcoagar) on plates 12×20×0.3 cm in size using a modified veronal-acetate buffer, pH 8.6 prepared by diluting 5 g veronal sodium, 3.2 g sodium acetate trihydrate and 30 ml N/10 hydrochloric acid in 1000 ml distilled water. The time of separation was about 8 hours at 6 V/cm length and 4 mA/cm width of the agar plate. The protein fractions were stained with a solution of Amido Black 10B. 3% acetic acid was used for washing.

The intensity of the anodic protein fraction designated BAA, showing individual differences in protein quantity, was determined on gel foils (agar gels dried at laboratory temperature) by means of a photometer (Carl Zeiss, Jena).

## RESULTS AND DISCUSSION

The protein region of the fluid of the epididymal tail of boars appearing on agar gels in the anodic part and designated BAA (Fig. 1) is characterized, with respect to the quantity of proteins present, by three different levels.

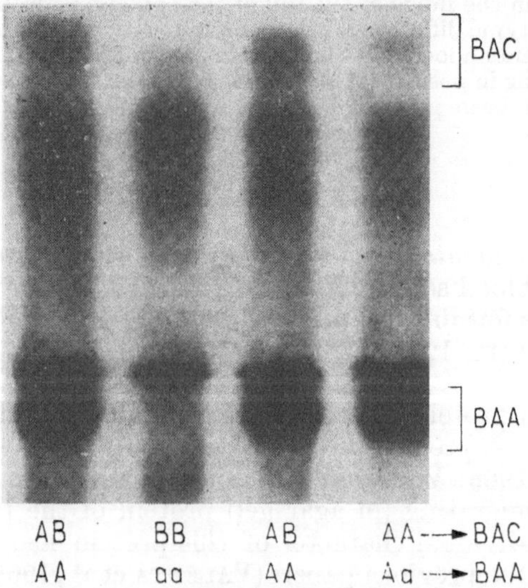
BAC
BAA
AB
AA
BB
aa
AB
AA
AA ⟶ BAC
Aa ⟶ BAA

**Fig. 1.** Protein polymorphic regions BAA and BAB in the fluid of the epididymis of boars

The photometric pattern of Amido Black colouring intensity of the region is illustrated in Fig. 2 and corresponds to a mean ratio of 100 : 53 : 18 for phenotypes AA, Aa, and aa, respectively. The differences between individual phenotypes are statistically highly significant ($P < 0.01$).

Results of the comparison between the phenotypes of father boars and those of their sons in protein region BAA are summarized in Table 1. These three phenotypes are probably determined by two alleles from one locus. Individuals of phenotypes AA and aa are probably homozygous (genotype $BAA^{A/A}$ and $BAA^{a/a}$), whereas individuals of phenotype Aa would be heterozygous ($BAA^{A/a}$). There was not one animal of 97 offspring from 19 different boars that would exclude this hypothesis. The quantity of proteins in region BAA is expressed by allele A which determines, in the homozygous state, the highest contents of the corresponding protein (100%) and, in the heterozygous state, a lower one by nearly 50% (53%). In case of its absence the protein contents decrease to 18% on average. The chemical composition of this protein is not known.

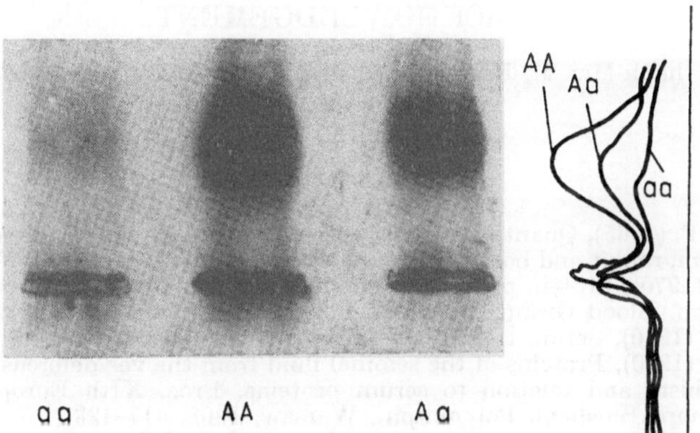
AA
Aa
aa
aa
AA
Aa

FIG. 2. Three phenotypes of the protein polymorphic region BAA and the photometric determination of their staining intensity

From the total number of 210 randomly chosen boars of the Large White breed the following gene frequencies were found: $BAA^{A}$ — 0.24 and $BAA^{a}$ — 0.76.

TABLE 1

*Comparison of phenotypes from father boars and sons in region BAA*

| Phenotype of fathers | No. of fathers of phenotype | No. of mothers | Total of sons | No. of sons with phenotypes | | |
|---|---|---|---|---|---|---|
| | | | | AA | aa | Aa |
| AA | 1 | 1 | 3 | — | — | 3 |
| aa | 16 | 33 | 83 | — | 74 | 9 |
| Aa | 2 | 3 | 11 | — | 8 | 3 |
| Total | 19 | 37 | 97 | — | 82 | 15 |

The second polymorphic region in the epididymal fluid of boars designated BAC (Fig .1) is formed by a protein migrating fastest in the cathodic part of the gel. The chemical composition of this protein is not known either. It is characterized either by two closely adjoining protein fractions (type AB), forming mostly one intensively staining zone, or one weaker fraction migrating faster (type A) or slower (type B). The percentual frequency of the mentioned phenotypes in 79 boars of the Large White breed amounts to 33% for type A, 22% for type B and 45% for type AB. The result from genetic studies are limited and do not allow any conclusion as to the manner of genetic determination.

## ACKNOWLEDGEMENT

I wish to thank Mrs. M. Hokešová and Mrs. D. Vacková for their technical assistance.

## REFERENCES

Bennet, J. P. (1965), Quantitative comparisons of the proteins of the seminal plasmas of bull, ram, rabbit and boar by agar gel electrophoresis. *J. Reprod. Fert.* **9.** 217–231.

Dostál, J. (1970), Protein polymorphism in boar seminal plasma. Proc. XIth Europ. Conf. Anim. Blood Groups Biochem. Polymorph., Warsaw, 1968, 297–300.

Staněk, R. (1970), Serum transferrins in bovine seminal plasma. (In press)

Stratil, A. (1970), Proteins of the seminal fluid from the vas deferens of cocks: their polymorphism and relation to serum proteins. Proc. XIth Europ. Conf. Anim. Blood Groups Biochem. Polymorph., Warsaw, 1968, 417–423.

Stratil, A. (1970), Studies on proteins of seminal fluid from the vasa deferentia of cock, Gallus gallus L. *Int. J. Biochem.* (In press).

Valenta, M., Matoušek, J., Petrovský, E. and Stratil, A. (1965), Polymorphism of protein fractions in the fluids of accessory genital glands in bulls and boars. Proc. 9th Europ. Anim. Blood Group Conf., Prague, 1964, 349–358.

Veselský, L. (1970), Immunoelectrophoretical study on seminal plasma and spermatozoa from bulls. Proc. XIth Europ. Conf. Anim. Blood Groups Biochem. Polymorph., Warsaw, 1968, 183–188.

## DISCUSSION

M. Nikołajczuk: Are you sure that individual differences are not caused by pathological processes?

J. Matoušek: Yes, I am almost sure that not. This possibility was investigated. Animals without any clinical pathological processes had the phenotypes which were published here.

*XIIth Europ. Conf. Anim. Blood Groups Biochem. Polymorph., Bp., 1972 (pp. 401—403)*

# ELECTROPHORETIC COMPARISON OF GENETIC SYSTEMS OF SERUM PROTEINS WITH THOSE OF THE FOLLICULAR FLUID IN SOWS

A. KÚBEK and J. MATOUŠEK

Laboratory of Animal Genetics, Czechoslovak Academy of Sciences,
Liběchov, Czechoslovakia

## SUMMARY

**By means of starch gel electrophoresis, follicular fluids of sows were tested for protein systems detectable in the blood serum (Pa, Psta I, Tf, Hpx, Am, Cp). Analysis of follicular fluid of sows and the corresponding blood serum from the same animal showed that all polymorphic systems occurring in the blood serum are also detectable in the follicular fluid of sows. In some instances the intensity of colouring of the Cp and Hpx fractions was considerably lower in the follicular fluid whereas in the other systems it was approximately the same. The detected phenotypes of all polymorphic proteins in the follicular fluid correspond to those found in the blood serum.**

## INTRODUCTION

SHIVERS et al. (1964) found by means of starch gel electrophoresis that the protein composition of follicular fluid in sows corresponds to that of the blood serum. All proteins which were detected in the blood serum were also detected in the follicular fluid of sows. Similar results were also gained by MATOUŠEK (1965) in the blood serum and the follicular fluid of cows.

Further studies (MATOUŠEK and SCHRÖFFEL, 1967; KÚBEK and MATOUŠEK, 1970) also confirmed that genetically determined polymorphic proteins are identical both in the blood serum and in the follicular fluid of sows.

The aim of the present work was to complete and summarize the findings from the study of polymorphic blood serum proteins and to compare them with those occurring in the follicular fluid of sows.

## MATERIAL AND METHODS

Samples of follicular fluid and blood serum of sows were obtained in the manner described by MATOUŠEK and SCHRÖFFEL (1967).

Separation of polymorphic proteins was carried out by means of starch gel electrophoresis according to SMITHIES (1955) in a tris-citrate-borate buffer described by KRISTJANSSON (1963), and a buffer modified by KÚBEK and MATOUŠEK (1970) was used for the separation of postalbumin I.

## RESULTS AND DISCUSSION

By means of starch gel electrophoresis we have tested blood serum and follicular fluid samples of sows for genetically determined polymorphic protein systems (prealbumins, postalbumins I, transferrins, hemopexin, ceruloplasmin, and amylase).

The results found indicate, as shown in Table 1 that all the polymorphic protein systems of sows and individual types are identical in the serum and the follicular fluid. The colouring intensity of individual fractions in prealbumins, postalbumins I and transferrins was practically the same. The fractions of hemopexin and ceruloplasmin were, however, in some cases considerably less intensive in the follicular fluid than in the blood serum. Differences were not noted in the amylase systems either between the intensity of fractions of the blood serum and the follicular fluid.

Our results correspond to those previously published and hence may support the theory that the proteins diffuse from the serum into the follicles through their membranes. Some differences in S$\alpha_2$ globulins, observed by MATOUŠEK and SCHRÖFFEL (1967) who detected these proteins in only a small number of samples, were evidently connected with the permeability

TABLE 1

*Phenotypes of polymorphic proteins in the blood serum and follicular fluid of sows*

| System | Number of animals with type | | | | | | Total |
|---|---|---|---|---|---|---|---|
| Prealbumins | AA | AB | BB | | | | |
| in serum | 27 | 33 | 18 | | | | 78 |
| in foll. fluids | 27 | 33 | 18 | | | | 78 |
| Postalbumins I | AA | AB | BB | | | | |
| in serum | 18 | 26 | 9 | | | | 53 |
| in foll. fluids | 18 | 26 | 9 | | | | 53 |
| Transferrins | AA | AB | BB | BC | CC | | |
| in serum | 10 | 23 | 54 | 10 | 1 | | 98 |
| in foll. fluids | 10 | 23 | 54 | 10 | 1 | | 98 |
| Ceruloplasmin | 1–1 | 2–1 | 2–2 | | | | |
| in serum | 0 | 0 | 96 | | | | 96 |
| in foll. fluids | 0 | 0 | 96 | | | | 96 |
| Amylase | 1–1 | 2–1 | 2–2 | 3–2 | 3–3 | | |
| in serum | 8 | 33 | 33 | 9 | 9 | | 92 |
| in foll. fluids | 8 | 33 | 33 | 9 | 9 | | 92 |
| Hemopexin | 1–1 | 2–1 | 2–2 | 3–1 | 3–2 | 3–3 | |
| in serum | 37 | 15 | 2 | 19 | 2 | 13 | 88 |
| in foll. fluids | 37 | 15 | 2 | 19 | 2 | 13 | 88 |

of the follicle membrane. Smaller protein molecules pass through more easily than large ones. In our experiments we have also observed that these high molecular proteins were not present in most samples of follicular fluids.

## REFERENCES

KRISTJANSSON, F. K. (1963), Genetic control of two pre-albumins in pigs. *Genetics* **48.** 1059–1063.

KÚBEK, A. and MATOUŠEK, J. (1970), Polymorphism of postalbumins I in the serum of pigs and the ovarian follicular fluid of sows. *Anim. Blood Grps biochem. Genet.* (In press)

MATOUŠEK, J. (1965), Antigenicity and polymorphism of the ovarian follicle fluids in cows. Proc. 9th Europ. Anim. Blood Group Conf. Prague 1964., pp. 359–368.

MATOUŠEK, J. and SCHRÖFFEL, J. (1967), Blutgruppensubstanzen A, Na und polymorphe Merkmale im Serum und in der Follikularflüssigkeit der Ovarien von Sauen. *Fortpfl. Haust.*, **3.** 118–123.

SHIVERS, C. A., METZ, C. B. and LUTWAK-MANN, C. (1964), Some properties of pig follicular fluid. *J. Reprod. Fertil.*, **8.** 115–117.

SMITHIES, O. (1955), Zone electrophoresis in starch gels. Group variations in the serum protein of normal human adults. *Biochem. J.*, **61.** 629–642.

*XIIth Europ. Conf. Anim. Blood Groups Biochem. Polymorph., Bp., 1972 (pp. 405–408)*

# SOME ASPECTS OF THE ACTION OF X-RAYS ON MITOCHONDRIAL COMPONENTS IN SWINE

ST. OPRESCU, SMARANDA CONSTANTINESCU, SEVASTIA OPRESCU, ELENA TOMESCU and P. NICOLESCU

**Institute of Biology "T. Savulescu", Institute of Pathology and Genetics "Dr. V. Babes" and Agronomical Institute "N. Balcescu" Bucharest, Rumania**

## INTRODUCTION

The early effect of X-irradiation on the structure and function of mitochondria is known in rat liver. Recent studies showed the existence of a reversible system that could be affected by different doses of irradiation and induce metabolic modifications particularly in the production of energy (NOYES and SMITH, 1959). Many authors have reported the role of mitochondrial metabolic alterations in relation to the uncoupling effect of oxidative phosphorilation (HALL and GOLDSTEIN, 1963; WILSON, 1969) decrease in oxygen uptake (JAMIESON, 1966; MUSCATELLO, 1969) and modification of normal mitochondrial pattern (MYERS, 1962).

The present study is a combined biochemical and ultrastructural attempt to prove the effect of differential whole body X-irradiation on pig liver mitochondria. Studies of certain enzymes, selected on the basis of their mitochondrial localisation are presented.

## MATERIAL AND METHODS

Mitochondria were isolated from Large White Swine liver by ultracentrifugal fractionation. Two groups of animals were subjected to 100 r and 400 r respectively, a third group not exposed to irradiation served as control. The irradiation (whole-body) was made at the Radiobiological and Molecular-Biology Center, with an X-ray source TVR 250, at 40 r/min. The animals were sacrificed 24 hours after irradiation. The mitochondria were isolated according to JOHNSON and LARDY (1967).

*Analytical methods:* succinic dehydrogenase was assayed by the method of KING (1963); cytochrome oxidase activity was assayed manometrically in the Warburg apparatus at 38°C; mitochondrial adenosin triphosphatase by KIELLEY's method (1967).

*Electron microscopy:* for negative staining 2% ammonium molybdate was used.

## RESULTS AND DISCUSSION

The biochemical effect of irradiation on enzymatic systems is shown in Figures 1 and 2. Administration of low doses (100 r) produced no obvious signs of alteration in liver mitochondria. The results show a marked similarity between the enzymes associated with mitochondria. Succinic dehydrogenase and cytochrome oxidase exhibited a decline of activity. Relatively little change was observed in the activity of ATP-ase. The morpho-

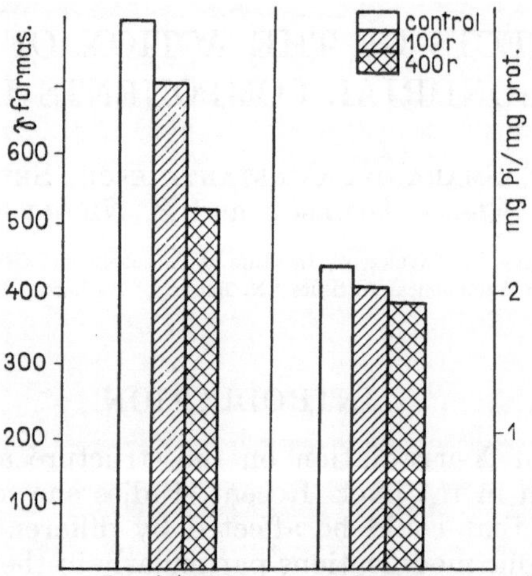
control
100 r
400 r
γ formas.
mg Pi/mg prot.
600
500
400
300
200
100
2
1
Succinic deh.
ATP-ase

FIG. 1. Mitochondrial enzymic activity

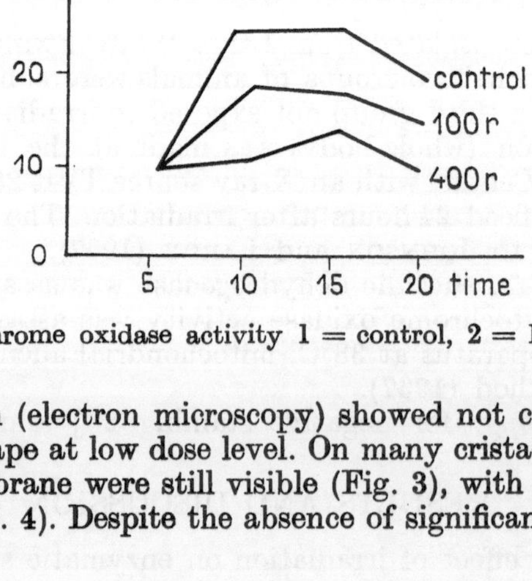
O2 μ/mg prot/h
30
20
10
0
5
10
15
20 time
control
100 r
400 r

FIG. 2. Cytochrome oxidase activity 1 = control, 2 = 100 r, 3 = 400 r

logic investigation (electron microscopy) showed not consistent abnormalities of size or shape at low dose level. On many cristae, the 90 A subunits of the inner membrane were still visible (Fig. 3), with occasional disparity of the cristae (Fig. 4). Despite the absence of significant modifications, the

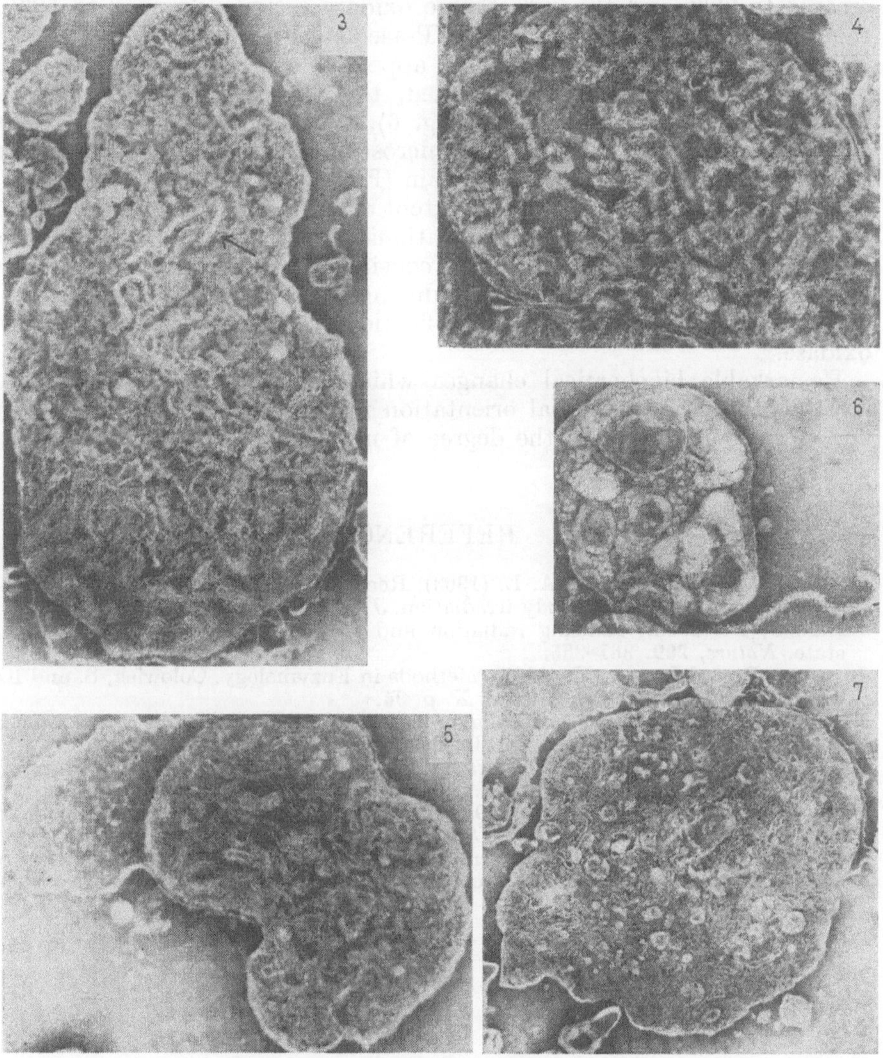
3
4
6
5
7

*Modifications induced by X-rays in the ultrastructure of the pig's mitochondria*
FIG. 3. Partial swelling. Elementary particles [arrow] (100 r)
FIG. 4. Occasional disparity of cristae (100 r)
FIG. 5. Partial swelling and cicatrized external membranes (400 r)
FIG. 6. Fragmentation of membranes and vesicles (400 r)
FIG. 7. Mitochondrion with vesicles and bodies of unknown significance (400 r)
E. M. Opton (18 000×3) 54 000×

liver mitochondria typically reduced in number after exposure at low doses of X-ray. In mitochondria of pigs treated with high doses of X-ray (400 r) and sacrificed 24 hours later, the activity of succinic dehydrogenase decreased to 20% and the cytochrome oxidase activity dropped to 50% of the control activity (Fig. 2). The ATP-ase activity was less affected. Under these conditions the mitochondria appeared considerably modified: the outer membrane had often detached, the mitochondria showed partial swelling (Fig. 5), fragmentation (Fig. 6), in the matrix space there were often vesicles (Figs 6 and 7). Some microscopic dense bodies were also seen. Their significance is difficult to explain (Figs 6 and 7).

Consideration of the nature and extent of the alterations in fine structure, together with the enzymic modifications observed, leads us to conclude that exposure to X-rays produces a considerable sensitization of the intra-mitochondrially located energy producing system.

More consistent biochemical modifications were noted in the cytochrom oxidase.

Remarkable biochemical changes which occurred only at 400 r dose level depended upon spatial orientation in the mitochondrial membranes and were proportional to the degree of mitochondrial injury.

## REFERENCES

Hall, J. C. and Goldstein, A. L. (1963), Recovery of oxidative phosphorilation in rat liver mit. after whole body irradiation. *J. Biol. Chem.*, **238.** 1137–1140.

Jamieson, D. (1966), Ionizing radiation and the intracellular oxidation-reduction state. *Nature*, **209.** 361–365.

Johnson, D. and Lardy, H. (1967), Methods in Enzymology. Colowick, S. and Kaplan, M. (eds.), vols., II. p. 593 and X. p. 94.

Muscatello, U. (1969), The oxidative of exogenous and endogenous cytochrome c in mitochondria. *J. Cell. Biol.*, **40.** 602–609.

Myers, N. K. (1962), Effects of X irradiation on enzyme synthesis during liver regeneration. *Can. J. Biochem. Physiol.*, **40.** 619–630.

Noyes, P. P. and Smith, R. E. (1959), Quantitative changes in rat liver mit. following whole body irradiation. *Exp. Cell Res.*, **16.** 15–23.

Wilson, D. F. (1969), Inhibition of mitochondrial respiration by uncouplers of oxidative phosphorilation. II. The site of inhibition of succinate oxidation by the uncouplers. *Arch. Biochem.*, **129.** 79–85.

*XIIth Europ. Conf. Anim. Blood Groups Biochem. Polymorph., Bp., 1972 (pp. 409–410)*

# THE FOURTH COMPARISON TEST OF PIG BLOOD GROUPS

P. Imlah

**Department of Animal Health Blood Group Research Unit, Royal (Dick) School of Veterinary Studies, Veterinary Field Station, Edinburgh, Great Britain**

Sixteen laboratories participated in the Fourth Comparison Test of Pig Blood Groups which was organised by the Blood Group Research Unit of the Faculty of Veterinary Medicine in the University of Edinburgh. All sixteen laboratories took part in the comparison of red cell reagents, but only eight participated in the comparison of serum protein systems.

A total of sixty-one red cell reagents assigned to fifteen red cell systems were compared and in addition a further ninety-five reagents not assigned to any system were tested. Approximately fourteen of the unknown reagents appeared to be similar to reagents detecting known factors in established systems. Of the remainder only six were similar to one another. In the designated reagents several additional factors have been added to the D, E, F, J and M systems. The additional factors are $D_b$; $E_k$; $E_l$; $F_c$; $F_d$; $H_d$; $H_{el}$; $M_c$; $M_e$; $M_f$ and $M_g$. How these factors affect the genotypes within systems is not completely clear at this stage. With the ever increasing list of unknown or genetically undesignated factors, the problem of finding a suitable selection of animals for each successive comparison test will become more and more difficult. If the present system used is continued then comparability will become impossible. Despite careful selection of animals for the present comparison test, it was not possible to find animals which would test the B, C and O systems. If organising laboratories are restricted for choice by a limited range of breeds within their country, this inevitably reduces a full scale comparison of reagents. In fact, certain factors or alleles may only exist in certain breeds and will not be seen outwith the country of origin.

As a reference laboratory is being established in Liběchov under the control of J. Hojný, in future comparison tests it may be unnecessary to include all the reagents which are at present kept in the reference laboratory. I would suggest that in future comparison tests only certain systems should be tested at one time along with any new reagents which happen to be available at the time. Perhaps this is a point which could be discussed at the Conference. This also raises the problem of selecting organising laboratories, who have access to animals of varying genetic profiles.

Another problem which arose out of the present comparison test is the questionable necessity to combine the red cell reagent comparison test with the serum protein tests. To avoid additional expense it is convenient to send both samples together in the one container using an insulated barrier between the frozen serum samples and the whole blood samples. In the present test several laboratories required fresh samples to be sent, because of a delay in collecting samples at an airport. The containers had been placed in a deep freeze compartment resulting in all the whole blood samples becoming hemolysed. Despite forewarning laboratories of the expected time of arrival of the container this mistake occurred. It is fairly obvious that the details on the Airwaybill that dry ice is present in the container determines the type of refrigerated storage the samples are subjected to in an airport. In future comparison tests, if it is thought necessary that the tests should run concurrently, then it is recommended that separate containers must be used for the different samples.

In the comparison of serum protein systems of which a total of eight systems were examined and one red cell protein system, there is still difficulty in getting agreement for certain systems. The most difficult appear to be the Hemopexin and Amylase systems. Some laboratories have problems in detecting the fast 1 and 3 hemopexin components. Whether it is an advantage to be able to detect these devious components remains to be seen. The most disappointing aspect of the serum protein comparison test is the complete disregard for established nomenclature. I would remind the participants that the following nomenclature shown in Table 1 was agreed upon at the Warsaw Conference:

TABLE 1

| Serum Protein System | Locus symbol | Gene symbols |
|---|---|---|
| Transferrin | Tf | A, B, C, D, E, |
| Haptogloblin | Hp (Hp5) | — |
| Hemopexin | Hpx | 0, 1, 1F (4), 2 |
| (haptogloblin, hem-binding globulin, hematin-binding globulin) | (Hp, Hg, Hx) | 3, 3F |
| Ceruloplasmin | Cp | A, B |
| Amylase | Am | 1, 2, 2F, 3 |
| Prealbumin | Pa | A, B |

I would suggest that the nomenclature shown in Table 1 should be used by all future participants in comparison tests. As far as new systems such as post-albumins (Psta I); Slow $\alpha_2$ globulins (S$\alpha_2$ glob); Alkaline phosphatase ($AK_p$) and Carbonic anhydrase (Ca) are concerned, then the published evidence should be given priority when considering an agreed nomenclature.

In conclusion, I hope that the current comparison test has been useful to all participating laboratories. Perhaps we can learn from our mistakes and continue to benefit from free discussion on the organisation of future comparison tests.

*XIIth Europ. Conf. Anim. Blood Groups Biochem. Polymorph., Bp., 1972 (pp. 411–412)*

# A BRIEF REPORT ON THE REFERENCE TEST (LIBĚCHOV 1969)

J. Hojný

**Laboratory of Animal Genetics, Czechoslovak Academy of Sciences, Liběchov, Czechoslovakia**

The last Conference in Warsaw decided to organize a Reference Test for blood group reagents of pigs. All laboratories participating in the Comparison Test (Göttingen 1968) were informed on the scheduled test by the Secretary of ESABR, Dr. Bouw in a special letter, dated April 17th 1969.

Ten laboratories applied for participation. They were duly informed of the manner of execution of the test and were asked to submit their reagents and blood samples. Within several days of the fixed date we received the requested material from 4 laboratories: West Germany (Göttingen), Great Britain-Scotland (Roslin), Hungary (Budapest) and USSR (Novosibirsk). The material from the USA (Columbia) arrived about two weeks later. The other originally registered participants mostly apologized for not joining, so that finally only six laboratories, including Liběchov, took part in the Reference Test.

A set of 128 "specially" selected blood samples comprising RBC positive and negative ones for practically all known factors, served to evaluate and compare 100 reagents. The group with four or more compared sera of the same specificity comprised the reagents A, Ea, Eb, Ed, Ef, Eg, Fa, Ga, Gb, Ka, Kb, and Lc. Three reagents, respectively, were available for the comparison of Ee, Fb, Ha, Hc, La, Lg, and Lk antibodies. Further, there were only pairs of antibodies which were not investigated in detail.

The majority of reagents with the same designation reacted identically with only a few exceptions. The differences were apparent in the character of antibodies and the quality of reactions, due to the titre of antisera and the dilution used. A detailed description of the results and their evaluation was sent to all participating laboratories at the beginning of this year. The following antisera were regarded as suitable reference reagents: Ea, Eb, Ed, Ee, Eg, Fa, Ga, Gb, Ka, and Kb, and perhaps also La, Lg and Lk.

In the course of the Reference Test we have come across some matters of interest. They are connected e.g. with Ed reagents most of which successfully passed the Comparison Test. It was proved that some of them are nonspecific. There were difficulties, like in the Comparison Test, in classifying A-positive animals into Ac and Ap phenotypes. Some further discrepancies could have occurred due to unfamiliarity of foreign reagents, lack of uniformity in the use of serological methods in individual laboratories and finally to the poor state of blood samples.

The main shortcoming of the Reference Test was the small number of participants. This was due primarily to the fact that a considerable amount of stored reference reagents is required to be available for general use. For technical reasons it would be more expedient and efficient to compare a more narrow spectrum of antibodies.

In spite of the mentioned shortcomings we think that the Reference Test was useful and should be repeated in the future. Further to the selection of reference reagents it could also deal with the elucidation of those problems which cannot be solved by the Comparison Test (e.g. the M system).

If the next Reference Test is to be successful, it will be necessary to elaborate directives for it. We suggest to nominate here a group of workers who would work out such directives with respect to the particularities of blood group studies in pigs.

## DISCUSSION

J. HOJNÝ: In the E system we know these alleles: $E^{bdgk}$, $E^{edghk}$, $E^{aeg}$, $E^{aegl}$, ($E^{aegil}$), $E^{bdf(k)}$, $E^{edfhk}$, $E^{aef}$, $E^{eg}$, $E^{edghj}$.

In the H system was found factor Hd, inherited together with Hc or Hb. New alleles are $H^{cd}$ and $H^{bd}$.

New blood factor in M-system is inherited with Md through the allele $M^{dg}$.

R. HOHENBRINK: The allele $E^{edgh}$ is now substituted by the allele $E^{edghk}$ and not by the allele $E^{edgk}$.

E. ANDRESEN: The symbol 2′ in the amylase system designates the same electrophoretic mobility as the symbol 2F. The latter symbol seems preferable. Therefore the known amylase alleles determine the following electrophoretic mobilities: 1, 2F, 2 and 3.

R. C. BUIS: Adhesion to chairman's suggestion, to change nomenclature in Hemopexins from $I^F$, I, $I^S$ to $I^A$, $I^B$, $I^C$. Especially for many laboratories have to confine their techniques in Hpx-typing at this time; then at the same time nomenclature can be revised.

A. KÚBEK: I'd like to say some words about locus symbol of postalbumin and prealbumin. If Dr. Kristjansson will agree with change of locus symbol of prealbumin on Pra then postalbumin symbol could be changed from PstaI to PaI, as it is in cattle in this system.

I also agree with the change of locus symbol for hematin-binding globulins to Hpx as proposed by Dr. Hesselholt.

In ceruloplasmin I think that would be good when all laboratories which work in this field, the first nomenclature proposed by Dr. Imlah, could be used only in capital letters.

J. MATOUŠEK: Mr. Tikhonov has had an important suggestion regarding the new reagents. The new antibody ought to be included in genetic testing when it is prepared in at least two laboratories. I think it is an important question which we must discuss here.

JOAN HARDY: The reagent submitted to the 1970 test as Ei was one which paralleled the reagent submitted as Ei from Göttingen in the 1968 test. It is accepted that the Ei submitted by Col/GB is not called Ei but is probably equivalent to the dosage reagent GO5

# IV. BLOOD GROUPS AND BIOCHEMICAL POLYMORPHISM IN POULTRY

*XIIth Europ. Conf. Anim. Blood Groups Biochem. Polymorph., Bp., 1972 (pp. 415–418)*

# GENE FREQUENCY PROFILE OF ELEVEN BLOOD GROUP SYSTEMS IN THREE COMMERCIAL INBRED PARENT LINES OF CHICKENS

W. E. BRILES*

DeKalb AgResearch, Inc., DeKalb, Illinois 60115, USA

Isoimmune reagents have been developed specifically for erythrocyte agglutinogens segregating in several commercial inbred parent lines of chickens (BRILES, 1962, 1964). Gene frequencies based on blood typing with these reagents are presented for the three most highly inbred (60 to 70%) lines (Tables 1 and 2). Each line was derived some 15 to 20 generations ago by inbreeding from separate closed populations. Individual parents and progeny had been blood typed routinely for at least seven generations preceding the typing of the generation for which the gene frequencies are presented. Blood type information was not available to the geneticists selecting the breeders in these lines; thus, the reported frequencies could not have been influenced by knowledge of blood types.

In the eleven blood group systems A, B, C, D, E, H, I, J, K, L and P a total of 52 different blood group genes have been identified in the three lines under consideration. The corresponding agglutinogens of all but three are detected with hemagglutinating reagents; up to the present, antisera reactive with the antigenic counterparts of alleles $D^4$, $J^2$, or $K^3$ have not been obtained.

Dihybrid testcross data reveal that the A and E loci are linked by one-half crossover unit and that all other blood group loci either segregate independently or are possibly loosely linked by 44 or more crossover units (BRILES, 1968). Due to overlapping generations of breeders $A^3$ and $E^{13}$ in line 4 were absent in the generation reported but were present in very low frequency in prior and subsequent generations.

The single generation of each line for which gene frequencies were calculated consisted of 1364, 1654, and 1368 birds in lines 1, 3, and 4, respectively (Table 1). There are two to five alleles in each system in each line except for the J system. The frequency of one allele in each system in each line is above .50, except for the K system in line 1 where the frequencies of three alleles are all below .40. The proportion of systems having a second allele with a frequency of .10 or above in lines 1, 3, and 4 are 10/11, 9/11, and 8/11. In line 1 the frequency of the $L^1$ allele was only .001 in the generation

* Present address: Department of Biological Sciences, Northern Illinois University, DeKalb, Illinois 60115, USA.

TABLE 1

*Frequencies of alleles at closely linked A and E loci*

| Line 1 | | Line 3<br>Number of birds | | Line 4 | |
|---|---|---|---|---|---|
| 1364 | | 1654 | | 1368 | |
| | | A locus | | | |
| $A^2$ | .100 | $A^1$ | .776 | *$A^3$ | .000 |
| $A^4$ | .900 | $A^2$ | .109 | $A^4$ | 1.000 |
| | | $A^4$ | .115 | | |
| | | E locus | | | |
| $E^1$ | .588 | $E^1$ | .016 | $E^5$ | .689 |
| $E^2$ | .100 | $E^2$ | .109 | $E^7$ | .288 |
| $E^4$ | .050 | $E^4$ | .005 | $E^{11}$ | .023 |
| $E^5$ | .262 | $E^6$ | .016 | *$E^{13}$ | .000 |
| | | $E^7$ | .854 | | |
| | | A—E allele combinations | | | |
| $A^4E^1$ | .588 | $A^4E^1$ | 016 | $A^4E^5$ | .689 |
| $A^2E^2$ | .100 | $A^2E^2$ | .109 | $A^4E^7$ | .288 |
| $A^4E^4$ | .050 | $A^4E^4$ | .005 | $A^4E^{11}$ | .023 |
| $A^4E^5$ | .262 | $A^4E^6$ | .016 | *$A^3E^{13}$ | .000 |
| | | $A^4E^7$ | .078 | | |
| | | $A^1E^7$ | .776 | | |
| | | *$A^2E^7$ | .000 | | |

* Present in very low frequency in prior and subsequent generations (see text).

reported and has since been lost; in line 3 the frequencies of two of the three B alleles are .055 ($B^1$) and .043 ($B^3$), and the frequency of one of two K alleles is .045 ($K^1$); in line 4 one of two A alleles is rare ($A^3$), and the J locus appears to be represented by only one allele ($J^2$). The proportion of systems in which the second most frequent allele reached a value of .20 or higher in lines 1, 3, and 4 are 5/11 (E, D, H, I, K), 6/11 (C, D, H, I, J, L), and 6/11 (E, B, C, I, L, P), respectively. The frequent occurrence of two or more alleles at moderate frequencies in these long-established populations indicates that these loci (perhaps including non-blood group loci in contiguous chromosome segments) are optimally beneficial to the population when maintained in a polymorphic state.

## REFERENCES

BRILES, W. E. (1962), Additional blood group systems in the chicken. *Ann. N. Y. Acad. Sci.* **97.** 173–183.

BRILES, W. E. (1964), Current status of blood groups in domestic birds. *Zschr. Tierzüchtung und Züchtungsbiologie,* **79.** 371–391.

BRILES, W. E. (1968), New evidence for close linkage between the *A* and *E* blood group loci of the chicken (abstract). *Genetics,* **60.** 164.

TABLE 2
*Gene frequencies*

| Line 1 | | Line 3 | | Line 4 | |
|---|---|---|---|---|---|
| $B^1$ | .165 | $B^1$ | .055 | $B^1$ | .288 |
| $B^2$ | .680 | $B^2$ | .902 | $B^2$ | .541 |
| $B^{10}$ | .102 | $B^3$ | .043 | $B^5$ | .165 |
| $B^{12}$ | .053 | | | $B^9$ | .004 |
| | | | | $B^{13}$ | .003 |
| $C^1$ | .083 | | | | |
| $C^2$ | .717 | $C^2$ | .301 | $C^2$ | .447 |
| $C^3$ | .087 | $C^4$ | .699 | $C^5$ | .553 |
| $C^5$ | .113 | | | | |
| $D^1$ | .333 | $D^1$ | .531 | $D^1$ | .919 |
| $D^2$ | .045 | $D^2$ | .180 | $D^4$ | .001 |
| $D^3$ | .622 | $D^3$ | .284 | $D^5$ | .080 |
| | | $D^4$ | .005 | | |
| $H^1$ | .452 | $H^1$ | .543 | $H^1$ | .164 |
| $H^2$ | .532 | $H^2$ | .457 | $H^2$ | .836 |
| $H^3$ | .016 | | | | |
| $I^2$ | .676 | $I^2$ | .436 | $I^1$ | .013 |
| $I^3$ | .029 | $I^3$ | .035 | $I^2$ | .705 |
| $I^4$ | .042 | $I^4$ | .529 | $I^3$ | .281 |
| $I^5$ | .253 | | | | |
| $J^1$ | .176 | $J^1$ | .477 | | |
| $J^2$ | .824 | $J^2$ | .523 | $J^2$ | 1.000 |
| $K^1$ | .280 | $K^1$ | .045 | $K^1$ | .083 |
| $K^2$ | .396 | $K^2$ | .955 | $K^2$ | .167 |
| $K^3$ | .324 | | | $K^3$ | .751 |
| $L^1$ | .999 | $L^1$ | .302 | $L^1$ | .626 |
| $L^2$ | .001 | $L^2$ | .698 | $L^2$ | .374 |
| | | $P^1$ | .188 | | |
| | | $P^2$ | .721 | | |
| $P^6$ | .118 | $P^3$ | .046 | $P^5$ | .694 |
| $P^{10}$ | .882 | $P^4$ | .045 | $P^{10}$ | .306 |

# DISCUSSION

**F. PIRCHNER: What is the inbreeding coefficient in the lines? Since there are many loci polymorphic there must be a considerable genetic load in these lines? Are the lines selected for crossline performance?**

**W. E. BRILES: The inbreeding coefficients are estimated to be between 60 and 70%. Inbreeding has actually been avoided since the original three to four generations of intense inbreeding.**

Apparently the theoretical basis for anticipating intolerable segregation loads in such systems must be inadequate, evidence is rapidly mounting that many balanced polymorphisms do indeed exist in almost every species that has been adequately examined. This apparent contradiction was discussed by Milkman and others in the papers appearing in a single issue of *Genetics* in either 1967 or 68. The essential point being that selection acts directly on individuals and only secondarily on the population as a whole.

These lines had been selected only in one or two generations for economical performance at the time of testing for the gene frequencies. In fact, the frequencies in the previous generation were essentially the same as those reported here.

J. Bouw: The alleles presented in your tables are detected with the reagents produced in your lines. In cattle and pig blood grouping we are nowadays finding many new specificities of reagents leading to the detection of many new alleles (actually subdividing the known alleles).

Are you sure that your reagents cannot be divided into subspecificities? If this should be so the frequencies presented in your tables could show quite a different picture.

W. E. Briles: Yes, most of the B and E reagents would likely show additional within locus specificities when used on cells of outside populations. This would not change the frequencies presented, because in mounting many antisera over the years the alleles common between lines invariably show parallel reactions with all locus-specific antisera.

A. Jamieson: If you forget about monitoring existing commercial lines, and simply select for homozygosity at all the known blood group loci, do the birds survive? Has this been attempted experimentally?

W. E. Briles: I have not done this but since occasional multiple homozygotes do occur, I have no reason to suspect that multiple homozygous extracted lines would not survive; however I would expect their general vigor to be somewhat adversely affected. Others in producing histocompatible lines in general experience observed considerable depression of vigor, yet I have found such lines to be segregating for at least four of the blood group loci.

*XIIth Europ. Conf. Anim. Blood Groups Biochem. Polymorph., Bp., 1972 (pp. 419—423)*

# DEVELOPMENT IN COMPARISON OF CHICKEN BLOOD TYPING REAGENTS

E. M. McDermid and C. C. Oosterlee*

Human Genetics Group, John Curtin School of Medical Research,
The Australian National University, Canberra, Australia
* Laboratorium voor Bloedgroepen Onderzoek, Landbouwhogeschool Wageningen, The Netherlands

## SUMMARY

During the period 1964--1968 four comparison tests of chicken blood typing reagents were performed. Subsequent tests have been designed to perform the comparisons differently. Two such tests were performed, the first in Autumn 1968 at The Charles Salt Research Centre, Oswestry, Great Britain and the second in Spring 1969 at the Stichting Bloedgroepen Onderzoek, Wageningen, The Netherlands. In the tests antisera supplied by eleven laboratories in eight European countries, one in the USA and one in Japan, were tested against erythrocytes from 11 different strains of chicken.
Reagents reacting with antigens of the A, B, E, C and D blood group systems were compared. With successive comparisons antisera from the various laboratories have shown an increasing agreement of test results. These developments are surveyed.

## INTRODUCTION

McDermid and Oosterlee (1966) presented the results of the first and second comparison tests. These tests concentrated attention on the A system reagents. Some of the discrepancies between the reagents which were shown up were assumed to occur because of the series of multiple alleles comprising the A system. Interpretation of the results is complicated further by the close linkage existing between the A and E systems. It is possible because of this linkage that antisera to antigens of the A system produced within a strain will contain antibodies to both anti-A and anti-E specificities. Such combined specificity makes it essential to study family groups in comparison tests.

A small number of antisera of B system specificity were also compared in these tests. B system antisera are usually produced by immunization within a strain. When such antisera are tested against erythrocytes of other strains and breeds they show characteristic cross reaction with antigens of the B system other than that used to provoke response (Briles et al., 1950; Gilmour, 1959; McDermid, 1963; Borel, 1964). The results of these tests and others performed by individual workers (Okada and McDermid, 1969) suggest that a number of the B system antigens may be common to many strains and breeds.

At the Eleventh European Animal Blood Group Conference held in Warsaw, Poland in 1968, it was decided to adopt a different procedure in performing the blood group comparison test. Prior to the Conference four comparison tests had been performed as follows by:

1. Thornbers Blood Research Laboratory, Halifax, Gt. Britain.

2 and 3. Stichting Bloedgroepen Onderzoek, Wageningen, The Netherlands.

4. The Department of Genetics, University of Agriculture, Brno, Czechoslovakia.

The organizing laboratory in these tests, tested all antisera sent in by participating laboratories. The introduction into the comparison of antisera other than those reacting with A system antigens created a burden of testing for the organizing laboratory which became a physical impossibility.

The necessary change, decided upon at the Warsaw Conference, was that each participating laboratory should test its own antisera against erythrocytes supplied by the organizing laboratory. A programme of future comparisons was planned. It turned out, however, that some changes had to be made, resulting in the following programme:

5. Autumn 1968 — McDermid — The Charles Salt Research Centre, Oswestry, Great Britain.

6. Spring 1969 — Oosterlee — Stichting Bloedgroepen Onderzoek, Wageningen, The Netherlands.

7. Autumn 1969 — Hála — Institute of Experimental Biology and Genetics, Prague, Czechoslovakia.

8. Spring 1970 — Okerman — Rijksstation voor Kleinveeteelt, Gontrode, Belgium.

9. Workshop, Summer 1970 — Papp — Bloodgrouping Laboratory, University of Veterinary Science, Budapest, Hungary.

Comparison tests 5 and 6 have been performed and interim reports sent to each participant.

## MATERIAL AND METHODS

The organizing laboratory distributed whole blood samples in anticoagulant to each participating laboratory. For test 5, samples from 50 birds belonging to four different strains of chicken and composed of groups of family material within each strain were distributed. For test 6, samples from 50 birds belonging to seven different strains of chicken and composed of sire and dam and five or six offspring of each strain were distributed.

Testing was carried out by each participant and the results obtained compared at the organising laboratory. Table 1 gives details of the laboratories receiving blood samples.

The authors have independently compared a number of reagents from Dr. W. E. Briles, DeKalb Laboratories, Illinois, USA, who did the same with a number of antisera sent to him. The nomenclature adopted has conformed as closely as possible to that of Dr. Briles.

## RESULTS

*A system.* The results obtained from comparison of A system antisera in tests 5 and 6 are summarized in Table 2. Those laboratories possessing antisera of apparently similar specificity are indicated. The anti Al antisera

TABLE 1

*Participants in the comparison tests*

| Worker | Laboratory and Code | | Test 5 | Test 6 |
|---|---|---|---|---|
| Ermencova | Sofia, Bulgaria | BG | — | + |
| Okerman | Lemberge, Belgium | B/L | + | + |
| Hála | Pragué, Czechoslovakia | CS/P | + | + |
| Schmid | Munich, Germany | D/M | + | + |
| McDermid | Oswestry, Great Britain | GB/O | + | +° |
| Papp | Budapest, Hungary | H/Hv | + | + |
| Okada | Sapporo, Japan | J/HO | + | + |
| Oosterlee | Wageningen, The Netherlands | NL | + | + |
| Muraviev | Moscow, U.S.S.R. | SU/MA | — | + |
| Kushner | Moscow, U.S.S.R. | SU/MG | — | + |
| Briles, W. E. | Illinois, U.S.A. | USA/1 | — | +° |

+° a number of reagents tested at Wageningen.

all reacted with the same cell samples in each strain. The antisera labelled A1, A2 react with both the A1 and A2 antigens. Three laboratories were able to identify the A5 antigen which, to date, has been found only in some of the mediumheavy breeds of chickens such as Rhode Island Red.

*E system.* Antisera identifying three E system antigens reacted uniformly in the tests. Anti-E1 has been produced in three laboratories, anti-E2 in six laboratories and anti-E5 in eight laboratories.

A number of other anti-E antisera have been produced in one or two laboratories only. They should be subjected to further comparison.

*C system.* The antigenicity of this system is rather weak. Few laboratories have produced anti-C antisera, but amongst those tested there were antisera showing similar specificities.

TABLE 2

*Results of the comparison of anti A sera*

| Antisera specificity | Antigens | | Participating laboratories | | | | | | | | | | |
|---|---|---|---|---|---|---|---|---|---|---|---|---|---|
| | test 5 | test 6 | BG | B/L | Cs/P | D/M | GB/O | H/Hv | J/Ho | NL | SU/ MA | SU' MG | USA/ I |
| Anti $A_1$ | + | + | + | + | | | + | + | | + | + | | + |
| Anti $(A_1)$ $A_2$ | + | + | | + | + | + | + | | + | + | | | + |
| Anti $A_3$ | | + | | | | | | + | + | + | | | + |
| Anti $A_4$ $(A_3, A_5)$ | + | + | | + | + | | + | | | + | | | + |
| *Anti $A_5$ | | + | | + | + | | | | + | + | | | |
| *Anti $AG_{B/0}$ | + | | | | | | + | | | | | | |
| *Anti $A_x$ | | + | | + | | | | + | | + | + | | |

* Experimental, probably identical antigens.

*D system.* Results suggest that the system may be less complex than the others examined. Six laboratories have produced anti-D2 antisera and four labboratories anti-D4 antisera.

*B system.* The B system is a system of multiple allelic alternatives. A large number of antigens co-exist. Antisera against these antigens show cross reactivity when tested against erythrocytes of different strains. However, the two tests demonstrate clearly that a number of B system antigens seem common to several strains and that antisera to these antigens can be produced in many of these strains. Table 3 presents a survey of the agreement in identification of the B system antigens. The information in the table is derived from these tests and also from comparison studies carried out by the authors with other workers.

TABLE 3

*Survey of the agreement in identification of B system alleles*

| Antisera specificity | Antigens | | Participating laboratories | | | | | | | | | | |
|---|---|---|---|---|---|---|---|---|---|---|---|---|---|
| | test 5 | test 6 | BG | B/L | Cs/P | D/M | GB/ O | H/ Hv | J/ Ho | NL | Su/ MA | Su/ MG | USA/ I |
| Anti $B_1$ | | | | | + | + | | | | + | | | + |
| Anti $B_2$ | + | + | + | + | + | | + | | + | + | | + | + |
| Anti $B_3$ | + | | | | | + | + | + | + | + | | | + |
| Anti $B_4$ | + | + | | | | + | + | + | + | + | | + | + |
| Anti $B_6$ | + | + | | + | + | | + | | | + | | | + |
| Anti $B_7$ | + | + | | | + | | + | + | | + | | + | + |
| Anti $B_8$ | | | | + | | | | | | | | | + |
| Anti $B_9$ | | | | | | + | + | | + | + | | | + |
| Anti $B_{10}$ | + | | | | | + | + | | + | | | | + |
| Anti $B_{21}$ | | + | | | | | + | | | + | | | |
| Anti $B_{23}$ | + | | | + | + | | + | | | | | | |
| Anti $B_{24}$ | + | + | + | | | | + | + | + | + | | | |

A number of antisera have been compared independently by several research workers.

*Unknown systems.* Several laboratories possess antisera which appear to identify alleles of systems other than those discussed. It is possible that a number of these antisera react with antigens of the I system but conclusive comparisons could not be made.

## DISCUSSION AND CONCLUSION

Several of the laboratories taking part in the 5th and 6th comparison tests have started work on chicken blood groups within the last three years. It is important to bear this in mind when studying the results obtained.

The specificity of antisera depends on the antigenic complement of the erythrocytes of the experimental strains of chicken used for immunization. Thus some laboratories may not be able to produce certain antisera because their chicken flocks lack diversity. The anti-A5 antisera can be taken as an example of this in that this antigen has not been demonstrated in White Leghorn strains.

The immediate purpose of these comparison tests has been to compare reaction patterns obtained with antisera from different laboratories. It can be seen in Table 2 that anti-A1,A2 antisera react with both A1 and A2 antigens. To genotype for the A1 and A2 alleles it is necessary to have information about the reaction of anti-A1 and anti-A1,A2 antisera. It is true also for other systems, and especially the B system, that to identify alleles it is necessary to have information about the reaction pattern of several antisera reacting with the antigens of the particular system.

As a first step towards the description of the antigenic specificity of blood group alleles on an international scale agreement about the specificity of antisera must be achieved. We believe that those comparison tests which have already been performed enable chicken blood group workers to agree about the A system, some of the antigens of the B system and also a number of antigens of the E and D system. It is essential that future and more detailed comparison studies build upon the basis which has been established so that a viable and internationally acceptable nomenclature for chicken blood groups can be constructed.

## ACKNOWLEDGEMENTS

E. M. McDermid carried out the work at Oswestry under the auspices of a Medical Research Council of Great Britain Grant. The receipt of a Welcome Travel Grant enabled him to attend the Warsaw Congress and assist in planning this work.

## REFERENCES

Borel, J. F. (1964), Recherches immuno-génétiques sur des substances spécifiques des groupes chez la poule et sur leur utilisation comme marqueurs de gènes dans l'élevage. Juris-Verlay Zurich, 66–75.

Briles, W. E., McGibbon, W. H. and Irwin, M. R. (1950), On multiple alleles affecting cellular antigens in the chicken. *Genetics*, **35.** 633–652.

Gilmour, D. G. (1959), Segregation of genes determining red cell antigens at high levels of inbreeding in chicken. *Genetics*, **44.** 14–34.

McDermid, E. M. (1963), Immunogenetics of the chicken. *Vox Sang.*, **9.** 249–287.

McDermid, E. M. and Oosterlee, C. C. (1966), Results of the first and second European comparison tests for blood typing of chicken, Proc. Xth. Europ. Conf. Anim. Blood Groups Biochem. Polymorph. Paris, 1966,253–259.

Okada, I. and McDermid, E. M. (1969), Results of international comparison tests. Proc. 1969. Spring. Meet. Japan Poultry Sci. Assoc., 15.

*XIIth Europ. Conf. Anim. Blood Groups Biochem. Polymorph., Bp., 1972 (pp. 425–428)*

# STRENGTH OF ERYTHROCYTE ANTIGENS IN CHICKEN ESTABLISHED BY MEANS OF ANTIBODY PRODUCTION

K. HÁLA

Institute of Experimental Biology and Genetics, Czechoslovak Academy of Sciences, Prague, Czechoslovakia

## SUMMARY

In the experiments inbred chicken lines were used. Such pairs of sublines were chosen which differed in one antigen only. Antigenic strength was measured by means of the titre of antibodies produced in immunized recipients. It was found that in t he lines used, the A antigen was able to induce a higher hemagglutinin titre than antigens of the B system.

## INTRODUCTION

All workers active in the immunogenetic problems and working on the preparation of antisera or other material involving great many antigens, realise that antigens differ considerably in their immunogenicity.

We studied the relative strength of A and B antigen systems. Their capacity to form hemagglutinating antibodies was selected as criterion for comparison.

## MATERIAL AND METHODS

Animals of the inbred lines C, I and W were used in the experiments Their genotype in the blood group loci is given in Table 1. Combinations of sublines of the inbred lines were chosen so as to differ in one antigen only (HAŠEK et al., 1966; HÁLA et al. 1966).

TABLE 1

*Blood type of animals*

| Genotype of donor | Genotype of recipient | Dif. antigen |
|---|---|---|
| IA – $A^{13}/A^{13}$; $E^{13}/E^{13}$; $B^{13}/B^{13}$; $C^{13}/C^{13}$ | IC – $A^{14}/A^{14}$; $E^{13}/E^{13}$; $B^{13}/B^{13}$; $C^{13}/C^{13}$ | A13 |
| WB – $A^{14}/A^{14}$; $E^{14}/E^{14}$; $B^{10}/B^{10}$; $C^{9}/C^{9}$; $D^{1}/D^{1}$ | WA – $A^{14}/A^{14}$; $E^{14}/E^{14}$; $B^{9}/B^{9}$; $C^{9}/C^{9}$; $D^{1}/D^{1}$ | B10 |
| CC – $A^{14}/A^{14}$; $E^{13}/E^{13}$; $B^{2}/B^{2}$; $C^{1}/C^{1}$; $D^{1}/D^{1}$ | CB – $A^{14}/A^{14}$; $E^{13}/E^{13}$; $B^{1}/B^{1}$; $C^{1}/C^{1}$; $D^{1}/D^{1}$ | B2 |

Both the recipients and donors were mostly cocks which had never been immunized. The age of the birds was 6 to 12 months. They were immunized with 40% suspension of unwashed erythrocytes (RBC), resuspended in citrated plasma. One immunizing dose of RBC was as a rule 0.1 ml to 16 ml intravenously. No correction was made for individual weight of the recipients. Blood samples were taken before the injection and on day 3 to 15 after it and the agglutination antibody titre was determined with the separated plasma. Titration was performed with a standard method, using TAKÁTSY's microtitrator. As antibody titre of the recipient, the maximum titre obtained on any day within the given range was considered. Normally the maximum titre appeared on day 4 to 7 following the injection.

## RESULTS

The antibody titres anti-B2, B10 and A13 formed after injection are shown in Table 2. These results were obtained after immunization of several bird groups. For each group one donor was selected, and we simultaneously

TABLE 2

*Antibody titre in chickens immunized with erythrocytes*

| Volume of immunizing RBC (in ml) | Anti-A 13 | | | Anti-B 10 | | | Anti-B 2 | | |
|---|---|---|---|---|---|---|---|---|---|
| | No. of birds | Titre* | | No. of birds | Titre* | | No. of birds | Titre* | |
| 0.1 | 6 | 44 | (0–128) | 2 | 0 | | 3 | 0 | |
| 0.5 | 4 | 58 | (8–128) | 3 | 1.3 | (0–2) | 4 | 0.25 | (0–1) |
| 1 | 2 | 64 | (64) | 2 | 3 | (2–4) | 4 | 0.5 | (0–1) |
| 2 | 7 | 384 | (256–1028) | 2 | 12 | (8–16) | 5 | 3.2 | (0–8) |
| 4 | 5 | 73 | (16–128) | 3 | 50.6 | (8–128) | 4 | 3.5 | (0–4) |
| 8 | | | | 3 | 32 | (32) | 3 | 4 | (0–8) |
| 16 | | | | 3 | 18 | (16–32) | 3 | 8 | (0–8) |

* Arithmetic mean and range of reciprocal titres.

injected all used amounts. Sex of the donor or recipient had no influence on the level of the titre.

The antibodies A13 showed a maximum titre after injection of 2 ml. After injecting a minimum amount 0.1 ml, there were only antibodies A13, whereas anti-B antibodies were not formed in any recipient. The results show that with the used combinations of the inbred sublines the formation of antibodies of anti-B specificity required a far higher amount of RBC administered in one run than that of the A antibodies.

We found also difference between anti-B2 and B10 antibodies. The latter occured in a higher titre and more frequently than the B2 antibodies.

Evaluating not only the titre but also the regularity of antibody formation (when 100% of the animals produce antibodies), it was found that anti-A13 required 0.5 ml RBC, while anti-B10 1 ml of blood and to induce anti-

B2 antibodies in all birds as much as 16 ml of 40% RBC suspension was necessary.

## DISCUSSION

As optimal immunization scheme for the formation of allo-antibodies in chickens, 3 injections of 3 ml blood applied at three-day intervals are reported (FANGUY, 1961). It was found in the inbred mouse line that repeated transplantation in certain combinations gradually lessened the difference between a stronger and a weaker antigen (HILDEMANN, 1970). In our experiment we used, therefore, such an immunization scheme which excluded the possibility of obtaining optimal immunization by repeated injections of RBC. That is why we selected the method of one immunization at a time.

The B system antigens are regarded as the strongest ones, since they induce an antibody formation in 100% of the cases (McDERMID, 1964). It can be suggested that this strong immunogenicity is in close relation with the complexity of the antigens of this system. In the previous works we found, however, that A13 antigen differed from the A14 antigen only in two antigenic factors (KNÍŽETOVÁ and HÁLA, 1970), while B10 antigen differed from B9, and B2 antigen from B1 each in four antigenic factors (HÁLA and KNÍŽETOVÁ, 1970). In spite of this, in our lines the A antigen is capable of inducing a stronger response. It seems that the difference found in the number of antigenic factors is not final. Another explanation may be a qualitative difference between the individual antigenic factors. There is evidence that in the case of B system the antigens present both in RBC and lymphocytes may be involved, whereas the antigens of the A system occur in RBC only. Last, but not least, the antigenic strength may be due to a number of immunologically competent cells which detect the antigen as alien, and react against it (SIMONSEN, 1962).

We are of course well aware of the fact that our results cannot be of general validity. In another line the A antigen has been described as a weak immunogen (SCHIERMAN and McBRIDE, 1967). The question remains whether we are able to distinguish exactly and type the antigens of the A and E system where 1% of crossovers can be found (BRILES, 1958).

Our results clearly indicate that in the determination of the immunogenicity of antigens both in mice (HILDEMANN, 1970) and in chickens a far greater role may be ascribed to alleles than to the loci.

## ACKNOWLEDGEMENT

The author is indebted to Miss LENKA NESLEROVÁ for her valuable technical help.

## REFERENCES

BRILES, W. E. (1958), A new blood group system E, closely linked with the A system in chickens. *Poult. Sci.*, **37**. 1189.

FANGUY, R. C. (1961), Blood typing technique in poultry. College Station, Texas.

HÁLA, K., HAŠEK, M., HLOŽÁNEK, I., HORT, J., KNÍŽETOVÁ, F. and MERVARTOVÁ, H. (1966), Syngeneic lines of chickens. II. Inbreeding and selection within the M, W and I lines and crosses between the C, M and W lines. *Fol. biol.* (Praha) **12.** 407.
HÁLA, K. and KNÍŽETOVÁ, F. (1970), Complex antigens of the B system in inbred lines of chickens. Proc. XIth Europ. Conf. Anim. Blood Groups Biochem. Polymorph. Warsaw, 1968, 385–387.
HAŠEK, M., KNÍŽETOVÁ, F. and MERVARTOVÁ, H. (1966), Syngeneic lines of chickens. I. Inbreeding and selection by means of skin grafts in C line chickens. *Fol. Biol.* (Praha), **12.** 135.
HILDEMANN, W. H. (1970), Components and concepts of antigenic strength. *Transplantation Rev.*, **3.** 5.
KNÍŽETOVÁ, F. and HÁLA, K. (1970), Complex antigens of the A system in chickens. Proc. XIth Europ. Conf. Anim. Blood Groups Biochem. Polymorph. Warsaw, 1968, 381–384.
MCDERMID, E. M., (1964), Immunogenetics of the chicken. *Vox Sang.*, **9.** 249.
SCHIERMAN, L. W. and MCBRIDE, A. R. (1967), Adjuvant activity of erythrocyte isoantigens. *Science*, **156.** 658.
SIMONSEN, M. (1962), The factor of immunization: clonal selection theory investigated by spleen assays of graft-versus-host reaction. In: Ciba Found. Symp. on Transplantation. Ed. G. E. Wolstenholme & M. P. Cameron & A. Churchill Ltd., London, p. 185.

## DISCUSSION

M. PAPP: How can you explain that you had better result with 2 ml than with 4 ml red cell suspension. In our earlier experiment carried out with 20 ml injected once and 2 ml red cell suspension given four times, the former quantity gave a surprisingly high titre which was not too much lower than when applying the 4 small doses.

K. HÁLA: I think that the antibody titre depends on the amount of antigen as is described in other experimental animals, too. One explanation for lower titre after higher dose of antigen is inhibition with the large doses.

E. PETROVSKÝ: There was marked difference in the age of recipients used in your experiment (6–12 months). Could not this influence the titre of antibodies in immune sera?

K. HÁLA: From results of other authors, it is evident, that chickens are immunologically fully competent from the 8th day after hatching and I did not find in the literature any evidence of differences in antibody formation between 6 and 12 month-old chickens.

*XIIth Europ. Conf. Anim. Blood Groups Biochem. Polymorph., Bp., 1972 (pp. 429–435)*

# THE STUDY OF LYTIC REACTIONS OF CHICKEN ERYTHROCYTES TO BOVINE SEMINAL PLASMA

E. PETROVSKÝ

**Department of Animal Biology and Biotechnology, Faculty of Agronomy, University of Agriculture, Brno**

## SUMMARY

**Chicken erythrocytes are lysed by bovine seminal plasma. Testing of erythrocytes of individuals of both sexes disclosed great differences in the lytic reaction.**

**A regular relationship has been found between egg-laying and lysability of erythrocytes. Non-laying birds had erythrocytes all resistant to spermatic lysin, while laying hens showed considerable lysability of red cells.**

**Testing of hormonally-treated individuals revealed that the lysability of chicken erythrocytes by bovine spermatic fluids was markedly dependent on the action of estrogenic hormones.**

## INTRODUCTION

In the past few years it has been found that certain fluids of the male genital tract of cattle and pigs display a special biological activity, manifesting itself both in vivo and in vitro.

FULKA et al. (1965) and MATOUŠEK (1966) have demonstrated the ability of bovine seminal vesicle and ampulla fluids to sensitize sperms to quick cooling of the ejaculate. The sensitizing substance has probably an antigenic character, as in 60 per cent of the related tests it was possible to block it with antisera to seminal vesicle fluid.

MATOUŠEK et al. (1965) found intramuscular immunization by hog seminal vesicle fluid to produce a diminution of sperm concentration in the ejaculates of some immunized bulls, progressing later on to complete aspermia. A much more reliable method of producing aspermatogenesis by heterologous or also isologous seminal vesicle fluid is its injection directly into the testes (MATOUŠEK, 1966; MATOUŠEK et al. 1966; PETROVSKÁ, 1968; PETROVSKÁ et al. 1968).

By polarographic study of the properties of bovine spermatic fluids PETROVSKÁ (1968) found that higher polarographically active protein level in the seminal plasma may be accompanied by a deterioration of certain properties of the germ cells. PETROVSKÁ (1968) and MATOUŠEK and PETROVSKÁ (1969) demonstrated that in mice and hens bovine seminal vesicle fluids produced an antifertile effect, viz. reduction of the number of embryos in the former species, while reduction of egg production in the latter species.

MILLAR (1956) proved bovine seminal plasma to be able to lyse cattle erythrocytes. On the basis of his observations, this investigator attached a certain importance to seminal plasma hemolysis in the evaluation of the state of health of the bovine genital tract or for the treatment of seminal samples intended for storage in insemination stations. MITSCHERLICH and PAUFLER (1960) and ROMANIUK (1960) noted that the hemolytic substance is of no importance from the viewpoint of fertility, for it occurs in fertile and less fertile bulls alike. Of interest are the observations made by ROMANIUK showing that the hemolytic activity is accompanied by a marked deterioration of the sperm quality occurring periodically in some bulls.

In a previous paper (PETROVSKÝ, 1964) we demonstrated that the lytic factor is also present in boar seminal plasma and that the result of the reaction of erythrocytes to the reacting spermatic fluid (seminal plasma and seminal vesicle fluid of bulls and boars, bovine ampulla fluid, boar supratesticle tail fluid) depends on the interrelationship between two factors: first, the resistance of the erythrocyte membrane; secondly, the activity of the hemolytic substance. We have also found individual differences in the resistance of red blood cells from cattle and pigs, but we did not pursue this topic in more detail. MATOUŠEK (1970) demonstrated that in cattle the lysability of erythrocytes is genetically controlled by two alleles of one locus.

The present paper deals with the results of the study of the factors responsible for differences in the lysability of chicken erythrocytes.

## MATERIAL AND METHODS

For the study of the lysability of chicken erythrocytes by spermatic fluids we used blood samples of cocks and hens from inbred lines of LW (5 lines), RIR (1 line) breeds, and of their hybrids from the Experimental Station for Poultry, University of Agriculture, at Drásov.

Lysis of chicken erythrocytes was produced either by seminal plasma from bull ejaculates from insemination stations or by seminal vesicle fluid of slaughtered bulls which we obtained immediately after their slaughtering by means of careful squeezing of the seminal vesicles. The lysing fluid was kept at a temperature of −18°C.

The lytic test was performed either on plexiglass plates or in test-tubes.

In the first procedure we mixed two volumes of various dilutions of spermatic fluid in saline with one volume of a 4% suspension of erythrocytes washed with saline. The mixture was thoroughly shaken and the plates were incubated at 38°C for 45 to 120 minutes. After incubation we read the reaction by means of a 0, ±1, 1–4 scale. The intensity of the hemolytic reaction was expressed as the degree of lysis in per cent, calculated from the number of dilutions and the sum of all values of lysis; the reaction was considered 100% when it reached the value of 4 in all dilutions (Table 1).

For reactions in test-tubes we mixed the spermatic fluid diluted to 1 : 50 with appropriately diluted blood (as a rule 1 : 400) at the ration 1 : 1, in six test-tubes. Three test-tubes were incubated at 38°C for 5–10 minutes, followed by centrifugation. The supernatants were collected in clean test-tubes. In the remaining test-tubes we produced complete hemolysis by alternate freezing and thawing followed by centrifugation, and collection of the supernatant over in clean test-tubes. In all test-tubes we assessed

TABLE 1

*Illustration of lytic test of blood cells of hens and cocks of the same parents*

| Line | Individual No. | Sex | Titre of seminal plasma | | | | | Control | Degree of lysis % |
|---|---|---|---|---|---|---|---|---|---|
| | | | 25 | 50 | 100 | 200 | 400 | | |
| $A_2$ (LB) | 1 | cock | — | — | — | — | — | — | 0.0 |
| | 2 | hen | 4 | 4 | 4 | 4 | 4 | — | 100.0 |
| | 3 | cock | — | — | — | — | — | — | 0.0 |
| | 4 | hen | 4 | 4 | 3 | 3 | 2 | — | 80.0 |
| | 5 | cock | 2 | 2 | 1 | — | — | — | 25.0 |
| | 6 | hen | 3 | 4 | 3 | 3 | 2 | — | 75.0 |
| H (LB) | 7 | cock | 3 | 2 | — | — | — | — | 25.0 |
| | 8 | hen | 4 | 4 | 4 | 1 | — | — | 65.0 |
| | 9 | cock | 1 | 2 | — | — | — | — | 15.0 |
| | 10 | hen | 4 | 4 | 4 | 3 | — | — | 75.0 |
| | 11 | cock | — | 1 | — | — | — | — | 5.0 |
| | 12 | hen | 4 | 4 | 4 | 4 | 3 | — | 95.0 |

the amount of released hemoglobin by photometric technique, at a wavelength of 414 nm. We expressed the intensity of the lytic reaction, per cent extinction of specific lysis related to the extinction of general lysis.

The dependence of the lysability of chicken erythrocytes on the level of estrogens was established in two experiments. In the first we followed the lysability of erythrocytes in hens after stopping egg production by a single intramuscular injection of 50 mg progesterone, while in the second the reaction of erythrocytes in cocks after two subcutaneous injections of 20 mg diethylstilbestrol.

## RESULTS AND DISCUSSION

The values of the lytic reactions of erythrocytes of sexually mature cocks and hens to bovine seminal plasma are presented in Table 1. Most striking are the differences between the strong reactions displayed by hen erythrocytes, on the one hand, and the resistance of cock erythrocytes, on the other hand, even though in both sexes individual variations manifest themselves as well.

Tests performed on hens and cocks of different age showed that over a longer period a much lower variation of lytic reactions was observed in cocks than in hens, the change — growth — of lysability in hens being associated with egg-laying (Table 2). The less marked differences in the lytic reaction values of cocks and their relation with egg-laying are also shown byt he fact that for the time being a statistically significant difference has been found in this reaction between individuals of genetically different populations only in hens in the production phase, but not in cocks. This is evident from Tables 3 and 4, showing clearly that elevated lysability of erythrocytes in egg-laying hens is responsible for the intersexual differences; the differences in the lysability of erythrocytes between hens of different LW lines and between hens of LW and RIR breeds and LW and LW × RIR

TABLE 2

*Intersexual differences in lysability of erythrocytes*

| Age of tested individuals in days | Stage of egg production (in hens) | Cocks | | Hens | | Significance of differences F-test P |
|---|---|---|---|---|---|---|
| | | Number | Degree of lysis % | Number | Degree of lysis % | |
| 150 | Before egg production | 15 | 12.2 | 24 | 10.4 | N |
| 250 | Egg production | 9 | 21.8 | 9 | 54.8 | 0.01 |
| 385 | Egg production | 21 | 21.2 | 21 | 69.6 | 0.01 |
| 500 | Beginning of moulting | 26 | 10.0 | 22 | 65.0 | 0.01 |

TABLE 3

*Interline differences in lysability of erythrocytes of hens (age 190 days)*

| Statistical value | C (LW) | H (LW) | Population | | R (RIR) | $A_1 \times R$ |
|---|---|---|---|---|---|---|
| | | | C×H | $A_1$ (LW) | | |
| $\bar{x}$ % | 13.2 | 25.3 | 18.7 | 19.2 | 38.1 | 42.7 |
| n | 8 | 8 | 8 | 35 | 35 | 35 |
| Significance of differences t-test P | all differences insignificant | | | | $P_{A_1}: R = 0.01$<br>$P_{A_1}: A_1 \times R = 0.01$<br>$R: A_1 \times R$ insignificant | |

TABLE 4

*Interline differences in lysability of erythrocytes of cocks and hens of the same parents (age 500 days)*

| Statistical value | Cocks | | | | F | Hens | | KR |
|---|---|---|---|---|---|---|---|---|
| | F | $A_1$ | $A_2$ | Line KR | | $A_1$ | $A_2$ | |
| x % | 12.9 | 7.5 | 19.4 | 15.5 | 39.8 | 62.0 | 86.6 | 63.3 |
| s$\bar{x}$ % | 10.11 | 0.39 | 10.94 | 1.89 | 20.72 | 5.03 | 6.78 | 8.72 |
| n | 3 | 11 | 7 | 5 | 3 | 9 | 5 | 5 |
| Significance of differences F-test P | | N | | 0.01 | | | 0.05 | |

hybrids manifest themselves both at the beginning of egg production and towards its end.

The relationship between egg production and lysability has been demonstrated by tests on hens — sisters of different ages, of which the elder were already laying eggs, while the younger either had not yet laid eggs or just began to do so (Table 5). The erythrocytes of egg-laying hens were actually more sensitive to the lytic action of spermatic fluids than the erythrocytes of their younger, not yet egg-laying sisters. Blocking of egg-laying through the application of progesterone manifested itself by a marked dim-

TABLE 5

*Lysability of erythrocytes of hens — sisters of different age*

| Line | Hen No. | Date of hatching | Degree of lysability on Nov. 10 in % | Weight on October 1 in kg | Egg production in October in pieces |
|---|---|---|---|---|---|
| F (LW) | 475 | March 22 | 46.9 | 1.60 | 15 |
| | 482 | May 10 | 21.9 | 1.00 | 0 |
| $A_1$ (LW) | 495 | March 22 | 71.9 | 1.70 | 28 |
| | 503 | May 10 | 12.5 | 1.20 | 0 |
| $A_2$ (LW) | 513 | March 22 | 37.5 | 1.45 | 13 |
| | 524 | May 10 | 18.7 | 1.15 | 1 |

inution of the lysability of erythrocytes. This follows quite clearly from Fig. 1. On the other hand, the application of diethylstilbestrol to cocks caused their erythrocytes to become equally sensitive to the lytic factor of the spermatic fluids as the erythrocytes of the egg-laying hens were (Table 6).

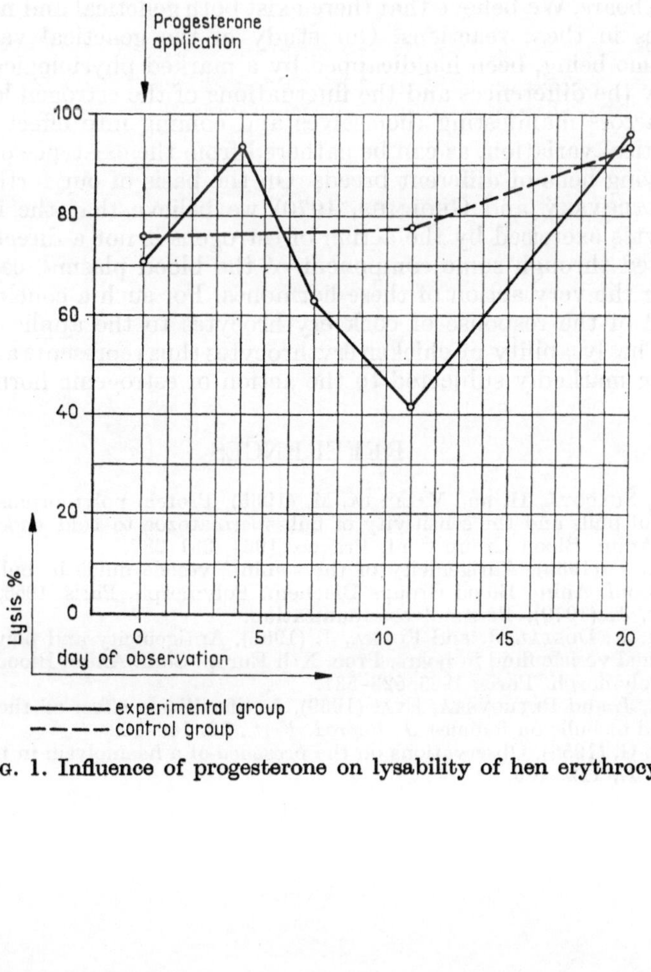
Progesterone
application
100
80
60
40
20
0
Lysis %
0
5
10
15
20
day of observation
experimental group
control group

FIG. 1. Influence of progesterone on lysability of hen erythrocytes

TABLE 6

*Lysability of cock erythrocytes 10 days after application of 20 mg of diethylstilbestrol*

| Group | Cock No. | Degree of lysibility in % |
|---|---|---|
| Experimental | 1 | 62.1 |
| | 2 | 78.5 |
| | 3 | 43.7 |
| | 4 | 31.2 |
| | 5 | 37.5 |
| Total | 1–5 | 50.6 |
| Control | 6 | 3.1 |
| | 7 | 3.1 |
| | 8 | 6.2 |
| | 9 | 40.6 |
| Total | 6–9 | 13.3 |

From all what has been said before it appears that the red cells of fowl are in certain cases sensitive to the lytic action of the spermatic fluids of bulls and boars. We believe that there exist both genetical and non-genetical differences in these reactions. Our study of the genetical variation has, for the time being, been handicapped by a marked physiological variation, caused by the differences and the fluctuations of the estrogen level, the genetical factors manifesting themselves and coming into effect also in this non-genetical variation, as can be gathered from the existence of differences in egg-laying hens of different breeds. On the basis of our further observations (PETROVSKÝ and GROLMUS, 1970) we believe that the influence on erythrocytes exercised by the action of estrogens is not a direct one, but it is mediated through some component of the blood plasma, causing or intensifying the very action of these hormones. For such a conclusion speaks the speed of the response of cock erythrocytes to the application of stilbestrol. The lysability of chicken erythrocytes thus represents another property being markedly subjected to the action of estrogenic hormones.

## REFERENCES

FULKA, J., ŠULCOVÁ, H. and VALENTA, M. (1965), Protein polymorphism of seminal vesicles of bulls and the sensitivity of bull spermatozoa to cold shock. Proc. 9th Europ. Anim. Blood Group Conf. Prague, 1964, 381–387.

MATOUŠEK, J. (1966), Antigenicity of the seminal vesicle fluids in bulls. Proc. Xth Europ. Conf. Anim. Blood Groups Biochem. Polymorph. Paris. 1966, 513–522.

MATOUŠEK, J. (1970), Personal communication.

MATOUŠEK, J., DOSTÁL, J. and FULKA, J. (1966), Antigenicity and polymorphism of the seminal vesicle fluid in boars. Proc. Xth Europ. Conf. Anim. Blood Groups Biochem. Polymorph. Paris, 1966, 523–531.

MATOUŠEK, J. and PETROVSKÁ, EVA, (1969), Antifertilizing effect of the seminal vesicle fluid of bulls on females. *J. Reprod. Fert.*, **20.** 189–192.

MILLAR, P. G. (1956), Observations on the presence of a haemolysin in bovine semen. *Br. Vet. J.*, **112.** 106.

MITSCHERLICH, E. and PAUFLER, S. (1960), Über eine hemolysierende Substanz im Bullensamen *Dt. Tierärztl. Wschr.*, **67.** 614.

PETROVSKÁ, EVA, (1968), Studium některých vlastností spermatickych tekutin hospodářských zvirat. *Thesis*, Liběchov.

PETROVSKÁ, EVA, ZEMÁNEK, F., PETROVSKÝ, E. and MÁCHA, J. (1968), The effect of bull seminal vesicle fluid in its isological intratesticular application. *Acta Univ. Agric.*, Brno, **16.** 673.

PETROVSKÁ, EVA (1964), Hemolytická aktivita semenné plasmy býků a její vztah k chladovému šoku spermií. (Haemolytic activity of bovine seminal plasma and its relation to cold shock of spermatozoa.) *Živ. výr.* **9.** 405.

PETROVSKÝ, E. and GROLMUS, J. (1970), Hormonal dependence of lysability of chicken erythrocytes by bovine seminal plasma. *Fol. biol.* (Praha), (In press).

ROMANIUK, J. (1961), Wystepowanie czynnika hemolytycznego w osoczu buhajów. Medycina Wet., **17,** 395.

## DISCUSSION

C. C. OSTERLEE: Did you take into consideration the marked difference in amount of red cells between cocks and hens?

E. PETROVSKÝ: The differences of hematocrit values could not influence our results. We were working with 4% suspension of red cells or with properly diluted blood.

A. PERRAMON: What about a possible relation between that lytic reaction and agglutination by lectins?

E. PETROVSKÝ: There is no relation between the Hi agglutinogen of Scheinberg and Reckel and the lytic reaction described here.

K. HÁLA: I think, when one is speaking about inbred lines, it is necessary to explain more precisely how this is understood.

E. PETROVSKÝ: Our experimental birds were of inbred lines, produced by brother × sister mating, during 3–4 generations.

V. BENDA: What influences has the author noted after the inactivation of seminal fluid and addition of complement?

E. PETROVSKÝ: Inactivation of the seminal fluid has no effect on its lytic activity. The addition of complement has an inhibitory effect on the reaction of erythrocytes caused by seminal lysin.

*XIIth Europ. Conf. Anim. Blood Groups Biochem. Polymorph., Bp., 1972 (pp. 437—440)*

# DETECTION OF ANTIGENS ON THE SURFACE OF CHICK FIBROBLASTS

V. BENDA, K. HÁLA and I. HLOŽÁNEK

Institute of Experimental Biology and Genetics, Czechoslovak Academy of Sciences, Prague, Czechoslovakia

## SUMMARY

The chickens of inbred lines C and W immunized with 12-day-old chick fibroblasts produced specific antibodies anti-B1 and anti-B9. The specificity of the reactions was tested by erythrocyte and lymphocyte hemagglutination both in homozygous and heterozygous animals, and by mixed agglutination. The antisera obtained from identically immunized rabbits reacted non-specifically even after-absorption.

Antigens corresponding to the adolescent tissue can be found on the cell surface already in the early stage of embryonic development. It can be presumed that the order of their demonstration is closely related to their significance for the ontogenesis of the organism. The surface antigens are important not only from the point of view of the genetic identification but also as a component of the cell surface participating in virus adsorption or penetration into the cell. A tissue culture of known composition of surface antigens enables, therefore, the study of genetically induced resistance or susceptibility to virus infection, e.g. on the model of chick embryonic cells and viruses of the avian leukosis and sarcoma complex.

The surface antigens of fibroblasts were detected by anti-fibroblast (PIOUS and MILLS, 1963) and anti-lymphocyte sera (ENGELFRIET et al., 1967; THORSBY and LIE, 1968). In previous work (HLOŽÁNEK and HÁLA, 1969) we found B antigen to be present on the fibroblasts of the inbred Iowa line. The aim of this paper is to confirm our previous findings also along other lines.

Chickens of the inbred lines C, I and W were used both in pure form and in the first backcross generation (HAŠEK et al., 1966; HÁLA et al., 1966). The heteroimmune sera were prepared in rabbits.

Primary chick embryo fibroblast cultures were prepared from 12-day embryos of C, I and W lines as described by HLOŽÁNEK and HÁLA (1969).

Anti-fibroblast sera (AFS) were prepared in chickens and rabbits on several immunization schemes. The optimal method proved to be the administration of $6 \times 10^6$ cells resuspended in 0.5 ml complete Freund adjuvant (Difco) twice a week. For immunization the cells were harvested by scraping, washed 5 times with phosphate buffered saline (DULBECCO and VOGT, 1954) by centrifugation at 800 r.p.m. for 10 min and finally resuspended in adjuvant. Whereas the hemagglutinin-significant titre appeared in rabbit normally after five injections, the preparation of the alloimmune sera required on the average ten immunizing doses.

The lymphocyte suspensions were prepared from citrated blood by careful centrifugation and the cells were resuspended in Hanks' solution with 1% EDTA (Chelaton I, Lachema) of approx. $5 \times 10^7$ lymphocyte/ml concentration.

The hemolysin titre was determined in inactivated rabbit AFS in the presence of guinea-pig complement.

The mixed agglutination test was performed as described by HLOŽÁNEK and HÁLA (1969).

Four alloimmune AFS agglutinated the erythrocytes from the individual inbred lines in a minimal titre of 8. Three of these sera were tested also by a mixed agglutination method (Table 1). Beside a pronounced reaction with the antigen donor erythrocytes the sera displayed in both tests similar cross reactions the characteristics of which coincided with the reactions obtained in alloimmune anti-erythrocyte sera. This phenomenon can be explained by the overlapping pattern of the antigenic factors (HÁLA and KNIŽETOVÁ, 1970). Identically, though in a lower titre, they agglutinated also the lymphocytes of the individual lines. The difficulties involved in the preparation of the fibroblast suspension did not allow for their direct agglutination.

In view of the fact that the antigens of the lines examined did not segregate, it was necessary to use a segregating first backcross generation for the determination of the specificity of the produced AFS. Consistently with our previous findings (HLOŽÁNEK and HÁLA, 1969) we found that the antibodies reacted only with the product of the allele B. For the AFS CA, CB erythrocytes from IA×(IA×CA) birds were used, for AFS WA were used erythrocytes CA× (CA × WA) (Table 2). The lymphocyte agglutination

TABLE 1

*Hemagglutination and mixed agglutination tests with anti-fibroblast sera*

| | Inbred lines | Genotype at B locus | Anti-fibroblast sera Donor/Recipient | | | |
|---|---|---|---|---|---|---|
| | | | CA—CB | | WA | I |
| | | | 6001—IA | 1750—WA | 5647—CB | 803—CB |
| Hemagglutination | CA | $B^1/B^1$ | + | + | — | — |
| | CB | $B^1/B^1$ | + | + | — | — |
| | CC | $B^2/B^2$ | + | — | — | — |
| | WA | $B^9/B^9$ | + | — | + | — |
| | WB | $B^{10}/B^{10}$ | + | — | + | + |
| | IA | $B^{13}/B^{13}$ | — | — | — | + |
| (RBC) | IB | $B^{13}/B^{13}$ | — | — | — | + |
| | IC | $B^{13}/B^{13}$ | — | — | — | + |
| Mixed Agglutination | CB | $B^1/B^1$ | + | + | — | |
| (CEF) | WA | $B^9/B^9$ | + | — | + | |
| | IA | $B^{13}/B^{13}$ | — | — | — | |

TABLE 2

*Hemagglutination tests with anti-fibroblast sera*

| CA—CB anti-fibroblast sera | | | | | WA anti-fibroblast serum | | | |
|---|---|---|---|---|---|---|---|---|
| | Genotype at B locus | | Recipient 6001—IA | Recipient 1750—WA | | Genotype at B locus | | Recipient 5647—CB |
| | 10322 | $B^{13}/B^{13}$ | — | — | | 10102 | $B^{1}/B^{1}$ | — |
| | 10324 | $B^{13}/B^{13}$ | — | — | | 10103 | $B^{1}/B^{1}$ | — |
| | 10328 | $B^{13}/B^{13}$ | — | — | | 10106 | $B^{1}/B^{1}$ | — |
| | 10327 | $B^{13}/B^{13}$ | — | — | | 10123 | $B^{1}/B^{1}$ | — |
| First | 10340 | $B^{13}/B^{13}$ | — | — | First | 10139 | $B^{1}/B^{1}$ | — |
| backcross | 10346 | $B^{13}/B^{13}$ | — | — | backcross | 10140 | $B^{1}/B^{1}$ | — |
| generation | 10351 | $B^{13}/B^{13}$ | — | — | generation | 10141 | $B^{1}/B^{1}$ | — |
| IA×(IA× | 10307 | $B^{1}/B^{13}$ | + | + | CA×(CA× | 10163 | $B^{1}/B^{1}$ | — |
| CA) | 10341 | $B^{1}/B^{13}$ | + | + | WA) | 10166 | $B^{1}/B^{1}$ | — |
| RBC | 10344 | $B^{1}/B^{13}$ | + | + | RBC | 10112 | $B^{1}/B^{9}$ | + |
| | 10350 | $B^{1}/B^{13}$ | + | + | | 10114 | $B^{1}/B^{9}$ | + |
| | 10354 | $B^{1}/B^{13}$ | + | + | | 10124 | $B^{1}/B^{9}$ | + |
| | | | | | | 10143 | $B^{1}/B^{9}$ | + |
| | | | | | | 10144 | $B^{1}/B^{9}$ | + |

of these heterozygotes was not distinct enough and therefore another test run with stronger AFS is recommended.

No antigens other than that of the B system, which is also a histocompatible system in chicken could be found on the studied fibroblasts (SCHIERMAN and NORDSKOG, 1961). Evidence of other antigens on the chick fibroblast surface (FRANKS and MCDERMID, 1962) is explicable by different techniques used.

Two heteroimmune AFS produced in rabbits showed nevertheless distinct though exclusively species-specific hemagglutination and hemolytic reactions. After absorption of species-specific antibodies both the agglutination and lytic activities of these sera disappeared completely. This result deviates from that of other authors who succeeded in preparing an anti-B antibody, following immunization of rabbits with chicken erythrocytes (BROWN, 1962).

## REFERENCES

BROWN, R. V. (1962), Use of rabbit antisera in blood studies of chickens. *Immunogenetics Letter*, **2.** 65.

DULBECCO, R. and VOGT, M. (1956), Plaque formation and isolation of pure lines with poliomyelitis virus. *J. Exp. Med.*, **99.** 167.

ENGELFRIET, C., MEERSCHE, J., EIJSVOOGEL, V. and VAN LOGHEM, J. (1967), Demonstration of leucocyte iso-antigens on skin fibroblasts by means of the cytotoxic antibody test. *Vox Sang.*, **11.** 625.

FRANKS, D. and MCDERMID, E. M. (1962), Antigens on chicken tissue culture cells. Rept. 8th Animal Blood Group Conference, Ljubljana, 1962 (Mimeo).

HÁLA, K., HAŠEK, M., HLOŽÁNEK, I., HORT, J., KNIŽETOVÁ, F. and MERVARTOVÁ, H. (1966), Syngeneic lines of chickens. II. Inbreeding and selection within the M, W. and I lines and crosses between the C, M, and W lines. *Fol. biol.* (Praha), **12.** 407.

HÁLA, K. and KNIŽETOVÁ, F. (1970), Complex antigens of the B system in inbred lines of chickens. Proc. XIth Europ. Conf. Anim. Blood Groups Biochem. Polymorph., Warsaw, 1968, 385–387.

HAŠEK, M., KNIŽETOVÁ, F. and MERVARTOVÁ, H. (1966), Syngeneic lines of chickens. I. Inbreeding and selection by means of skin grafts and tests for erythrocyte antigens in C line chickens. *Fol. biol.* (Praha), **12.** 335.

HLOŽÁNEK, I. and HÁLA, K. (1969), The detection of B red blood cell allele on chick embryo fibroblasts. *Fol. biol.* (Praha), **15.** 474.

PIOUS, D. A. and MILLS, S. E. (1963), Killing of cultured fibroblasts with isoimmune serum. *Science*, **142.** 52.

SCHIERMAN, L. W. and NORDSKOG, A. W. (1961), Relationship of blood type to histocompatibility in chickens. *Science*, **134.** 1008.

THORSBY, E. and LIE, S. (1968), Antigens on human fibroblasts demonstrated with HL-A antisera and anti-human lymphocytic sera. *Vox Sang.*, **15.** 44.

*XIIth Europ. Conf. Anim. Blood Groups Biochem. Polymorph., Bp., 1972 (pp. 441–446)*

# EFFECT OF BURSECTOMY ON NATURAL ANTIBODY TITRES AND PLASMA PROTEINS IN CHICKS AND GEESE

G. PETHES*, G. KOVÁCS** and S. LOSONCZY***

* Radioisotope Laboratory, Department of Physiology
** Department of Animal Husbandry
*** Bloodgrouping Laboratory
University of Veterinary Science, Budapest, Hungary

## SUMMARY

1. Natural antibodies occur in the chicken and goose from the 2nd–3rd week of life on.
2. Early bursectomy will delay the appearance of natural antibodies and also quantitatively diminish their level.
3. Level of γ-globulin is higher in the goose than in the chicken. This may be related with the lower capability of the goose for specific immune-response and higher capability of the chicken for heteroimmune responses.
4. The serum γ-globulin ratio diminishes simultaneously with the appearance of natural antibodies in the circulation. This may suggest that production of natural antibodies and/or γ-globulins is controlled by independent regulatory mechanisms.

## INTRODUCTION

The appearance of natural antibodies has been examined in chicks and geese. Red blood cells (RBC) from cattle, horse, rabbit, duck and pheasant were used for their detection. Adult animals showed strong antibody response to blood cells from certain species (e.g. horse), while the antiserum of the same animal gave weaker reaction with blood cells of other species (e.g. cattle). Inherited polymorphism, however, could not be detected.

It has been established (LOSONCZY, 1969, unpublished), that no natural antibody is present immediately after hatching. Its level increases gradually and a considerable antibody level can be demonstrated only during the third week of age. The adult level is reached by the 8th–10th week.

It is known that the formation of specific antibodies is delayed and diminished by the removal of the bursa of Fabricius (GLICK, 1964; PETHES and KEMENES, 1967). In the present study it was examined whether also the non-specific antibodies were influenced by bursectomy. The data suggest a delay also in the formation of natural antibodies.

Literary data have been contradictory concerning the natural antibodies. In general, one may say that they appear early at the beginning of ontogenesis, mostly without detectable stimuli.

Recently it became known that in birds the sack-like lympho-epithel organ on the dorsal wall of the cloaca, the bursa of Fabricius, plays an important role in specific immune response. After bursectomy or in case of diminished function no or significantly lower antibody response can be provoked. The role of the bursa in humoral immune-response has been mostly studied in the chicken, but some data are available also on the pheasants (PETHES, KEMENES and PÁSZTOR, 1970) and ducks (GLICK, 1964). There have been however, no similar data on geese.

The natural antibodies are chemically non-uniform substances, defined as precipitins, hemagglutinins, hemolysins, etc. The subject of the present study were hemolysins and, using RBC of many species, the majority of the natural antibodies could be detected.

Aim of the work was to obtain information on the following problems:

1. When will natural antibodies appear in the circulation of chickens and geese?

2. Does the early removal of bursa influence the appearance of natural antibodies and/or their serum levels in a similar way as that of the specific immune response?

3. Percentual serum protein fractions were measured and the results were expressed in relation of the chicken and goose and with reference to bursectomy.

## MATERIAL AND METHODS

20 Leghorn chickens and White German (Rhine) geese were used. The optimal serum sampling period was determined in preliminary experiments. Surgical bursectomy was carried out on the 1st or 2nd posthaching days. Postoperative losses were less than 2%.

Sera of the experimental animals were tested with 2.5% washed duck, pheasant, pintado, rabbit, cattle, horse and sheep RBC.

## RESULTS AND CONCLUSIONS

Using duck RBC, natural antibodies can be detected in the chicken from the 10–15th day on (Fig. 1). The bursectomized birds showed some 3 weeks'

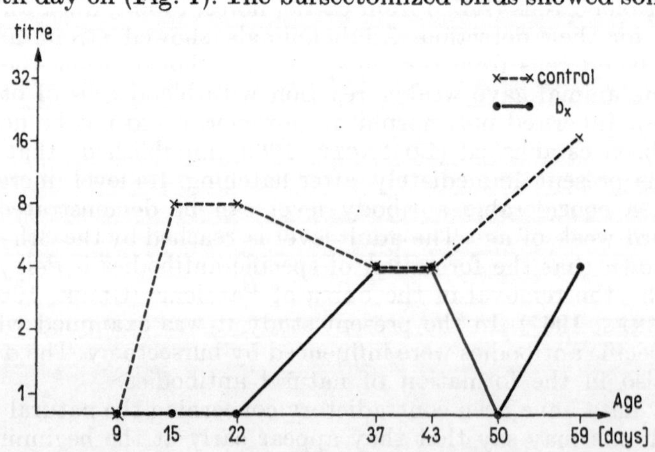
titre
32
16
8
4
2
1
control
$b_x$
Age
9
15
22
37
43
50
59 [days]

**FIG. 1. Formation of the level of natural antibodies in chicken against duck red cells**

delay in antibody appearance. Very similar results were found with pheasants (Fig. 2) and pintado RBC (Fig. 3). It is worthwhile to mention that in bursectomized birds the level of natural antibodies was much lower throughout the whole period of investigation.

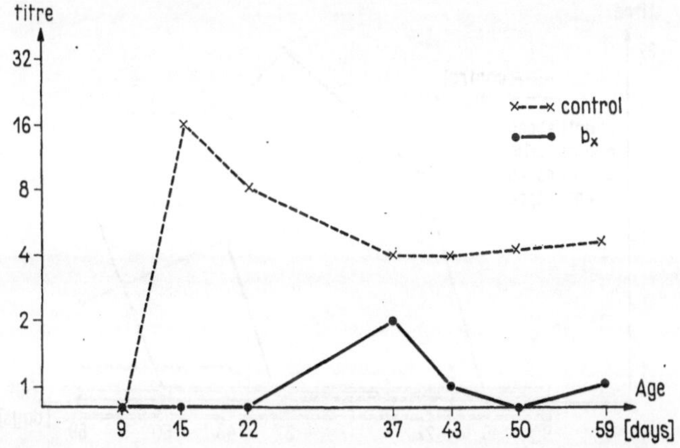
titre
32
16
8
4
2
1
control
$b_x$
Age
9
15
22
37
43
50
59 [days]

FIG. 2. Formation of the level of natural antibodies in chicken against pheasant red cells

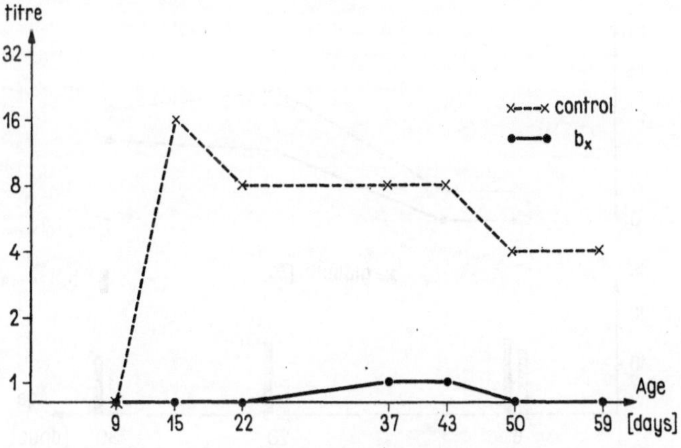
titre
32
16
8
4
2
1
control
$b_x$
Age
9
15
22
37
43
50
59 [days]

FIG. 3. Formation of the level of natural antibodies in chicken against guinea-fowl red cells

Actually the same was found (Fig. 4) with rabbit RBC. With cattle RBC there was a slighter response beginning earliest at 5 weeks of age, and with sheep RBC there was a delayed response (60 days old). With horse RBC no natural antibodies could be demonstrated in the first two months.

In the goose, natural antibodies appeared essentially on the same scheme (Fig. 5). Bursectomy had a depressing and delaying influence also on this species. The level of natural antibodies was, however, much lower.

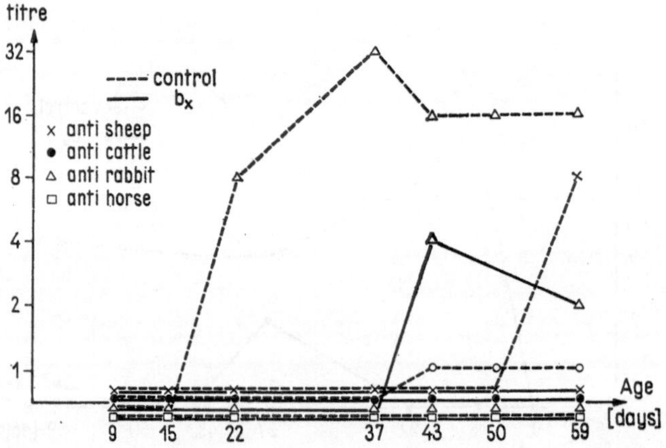
titre
32
16
8
4
2
1
---- control
—— $b_x$
× anti sheep
● anti cattle
△ anti rabbit
□ anti horse
Age
[days]
9
15
22
37
43
50
59

FIG. 4. Formation of the level of natural antibodies in chicken against sheep, rabbit, cattle and horse red cells

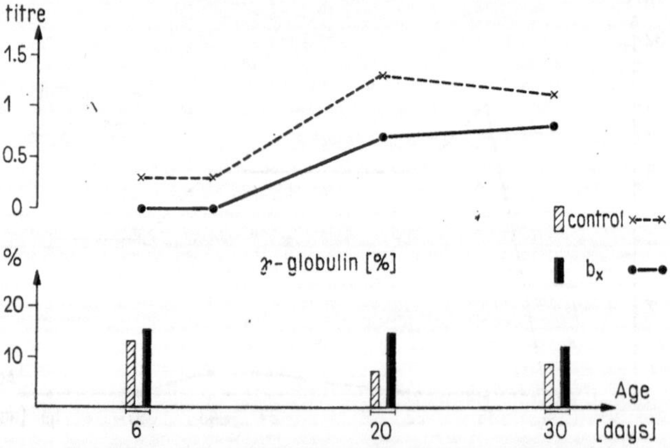
titre
1.5
1
0.5
0
control ×--×
$b_x$ ●—●
%
γ- globulin [%]
20
10
Age
6
20
30
[days]

FIG. 5. The relationship between the level of natural antibodies and the percentage of γ-globulin in goose examining with cattle red cells

Concerning serum-protein spectra of the two species, it is remarkable that the percentage of γ-globulin was higher in the goose than in the chicken (Fig. 5). Also, the percentual quantity of γ-globulin was higher than in the controls.

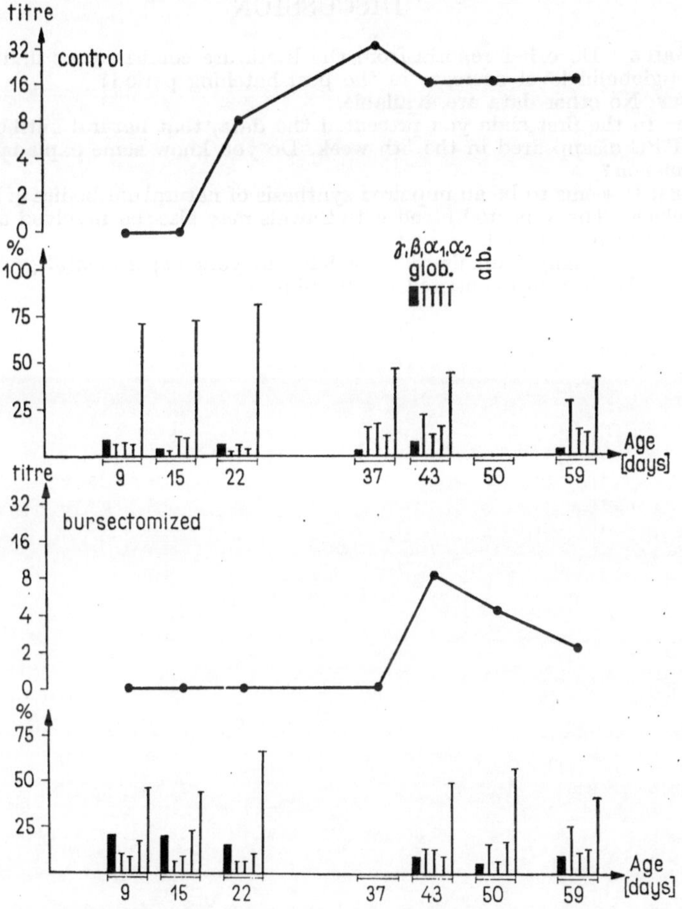
titre
32
16
8
4
2
0
control
%
100
75
50
25
glob.
alb.
9
15
22
37
43
50
59
Age
[days]
titre
32
16
8
4
2
0
bursectomized
%
75
50
25
9
15
22
37
43
50
59
Age
[days]

FIG. 6. Formation of the level of natural antibodies against rabbit red cells and the spectrum of serum protein in chicken

When natural antibodies appear in the circulation, the percentual amount of $\gamma$-globulins will diminish. The same can be observed also in the chickens (Fig. 6).

## REFERENCES

GLICK, B. (1964), The bursa of Fabricius and the developments of immunologic competence. The Thymus in Immunology. (R. A. Good and A. Gabielsen eds.) Harper and Row, N. Y., 343–358.

GLICK, B. (1964), Effect of surgical and chemical bursectomy in duck. *Poult. Sci.*, **43.** 1107.

PETHES, G. and KEMENES, F. (1967), *Acta Univ. Agricultural Brnensis* **36.** 439–448.

PETHES, G., KEMENES, F. and PÁSZTOR, L. (1970), XII. Internationales Symposium über die Erkrankungen der Zootiere. Budapest, pp. 101–102.

## DISCUSSION

W. E. BRILES: Do other reports from the literature confirm your finding that the gamma-globulin level decrease in the post-hatching period?

G. PETHES: No other data are available.

K. HÁLA: In the first slide you presented the data, that natural antibodies against chick RBC disappeared in the 5th week. Do you know some explanation for this phenomenon?

G. PETHES: It seems to be an impaired synthesis of natural antibodies in bursectomized chickens. The repeated blood withdrawals may also be involved at the end of the experimental period.

K. HÁLA: How many animals did you have in your experimental groups?

G. PETHES: Each group consisted of 20 animals.

*XIIth Europ. Conf. Anim. Blood Groups Biochem. Polymorph., Bp., 1972 (pp. 447–450)*

# COMPARATIVE ANALYSIS OF SERA FROM RHODE ISLAND RED, GUINEA-FOWL AND THEIR HYBRID

P. LEROY, J. MORETTI, YOLAINE BARBIER and R. DONATI[1]

Laboratoire du C.S.A.L. 91, Gif-sur-Yvette, France, Faculté des Sciences, Université de Montpellier, France

## INTRODUCTION

It has been reported that protein constituents of serum of cock, and guinea-fowl present small differences. What would be the protein phenotype of a hybrid derived from crossing between cock and guinea-fowl through artificial insemination? In such a crossing in which the pairing of chromosomes is difficult should the genes act differently and give other protein phenotypes by inhibition or formation of new proteins, should the existing ones be modified, or should they repeat the parental phenotype with the same constituents in equal or different number?

## MATERIAL AND METHODS

The external appearance of the hybrid (Numigall) does not resemble the phenotype of the cock nor of the guinea-fowl; head, legs, bill, colour pattern of plumage, body-weight are different (body weight: Numigall 2200 g, Cock 3000 g, Guinea-fowl 1600 g). Hybrid males only, are known; whatever may be the age of the hybrid the testes remain small and are deprived of any sexual activity. Females do not survive after hatching; most of the time they die at the end of the incubating period.

12 domestic cocks (Rhode Island Red); 6 guinea-fowls *(Numida meleagris)* and 6 Numigalls were used for our present study.

All the animals were of the same age (8–10 months old). The blood collected from the wing vein was kept at room temperature during 24 hours. 0.5 ml of serum mixed with 0.5 ml of Freund's adjuvant were injected subcutaneously in the neck of 2 rabbits for each category. Every 3 weeks, one injection was given; in total 3 injections in 9 weeks. Another injection of the antigen mixed with saline (0.9%) and without adjuvant, was given intravenously as needed. The dose in that case was 0.25 ml of serum and 0.25 ml of physiological solution. Blood was collected 15 days after the last injection[2].

[1] Supported in part by a grant from Schneider and Co. (Paris — Le Creusot). Contract no 659952. C.N.R.S. Paris.

[2] *Technical material:* Immuno-electrophoresis: Hirschfeld technique modified by Jan. Trough cutter LKB — Gilson. Supporting medium: agerose (without pectin), and agar-agar. Buffer: sodium Veronal(diethylmalonylurea); pH 8.6. Voltage 50 per microscope plate (75 mm). Time 90 minutes.

All the immunsera were first tested, antisera giving the maximum of precipitin lines were used. Sera of same species were compared in order to detect individual variations. Cross reactions have been made, the same immunserum revealing different types of sera, or conversely the same sera being revealed by different immunsera.

## RESULTS

A diagramm of repartition of precipitin lines (Fig. 1) gives the illustration of all the analysed sera. Since a more detailed study is necessary to identify all the proteins, arbitrary signs have been selected to distinguish the differ-

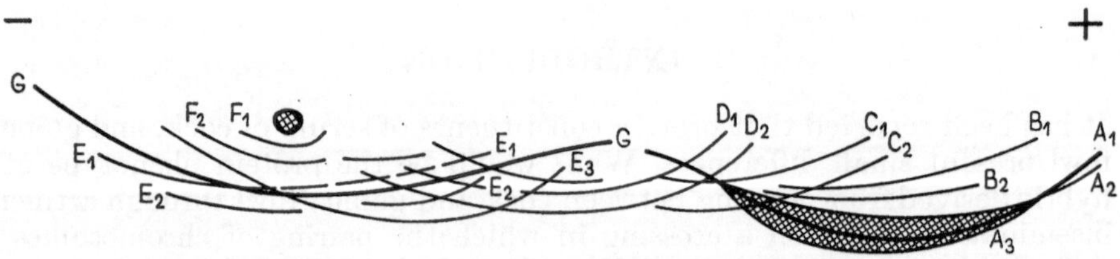

−
+
G
F2
F1
E1
E2
E1
G
E3
E2
D1
D2
C1
C2
B1
A1
B2
A2
A3


FIG. 1. General diagramm of repartition of precipitin lines

ent lines in the classical repartition region: alpha-1, alpha-2, beta-1, beta-2 and gamma-globulins, without presuming the genuine name of the proteins.

The albumin forms a very thick precipitin line $A_2$ in which another important line $A_1$ and sometimes $A_3$ are distinguishable.

In the concavity of albumin $A_1$, there are 2 precipitin lines which can go further than the albumin towards the anode; we call them pre-albumins.

From the anode to the cathode, behind the precipitin lines A and B, there are 2 groups $C_1$, $C_2$ (alpha-1 globulin), $D_1$, $D_2$ (alpha-2 globulin).

Beta-globulins are revealed by E and F corresponding most probably to beta-1 and beta-2.

Gamma-globulin is easy to recognize between E and F in beta-globulin region. We call it G.

Rhode serum with anti-Rhode 2, gives 4 precipitin lines in the albumin: $A_1$, $A_2$, $B_1$, $B_2$; $5 \pm 1$ lines in the alpha-globulin zone: $C_1$, $C_2$, $D_1$, $D_2$, $D_3$, $D_4$; $5 \pm 1$ lines in the beta-globulin: $E_1$, $E_2$, $E_3$, $F_1$, $F_2$.

Guinea-fowl serum with anti-guinea-fowl reveals 4 or 5 lines in the albumin: $A_1$, $A_2$, $B_1$, $B_2$; 5 or 6 lines in the alpha-globulin: $C_1$, $C_2$, $D_1$, $D_2$, $D_3$, $D_4$; 3 lines in the beta-globulin: $E_1$, $E_2$, $F_1$.

Numigall serum with anti-numigall 17 reveals $4 \pm 1$ precipitin lines in the albumin region: $A_1$, $A_2$, $B_1$, $B_2$; $4 \pm 1$ lines in the alpha-globulin: $C_1$, $C_2$, $D_1$, $D_2$; $4 \pm 1$ lines in the beta-globulin: $E_1$, $E_2$, $F_1$, $F_2$.

Identical Numigall sera with anti-Rhode or anti guinea-fowl reveal the same amount of proteins in both cases (Fig. 2).

Specific anti-numigall serum reveals therefore less proteins in the numigall than unspecific antisera. A similar apparent contradiction has been published in a previous publication (LEROY et al., 1968).

After this succinct enumeration of numerous precipitin lines and despite the fact that electrophoresis has been repeated several times, it does not appear that definite conclusion can be drawn from the present study; this is partly due to individual variations of differences in immunsera as well as accidental modifications in the course of electrophoresis.

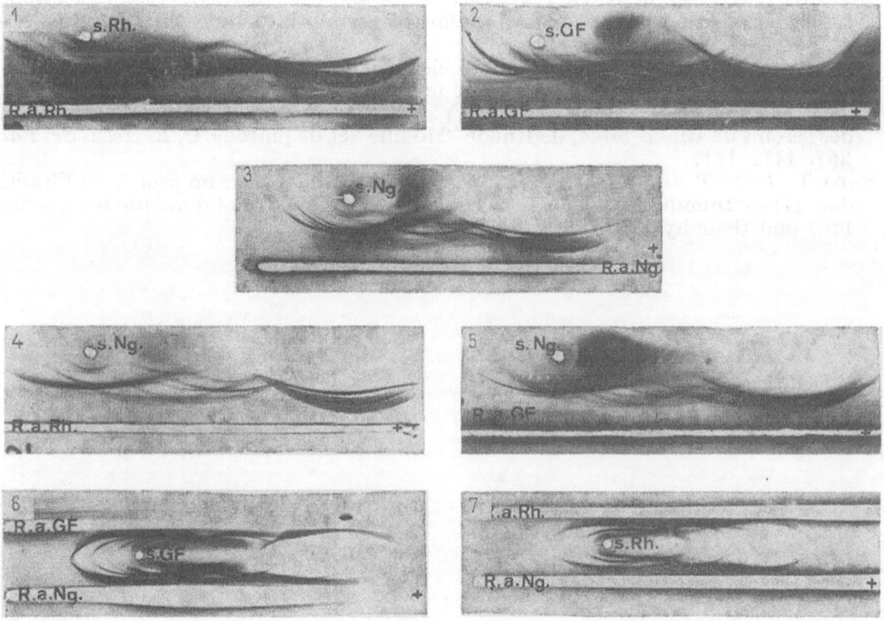
1
s.Rh.
R.a.Rh.
+
2
s.GF
R.a.GF
+
3
s.Ng
+
R.a.Ng.
4
s.Ng.
R.a.Rh.
+
5
s. Ng
R.a.GF
+
6
R.a.GF
s.GF
R.a.Ng.
+
7
R.a.Rh.
s.Rh.
R.a.Ng.
+

FIG. 2. Immuno-electrophoretic analysis. 1 — RIR serum (s. Rh. and rabbit anti-Rh) 2 — Guinea-fowl serum (s. GF and r. anti-GF) 3 — Numigall serum (s. Ng and r. anti-Ng)
— Cross-reaction (immuno-electrophoresis). 4 — between same Numigall serum (s. Ng) and Rh immunserum (r. anti-Rh) 5 — between same Numigall serum (s. Ng and guinea-fowl immunserum (r. anti-GF)
— Immuno-electrophoretic analysis.
6 — Guinea-fowl serum (s. GF) by 2 immunsera: anti-guinea-fowl (r. anti-GF) and anti-Numigall (r. anti-Ng) 7 — Rhode serum (s. Rh) by 2 immunsera: anti-Rh (r. anti-Rh) and anti-Numigall (r. anti-Ng)

However it seems probable that Numigall serum has inherited all the plasmatic proteins of both genitors, cock and guinea-fowl, and not less. Since sera are revealed by unspecific immunsera as shown by cross-reaction the same antigenicity has been preserved.

## CONCLUSION

Numigalls seem to possess all the proteins of the cock and all the proteins of the guinea-fowl. This study is in progress and will be the subject of a subsequent report.

## REFERENCES

BECKMAN, L. D., CONTERIO, and MAINARDI, D. (1963), Serum protein variations in bird species and hybrids. *Bull. Serol. Mus.*, **29.** 5–8.

BRANDT, I. W., SMITH, H. D., ANDREWS, A. C. and CLEGG, R. E. (1952), Electrophoretic investigation of the serum proteins of certain birds and their hybrids. *Arch. Biochem. Biophys.*, **36.** 11–17.

HIRSCHFELD and JAN. (1960), Immuno electrophoresis procedure and application to the study of group specific variations in sera. *Science Tools*, **7.** 2–18.

LEROY, P., MORETTI, J. and BARBIER, Y. (1968), Immunoélectrophorèses comparées des sérums de Rhode M-44, de Rhode "Modifié" et de pintade. *C. R. Acad. Sc. Paris*, **267.** 1412–1413.

SATO, T., ISHII, T. and HIRAI, Y. (1967), Genetic studies on serum protein in Phasianidae. III — Immunoélectrophoretic comparison of the sera of domestic fowl, guinea-fowl and their hybrid. *Jap. J. Genet.* **42.** 51–59.

*XIIth Europ. Conf. Anim. Blood Groups Biochem. Polymorph., Bp., 1972 (pp. 451–454)*

# TIME OF APPEARANCE OF THE B SYSTEM RED CELL ANTIGENS IN THE CHICKEN

C. O. BRILES* and K. E. LEE**

Department of Animal Science, Macdonald College of McGill University, Montreal, P. Q. Canada

## SUMMARY

The development of red cell antigens of the *B* blood group system of the domestic chicken were studied at various embryonic stages and young chick development. Antigens of this system gave weak agglutination, reactions, but increased slowly with increasing age of embryos and newly hatched chicks. The agglutinability of 30-day old chicks' red cells was equivalent to those of adult birds in serial dilution tests (titers) with blood-typing reagents. Homozygous chicks' cells acquired detectable antigens earlier, and reacted stronger than heterozygotes. Cellular antigens appeared on cells of heterozygous embryos and young chicks later than on those of homozygotes.

## INTRODUCTION

The appearance of cellular antigens during embryonic stages and earlier life have been studied in humans, other mammalian and avian species. Experiments have been conducted on blood group systems in man (BORNSTEIN and ISRAEL, 1942; INO, 1950), and a small number of reports have been published in cattle (STORMONT et al., 1958), in rabbits, (KEELER and CASTLE, 1934), and in pigeons (MILLER, 1953).

The appearance of cellular antigens in developing embryos and newly hatched chicks was first studied by BRILES et al., (1948). These workers observed three A locus antigens on the fifth day of incubation. One B system antigen was found on three-day-old embryos, while three others were not detectable until the sixth day post-hatching.

JOHNSON (1956) continued this work and found that the A and B antigens in developing embryos were detectable on the second day of incubation. Three B antigens did not appear in embryos until the seventh day. He also observed that red cells from homozygous embryos were more agglutinable than those from heterozygotes.

DROBNA (1966) found that one cellular antigen of the B system could be observed at early stages of embryonic development. One other was not detectable until ten days post-hatching.

This research was conducted in order to obtain further information on the appearance of red cell antigens of the B system in embryos and young chickens in a high producing egg strain.

Present addresses:
* Carver Research Foundation, Tuskegee Institute, Alabama 36088, USA.
** Department of Poultry Science, University of Sackatchewan, Saskatoon, Canada.

## MATERIAL AND METHODS

Single Comb White Leghorns (Line M1) were used throughout this study. Birds of known genotype of the B system were mated to produce progeny for collecting the data in these experiments.

The reagents utilized were prepared by iso-immunizations and differential absorption methods as described by BRILES et al. (1959).

Blood samples from embryos were obtained by puncturing the heart and extra-embryonic blood vessels. Blood samples from young chicks were collected by making a diminutive incision in the brachial vein with a scapel.

The serological tests and designation of blood group genes, antigens and test-fluids were described by FANGUY (1961). Control (adult) cells with the same genotype as embryos were utilized in all tests.

## RESULTS

Embryos possessing homozygous and heterozygous genotypes consisting of $B^1$, $B^2$, and $B^3$ alleles were sacrificed from 5 to 19 days of incubation every second day. Another group was allowed to hatch in which the chicks' cells were blood-typed at two-day intervals from day-old to 15 days of age and thereafter at 20, 30, and 40 days. The agglutination strength of cells of embryos with $B_1$ antigen increased more rapidly than embryos possessing the $B_2$ and $B_3$ antigens.

Red cells of embryos and chicks with known B system genotypes were tested with serial dilutions of B2 and B3 reagents, in which the reactions of homozygous embryo and chick cells produced greater agglutination scores than those genotypes heterozygous. When chicks were maintained beyond 20 days of age, cells of homozygous and heterozygous birds reacted approximately equal at the same titer.

Cells from homozygous $B^2B^2$ and heterozygous $B^2B^3$ embryos at the fifth and seventh day of incubation were tested with undiluted B2 reagent; only cells from the 7-day-old homozygous $B^2B^2$ embryos reacted weakly (Fig. 1).

The titer of B2 reagents used to detect the $B_2$ antigen of the homozygous $B^2B^2$ embryos from 7 to 13 days of incubation increased from 1 : 1 to 1 : 16 (Fig. 1), and the titer declined to some extent with cells of 15-day-old embryos to the termination of incubation. The detectable agglutination in the titer of 19-day embryo cells was reduced to 1 : 4. However, for cells from day-old chicks the titer increased. The reason for a decrease in reactivity of the $B_2$ antigens of cells from 13-day embryos is unknown.

The $B_2$ antigens from homozygous $B^2B^2$ chicks' cells appeared to reach almost complete expression at 15 days of age, as did the control cells (adults). For the heterozygous $B^2B^3$ cells the age was about 30 days (Fig. 1).

Cells of the homozygous $B^3B^3$ genotype from 5 and 7 day-old embryos did not reveal any agglutination with the B3 reagent diluted 1 : 6 (Fig. 2). Cells from 9-day homozygous $B^3B^3$ and 15 day heterozygous $B^2B^3$ embryos reacted consistently with B3 at this dilution. The cells reacted only at the 1 : 6 dilution or lower throughout the incubation period. The same level

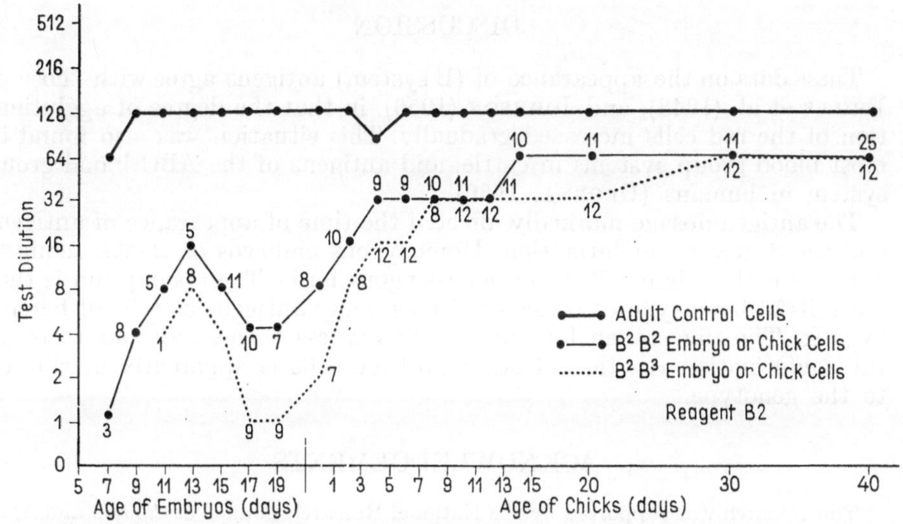
512
216
128
64
32
16
8
4
2
1
0
Test Dilution
5 7 9 11 13 15 17 19
1 3 5 7 9 11 13 15 20 30 40
Age of Embryos (days)
Age of Chicks (days)
Adult Control Cells
B² B² Embryo or Chick Cells
B² B³ Embryo or Chick Cells
Reagent B2

FIG. 1. Time of appearance of red cell antigens of embryos and post-hatched chicks. Reagent B2

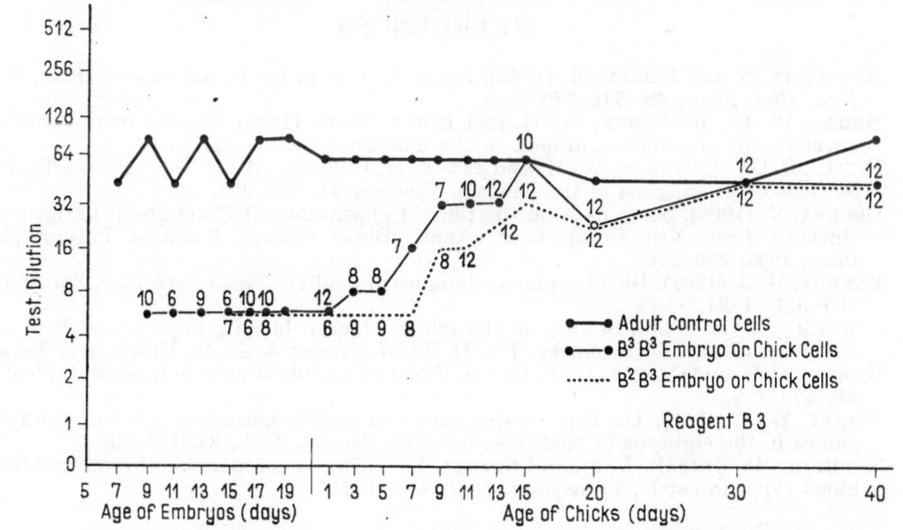
512
256
128
64
32
16
8
4
2
1
0
Test Dilution
5 7 9 11 13 15 17 19
1 3 5 7 9 11 13 15 20 30 40
Age of Embryos (days)
Age of Chicks (days)
Adult Control Cells
B³ B³ Embryo or Chick Cells
B² B³ Embryo or Chick Cells
Reagent B3

FIG. 2. Time of appearance of red cell antigens of embryos and post-hatched chicks. Reagent B3

of reactivity was maintained throughout the embryonic stages and began to increase gradually after hatching as shown in Fig. 2.

The titer of the B3 reagent with cells of heterozygous $B^2B^3$ chicks, as well as embryos, remained at 1 : 6 for the first 7 days of age, but increased gradually thereafter. At about 30 days of age, cells of homozygous and heterozygous birds reacted at the same dilutions as the control cells of adults.

## DISCUSSION

These data on the appearance of (B system) antigens agree with those of BRILES et al. (1948), and JOHNSON (1956), in that the degree of agglutination of the red cells increased gradually. This situation was also found in most blood group systems in cattle, and antigens of the ABO blood group system in humans (DROBNA, 1966).

The antigen dosage markedly affected the time of appearance of antigens and the degree of agglutination. Homozygous embryos or chicks acquired antigens earlier than cells from heterozygous birds. The assumption is that red cells of homozygotes possess twice as many antigens as cells of heterozygotes. The time taken for antigens to express themselves and to reach the agglutination reaction of adult positive cells is apparently attributed to the genotype.

## ACKNOWLEDGEMENTS

This research was supported by the National Research Council of Canada and Agriculture Research Council of Quebec.

## REFERENCES

BORNSTEIN, S. and ISRAEL, M. (1942), Agglutinogens in fetal erythrocytes. *Proc. Soc. Exp. Biol. Med.*, **49.** 718–720.

BRILES, W. E., MCGIBBON, W. H. and IRWIN, M. R. (1948), Studies of the time of development of cellular antigens in the chicken. *Genetics*, **33.** 96–97.

BRILES, C. O., MCGIBBON, W. H. and IRWIN, M. R. (1959), Additional alleles affecting red blood cell antigens in the chicken. *Genetics*, **44.** 955–965.

DROBNA, V. (1966), Some data on the time of appearance of blood group antigens in chickens. Proc. Xth Europ. Conf. Anim. Blood Groups Biochem. Polymorph., Paris, 1966, 235–239.

FANGUY, R. C. (1961), Blood typing techniques in poultry. *Texas Agri. Exp. Sta. Bull.*, M P-551, 1–24.

JOHNSON, L. W. (1956), Studies on the role of chicken blood groups A and B in the induction of anemia in chicks. Ph. D. *Thesis*, Texas A. &. M. University, Texas.

KEELER, C. E. and CASTLE, W. E. (1934), Blood group inheritance in rabbits. *J. Hered.*, **25.** 433–439.

MILLER, W. J. (1953), The time of appearance of species-specific antigens of Columba guinea in the embryos of backcross hybrids. *Physiol. Zool.*, **26.** 124–131.

STORMONT, C., JULIAN, L. M. and SUZUKI, Y. (1958), A comparison of maternal-fetal blood types in cattle. *Transplant. Bull.*, **5.** 133–137.

*XIIth Europ. Conf. Anim. Blood Groups Biochem. Polymorph., Bp., 1972 (pp. 455–458)*

# BASIC ASPECTS OF THE UTILIZATION OF BLOOD GROUPS IN POULTRY BREEDING

M. Papp

**Bloodgrouping Laboratory, University of Veterinary Science, Budapest, Hungary**

## INTRODUCTION

Apart from the papers dealing with the relationships between blood group alleles and economical traits, only a few publications dealt with other possibilities which also can be utilized in practical poultry breeding.

In his work entitled "Blood groups of chicken, their nature and utilization" W. E. Briles (1960) discussed chiefly questions of correlation and McDermid (1964) only mentions that some commercial poultry breeders have studied the effect of various methods of selection (inbreeding or recurrent selection) on the changes taking place in the frequencies of blood group genes from generation to generation, together with the changes of economical traits.

In connection with my present work — collaboration in the new poultry breeding programme of the Bábolna State Farm, — I have tried to review all the possibilities by which the blood group work can contribute to the poultry breeding. As can be seen below, a number of fields of examination generally used in the blood group work of other species can be of greater importance in chicken in the future than it has been up to the present.

## A) CONTROL OF ORIGIN

In cattle, pig and sheep breeding the parentage control is of prime necessity. According to Kovács (1965) the parentage of 21.92% of the home cattle population tested proved to be incorrect on the basis of blood group, transferrin and hemoglobin examinations. Buschman (1964) found 9.6% of pig progeny groups, while Larsen and Bräuner-Nielsen (1964) 8% of them to be erroneous. In Bräuner-Nielsen's (1960) opinion in 30–40% of the pig species erroneous parentage cases can be detected by blood typing. This percentage was 45.82 in sheep according to Fésüs' (1969) calculations.

Poultry breeders held the erroneous belief that the selflocking laying box eliminates incorrect pedigree almost completely. On the contrary we know very well that apart from the handling failures in egg collection, errors can occur also in the pedigree hatching. Nevertheless we cannot find any literary data on the percentage of incorrect pedigree in poultry detectable by blood typing.

In 1968, 625 individuals (50 families) were blood typed in a Yellow Hungarian stock at Mosonmagyaróvár. 50 of 562 offspring — the rest (63) were parents — had incorrect pedigree. In the case of one family the sire was wrongly designated and as a result of this, 8.89% of the offspring examined proved to be of erroneous parentage in the Farm of the College of Agricultural Science in Mosonmagyaróvár where otherwise the registration work has been very accurate. Presumably in an average poultry farm, the percentage of erroneous parentage ranges between 12–15%. It is a considerable ratio especially if it occurs in the parental or grandparental lines of a hybrid-producing farm.

In the farms having pedigree and replacement populations it is usual to keep more than one cock in a test pen. (Generally 5 cocks are assigned to 40 pullets.) In this case, the breeder disregards the differences existing in the breeding value of the individual cocks, although their knowledge would be advantageous. This problem can also be solved technically by blood typing.

Finally blood typing is of decisive importance in the control of the purity of the parental and grandparental lines i.e. in the maintenance of their genetical homogenity.

In the course of business transaction that is, buying or selling poultry, chiefly on marketing a great number of baby chicks, forensic affairs can frequently occur. Particularly when chicks are hatched on contract basis, from eggs derived from several farms, the verification of the origin of diseased chicks is very important. The blood typing can give a considerable support in this case, too.

## B) EXAMINATIONS OF POPULATION GENETICS

One of the most important fields of the blood group work is the detection of blood-group-genetic status (homogeneity, heterogeneity) of the populations and its influencing in favourable direction.

The breeding methods used earlier, as closed breeding or inbreeding, are more and more replaced by "linebreeding" and crossbreeding for hybrid production. Besides the crossbreedings the recurrent or reciprocal recurrent selections are also carried out by using two pure lines. There is no doubt, that to get a standard final product, homogenous, constant, pure lines are needed both in the grandparental and parental generations.

The blood typing accelerates the development of the homogenous lines, as lines homogenous in respect of blood group alleles are supposedly also genetically homogenous at the same time. On the other hand, before making a line homogenous, we should examine the value of every single allele in view of all the essential economical traits not only in the respective line in itself but also in the respective crossbred generation i.e. in the hybrids (see later). Accordingly, not always the most frequent alleles are the most favourable from the breeding point of view. Having found the most advantageous or rather most compatible blood group genotype for the line in question, we increase their occurrence from generation to generation through purposeful selection.

In the case of single cross, the maternal and paternal lines should be homozygous for the essential B or/and A alleles in order to increase the heterosis effect. It is well known that the heterozygous animals are superior to the homozygous ones in respect of some productive characteristics. The selection of the most suitable genotypes both in the grandparental and parental populations is a much more difficult task in double cross than in single cross, but the superiority of the heterozygous blood type should be considered also in this context.

Of course, the close inbreeding, especially the brother–sister matings, accelerate the development of a genetically homogenous line, but this method involves considerable losses through low fertility and hatchability. It seems self-evident that by mating birds of the same blood type without being closely related, such losses can be decreased.

Having done all these, the task remains to maintain the homogenous lines by controlling their purity.

## C) EXAMINATIONS OF RELATIONSHIP BETWEEN BLOOD GROUP ALLELES AND ECONOMICAL TRAITS

This is the question most often dealt with in publications, especially as regards the superiority of heterozygous genotypes to homozygous ones and some heterozygous combinations to other ones. Therefore, only a few points of view are referred to here.

Considering that by a possible linkage a single blood group allele might have a favourable or unfavourable effect on a certain productive characteristic we should examine the relationships existing in the case of its presence or absence in the population.

We must also observe the relative advantage between each pair of alleles within one system to be able to select the allele pairs most advantageous in a productional respect.

Relationship might exist between some blood group alleles or genotypes and certain diseases heritable or hereditary character. Some unpublished data are known concerning the Marek disease.

## REFERENCES

BRILES, W. E. (1960), Blood groups in chickens, their nature and utilization *Wld. Poult. Sci. J.*, **16.** 223–242.

IRWIN, M. R. and STONE, W. H. (1961), Immunogenetics and its application to livestock improvement *Germ. Plasm. Resources.*

McDERMID, E. M. (1964), Immunogenetics of the chicken. *Vox Sang.*, **9.** 249–267.

PAPP, M. (1968), Relationships between blood group factors and economical traits in various breeds of chicken. *Magy. Áo. Lapja*, **23.** 580–583.

SCHULTZ, F. T. and BRILES, W. E. (1952), The adaptive value of blood group genes in chickens. *Genetics*, **38.** 34–50.

# DISCUSSION

C. C. OOSTERLEE: How many cocks do you use to maintain your lines? If these are limited it is essential to control your lines on random drift otherwise it will be possible that after some years you will have lost a number of genes and your line will be completely an other line.

M. PAPP: I do not know still the exact breeding method of Bábolna State Farm. I think a limited number of cocks is used — perhaps 20 — to maintain a line. One of the main purpose of our blood typing work in this Farm is the control of genetic state of the different lines. Actually this work is now in progress.

*XIIth Europ. Conf. Anim. Blood Groups Biochem. Polymorph., Bp., 1972 (pp. 459–466)*

# GENETIC VARIABILITY OF SERUM ALKALINE PHOSPHATASE, LEUCINE AMINOPEPTIDASE AND ACID PHOSPHATASE IN CHICKENS

J. CSUKA and E. PETROVSKÝ*

**Department of Genetics, Agricultural University, Nitra,**
*** Research Institute of Animal Sciences, University of Agriculture of Brno**

## SUMMARY

The genetic polymorphism of the enzyme systems of serum alkaline phosphatase and leucine aminopeptidase was confirmed by starch gel electrophoresis studies.

The genetic variability of acid phosphatase was also disclosed.

All the three enzymes mentioned above had two types which were controlled by two alleles at the same locus (pleiotropy).

The relation of these to economic characters was also studied.

## INTRODUCTION

Recently attention has been increasingly focused on the genetic variability of enzymes of various body fluids in all the species. The variations of the following enzymes were described in chicken: serum esterase I (BOREL, 1964; CSUKA and PETROVSKÝ, 1968) serum esterase II (KIMURA, 1969) alkaline phosphatase (LAW and MUNRO, 1965; WILCOX, 1966) leucine aminopeptidase (LAW, 1967) and liver acid phosphatase (OKADA and HASHINOHE, 1968).

Our studies on the genetic and non-genetic variability of the series of serum enzymes included also phosphatases and leucine aminopeptidase. A summary of the results is presented in this communication.

## MATERIAL AND METHODS

For the study of enzyme polymorphism, plasma or serum samples were withdrawn first from hens of White Leghorn (WL) White Plymouth (WP) and White Cornish (WC) breeds raised at the Enterprise for the Poultry Breeding at Chrustenice. Additional samples from WL, Rhode Islande Red (RIR), Sussex (SU) and WC hens bred at the Institute for Raising Small Animals at Lemberge-Merelbeke were kindly supplied by Dr. OKERMAN. The genetic control of enzyme phenotypes was investigated with the material of the Experimental Poultry Station of the Research Institute of Animal Sciences of Agronomical Faculty, Agricultural College in Brno.

Analysis was performed by horizontal electrophoresis in starch gel. Potato starch hydrolyzed in our laboratory was used. The starch concentration was 12%. The discontinual POULIK (1957) buffer system modified by

ASHTON and BRADEN (1961) was used. Blood plasma samples were placed into gel using the filter paper Whatman 3. Electrophoresis was carried out under 500–600 V, 50–70 mA with simultaneous division on two gels. Separation was allowed to take place for 6–7 hours and was finished when the borate line was 17 cm distant to the start. With the electrophoretic separation completed, the gels were cut into halves and stained. Alkaline phosphatase and leucine aminopeptidase were detected with the method described by LAWRENCE et al. (1960). Acid phosphatase was stained with the technique of ALLEN et al. (1963).

## RESULTS AND DISCUSSION

The overwhelming majority of the chickens tested had two phenotypes of alkaline phosphatase. Phenotype F represented a rapidly migrating fraction between the albumin and transferrin regions. Phenotype S also had

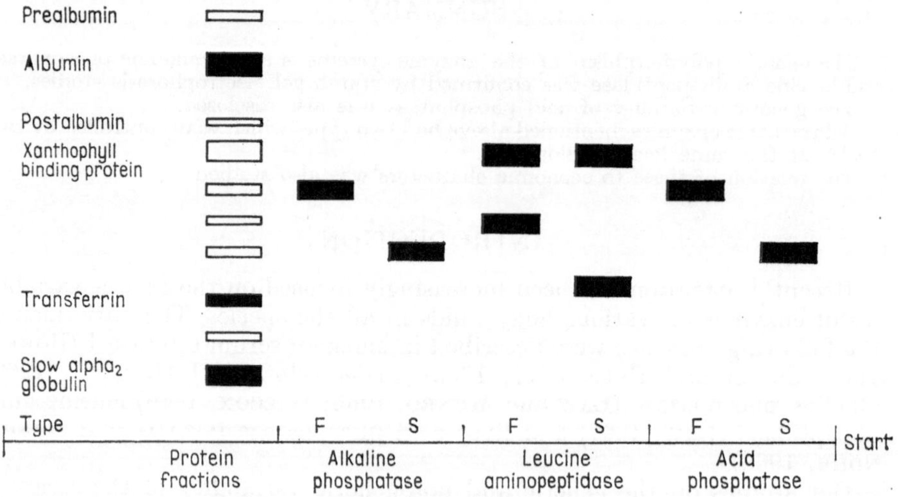
Prealbumin
Albumin
Postalbumin
Xanthophyll binding protein
Transferrin
Slow alpha₂ globulin
Type
F
S
F
S
F
S
Start
Protein fractions
Alkaline phosphatase
Leucine aminopeptidase
Acid phosphatase

FIG. 1. Phenotypes of alkaline phosphatase, leucine aminopeptidase and acid phosphatase

only a single fraction, but migrated more slowly and appeared in the transferrin region (Fig. 1). In three cases we also encountered sera without any phosphatase activity. In gels stained for leucine aminopeptidase by the technique of LAWRENCE's et al. (1960) only one slight fraction appeared. With 5-fold increase of the substrate's concentration two fractions originated giving two phenotypes. Phenotype F was represented by one fraction lying between transferrins and albumins and by another weaker fraction appearing on the catode side of the strong fraction. In phenotype S there were also two fractions, the first corresponding to the strong fraction of phenotype F, the second appearing in the transferrin region.

Acid phosphatase of hen blood plasma also was of two phenotypes, F and S. Phenotype F was represented by one fraction appearing between the albumin and transferrin zones, whereas phenotype S by one fraction migrating slowly and appearing in the transferrin zone. Acid phosphatase compared to alkaline phosphatase was of lower activity and consequently it was difficult to determine its phenotypes in older animals.

Individuals with F phenotype in acid phosphatase had the F type also in alkaline phosphatase and leucine aminopeptidase and vice versa. This rule had no exception though we identified the types of all three enzymes in 524 hens and cocks of different breeds.

## THE HEREDITY OF PHENOTYPES

We considered necessary to confirm whether the differences detected in alkaline phosphatase, leucine aminopeptidase and acid phosphatase had the same way of genetic control as those described by LAW and MUNRO (1965) and WILCOX (1966). For this purpose we tested the offspring of parents arranged into the suitable mating combinations. It is evident (Table 1)

TABLE 1

*Distribution of phenotypes among the progeny from various matings*

| Mating types | | Number of | | Phenotypes of alkaline phos. leucine amino-acid phosphatase observed/expected | | Agreement of observed and expected numbers | |
|---|---|---|---|---|---|---|---|
| Phenotypes | Genotypes | Dam families | Progeny | F | S | $\chi^2$ | P |
| F×F | A/A×A/A | 4 | 24 | 24/24 | — | — | — |
| S×S | a/a×a/a | 20 | 156 | — | 156/156 | — | — |
| F×S | A/A×a/a | 10 | 35 | 35/35 | — | — | — |
| F×F | A/a×A/A | 2 | 7 | 7/7 | — | — | — |
| F×S | A/a×a/a | 13 | 52 | 22/26 | 30/26 | 0.306 | 0.50 |

that all enzymes studied had the same way of genetic control: at one locus two alleles were segregating, one of them dominant, the other recessive. This phenomenon has two hypothetic explanations, one being that the genetic control took place from one locus (pleiotropic effect), the other that all the three different activities appeared on one protein carrier so that the polymorphism showed apoenzyme rather than coenzyme. Based on the finding that both phosphatases migrated differently as compared to leucine aminopeptidase fractions, as well as on the similarity of acid and alkaline phosphatase zymograms we suppose that in the phosphatase-leucine aminopeptidase relation the first hypothesis is more feasible, while in the relation of both phosphatases the second one. LAW (1967) arrived at similar conclusions in respect of alkaline phosphatase and leucine aminopeptidase.

TABLE 2

*Estimates of the frequencies of phenotypes and alleles in different populations*

| | Breed | Origin | Line | Sex | Number of tested chickens | Frequency of phenotypes (No.) % F | Frequency of phenotypes (No.) % S | Frequency of alleles A | Frequency of alleles a |
|---|---|---|---|---|---|---|---|---|---|
| Closed populations | WL | ČSSR | 31 | female | 164 | 14/ 8.5 | 150/91.5 | 0.044 | 0.956 |
| | WL | ČSSR | 33 | female | 605 | 242/40.0 | 363/60.0 | 0.226 | 0.774 |
| | WL | ČSSR | 36 | female | 752 | 313/41.6 | 339/58.4 | 0.236 | 0.764 |
| | WL | ČSSR | 37 | female | 440 | 141/32.0 | 299/68.0 | 0.176 | 0.824 |
| | WL | ČSSR | 31–37 total | female | 1961 | 710/56.2 | 1251/63.8 | 0.202 | 0.798 |
| | WL | ČSSR | 31–37 total | male | 96 | 37/38.5 | 59/61.5 | 0.216 | 0.784 |
| | WL | ČSSR | S-1968 | female | 1271 | 576/45.3 | 695/54.7 | 0.261 | 0.739 |
| | WL | ČSSR | S-1968 | male | 206 | 93/45.2 | 113/54.8 | 0.260 | 0.740 |
| | WL | Belgium | — | female | 252 | 175/69.4 | 77/30.6 | 0.447 | 0.553 |
| | RIR | ČSSR | A | female | 142 | 23/16.2 | 119/83.8 | 0.085 | 0.915 |
| | RIR | Belgium | — | female | 238 | 44/18.5 | 194/81.5 | 0.100 | 0.900 |
| | WC | ČSSR | 1 | female | 286 | 160/55.9 | 126/44.1 | 0.337 | 0.663 |
| | WC | Belgium | — | female | 130 | 112/86.1 | 18/13.9 | 0.628 | 0.372 |
| | WP | ČSSR | 1 | female | 284 | 186/65.5 | 98/34.5 | 0.414 | 0.586 |
| | SU | Belgium | — | female | 231 | 115/49.8 | 116/50.2 | 0.292 | 0.708 |
| Inbred populations | WL | ČSSR | F | female | 13 | 4 | 9 | | |
| | WL | ČSSR | $A_1$ | female | 36 | 0 | 36 | | |
| | WL | ČSSR | $A_2$ | female | 35 | 1 | 34 | | |
| | WL | ČSSR | He | female | 23 | 2 | 21 | | |
| | WL | ČSSR | Hb | female | 42 | 5 | 37 | | |
| | WL | ČSSR | Hy | female | 27 | 11 | 16 | | |
| | RIR | ČSSR | R | female | 74 | 1 | 73 | 0.007 | 0.993 |

## PHENOTYPIC AND GENOTYPIC FREQUENCIES

Breeds and lines within one breed differ considerably in the frequencies of enzyme phenotypes. The low number of populations tested from various breeds as well as the inadequate knowledge of their origin and development do not permit any definite conclusions on interbreed differences, the inter-line differences within one breed being sometimes greater than those existing between different breeds.

In the majority of the populations tested, the recessive allele and phenotype S (Table 2) was predominant. Inbreeding is likely to increase the difference in frequencies. This is confirmed by the high frequency of the recessive allele in inbred populations and in line 31 established by this breeding technique.

## RELATIONSHIP BETWEEN POLYMORPHIC TYPES AND SEVERAL QUANTITATIVE PROPERTIES

The relation of phenotypes of the three enzymes studied with egg laying was followed up in certain populations for two years. Table 3 shows changes in enzyme phenotype frequency after the selection of 1842 hens from four lines egg-laying, egg weight and health state. In the lines 31 and 36, the group of hens selected for further breeding had significantly higher frequency of phenotype S. The two other lines showed negligible difference between hens selected for further breeding and those excluded from the breeding.

TABLE 3

*Frequencies of phenotypes in laying hens with higher and lower performance*

| Line | Group | Number of tested hens | Phenotype frequency in % Alkaline phosphatase Leucine aminopeptidase Acid phosphatase F | S | Significance of differences P |
|---|---|---|---|---|---|
| 31 | A | 95 | 3.16 | 96.84 | 0.02 |
| | B | 52 | 13.46 | 86.54 | |
| | Total | 147 | 6.80 | 93.20 | |
| 33 | A | 375 | 39.20 | 60.80 | — |
| | B | 175 | 40.57 | 59.43 | |
| | Total | 550 | 39.64 | 60.36 | |
| 36 | A | 379 | 37.47 | 62.53 | 0.01 |
| | B | 341 | 47.51 | 52.49 | |
| | Total | 720 | 42.22 | 57.78 | |
| | A | 127 | 30.71 | 69.29 | — |
| | B | 298 | 31.54 | 68.46 | |
| | Total | 425 | 31.29 | 68.71 | |

A = Hens with higher level of efficiency (included in further breeding)
B = Hens with lower level of efficiency

Estimation of egg production of hens of various types in these populations indicated that the hens of S type belonging to the line 33 had higher egg yields. In the remaining lines there was no significant difference in egg production between individuals of various phenotypes. The results are summarized in Table 4.

The repetition of this study on WL hybrids in the next year suggested that hens of type S may have higher egg yields at least in some populations (Table 5).

The relation between the types of enzymes and body weight was studied in individuals of WL, SU and WC breeds. No significant difference was found in these populations (Table 5).

The present findings correlate well with those of other authors. Consistently with LAW and MUNRO (1965) and WILCOX (1966) we identified two phenotypes controlled by a dominant and a recessive allele of one locus. The allele frequencies calculated by us for WL hens were comparable with WILCOX's (1966) results. Our results disagreed with those of WILCOX (1963, 1966) only in respect of the relation between enzyme types and egg production.

## REFERENCES

ALLEN, S. L., MISCH, M. S. and MORRISON, B. M. (1963), Genetic control of an acid phosphatase in Tetrahymena: Formation of a hybrid enzyme. *Genetics*, **48.** 1635.

ASHTON, G. C. and BRADEN, A. W. H. (1961), Serum beta-globulin polymorphism in mice. *Austr. J. Biol. Sci.* **14.** 248.

BOREL, J. F. (1964), Recherches immuno-génétiques sur les substances spécifiques de groupes chez la poule et sur leur utilisation comme marquers de génes dans l'élevage. *Thesis.* Zürich.

CSUKA, J. and PETROVSKÝ, E. (1968), Study of polymorphism of esterase of chicken egg white and blood serum. *Fol. biol.* **14.** 165.

KIMURA, M. (1969), Genetic studies on plasma esterase isozymes in chicken. *Jap. Poult. Sci.*, **6.** 68.

LAW, G. R. J. (1967), Alkaline phosphatase and leucine aminopeptidase association in plasma of the chicken. *Science*, **156.** 1106.

LAW, G. R. J. and MUNRO, S. S. (1965), Genetically determined variations in alkaline phosphatase of White Leghorn chickens revealed by starch-gel electrophoresis. *Genetics*, **52.** 454.

LAWRENCE, S. H., MELNICK, P. J. and WEIMER, H. E. (1960), A species comparison of serum proteins and enzymes by starch gel electrophoresis. *Proc. Soc. Exp. Biol. Med.*, **105.** 572.

OKADA, I. and HACHINOHE, Y. (1968), Inheritance of liver acid phosphatase isozymes of chickens. From abstr., Proc. XIIth Int. Cong. Genet. Tokyo, 1968, *I.* 127.

POULIK, M. D. (1957), Starch gel electrophoresis in a discontinuous system of buffers. *Nature*, **180.** 1477.

WILCOX, F. H. (1963), Genetic control of serum alkaline phosphatase in the chicken. *J. Exp. Zool.*, **152.** 195.

WILCOX, F. H. (1966), A recessively inherited electrophoretic variant of alkaline phosphatase in chicken serum. *Genetics*, **53.** 799.

TABLE 4

*The egg production of hens with different phenotypes (Dobřenice WL, 1967)*

| | Alkaline phosphatase, leucine aminopeptidase and acid phosphatase | | | | | | | |
|---|---|---|---|---|---|---|---|---|
| | Line 31 | | Line 33 | | Line 36 | | Line 37 | |
| | F | S | F | S | F | S | F | S |
| $\overline{X}$ | 131.76 | 125.69 | 114.84 | 121.79 | 113.34 | 115.04 | 126.84 | 126.70 |
| s | 24.96 | 31.46 | 25.22 | 24.31 | 20.16 | 27.16 | 21.01 | 21.70 |
| Number of tested chickens | 9 | 139 | 213 | 327 | 281 | 393 | 136 | 295 |
| Significance of differences P | — | | 0.01 | | — | | — | |

TABLE 5

*The egg production and body weight of hens with different phenotypes*

| | Egg production | | Body weight (g) | | | | | |
|---|---|---|---|---|---|---|---|---|
| Breed | W L, "S" | | W L "S" (150 days) | | W C (140 days) | | S U (140 days) | |
| Phenotype | F | S | F | S | F | S | F | S |
| $\bar{x}$ | 35.50 | 37.32 | 1534.73 | 1534.25 | 2702.12 | 2611.23 | 2627.80 | 2634.90 |
| s $\bar{x}$ | 0.69 | 0.64 | 7.57 | 6.95 | 21.38 | 62.44 | 20.81 | 16.68 |
| Number of tested chickens | 427 | 510 | 430 | 522 | 105 | 17 | 115 | 116 |
| Significance of differences P | 0.01 | | — | | — | | — | |

## DISCUSSION

W. E. Briles: Did you make your comparison between types existing within families? If not, perhaps family differences being confounded with the genotype classifications could yield spurious significant associations with egg production, etc. Also of course, the existence of such confounding with family structure in the populations could very well obscure the relationships that may exist between your polymorphic classifications and economic traits.

J. Csuka: The "hybrid" population was a group produced by hybridization of 4 lines of WL breed. It was the basic material for breeding a new "synthetic" strain. — The material was not divided according to the families.

M. Papp: Concerning your study, I just mention that I also found relationships between blood group alleles and egg production rather than between those and body weight.

*XIIth Europ. Conf. Anim. Blood Groups Biochem. Polymorph., Bp., 1972 (pp. 467—472)*

# QUANTITATIVE STUDY OF GENETIC AND PHYSIOLOGICAL VARIATION OF SERUM ESTERASE IN CHICKENS

E. PETROVSKÝ and JANA MUZIKANTOVÁ*

Research Institute of Animal Sciences, University of Agriculture, Brno

* Laboratory of Animal Genetics, Czechoslovak Academy of Sciences, Liběchov, Czechoslovakia

## SUMMARY

**Studying the genetic variation of serum esterase in the postalbumin polymorphic region Es-1, we actually observed four phenotypes: Es A, B, AB, and O, described earlier in our laboratory. But in repeated tests of the sera of zero type individuals it was as a rule possible to classify them with some active types. Explanation was sought for the genetically conditioned low activity of esterase in this region in individuals formerly ranged among phenotype Es O.**

**In the examined LW individuals we noted a rather high frequency of type Es A, while in the RIR animals type Es B prevailed. Type Es O occurred only in the LW individuals.**

**Between the two breeds there was a marked difference in the strength of the fractions in the region Es-1 and in the overall activity of eserine-resistant esterase: the RIR breed was considerably superior to the LW breed. But we did not observe any intersexual differences.**

**In the study of the physiological variation of esterase in the region Es-1 we confirmed our previous observations as well as the connections between the manifestation of esterase fractions in this region and the elevation of the level of estrogenic hormones, no matter whether they occurred during egg-laying or after the application of hormonal preparations.**

## INTRODUCTION

In the past few years great attention has been paid to the genetic variation of chicken blood serum esterase.

BOREL (1964) detected in chicken serum by electrophoresis either two esterase fractions, a slow and a fast one, or only one fraction, a slow one.

CSUKA and PETROVSKÝ (1968a) distinguished in starch gel electrophoretogram four regions, the fastest of which was polymorphic. In it occurred two variants of different mobility, forming three types: type Es A with one fast migrating fraction, type Es B with one more slowly migrating fraction, and type Es AB containing both variants. Besides active phenotypes, also a type without esterase activity in the polymorphic region — type Es O — occurred in certain individuals.

The same authors have demonstrated that the stainability of the polymorphic esterase fractions changes or sometimes disappears in connection with egg production; that was the reason why the more productive layer-hens did not show a higher representation of phenotype Es O than hens with a lower egg production at the beginning and towards the end of the egg-laying period (CSUKA and PETROVSKY, 1968a, 1968b, 1970). The same changes could be produced in cocks by the application of diethylstilbestrol (PETROVSKÝ et al., 1970).

Chicken serum esterase variation was studied also by GRUNDER (1968) and KIMURA (1969a, 1969b). The former author described polymorphism in the fastest of the three detected esterase regions, consisting of the occurrence of three phenotypes (A, B, AB), controlled by two codominant alleles — $Es^A$ and $Es^B$. At the same time he found that the stainability of esterase zones in the polymorphic region was in some association with the sex, for the mean intensity of colouring of these fractions was higher in cocks than in hens.

KIMURA (1969a) too, described two variants of the fastest migrating esterase region in chicken serum, Es-1 F and Es-1 S, which were genetically controlled by a pair of autosomal codominant alleles, $Es\text{-}1^F$ and $Es\text{-}1^S$. In another report he (KIMURA, 1969b) described hereditary variation in another esterase region, consisting of the presence or absence of one slowly migrating eserine-resistant fraction. This region is genetically controlled on the locus Es-2 by the dominant allele $Es\text{-}2^A$ (for activity of the fraction) and the recessive allele $Es\text{-}2^a$ (for its absence).

The above data agree on certain points and disagree others. All cited authors confirmed the existence of three active esterase phenotypes in the fastest migrating region and showed the influence on their activities of processes connected with egg production. The difference resides in the fact that the phenotype Es O described in our laboratory was not detected in other laboratories, although GRUNDER mentioned variation in the intensity of phenotype colouring in sexually immature individuals and the existence of a new variant. To elucidate these differences and to find solution to certain other problems, detailed studies were carried out on chicken serum esterase, both from the qualitative and from the quantitative point of view. Some of the related findings are reported in this paper.

## MATERIAL AND METHODS

Blood serum of heparinized plasma samples taken from individuals of the inbred lines of LW and RIR breeds (3–4 generations of brother × sister mating) and from their hybrids were used; for details see the section on Results and Discussion.

Detection of polymorphic esterase was made by starch gel electrophoresis, according to the earlier described technique (CSUKA and PETROVSKÝ, 1968a).

The quantitative determination of esterase was made by colorimetry, estimating the amount of naphthol released through enzymic hydrolysis of naphthol esters, to the analogy of the method employed for the quantitative determination of pig blood plasma arylesterase [KÚBEK and HESSELHOLT, 1969 (cit. KÚBEK, 1969)]. The method used by us is described in detail in another communication (MUZIKANTOVÁ and PETROVSKÝ, 1970).

## RESULTS AND DISCUSSION

### A) GENETIC VARIATION OF POLYMORPHIC ESTERASE

In a single test of sexually mature LW (18 individuals) and RIR (13 individuals) cocks, 12 cocks of the LW breed displayed type Es A, while RIR birds type Es O; in the latter breed we found 12 individuals of type Es B and 1 cock of phenotype Es AB (Fig. 1).

In tests on further 67 LW and RIR cocks of inbred lines and their hybrids, we found in the period before sexual maturity (age — about 5 months) four phenotypes described earlier in our laboratory. At the same time there was a marked difference in their representation in different populations. In RIR cocks we found only type Es B, in the LW individuals chiefly

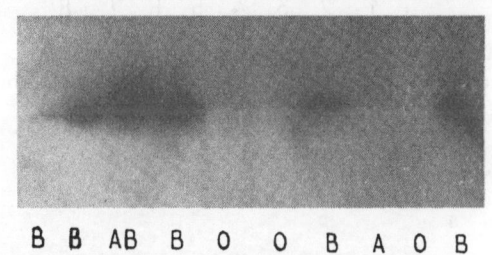

Type B B AB B O O B A O B

FIG. 1. Samples of esterase phenotypes

types Es A and Es O, while in the hybrids all esterase phenotypes were present. Repeated tests performed at monthly intervals until the age of twelve months showed as a rule a very weak A or B esterase fraction in the zero phenotype individuals.

On the basis of these observations the four esterase phenotypes appear explicable not only by the existence of two dominant alleles and one recessive allele, but also by quantitative differences genetically controlled by another locus (by other loci). This initiated further studies on the quantitative variation of chicken serum esterase.

A comparison of the activities of the electrophoretic fractions of the post-albumin polymorphic region, which we denoted with the symbol Es-1 after KIMURA (1969), showed marked differences in the LW and RIR individuals. The absolute majority of tested RIR individuals (20 cocks) had a strong fraction, while the LW birds (38 cocks) displayed, as a rule, a weak fraction. As already noted, type Es O occurred exclusively in the LW breed. Interbreed differences in the strength of the esterase fractions are shown in Fig. 2.

Preliminary tests of the eserine-resistant esterase level on 10 individuals confirmed the results obtained by electrophoresis. In the RIR breed, activity was 2.3 times higher than in the LW breed. A more detailed description of the results of quantitative observations is given elsewhere (MUZIKANTOVÁ and PETROVSKÝ, 1970).

In none of the breeds examined was found, any quantitative difference between sexually immature hens and cocks. But in the period of sexual maturation, marked differences came into display, owing to diminuation of esterase activity in the egg-laying individuals, as described earlier (PETROVSKÝ et al., 1970).

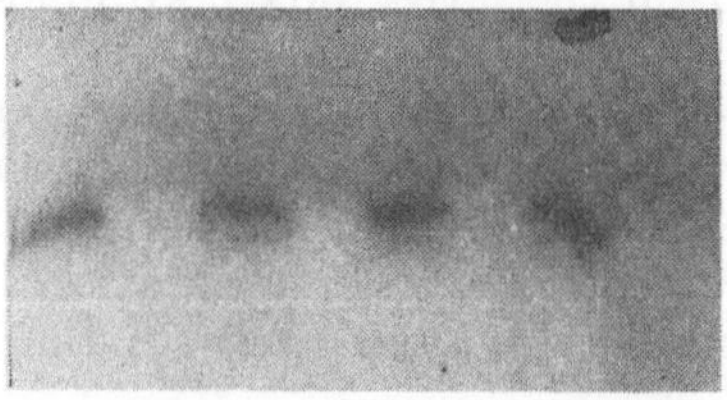
1 2 3 4 5 6 7 8

FIG. 2. Differences in the activity of fractions from WL and RIR cocks
RIR cocks — samples 1, 3, 5, 7; WL cocks — other samples

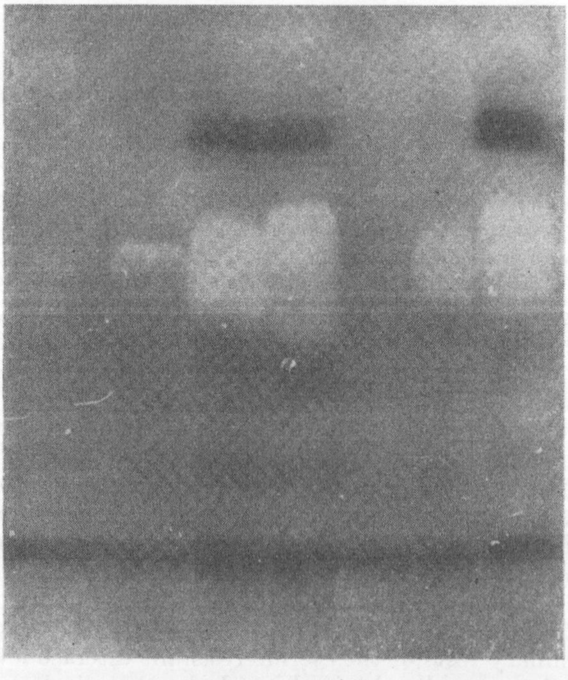
Age in weeks: 25 24 23 22 25 23 22
Hen No. 1102 1101

FIG. 3. Demonstration of changes in stainability of esterase fractions in hens at sexual maturity

## B) PHYSIOLOGICAL VARIATION OF POLYMORPHIC ESTERASE

The study of the physiological changes in the stainability of the fractions in the Es-1 region and of the eserine-resistant esterase level was performed on 5 LW hens, 5 RIR hens, and 10 LW × RIR hybrids. The birds were tested at weekly intervals in the period of sexual maturation.

Repeated electrophoretic examinations have shown the intensity of the esterase fractions to be influenced not only by genetic, but also by physiological factors. In the period of sexual maturation, the intensity of the fractions diminuted in all hens. Moreover a marked decline of the stainability

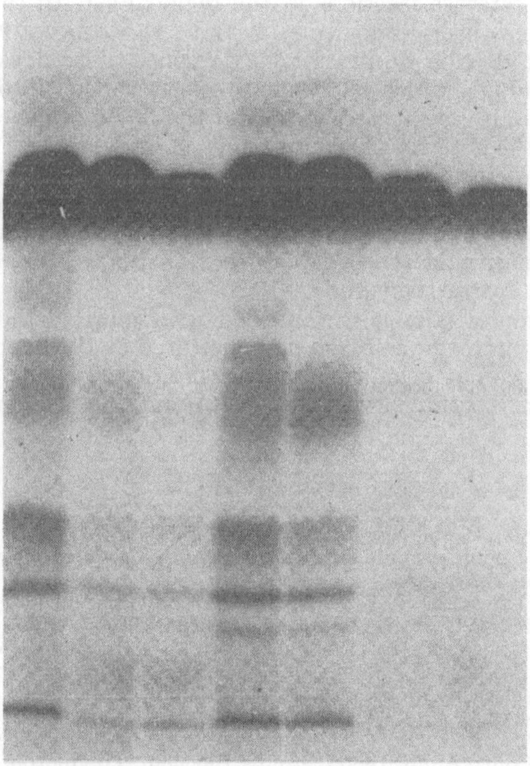

FIG. 4. Demonstration of changes of electrophoretic protein fractions in hens at sexual maturity (Second half of gel from Fig. 3. Evaluate the two pieces together)

of the esterase zones was noted 15–20 days before laying the first egg. Immediately before the beginning of egg production the fractions were hardly visible (Fig. 3).

In some cases there was a weakening of the polymorphic esterase region, despite staining for proteins. The changes in the protein fractions at the beginning of egg production are apparent from Fig. 4.

The quantitative study revealed a marked decrease also of eserine-resistant esterase at the beginning of egg production. The results so obtained are described in detail in another paper (MUZIKANTOVÁ and PETROVSKÝ, 1970).

The mentioned weakening of the esterase fractions occurred also after a single application of 7.5 mg estradiol benzoate to adult cocks.

From the results of the cited authors and those presented by us it can be safely concluded that in chicken serum esterase occurs partly genetical partly non-genetical variation.

Genetical variation manifests itself in two ways, by the presence of at least two alleles in the esterase loci Es-1 and Es-2 and by the different intensity of the esterase fractions in the region Es-1. In this context we revised our earlier observations (CSUKA and PETROVSKÝ, 1968a) in the sense that by repeated tests we proved the presence of weak esterase zones also in those individuals which had been formerly ranged into the group Es O. Thus, of course, it was necessary to find the mode of genetic control of the differences in the activity of esterase in such individuals. For the time being the existence of interbreed quantitative differences was demonstrated observing at the same time also individual variations within the examined breeds. As yet no sufficient information is available on the mode of the genetic conditionality of these differences. Therefore, this question remains to be clarified by further study.

As to non-genetic changes, a marked weakening of the postalbumin region of serum esterase was again demonstrated at the beginning of egg production. Thus we have confirmed our earlier observations (CSUKA and PETROVSKÝ, 1968a; PETROVSKÝ et al., 1970) and the results of GRUNDER as well (1968).

## REFERENCES


BOREL, J. F. (1964), Recherches immuno-genetiques sur les substances spécifiques des groupes chez la poule et sur leur utilisation comme marqueurs de génes dans l'elevage. *Thesis*, Zürich.

CSUKA, J. and PETROVSKÝ, E. (1968a), Study of polymorphism of esterase of chicken egg-white and blood serum. *Fol. Biol.* **14.** 165–168.

CSUKA, J. and PETROVSKÝ, E. (1968b), Genetics of immunological and physiological characters in poultry. IV. Changes in phenotype frequencies of serum esterase in hens due to selection for characters of economic importance. *Acta Univ. Agri.*, Brno **3.** 380–383.

CSUKA, J. and PETROVSKÝ, E. (1970), Serum esterase polymorphism in chicken. Proc. XIth Europ. Conf., Anim. Blood Groups Biochem. Polymorph. Warsaw, 1968.

GRUNDER, A. A. (1968), Inheritance of electrophoretic variants of serum esterases in domestic fowl. *Can. J. Genet. Cytol.*, **10.** 961–967.

KIMURA, M. (1969a), The hereditary variation of eserine-resistant esterases in chickens. *Jap. J. of Genetics.*, **44.** 107–108.

KIMURA, M. (1969b), Genetic studies on plasma esterase isozymes in chicken. *Jap. Poultry Sci.*, **6.** 68–72.

KÚBEK, A. (1969), Študium polymorfných proteinov a enzymov krvného séra a niekterých pohlavných tekutin ošipaných. Czechoslovak Academy of Sciences, Laboratory of Animal Physiology and Genetics, Liběchov. *Thesis.*

PETROVSKÝ, E., CSUKA, J., MUZIKANTOVÁ, J., SMUTNÁ, I. (1970), Hormonálni ovlivnění manifestace polymorfní esterasy krevní plasmy slepic. *Fol. biol.* (Praha) (in press).

*XIIth Europ. Conf. Anim. Blood Groups Biochem. Polymorph., Bp., 1972 (pp. 473–480)*

# ESTERASE IN SERA OF RHODE ISLAND RED, GUINEA-FOWL AND THEIR HYBRIDS OBTAINED THROUGH ARTIFICIAL INSEMINATION

MARIE KAMINSKI, P. LEROY and MICHÈLE SYKIOTIS

**Laboratoire d'Enzymologie CNRS, 91, Gif-sur-Yvette, France**

The esterases demonstrable in sera of hen, guinea-fowl and their hybrids (LEROY et al., 1968) were compared by starch gel electrophoresis and agar gel immunoelectrophoresis, using two chromogenic substrates: $\alpha$- and $\beta$-naphthyl·acetates. The techniques were described previously (KAMINSKI, 1966, 1969).

33 hybrids, 25 RIR, chosen among the progeny of the brothers and sisters of the donor cocks, and 20 GF, chosen among the progeny of the brothers and sisters of the layers, were examined. All the birds were aged 8 to 10 months. The RIR (7 hens and 18 cocks) comprised 2 groups of 8 and 9 brothers and sisters; the others were related through one of the parents. The GF came from a highly inbred flock, but neither sex nor the precise parentage could be determined. The hybrids were closely related between them, as 20 were produced by a pool of sperm from 3 brother-cocks and the remaining 13 by an other similar pool; besides, they were laid, by groups of 2 or 3, by the same GF layer.

A) The starch gel comparison was carried out taking account of the esterase polymorphism of the parental species.

1. The RIR were bled twice, before and during the laying season: in this way the phenotypes could be determined for all animals. We observed more variable bands than reported (CSUKA and PETROVSKÝ, 1968); consequently, we assumed a greater number of alleles. The following phenotypes of 2, 3 or 4 bands were encountered (Fig. 1):

| | | | | |
|---|---|---|---|---|
| AA–O | | | | |
| AB–O | BB–O | | | |
| AC–1 | BC–5 | CC–9 | | |
| AD–O | BD–O | CD–3 | DD–O | |
| AE–O | BE–O | CE–7 | DE–O | EE–O |

| Family distribution: | group I | group II |
|---|---|---|
| | BC–2 | BC–2 |
| | CC–2 | CC–3 |
| | CE–4 | CD–2 |
| | | CE–2 |

2. The patterns obtained for GF differ markedly from those of RIR: in addition to similar bands, there is a more anodic series of bands, grouped by 4, more or less intensely stained, and migrating slightly slower or faster (Fig. 2).

In 2 GF sera this second system was lacking, and the resulting zymogram resembled that of RIR; such absence could be due either to seasonal variation or to a true genotypic difference; at any rate, this indicates that the

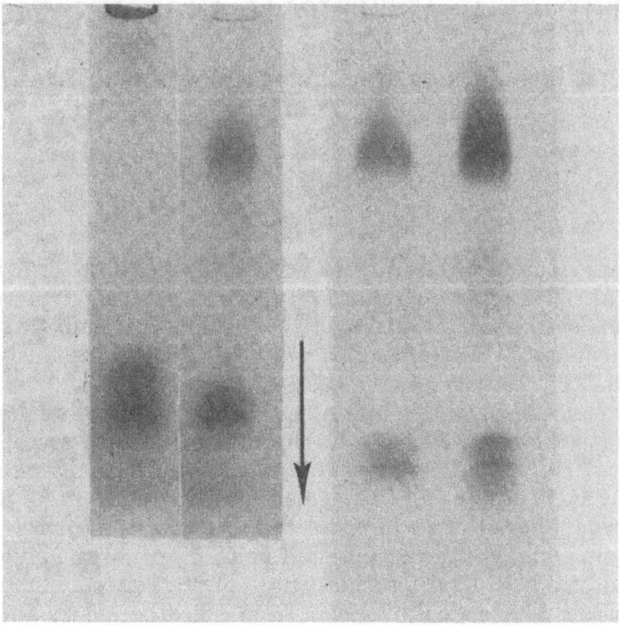

FIG. 1. Starch gel phenotypes of esterases ($\beta$-NaAc) of RIR

two systems are not linked. The differentiation by chemical properties was unsuccessful, as both react with the same substrates and neither of them is inhibited by prostigmin $10^{-3}$ M or DFP $10^{-5}$ M.

Because of the superimposition of bands, the phenotypes of the first system were not determined, except for the 2 sera lacking the second system: they had 2 and 3 bands, respectively analogous to the BB and BC phenotypes of RIR. The second esterase system showed following variation:

absence—2
intense—12
weak—6 (out of which 3 slow).

3. Hybrid zymograms appeared very similar to those of GF, but seemed more subjected to individual variation (Fig. 2). The slower system yielded two phenotypes: 2 bands, slow weak and fast intense, or 4 bands, of equal

intensity. In some sera the distinction was troublesome because of the superimposition of the second system. It was tentatively assumed that 16 sera showed a 2-band pattern and 17 the 4-band. The second system showed more marked variation than in GF, but only 2 slow types were found for 15 fast, 15 intermediate and 1 absence.

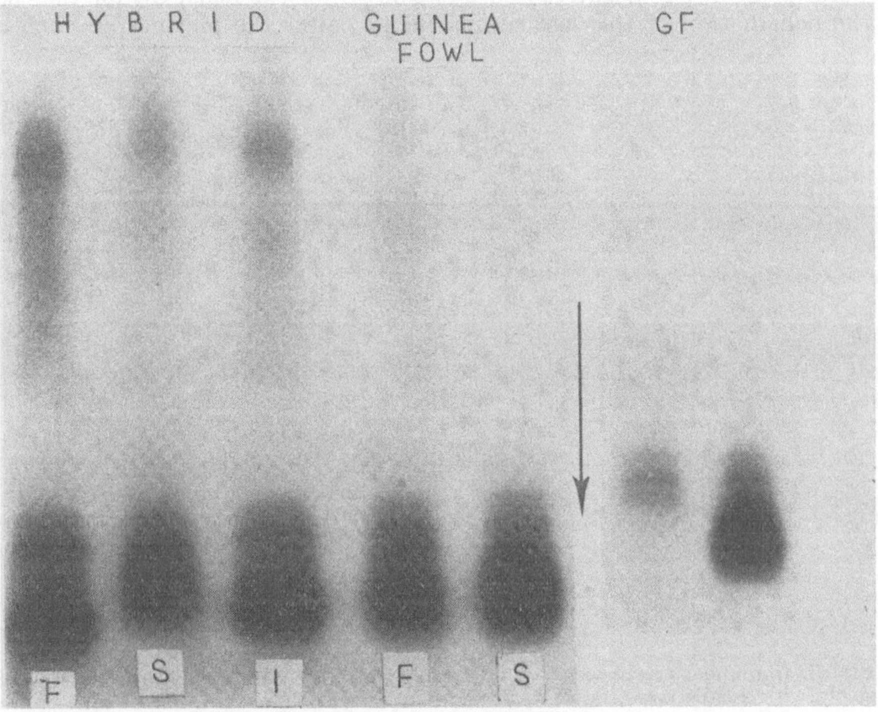
HYBRID
GUINEA FOWL
GF
F
S
I
F
S

FIG. 2. Starch gel phenotypes of Guinea-fowl and hybrids

B) To ensure a correct immunoelectrophoretic comparison of analogous components in sera of the studied species, several anti-RIR, anti-GF and anti-hybrid antisera were used. Moreover, the use of 2 substrates enabled a distinction between two groups of esterases; a) $\alpha$ NaAc revealed 2 lines: 1 in $\beta_1$ and 1 in $\alpha_1$ zone (the latter was also active on $\beta$ NaAc); b) $\beta$ NaAc revealed 5 lines: 1 $\beta_1$, different from the previous $\beta_1$ (Fig. 4); 2 lines of $\alpha_2$ and $\varrho$ location, corresponding to the two forms of the $\alpha$-lipoprotein (KAMINSKI and PODLIACHOUK, 1970), a line of $\beta$-lipoprotein, very variable in location, and the above mentioned $\alpha_1$. As the antisera differed by the presence or absence of antibodies specific of different esterases, the examined sera had to be checked with the whole panel of antisera.

1. RIR (Figs 3 and 4). The $\beta_1$ revealed with $\alpha$ NaAc was found in all sera but varied in amount, the other $\beta_1$ was rare; $\alpha_1$ esterase lacked in 6 sera and

was very faint in others; $\alpha_2$ was found in 5 cases and lacked in 6; the faster component of $\alpha$-lipoprotein was present in all sera. However, 3 sera, containing only $\beta_1$ and $\alpha$ Lp esterases, displayed a line extending over $\beta_1$-$\alpha_1$ zone instead of being faster than albumin (Fig. 4); such pattern was observed for fresh serum while the stored samples showed the usual pattern. Two of these animals were hens, the third was a cock. Their starch phenotypes were: weak CC; 1° negative and 2° weak CC for the hens; CE for the cock; the delipidation of this last serum did not alter the phenotype. Such ab-

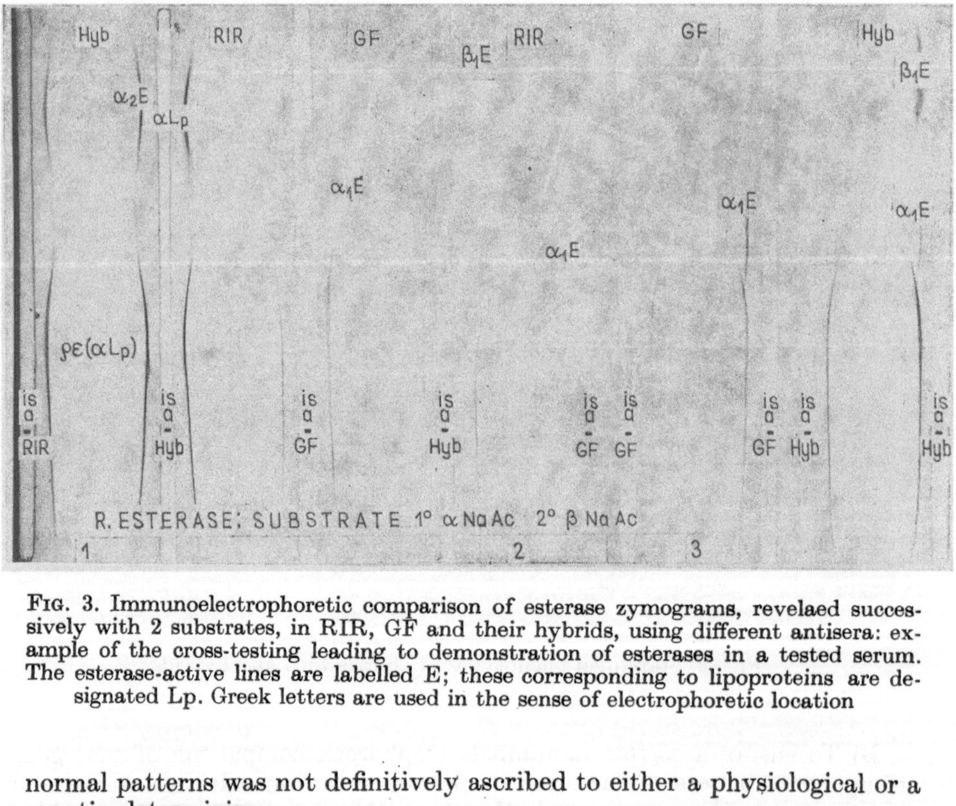
Hyb
RIR
GF
$\beta_1$E
RIR
GF
Hyb
$\beta_1$E
$\alpha_2$E
$\alpha$Lp
$\alpha_1$E
$\alpha_1$E
$\alpha_1$E
$\alpha_1$E
$\rho$E($\alpha$Lp)
is a RIR
is a Hyb
is a GF
is a Hyb
is a GF
is a GF
is a GF
is a Hyb
is a Hyb
R. ESTERASE; SUBSTRATE 1° $\alpha$ NaAc 2° $\beta$ NaAc
1
2
3

FIG. 3. Immunoelectrophoretic comparison of esterase zymograms, revelaed successively with 2 substrates, in RIR, GF and their hybrids, using different antisera: example of the cross-testing leading to demonstration of esterases in a tested serum. The esterase-active lines are labelled E; these corresponding to lipoproteins are designated Lp. Greek letters are used in the sense of electrophoretic location

normal patterns was not definitively ascribed to either a physiological or a genetic determinism.

2. GF (Figs 3 and 5). The $\beta_1$ esterase ($\alpha$ NaAc) was present in 6 sera, absent in 6 and not checked in 8; $\alpha_2$ and $\alpha_1$ lines, very intense, were present, except in 1 and 2 sera respectively; $\varrho$ was uniform in all sera.

As no components differing from those of RIR were found, the immunoelectrophoretical counterpart of the most anodic esterase of GF, detected on starch, could not be directly established. Therefore an identification was attempted by cutting off the active area of starch and inserting the piece into an agar gel, which led to demonstration of $\alpha_1$ esterase. Such reverse electrophoretic location between starch and agar gels was pointed out in the case of two analogous esterase components in horse serum (KAMINSKI,

1969). The $\alpha_1$ esterase, even though not detected in RIR on starch, has been found by immunoelectrophoresis; truly, it was in very low amounts. Such apparent contradiction proceeds from the superior power of resolution of immunoelectrophoresis compared to the direct revelation on starch. Similar case was recently described for horse (KAMINSKI and PODLIACHOUK, 1970).

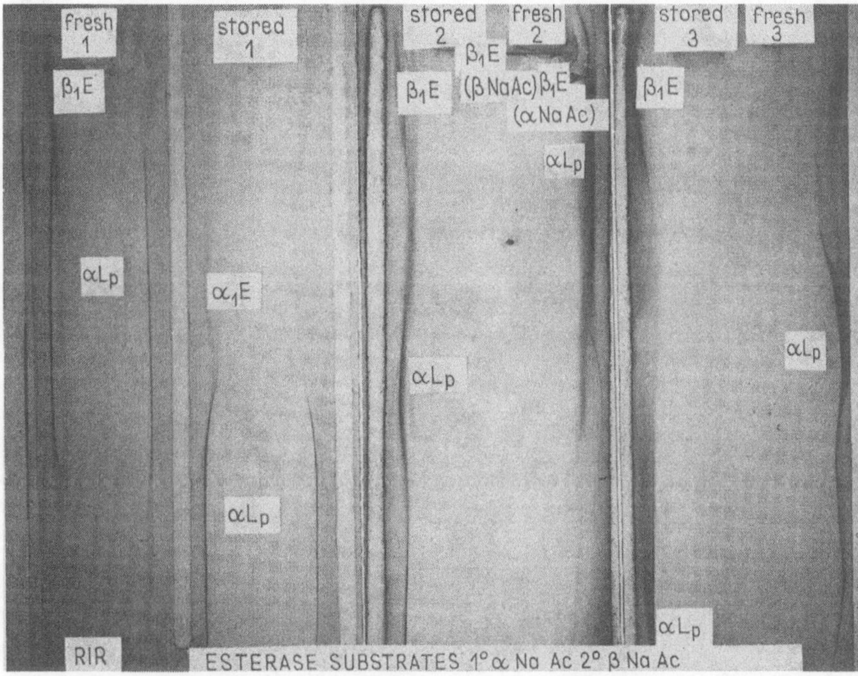
fresh 1
stored 1
stored 2
fresh 2
stored 3
fresh 3
β1E
(βNaAc)β1E
(αNaAc)
αLp
α1E
RIR
ESTERASE SUBSTRATES 1° α Na Ac 2° β Na Ac

FIG. 4. Immunoelectrophoretic comparison of sera of RIR, fresh and stored, showing the variable location of the αLp esterase. 1 and 2 correspond to abnormal patterns, 3 is a control sample of a normal alteration due to storage. The serum 2 contains both $\beta_1$ esterases, revealable each with one of the substrates

3. Hybrids (Figs 3 and 6). The overall pattern resembles very much those of both parental species. The occurrence of $\beta_1$ seems slightly lower than in GF 4 positive sera and 7 negative; $\alpha_2$ was present in 18 sera and absent in 12; $\alpha_1$ present in 31 and absent in 2.

In conclusion, the hybrids appear to have inherited all the same esterase components of both parents, with probably a similar extent of allelic variation.

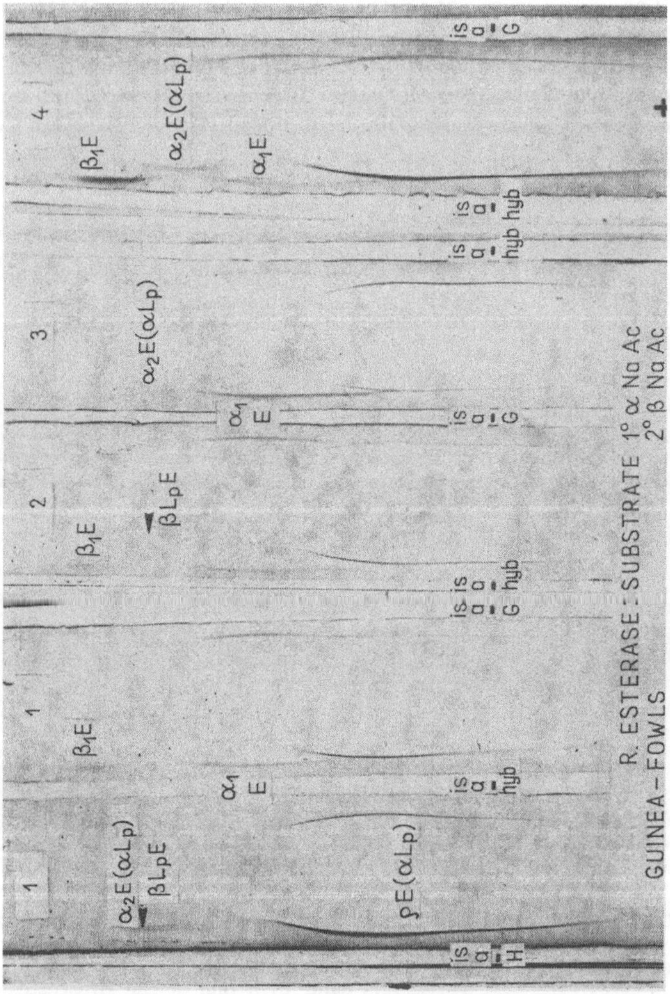
GUINEA - FOWLS
R. ESTERASE : SUBSTRATE 1° α Na Ac
2° β Na Ac
1
2
3
4
β1E
βLpE
α2E(αLp)
ρE(αLp)
α1 E
is a G
is a H
is a hyb
hyb

FIG. 5. Immunoelectrophoretic zymograms of individual GF; cross testing with different antisera, showing presence or absence of esterases and their independence

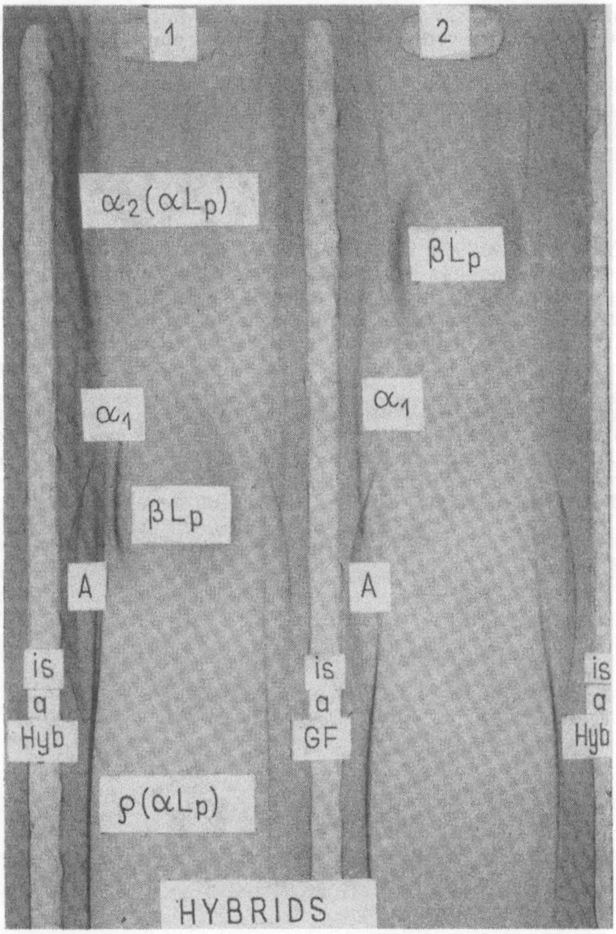
1
2
α2(αLp)
βLp
α1
α1
βLp
A
A
is
a
Hyb
is
a
GF
is
a
Hyb
ρ(αLp)
HYBRIDS

FIG. 6. Immunoelectrophoretic identification of components of sera of hybrids, revealed with a single substrate (β-NaAc); the variable location of the β-lipoprotein-esterase does not interfere with other lines

## REFERENCES

LEROY, P., BARBIER, Y. and AUZANNEAU, C. (1968), L'Hybride de Gallus domesticus and Numida meleagris. Histologie du testicule — comparaison avec le coq et le mâle Pintade. *Arch. Anat. Hist. Embr.*, **51.** 401–406.

KAMINSKI, MARIE (1966), Immunoelectrophoretic examination of avian serum esterases *Nature*, **209.** 723–725.

KAMINSKI, MARIE (1969), Common and species-specific serum esterases of Equidae. Horse and Donkey. *Biochim. Biophys. Acta*, **191.** 611–620.

CSUKA, J. and PETROVSKÝ, E. (1968), Study of polymorphism of esterase of chicken egg-white and blood serum. *Folia Biol.* **14.** 165–168.

KAMINSKI, MARIE and GUÉRIN, MARIE (1970), Lipid and esterase reactions of the duck serum HDL lipoproteins, the apoprotein and their immunoprecipitates. *Prot. Biol. Fluids* (in press).

KAMINSKI, MARIE and PODLIACHOUK, LUBA (1970), Serum esterases of Equidae. Truly or apparently negative phenotypes. *Comp. Biochem. Physiol.* (in press)

## DISCUSSION

J. DOSTÁL: You have described changes in electrophoretic mobility of esterase aftes storage in freezer using agar-gel electrophoresis. Did you observe also some changer of esterase fractions using starch-gel electrophoresis? Does the antigenicity of the esterase change also after storage in freezer?

M. KAMINSKI: As to the changes of mobility related to storage; there was no change in starch-gel but there was change in agar gel. As to the antigenicity; there were no changes.

A. KÚBEK: Do you know, Dr. Kaminski, the kind of esterase which is polymorphic in chicken serum? (I mean in which group of esterases can it be included?)

M. KAMINSKI: As to the nomenclature of esterases: Personally I do not classify the esterases into ali-or arylesterases; until a general nomenclature is established I simply indicate the substrate used and the inhibitors acting on the activity.

*XIIth Europ. Conf. Anim. Blood Groups Biochem. Polymorph., Bp., 1972 (pp. 481–483)*

# GENETIC POLYMORPHISM OF BLOOD CATALASE IN FOWLS

ALBINA. T. SHABALINA

**Institute of Animal Breeding, Kostinbrod, Sofia, Bulgaria**

The character of inheritance of blood catalase was studied in man (TAKAHARA, 1952, 1960; NISHIMURA et al., 1959; AEBI et al., 1961); mice (FEINSTEIN R. et al., 1964); dogs (ALLISON A. et al., 1957); guinea pigs (KOLZOV, 1927; RADEV, 1960) etc. One object of our studies was to establish the character of catalase inheritance in fowls (SHABALINA and RUSEV, 1968; SHABALINA, 1970). Analyses of hybrids were made by crossing $F_1 \times F_1$ and an analysing crossing of 643 fowls. It was found that hypocatalasemia in fowls has a recessive autosome character of inheritance.

It was of interest to verify whether there exists a dependence between the activity of various catalase forms and their electrophoretic mobility. To this end in our joint work with SEROV (1970) we made use of vertical electrophoresis on starch gel after the method of THORUP et al. (1961) with slight modifications, described earlier, (SHABALINA and SEROV, 1970). Electrophoresis showed (Fig. 1) blood catalase in fowls to have two fractions differing in respect of mobility. Fowls with low catalase activity have a slightly mobile fraction, B (1,3,5 — Fig. 1), and those with high catalase activity the fraction A (6). The simultaneous presence of two fractions was observed (2,4 — Fig. 1) in the electrophoretograms of heterozygotes.

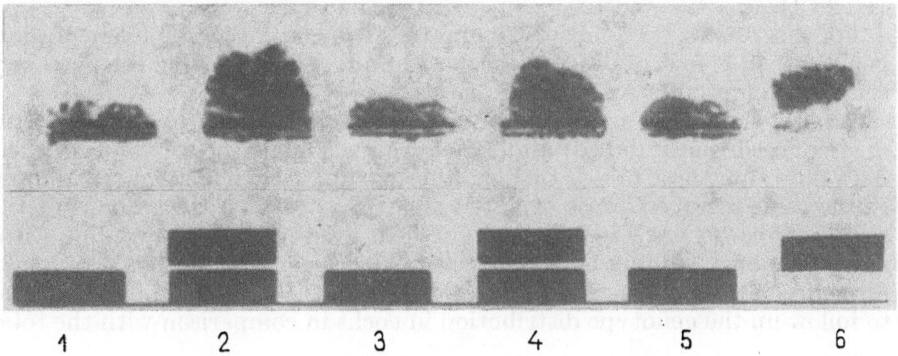
1
2
3
4
5
6

FIG. 1. A photograph of starch gel electrophoretogram and diagram with different types of blood catalase. Type A-6, B-1, 2, 5 and AB-2,4

TABLE 1

*Gene frequencies of the alleles $Ct^A$ and $Ct^B$ controlling blood catalase of the fowls*

| Breed | | Gene frequency | | Frequency of genotype | | |
|---|---|---|---|---|---|---|
| | | CtA | CtB | AA | AB | BB |
| Leghorn | 339 | 0.7396 | 0.2604 | 0.55 | 0.38 | 0.07 |
| New Hampshire | 268 | 0.5595 | 0.4405 | 0.31 | 0.50 | 0.19 |
| Rhode Island | 145 | 0.4615 | 0.5385 | 0.21 | 0.50 | 0.29 |
| Plymouth Rock | 629 | 0.3849 | 0.6151 | 0.15 | 0.47 | 0.38 |
| Faverolle | 94 | 0.3326 | 0.6684 | 0.11 | 0.44 | 0.45 |

Extensive biochemical investigations were undertaken with the aim to demonstrate the isoenzyme nature of these fractions. It was checked how the activity of these fractions was modified by changes in temperature, pH of the medium and time. Fraction B was found less thermostable than fractions A and AB. The activity of this fraction dropped to 29.1 per cent at 50°C as compared its initial activity; at the same time the activities of A and AB remained at the level of 41.7 per cent ($p < 0.001$). Furthermore, in our previous experiments (SHABALINA and RUSEV, 1968) with isotopes $Fe^{59}$ and $^{14}C$-valine this fraction was found to differ in respect of iron content in the active center and inducibility. All this gives us justification to consider the fractions obtained as catalase isoenzymes. The gene for hypocatalasemia was designated by us as $Ct^B$ and its allele, controlling the normal blood catalase activity in fowls, as $Ct^A$ (SHABALINA, 1970). The distribution of the genotypes Ct AA, Ct AB, Ct BB in various fowl populations was studied in the same work.

Table 1 shows that the frequency of these genotypes differs substantially in various breeds. Genotype Ct BB is more often encountered in broiler breeds — Faverolle (0.45) and Plymouth Rock (0.38), while almost absent in the Leghorn breed (0.07). But we also noted that it is associated not so much with the breed as with the level of breeding work in this breed. In the absence of thorough breeding work we observed a substantial change in the ratio of the genotypes Ct AA, Ct AB, Ct BB in the Plymouth Rock and Leghorn breeds with a tendency toward recovery of the gene balance (1 : 2 : 1).

For this reason, it seemed to be of interest to check the frequency of these genotypes in breeds of different purpose and in a flock lacking the necessary selection for productive characteristics.

Table 2 shows the results of these investigations. Two breeds were studied: Cornish (meat breed) and Black Shumen breed of a light type. The data show that these breeds though differing greatly in live weight, had nearly the same genotype ratio AA, AB and BB (1 : 2 : 1). On account of the small number of fowls in this flock, all hens are retained for breeding without preliminary culling. Cocks are selected according to the egg laying performance of their mothers and sisters. For that reason it would be of interest to follow up the genotype distribution in cocks in comparison with the total

TABLE 2

*Frequency changes of $Ct^A$ and $Ct^B$ in a flock of non-selected chickens*

| Breed | | Gene frequency | | Frequency of genotype | | |
|---|---|---|---|---|---|---|
| | | CtA | CtB | AA | AB | BB |
| Black Shumenska | 91 | 0.4868 | 0.5132 | 0.24 | 0.50 | 0.26 |
| Cornish | 107 | 0.4666 | 0.5334 | 0.21 | 0.51 | 0.28 |
| Black Shumenska (cocks) | 43 | 0.5965 | 0.4035 | 0.36 | 0.48 | 0.16 |

flock of the Black Shumen breed (Table 2). It was found that genotype Ct AA increased and genotype Ct BB diminished in the group of cocks under the influence of breeding for egg laying performance (of mothers and sisters). In all probability there exists a connection between $Ct^A$ and some characters controlling the egg production of breeds.

## REFERENCES


AEBI, H., HEIHIGER, J., BÜTLER, R. (1961), Two cases of Acatalasia in Switzerland., *Experientia*, **17**, 466.

ALLISON, A. and REES, G. (1957), Genetically controlled differences in catalase activity of dog erythrocytes, *Nature*, **180**, 649.

FEINSTEIN, R., SEAHOEM, J. and NOWARD, J. (1964), Acatalasemic mice, *Proc. Natl. Acad. Sci.*, **52**, 661.

KOLZOV, N. (1927), Catalase content in blood of vertebrates as a hereditary characteristic) Soderjanie katalasi v krovi pozvonochnih kak nasledstvenii priznak, *Jurnal Experimentalnoi biologii i medizini*, T. V., **15**, 303.

NISHIMURA, E., HAMILTON, T. and KOBARA, T. (1959), Carrier state in human acatalasemia, *Science*, **130**, 339.

RADEV, T. (1960), Inheritance of hypocatalasemia in guinea-pigs, *J. Genet.*, **57.** 169.

SHABALINA, ALBINA T. and RUSEV, G. K. (1968), Nasledovanie katalasi krovi u Gallus domesticus (Heredity of blood catalase in Gallus domesticus), *Genetica* (USSR), **4**, 81.

SHABALINA, ALBINA T. (1968), Varhu charaktera na nasledjavaneto na katalasata pri kokoshkite (On the characterof inherita nce of the catalase in fowls), *Genetika i selekzia* (Sofia), **6**, 449.

SHABALINA, ALBINA T. (1970), Opredelenie genotipa razlichnih populjazii Gallus domesticus po priznaku katalazi krovi (Determination of the genotype in different populations of Gallus domesticus according to the level of catalase in the blood), *Genetika* (USSR), **6**, 48.

SHABALINA, ALBINA T. and SEROV, O. (In press) Genetics of blood catalase isozymes in Gallus domesticus.

TAKAHARA, S. and HAMILTON, H. (1960), Hypocatalasemia. A new genetic carrier state, *J. Clin. Invest.*, **39**, 610.

THORUP, O. A., STROLL, W. and LEAVELL, B. (1961), *J. Labor. Clin. Med.*, **58**, 122.

*XIIth Europ. Conf. Anim. Blood Groups Biochem. Polymorph., Bp., 1972 (pp. 485–489)*

# BLOOD GROUP COMPOSITION IN FOWLS OF DIFFERENT POULTRY BREEDS AT DIFFERENT LEVELS OF CATALASE ACTIVITY

LIDIA ERMENCOVA and ALBINA T. SHABALINA

Institute of Animal Breeding, Kostinbrod, Academy of Agricultural Sciences, Bulgaria

## INTRODUCTION

The studies of BRILES (1957) and GILMOUR (1959) suggested B-blood-group system to be related to hatchability, viability, survival and egg-production of birds. Similar relationships regarding fertility and hatchability were established by BRILES and ALLEN (1961, 1962) in various lines during different age periods. MCDERMID (1964) has made it clear that the B-system is in some way connected with the mortality in crossbreds and their egg-production.

The related studies on the blood group system are still insufficient. Using normal bovine serum, OLSEN (1943) and BRILES (1952) succeeded in identifying the $A_6$ blood group antigen in hens. Antibodies to this antigen produce a very pronounced reaction with the homologous antigen of Rhode Island and Sussex broilers.

HILFIKER–HANGARTNER (1967) found a reliable difference in the egg weight related to the gene frequency in $A_1$ and $A_2$ blood group antigens.

Breed-dependent gene frequencies of the hypocatalase activity of blood (CtBB) in hens were described earlier (SHABALINA, 1970). From this viewpoint it is interesting to find a possible relationship between the $Ct^B$ antigen and some antigens of different blood group systems.

The present study was aimed at establishing the relationship between blood types and different blood catalase activities of birds.

## MATERIAL AND METHODS

A total of 569 cocks and hens of the Leghorn, Rhode Island, New Hampshire, Plymouth Rock and Cornish breeds were used to study the dependence of blood type composition on the breed and catalase activity of blood.

The determination of the blood groups was performed by the classical agglutination method, using 20 monospecific sera obtained at the laboratories taking part in an international comparison test.

The catalase activity was determined by the method of BACH and ZOUBAKOVA. The Rhode Island and New Hampshire hens used by us were the result of long-term selection and inbreeding work with special regard to the catalase activity of blood (SHABALINA et al., 1968; SHABALINA, 1970). The

TABLE 1

*Distribution of hens tested according to catalase activity of blood (catalase number)*

| Breed | Number of birds tested | Catalase activity m± | | |
|---|---|---|---|---|
| | | low | high | intermediate |
| Leghorn | 175 | 0.24±0.017 | 0.92±0.032 | 0.56±0.009 |
| Rhode Island | 85 | 0.18±0.012 | 1.03±0.130 | 0.53±0.022 |
| New Hampshire | 159 | 0.19±0.007 | 1.00±0.016 | 0.59±0.014 |
| Plymouth Rock | 106 | 0.17±0.007 | 0.97±0.025 | 0.60±0.021 |
| Cornish | 44 | 0.26±0.015 | 1.04±0.081 | 0.60±0.026 |

birds were divided into three main groups of low, high and medium catalase activity respectively (Table 1).

## RESULTS AND DISCUSSION

Table 2 shows the breed differences in blood type.

Certain differences were established by analyzing the frequencies of blood group antigens. The $A_1$ antigen proved to be of particular interest. Its gene frequency (.009) was considerably lower in the Leghorn breed as compared to other breeds, such as Cornish (.248) and Plymouth Rock (.210). Breed differences of similar type have been reported also by other authors. For instance, GASPARSKA (1963) did not find this antigen in Leghorn and Plymouth Rock, while it appeared, though rarely, in Rhode Island. The absence of this antigen in some of the breeds studied by GASPARSKA was probably due to the relatively small number of the birds tested. OLSEN (1943) and BRILES et al. (1952) established certain relationship between the $A_6$ antigen and meat performance in Rhode Island and Sussex. This antigen was detected in the same way as our $A_1$ antigen i.e. using normal bovine serum. Weassume that $A_1$ in our case is also related to the meat productive characteristics of the birds, as the highest antigen frequency was observed primarily in broiler breeds, such as Cornish and Plymouth Rock. Gene frequency was lowest in the Leghorn breed. New Hampshire and Rhode Island were intermediate between the two.

Our findings correlate with those of GASPARSKA (1963) as far as the absence of $A_3$ in Leghorn and Plymouth Rock is concerned.

The $E_4$ antigen frequency was high in the Leghorn (.228), Rhode Island (.321), and New Hampshire (.267) breeds, while it was low in the Cornish breed (.056).

It should be noted that the $B_1$ antigen showed highest frequency in Cornish (.219) and Plymouth Rock (.292) followed in succession by New Hampshire (.163) and Rhode Island (.157), while it was 8 times lower in Leghorn (.037).

TABLE 2

*Gene frequencies in hens of different breeds*

| | Leghorn | Rhode Island | New Hampshire | Plymouth Rock | Cornish |
|---|---|---|---|---|---|
| | | A locus | | | |
| $A^1$ | .009 | .073 | .061 | .210 | .248 |
| $A^2$ | .273 | .374 | .350 | .294 | .247 |
| $A^3$ | — | .020 | .006 | — | .061 |
| $A^4$ | .347 | .344 | .381 | .328 | .290 |
| $A^x$ | .371 | .189 | .202 | .168 | .154 |
| | | E locus | | | |
| $E^1$ | .273 | .315 | .374 | .170 | .264 |
| $E^2$ | .254 | .283 | .356 | .471 | .566 |
| $E^3$ | .179 | .081 | — | .235 | — |
| $E^4$ | .228 | .321 | .267 | .110 | .056 |
| $E^5$ | .045 | — | .003 | .014 | .013 |
| $E^6$ | .021 | — | — | — | — |
| | | B locus | | | |
| $B^1$ | .037 | .157 | .163 | .292 | .219 |
| $B^2$ | .071 | .141 | .063 | .132 | .069 |
| $B^4$ | .466 | .285 | .351 | .092 | .267 |
| $B^8$ | .008 | .020 | .080 | .014 | — |
| $B^9$ | .003 | .057 | .006 | .024 | .056 |
| $M^3$ | .008 | .040 | .031 | .071 | .072 |
| $M^0$ | .219 | .237 | .207 | .241 | .187 |
| $B^{21}$ | .188 | .063 | .099 | .134 | .130 |

Analysis of data on gene frequencies and blood catalase activities in the four breeds also revealed many dissimilarities (Table 3). There was no evidence for the presence of $A_1$ antigen in Leghorn birds with high catalase activity. At lower catalase activity, however, the antigen frequency of $A_1$ proved to be higher (.066). The same was observed also in New Hampshire hens. In Rhode Island, $A_1$ was encountered in birds with high catalase activity in a frequency of .029, being almost three times less as compared to that of the birds with low catalase activity (.071). The higher antigen frequency of $A_1$ in hens with low catalase activity points to the relation of this antigen with the meat productive characteristics of the fowls. This confirmed our assumption (1970) of the improved meat performance of hens showing low catalase activity.

Plymouth Rock birds are distinguished by the higher frequency of $A_1$, regardless of blood catalase activity.

There is evidence of the higher antigen frequency of $B_1$ in groups of birds with low catalase activity of the New Hampshire breed.

High $B_2$ antigen frequency in Leghorn and Rhode Island breeds was found in birds with high catalase activity. In Plymouth Rock it was higher in hens showing low catalase activity.

TABLE 3

*Gene frequencies and catalase activities of blood in hens*

| | Breeds | | | | | | | | | | | |
|---|---|---|---|---|---|---|---|---|---|---|---|---|
| | Leghorn (catalase) | | | Rhode Island (catalase) | | | New Hampshire (catalase) | | | Plymouth Rock (catalase) | | |
| | high | low | interm. | high | low | interm. | high | low | interm. | high | low | interm. |
| $A^1$ | — | .066 | .004 | .029 | .071 | .089 | — | .136 | ,049 | .263 | .247 | .238 |
| $A^2$ | .260 | .267 | .251 | .371 | .357 | .354 | .339 | .296 | .378 | .276 | .297 | .309 |
| $A^3$ | — | — | — | — | — | .025 | — | .012 | .005 | — | — | — |
| $A^4$ | .391 | .267 | .284 | .286 | .309 | .278 | .415 | .370 | .373 | .303 | .317 | .321 |
| $A^x$ | .349 | .400 | .461 | .314 | .263 | .254 | .246 | .185 | .195 | .158 | .139 | .132 |
| $E^1$ | .216 | .208 | .256 | .309 | .297 | .313 | .400 | .397 | .356 | .111 | .211 | .125 |
| $E^2$ | .270 | .250 | .273 | .236 | .281 | .242 | .400 | .383 | .330 | .528 | .465 | .464 |
| $E^3$ | .189 | .292 | .177 | .164 | .125 | .132 | — | — | — | .194 | .127 | .143 |
| $E^4$ | .243 | .250 | .263 | .291 | .297 | .313 | .186 | .219 | .314 | .167 | .155 | .232 |
| $E^5$ | .054 | — | .015 | — | — | — | .014 | — | — | — | .042 | .036 |
| $E^6$ | .027 | — | .016 | — | — | — | — | — | — | — | — | — |
| $B^1$ | .080 | .059 | .021 | .104 | .113 | .136 | .171 | .257 | .129 | .208 | .239 | .204 |
| $B^2$ | .120 | .059 | .042 | .250 | .169 | .212 | .061 | .043 | .071 | .078 | .130 | .102 |
| $B^4$ | .320 | .294 | .481 | .333 | .239 | .271 | .341 | .430 | .329 | .208 | .261 | .284 |
| $B^8$ | — | — | — | .021 | .028 | .017 | .122 | .056 | .071 | — | — | — |
| $B^9$ | .040 | — | .008 | — | .056 | .017 | — | — | .009 | .026 | — | .023 |
| $M^3$ | — | — | — | — | .099 | .042 | — | — | .053 | .078 | .033 | .045 |
| $M^6$ | .240 | .294 | .299 | .292 | .211 | .237 | .195 | .186 | .219 | .247 | .228 | .227 |
| $B^{21}$ | .200 | .294 | .148 | — | .085 | .068 | .110 | .028 | .119 | .157 | .109 | .115 |

## CONCLUSION

Certain breed differences were found concerning the antigen frequency of $A_1$, $A_3$, $E_4$ and $B_1$. It is interesting to note that the highest antigen frequency of $A_1$ was observed in meat type breeds (Cornish — .248), whereas the lowest frequency in breeds of higher egg productivity (Leghorn — .009). In Leghorn, Rhode Island and New Hampshire the $A_1$ antigen frequency was higher in groups of hens showing low catalase activity of the blood. In Leghorn and Rhode Island, $B_2$ antigen frequency was higher in bird groups with high catalase activity, while in Plymouth Rock it occurred in groups of birds with lower catalase activity.

## REFERENCES

ALLEN, C. P. (1962), The effect of parental B locus genotypes on multiple cross performance in chickens. *Ann. N. J. Acad. Sci.*, **97.** 184–193.

BACH and ZUBKOVÁ (po Predtechenski) (1960), Rukovodstvo po klinicheskim i laboratornim issledovaniam (Methods in clinical anal. laboratory research), Moscow, 260.

Briles, W. E., Briles, R. W. and Irwin, M. R. (1952), Differences in specificity of the antigenic products of a series of alleles in the chickens, *Genetics*, **37.** 359–368.

Briles, W. E., Allen, C. P. and Millen, T. W. (1957), The B blood group system of chickens. I. Heterozygosity in closed populations, *Genetics*, **42.** 631–648.

Briles, W. E. (1960), Blood groups in chickens, their nature and utilisation, *Wld. Poult. Sci.*, **16.** 223–242.

Briles, W. E. and Allen, C. P. (1961), The blood group system of chickens II, The effects of genotype on livability and egg production in seven commercial inbred lines. *Genetics*, **46.** 1273–1293.

Gasparska, Jolanta and Rybiska, Ursula (1962), Wykrywanie antygenow krwinnowych ukladu A u kur za pmocaprzeciwcial normalnych surowicy bydlecej. *Roczn. Nauk, Roln.*, **80**–B–2, 211–215.

Gasparska, Jolanta (1963), The blood groups in red cells of Rhode Island red chickens. *Roczn. Nauk, Roln.*, **83**–B–2, 343–347.

Gasparska, Jolanta and Gasparski, J. (1966), Comparison of red cell antigenic factor frequencies in four chicken breeds reared in Poland. *Roczn. Nauk Roln.*, **88**–B–1, 11–17.

Gilmour, D. G. (1959), Segregation of genes determining red cell antigens at high levels of inbreeding in chickens, *Genetics*, **44.** 14–33.

Hilfiker-Hangartner, H. (1967), Die Veränderung der Blutgruppenfrequenzen im Laufe eines Selektionsexperimentes nach Anfangseigewicht, *Z. Tierzücht Zücht.* **84.** 73–79.

McDermid, E. M. (1965), The effect of blood group genotypes of the B system on the performance of hybrid chickens, Proc. 9th Europ. Anim. Blood Group Conf. Prague 1964, 173–179.

Olsen, C. (1943), The inheritance of an agglutinogen of the chicken erythrocyte, *J. Immun.*, **47.** 149–154.

Oosterlee, C. C. (1965), Some aspects of studies on the relationship between blood groups and economic characters in farm animals, *World Rev. Anim. Prod.*, **2.** 21–26.

Schmid, D. O., Thein, P., Huber, E. (1966), Erfahrungen bei der Gewinnung von Iso-Immunseren zur Feststellung von Blutgruppenmerkmalen beim Huhn, *Arch. Geflügel*, **30.** 392–400.

Shabalina, Albina T. and Rusev, G. K. (1968), Nasledovanie katalasi krovi u Gallus domestica, (Heredity of blood catalase in Gallus domestica), *Genetika* (USSR), **1.** 81–89.

Shabalina, Albina T. (1970), Opredelenie genotipa raslichnikh populiatsei Gallus domesticus (Determination of the genotype in different populations of Gallus domesticus according to the level of catalase in the blood), *Genetika* (USSR) **6.** 48–52.

*XIIth Europ. Conf. Anim. Blood Groups Biochem. Polymorph., Bp., 1972 (pp. 491 - 498)*

# THE EFFECT OF BLOOD GROUP ALLELES ON EGG PRODUCTION, EGG WEIGHT AND BODY WEIGHT IN A CLOSED YELLOW HUNGARIAN BREED

M. PAPP, L. SZAJKÓ* and J. SCHMIDT*

Bloodgrouping Laboratory, University of Veterinary Science, Budape

* College of Agricultural Sciences, Mosonmagyaróvár, Hungary

## SUMMARY

A total of 1284 Yellow Hungarian birds were blood typed and examined by variance analysis for three economical traits — egg production, egg weight and body weight. The effect of 4 alleles in A system ($A^1$, $A^2$, $A^3$ and $A^6$) and that of 8 alleles in B system ($B^1$, $B^2$, $B^3$, $B^4$, $B^5$, $B^7$, $B^{10}$, and $B^{13}$) were analyzed.

The interaction of the alleles and the three characteristics were studied by two methods. The effect of the presence or absence of the respective antigen on the productive traits was examined (a) and comparisons were made between the effects of each antigen belonging to the same system (b). The interaction of the egg production and tho different alleles was examined on the basis of the performance for 3 1/2, 6, and 11 months.

The presence of $A_2$ and $A_6$ and the absence of $A_1$ associated with higher egg production over all the three periods. (All interactions were significant except the effect of $A_1$ in the 11 month period.) In the B system, $B_5$ correlated positively with the egg production over all the three periods ($P < 0.01$ and $P < 0.05$ respectively), while $B_1$ and $B_2$ correlated negatively with the egg production, though the interactions were significant in the first two periods only. The egg weights of birds possessing $A_2$ and $B_4$ antigens were significantly higher than that of those not possessing these factors. Conversely, $A_3$ and $B_{10}$ had negative effect on the egg weight. — Considering the body weight, two positive interactions were found. Birds having $A_2$ and $A_6$ antigens were significantly heavier, than those not possessing these factors.

The comparisons made between each antigen within both systems corresponded to the above findings.

## INTRODUCTION

Ever since the first blood group systems were detected (BRILES et al., 1950, 1953) a great interest has been shown in the relationships of blood group alleles and productive characteristics. From the beginning of the early fifties, many examinations were carried out along this line, chiefly in pure lines and crossbred poultry stock. Concerning closed populations relatively few data of this kind are available.

Present study is dealing with the effect of some blood group alleles on the productive traits, examining the consequence of their presence or absence and their comparative values in this respect.

## MATERIAL AND METHODS

The experiment was carried out in a closed Yellow Hungarian population at Mosonmagyaróvár. This breed derived from the native Hungarian Barnyard Fowl which was improved by the Yellow Orpington, Langsan and Velsum before World War II and by New Hampshire and Rhode Island Red immediately after it. During the last 15–20 years it has been kept in closed breed. The population in Mosonmagyaróvár is now the only Yellow Hungarian stock in the country serving as reserve. Owing to the various kinds of ancestors, the gene-pool of this breed is also varied.

Blood samples were taken from 1284 hens 9 months old, immediately after housing. They were examined in agglutination test according to BRILES et al. (1960, 1961, 1962 and 1963) and FANGUY (1961), but using plexiplates and shaking machine.

First of all, reagents belonging to the A and B system were used, which are A1, A2, A3, A6 and B1, B2, B3, B4, B5, B7, B10, B13 respectively.

Egg production was calculated on the basis of 3 1/2, 6 and 11 monht performance of the birds. Egg weight used in the calculation was the average of the measurements made during one month (April). Body weighing was performed at the age of 12 months.

The interaction of the alleles and the characteristics mentioned was studied by two methods, with regard to the effect of the presence or absence of the respective antigen on the productive traits (a) and by making a comparison between the effects of each antigen belonging to the same system (b).

## RESULTS

### EGG PRODUCTION

Positive and negative significant interactions were observed between the egg production and some alleles of both A and B system over all the three periods (Table 1).

(a) In the A system the presence of $A^1$, had an adverse effect on egg production, which was significant during 3 1/2 and 6 months ($P < 0.05$) and non-significant in 11 months. $A^6$ allele was significantly favourable for the egg production all over the three periods ($P < 0.05$). The non-significant favourable effect of the $A^2$ allele in the period of 3 1/2 months became significant ($P < 0.05$) in the next two periods.

In the B system, $B^1$ was unfavourable throughout, in the first two periods significantly ($P < 0.01$ and $P < 0.05$) and in the last period non-significantly. $B^2$ and $B^{10}$ alleles were also unfavourable considering the whole period, although this effect was significant in one period only. On the other hand $B^5$ had a remarkably advantageous effect on the egg production in all the three periods of the year. ($P < 0.01$ in the first two periods and $P < 0.05$ in last one.)

(b) Comparisons made between each of the antigens within a system confirmed the above findings (Table 2). $A^1$ compared with $A^2$, $A^3$ and $A^6$, re-

TABLE 1

*Relationship of blood group antigens and average egg production for 3 1/2, 6 and 11 months in the closed Yellow Hungarian breed in Mosonmagyaróvár*

| Antigens | 3 1/2 months (1279) pullets | | | | | 6 months (1284 pullets) | | | | | 11 months (754 pullets) | | | | |
|---|---|---|---|---|---|---|---|---|---|---|---|---|---|---|---|
| | No. of pullets | | Egg production of pullets | | $t$ value | No. of pullets | | Egg production of pullets | | $t$ value | No. of pullets | | Egg production of pullets | | $t$ value |
| | reacting (+) | non-reacting (—) | reacting (+) | non reacting (—) | | reacting (+) | non reacting (—) | reacting (+) | non reacting (—) | | reacting (+) | non reacting (—) | reacting (+) | non reacting (—) | |
| $A_1$ | 240 | 1039 | 26.71 | 29.66 | —2.28* | 240 | 1044 | 72.57 | 76.21 | —2.19* | 134 | 620 | 144.50 | 149.41 | —1.82 |
| $A_2$ | 546 | 733 | 29.76 | 28.62 | 1.15 | 542 | 742 | 77.35 | 74.20 | 2.37* | 310 | 444 | 151.74 | 146.32 | 2.51* |
| $A_3$ | 490 | 789 | 29.33 | 28.96 | 0.36 | 494 | 790 | 77.22 | 74.47 | 2.02* | 296 | 458 | 148.76 | 148.40 | 0.17 |
| $A_6$ | 223 | 1056 | 31.55 | 28.59 | 2.24* | 224 | 1060 | 79.65 | 74.66 | 2.76** | 131 | 623 | 154.21 | 147.36 | 2.38* |
| $B_1$ | 562 | 717 | 27.20 | 30.59 | —3.45*** | 563 | 721 | 73.63 | 77.01 | —2.57* | 338 | 416 | 147.33 | 149.53 | —1.04 |
| $B_2$ | 494 | 785 | 27.93 | 29.84 | —1.94 | 499 | 785 | 73.31 | 76.94 | —2.66** | 289 | 465 | 146.36 | 149.90 | —1.63 |
| $B_3$ | 265 | 1014 | 28.05 | 29.38 | —1.06 | 271 | 1013 | 74.49 | 75.81 | —0.81 | 169 | 585 | 146.69 | 149.08 | —0.97 |
| $B_4$ | 297 | 982 | 27.36 | 29.63 | —2.07* | 298 | 986 | 73.88 | 76.03 | —1.43 | 183 | 571 | 149.40 | 148.27 | 0.46 |
| $B_5$ | 281 | 998 | 32.23 | 28.23 | 3.24** | 281 | 1003 | 79.56 | 74.40 | 3.11** | 162 | 592 | 153.07 | 147.31 | 2.16* |
| $B_7$ | 684 | 595 | 29.06 | 29.16 | —0.10 | 689 | 595 | 75.72 | 75.31 | 0.31 | 417 | 337 | 147.83 | 149.42 | —0.76 |
| $B_{10}$ | 123 | 1156 | 25.79 | 29.46 | —2.33* | 124 | 1160 | 72.95 | 75.81 | —1.30 | 67 | 687 | 143.76 | 149.01 | —1.51 |
| $B_{13}$ | 524 | 755 | 29.30 | 28.97 | 0.33 | 529 | 755 | 75.52 | 75.54 | —0.02 | 319 | 435 | 148.15 | 148.83 | —0.32 |

* Significant at 0.05 level of probability
** Significant at 0.01 level of probability
*** Significant at 0.001 level of probability

TABLE 2

*Significant relationship between some blood group antigens belonging to the same system concerning the average egg production for 3 1/2, 6 and 11 months*

| Antigens (1) | Antigens (2) | 3 1/2 months d.f. | 3 1/2 months No. of antigen (1) | 3 1/2 months Egg production of pullets having antigen (1) | 3 1/2 months Egg production of pullets having antigen (2) | 3 1/2 months t value | 6 months d.f. | 6 months No. of antigen (1) | 6 months Egg production of pullets having antigen (1) | 6 months Egg production of pullets having antigen (2) | 6 months t value | 11 months d.f. | 11 months No. of antigen (1) | 11 months Egg production of pullets having antigen (1) | 11 months Egg production of pullets having antigen (2) | 11 months |
|---|---|---|---|---|---|---|---|---|---|---|---|---|---|---|---|---|
| $A_1$ | $A_2$ | 784 | 240 | 26.71 | 29.76 | —2.17* | 780 | 240 | 72.57 | 77.35 | —2.66** | 442 | 134 | 144.50 | 151.74 | —2.43* |
| $A_1$ | $A_3$ | 728 | 240 | 26.71 | 29.33 | —1.83 | 732 | 240 | 72.57 | 77.22 | —2.52* | 428 | 134 | 144.50 | 148.76 | —1.48 |
| $A_1$ | $A_6$ | 461 | 240 | 26.71 | 31.55 | —2.87** | 462 | 240 | 72.57 | 79.65 | —3.17** | 263 | 134 | 144.50 | 154.21 | —2.70** |
| $B_1$ | $B_5$ | 841 | 562 | 27.20 | 32.23 | —3.78*** | 842 | 563 | 73.63 | 79.56 | —3.35*** | 498 | 338 | 147.33 | 153.07 | —2.02* |
| $B_2$ | $B_5$ | 773 | 494 | 27.93 | 32.23 | —3.21** | 778 | 499 | 73.31 | 79.56 | —3.40*** | 449 | 289 | 146.36 | 153.07 | —2.28* |
| $B_3$ | $B_5$ | 544 | 265 | 28.05 | 32.23 | —2.64** | 550 | 271 | 74.49 | 79.56 | —2.44** | 329 | 169 | 146.69 | 153.07 | —1.98* |
| $B_4$ | $B_5$ | 576 | 297 | 27.36 | 32.23 | —3.36*** | 577 | 298 | 73.88 | 79.56 | —2.88** | 343 | 183 | 149.40 | 153.07 | —1.14 |
| $B_7$ | $B_5$ | 963 | 682 | 29.06 | 32.23 | —2.44* | 968 | 687 | 75.72 | 79.56 | —2.21* | 577 | 415 | 147.83 | 153.07 | —1.85 |
| $B_{10}$ | $B_5$ | 402 | 121 | 25.79 | 32.23 | —3.47*** | 403 | 122 | 72.95 | 79.56 | —2.57* | 227 | 65 | 143.76 | 153.07 | —2.28* |
| $B_{13}$ | $B_5$ | 803 | 522 | 29.30 | 32.23 | —2.14* | 808 | 527 | 75.52 | 79.56 | —2.23* | 479 | 317 | 148.15 | 153.07 | —1.68 |

* Significant at 0.05 level of probability
** Significant at 0.01 level of probability
*** Significant at 0.001 level of probability

spectively proved to be consistently unfavourable for egg production in all the three periods. (Among 9 comparisons only 2 were non-significant). On the contrary, $B^5$ was significantly superior to the other 7 B alleles, with the exception of three cases, from the 21 comparisons.

## EGG WEIGHT

(a) Considerable relationships showed in the egg weight, too (Table 3). The unfavourable effect of $A^3$ proved to be highly significant ($P < 0.001$). At the same time $A^2$ had a significantly advantageous effect on the egg weight ($P < 0.01$). In the B system, $B^{10}$ effected adversely, while $B^4$ favourably, these characteristics (in both cases $P < 0.05$).

(b) Comparing each of the alleles in respect of egg weight (Table 4), $A^3$ was significantly ($P < 0.001$, $P < 0.01$, and $P < 0.05$ respectively) inferior to the other three A alleles. In the B system birds having $B^{10}$ allele had consistently lower egg weight than those having other B alleles examined. In the comparison made with $B^1$, $B^2$, $B^4$ and $B^{13}$ alleles these differences were significant.

## BODY WEIGHT

Concerning the body weight, in the A system two significant interactions were found (Table 5). Both were positive, i.e. birds having $A^6$ and $A^2$ alleles were significantly heavier, than those not possessing these factors. In the B system no significant connection was found with body weight. Comparing each of the A alleles (Table 6), $A^6$ was significantly superior to $A^1$ and $A^3$ and non-significantly to $A^2$.

TABLE 3

*Relationship of blood group antigens and average egg weight in the closed Yellow Hungarian breed at Mosonmagyaróvár (1281 pullets)*

| Antigens | No. of pullets | | Egg wt. (g) of pullets | | $t$ value |
|---|---|---|---|---|---|
| | reacting (+) | non reacting (−) | reacting (+) | non reacting (−) | |
| $A_1$ | 240 | 1041 | 55.14 | 55.13 | 0.03 |
| $A_1$ | 544 | 737 | 55.71 | 54.70 | 4.57*** |
| $A_3$ | 491 | 790 | 53.97 | 55.85 | −8.58*** |
| $A_6$ | 223 | 1058 | 54.78 | 55.20 | −1.37 |
| $B_1$ | 561 | 720 | 55.30 | 55.00 | 1.33 |
| $B_2$ | 495 | 786 | 55.38 | 54.97 | 1.79 |
| $B_3$ | 268 | 1013 | 55.05 | 55.15 | −0.35 |
| $B_4$ | 298 | 983 | 55.77 | 54.93 | 3.04** |
| $B_5$ | 278 | 1003 | 54.94 | 55.18 | −0.85 |
| $B_7$ | 686 | 595 | 55.25 | 54.99 | 1.18 |
| $B_{10}$ | 124 | 1157 | 54.27 | 55.22 | −2.31* |
| $B_{13}$ | 529 | 752 | 55.34 | 54.98 | 1.65 |

* Significant at 0.05 level of probability
** Significant at 0.01 level of probability
*** Significant at 0.001 level of probability

TABLE 4

*Comparisons between some blood group antigens belonging to the same system concerning the average egg weight in the closed Yellow Hungarian breed at Mosonmagyaróvár*

| Antigen (1) | Antigen (2) | d.f. | No. of antigen (1) | Egg wt. (g) of pullets having antigen (1) | Egg wt. (g) of pullets having antigen (2) | $t$ value |
|---|---|---|---|---|---|---|
| $A_1$ | $A_2$ | 782 | 240 | 55.14 | 55.71 | −1.92 |
| $A_1$ | $A_3$ | 730 | 240 | 55.14 | 53.98 | 3.87*** |
| $A_1$ | $A_6$ | 460 | 240 | 55.14 | 54.77 | 0.95 |
| $A_2$ | $A_3$ | 1034 | 544 | 55.71 | 53.98 | 7.19*** |
| $A_2$ | $A_6$ | 764 | 544 | 55.71 | 54.77 | 2.80** |
| $A_3$ | $A_6$ | 712 | 492 | 53.98 | 54.77 | −2.39* |
| $B_1$ | $B_{10}$ | 684 | 562 | 55.30 | 54.27 | 2.37* |
| $B_2$ | $B_{10}$ | 616 | 494 | 55.38 | 54.27 | 2.54* |
| $B_3$ | $B_{10}$ | 390 | 268 | 55.05 | 54.27 | 1.65 |
| $B_4$ | $B_{10}$ | 420 | 298 | 55.77 | 54.27 | 3.21** |
| $B_5$ | $B_{10}$ | 400 | 278 | 54.94 | 54.27 | 1.45 |
| $B_7$ | $B_{10}$ | 450 | 285 | 55.25 | 54.27 | 2.28* |
| $B_{13}$ | $B_{10}$ | 651 | 527 | 55.34 | 54.27 | 2.49 |

* Significant at 0.05 level of probability
** Significant at 0.01 level of probability
*** Significant at 0.001 level of probability

TABLE 5

*Relationship of blood group antigens and average body weight in the closed Yellow Hungarian breed at Mosonmagyaróvár (213 pullets)*

| Antigens | No. of pullets reacting (+) | No. of pullets non reacting (−) | Body wt. (kg) of pullets reacting (+) | Body wt. (kg) of pullets non reacting (−) | $t$ value |
|---|---|---|---|---|---|
| $A_1$ | 36 | 177 | 1.8833 | 1.9851 | −1.33 |
| $A_2$ | 89 | 124 | 2.0347 | 1.9210 | 2.28* |
| $A_3$ | 76 | 137 | 1.9768 | 1.9632 | 0.27 |
| $A_6$ | 37 | 176 | 2.1038 | 1.9398 | 2.85** |
| $B_1$ | 81 | 132 | 1.9679 | 1.9681 | 0.00 |
| $B_2$ | 82 | 131 | 1.9352 | 1.9883 | −1.00 |
| $B_3$ | 51 | 162 | 1.9024 | 1.9885 | −1.33 |
| $B_4$ | 51 | 162 | 2.0084 | 1.9555 | 0.97 |
| $B_5$ | 55 | 158 | 1.9915 | 1.9600 | 0.53 |
| $B_7$ | 110 | 103 | 1.9866 | 1.9486 | 0.75 |
| $B_{10}$ | 15 | 198 | 2.0367 | 1.9629 | 0.71 |
| $B_{13}$ | 92 | 121 | 1.9597 | 1.9743 | −0.28 |

* Significant at 0.05 level of probability
** Significant at 0.01 level of probability

TABLE 6

*Comparisons between each of the blood group antigens belonging to the A system concerning the average body weight in a closed Yellow Hungarian breed at Mosonmagyaróvár*

| Antigens (1) | Antigens (2) | d.f. | No. of antigens (1) | Body wt. (kg) of pullets having antigen (1) | Body wt. (kg) of pullets having antigen (2) | $t$ value |
|---|---|---|---|---|---|---|
| $A_1$ | $A_2$ | 119 | 36 | 1.8833 | 2.0175 | −1.64 |
| $A_1$ | $A_3$ | 110 | 36 | 1.8833 | 1.9768 | −1.15 |
| $A_1$ | $A_6$ | 71 | 36 | 1.8833 | 2.1038 | −2.52* |
| $A_2$ | $A_3$ | 159 | 85 | 2.0175 | 1.9768 | 0.74 |
| $A_2$ | $A_6$ | 120 | 85 | 2.015 | 2.1038 | −1.36 |
| $A_3$ | $A_6$ | 111 | 76 | 1.9768 | 2.1038 | −2.00* |

* Significant at 0.05 level of probability

## DISCUSSION

As it is well known, the genetical explanation of the relationships between blood group allele and productive characteristics can be linkage, pleiotropy and heterosis. The first two ways can be examined by a single allele effect, while the latter one on the basis of those connected with heterozygous combinations.

Although a number of genic and environmental effects can simultaneously be manifested in the organism during its lifetime, even several blood group alleles can exert favourable or unfavourable effect at the same time and they perhaps compensate each other, these kinds of investigations might, however, clear up those relationships which surpass the other ones.

According to the present findings and considering the previous one (PAPP, 1968) it appears that there can exist relationships of the same trend between blood group alleles and productive characteristics, in the subsequent seasons of a single generation rather than in the generations following upon each other. Obviously it is due to the year interaction.

Alleles of one system generally have a consistent rank of value concerning a certain productive trait, at least within one generation.

## REFERENCES


ALLEN, C. P. and GILMOUR, D. G. (1962), The B blood group system of chickens III. The effects of two heterozygous genotypes on the survival and egg production of multiple crosses. *Genetics*, **47.** 1711–1718.

BRILES, W. E. (1956), The relationship between B blood group genotypes and adult performance in two white Leghorn inbred lines. *Poult. Sci.*, **35.** 1134–1135.

BRILES, W. E. (1957), The effect of B blood group system on 10 week weights of chicks resulting from a cross between inbred lines. *Poult. Sci.*, **36.** 1106.

BRILES, W. E. and ALLEN, C. P. (1961), The B blood group system of chickens II. The effects of genotype on livability and egg production in seven commercial inbred lines. *Genetics*, **46.** 1273–1293.

BRILES, W. E. and KRUEGER, W. F. (1955), The effect of parental B blood group genotypes on hatchability and livability in Leghorn inbred lines. *Poult. Sci.*, **34** 1182.

GILMOUR, D. G. (1954), Selective advantage of heterozygosis for blood group genes among inbred chickens. *Heredity*, **8.** 291.

McDERMID, E. M. (1965), The effect of blood group genotypes of the B system on the performance of hybrid chickens, Proc. 9th Europ Anim. Blood Group Conf. Prague, 1964, 173–178.

McDERMID, E. M. (1966), Further experiments on the relationship of the B-system blood group genotype to production characters in the chicken. Proc. Xth Europ. Conf. Anim. Blood Groups Biochem. Polymorph. Paris, 1966, 223–230.

OKADA, I. and MATSUMOTO, K. (1962), Fitness of the genotypes at the B locus determining the blood, group of chickens. *Jap. J. Genet.*, **4.** 267–275.

OKADA, I., HASEGAWA, T., SEKIDERA, S., SHIMIZU, H. and HACHINOHE, Y. (1966), Association of the B blood group alleles with production characters in chickens. *Jap. J. Zootechn. Sci.*, **37.** 302–311.

PAPP, M. (1968), Relationships between blood group factors and economical traits in various breeds of chicken. *Magy. Áo. Lapja*, **23.** 580–583.

## DISCUSSION

C. C. OOSTERLEE: You have mentioned that you have $A^1$, $A^2$, $A^3$ and $A^0$ alleles, but in the Table 1, for instance, you speak about positive and negative reactions. Have you genotyped your birds?

M. PAPP: When this population was tested no family material was available to determine the genotypes. In the following generation — as I mentioned in my previous report — on the basis of 50 families i.e. 625 birds, the genotyping was carried out. These allele designations were applied in this report. As we failed to detect all alleles in the population, for statistical reasons it seemed to be practical to calculate with antigens, actually detected by testing. Therefore we used positive and negative reactions in the Tables 1, 3 and 5. We thought that by these means it would be easy to survey the effect of the presence or absenceof a certain allele on the productive characteristics studied, on population level.

K. HÁLA: What do you mean by allele?

You had 1284 animals in experiment. We know, that every animal has 2 alleles in one system — it represents 2568 alleles. But in B system you found a much higher number of alleles, than 2568.

M. PAPP: I mean, one allele corresponds to one antigen, probably the B system is the exception. This is the reason that you have seen more antigens — or rather antigenic factors — in the B system in my tables. Namely, statistical calculations — as I said to Dr. Oosterlee — were carried out with antigens.

As I see that the term allele — especially in the tables — may cause confusion I will use antigens instead of alleles in all tables.

*XIIth Europ. Conf. Anim. Blood Groups Biochem. Polymorph., Bp., 1972 (pp. 499 – 500)*

# RELATIONSHIP OF B LOCUS GENOTYPE WITH EGG PRODUCTION IN HENS

V. I. Muraviev, Svetlana Samodelkina and Iraida Sovetova

**All-Union Poultry Research and Technological Institute, Moscow-region, USSR**

The red cell antigens forming the basis of blood groups are controlled by certain allele genes.

Neither environmental conditions nor physiological state of the animal can change the specificity of their albumin structures formed on the basis of DNA code.

Immunological and serological methods and genetic analysis of red cell antigens allow to establish their specificity and to identify the genes controlling them where hereditary alleles of one and the same blood group system are interacting following the pattern of codominance.

Evidence to the hereditary nature of a number of specific albumin components, i.e. red cell antigens, easily determined by the laboratory tests makes it possible to decode the working mechanism of genes in the synthesis of those albumins which have a direct or indirect bearing on the change of the performance indices of the animals. These data can be used for determining genetic correlations both through the gene linkage and possibly through pleiotropic effects on different traits.

Such correlations were established between the blood group genotype of hens and their reproductive and performance qualities (Briles, 1952, 1957; Briles and Allen, 1962; Morton, et al., 1965).

It is known that hens of different populations and lines are characterized by a certain specific blood group genotype which is in close correlation with their productivity. It has been also noted that optimal productivity often correlates with the presence of a certain allele in the genotype.

Having taken all this into consideration we tried to define B locus genotype of blood group in hens of our selection stock represented by the breeds Russian White — line G and White Leghorn — lines Katman 18 and 63.

Monospecific sera obtained by isoimmunization of the experimental hens of the closed high inbred stock were used for testing hens.

Having examined over 400 hens with egg production recorded individually for the period of nine months, we compared the B locus genotypes of blood groups in these hens with the level of their egg production.

As a result it was established that hens of the Russian White breed, line G with blood group genotype $B^1B^2$ and $B^1B^x$, the White Leghorn hens, line Katman-63 with blood group genotype $B^1B^8$ and $B^1B^x$ had a higher egg production as compared to hens of the same lines with dissimilar blood group genotype of locus B. The difference in the level of egg production was statistically significant ($P < 0.5$).

In hens of the White Leghorn breed, line Katman-18, no statistically significant differences were found in egg production in relation to the individual B locus genotype of blood groups. However, in this case hens with genotype $B^1B^x$ laid by 5–14 eggs more than their contemporaries with different blood group genotypes.

The above data show that the relationship of certain B locus genotypes of blood groups with egg production has certain particularities for each line of hens. In the case studied, the presence of allele $B^x$ correlates in a greater degree with egg production of hens of all lines.

Taking into account that the egg production of hens has a polygenic foundation these data can be considered as preliminary. Comprehensive and detailed studies remain to be carried out along this line, which seems very promising and important from both the theoretical and practical point of view.

*XIIth Europ. Conf. Anim. Blood Groups Biochem. Polymorph., Bp., 1972 (pp.501 – 507)*

# ON THE BLOOD GROUPS AND THE APPEARANCE OF NATURAL ANTIBODIES IN THE GOOSE

S. LOSONCZY

Bloodgrouping Laboratory, University of Veterinary Science, Budapest, Hungary

## INTRODUCTION

In the goose, similarly to other species, the blood group antigens show such a wide variety that they can be helpful both in parentage control tests and in other aspects of breeding. We studied the blood group antigens of the goose and the properties of non-specific (natural) antibodies in the goose serum. The time of their appearance and their level were determined in different seasons.

## MATERIAL AND METHODS

For the detection of antibodies 2.5% suspensions of washed red blood cells (RBC) from different species were used.

To examine the natural antibodies, heteroimmunizations were carried out with 1 ml of 30% suspensions of cattle, horse and human washed RBC.

## RESULTS

### THE BLOOD GROUP ANTIGENS

The production of isoimmune antibodies is more difficult in the goose than in other species, but with successful induction of antibodies the diagram of the specific antibody level resembles that of other species, as shown in Fig. 1.

The methods of antibody production were elaborated earlier (LOSONCZY, 1967; 1968a; 1968b).

Table 1 shows the immune response of the goose to various modes of immunization. To express the degree of the immune response, percentages of successful immunizations and average titres are presented. Intravenous immunization evoked an intensive immune response, when the immunizing dose was at least 10 ml blood. More blood was not necessary. There was no immune response to intramuscular immunization alone, but when combined with intravenous immunization, the immune response was sometimes more pronounced. The present results and several hundreds of examinations carried out suggest that more than one immunizations do not result in higher levels of antibodies than those carried out on a single occasion.

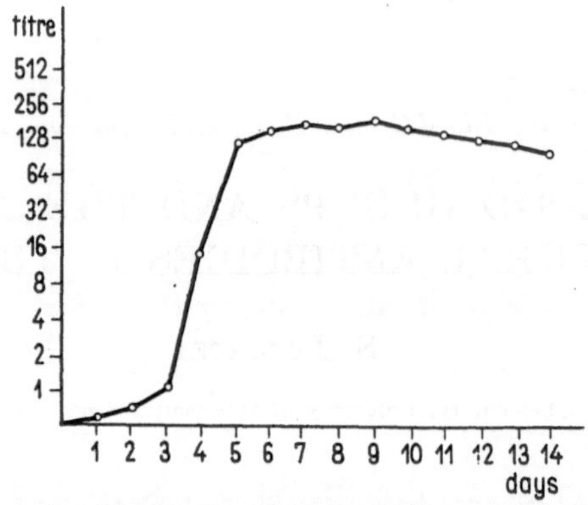
titre
512
256
128
64
32
16
8
4
2
1
1 2 3 4 5 6 7 8 9 10 11 12 13 14
days

FIG. 1. The formation of antibody level after immunization. Geese were given a single intravenous injection of 40 ml 30% RBC suspension

40 different monospecific antisera (blood group reagents) were prepared and used in parentage control tests. To clarify the genetic system of the red cell antigen in the goose, family material was used. Figure 2 shows the test results obtained by using 17 monospecific sera in 6 families.

Recently, a family material consisting of 1200 geese was bloodtyped with 40 reagents. Determinations of the blood group systems from the test results are now in progress.

## NATURAL ANTIBODIES IN THE GOOSE

Figure 3 shows the age of the birds at the time of appearance of the natural antibodies (hemolysins). Antibodies generally appear by the third week of life. They hardly can be detected before this time. Their level increases intensively from the third to the fifth week. This observation is in accordance with SETO and HENDERSON'S (1969) results concerning chicken. The level of antibodies becomes constant at the age of 2 months. Natural antibodies appeared earlier than usually if the birds were immunized at young age. In young birds the antigenic stimulus evoked an immune response already by the end of the first week of life, although the antibody level was lower than in adults. This is in accordance with SOLOMON'S (1965) results obtained in chickens.

Supposing that the hormonal activity and character of metabolism depended also on the season the examination of the formation of natural antibodies was continued during the full period of the year. The results indicated a seasonal change of the natural antibody level (Fig. 4), but no significant variation of the titre of natural antibodies was found in the different seasons.

Natural antibodies from adult goose sera were tested with RBC from different species. The extent of the reactions before immunization (basic level of natural antibodies) was in decreasing sequence as follows: horse, cattle, chicken (Fig. 5). As a result of immunization with cattle RBC,

**TABLE 1**

*Effect of the different modes of immunization on isoimmune antibody titres*

| | Mode of immunization | | No. of animals | Average titre and percentual number of the effective immunizations on post-injection days | | | | | | | | | |
|---|---|---|---|---|---|---|---|---|---|---|---|---|---|
| | | | | 3rd day | | 4th day | | 5th day | | 6th day | | 7th day | |
| | | | | t | % | t | % | t | % | t | % | t | % |
| Intravenous | 1×1 ml | | 30 | — | — | | | 0.53 | 6.6 | 13.03 | 23.3 | 9.0 | 16.7 |
| | 1×5 ml | | 30 | — | — | | | 0.7 | 13.3 | 2.56 | 20.0 | 0.86 | 23.33 |
| | 1×10 ml | | 30 | — | — | | | 4.36 | 26.6 | 16.0 | 40.0 | 14.36 | 26.6 |
| | 1×40 ml | | 22 | — | — | | | 11.7 | 13.6 | 23.4 | 13.6 | 1.5 | 13.6 |
| | 2×5 ml | one-day interval | 13 | 0.51 | 15.0 | | | | | | | | |
| | 2×10 ml | two-day interval | 28 | 8.9 | 32.0 | | | | | | | | |
| | 3×5 ml | one-day interval | 20 | — | — | 0.15 | 10.0 | — | — | 0.45 | 30.0 | | |
| Intramuscular | 1×1 ml | | 30 | — | — | | | — | — | — | — | | |
| | 1×5 ml | | 16 | — | — | | | — | — | — | — | | |
| | 1×10 ml | | 22 | — | — | | | | | | | | |
| | 2×5 ml | one-day interval | 25 | 0.08 | 0.04 | | | — | — | | | | |
| | 2×10 ml | two-day interval | 22 | — | — | — | — | | | | | — | — |
| | 3×5 ml | one-day interval | 18 | | | — | — | | | | | — | — |
| Iv. + im. | 1×10 ml<br>5 ml | iv.<br>im. | 20 | | | 20.15 | 55.0 | | | | | | |
| | 1×15 ml<br>5 ml | iv.<br>im. | 30 | — | — | 2.5 | 23.3 | | | | | | |
| | 1×8 ml<br>2 ml | iv.<br>im. | 26 | — | — | | | 3.4 | 27.0 | | | | |

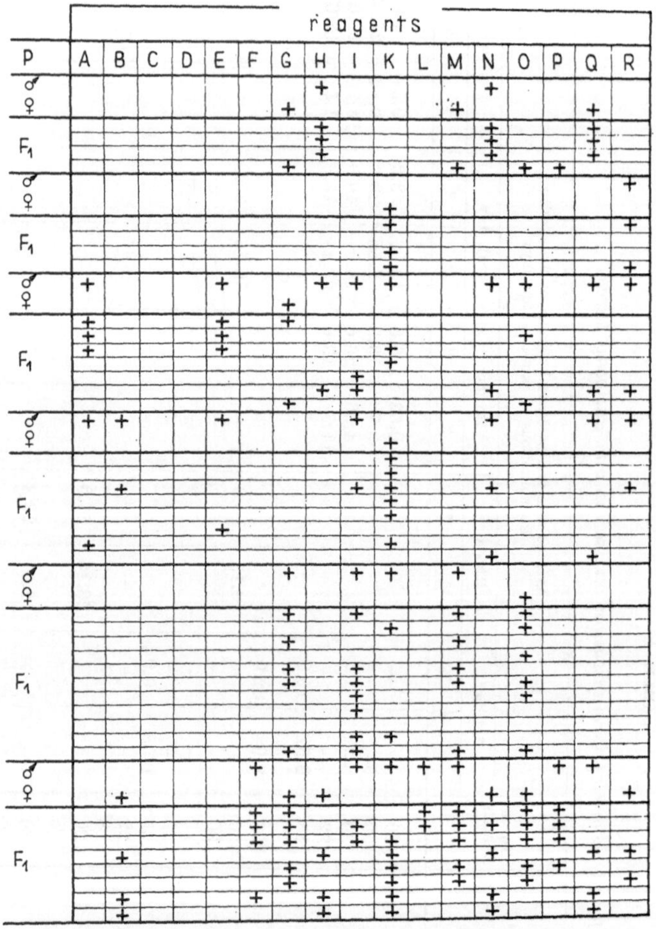
reagents
P
A B C D E F G H I K L M N O P Q R
♂
♀
F1

FIG. 2. Reaction scheme of six families with monospecific antisera

antibody levels increased and the hemolysis of RBC from all the three species tested was proportional with their basic levels. Thus it can be established that the non-specific immune response is proportional with the basic level.

Increasing natural antibody level by induction very interesting results have been obtained concerning the immunity of geese. Having considerably increased the titre of antibodies, it was found, that following the infection of the birds with various Pasteurella strains the time of death was prolonged (Fig. 6).

This is the so-called limited anti-bacterial effect of natural antibodies. This property has a great significance particularly for the goose, considering

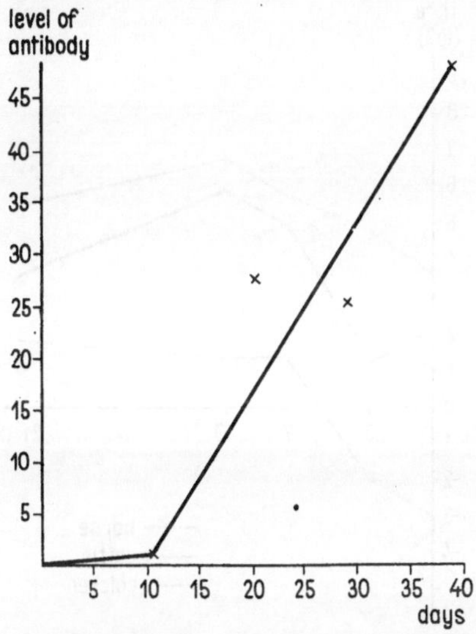
level of
antibody
45
40
35
30
25
20
15
10
5
5 10 15 20 25 30 35 40
days

FIG. 3. Appearance of natural antibodies in goose

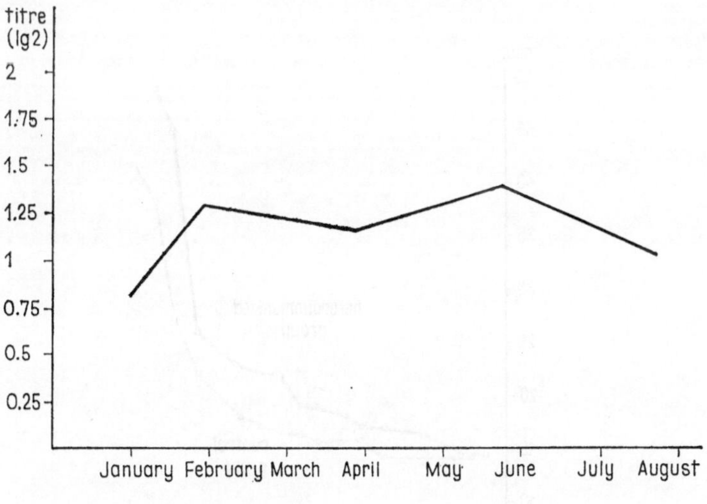
titre
(lg2)
2
1.75
1.5
1.25
1
0.75
0.5
0.25
January February March April May June July August

FIG. 4. Seasonal fluctuation of the natural antibody level

its increased susceptibility to infection in consequence of the special technology of goose breeding. The increase of the level of the natural antibodies is a possibility of protecting the animals against infectious diseases under critical conditions.

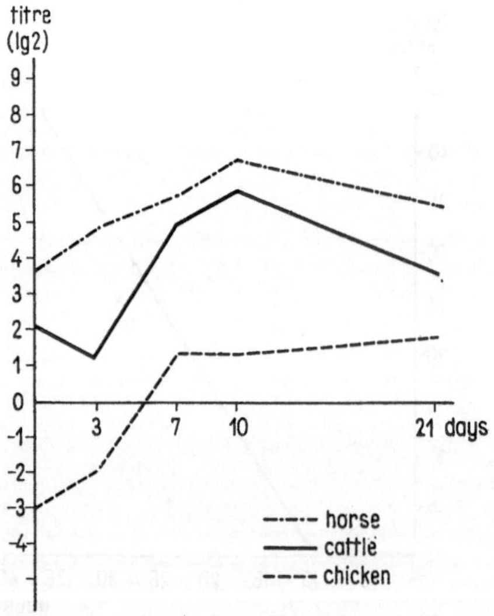
titre
(lg2)
9
8
7
6
5
4
3
2
1
0
-1
-2
-3
-4
3
7
10
21 days
horse
cattle
chicken

FIG. 5. Titre changes of natural (antibody) hemolysin after heteroimmunization with cattle erythrocytes

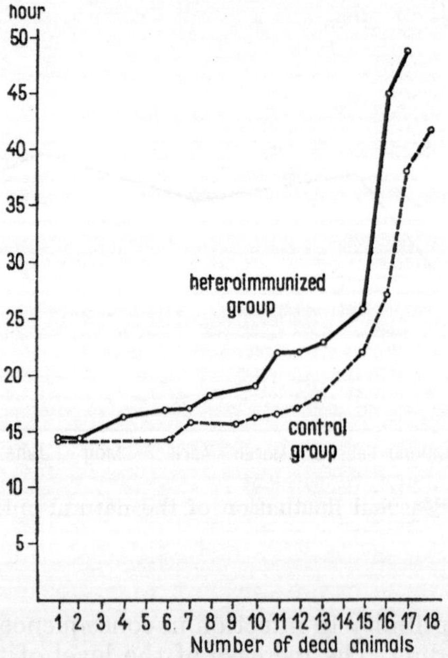
hour
50
45
40
35
30
25
20
15
10
5
heteroimmunized
group
control
group
1 2 3 4 5 6 7 8 9 10 11 12 13 14 15 16 17 18
Number of dead animals

FIG. 6. Deaths of the heteroimmunized and control animals after infection with V cholerae

## REFERENCES

LOSONCZY, S. (1967), The methods and results of goose blood group investigations. *Thesis.*

LOSONCZY, S. (1968a), Initial results of investigations into the blood groups of the goose. *Acta Vet. Acad. Sci. Hung.* **18**, 287–293.

LOSONCZY, S. (1968b), Data on the blood group properties of geese. *Magy. Áo. Lapja*, **23.** 584–586.

SETO, F. and HENDERSON, W. G. (1969), Natural and immune hemagglutinin forming capacity of immature chickens. *J. Exp. Zool.*, **169.** 501–512.

SOLOMON, J. B. (1965), Onset and nature of the immune response to heterologous erythrocytes in embryo and young chickens. pp. 371–379. In: The Molecular and Cellular Basis of Antibody Formation. Publishing House of the Czech. Acad. Sci. Prague.

## DISCUSSION

W. E. BRILES: How do you account for the appearance of the normal antibodies? You may be interested in a possible third explanation suggested by the work of Dr. George F. Springer. He exposed germ-free chicks to cultures of certain bacteria and found that he could engender antibodies that agglutinated human B blood type cells. Serum from chickens normally contain such anti-B agglutinins and it is likely that they are accounted for bacteria ingested by chickens in their normal environment. Thus, since your normal antibodies are found after the chicks are old enough to be immunocompetent it could be quite possible that the normal antibodies observed for goose, sheep, cows, etc. are in reality cross-reactive antibodies produced against ingested bacteria.

S. LOSONCZY: Thank you for your comment and completion. The explanation mentioned by you is quite possible also in our practice.

E. PETROWSKÝ: Is there a correlation between the titre of natural and immune antibodies?

S. LOSONCZY: In most cases there is positive correlation between the basic level of natural antibodies and that of the heteroimmune antibodies induced by immunization with cattle red blood cells.

*XIIth Europ. Conf. Anim. Blood Groups Biochem. Polymorph., Bp., 1972 (pp. 509–512)*

# PHYSIOLOGICAL PHENOMENA ACCOMPANYING CHANGES IN THE SERUM AMYLASE ACTIVITY IN GEESE

MÁRIA LOSONCZY*, S. LOSONCZY** and ERZSÉBET TAKÁCS**

* Hungarian Meat Research Institute, Budapest, Hungary
** Bloodgrouping Laboratory, University of Veterinary Science, Budapest, Hungary

## INTRODUCTION

The liver plays an important role in producing and controlling the serum proteins out of which the amylase enzyme occurs also in goose serum.

Synthesis of amylase is connected with the microsome fraction of liver cells. It is, however, unknown, how the amylase enzyme is released and how it enters the blood circulation. Depending on species and organ, the amylases consist of several izoenzyme fractions, but we have failed to find any fractions in goose serum by means of starch gel electrophoresis.

It is important, especially in the case of the goose, that the level of serum amylase activity is connected to a certain extent with the liver metabolism, its level being controlled by the liver. Hepatic glycogen is effected by both the liver and serum amylases, whereby it can be taken for an indicator of carbonhydrate metabolism. It is readily available for testing in serum samples.

With these facts in mind, attempt was made to elaborate a biochemical method based on serum amylase activity for the practical purpose of estimating the liver increase of geese masted by stuffing.

The method of HOREJSI and SLAVIK (Laboratoriumstechnik für Biochemiker, eds. Keil and Sormova, 1965) was used for the determination of amylase activity.

## RESULTS

First it was examined, how increase of the liver affects serum amylase activity during a 5-week stuffing period. From a number of amylase test in four experimental groups, the curves showed in Fig. 1 have been obtained.

A steep increase occurred during the first week, which corresponded to the adaptation of the liver to the increased carbohydrate rations through activation of protein synthesis and multiplication of liver cells. The metabolism altered toward a more intensive reserve accumulation. From the 2nd week on the activity decreased. As amylase plays a role in the breakdown of glycogen, the decrease of its activity may be related to the de-

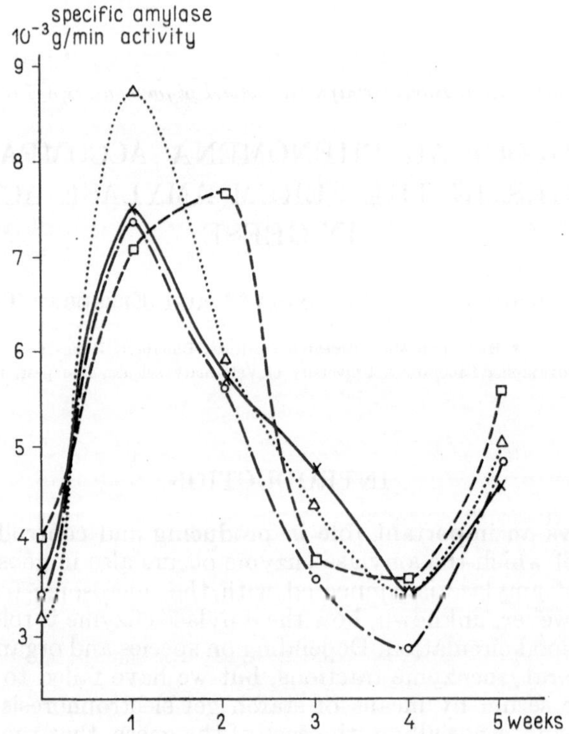
specific amylase
$10^{-3}$ g/min activity
9
8
7
6
5
4
3
1
2
3
4
5 weeks

FIG. 1. Changes of amylase activity during fattening

crease of liver glycogen. The serum amylase activity may be supposed to indicate this change in liver metabolism.

From the 2nd week on the breakdown of glycogen tended to increase, and fat synthesis became predominant. There is histological evidence (LABIE, 1967) supporting the depletion of glycogen from the second week, replaced later on by fat droplets depositing in microvacuoles. Minimum activity was noted at the end of the 4th week, and increasing tendency during the 5th week, as observed also in the present study.

There were differences in the time and extent of maximum and minimum activities.

Characteristically of the goose species, most of the enzyme activity levels strongly depended on seasonal changes in sexual hormon activity.

In spring and early summer, the amylase activity depends to a greater degree on sexual hormone activity than on the increasing fat-stage of the liver. Therefore, more precise examinations were carried out in autumn and late summer.

In the 4th and 5th weeks of the autumn stuffing period, we determined the individual amylase activities as well as the individual liver weights after slaughter.

Both weights and amylase activity values varied widely. To support the supposition, that the greater the liver weight the less the amylase activity, rank correlation coefficients were computed. The rank correlation coefficient was 0.40 after the 4th week, and −0.52 (+) after the 5th week.

These results are demonstrated in Fig. 2. Individual activities measured during the 4th and 5th week depended on the liver weight.

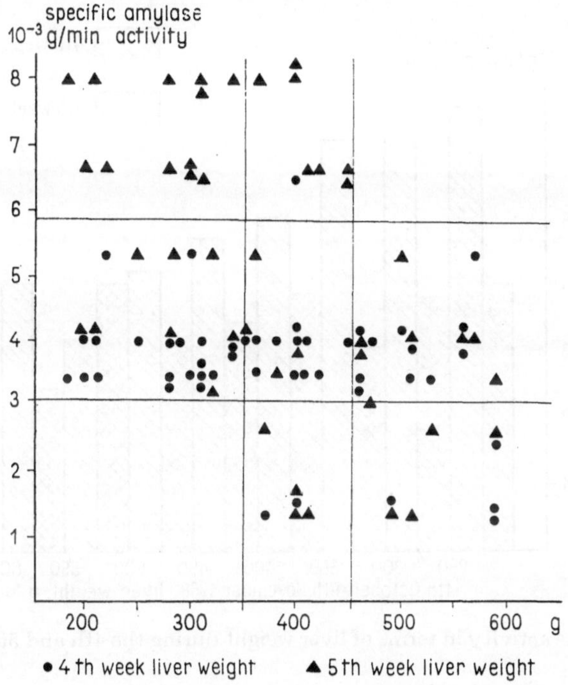
specific amylase
$10^{-3}$ g/min activity
8
7
6
5
4
3
2
1
200
300
400
500
600 g
● 4th week liver weight
▲ 5th week liver weight

FIG. 2. Distribution of individual liver weights in terms of amylase activity in the 4th and 5th week of fattening

Low correlation coefficients were explicable by the following finding. Although there was no correlation between activities and liver weights in the case of medium-weight livers, there was a considerable difference between the activities of small and large livers. This may have been due to the fact, that when glycogen diminishes, activity decreases, but there had been just a small fat deposition yet, so the liver weight had not increased too much. Apparently the depletion of glycogen and deposition of fat do not take place simultaneously in each liver cell.

As shown in Fig. 3 the activities measured during the 4th and 5th weeks depended on the liver weight. The diagram clearly shows that (1) simultaneously with the increase of liver, both of 4-week and 5-week activities decreased, (2) 5-week activities were higher than 4-week activities, (3) the difference between 4-week and 5-week activities decreased with the increase of liver weight.

From the above findings the following conclusions can be drawn: In autumn and late summer, viz. by the end of the stuffing period, the lower amylase activity is a favourable sign of obtaining a larger liver. Contrarily a liver of small size is expected, if the amylase activity increases again after having reached its minimum level.

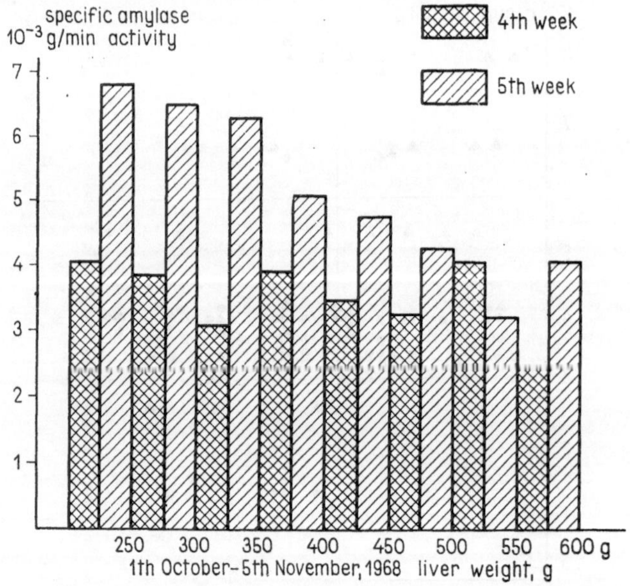
specific amylase
$10^{-3}$ g/min activity
4th week
5th week
7
6
5
4
3
2
1
250 300 350 400 450 500 550 600 g
1th October–5th November, 1968 liver weight, g

Fig. 3. Amylase activity in terms of liver weight during the 4th and 5th week of stuffing

## REFERENCES

Keil, B. and Sormova, Z. (1965), Laboratoriumstechnik für Biochemiker. Akademische Verlagsgesellschaft, Leipzig.

Labie, C. H. (1967), Quelques remarques sur l'élevage et la production de l'oie. Jouy-en-Josas.

*XIIth Europ. Conf. Anim. Blood Groups Biochem. Polymorph., Bp., 1972 (pp. 513–515)*

# GAMMA-GLOBULIN ALLOTYPIC SPECIFICITIES IN TURKEYS (*Meleagris gallopavo*)

J. M. GASPARSKI and R. W. C. STEVENS

Department of Biomedical Sciences, University of Guelph, Canada

## SUMMARY

This report describes allotypic specificities in the domestic turkeys. Isoimmune antisera to turkey serum proteins were obtained indirectly by intravenous injections of turkeys with anti-Brucella abortus and anti-Salmonella typhimurium antisera with the homologous antigens. The precipitation reactions of antigen-antibody were investigated in agar gel. The allotypic specificities were identified as gamma-globulins by immunoelectrophoresis. In tested turkeys the frequencies of particular antigens were from 0.3536 (lowest) to 0.7234. The genetic relationships of the five different antigens have not been determined.

## INTRODUCTION

Among the investigations on the blood proteins of man and many different animal species, allotypic specificities of serum proteins merit some position according to their structure, as well as genetical importance.

GRUBB and LAURELL (1956) reported on group differences in the gamma-globulin component of human serum proteins. Since that time many scientists investigated and are still investigating the allotypes in different species of animals e.g. in rabbits (DRAY et al., 1962, 1963; DUBISKI et al., 1959; DUBISKI and MULLER, 1967; OUDIN, 1960), in hens (DAVID et al., 1965: SKALBA, 1966), in mice (LIBERMAN and DRAY, 1964a, 1964b) in pigs (RASMUSEN, 1965) in cattle (RAPACZ et al., 1968).

This paper informs of allotypes in domestic turkeys (*Meleagris gallopavo*) in which five allotypes were observed.

## MATERIAL AND METHODS

To obtain isoimmune antisera isoimmunizations were performed indirectly, according to the method published in 1968 (GASPARSKI and STEVENS). Donors received intraveously ten 1 ml injections of the suspension of $10^9$ cells of killed bacteria, *Brucella abortus* or *Salmonella typhimurium*. Recipients were injected also intravenously with anti-Brucella or anti-Salmonella antisera with the homologous antigens, respectively. The inoculum was a suspension of $10^9$ killed bacterial cells agglutinated with anti-bacterial serum. It was unknown which differences of antigens existed between donors and recipients. The number of injections varied from fifteen to thirty-five per bird.

Precipitation reactions of antigen-antibody were investigated in agar gel, according to the modified Ouchterlony method. For agar solution the barbital buffer with eight per cent sodium chloride was used. In one Petri dish forty-two serum samples were tested. The results of precipitation reactions were read after ten–twenty hours of incubation at room temperature.

The characterization of allotypic serum proteins was studied by immunoelectrophoretic method. The barbital buffer (pH 8.6 ionic strength 0.1) was used to prepare on agar gel and to vessels. The time of separation of the antisera was 60 minutes at 10 V/cm. For protein staining Amido Black solution was used.

## RESULTS

After isoimmunizations all normal sera of donors and recipients were tested with different isoimmune antisera.

The first group of eight recipients received fifteen injections over three months' period in three series (9, 3, and 3 injections, respectively). In this group the serum of only one recipient was active. Second group of eight recipients received thirty-five injections over thirteen months' period in seven series (9, 6, 3, 3, 4, 5, and 4 injections, respectively). In this group the sera of three recipients were active in antigen-antibody precipitation reactions.

The specific precipitating antisera were used to test normal sera of turkeys, which originated mostly from the Poultry Science Experimental Farm, University of Guelph. Altogether 246 turkeys were tested. Table 1 illustrates the frequencies of serum antigens of random adult individuals of different lines: A, B, C and D.

TABLE 1

*The frequencies of serum antigens in turkeys*

| Antigen | Number of individuals | Frequencies of antigens | | |
|---|---|---|---|---|
| | | General group | Line A | Line B |
| $A_1$ | 246 | 0.6423 | 0.6800 | 0.5851 |
| $A_2$ | 246 | 0.3536 | 0.4200 | 0.3851 |
| $A_3$ | 229 | 0.4148 | 0.6321 | 0.3939 |
| $A_4$ | 41 | 0.7234 | — | — |
| $A_5$ | 246 | 0.4878 | 0.6500 | 0.4571 |

The frequencies of antigens $A_1$, $A_2$, $A_3$, $A_4$ and $A_5$ were calculated for general group and for random individuals of line A and line B. In line C and line D the number of tested individuals were too small.

In addition, the comparison between line A and line B was performed to estimate differences according to occurrence of the particular antigens. The results are illustrated in Table 2.

TABLE 2

*Comparison test between line A and line B*

| Antigen | Line | Frequencies + observ. | + expect. | — observ. | — expect. | $\chi^2$ | d.f. | $P$ |
|---|---|---|---|---|---|---|---|---|
| $A_1$ | A | 68 | 64.15 | 32 | 35.85 | 1.5244 | 1 | $0.5 > P > 0.3$ |
| $A_1$ | B | 41 | 44.85 | 29 | 25.15 | | | |
| $A_2$ | A | 42 | 40.59 | 58 | 59.41 | 0.2101 | 1 | $0.7 > P > 0.5$ |
| $A_2$ | B | 27 | 28.41 | 43 | 41.59 | | | |
| $A_3$ | A | 55 | 46.06 | 32 | 40.94 | 8.5421 | 1 | $0.01 > P > 0.001$ |
| $A_3$ | B | 26 | 34.94 | 40 | 31.06 | | | |
| $A_5$ | A | 65 | 57.06 | 35 | 42.94 | 6.2498 | 1 | $0.02 > P > 0.01$ |
| $A_5$ | B | 32 | 39.94 | 38 | 30.66 | | | |

## DISCUSSION

The reported investigations on allotypic specificities in domestic turkeys showed that the polymorphism of these specificities exists.

The indirect method for obtaining the specific isoprecipitating sera appeared satisfactory but required a long period of time and a large number of injections as well as a large number of recipient-donor sets to obtain specific antibodies against different allotypic antigens.

The results of investigation of experimental farm turkey sera proved that the frequencies of antigens were different. In general group the range was from 0.3536 to 0.7234 for five antigens: $A_1$, $A_2$, $A_3$, $A_4$ and $A_5$. The differences of antigen frequencies were also observed between subgroups A and B. These differences were significant for antigens $A_3$ and $A_5$, as $P$ was 0.001 and 0.01 respectively.

Since the reported allotypic specificities are polymorphic and the differences can be observed among several groups of turkeys this system was found useful for further investigations as well as to include them in studies on the different blood traits and other traits in domestic turkeys.

## REFERENCES

DAVID, C. S., KAEBERLE, M. L. and NORDSKOG, A. W. (1965), *Immunogenetics Letter* 4,

DRAY, S., DUBISKI, S., KELUS, A., LENNOX, E. S. and OUDIN, J. (1962), *Nature*, **195**. 785–786.

DRAY, S., YOUNG, G. O. and GERALD, L. (1963), *J. Immun.*, **91**, 403–415.

DUBISKI, S., DUDRIAK, Z., SKALBA, D. and DUBISKA, A. (1959), *Immunology*, **11**, 84–92.

DUBISKI, S. (1963), Prace Wroctawskiego Towaszystwa Nauk., Ser. B. **113**, 1–66.

DUBISKI, S. and MULLER, P. J. (1967), *Nature*, **214**, 696–697.

GASPARSKI, J. M. and STEVENS, R. W. C. (1968), *Immunogenetics Letter*, **5**, 148–149.

GRUBB, R. and LAURELL, A. B. (1956), *Acta Path. Micro. Scand.*, **39**, 390–398.

LIBERMAN, R. and DRAY, S. (1964a), *Proc. Soc. Exp. Biol. Med.*, **116**, 1069–1074.

LIBERMAN, R. and DRAY, S. (1964b), *J. Immun.*, **93**, 584.

OUCHTERLONY, O. (1949), *Acta Path. Micro. Scand.*, **26**, 507.

OUDIN, J. (1960), *J. Exp. Med.*, **122**, 125–142.

RAPACZ, J., KORDA, N. and STONE, W. H. (1968), *Genetics*, **58**. 387–398.

RASMUSEN, B. A. (1965), *Science*, **148**, 1742–1743.

SKALBA, D. (1966), *Bulletin De Lacademic Pol. Ses.*, XIV-3, 159–162.

*XIIth Europ. Conf. Anim. Blood Groups Biochem. Polymorph., Bp., 1972 (pp. 517 – 521)*

# ARGUMENTS IN FAVOUR OF A HYPOTHESIS OF GAMETIC INCOMPATIBILITY CONCERNING BLOOD GROUPS IN DOMESTIC QUAIL*

A. PERRAMON

**Department of Animal Genetics, I.N.R.A.-C.N.R.Z., 78-Jouy-en-Josas**
**with the technical collaboration of J. L. Monvoisin**

In earlier publications (PERRAMON and MERAT, 1968a, 1968b, 1970a; PERRAMON, 1969; PERRAMON and MERAT, 1970) we presented results on domestic birds suggesting that when the female is heterozygous for a pair of close-linked blood group genes, a mitotic recombination mechanism is present in the gonads (premeiotic recombination). Using the species, *Coturnix coturnix japonica*, we continued our research to try to prove that such a mechanism is present in the male.

The possibility of a mitotic recombination itself a rare occurrence) producing a homozygous clone (gonadal mosaic ) in heterozygous males is obviously more difficult to demonstrate becau of the persistance of a large number of germ cells which become functional. (In the female, only the left ovary persists during post-embryonic life.)

We observed an apparent absence of B1, significant at the 5% threshold, in the offspring of a male (heterozygous for the B1 factor) crossed with a female not having that factor (Table 1). Naturally, the possibility was not excluded that such a deviation might appear at random among 50 observed sibship and we considered it from the beginning as a simple indication which should be confirmed.

TABLE 1

| Parental genotype | ♂ 52757 B1/— × ♀ 52805— /— | |
|---|---|---|
| Proportions and genotypes of offspring | 5 B1/— | 16 —/— |

Because of the small number of offspring obtained due to the premature death of the female, we replaced that female by another one also negative

* The work presented in this publication is part of the R.C.P. 224 research plan. We wish to thank M. E. BOESIGER, C.N.R.S. Research Director, for making available the animals used in our research and carefully observing all the crosses necessary to the experiments.

for the B1 factor, but having produced a significantly high number of offspring carrying the B1 factor in a previous crossing with a male genotypically analogous to male 52757 (Table 2).

TABLE 2

| Parental genotypes | ♂ 54460 B1/— × ♀ 54436 —/— | |
|---|---|---|
| Proportions and genotypes of offspring | 16 B1/— | 5 —/— |

Previous crossing of female 54436 used in a 2nd crossing with male 52757

Using the hypothesis of a gonadal mosaicism in male 52757, we should have obtained a result similar to that in Table 1 in the new crossing of male 52757 with female 54436 (Table 3), supposing that the same cellular clone persists as before in a functional state. On the contrary, we obtained proportions which agree more closely with those of Table 2.

TABLE 3

| Parental genotypes | 52757 B1/— × 54436 —/— | |
|---|---|---|
| Proportions and genotypes of offspring | 18 B1/— | 7 —/— |

These unexpected results lead us to think that, depending on the genotype of the egg, gametes are selected which are capable of fertilizing it. Moreover, the well-known gallinacean polyspermy seems to suppose that there is an *in ovo* choice of spermatozoa.

We have tried to find similar cases in other crosses between partners which are respectively heterozygous and negative for the B1 factor. We saw one such case between a female carrying the same B1 factor in a heterozygous state and a male negative for that same factor. However, that time the male was heterozygous for a new factor which we shall call X for the moment because we still do not understand its genetic dependence relations with present-known factors. Table 4 summarizes the actual results obtained and the expected theoretical ones in such a cross as well as the genotypes of parents and offspring. The A1 locus controlling the synthesis of A1 and A1′ alleles is closely related to the two loci determining the A2 and A3 factors forming a "supergene", which is situated about 25 recombination units away from the B1 locus. In the Table 4 cross, the sole genotype is $\frac{A1A2A3}{A1'A2-}$ (— signifies the absence of A3 but corresponds to an antigene produced by an A3 gene allele), and no vanability in the offspring should be found because the "supergene" factors are homozygous in the two parents.

We shall see these factors act in future crosses, and that is why they are mentioned here.

No matter what the genetic dependence relation between B1 and X factors, we expect to find the four possible categories, B1, X; B1, –; X, –; and –,– in equal proportion in the offspring.

The results in Table 4 show an almost total absence of B1, X, and a complete absence of –,–.

TABLE 4

Parental genotypes ♂ 41779 $\frac{\text{A1}'\ \text{A2}\ —}{\text{A1}'\ \text{A2}\ —}$ — — ? — $\frac{\text{X}}{—}$ × ♀ 42125 — $\frac{\text{A1A2A3}\ \ \text{B1}}{\text{A1A2A3}\ \ —}$

| | | | | | |
|---|---|---|---|---|---|
| offspring | Expected categories and proportions | 1 B1, X | 1 B1, — | 1 X, — | 1 —, — |
| | Observed categories and proportions | 1 | 39 | 28 | 0 |

We know that this very significant disproportion of offspring categories is not explainable by differential mortality. As a matter of fact, embryonic and post-embryonic death in all the crosses described here do not significantly deviate from the average death rate found in crosses in the general population of this laboratory species.

Thus, only two hypotheses remain which may explain the results in Table 4: interaction in factor determinism (not usual in blood groups), and gamete incompatibility in conformity with haploid genotype.

Of the two female pronuclei which can form at the end of meiosis in a cytoplasm having several spermatozoa (there being no doubt about polyspermy in this species), the pronucleus B1 carrier should produce amphimixia more easily with a spermatozoon not carrying X, and vice-versa, the pronucleus not carrying B1 should produce it with a spermatozoon carrying X.

Since several physiological conditions are requisite, the mechanism does not necessarily follow the law of all or nothing, and this would explain some exceptions to the reciprocal choice of gametes.

At the present time we have no direct, physiological reason for defending this hypothesis rather the alternative one which proposes the interaction of several factors to explain the determinism of the blood groups studied.

However, some later crosses give results which rather disagree with the second hypothesis. Thus, we repeated the Table 4 cross with different offspring of that same family, taking care to always select females heterozygous for B1 and males heterozygous for X, thus reproducing crosses between the same genotypes as the parental ones (at least for these two factors). The results are shown in Table 5.

TABLE 5

*Crosses of male type X/– × female B1/–*

| Crosses between offspring of mother 42125 | Categories and proportions of offspring | | | |
|---|---|---|---|---|
| | B1, X | B1, – | X, – | –, – |
| 1 | 0 | 8 | 6 | 0 |
| 2 | 0 | 9 | 7 | 1 |
| 3 | 0 | 12 | 9 | 2 |
| 4 | 2 | 11 | 9 | 8 |

We know that of these four crosses, the first is analogous to the parental cross, two are missing one category, and the last has the theoretical proportions of a cross with random meeting of gametes.

According to our knowledge of other cases, we think that determinism based on genic interaction should almost always be manifested in the same way, and especially in a group of crosses in which the partners belong to the same family. On the contrary, more flexible conditions controlling relative incompatibility might explain the absence of incompatibility in the last cross and the lessened incompatibility in crosses 2 and 3 (Table 5).

Moreover, in order to postulate an interactive mechanism between the two factors in Table 4 under discussion, and to explain the results obtained, one and only one of the two factors (either B1 or X) must be heterozygous in the parents. It would be necessary to hypothesize that B1, for example, was really homozygous in the mother, but that it is only manifested in the offspring when it coincides with the X factor allele (hterozygous in the father, one-half passing into the offspring). Interaction may no longer be considered if both factors are presumed to be homozygous in the parents. The same is true for the interaction hypothesis, if B1 and X are both heterozygous in the parents, because the absence of –,– (very significant) in the Table 4 cross and the first cross of Table 5 cannot be explained.

As a result, we wished to be sure that the two parents, 41779 and 42125 respectively, were heterozygous for X and B1 factors, and we re-crossed them with partners lacking both X and B1.

In Table 6, female 42125, in keeping with the postulated heterozygous hypothesis, produced offspring in which one-half lacked B1.

Table 7 shows the result observed when 41779 is crossed a second time with an A1A2A3 genotype female.

TABLE 6

| Parental genotypes | ♂ 55712 –/– × ♀ 42125 B1/– | |
|---|---|---|
| Proportions and genotypes of offspring | 14 –/– | 12 B1/– |

TABLE 7

| Parental genotypes | ♂ 41779 $\frac{A1'A2 -}{A1'A2 -}$ — ? $\frac{X}{-}$ × ♀ 53433 $\frac{A1A2A3\ -}{-\ -\ -\ -}$ | | | |
|---|---|---|---|---|
| Proportions and genotypes of offspring | $\frac{-\ -\ -\ \ -}{A1'A2 -\ \ -}$ | $\frac{-\ -\ -\ \ X}{A1'A2 -\ \ -}$ | $\frac{A1A2A3\ \ -}{A1'A2 -\ \ -}$ | $\frac{A1A2A3\ \ X}{A1'A2 -\ \ -}$ |
| | 0 | 23 | 24 | 0 |

When male 41779 is checked for heterozygote factor X, the genic interaction hypothesis may be discarded. Moreover, Table 7 shows in a significant way as Table 4 a complete absence among the offspring of the two categories, $\frac{A1A2A3\ \ X}{A1'A2-\ \ -}$ and $\frac{-\ -\ -\ \ -}{A1'A2-\ \ -}$:

This means that every time the female gamete is A1A2A3, X spermatozoa do not fertilize, and can only effect amphimixia with - - - female pronuclei.

Knowing that the A1A2A3 complex is in the same chromosome as the B1 locus, it would appear that apparent incompatibility occurs between the X factor and a long genic complex, which makes the gametic incompatibility hypothesis we proposed more likely. Of course we need to discover more objective data concerning this phenomenon. We hope to obtain these data in the future and to reproduce analogous results to confirm previous observations.

## REFERENCES

PERRAMON, A. and MERAT, P. (1968a), *Ann. Biol. anim. Biochim. Biophys.*, **8** (4). 485–500.

PERRAMON, A. and MERAT, P. (1968b), *Proc. XIIth Int. Congr. Genet., Tokyo*, **1**, 277.

PERRAMON, A. and MERAT, P. (1970a), *XIth. Europ. Conf. Anim. Blood Groups Biochem. Polymorph.*, Warsaw, 1968, 373.

PERRAMON, A. (1969), *Bull. Biol. de la France et de la Belgique*, **3–4.** 489.

PERRAMON, A. and MERAT, P. (1970), *Ann. Génét. Sél. anim.*, **2** (in press).

## DISCUSSION

C. C. OOSTERLEE: Do you know if the phenomena you have described has been found in other species.

I do know that in the laboratory of Prof. van Laggtem in Amsterdam families with lethal factors are studied.

A. PERRAMON: I think also that in Japan, HIRARUMI is working in the same problem in humans, but I think that in humans it is very difficult to be sure that it is not very early zygotic mortality to explain absence of some categories.

# V. BLOOD GROUPS AND BIOCHEMICAL POLYMORPHISM IN HORSES

*XIIth Europ. Conf. Anim. Blood Groups Biochem. Polymorph., Bp., 1972 (pp. 525–526)*

# GENETIC CONTROL OF AN *IN VITRO* AUTOLYTIC FACTOR IN HORSE RED CELLS

YOSHIKO SUZUKI and C. STORMONT

Serology Laboratory, Deparment of Veterinary Microbiology
University of California, Davis, California 95616

Some years ago, when we began blood typing studies in an inbred herd of Arabians, we found that about half of the samples could not be completely typed because of extensive lysis in certain sections of the tests. In later samplings, which included many repeats, the frequency of untypeable cells remained high. However, the resamplings did provide us with an opportunity to relate this problem to the existence of an *in vitro* autolytic factor, which we will describe here.

The testing procedures used in these studies have been described elsewhere (STORMONT et al., 1963). We modified the .9% saline by adding 2% dextrose, which appeared to stabilize horse red cells and was particularly useful when blood samples had been stored longer than one or two weeks.

The repeated samplings showed that the untestable cells always originated from the same individuals. Apparently, some intrinsic property of the red cells of such individuals rendered the cells susceptible to lysis. Once the individual specificity of the lytic factor (LF) was established, our attention was directed towards the agent which triggered the lysis of the LF cells. Characteristically, LF cell samples had the appearance of being in poor condition, yet the cells washed up into stable suspensions were similar to those of normal cells. In the blood typing tests, the cells were stable in the agglutination tests and in the lytic tests mediated by guinea-pig complement, but not in the tests mediated by absorbed rabbit complement. Reabsorption of the rabbit complement with LF cells did not prevent the extensive hemolysis in these tests, thereby, ruling out rabbit complement as the sole lytic agent. The solution came unexpectedly in a cross match test involving a mare with normal cell type and a colt of LF cell type. Their sera were heated at 56°C for 30 minutes and the tests were run without complement; yet, the colt's cells (LF) were lysed by the mare's serum and by the autogenous serum serving as negative control. In subsequent tests of 7 LF cells and 25 normal cells (20 from outside the herd) only the LF cells reacted strongly with autogenous serum. Furthermore, the non-complement mediated lysis of LF cells was effected by all normal sera tested: horses (21), cattle (10), sheep (4), dogs (28), rabbits (pooled), guinea-pig (pooled) and chicken (pooled).

The heat stable lytic agent, which rarely exceeded titers of 1/8, was absorbable, but of low avidity as shown by the need for three serial absorptions before all reactivity was removed. Surprisingly, not only LF cells, but cells from normal horses as well as from all representative species tested, absorbed the lytic agent. These absorption results would suggest that the lytic agent may be related to the so called cold autoagglutinins which have been long recognized in man and animals.

Referring back to the blood typing results on the LF cells, the cells were stable in tests with reagents prepared by absorbing heteroimmune antisera because the lytic agent was completely removed. In contrast, reagents prepared from isoimmune antisera were usually under-absorbed with respect to the lytic agent. The same was also true for the absorbed rabbit complement which was used undiluted.

The genetic control of this trait was evident from the well documented pedigrees. Data from the pedigrees are summarized in Table 1. They provide evidence indicating that the autolytic trait (LF) is inherited as an autosomal recessive. Every animal of normal phenotype in the herd was found to be heterozygous for the trait.

TABLE 1

*Inheritance of an in vitro autolytic factor in horses*

| Number of matings | *Types of matings | Offspring of phenotype | |
|---|---|---|---|
| | | N | LF |
| 1 | N×N | 1 | 0 |
| 39 | N×LF | 20 | 19 |
| 23 | LF×LF | 0 | 23 |

* N, normal individual, LF, individual with autolytic factor

There were five presumptive cases of hemolytic disease of the newborn foal recorded in the herd, but these could not be directly associated with the autolytic trait. In fact, there appears to be no deleterious effect due to this trait, since the herd as a whole is reported to be quite normal.

Further studies in our laboratory indicate that a similar if not identical autolytic factor exists in dogs. However, in dogs the factor appears to be a species attribute, since every individual in a group of 28 experimental dogs of diverse origins possessed this property. Present efforts, directed towards the isolation of the lytic agent from normal serum, are expected to define the serological relationship between autolytic factors found in dogs and horses as well as the relationship — if it exists — between the lytic agent and normal auto-antibodies.

## REFERENCE

STORMONT, C. J., SUZUKI, YOSHIKO and RHODE, E. A. (1964), Serology of horse blood groups. *Cornell Vet.* **54**. 400–452.

*XIIth Europ. Conf. Anim. Blood Groups Biochem. Polymorph., Bp., 1972 (pp. 527–531)*

# POLYMORPHISM OF HEMOGLOBIN AND 6-PHOSPHOGLUCONATE DEHYDROGENASE IN HORSE ERYTHROCYTES

K. SANDBERG and S. BENGTSSON

Department of Animal Breeding, Agricultural College, 75007 Uppsala, Sweden

## SUMMARY

Hemoglobin and 6-PGD polymorphism in horse erythrocytes was studied by means of starch gel electrophoresis. Three hemoglobin phenotypes were observed. Extensive family data supported the genetic theory that the alleles $Hb^{m+}$ and $Hb^{m-}$ controlled the variation (BRAEND, 1967).

Six 6-PGD phenotypes were found. Presented family data were consistent with the interpretation that the 6-PGD phenotypes were controlled by three codominant, autosomal alleles, designated $PGD^D$, $PGD^F$ and $PGD^S$. Hemoglobin and 6-PGD gene frequencies in two Swedish horse breeds were calculated.

## INTRODUCTION

Since the electrophoretically different variants of hemoglobin were first observed (CABANNES and SERAIN, 1955) several reports on the genetics of horse hemoglobin have been published (BRAEND and EFREMOV, 1965; SCHMID, 1965; BRAEND, 1967). SCHMID (1965) suggested that the three phenotypes observed were controlled by a genetical system in which the two alleles $HB^A$ and $Hb^a$ were active. BRAEND (1967) proposed the existence of a modulating locus with the alleles $Hb^{m+}$ and $Hb^{m-}$ being responsible for the variation.

Genetic polymorphisms in the enzyme 6-phosphogluconate dehydrogenase (6-PGD) from red cells have been described in humans (FILDES and PARR, 1963), pigs (SAISON and GIBLETT, 1969), horses (BERGMAN and GUSTAVSSON, 1970) and also in other species. In most of the species studied three electrophoretic phenotypes have been observed; one with a single fast band, one with a single slow band and one composed of a fast, a slow and an intermediate band. This typical variation pattern is supposed to be caused by a dimeric structure of the enzyme (KAZAZIAN et al., 1965, CARTER et al., 1968). The 6-PGD variation has proved to be controlled by autosomal genes in all examined species except Drosophila, where the 6-PGD locus is X-linked (YOUNG, 1966).

The present report deals with a genetic study on hemoglobin and 6-PGD from horse erythrocytes.

## MATERIAL AND METHODS

Undiluted hemolysates were prepared by freezing and thawing fresh, packed horse erythrocytes, which had been washed three times. The lysates were centrifuged at about 3000 r.p.m. for 15 minutes to get unlysed cells and cell debris to settle. The samples were inserted in the gel on pieces of heavy filter paper. Starch gel electrophoresis was performed mainly according to GAHNE (1963). For the hemoglobin separation a slight modification of the discontinuous buffer system (pH 8.5) described by GAHNE (1966) was used. This system made it possible to determine the carbonic anhydrases (CA) simultaneously (SABDNERG, 1968). The hemoglobins were preferably classified without staining the gel. For the 6-PGD separation a continous phosphate buffer system (pH 6.8) was used (FILDES and PARR, 1963). After slicing the gel the 6-PGD activity was visualized according to the paint brush method of FITCH and PARR (1966).

The horse families in the hemoglobin study comprised 38 stallions and 552 mares with 863 offspring of the North-Swedish Horse breed. The inheritance of the 6-PGD types was studied on 37 stallions and 420 mares with 603 offspring also included in the hemoglobin study, and 44 stallions and 69 mares with 69 offspring of the Swedish Trotter breed, which mainly derives its origin from the American Trotter. Gene frequencies of the North-Swedish Horse were estimated from all parental animals in the inheritance studies and for the Swedish Trotter from 186 stallions and 78 mares, including the parental animals in the 6-PGD inheritance study. All blood samples were sent to our laboratory for routine bloodgrouping in connection with the pedigree control of trotters which is demanded by The Swedish Trotting Association.

For the hemoglobins the nomenclature suggested by BRAEND (1967) was used.

## RESULTS AND DISCUSSION

Three hemoglobin phenotypes were observed. They seemed to be identical with those reported by SCHMID (1965), BRAEND and EFREMOV (1965) and BRAEND (1967). The distribution of hemoglobin phenotypes among the progeny after 863 matings, representing the six possible mating types, is shown in Table 1.

The segregation ratios were, with one exception, in good agreement with the theory about a modulating locus as postulated by BRAEND (1967). They also fit the genetic theory advanced by SCHMID (1965), as m+ corresponds to A and m− corresponds to a. The matings between heterozygotes (m+m− × m+m−) gave a significant ($\chi^2$ = 8.61; 1 d.f.) deviation from expectation. So far we have not found any reasonable explanation of this fact. However, as the deviation decreased as the number of tested animals increased, we believe that the deviation was due to chance. There is nothing in the data published by other authors (SCHMID, 1966; BRAEND, 1967) that points to a similar deviation from expected segregation ratios. Five individual exceptions to the genetic theory were observed. They could all be ascribed to

TABLE 1

*Inheritance of hemoglobin phenotypes in 863 matings*

| Mating type | No. and phenotype of offspring | | |
|---|---|---|---|
| | m+m+ | m+m− | m−m− |
| m+m+×m+m+ | 183 | | |
| m+m+×m+m− | 163 | 172 | |
| m+m+×m−m− | | 72 | |
| m+m−×m+m− | 53 | 101 | 29 |
| m+m−×m−m− | | 40 | 39 |
| m−m−×m−m− | | | 11 |

erroneous registration on the basis of blood group factors or serum protein systems.

The hemoglobin gene frequencies of the two studied Swedish breeds are given in Table 2. There was a good agreement between observed genotypic frequencies and those expected on the basis of the Hardy-Weinberg law (Table 3).

TABLE 2

*Gene frequencies of hemoglobin and 6-PGD in two Swedish horse breeds*

| Breed | Hemoglobin | | | 6-PGD | | | |
|---|---|---|---|---|---|---|---|
| | $Hb^{m+}$ | $Hb^{m-}$ | *n* | $PGD^D$ | $PGD^F$ | $PGD^S$ | *n* |
| North-Swedish Horse | 0.678 | 0.322 | 590 | 0.092 | 0.907 | 0.001 | 457 |
| Swedish Trotter | 0.930 | 0.070 | 264 | 0.002 | 0.794 | 0.204 | 264 |

TABLE 3

*Observed and expected distribution of hemoglobin and 6-PGD types in two Swedish horse breeds*

| Breed | | Hemoglobin types | | | 6-PGD types | | | | | |
|---|---|---|---|---|---|---|---|---|---|---|
| | | m+m+ | m+m− | m−m− | DD | DF | DS | FF | FS | SS |
| North-Swedish Horse | Obs. | 268 | 264 | 58 | 4 | 76 | 0 | 376 | 1 | 0 |
| | Exp. | 271.2 | 257.6 | 61.2 | 3.9 | 76.2 | 0.1 | 376.0 | 0.9 | 0 |
| | | $\chi^2 = 0.361$ 1 d.f. | | | $\chi^2 = 0.104$ 1 d.f. | | | | | |
| Swedish Trotter | Obs. | 228 | 25 | 1 | 0 | 0 | 1 | 166 | 87 | 10 |
| | Exp. | 228.3 | 34.4 | 1.3 | 0 | 0.8 | 0.2 | 166.3 | 85.7 | 11.0 |
| | | $\chi^2 = 0.003$ 1 d.f. | | | $\chi^2 = 0.112$ 1 d.f. | | | | | |

Six different 6-PGD phenotypes, five of which are represented in Fig. 1 were observed. Three of them, Nos 3, 4 and 5 in Fig. 1, have been found by Bergman and Gustavsson (1970). They designated them F, FS and S respectively. Two new phenotypes appeared in the North-Swedish Horse breed. One of them (No. 1 in Fig. 1) consisted of a single band migrating faster than the F band and staining with the same intensity as the F and S types. We tentatively decided to designate this new type D. The second new type (No. 2 in Fig. 1) was composed of a D band, a F band and a third band between the other two but closer to the D band. The middle band stained slightly weaker than the other two bands. We decided to call this type DF. It should be pointed out that the DF type pattern is divergent from the typical heterozygous 6-PGD pattern. However, the appearence of the DF type is not contradictory to the theory that the enzyme has a di-

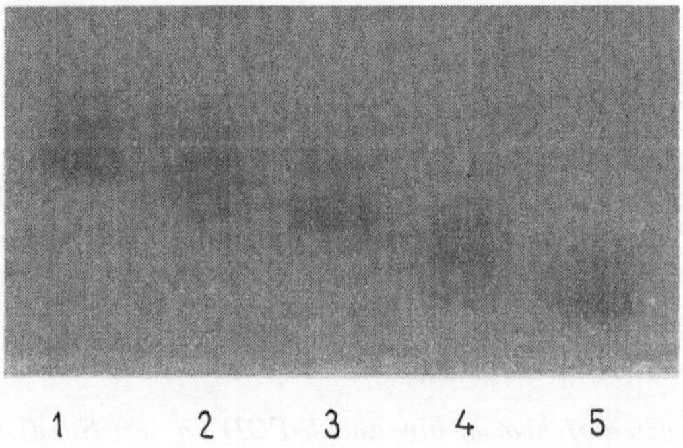
1
2
3
4
5

Fig. 1. A photograph showing five different 6-PGD types

meric structure. A third new 6-PGD phenotype, designated DS by us, was observed in one individual. It showed a triple band pattern composed of a D band, a S band and a third band midway between the other two. The middle band stained more intensely than the D and S bands.

Data on the inheritance of the 6-PGD phenotypes are compiled in Table 4. The observed types of offspring were consistent with the interpretation that the 6-PGD phenotypes were controlled by three codominant, autosomal alleles, $PGD^D$, $PGD^F$ and $PGD^S$. Two exceptions to this genetic theory were observed. On the basis of blood group antigens both of them proved to be due to erroneous registration.

The frequencies of 6-PGD genes in two investigated breeds are given in Table 2. The $PGD^S$ gene seemed to be very rare in the North-Swedish Horse as it was observed in a single individual only. The same thing was valid for the $PGD^D$ gene in respect to the Swedish Trotter breed. As one of the main purposes of the pedigree control of trotters in Sweden is to prevent crossing of the two breeds, the 6-PGD system will be of great value in these tests.

The observed number of animals in each 6-PGD genotype group agreed very well with the Hardy–Weinberg expectations (Table 3).

TABLE 4

*Inheritance of 6-PGD phenotypes in 672 matings*

| Mating type | No. and phenotype of offspring | | | | |
|---|---|---|---|---|---|
| | DD | DF | FF | FS | SS |
| DD×DF | | 1 | | | |
| DD×FF | | 3 | | | |
| DF×DF | 2 | 3 | 1 | | |
| DF×FF | | 74 | 59 | | |
| DF×FS | | | | 1 | |
| FF×FF | | | 480 | | |
| FF×FS | | | 16 | 19 | |
| FF×SS | | | | 4 | |
| FS×FS | | | 2 | 5 | 1 |
| FS×SS | | | | 1 | |

## REFERENCES

BERGMAN, H. and GUSTAVSSON, I. (1970), (In press).

BRAEND, M. and EFREMOV, G. (1965), Hemoglobins, haptoglobins and albumins of horses. Proc. 9th Europ. Anim. Blood Group Conf. Prague, 1964, 253–259.

BRAEND, M. (1967), Genetic variation of horse hemoglobin. *Hereditas*, **58**, 385–392.

CABANNES, R. and SERAIN, C. (1955), Etude électrophorétique des hémoglobines des mammifères domestiques d'Algérie. *C. R. Soc. Biol.* (Paris) **149**, 1193–1197.

CARTER, N. D., FILDES, R. A., FITCH, L. I. and PARR, C. W. (1968), Genetically determined electrophoretic variations of human phosphogluconate dehydrogenase. *Acta genet.* (Basel), **18**, 109–122.

FILDES, R. A. and PARR, C. W. (1963), Human red cell phosphogluconate dehydrogenase. *Nature*, **200**, 890–891.

FITCH, L. I. and PARR, C. W. (1966), Development of zymograms by the paint brush technique. *Biochem. J.* **99**, 20.

GAHNE, B. (1963), Inherited variations in the postalbumins of cattle serum. *Hereditas* **50**, 126–135.

GAHNE, B. (1966), Studies on the inheritance of electrophoretic forms of transferrins, albumins, prealbumins and plasma esterases of horses. *Genetics*, **53**, 681–694.

KAZAZIAN, H. H. JR., YOUNG, W. J. and CHILDS, B. (1965), X-linked phosphogluconate dehydrogenase in Drosophila: Subunit associations. *Science*, **150**, 1601–1602.

SAISON, RUTH and GIBLETT, E. R. (1969), 6-Phosphogluconic dehydrogenase polymorphism in the pig. *Vox. Sang.*, **16**, 514–516.

SANDBERG, K. (1968), Genetic polymorphism in carbonic anhydrase from horse erythrocytes. *Hereditas*, **60**, 411–412.

SCHMID, D. O. (1965), Über den Hämoglobin-Polymorphismus beim Pferd. *Z. Immunforsch.* **128**, 499–503.

SCHMID, D. O. (1966), Further progress in serogenetics in horses. Proc. Xth Europ. Conf. Anim. Blood Groups and Biochem. Polymorph. Paris, 1966, 339–343.

YOUNG, W. J. (1966), X-linked electrophoretic variation in 6-phosphogluconate dehydrogenase. *J. Hered.*, **57**, 58–60.

*XIIth Europ. Conf. Anim. Blood Groups Biochem. Polymorph., Bp., 1972 (pp. 533 – 536)*

# IMMUNOGENETIC STUDY OF THE MUR-INSULAN HORSES

LUBA PODLIACHOUK,* H. BALBIERZ,** MARIE KAMINSKI,*** MARIA NIKOŁAJCZUK** and ANNA STRZELECKA**

* Institut Pasteur, Paris
** Immunopathologic Laboratory Chair of Obsterics, Wroclaw, Poland
*** Laboratorie d'Enzymologie CNRS, 91 Gif-sur-Yvette, France

The Mur-Insulan horse, originâting from North-Western Yugoslavia, was a product of cross-breeding between local warm-blooded horses with alpine Noryks and Hungarian breeds. In the second half of the 19th century, they were in turn cross-bred with Ardennes and Belgian horses, which resulted in production of a robust horse, employed for tramways and coaches.

The breeding of Mur-Insulans in Poland started in 1940 (it is actually located in Moszna). These horses are easy to breed, hard working and resistant. They are mainly employed in agriculture (Fig. 1).

We determined the blood groups, the naturally occurring isohemagglutinins, the phenotypes of transferrins, erythrocyte acid phosphatases and esterases of 82 horses of known pedigree.

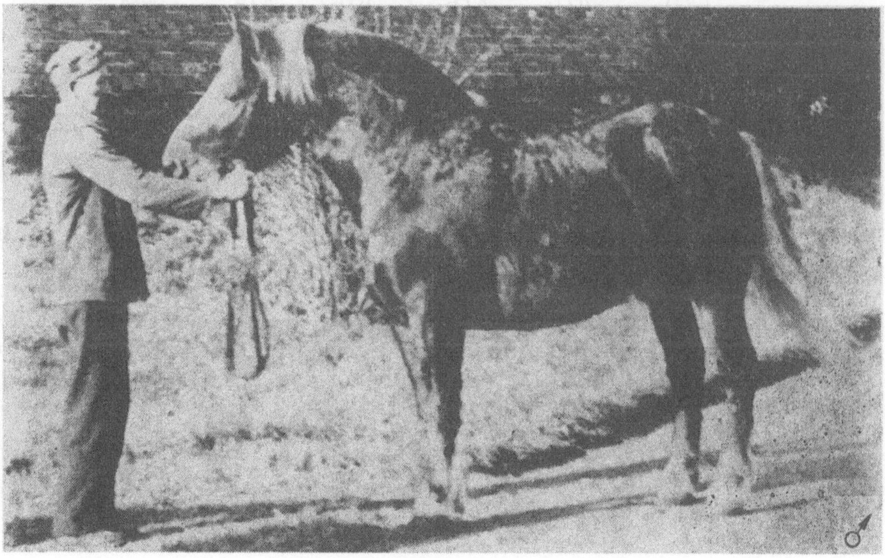

FIG. 1. Stallion Mur-Insulan "Kroat"

*The blood groups* have been determined in 80 animals, using 14 blood typing reagents: anti-$A_1$, C, $D_1$, E, F, G, $H_2$, $J_1$, $J_2$, K, $Fr_1$, $Fr_3$, $Fr_4$ and $Fr_5$. The anti-$A_1$, $D_1$, F, $J_1$ and $J_2$ are naturally occurring isoagglutinins; the anti-E, G, $H_2$, K, $Fr_3$, $Fr_4$ and $Fr_5$ are iso-immunantibodies (either agglutinating or hemolysing), the anti-C is a heteroagglutinin and anti-$Fr_1$ a hetero-immunagglutinin.

The frequencies of the erythrocyte antigens are as follows:

| | | | | |
|---|---|---|---|---|
| $A_1$ = 0.863 | C = 0.925 | $D_1$ = 0.370 | E = 0.062 | F = 0.963 |
| G = 0.000 | $H_2$ = 0.259 | $J_1$ = 0.000 | $J_2$ = 0.173 | K = 0.000 |
| $Fr_1$ = 0.654 | $Fr_3$ = 0.812 | $Fr_4$ = 0.037 | $Fr_5$ = 0.650 | |

Acid phosphatases

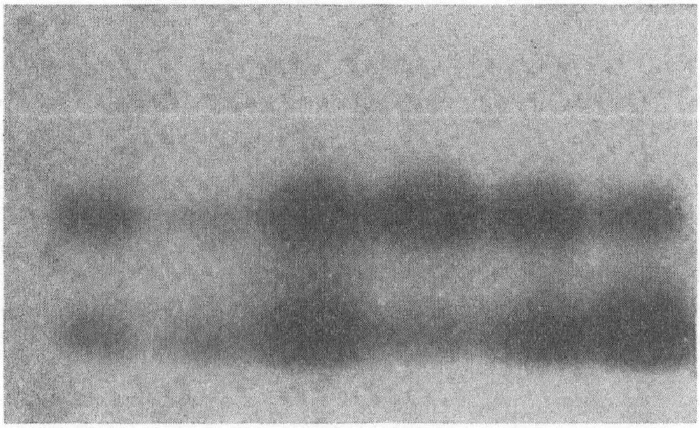

A/B A/B A/B B/B A/B A/A

FIG. 2. Acid phosphatases

Thus the most frequent factors in Mur-Insulans are $A_1$, C and F; the less frequent are E and $Fr_4$; the factors G, $J_1$ and K have not been encountered.

The genetic study of erythrocyte antigens was carried out on 48 complete families: 25 mares sired by 6 stallions gave birth to 48 offspring: 8 mares had 1 offspring each, 12 mares had 2 offspring each, 4 mares had 3 offspring each and 1 mare had 4 offspring. As to the 6 stallions, Petehöza had 14 offspring, Pamir 13, Zalawar 13, Kroat 5, Haber 2, and Kafr had 1 offspring.

*Naturally occurring isoagglutinins* have been found in sera of 14 horses out of 82 examined (16.9%). These were 9 anti-$A_1$, 3 anti-C and 2 anti-$Fr_6$.

*Transferrins:* in 84 sera 11 different phenotypes have been observed out of 21 theoretically possible phenotypes:

| | DF | DH | DO | FF | FH | FM | FO | HM | MO | HO | OO |
|---|---|---|---|---|---|---|---|---|---|---|---|
| number of cases: | 2 | 1 | 1 | 18 | 5 | 1 | 24 | 1 | 4 | 10 | 14 |

As shown above, the most frequent phenotypes are FO, FF and OO. The observed frequencies of 6 alleles are:

$$Tf^{D} = 0.0247 \quad Tf^{F} = 0.4197 \quad Tf^{H} = 0.1050$$
$$Tf^{M} = 0.0370 \quad Tf^{O} = 0.4136 \quad Tf^{R} = 0.0000$$

*Erythrocyte acid phosphatases* have been determined in 80 horses, using the same method as in man (HOPKINSON et al., 1963). Family study suggests that the erythrocyte acid phosphatases are controlled by 2 codominant alleles, named A and B.

Each allele produces a zymogram of two bands differing in intensity: the type A is composed of one weak fast band and one intense slow band. The type B is composed of one intense fast band and one weak slow band. The heterozygote AB has either 2 intense or 2 weak bands (Fig. 2).

The system of horse erythrocyte acid phosphatases closely resembles that of man, except the absence of allele C. In 80 examined samples we found 9 A, 28 B and 43 AB types.

*Esterases:* All examined sera contained $E_2$ and $E_4$ esterases. Concerning the $E_5$, it was present in trace amounts in two sera (one mare and her foal) contrasting with high amounts in all other samples (Fig. 3).

The blood groups of all the offspring are in agreement with the laws of Mendel. False filiation based on the determination of transferrins and erythrocyte acid phosphatases was found in one case; in a second case the determination of erythrocyte acid phosphatase alone permitted the conclusion.

The most frequent alleles of transferrin in the 82 examined Mur-Insulan horses are $Tf^{F}$ and $Tf^{O}$. This can be explained by in-breeding: among the stallions, 4 are FO, 1 is FF and 1 HO. 23 out of 25 mares have alleles $Tf^{F}$ and $Tf^{O}$ (9 are homozygotes and 14 heterozygotes).

In other breeds examined (Table 1) (Thoroughbred, Fjörding, Shetland ponies, Boulonnais) the frequency of $Tf^{F}$ is very close to that of Mur-Insulans. On the other hand, $Tf^{O}$ is either entirely absent, or very rare in the former breeds. $Tf^{R}$ absent in Mur-Insulans, has a low frequency in Thoroughbred and in Salernitan breeds, but a very high in Döle horse. Finally, $Tf^{M}$ rare in Mur-Insulans is also rare or absent in the other breeds examined.

TABLE 1

*Frequencies of transferrin alleles in various breeds*

| Alleles | Pony Shetland (6) | Thoroughbred (6) | Thoroughbred (5) | Boulonnais (5) | Döle (1) | Fjord (1) | Iceland (3) | Salernitain (2) |
|---|---|---|---|---|---|---|---|---|
| $Tf^{D}$ | 0.17 | 0.27 | 0.28 | 0.11 | 0.00 | 0.15 | 0.20 | 0.41 |
| $Tf^{F}$ | 0.46 | 0.56 | 0.53 | 0.44 | 0.23 | 0.62 | 0.27 | 0.39 |
| $Tf^{H}$ | 0.03 | 0.03 | 0.05 | 0.17 | 0.07 | 0.00 | 0.07 | 0.03 |
| $Tf^{M}$ | 0.03 | 0.00 | 0.00 | 0.01 | 0.00 | 0.00 | 0.01 | 0.00 |
| $Tf^{O}$ | 0.11 | 0.09 | 0.00 | 0.00 | 0.02 | 0.04 | 0.25 | 0.16 |
| $Tf^{R}$ | 0.20 | 0.05 | 0.13 | 0.28 | 0.67 | 0.19 | 0.20 | 0.01 |

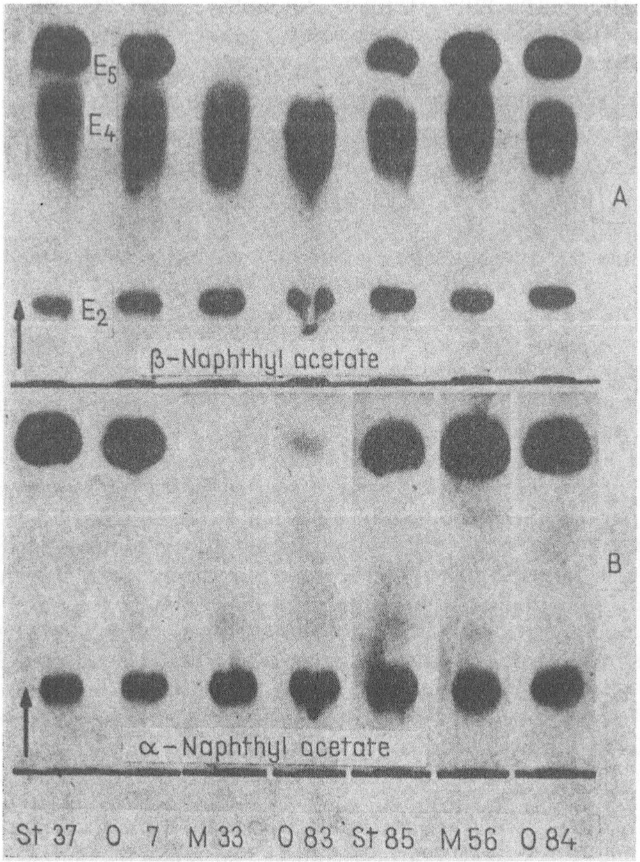
E5
E4
A
E2
β-Naphthyl acetate
B
α-Naphthyl acetate
St 37 O 7 M 33 O 83 St 85 M 56 O 84

FIG. 3. Family study of esterases:
offspring 7 from mare 33 sired by stallion 37
offspring 83 from mare 33 sired by stallion 85
offspring 84 from mare 56 sired by stallion 85

## REFERENCES

BRAEND, M. (1964), Serum types of Norvegian horses, *Nord. vet. Med.*, **16,** 363–373.

GAHNE, B. (1966), Studies on the inheritance of electrophoretic forms of transferrins, albumins, prealbumins and plasma esterases of horses. *Genetics*, **53,** 681–694.

HESSELHOLT, M. (1966), A study of blood groups and serum types of the icelandic horse. Proc. Xth Europ. Conf. Anim. Blood Groups Biochem. Polymorph., Paris, 1966, 325–331.

HOPKINSON, D. A., SPENCER, N. and HARRIS, H. (1963), Red cell acid phosphatase variants: a new human polymorphism. *Nature*, **199,** 969.

PODLIACHOUK, LUBA, MADEYSKA, ANNA and DEGUINES, D. (1966), Etude immunogénétique du cheval boulonnais. Proc. Xth Europ. Conf. Anim. Blood Groups Biochem. Polymorph., Paris, 1966, 333–338.

STORMONT, C. and SUZUKI, Y. (1965), Paternity tests in horses. *Cornell Vet.*, **55,** 365–277.

*XIIth Europ. Conf. Anim. Blood Groups Biochem. Polymorph., Bp., 1972 (pp. 537—540)*

# DETERMINATION OF BLOOD GROUPS AND TRANSFERRINS IN THREE BREEDS OF HORSES IN BULGARIA

D. DOBREV, LIDIA ERMENCOVA, R. KARAIVANOV and TS. TSANCOV

Research Institute of Animal Breeding, Kostinbrod, Academy of Agricultural Sciences
Higher Agricultural Institute "G. Dimitrov" Sofia, Bulgaria

## SUMMARY

The frequencies of blood group antigens and transferrins were determined in 280 horses of the Native Mountain, Pleven and Danube breeds.

The distribution of antigens amongst the three breeds is characterised for O and $J_1$ antigens which are most common in the Native Mountain breed, with a frequency of 0.7143 and 0.7714 whereas in the Danube breed the respective frequencies were 0.0851 and 0.2340. Factor E is more common in the Danube breed (0.3404) as compared to the other two breeds.

As far as transferrins are concerned the authors detected 6 alleles and all of the known 11 phenotypes were found exclusively in the Native Mountain breed. In the Pleven breed $Tf^O$ and in the Danube breed transferrin $Tf^M$ and $Tf^O$ were absent.

## INTRODUCTION

Thorough studies on the antigenic structure of erythrocytes in horses were undertaken by PODLIACHOUK (1957, 1965). PODLIACHOUK et al. (1960, 1962, 1966, 1967), SCHMID (1965, 1966) STORMONT et al. (1965) and others. Protein polymorphism and its genetic control have been investigated by BENGTSSON et al. (1968), BAER (1968), BUSCHMAN and SCHMID (1968), BRAEND and EFREMOV (1968), GRAETZER et al. (1965), GAHNE (1966), HESSELHOLT (1966) KRISTJANSSON (1962) and others but no such studies in horses have been carried out previously in Bulgaria.

Our aim was to determine the frequencies of blood group antigens and of serum transferrins in common horse breeds of the country, viz. native Mountain, Pleven and Danube breeds.

## MATERIAL AND METHODS

The study covered 280 horses 70 of which were of Native Mountain breed, 117 of Pleven breed and 94 of Danube breed.

The antigenic factors of erythrocytes were determined by 14 monospecific reagents (A, F, I, C, F, D, E, G, $J_1$, $J_2$, K, O, $Fr_1$, $Fr_3$ and BB). Isoimmunization of the horses, cross reactions, absorption and identification of reagents were performed according to PODLIACHOUK's technique (1957) with some slight modifications related to local conditions.

Electrophoretic investigations of serum transferrins were carried out by means of the horizontal electrophoresis technique of KRISTJANSSON (1962).

## RESULTS

The frequencies of blood group antigens in the three breeds of horses are shown in Table 1.

TABLE 1

*Frequency of antigenic factors in the erythrocytes of three horse breeds* (%)

| Antigens | Breeds | | |
|---|---|---|---|
| | Native mountain | Pleven | Danube |
| A | 88.5 | 93.6 | 81.9 |
| F | 88.5 | 92.6 | 81.1 |
| I | 92.6 | 97.4 | 81.2 |
| C | 55.7 | 88.7 | 79.8 |
| D | 71.4 | 18.9 | 8.5 |
| E | 15.7 | 13.8 | 34.0 |
| G | 70.0 | 70.7 | 32.9 |
| $J_1$ | 77.1 | 35.2 | 23.4 |
| $J_2$ | 75.7 | 64.6 | 41.5 |
| K | 60.0 | 53.4 | 81.9 |
| O | 47.1 | 22.4 | 39.3 |
| $Fr_1$ | 42.8 | 54.3 | 56.4 |
| $Fr_3$ | 32.8 | 39.6 | 36.2 |
| B | 87.3 | 84.5 | 57.7 |

Table 1 clearly shows that there were no breed differences in antigens A, F and I. There were differences in D, G and $J_1$ the highest percentage of frequency being established in the Native Mountain breed and the relatively lowest in the Danube breed. There were also some differences in antigen C found least often in the Native Mountain breed. Differences concerning the remaining antigenic factors were low and statistically not significant.

Data on distribution of transferrins phenotypes are shown in Table 2

TABLE 2

*Distribution of transferrin phenotypes in the three horse breeds*

| Breeds | Phenotypes in % | | | | | | | | | | | |
|---|---|---|---|---|---|---|---|---|---|---|---|---|
| | DD | DF | DH | DO | DR | FF | FH | FM | FR | HH | HR | MR |
| Native Mountain | 7.2 | 10.1 | 7.2 | 1.4 | 8.7 | 18.8 | 17.4 | — | 13.0 | 7.2 | 7.2 | 1.4 |
| Pleven | 1.8 | 23.2 | 6.2 | — | 7.1 | 33.9 | 6.2 | 0.9 | 17.8 | 0.9 | 0.9 | 0.9 |
| Danube | 1.4 | 40.0 | 1.4 | — | 2.8 | 25.7 | 8.5 | — | 14.3 | 1.4 | 4.3 | — |

TABLE 3

*Gene frequency of the alleles*

| Allele | Breeds | | |
|---|---|---|---|
| | Native Mountain | Pleven | Danube |
| $Tf^D$ | 0.210 | .201 | .236 |
| $Tf^F$ | .391 | .580 | .571 |
| $Tf^H$ | .232 | .076 | .086 |
| $Tf^M$ | .007 | .009 | — |
| $Tf^O$ | .008 | — | — |
| $Tf^R$ | .152 | .134 | .107 |

As to the estimation of gene frequencies of transferrin alleles (Table 3) the allele $Tf^F$ is more common in the Pleven and Danube breeds and rare in the Native Mountain breed. While the allele $Tf^H$ is significant in the Native Mountain horse, it is rare in the other two breeds.

## CONCLUSIONS

Comparing our findings on the antigenic factors of erythrocytes in the three breeds of horse with those of PODLIACHOUK et al., a similarity of the results was found as far as factors A, I and C are concerned. It should be noted that factor D, which PODLIACHOUK et al. failed to identify in the English thoroughbreds found little of it in Arabian and Poznan horses and a high occurrence in the Primitive Polish horse, was mostly present in the Native Mountain breed in Bulgaria.

As to transferrin alleles, STORMONT et al. (1965) and GAHNE (1966) established 16 phenotypes out of which we detected so far 11 in the Native Mountain and Pleven breeds and 9 in the Danube breed. STORMONT et al. (1965) and others described 6 transferrin alleles and BENGTSSON et al. (1968) have described 5 (without $Tf^M$). We detected all the 6 alleles in the Native Mountain breed; in the Pleven breed $Tf^O$, and in Danube horse $Tf^M$ and $Tf^O$ are absent.

The following conclusions were drawn:

In all the three breeds of horses the blood group factors A, F and I occur in a comparatively wide range. The factors D and $J_1$ are more significant in the Native Mountain horses and comparatively insignificant in the Danube breed.

The Native Mountain horses, which are much more the product of natural selection, display a greater variety, particularly of transferrin phenotypes and alleles. The Pleven and Danube breeds which are the result of a more purposeful selection possess a smaller variety of transferrin alleles. This applies particularly to the Danube breed which is genetically more balanced.

## ACKNOWLEDGEMENTS

We wish to thank most cordially Dr. Luba PODLIACHOUK Pasteur Institute, Paris, for the typing of our reagents derived from 89 horses from the farm of the Higher Agricultural Institute "G. Dimitrov" in Sofia.

## REFERENCES

BENGTSSON, GAHNE, B. and RENDEL, J. (1968), Genetic Studies on Transferrins, Albumins, Prealbumins and Esterases in Swedish Horses, *Acta Agr. Scand.*, **18**, 60–64.

BAER, A. (1968), Lapo-transferrin du sérum équin; son importance dans la typisation phenotypique, *Schweiz. Arch. Tierheil.*, **110**, 463–466.

BUSCHMAN, H. and SCHMID, D. O. (1968), Serumgruppen bei Tieren, Berlin.

BRAEND, M. and EFREMOV, G. (1965), Haemoglobins, Haptoglobins and Albumins of Horses, Blood Groups of Animals, Proc. 9th Europ. Conf. Anim. Blood Groups, Prague, 1964, 253–259.

GRAETZER, MARY ALICE, HESSELHOLT, M., MOUSTGAARD, J. and THYMANN, MARY (1965), Studies on protein polymorphism in pigs, horses, and cattle. Proc. 9th Europ. Conf. Anim. Blood Groups, Prague, 1964, 279–293.

GAHNE, B. (1966), Studies on the inheritance of electrophoretic forms of transferrins, albumins, pre-albumins and plasma esterases of horses, *Genetics*, **53**, 681–694.

HESSELHOLT, M. (1966), Studies on blood serum types of the Icelandic horses. *Acta Vet. Skand.*, **7**, 206–225.

KRISTJANSSON, F. K. (1962), Recent research in serum protein polymorphism of livestock., VIIIth Europ. Blood Group Conf., Ljubljana (Mimeo).

OSTERHOFF, D. R., SCHMID, D. O. and WARD-COX, I. S. (1970), Blood group and serum type studies in Basuto Ponies. Proc. XIth Europ. Conf. Anim. Blood Groups, Biochem. Polymorph. Warsaw, 1968, 453–457.

PODLIACHOUK, LUBA (1957), Thèse de doctorat, Paris.

PODLIACHOUK, LUBA, KACZMAREK, A. and ZWOLINSKI, J. (1962), Les groupes. sanguins de chevaux de six races de Pologne, *Ann. de l'Inst. Pasteur*, **103**, 943–949.

PODLIACHOUK, LUBA (1965), Les groupes sanguins et sériques des équins. Paris.

PODLIACHOUK, LUBA, WADOWSKA, I. WADOWSKI, ST. (1967), Badania nad grupami krwi koni hodowlanijch w województwie Olstynskim, *Roczniki Nauk Rolniczych*, **89**, 477–489.

PODLIACHOUK, LUBA (1965), The blood groups of equidae, Proc. 9th Europ. Conf. Anim. Blood Groups, Prague, 1964, 229–235.

PODLIACHOUK, LUBA, SALERNO, A. and LABERT, D. (1966), Les groupes sanguins du cheval salernitain, *Ann. de l'Inst. Pasteur*, **110**,

PODLIACHOUK, LUBA, SIRBU, Z., KOWNACKI, MARIA and SREWONSKA, L. (1960), Les groupes sanguins des chevaux, *Ann. de l'Inst. Pasteur*, **98**,

SCHMID, D. O. (1965), Blood group studies in horses, Proc. 9th Europ. Conf. Anim. Blood Groups, Prague, 1964, 237–243.

SCHMID, D. O. and MANCI, D. (1966), Blutgruppenuntersuchungen beim Bosnischen Gebirgspferde, *Zentbl. Vetmed.*, **13**, 75–77.

STORMONT, C., SUZUKI, YOSHIKO and RENDEL, J. (1965), Application of blood typing and protein tests in horses, Proc. 9th Europ. Conf. Anim. Blood Groups, Prague, 1964, 221–228.

*XIIth Europ. Conf. Anim. Blood Groups Biochem. Polymorph., Bp., 1972 (pp. 541 – 546)*

# QUANTITATIVE STUDIES ON HORSE HEMOGLOBINS

D. R. OSTERHOFF and I. S. WARD-COX

Department of Zootechnology, Faculty of Veterinary Science, University of Pretoria
Onderstepoort, Rep. of South Africa

## SUMMARY

An attempt is made to explain the phenomenon of the double-banded appearance of horse hemoglobins in starch gel. Densitometer readings were taken at regular time intervals to study the possible changes in the relationship of the one hemoglobin band to the other. Influences of sex, performance, condition and diet were studied, and a new theory of the inheritance of the horse hemoglobin is put forward.

Oxygen equilibria curves of blood samples from horses belonging to different hemoglobin types are presented showing interesting differences between these animals. The importance of these findings is discussed.

## INTRODUCTION

Hemoglobin variants in animals are of great interest not only in clinical surveys pertaining to certain anaemias and hemopoetic abnormalities (NEETHLING et al., 1969) but also in phylogenetic surveys especially in wild animals (OSTERHOFF and KEEP, 1970; OSTERHOFF et al., 1970). While in most species the mode of inheritance can easily be deduced from the migrating bands, horse hemoglobins have provided some difficulties due to their double-banded appearance in the electrophoretogram. BRAEND (1968) and SCHMID (1965) reported on a single-banded horse hemoglobin type, but in many breeds this type is extremely rare or completely absent. In the present study previous work (OSTERHOFF, 1966; OSTERHOFF and WARD-COX-1967) was followed up by quantitative studies to verify the mode of inheritance suggested by various workers, whereby the horse hemoglobin classification in $A_1$, $A_1A_2$ and $A_1^+A_2^-$ was retained. In the later stage of the experimental work donkeys and mules were included as controls of the experiments performed on horse blood samples.

## MATERIAL AND METHODS

Three groups of horses were available from the various units at the Veterinary Research Institute, comprising 20 brood mares, 20 riding horses and 20 horses used for horse sickness challenging experiments. Blood specimens were taken in ACD solution for the purpose of electrophoretic examination and in heparin for biochemical analyses. The former were taken once a month over a period of twelve months while the latter were collected only twice during the last two months of the experiment.

*Electrophoresis.* Having prepared the hemolysates by adding four drops of three times washed packed cells to 2 ml of distilled water and spinning down the ghost cells for six minutes, they were subjected to electrophoresis using the buffers of BUSCHMANN (1963). Electrophoresis was stopped after 1 1/2 hour, when the two migrating bands had separated sufficiently to allow a clear space in between. After staining and complete destaining each gel was dehydrated over night in pure methanol and placed in a mixture containing six parts benzyl alcohol and one part glycerol. Being another night in this mixture the gels were sufficiently transparent for densitometry.

*Densitometry.* The gels were placed between two Kodak photofilm glass plates, ensuring that no air bubbles were trapped. They were then scanned manually on a Photovolt Densitometer, calibrating for zero between the two bands and recording the peak results only. Filters were chosen so as to give density readings between 2 and 7, depending on the degree of background staining. The concentration of the bands was calculated as percentage of the $A_2$ band, following the example of BRAEND (1968).

*Biochemistry.* Hematocrit, hemoglobin, blood sugar, blood urea nitrogen, glutamic oxalacetic transaminase, serum glutamic pyruvic transaminase and total plasma protein determinations were carried out on the samples collected over the last two months of the brood mare and riding horse groups.

*Blood oxygen dissociation.* Following the densitometric analyses samples were selected for the determinations of the blood oxygen dissociation curves. Making use of the Photometer Acid-Base Analysis System, incorporating the Blood Micro System, Acid-Base Analyser and the $pO_2$ electrode, six points of the dissociation curve were determined at oxygen concentrations of 0.95%, 1.90%, 3.80%, 5.70%, 7.60% and 9.50% corresponding to the oxytension at 38°C and 760 mm Hg of 6.8 mm, 13.5 mm, 27.0 mm, 40.5 mm, 53.8 mm and 67.5 mm Hg respectively. Following every $pO_2$ determination, the oxygen saturation levels were read on a Photometer Oximeter after hemolysing the specimens with Triton X-100 in capillary tubes sealed at either end. Using the $pO_2$ and oxygen saturation values, oxygen dissociation curves were drawn not only for horse, but also for mule and donkey plus human control samples, taking note of the actual blood pH in every case.

## RESULTS

From the 60 animals only four — two brood mares and two riding horses — belonged to the $A_1^+A_2^-$ hemoglobin type. Using the results of other horses tested earlier (OSTERHOFF and WARD-COX, 1967), it is obvious that so far not a single horse has been found in South Africa belonging to the $A_1$ type (Table 1).

The correct typing of the horse possessing the extremely rare type $A_1^-A_2^+$ was confirmed earlier (OSTERHOFF, 1966).

TABLE I

*Percentage of hemoglobin types in different horse breeds*

| Breed | No. tested | Hemoglobin type | | | |
|---|---|---|---|---|---|
| | | $A_1$ | $A_1A_2$ | $A_1{}^+A_2{}^-$ | $A_1{}^-A_2{}^+$ |
| Common horses | 228 | 0 | 97.4 | 2.2 | 0.4 |
| Arabs | 46 | 0 | 89.1 | 10.9 | 0 |
| Percheron | 45 | 0 | 77.8 | 22.2 | 0 |
| Thoroughbred | 54 | 0 | 100.0 | 0 | 0 |
| Welsh and Welsh cross | 18 | 0 | 100.0 | 0 | 0 |
| Experimental group | 60 | 0 | 93.4 | 6.6 | 0 |

Hemoglobin typing was performed on all horses and densitometric readings were taken on the transparent gels every month throughout the period of one year. The percentage of $A_2$ varied between 34 and 40 with an average of 38% in the horses belonging to the $A_1A_2$ group and between 15 and 23 with an average of 18% in the four animals belonging to the $A_1{}^+A_2{}^-$ hemoglobin type. These percentages were never exactly the same but showed certainly no seasonal variation as can be seen in Fig. 1 on four typical horse samples, two belonging to each type. The readings on mule samples belonging to the $A_1{}^-A_2{}^+$ type taken from January onwards are also depicted in Fig. 1.

The deviations obtained are apparently due to errors in the technique, and it might be stated that the percentage of $A_2$ is very constant and not influenced by season. No differences between the average $A_2$ values in the $A_1A_2$ types of brood mares, riding horses and experimental horses could be established. Age, sex and feeding did not influence the percentage of $A_2$

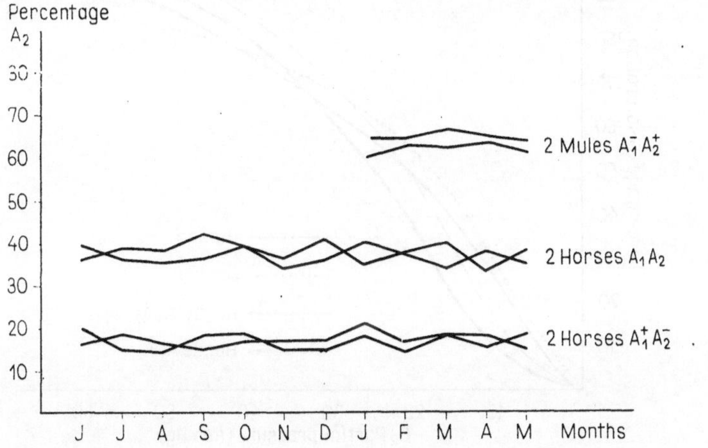
Percentage
A2
80
70
60
50
40
30
20
10
2 Mules A1- A2+
2 Horses A1 A2
2 Horses A1+ A2-
J J A S O N D J F M A M Months

FIG. 1. Quantitative studies of the equine $A_2$ hemoglobin band

TABLE 2

*Average hematological values in horses of different hemoglobin type*

| Hb type | No. of animals | Hb g % | Ht % | Bl. Sugar mg % | B. U. N. mg % | SGOT K. U. | SGPT K. U. | TPP g % |
|---|---|---|---|---|---|---|---|---|
| $A_1^+A_2^-$ | 4 | 11.2 | 37 | 77.3 | 10.8 | 199 | 25 | 5.8 |
| $A_1A_2$ | 36 | 11.3 | 36 | 68.0 | 12.4 | 190 | 22 | 5.5 |

either and samples taken before and after exercise of horses for the determination of $pO_2$ values, as expected did not differ in the percentage of $A_2$. In six foals from the mares mentioned the percentage of $A_2$ seemed to be somewhat higher, but the hemoglobin type inherited was certainly the same as that of their parents.

The average hematological values for 36 horses belonging to the $A_1A_2$ and the four animals belonging to the $A_1^+A_2^-$ hemoglobin types, are given in Table 2.

It is almost impossible to compare the values of four animals with those of 36 and no significant differences could be established.

HUISMAN (1966) showed considerable differences in the oxygen affinity of adult and fetal blood samples of cattle, sheep and goat and also of animals belonging to different hemoglobin types. Since the horse was excluded in his studies, attempts were made to compare the oxygen affinity of the two hemoglobin types under discussion. The technical difficulties experienced during this part of the work will not be reported here, but the reliable

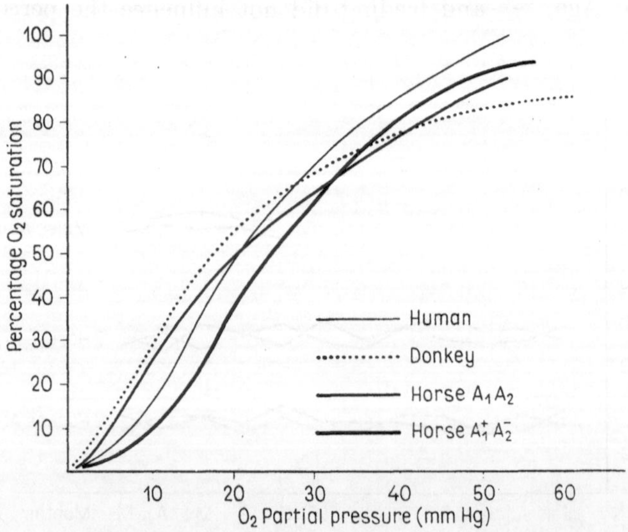
100
90
80
70
60
50
40
30
20
10
Percentage O2 saturation
10
20
30
40
50
60
O2 Partial pressure (mm Hg)
Human
Donkey
Horse A1A2
Horse A1+ A2-

FIG. 2. Oxygen equilibria curves for equine blood

repeatability of results was not easily obtained. Typical blood oxygen dissociation curves of horses of different hemoglobin types are presented in Fig. 2 together with the curves of a human blood sample and donkey blood belonging to the equine hemoglobin type $A_2$.

One should expect that all curves would run parallel but interesting differences could be established, since in the region of lower partial oxygen pressure the blood samples of the hemoglobin type $A_1A_2$ and the donkey control showed a higher oxygen affinity than that of human blood which was reversed in the region of higher partial oxygen pressure. The oxygen affinity curves of blood belonging to the hemoglobin type $A_1^+A_2^-$ paralleled that of human blood, but provided lower values throughout.

## DISCUSSION AND CONCLUSIONS

In an attempt to correlate the two hemoglobin components with any physiological character it was necessary to ensure that all possible environmental influences could be excluded. It can be stated that the two bands appearing in the great majority of all horses are inherited and not influenced by age, condition, sex, stage of pregnancy, feeding or season, to the conclusion partly also come by BRAEND (1968).

The correlation between hemoglobin types and hematological values could not give any conclusive results due to the unbalanced numbers in the groups. This part of the study will be followed up by comparing the hematological values of all possible equine hemoglobin types, i.e. types $A_1$, $A_1{}^+A_2{}^-$, $A_1{}^-A_2{}^+$ and $A_2$ in reasonably well represented numbers, since interesting preliminary results were obtained from mules with type $A_1{}^-A_2{}^+$ and donkeys with type $A_2$ showing lower Hb, lower hematocrit, higher blood sugar, higher B.U.N. and clearly lower total plasma protein values in these animals, indicating lower metabolic rates and diminished levels of physiological functions. It is hoped to find horses with the type $A_1$ to have the complete range of all types for these studies.

Future amino-acid replacement studies in hemoglobins, similar to those performed by KILMARTIN and CLEGG (1967) should also include the mule and donkey hemoglobin to come to clearer conclusions regarding the genetic evolutionary theories.

The same applies for the studies of oxygen affinity; also here all five hemoglobin types should be included since interesting differences could be established not only in the oxygen equilibria curves (see Fig. 2) but also in the direct $pO_2$ measurements. Twelve horses of hemoglobin type $A_1A_2$ at rest averaged a partial pressure of 39 mm Hg $pO_2$, six mules 32 and six donkeys only 25 mm Hg. The same horses directly after exercise averaged a value of 57, and the $pO_2$ measurements of arterial blood averaged 86 mm Hg. The field is wide open and a direct relationship between hemoglobin types and oxygen pressure and binding affinity could possibly be verified.

Regarding the mode of inheritance of equine hemoglobins, the different suggestions made by BRAEND and EFREMOV (1965), SCHMID (1965) and BRAEND (1968) do not cover the hemoglobin type $A_1{}^-A_2{}^+$ and therefore, it is suggested that the hemoglobins could be inherited in the simplest form of multiple-gene heredity. Two gene pairs might act in a cumulative way, whereby each additional gene adds a certain percentage of hemoglobin to the $A_2$ band. The cross between $A_1A_2$ and $A_1A_2$ could be demonstrated by using the $h_1h_1H_2H_2 \times H_1H_1h_2h_2$-combination whereby each capital letter stands for a specific percentage of hemoglobin $A_2$. In this way all possible combinations in the equine hemoglobins could be explained. Further work, however, is necessary to explain the overwhelming frequency of the $A_1A_2$ type in all horse breeds.

## REFERENCES


BRAEND, M. (1968), *Hereditas*, **58,** 385.
BRAEND, M. and EFREMOV, G. (1965), Proc. 9th Europ. Conf. Anim. Blood Groups, Pague, 1964, 253.
BUSCHMANN, H. (1963), *Zbl. Vet. Med.*, B **10,** 49.
HUISMAN, T. H. J. (1966), Proc. Xth Europ. Conf. Anim. Blood Groups Biochem. Polymorph., Paris, 1966, 61.
KILMARTIN, J. V. and CLEGG, J. B. (1967), *Nature*, **213,** 318.
NEETHLING, L. P., BROWN, J. M. M., OSTERHOFF, D. R., DE WET, P. J. and WARD-COX, I. S. (1969), *Jl. S. Afr. Vet. Med. Ass.*, **40,** 121.
OSTERHOFF, D. R. (1966), Proc. Xth Europ. Conf., Anim. Blood Groups Biochem. Polymorph., Paris, 1966, 345.
OSTERHOFF, D. R. and WARD-COX, I. S. (1967), Proc. S. Afr. Soc. Anim. Prod., 218.
OSTERHOFF, D. R. and KEEP, M. R. (1970), *The Lammergeyer* **10,** 50.
OSTERHOFF, D. R., YOUNG, E. and WARD-COX, I. S. (1972), Proc. XIIth Europ. Conf. Anim. Blood Groups Biochem. Polymorph., Budapest, 1970.
SCHMID, D. O. (1965), *Z. Immun.-Allergieforsch.*, **128,** 499.

*XIIth Europ. Conf. Anim. Blood Groups Biochem. Polymorph., Bp., 1972 (pp. 547–550)*

# INFLUENCE OF BREEDING ASPECTS ON SERUM ALBUMIN AND TRANSFERRIN GENE FREQUENCIES IN HUNGARIAN THOROUGHBRED HORSES

PÁL SOÓS, J. STUKOVSZKY and PÉTER SOOS

Center of Artificial Insemination, Laboratory of Immunogenetics, Budapest, Hungary

## SUMMARY

208 blood samples were examined for transferrin and albumin types. 12 transferrin phenotypes, controlled by 5 alleles ($Tf^D$, $Tf^F$, $Tf^H$, $Tf^O$ and $Tf^R$) and 3 albumin phenotypes, controlled by two alleles ($Al^F$ and $Al^S$) were found.

Significant deviation was observed ($P < 0.01$) from the Hardy–Weinberg equilibrium.

The possibility of erroneous paternity exclusion was 40.6% by transferrin and 14.1% by albumin examination, respectively.

The $Tf^D$ and $Tf^O$ deviations from other Thoroughbreds are explained by the breeding policy used.

## INTRODUCTON

Several genetic variations have been demonstrated in horse serum proteins. BRAEND and STORMONT (1964) found 16 different transferrin types, which were controlled by six codominant autosomal alleles, as follows: $Tf^D$, $Tf^F$, $Tf^H$, $Tf^M$ $Tf^O$ and $Tf^R$. This theory was confirmed by many authors.

STORMONT and SUZUKI (1963) studied the horse albumins and described a pair of codominant alleles, designated by $a^A$ and $a^B$ respectively. The same variations were observed by BRAEND (1964) and GAHNE (1966), who used the symbols $Al^F$ and $Al^S$. Since then the latter symbols have been generally used.

In this report the transferrin and albumin variations observed in the Hungarian Thoroughbred horses are presented and compared with the results published by other authors.

## MATERIAL AND METHODS

Blood samples were collected from 208 Hungarian Thorough bred mares covering about 30% of the total number of mares in the breed; from the stallions being used and from the offspring. Most of the animals examined are stationing at the State Farm of Bábolna. Blood samples were withdrawn from the jugular vein and after 24-hour storage they were centrifuged and the sera — after pipetting — were stored at —20°C until used.

The samples were examined with horizontal starch-gel electrophoresis, partly with the method of MAKARECHIAN and HOWELL (1966), partly using a tris-citric acid gel-buffer of pH 6.8 and a vessel-buffer of pH 8.7 as described by KRISTJANSSON (1963), which makes possible the simultaneous determination of transferrins and albumins. Albumins were examined with the method of EFREMOV et al. (1968).

## RESULTS AND DISCUSSION

In the population examined we found 12 transferrin phenotypes, controlled by 5 different alleles. We did not find the $Tf^M$ allele, which is in good agreement with other authors' observations (BENGTSSON et al., 1968; GAHNE, 1966; OSTERHOFF and WARD-COX, 1967; and STORMONT et al., 1965).

Table 1 shows the gene frequencies of transferrin alleles compared with the observations of others. The high frequency of the allele $Tf^D$ and $Tf^O$ in Hungarian Thoroughbreds is obvious. Table 2 shows the result of the test

TABLE 1

*Transferrin and albumin gene frequencies of Thoroughbred horses, observed by different authors*

| Authors | Transferrins | | | | | Albumins | |
|---|---|---|---|---|---|---|---|
| | $Tf^D$ | $Tf^F$ | $Tf^H$ | $Tf^O$ | $Tf^R$ | $Al^F$ | $Al^S$ |
| BENGTSSON et. al. (1968) | 0.231 | 0.639 | — | 0.076 | 0.054 | 0.408 | 0.592 |
| OSTERHOFF et. al. (1967) | 0.167 | 0.648 | 0.009 | 0.046 | 0.130 | 0.278 | 0.722 |
| STORMONT et. al. (1965) | 0.266 | 0.564 | 0.027 | 0.090 | 0.053 | 0.210 | 0.790 |
| GAHNE (1966) | 0.410 | 0.390 | 0.030 | 0.160 | 0.010 | 0.340 | 0.660 |
| In present study | 0.313 | 0.471 | 0.019 | 0.101 | 0.096 | 0.216 | 0.784 |

TABLE 2

*The observed and expected distribution of transferrin and albumin phenotypes among 208 Thoroughbred horses*

| | Transferrin | | | | | | | | |
|---|---|---|---|---|---|---|---|---|---|
| | DD | DF | DH | DO | DR | FF | FH | FO | FR |
| Obs. | 20 | 63 | 5 | 11 | 11 | 45 | 1 | 20 | 22 |
| Exp. | 20.32 | 61.26 | 2.51 | 13.14 | 12.50 | 46.18 | 3.77 | 19.80 | 18.74 |

| | Transferrin | | | | | | Albumin | | |
|---|---|---|---|---|---|---|---|---|---|
| | HH | HO | HR | OO | OR | RR | FF | FS | SS |
| Obs. | 0 | 0 | 2 | 3 | 5 | 0 | 8 | 74 | 126 |
| Exp. | 0.09 | 0.81 | 0.78 | 2.13 | 4.04 | 1.93 | 9.73 | 70.52 | 127.75 |

$\chi^2$ for transferrins 10.86 ($P < 0.01$), $\chi^2$ for albumins 0.48

for the Hardy–Weinberg equilibrium. There is a significant deviation between the observed and expected values.

In Table 3 the theoretical detectability of the erroneous paternity cases was calculated with the method of JAMIESON (1965). Using combined systems, a considerably high exclusion percentage is possible in the population. The transferrin system, however, proved to be more efficient than any other protein system.

TABLE 3

*The expected percentage of paternity exclusions using various protein systems by different authors*

| Author | | Breed | System | | | All systems combined |
|---|---|---|---|---|---|---|
| | | | Transferrin | Albumin | Esterase | |
| BENGTSSON et al. | 1968 | Gotland ponies | 32.0 | 11.1 | 14.6 | 48.1 |
| BENGTSSON et al. | 1968 | North-Swedish | 48.0 | 17.1 | 19.4 | 65.2 |
| OSTERHOFF et al. | 1967 | Thoroughbred | 24.6 | 16.0 | 5.0 | 39.9 |
| STORMONT et al. | 1965 | Thoroughbred | + | + | — | 44.1 |
| STORMONT et al. | 1965 | Shetland ponies | + | + | — | 60.0 |
| In present study | | Thoroughbred | 40.6 | 14.1 | — | 49.0 |

All three possible albumin phenotypes were present in our material. The results are in agreement with corresponding findings of other authors.

The deviations observed in Tables 1 and 2 can be explained by the breeding tendency. It has been established that the following stallions were preferred during the last 10 years:

| Name | Transferrin type | Albumin type |
|---|---|---|
| Imi | DF | SS |
| Imperial | DO | SS |
| Nostradamus | DO | FS |

all of which were carrying the $Tf^D$ and $Al^S$ alleles. In the first five years *Imi* was used very frequently, which is in good agreement with the observation that 50% of the examined animals possess Tf FF and Tf DF phenotypes.

## REFERENCES

BENGTSSON, S., GAHNE, B. and RENDEL, J. (1968), Genetic studies on transferrins, albumins, prealbumins, and esterases in Swedish horses. *Acta Agric. Scand.*, **18**, 60.
BRAEND, M. (1964), Serum types of Norwegian horses. *Nord. Vet. Med.*, **16**, 353.
BRAEND, M. and STORMONT, C. (1964), Studies on haemoglobin and transferrin types of horses. *Nord. Vet. med.*, **16**, 31.
EFREMOV, G., VASKOV, B. and HRISOHO, R. (1970), Inherited variations in the prealbumins of sheep serum. Proc. XIth Europ. Conf. Anim. Blood Groups Biochem Polymorph., Warsaw, 1968, 505–511.

GAHNE, B. (1966), Studies on the inheritance of electrophoretic forms of transferrins, albumins, prealbumins and plasma esterases of horses. *Genetics*, **53**, 681.
JAMIESON, A. (1965), The genetics of transferrins in cattle. *Heredity* (Lond.), **20**, 419.
KRISTJANSSON, F. K. (1963), Genetic control of two prealbumins in pigs. *Genetics*, **48,** 1059.
MAKARECHIAN, M. and HOWELL, W. E. (1966), Improved technique for separation and identification of bovine beta globulins by starch-gel electrophoresis. *Canad. J. Biochem.*, **44,** 1089.
OSTERHOFF, D. R. and WARD-COX, F. S. (1967), A preliminary horse breed comparison with regard to haemoglobin and serum type polymorphism. *Proc. S. Afr. Soc. Anim. Prod.*, **6,** 218.
STORMONT, C. and SUZUKI, Y. (1963), Genetic control of albumin phenotypes in horses. *Proc. Soc. Exp. Biol. Med.*, **114,** 673.
STORMONT, C., SUZUKI, Y. and RENDEL, J. (1965), Application of blood protein tests in horses. Proc. 9th Europ. Conf. Anim. Blood Groups, Prague, 1964, 221–228.

*XIIth Europ. Conf. Anim. Blood Groups Biochem. Polymorph., Bp., 1972 (pp. 551—553)*

# IMPROVED SEPARATION OF POLYMORPHIC ESTERASES IN HORSES

A. M. Scott

Equine Research Station of the Animal Health Trust, Newmarket, Suffolk, England

## SUMMARY

Using a low ionic strength, low pH, Tris/Citric buffer, in starch gel, improved separation of the polymorphic esterases is obtained, which reveals the existence of two ne bands.

## INTRODUCTION

Previous reports on the separation of the polymorphic esterases of horses have shown the inheritance of these enzymes (Es system) to be controlled by four alleles F, I, S and O (Gahne, 1966; Bengtsson et al., 1968). Here, two buffer systems were used, one at pH 8.5, the other at pH 5.4; with this technique, the esterase bands run fairly close together. This preliminary, technical paper, describes a technique for improving the clarity of separation of these esterases.

## MATERIAL AND METHODS

The electrophoresis was carried out as described by Kristjansson (1963). Stock solutions "A" (10.5 g citric acid per litre) and "B" (23 g Tris per litre) were used, as described by Tucker et al. (1967) for preparing the gel buffer. This consisted of 12.0 ml of A and 4.0 ml of B made up to 250 ml with distilled water. The pH of this buffer is 4.05; a starch concentration of 10% is used (Connaught Medical Research; batch No. 273.1). The tank buffer contained 18.55 g of boric acid and 4 g of NaOH per litre (pH 8.70).

Gel dimensions were 21.7 cm × 12.8 cm × 0.45 cm and the insert line was 4.0 cm from the cathode end; 0.8 cm of gel were exposed at each end for contact with sponge rubber wicks.

A constant current of 16.0 mA was applied for 5 minutes, after which the inserts were removed.

A constant current of 8.0 mA was then applied until the borate boundary had migrated 9.0 cm from the insert line (this takes approximately 3 hours).

The gels were sliced and stained as described by Gahne (1966), except that methanol was not used in the reaction mixture. The bands were allowed to stain lightly (approximately one minute), and then the gel was fixed in methanol.

## RESULTS

The photo (Fig. 1) shows the results obtained using these conditions. Each allele appears to instruct for one main band and two subsidiaries, one faster and one slower (in fact, if the staining is prolonged, two additional bands appear, in front of, and behind the last two mentioned bands). There are clearly two types of bands in the F region and two in the I region. I have tentatively labelled the band slower than F, the "G" band, and the band faster than I, the "H" band.

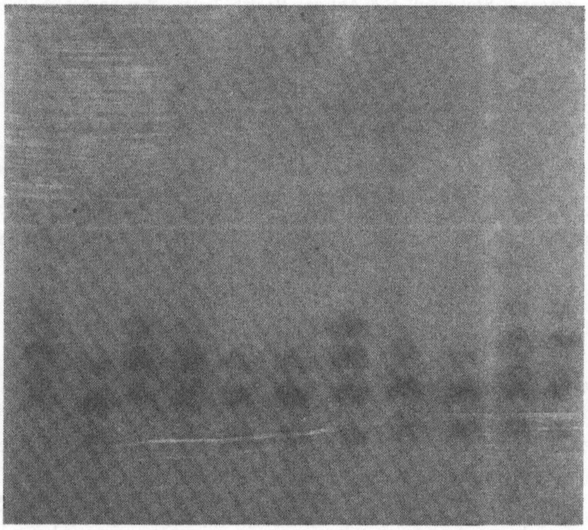
FI II FG FG II II GI HI II FI FI

FIG. 1. Starch gel; pH 4.05, stained with α-naphthyl acetate + fast Garnet GBC

## DISCUSSION

Some phenotypes have been shown diagrammatically in Fig. 2 and a study of the patterns suggests that each main band corresponds in position to a subsidiary of another band type. For instance, the band labelled "H" corresponds to the slow "F" band, "I" corresponds to the slow subsidiary of the band labelled "G" and so on. Heterozygote patterns involve six bands (two main and four subsidiary), unless any two bands occupy the same position. For example GI is a four band pattern because the slow G corresponds to I and the fast I to G.

If the observations of band patterns are correct, there should be a main band in the position of the slow band labelled "H". In fact, SANDBERG (1968) described to me and sent me a sample of serum with a band controlled by an allele he has called "X". This corresponds to the slow subsidiary of the band labelled "H". An FX heterozygote has a five band pattern because the slow F and the fast X subsidiaries fall in the same position (Fig. 2).

Limited family data involving offspring of three stallions of types Fl, Gl and Hl are consistent with the interpretation that the new bands described are controlled by separate alleles to those hitherto reported.

A full investigation to ascertain the frequencies of the alleles is being undertaken.

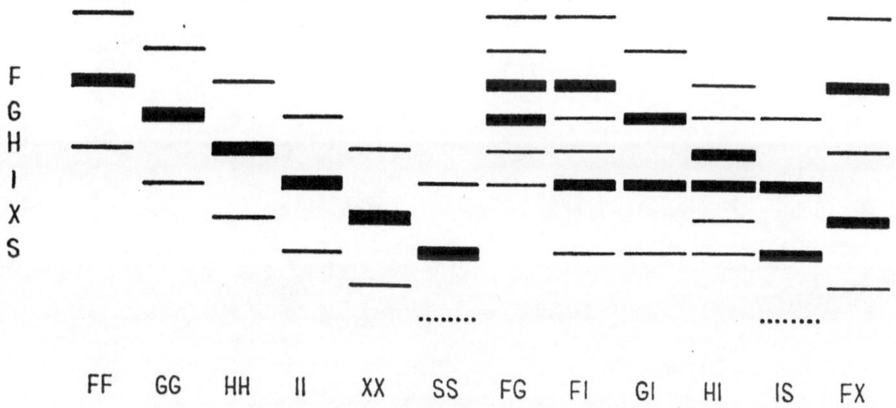
F
G
H
I
X
S
FF
GG
HH
II
XX
SS
FG
FI
GI
HI
IS
FX

FIG. 2. Diagram to illustrate various esterase phenotypes

## ACKNOWLEDGEMENTS

I should like to thank Dr. E. M. TUCKER for helpful advice.

## REFERENCES

BENGTSSON, S., GAHNE, B. and RENDEL, J. (1968), Genetic Studies on Transferrins, Albumins, Prealbumins and Esterases in Swedish Horses. *Acta Agric. Scand.*, **18**, 60–64.

GAHNE, B. (1966), Studies on the Inheritance of Electrophoretic Forms of Transferrins, Albumins, Prealbumins and Plasma Esterases of Horses. *Genetics*, **53**, 681–694.

KRISTJANSSON, F. K. (1963), Genetic Control of Two Prealbumins in Pigs. *Genetics*, **48**, 1059–1063.

SANDBERG, K. (1968), Personal communication.

TUCKER, ELIZABETH, M., SUZUKI, Y. and STORMONT, C. (1967), Three New Phenotypic Systems in the Blood of Sheep. *Vox Sang.*, **13**, 246–262.

# VI. BLOOD GROUPS AND BIOCHEMICAL POLYMORPHISM IN SHEEP AND GOATS

*XIIth Europ. Conf. Anim. Blood Groups Biochem. Polymorph., Bp., 1972 (pp. 557–559)*

# THE INFLUENCE OF ANTIGEN TYPE ON ACTIVE POTASSIUM TRANSPORT IN THE RED CELLS OF SHEEP AND GOATS

ELIZABETH M. TUCKER and J. C. ELLORY

A.R.C. Institute of Animal Physiology, Babraham, Cambridge, Great Britain

## INTRODUCTION

There is now good evidence that the M-L blood group system is involved in the active potassium transport mechanism in sheep red cells (RASMUSEN and HALL, 1966; ELLORY and TUCKER, 1969a; TUCKER and ELLORY, 1970). Sheep which are homozygous for the gene controlling the presence of the red cell antigen M are also homozygous for the gene controlling high potassium (HK) levels, whereas those sheep which are homozygous for the gene for the L antigen are also homozygous for the gene for low potassium (LK) levels. Heterozygous sheep have LK type red cells because the gene for LK is almost completely dominant to that for HK. Although M and L antigens are present on these heterozygous red cells, the serological reactivity of the cells is much stronger with the M reagent than with the L reagent, and the latter reaction may be so weak as to be barely detectable.

HK type sheep red cells have a 3–4 times greater ($Na^+$-$K^+$) activated ATPase activity (TOSTESON, MOULTON and BLAUSTEIN, 1960), a 3–4 times increase in active $K^+$ transport (TOSTESON and HOFFMAN, 1961), and a larger number ouabain-binding sites (DUNHAM and HOFFMAN, 1969) than LK type red cells. ELLORY and TUCKER (1969a) showed that when LK type sheep red cells were sensitized by incubating *in vitro* with anti-L, the active $K^+$ transport rate and ATPase activity were stimulated to the same level or above that found in HK type red cells. In addition, more $^3H$-ouabain binding sites were exposed (ELLORY and TUCKER, 1969b, 1970a) whilst the apparent affinity of the transport system for $K^+$ remained unchanged. It was therefore postulated that the L antigen acts as an inhibitor of active potassium transport in LK type red cells and that the effect of anti-L is to unmask more pump sites on the cell and restore a high rate of active cation transport. RASMUSEN and HALL (1966) described the existence of HK and LK type red cells in goats and indicated that the M antigen system may also be connected with potassium types in this species. This paper presents the results of a comparative study of the effect of anti-L on sheep and goat red cells.

## METHODS

These have been described elsewhere (ELLORY and TUCKER, 1970b; TUCKER and ELLORY, 1970).

## RESULTS AND CONCLUSIONS

We tested red cells from 246 goats and found that potassium values ranged from 25–110 mmole/litre red cells and the distribution was consistent with there being two potassium types, HK type (227 of the goats) with levels greater than 65 mM and LK type (19 goats) with less than 60 mM. The red cells of 132 of these goats, including the 19 LK individuals, were tested for the presence of the L and M antigens. Red cells from 116 of the goats, including some HK and LK individuals were hemolysed by the M reagent. The L reagent however did not hemolyse any of the red cells, not even those of the LK type. Ten LK and 9 HK goats were subsequently selected for more detailed observations on their red cells.

The rate of active potassium transport was found to be higher in the HK type goat red cells (0.733 $\pm$ 0.067 m-moles/l p. cells/hour) than in the LK type cells (0.311 $\pm$ 0.035 m-moles/l p. cells/hour), and also ($Na^+$-$K^+$) activated ATPase activity was higher (0.02 for LK and 0.067 $\mu$mole Pi/mg protein/hour for HK type cells).

Treatment of LK type goat red cells with anti-L stimulated active potassium transport and ATPase activity in these cells, but the degree of stimulation was more variable than in LK sheep red cells treated with anti-L. The binding of $^3$H-ouabain was also stimulated by treatment of LK goat red cells with anti-L (Table 1). It was of considerable interest that anti-L had this effect on these goat red cells because we could find no evidence serologically that this antibody was in fact bound to the cells.

Absorption of L-reagent with LK sheep red cells removed both the serological activity for sheep red cells and the ability to stimulate the potassium pump of LK sheep and goat red cells. Absorption of L reagent with LK goat red cells removed the pump stimulatory activity for goat red cells, but did not affect the serological activity of anti-L against sheep LK red cells. This absorbed reagent still had the power to slightly stimulate active $K^+$ transport in LK sheep red cells, but the major part of this activity was lost (Table 2).

Red cell membranes from LK sheep and goat red cells were treated with anti-L and the antibody was then eluted by ether extraction. The eluates

TABLE 1

*The effect of anti-L on the active potassium uptake, $Na^+$—$K^+$ activated ATPase activity and ouabain binding capacity of goat and sheep red cells. Results for one individual of each type*

| Red cell type | Ouabain sensitive $K^+$ uptake (m-moles/l packed cells/hour) | | $Na^+$—$K^+$ activated ATPase activity ($\mu$ moles Pi-mg protein/hour) | | Bound ouabain (molecules/cell) | |
|---|---|---|---|---|---|---|
| | Control | anti-L treated | Control | anti-L treated | Control | anti-L treated |
| LK goat | 0.309 | 0.705 | 0.020 | 0.043 | 16.1 | 26.8 |
| HK goat | 0.798 | 0.763 | 0.067 | 0.069 | 31.2 | — |
| LK sheep | 0.123 | 0.890 | 0.016 | 0.072 | 8.4 | 25.4 |
| HK sheep | 0.442 | 0.446 | 0.181 | 0.174 | 41 | 42 |

TABLE 2

*The effect of anti-L after absorption with LK sheep and goat red cells or after elution from LK sheep and goat red cells, on active $K^+$ uptake and hemolysis of LK sheep and goat red cells*

| | Stimulation of active $K^+$ uptake | | Hemolytic Test | |
|---|---|---|---|---|
| | *LK sheep red cells | LK goat red cells | *LK sheep red cells | LK goat red cells |
| Anti-L absorbed with HK goat or sheep red cells | + | + | + | − |
| Anti-L absorbed with LK sheep red cells | − | − | − | − |
| Anti-L absorbed with LK goat red cells | ± | − | + | − |
| Anti-L eluted from LK sheep red cells | + | + | + | − |
| Anti-L eluted from LK goat red cells | + | + | ± | − |

Results scored non-parametrically
+ = Positive reaction
− = Negative reaction
± = Weak positive reaction
* = The LK sheep red cells were homozygous

were tested both for their serological activity and for their ability to stimulate active $K^+$ transport-in LK sheep and goat red cells. The eluates from LK goat red cells stimulated active potassium transport in both the sheep and goat red cells and also weakly hemolysed the sheep but not the goat red cells. The eluates from LK sheep red cells stimulated the active $K^+$ flux in both sheep and goat red cells, and hemolysed the sheep, but not the goat red cells.

To explain these results, we put forward the tentative hypothesis, that there are two specificities of the L antigen, Ls and Lp. Specificity Lp is concerned with active $K^+$ transport, and Ls with strong serological activity. Our anti-L reagent has both specificities. According to this hypothesis, LK sheep red cells have Lp and Ls and goat red cells only Lp. Support to this theory is given by the fact that different batches of anti-L obtained from the same sheep during several immunization courses, although increasing sequentially in serological titre when tested against LK sheep red cells, did not show a comparable increase in their ability to stimulate active $K^+$ transport. Presumably the antibody which was showing a rise in titre was anti-Ls and not anti-Lp.

## REFERENCES

DUNHAM, P. B. and HOFFMAN, J. F. (1969), *Fed. Proc.*, **28**, 339.

ELLORY, J. C. and TUCKER, ELIZABETH M. (1969a), *Nature*, **222**, 477.

ELLORY, J. C. and TUCKER, ELIZABETH M. (1969b), *Proc. Int. Conf. Biological Membranes Stresa.* North Holland (in press).

ELLORY, J. C. and TUCKER, ELIZABETH M. (1970a), *J. Physiol.*, **208**, 18.

ELLORY, C. and TUCKER, ELIZABETH M. (1970b), *Biochim. Biophys. Acta* (in press).

RASMUSEN, B. A. and HALL, J. G. (1966), Proc. Xth Europ. Conf. Anim. Blood Groups Biochem. Polymorph., Paris, 1966, 453.

TOSTESON, D. C., MOULTON, R. H. and BLAUSTEIN, M. (1960), *Fed. Proc.*, **19**, 128.

TOSTESON, D. C. and HOFFMAN, J. F. (1961), *J. Gen. Physiol.*, **44**, 169.

TUCKER, ELIZABETH M. and ELLORY, J. C. (1970), Animal Blood Groups and Biochemical Genetics, **1** (in press).

*XIIth Europ. Conf. Anim. Blood Groups Biochem. Polymorph., Bp., 1972 (pp. 561–566)*

# NEW ASPECTS OF SHEEP BLOOD GROUPS

D. O. SCHMID

**Institute of Animal Blood Group and Resistance Research, Livestock Breeding Research Organization, Munich, GFR**

In the Institute of Blood Group and Resistance Research in Munich research of sheep blood groups has been conducted for nearly four years. According to the experiences of STORMONT and RASMUSEN in sheep and our own experiences in cattle, we first tried to produce the necessary test reagents by isoimmunization of sheep. Valuable assistance was extended by RASMUSEN who typed 50 sheep with the reagents anti-A, B′, E, $I_x$, N, Y, $P_2$, $O'_1$, $O'_x$, M + $M_x$, $M_x$, R, O, Z and D, and gave us the reagents. Although these reagents reacted only very weakly with the blood of our experimental animals and did not include all features, we had not to immunize without any direct orientation. As experimental animals we disposed of two herds with a total number of 300 sheep belonging to the breeds Merino, Texel, Schwarzkopf (Black-face) and Bergschaf (Mountain-sheep). During the last years we immunized 160 sheep, 44 of which could be reimmunized once and 15 twice. For first immunization, we immunized weekly 8 times with 10.0 ml ACD-blood intramuscularly, or with the equivalent amount of blood as red cell-deepfrozen sample. Blood for serum samples was collected one week after the last injection. In average 180–220 ml serum was obtained from one adult sheep. Sampling could be repeated earliest after one week. Yearlings are not particularly suitable for serum production being on the one hand very bad antibody-producers, on the other hand, they respond often with collapse to withdrawal of larger amounts of blood. Reimmunization — 6 injections weekly — were carried out after 3 months. 44 sheep out of 160 formed no antibodies during the first immunization. Only two sheep failed to produce antibodies on repeated immunization as well. Already OSTERHOFF (1960) stated that the formation of antibodies in sheep was often unsatisfactory after the first immunization.

The antibody-titers of the unabsorbed raw sera varied between 1/2–1/1024 after the first immunization, in one case they even reached 1/4096. After reimmunization, a titer was 1/4–1/2048, being usually by 2–5 dilutions higher. Also in those sheep, which did not respond by hemolysin production to the first immunization, titers of 1/512–1/2048 were often assessed after the second immunization, which followed 3 months later. After 12 months, there was a persistent titer level of 1/2–1/32 (Table 1).

TABLE 1

*Production of blood group isoantibodies in sheep*

| Antibody-titer after the first immunization period (8 injections) | Antibody-titer after the re-immunization period (6 injections) | Antibody-persistence after 12 months |
|---|---|---|
| 1/2 | 1/512 | 1/8 |
| — | 1/256 | 1/8 |
| 1/4 | 1/1024 | 1/2 |
| 1/8 | 1/1024 | 1/32 |
| 1/4 | 1/32 | — |
| — | 1/2048 | ? |

In two sheep we found D-specific agglutinins with titers of 1/4 after the first immunization.

It was interesting to observe in isoimmune sera of sheep the frequent existence of prozone-phenomenon of the immune hemolysins. The undiluted serum reacted negatively, while in 1/4 dilution the same serum produced a complete lysis. As source of complement undiluted, absorbed rabbit-serum was used. The hemolysis-reaction was allowed to take place at 27°C, and was read after 3-hour incubation. After numerous absorption-experiments we succeeded in producing 36 monovalent test reagents. The classification was effected in accordance with the 1st International Sheep-reagent Comparison test, proposed by RASMUSEN (Table 2). We isolated anti-A, in the B-system anti-B, $I_1$, $I_2$, $I_{x1}$, $I_{x2}$, N, $N_x$, $P_1$, $P_2$, $P_3$, $P_4$, $P_x(S_{26})$, $Y_1$, $Y_2$, B′, I′, $O_1'$, $O_2'$, $O_x'$, Mü-16, Mü-42, Mü-68, Mü-70, Mü-72, Mü-73 and Mü-75. The B-bloodgroup system is similar to that of cattle, which is a system of multiple alleles. Up to now we identified 30 different B-phenogroups (Table 3). In the C system we isolated anti-$C_x$, in the M system anti-$M_1$, $M_2$, $M_3$ and $M_x$, in the D system anti-D and in the R-O-system anti-$R_1$,

TABLE 2

*Blood group reagents for sheep*

Comparison—Test 1967/68
Rasmusen—Urbana—USA

| | | | | | | | | | |
|---|---|---|---|---|---|---|---|---|---|
| A | USA-D | USA-U | | D-Mü | F | GB | IND | PL | |
| B′ | | USA-U | BG ? | D-Mü | F ? | GB | | | |
| E′ | USA-D | | BG | D-Mü | F | GB | | | |
| R | USA-D | USA-U | BG | D-Mü | F | GB | IND | PL | |
| O | USA-D | USA-U | | | F | GB | IND | | |
| S | USA-D | USA-U | | | | | | | $ZA_{21}$ |
| U | USA-D | | | | | GB | | | $ZA_{33, 60}$ |
| $I_x$ | USA-D | | | D-Mü | | | | | |
| C | USA-D | USA-U | $BG_9$ | | F | GB | IND | | $ZA_{47}$ |
| $C_x$ | USA-D | USA-U | | D-Mü | F | | IND | | |
| $S_{26}$ | USA-D | USA-U | BG | D-Mü | F | GB | IND | | ZA |

TABLE 3

*B-phenogroups in sheep*

| | |
|---|---|
| $BI_1NP_1B'O'_x$ 42 70 75 | $Y_2B'$ |
| $BI_1P_1B'O'_x$ 72 | $Y_2B'$ 70 |
| $BI_1P_1B'O'_x$ 42 | $Y_2O'_x$ |
| $BI_1P_1B'O'_x$ 42 70 75 | $Y_2$ 70 |
| $BNY_2O'_x$ | |
| $BY_2$ | $B'O'_x$ |
| $BY_2B'$ 70 | $B'O'_x$ 75 |
| $BB'O'_x$ 16 70 75 | |
| | |
| $I_1$ | $O'_x$ |
| $I_2$ | $O'_x$ 75 |
| $I_{x1}$ | |
| $I_{x1}N_x$ | 42 73 |
| $I_1NY_2B'O'_x$ 16 | 68 |
| | |
| $P_1O'_x$ | |
| $P_1$ 72 | $Y_1$ |
| $P_2$ | $Y_1$ 68 |
| $P_1B'O'_x$ | |
| $S_{26}Y_2B'$ 70 75 | |

a natural antibody which has been demonstrated in normal sera of many sheep, and anti-$R_x$. We were not able to isolate anti-X and anti-Z from sheep-isoimmune sera. Our anti-Mü-24 comprises an antigenic spectrum similar to the anti−Z. Studying the inheritance of blodd groups of known systems we found that some of our isoimmune sera prepared in sheep did not react with blood groups of these sheep but referred to blood groups of systems not yet known. The first one of these systems includes the blood group factors Mü-24 and Mü-29, which do not belong to a closed blood group system. Two other systems were up to now represented by the blood group factors Mü-10, Mü-76 and Mü-41, respectively. We isolated from cattle isoimmune sera anti-E′ (from anti-$E'_3$) $I_x$ (from $I_2$) $R_2$ (from K) and Y (from Y′).

We isolated from a sheep isoimmune serum an antibody for the detection of a bloodgroup factor, which can be identified also by anti-T′ of cattle. Therefore we called also this factor as T′.

Extensive research work was done to clarify to what extent sheep and cattle were antigenically related as stated by RASMUSEN (1966) (Table 4). The isohemolysin anti-$J_1$ of cattle contrary to all expectations, failed to detect the R-antigen of sheep in the hemolysis-reaction, but detected it in the agglutination reaction. In general, the R-antigen can be identified exclusively by the agglutination technique. With anti-R-sera of higher titres (1/8) it may be possible to detect the R-antigen by hemolysis. The same is true for anti-D. The anti-$E'_3$ of cattle occasionally corresponds to the anti-E′ of sheep in respect of hemolysis. Anti-Y′ of cattle enables detec-

TABLE 4

*Similarity of blood groups in sheep and cattle*

(RASMUSEN, 1966)

| Sheep | Cattle |
|---|---|
| $B'$ | $B'$ |
| $O_1'$ | $O_1'$ |
| $O_x'$ | $O_x'$ |
| $I'$ | $I_1$ |
| $I_x$ | $I_2$ |
| $E'$ | $E_2$ |
| $P_2$ | $P_3'$ |
| $Y$ | $Y'$ |
| S | ($I'$ ?) |
| C | R |
| $C_x$ | $RX_2$ |
| M | $S_2$ and $U_2$ |

tion of sheep of blood type Y. The results shown in Table 5 are based exclusively on internationally verified isoimmune sera of cattle, recommend a strict reservation in interpreting. No doubt, that there are cross-reactions between blood groups of sheep and cattle, especially in the blood group system B, but we know very little concerning the exact relation and the structure of partial antigens. These questions can only be clarified by means of extensive cross absorptions. This is a promising task for an immunogeneticist.

During the recent weeks we demonstrated by means of the extract of the protein gland of the edible snail *Helix pomatia*, a bloodgroup-antigen unknown up to now and present on the surface on sheep red cells (Table 6). The natural antibody in the secretion of the protein gland of the snails belongs to the group of protectins, the anti-$A_{hel}$ reacts in a very strong way with the erythrocytes of most of the sheep. Indications for this, that

TABLE 5

*Reactions between sheep red cells and cattle blood group isoimmune-antisera*

Positive hemolytic reactions with the red cells of some sheep

$A_2$ B $G_1$ $G_2$ K $I_1$ $I_2$ $O_1$ $O_3$ $P_2$ Q $T_1$ $T_2$ $Y_1$ $A_1'$ $B'$ $D'$ $E_1'$ $E_3'$ $F'$ $G'$ $I'$ $J_1'$ $J_2'$ $K'$ $O'$ $P'$ $Y'$ $G''$ Gő-10 Gő-12 $C_1$ $C_2$ E $R_1$ $R_2$ W $X_1$ $X_2$ $C'$ $L'$ F V $J_1$ L M $S_1$ $H'$ U $U'$ $U_2'$ $H''$ $S''$ Z $T'$ $A_2'$

Positive hemolytic reactions with the red cells of all sheep

$A_1$ $A_2$ H $Z'$ $G_1$ $G_3$ $I_1$ $O_1$ $O_2$ $Y_2$ $E_2'$ $S''$ $S'$

Negative reactions

$Z'$ $R_1$ $O_3$ $U''$ $A_2'$

Different reaction-patterns with different bloodtyping reagents of the same specificity

$A_2'$ $D'$ $O'$ $C_2$ E $R_1$ F L M

TABLE 6

*Helix-antiboides*

| | |
|---|---|
| **Helix pomatia** | **Anti – $A_{HP}$ (=Anti – αβ – N – Ac – Galm)** |
| **Anti – $A_{hel}$** | **Anti – $A_{HN}$ (= Anti – α – N – Ac – Galm)** |
| **Helis nomoralis** | **Anti – $A_{HN(HH)}$** |
| | **Anti – $M^{duc}_{HN}$ (= Anti – α – N – Ac – Neuramingl – )** |

is to say the quantitative differences in the determination of characteristics were obtained by proving an obviously feeble agglutination in a mother and her daughter. Since a short time we dispose of sheep the blood cells of which do not react with the antibody anti-$A_{hel}$. Family studies have shown that this antigen called Hel is genetically determined and of autosomal and codominant inheritance. From this point of view, sheep with strong features are homozygote positive Hel/Hel, and sheep with weak characters are heterozygote positive Hel/hel and all other sheep the red cells of which show no reaction are homozygote negative hel/hel (Table 7).

The assumption of HAFERLAND, KIM, UHLENBRUCK and NELSON (1967), that perhaps there may be an identity between the Hel-receptor and the blood group antigen R, can surely not be supported.

Blood group research is surely not only of theoretical interest, but also has a great practical importance in Germany.

Utilizing 47 different test reagents in order to prove the sheep blood groups we examined 2500 sheep. On the occasion of elite auction sales for rams in Bavaria only those animals which are progeny-tested for fattening and slaughtering performances are controlled by blood groups since more than one year. These controls are absolutely necessary. The verification of the descendance of sheep at our Munich institute showed that 16% of all controlled animals had wrong pedigrees. After it had become known, that we were able to control the submitted pedigrees serologically, the percentage of wrong pedigrees could be decreased by 6.5%. Bloodgrouping can, therefore, be considered an important educational factor for the breeders.

TABLE 7

*Hel-antigen in sheep*

| | Hel | hel |
|---|---|---|
| | + | – |
| Hel × Hel<br>+ + | + | + |
| Hel × hel<br>+ – | + | + |
| hel × hel<br>– – | – | + |

It is of great importance for bloodgroup-serologists, that blood cell mosaicism tends to become frequent, which is explicable by the high frequency of twins in modern sheep breeding. STORMONT, WEIR and LANE (1953) stressed already the different amounts of hemolysin in different sheep and showed the possibility of the existence of such a mosaic. Before excluding a sheep from the pedigree point of view it is absolutely necessary to effect an absorption test concerning all blood group factors important in respect of the exclusion, otherwise we run the risk of a negligent judgement. In Munich we could already observe parents of sheep, where some blood groups could only be detected by means of absorption.

## REFERENCES

HAFERLAND, W., KIM, Z., UHLENBRUCK, G. and NELSON, D. S. (1967), Zur Frage der Einheitlichkeit des Agglutinins Anti-$A_{hel}$, *Z. Immun. forsch.*, **132**, 93.
OSTERHOFF, D. R., Blood group studies in sheep, *Immunogenetica Edinburgensis* 1960–89.
PROKOP, O. (1969), Einige aktuelle Probleme der Blutgruppenkunde, *Klin. Wschr.*, **47**. 605.
PROKOP, O., UHLENBRUCK, G. and KÖHLER, W. (1968), Protectine eine neue Klasse antikörperähnlicher Verbindungen, *Das Deutsche Gesundheitswesen*, **23**, 318.
PROKOP, O. and UHLENBRUCK, G. (1969), N-Acetyl-D-Galaktosamin an Tumorzellmembranen: Nachweis mittels Helix-Agglutininen, *Med. Welt*, **20** (NF), 2515.
RASMUSEN, B. A. (1962), Bloodgroups in sheep, *Ann. NY. Acad. Sci.*, **97**, 306.
RASMUSEN, B. A. (1966), Personal communication.
SCHMID, D. O. and BUSCHMANN, H. (1966), A new bloodgroup antigen in pigs, Proc. Xth Europ. Conf. Anim. Blood Groups Biochem. Polymorph., Paris, 1966, 165.
SCHMID, D. O., BUSCHMANN, H., PROKOP, O. and UHLENBRUCK, G. (1967), Hel, ein neues Blutgruppen-Antigen beim Schwein, *Z. Immun. forsch.*, **133**, 54.
STORMONT, C., WEIR, W. C. and LANE, L. L. (1953), Erythrocyte mosaicism in a pair of sheep twins, *Science*, **118**, 695.
UHLENBRUCK, G., OTTEN, H., REHFELDT, U., REIFENBERG, U. and PROKOP, O. (1968), Enzymatischer Abbau der Erythrozytenmembran: Topochemie verschiedener $A_{hel}$-Rezeptoren sowie Nachweis inkompletter Antikörper durch Subtilisin-A-Behandlung, *Z. Immun-forsch.* **134**, 476.
UHLENBRUCK, G., SCHMID, D. O. and PROKOP, O. (1966), Über die Natur des $A_{he}$ Rezeptors an menschlichen und tierischen Blutkörperchen, *Acta biol. med. germanica*, **16**. K-9.

## DISCUSSION

DR. MILLOT: 1. You insist on absorption which is absolutely necessary with different reagents for parentage cases. I think it is because different subgroups are more marked in sheep than in cattle as I had previously observed in my works and also Mary Icas.

2. Have you studied inhibition reaction by normal sera of R-anti-R reaction and can you say if inhibition is exactly in agreement with the direct test on the cells which is not the case for J in cattle?

D. O. SCHMID: Absorption is very important.

R. HOHENBRINK: Did you try any absorption with helix and where did you get the extract from?

D. O. SCHMID: By means of cross absorption we should be able to determine the relationship between the Hel-system in sheep and in pigs.

*XIIth Europ. Conf. Anim. Blood Groups Biochem. Polymorph., Bp., 1972 (pp. 567–573)*

# APPARENT DISTURBED SEGREGATION AT THE HEMOGLOBIN AND TRANSFERRIN LOCI IN HUNGARIAN MERINO SHEEP

L. Fésüs

Bloodgrouping Laboratory, University of Veterinary Science, Budapest, Hungary

## SUMMARY

**The hemoglobin types of 2919 Hungarian Merino sheep (28 rams, 1383 ewes, and 1508 lambs) were determined and the frequency of the $Hb^A$ was found to be 0.1535. Comparisons were made between the expected and observed numbers of hemoglobin genotypes in offspring from parents of known hemoglobin genotypes. Statistically significant differences were found (A/B ♂ × B/B ♀, $\chi^2 = 5.84$*, d. f. = 1; B/B ♂ × A/B ♀, $\chi^2 = 5.34$*, d. f. = 1; A/B♂ × A/B♀, $\chi^2 = 9.30$**, d. f. = 2). An excess of B/B genotypes and a shortage of A/B genotypes was observed.**

**After determining the transferrin types of 2887 Hungarian Merino sheep (28 rams, 1385 ewes, and 1474 lambs) the following gene frequencies were found: $Tf^I$ 0.0124; $Tf^A$ 0.2406; $Tf^G$ 0.0282; $Tf^B$ 0.1238; $Tf^C$ 0.1368; $Tf^M$ 0.1863; $Tf^D$ 0.2178; $Tf^Q$ 0.0026; $Tf^E$ 0.0498; $Tf^P$ 0.0017. With the exception of three sire families in which the rams had I/A, B/D, and C/M genotypes, no significant departures were found from the expected 1 : 1 ratio among the heterozygous offspring. The results of these exceptional sire families suggest that there may be some maternal-foetal incompatibility, or linkage of the transferrin locus to a lethal factor.**

**After combining the data using the method of Khattab et al. (1964), statistically significant differences were found between the expected and observed numbers of transferrin genotypes amongst the offspring in four out of eleven cases.**

## INTRODUCTION

Using the method of starch-gel electrophoresis, nineteen transferrins and four hemoglobins have been found in the sheep blood (transferrins I, A, G, B′, $B'_{Hungary}$, B, C, $C_{Hungary}$, M, D, U, N, $N_{Hungary}$, Q, E, R, V, P, $E_2$ by Ashton, 1958; Ashton and Ferguson, 1962; Khattab et al., 1964; Fésüs, 1967a and b; Osman, 1967; Fésüs and Orbányi, 1968; Stormont et al., 1968; Fésüs, 1969; hemoglobins A, B, C, D by Harris and Warren, 1955; Blunt and Evans, 1963; Vaskov and Efremov, 1967).

Some investigators found apparently disturbed segregation at both the transferrin and hemoglobin loci; others, after calculating the expected numbers at the transferrin and hemoglobin loci by using the Hardy–Weinberg formula, found statistically significant differences between the expected and observed numbers of some transferrin and hemoglobin genotypes (Evans et al., 1956; Fechter and Myburgh, 1966; Meyer et al., 1967; Orbányi and Fésüs, 1968; Agar et al., 1969; Khattab et al., 1964;

* $0.02 < P < 0.01$
** $P > 0.01$

EFREMOV and BRAEND, 1964; KING and FECHTER, 1966; COOPER et al., 1967; NI et al., 1967; OSMAN, 1967; FÉSÜS and ORBÁNYI, 1968).

During the past two years, a large number of Hungarian Merino sheep were studied to see if the segregation ratios at either the transferrin or hemoglobin loci deviated from expectation and to aid in planning of experiments intended to determine the maternal-foetal incompatibility or selective fertilization which plays an important role in sheep reproduction.

## MATERIAL AND METHODS

The hemoglobin types of 28 rams, 1383 ewes and 1508 lambs were determined, and samples taken from 28 rams, 1385 ewes and 1474 lambs were tested for serum transferrin types.

The methods used for collecting and preparing samples as well as the technique of starch-gel electrophoresis are described elsewhere (FÉSÜS, 1965 and 1967a).

Calculating the gene frequencies at both the hemoglobin and transferrin loci, exclusively the parental genotypes were considered.

To analyze the segregation data the following methods were used:

1. Comparisons were made between the expected and observed numbers of transferrin and hemoglobin genotypes in offspring derived from parents of known transferrin and hemoglobin genotypes.
2. The method used by KHATTAB et al. (1964).
3. The progeny of each heterozygous ram was tested to see if the transferrin alleles were transmitted to the offspring at a ratio of 1 : 1.

## RESULTS

The following hemoglobin and transferrin gene frequencies were found:

| | |
|---|---|
| $Hb^A$ | 0.1535 |
| $Hb^B$ | 0.8465 |
| $Tf^I$ | 0.0124 |
| $Tf^A$ | 0.2406 |
| $Tf^G$ | 0.0282 |
| $Tf^B$ | 0.1238 |
| $Tf^C$ | 0.1368 |
| $Tf^M$ | 0.1863 |
| $Tf^D$ | 0.2178 |
| $Tf^Q$ | 0.0026 |
| $Tf^E$ | 0.0498 |
| $Tf^P$ | 0.0017 |

The expected numbers of each transferrin and hemoglobin genotype were determined by using the Hardy–Weinberg formula and calculations were made to reveal differences between the expected and observed numbers of each genotype (Tables 1 and 2).

Tables 3–8 show the cases in which statistically significant differences were found.

TABLE 1

*Comparisons between the expected and observed numbers of transferrin genotypes*

| Transferrin genotypes | Expected | Observed | Transferrin genotypes | Expected | Observed |
|---|---|---|---|---|---|
| I/I | (0.39) | — | G/C | 19.74 | 19 |
| A/A | 148.13 | 163 | G/M | 26.85 | 20 |
| G/G | (2.03) | 1 | G/D | 31.43 | 41 |
| B/B | 39.22 | 37 | G/Q | (0.37) | 3 |
| C/C | 47.88 | 43 | G/E | 7.18 | 4 |
| M/M | 88.81 | 74 | G/P | (0.24) | — |
| D/D | 121.39 | 116 | B/C | 87.72 | 89 |
| Q/Q | (0.01) | — | B/M | 118.04 | 118 |
| E/E | 6.34 | 7 | B/D | 137.99 | 140 |
| P/P | (0.007) | — | B/Q | (1.64) | — |
| I/A | 15.26 | 11 | B/E | 31.55 | 33 |
| I/G | (1.78) | 2 | B/P | (1.07) | 3 |
| I/B | 7.85 | 9 | C/M | 130.44 | 135 |
| I/C | 8.68 | 12 | C/D | 152.49 | 158 |
| I/M | 11.82 | 8 | C/Q | (1.82) | 3 |
| I/D | 13.82 | 15 | C/E | 43.86 | 39 |
| I/Q | (0.16) | — | C/P | (1.19) | — |
| I/E | (3.16) | 6 | M/D | 207.66 | 222 |
| I/P | (0.10) | — | M/Q | (2.47) | 1 |
| A/G | 34.72 | 34 | M/E | 47.48 | 57 |
| A/B | 152.44 | 149 | M/P | (1.62) | — |
| A/C | 168.45 | 159 | D/Q | (2.89) | 2 |
| A/M | 229.40 | 244 | D/E | 55.51 | 49 |
| A/D | 268.19 | 252 | D/P | (1.89) | 4 |
| A/Q | (3.20) | 3 | Q/E | (0.66) | 1 |
| A/E | 61.32 | 52 | Q/P | (0.02) | — |
| A/P | (2.09) | 2 | E/P | (0.43) | — |
| G/B | 17.86 | 19 | | $\chi^2 = 23.82$ | d.f. = 24 |

Values in parantheses indicate that classes have been pooled for $\chi^2$ analysis because of the too small numbers of certain samples

TABLE 2

*Comparisons between the observed and expected numbers of hemoglobin genotypes*

| Hemoglobin genotype | Observed | Expected | $\chi^2$ |
|---|---|---|---|
| A/A | 67 | 65.64 | 0.028 |
| A/B | 721 | 724.01 | 0.012 |
| B/B | 1998 | 1996.44 | 0.001 |
| | | | 0.041; d.f. 1 |

TABLE 3

*Comparisons between the expected and observed numbers of hemoglobin genotypes in offspring from parents of known genotypes*

| Ram | Ewe | Number of matings | Number of offspring | A/A exp. | A/A obs. | A/B exp. | A/B obs. | B/B exp. | B/B obs. | $\chi^2$ | d.f. |
|---|---|---|---|---|---|---|---|---|---|---|---|
| A/A × | A/A | — | — | — | — | — | — | — | — | — | |
| B/B × | A/A | 25 | 25 | — | — | 25 | 25 | — | — | | |
| A/B × | A/A | 4 | 5 | 2.5 | 1 | 2.5 | 4 | — | — | 1.80 | 1 |
| A/A × | B/B | 32 | 32 | — | — | 32 | 32 | — | — | | |
| B/B × | B/B | 578 | 659 | — | — | — | — | 659 | 659 | | |
| A/B × | B/B | 391 | 411 | — | — | 205.5 | 181 | 205.5 | 230 | 5.84* | 1 |
| A/A × | A/B | 10 | 10 | 5 | 7 | 5. | 3 | — | — | 1.60 | 1 |
| B/B × | A/B | 213 | 229 | — | — | 114.5 | 97 | 114.5 | 132 | 5.34* | 1 |
| A/B × | A/B | 130 | 137 | 34.25 | 23 | 68.5 | 66 | 34.25 | 48 | 9.30** | 2 |

* 0.02 < P < 0.01
** P > 0.01

TABLE 4

*Disturbed segregation ratios in two matings*

| Ram | Ewe | Number of matings | Number of offspring | Tf genotype of offspring | Expected | Observed |
|---|---|---|---|---|---|---|
| I/A × | A/D | 7 | 7 | I/A | 1.75 | 2 |
| | | | | I/D | 1.75 | — |
| | | | | A/A | 1.75 | — |
| | | | | A/D | 1.75 | 5 |
| | | | | $\chi^2 = 9.56^*$, d. f. = 3 | | |
| A/C × | A/M | 14 | 16 | A/A | 4 | 9 |
| | | | | A/C | 4 | 1 |
| | | | | A/M | 4 | 3 |
| | | | | C/M | 4 | 3 |
| | | | | $\chi^2 = 9.00^*$, d. f. = 3 | | |

* 0.05 < P < 0.02

TABLE 5

| Ram | Ewe | I/I exp. | I/I obs. | I/+ exp. | I/+ obs. | +/+ exp. | +/+ obs. | $\chi^2$ | d. f. |
|---|---|---|---|---|---|---|---|---|---|
| I/+ | × I/+ | 1.75 | — | 3.5 | 7 | 1.75 | — | 7.00* | 2 |
| I/+ | × +/+ | — | — | 95 | 92 | 95 | 98 | 0.18 | 1 |
| +/+ | × I/+ | — | — | 16 | 14 | 16 | 18 | 0.50 | 1 |

* 0.05 < P < 0.02

TABLE 6

| Ram | Ewe | A/A | | A/+ | | +/+ | | $\chi^2$ | d. f. |
|---|---|---|---|---|---|---|---|---|---|
| | | exp. | obs. | exp. | obs. | exp. | obs. | | |
| A/A × | A/+ | 15 | 17 | 15 | 13 | — | — | 0.52 | 1 |
| A/+ × | A/A | 9 | 13 | 9 | 5 | — | — | 3.44 | 1 |
| A/+ × | A/+ | 31.75 | 33 | 63.50 | 65 | 31.75 | 29 | 0.23 | 2 |
| A/+ × | +/+ | — | — | 111 | 133 | 111 | 89 | 8.72** | 1 |
| +/+ | A/+ | — | — | 172 | 194 | 172 | 150 | 5.62* | 1 |

* 0.02 < P < 0.01
** P > 0.01

TABLE 7

| Ram | Ewe | E/E | | E/+ | | +/+ | | $\chi^2$ | d. f. |
|---|---|---|---|---|---|---|---|---|---|
| | | exp. | obs. | exp. | obs. | exp. | obs. | | |
| E/+ × | E/+ | 3.25 | 3 | 6.5 | 10 | 3.25 | — | 5.14 | 2 |
| E/+ × | +/+ | — | — | 98 | 115 | 98 | 81 | 5.88* | 1 |
| +/+ × | E/+ | — | — | 64.5 | 55 | 64.5 | 74 | 2.78 | 1 |

* 0.02 < P < 0.01

## DISCUSSION

The calculated hemoglobin and transferrin gene frequency values are almost the same as those ascertained in earlier studies (FÉSÜS, 1965 and 1967b). The small differences found may be due to the continuous use of different Soviet Merino rams. There is a good agreement between the expected and observed numbers of both the hemoglobin and transferrin genotypes, indicating that the studied populations are in genetic equilibrium with respect to these loci (Tables 1 and 2).

In those matings in which there was a large number of offspring, a higher proportion of lambs with B/B genotype was found. The differences between the expected and observed numbers of genotypes were statistically significant in these matings (Table 3). The same results were found by MEYER et al. (1967). The explanation seems to be obvious that the hemoglobin B/B genotype is more advantageous to a sheep than either the hemoglobin A/A or the hemoglobin A/B.

Among offspring from two matings (Table 4), there was a statistically significant difference between the expected and observed numbers of offspring with certain transferrin genotypes, but the number of observations, especially in the first mating, was too small to permit any definitive conclusion. No reciprocal matings were available.

After combining the data using the method of KHATTAB et al., 1964 (Tables 5, 6 and 7), statistically significant differences were found between the expected and observed numbers of transferrin genotypes of offspirng in the case of transferrins I, A and E. On the basis of these differences there is no indication of a possible maternal-foetal incompatibility.

TABLE 8

*Ratio of heterozygous progeny of rams with I/A, B/D, and C/M transferrin genotype*

| Ram | Ratio of heterozygous progeny | | | |
|---|---|---|---|---|
| I/A | I/G : A/G | 1 : 1 | | |
| | I/B : A/B | 2 : 4 | | |
| | I/C : A/C | 3 : 5 | | |
| | I/M : A/M | 3 : 11 | | |
| | I/D : A/D | 1 : 12 | | |
| | I/E : A/E | — : — | | |
| | | 10 : 33 | $\chi^2 = 12.30^{**}$, d. f. 1 | |
| B/D | A/B : A/D | 6 : 4 | | |
| | G/B : G/D | 1 : — | | |
| | B/C : C/D | 2 : 1 | | |
| | B/M : M/D | 6 : 2 | | |
| | B/Q : D/Q | 1 : — | | |
| | B/E : D/E | 2 : — | | |
| | | 18 : 7 | $\chi^2 = 4.84^{*}$, d. f. 1 | |
| C/M | I/C : I/M | 3 : 2 | | |
| | A/C : A/M | 39 : 32 | | |
| | G/C : G/M | 3 : 2 | | |
| | B/C : B/M | 20 : 9 | | |
| | C/D : M/D | 32 : 28 | | |
| | C/Q : M/Q | — : 2 | | |
| | C/E : M/E | 12 : 5 | | |
| | C/P : M/P | 1 : 1 | | |
| | | 110 : 81 | $\chi^2 = 4.40^{*}$, d. f. 1 | |

* $0.05 < P < 0.02$
** $0.01 < P$

With the exception of three sire families, where the rams had I/A, B/D and C/M genotype, deviations were found from the expected 1 : 1 ratio among the heterozygous offspring. The results of these exceptional sire families (Table 8) indicate that there may have been some maternal-foetal incompatibility.

These results may be due to a recessive factor which is lethal in homozygous form and linked to the transferrin locus.

If $l$ is the allele which determines the above mentioned factor and the dominant allele is $L$, the following types of matings may occur:

| *Ram* | | *Ewe* | *Possible genotypes of offspring* | | |
|---|---|---|---|---|---|
| $L^LL^L$ | × | $L^LL^L$ | $L^LL^L$ | | |
| $L^LL^L$ | × | $L^LL^l$ | $L^LL^L$ | $L^LL^l$ | |
| $L^LL^l$ | × | $L^LL^L$ | $L^LL^L$ | $L^LL^l$ | |
| $L^LL^l$ | × | $L^LL^l$ | $L^LL^L$ | $L^LL^l$ | $L^lL^l$ (lethal) |

The frequency of $l$ is probably very low and since the selection differential of the rams, both natural and artificial, would be greater than that of the ewes, most of the rams used for breeding purposes would have genotype $L^L L^L$.

We may expect statistically significant deviations from the expected 1:1 ratio among offspring of a heterozygous ram only in those cases in which the ram has $L^L L^l$ genotype. The size of these deviations will depend on the number of $L^L L^l$ ewes mated to the $L^L L^l$ ram.

## REFERENCES


AGAR, N. S., RAWAT, J. S. and ROY, A. (1969), Haemoglobin variants in Indian sheep. *Anim. Prod.*, **11.** 247–250.

ASHTON, G. C. (1958), Polymorphism in the beta-globulins of sheep. *Nature*, **181.** 849–850.

ASHTON, G. C. and FERGUSON, K. A. (1962), Serum transferrins in merino sheep *Genet. Res.*, **4.** 240–247.

BLUNT, M. H. and EVANS, J. V. (1963), Changes in the concentration of potassium in the erythrocytes and in hemoglobin type in merino sheep under a severe anemic stress. *Nature*, **200.** 1215–1216.

COOPER, D. W., BAILEY, L. F. and MAYO, O. (1967), Population data for the transferrin variants in the Australian Merino. *Aust. J. Biol. Sci.*, **20.** 959–966.

EFREMOV, G. and BRAEND, M. (1965), Haemoglobins, transferrins and albumins of sheep and goats. Proc. 9th Europ. Anim. Blood Group Conf., Prague, 1964, 323–320.

EVANS, J. V., KING, J. W. B., COHEN, B. L., HARRIS, H. and WARREN, F. L. (1956), Genetics of haemoglobin and blood potassium differences in sheep. *Nature*, **178.** 849–850.

FECHTER, H. and MYBURGH, S. J. (1966), Haemoglobin and potassium types in South African sheep breeds. Proc. Xth Europ. Conf. Anim. Blood Groups. Biochem. Polymorph. Paris, 1966, 395–399.

FÉSÜS, L. (1965), The frequencies of haemoglobin genes observed in some sheep breeds in Hungary. *Z. Tierzücht., Zücht.*, **82.** 94–98.

FÉSÜS, L. (1967a), A new sheep transferrin allele: $Tf^I$. *Acta Vet. Acad. Sci. Hung.*, **17.** 95–98.

FÉSÜS, L. (1967b), Transferrin alleles in some sheep breeds in Hungary. *Acta Vet. Acad. Sci. Hung.*, **17.** 433–438.

FÉSÜS, L. (1970), Hogyan értékelhetjük a juh transzferrin meghatározások eredményeit? *Állattani Közlemények.* (Budapest) XVII. 1–4. 67–73.

FÉSÜS, L. and ORBÁNYI, I. (1968), On the occurrence of alleles $Tf^{N_{\text{Hungary}}}$ $Tf^U$ and $Tf^V$ in sheep. *Acta Vet. Acad. Sci. Hung.*, **18.** 415–422.

HARRIS, H. and WARREN, F. L. (1955), Occurrence of electrophoretically distinct haemoglobins in ruminants. *Biochem. J.*, **60.** xxix.

KHATTAB, A. G. H., WATTSON, J. H. and AXFORD, R. F. E. (1964), Associations between serum transferrin polymorphism and disturbed segregation ratios in Welsh Mountain sheep. *Anim. Prod.*, **6.** 207–213.

KING, P. and FECHTER, H. (1966), Transferrin polymorphism in South African sheep breeds. Proc. Xth Europ. Conf. Anim. Blood Groups Biochem. Polymorph. Paris, 1966, 307–312.

NI, G. V., EGOROV, E. A. and RIS, M. A. (1967), Inherited transferrin types in Karakul sheep. Dokl. VASzHNIL. Moscow. **2.** 32–34.

ORBÁNYI, I. and FÉSÜS, L. (1968), Hemoglobin types and production traits in sheep. *Állattenyésztés*, **17.** 81–87.

OSMAN, H. EL S. (1967), Serum transferrin polymorphism in the Desert Sheep of the Sudan. *Nature*, **215.** 162–163.

STORMONT, C., SUZUKI, Y., BRADFORD, G. E. and KING, P. (1968), A survey of hemoglobins, transferrins and certain red cell antigens in nine breeds of sheep. *Genetics*, **60.** 363–371.

VASKOV, B. and EFREMOV, G. (1967), Fourth hemoglobin type in sheep. *Nature*, **216.** 593–594.

*XIIth Europ. Conf. Anim. Blood Groups Biochem. Polymorph., Bp., 1972 (pp. 575—578)*

# POLYMORPHISM OF SOME SERUM PROTEIN SYSTEMS IN GOATS

S. TJANKOV

Chair of Sheep Breeding, Higher Agricultural Institute, Sofia, Bulgaria

## SUMMARY

The gene frequencies of $Tf^A$ and $Tf^B$ differed in two goat breeds (Togenburg and Bulgarian native goats) and their crosses. The highest frequency can be found in the Togenburg breed ($Tf^A$ 1000). All imported animals have only $Tf^A/Tf^A$. The native goats have a frequency of $Tf^A$ 0.77, while the cross-bred $F_1$ and $F_2$ 0.85 and 0.96. The cause of increasing the frequency in $Tf^A$ in the offspring generations is that all, male goats have only homozygous $Tf^A/Tf^A$.

From the albumin alleles allele $Al^S$ is of the highest frequency, 0.834 in the Togenburg, while 0.762 in the native goats. The crossbred first and second generations have 0.68 and 0.91. Here also male goats with homozygous phenotypes $Al^S/Al^S$ were used which increases the gene frequency of $Al^S$.

## INTRODUCTION

Transferrin polymorphism in goats was first described by ASHTON and MCDOUGALL (1958). Similar studies have been made by EFREMOV and BRAEND (1965), WATANABE (1965). Three phenotypes were established (AA, BB and AB) which are being controlled by two alleles.

Same albumin phenotypes as in sheep (FF, SS and FS) were found in goats by EFREMOV and BRAEND (1965).

Hemoglobin polymorphism in goats was first described by KHANOLKAR in 1963. According to BERNHARDT (1964) in goats two alleles are existing They correspond to three hemoglobin phenotypes.

There is no information for amylase and ceruloplasmin polymorphism in goats.

The object of this report is to follow the variability of gene frequency of transferrin, albumin and hemoglobin alleles of progeny cross-bred ($F_1$ and $F_2$) from Bulgarian native goats and German Togenburg male goats. Serum amylase and ceruloplasmin polymorphism were investigated.

## MATERIAL AND METHODS

Two breeds of goat were used in the present investigation:

a) *German Togenburg goats*. Male goats of this breed have been regularly imported since 1964 for crossing our native goats. The purpose is to produce Bulgarian milk goat for mountain regions. In 1969, 14 goats were imported for male goat production.

b) *Bulgarian native goats.* The investigated animals have been bred for more than 100 years in a flock, in the region of the Rila Monastery. During this period no male goats from other regions had been mated with the goats of this flock. The total number of the flock never exceeded 150–200 animals. They were crossed for the first time with Togenburg male goats in 1964.

c) *$F_1$ and $F_2$ generation of crosses* between Togenburg male goats and aboriginal goats.

A total of 241 goats were used in the investigations: 19 Togenburg goats, 24 native goats, 101 animals from the $F_1$ generation and 97 animals from the $F_2$ generation.

For determining goat transferrins, BRAEND's and EFREMOV's method (1964) with buffer system of KRISTJANSSON (1962) was used. For preparing gels 12 g starch suspended in 100 ml buffer (40 ml A, 38 ml B and distilled water 500 ml) wasused. The size of the gel was 14×18×0.5 cm. Voltage was 300 V and 20–60 mA for about 4 hours. Electrophoresis was finished when the boundary line moved 10–11 cm beyond the insertion line.

EFREMOV's and BRAEND's method was used for goat albumins (13 ml A, 9 ml B and distilled water 250 ml). The start line was at 4 cm distance to the cathode end.

The hemoglobin was fixed by using tris-borate buffer, pH 9.2.

Serum ceruloplasmin and amylase were determined by SCHRÖFFEL's method (1968).

## RESULTS

Table 1 shows the gene frequency of the transferrin alleles in the two pure-bred groups and their crosses. $Tf^A$ is the most frequent in the Togenburg

TABLE 1

*Observed and expected transferrin phenotypes in two goat breeds and their crosses. Gene frequencies*

| Breed | | Tf phenotypes | | | Total | Gene frequencies | | $\chi^2$ |
|---|---|---|---|---|---|---|---|---|
| | | AA | AB | BB | | $Tf^A$ | $Tf^B$ | |
| Togenbourg goat | obs. | 19 | — | — | 19 | 1.000 | 0.000 | — |
| | exp. | 19 | — | — | 19 | | | |
| Native goat | obs. | 14 | 9 | 1 | 24 | 0.771 | 0.229 | 0.091 |
| | exp. | 14.2 | 8.6 | 1.2 | 24.0 | | | |
| Cross-bred-$F_1$ | obs. | 72 | 29 | — | 101 | 0.856 | 0.144 | 1.760 |
| | exp. | 74.1 | 24.8 | 2.1 | 101 | | | |
| Cross-bred-$F_2$ | obs. | 90 | 7 | — | 97 | 0.963 | 0.037 | 1.140 |
| | exp. | 89.9 | 6.9 | 0.1 | 97.0 | | | |

TABLE 2

*Observed and expected albumin phenotypes in two goat breeds and their crosses. Gene frequencies*

| Breed | | Alb Phenotypes | | | Total | Gene frequencies | | $\chi^2$ |
|---|---|---|---|---|---|---|---|---|
| | | FF | FS | SS | | $Al^F$ | $Al^S$ | |
| Togenburg goat | obs. | — | 5 | 10 | 15 | 0.166 | 0.834 | 1.185 |
| | exp. | 0.4 | 4.2 | 10.4 | 15 | | | |
| Native goat | obs. | — | 10 | 11 | 21 | 0.238 | 0.762 | 1.850 |
| | exp. | 1.9 | 7.0 | 12.1 | 21.0 | | | |
| Cross-bred-$F_1$ | obs. | 14 | 26 | 46 | 86 | 0.314 | 0.686 | 7.64 |
| | exp. | 8.5 | 37.0 | 40.5 | 86 | | | |
| Cross-bred-$F_2$ | obs. | 4 | 9 | 83 | 96 | 0.088 | 0.912 | 16.9 |
| | exp. | 0.7 | 15.6 | 79.7 | 96.0 | | | |

breed (1000). The native goats have lower frequency of the $Tf^A$. In the crosses $F_1$ and $F_2$ the frequency of the allele $Tf^A$ increased progressively.

In the native goats and cross-bred animals, the observed numbers of the three transferrin types did not significantly differ from their numbers expected on the basis of the Hardy–Weinberg equilibrium.

The gene frequency of the albumin alleles is shown in Table 2. The highest frequency is in the allele $Al^S$ of all the groups. A tendency could be seen for an increase of the frequency of $Al^S$. In the two pure-bred groups as well as in the cross-breed of the first generation the observed and the expected numbers differed slightly. There was a statistically significant difference in relation to the $F_2$ generation ($\chi^2 = 16.9$, $P > 0.001$).

In all animals (240 in number) only a single type each of hemoglobin, amylase and ceruloplasmin was found.

## DISCUSSION

Table 1 clearly shows the difference in gene frequencies between the two pure-breeds and their crosses. While all goats imported from Germany have Tf AA phenotypes, gene frequency in the native goats is 0.771. An increase of gene frequencies $Tf^A$ (in second generation 0.963) was observed in the cross-bred goats. The cause of this phenomenon is that for some years, male goats with homozygous transferrin phenotypes-AA, have been used for service. It is natural to expect that if male goats with heterozygous phenotype AB or homozygous BB will not be procured, the homozygous type "AA" will gradually supersede the other two types.

For example, while the transferrin phenotype "AA" in pure-bred native goats had been only 58.3%, it is 71.3% in the $F_1$ generation, and in the second generation Tf AA is 92.8%. In contrast in native goats the homozygous phenotype "BB" had been only 4.1% and after that disappeared. Heterozygous type Tf AB decreased from 37.5% in native goats to 28.7% in first cross-bred generation. In the second generation it is only 7.2%.

Data of Table 2 show that the populations are in equilibrium.

In serum albumins higher frequencies are observed in slower allele ($Al^S$). In the Togenburg breed it is 0.834 and in the native goats 0.762. In both breeds homozygous type $Al^F/Al^F$ was not found. Heterozygous phenotype FS is 33.3% in Togenburg goats and 47.6% in the native goats.

The cross-bred $F_1$ generation has lower frequencies of the allele $Al^S$ (0.68) in comparison with purebred animals. The cause of this phenomenon is that two of the five male goats that serve during the first three years had Al FS. That is why homozygous forms FF were obtained in the $F_1$ generation in a proportion of 16.3%. In the $F_2$ generation the gene frequencies of $Al^S$ is highest, 0.91. This is due to the fact that the last male goats which acted in the flock had the homozygous phenotype SS.

Consequently if we want to increase the genetic variety of the new breed in connection with transferrin, albumin and hemoglobin phenotypes, which eventually may be in relation with other indications, male goats having transferrin phenotypes BB and AB or albumins FF and FS must be chosen.

## REFERENCES

Ashton, G. C. and McDougall, E. I. (1958), *Nature*, **182**, 945.
Bernhardt, D. (1964), *Dt. Tierärzt. Wschr.*, **71**, 461–462.
Efremov, G. and Braend, M. (1965), Proc. 9th Europ. Conf. Anim. Blood Groups, Prague, 1964.
Harris, H. and Warren, F. L. (1955), *Biochem. J.*, **60.** 29.
Kristjansson, F. K. (1962), VIIIth Europ. Conf. Anim. Blood Groups, Ljubljana, 1962 (Mimeo).
Salisbury, G. W. and Schreffler, D. C. (1957), *J. Dairy Sci.*, 1198–1199.
Schröffel, J., Kúbek, A. and Glasnák, V. (1970). Proc. XIth Europ. Conf. Anim. Blood Groups Biochem. Polymorph. Warsaw, 1968, 207–210.
Watanabe, S. et al. (1965), *Proc. Japan Acad.*, **41**, 326.

*XIIth Europ. Conf. Anim. Blood Groups Biochem. Polymorph., Bp., 1972 (pp. 579–582)*

# SERUM POLYMORPHISM IN THREE SOUTH AFRICAN GOAT BREEDS

D. R. OSTERHOFF and I. S. WARD-COX

Department of Zootechnology, Faculty of Veterinary Science, University of Pretoria
Onderstepoort, Rep. of South Africa

## SUMMARY

A survey is given of the various biochemical polymorphic systems of goats in South Africa viz. Angora, Indigenous and Boer goats, together with some preliminary results on the frequency with which the presently known blood group factors appear in each race. A comparison of the breeds is interesting especially from the phylogenetic points of view.

Speculation is made on the aetiology of the problem of spontaneous abortions in Angora goats, this being the major difficulty in the breeding and selection of these animals. The possibility of blood factor incompatible matings is discussed, and the further direction of these investigations is indicated by following primarily the comparison of the primitive (indigenous) breed with the highly improved Angora goat along the immunogenetical path and secondarily making use of modern concepts with regard to tissue rejection phenomena.

## INTRODUCTION

South Africa is not a recognized goat breeding country nor does its economy depend to any great extent on an advanced goat or related industry. However, the mohair production from the Angora goats amounts to about 10 million pounds annually keeping South Africa in the third position after the U.S.A. and Turkey.

The indigenous and improved Boer goats are present in a far greater number than the Angora goats (5 millions versus 2 millions) but these animals play a role only in controlling bush encroachment and in the production of lower grade meat. The improved Boer goats have been bred and systematically selected from the indigenous goats, providing thus the ideal material for phylogenetical studies.

In the Angora goat on the other hand immunogenetical studies could possibly provide some information to solve the most important problem of the Angora — the high abortion rate (VAN HEERDEN, 1963; VAN RENSBURG, 1970). Therefore a preliminary study was undertaken to provide some information on both the abortion problem of the Angoras and the breed comparison for phylogenetic analyses.

## MATERIAL AND METHODS

Using the technique of starch gel electrophoresis, analyses were carried out on different protein types including hemoglobin, transferrin and amylase variants. In the red cell identification eight newly produced and absorbed antisera were used applying standard hemolytic tests with rabbit complement absorbed with goat cells as complement.

Blood samples of Boer goats were collected at different farms, while the samples from indigenous goats were drawn from the Onderstepoort experimental flock. The Angora goat samples could be collected from two experimental flocks — Onderstepoort and Grootfontein — being kept to study the abortion problem. All the available evidences for the reason of the high abortion rate in these animals point to a genetical predisposition possibly related to isoimmune reactions in the does. It is certainly a more complex problem since a doe may abort the very first foetus and then having three to four normal births and another abortion again (VAN HEERDEN, 1963). The samples included in the study could be divided into samples from aborters and non-aborters, aborters being those does having aborted at least once at any stage of the reproductive life while non-aborters being those having produced normal kids throughout.

## RESULTS

The results of the starch gel analyses of hemoglobin, transferrin, albumin and amylase phenotypes are compiled and gene frequencies are given in Table 1.

From the albumin gene frequency only the clear difference between the Angora goat with its origin in Turkey and the other two breeds being indigenous to South Africa can be seen.

Since red cell antigens are usually better markers than albumins, for detecting breed differences, immunogenetical studies were initiated. No natural occurring antibodies could be established in any of the three breeds, therefore the production of immune antibodies was indicated. To accelerate the process initial immunizations were carried out using Angora donors and Dorper sheep recipients at random. In this way it was possible to iden-

TABLE 1

*Biochemical polymorphism in three breeds of goats*

| Breed | No. of animals | Gene frequencies | | | | | | | | | |
|---|---|---|---|---|---|---|---|---|---|---|---|
| | | $Hb^A$ | $Hb^B$ | $Tf^A$ | $Tf^B$ | $Tf^C$ | $Tf^D$ | $Alb^A$ | $Alb^B$ | $Am^A$ | $Am^B$ |
| Boergoat | 212 | .91 | .09 | .70 | .30 | .00 | .00 | .08 | .92 | 1.0 | .00 |
| Indigenous | 87 | .95 | .05 | .72 | .28 | .00 | .00 | .01 | .99 | 1.0 | .00 |
| Angora: | | | | | | | | | | | |
| Aborters | 110 | .93 | .07 | .76 | .23 | .00 | .01 | .32 | .68 | .90 | .10 |
| Non-aborters | 147 | .94 | .06 | .80 | .19 | .01 | .00 | .15 | .85 | .97 | .03 |

tify eight types of antibodies produced by twelve donor-recipient pairs, all eight being hemolysins. In the following three immunization programs using goat donors against goat recipients three more monospecific reagents could be produced.

Using only the first eight reagents, a preliminary survey was carried out on 62 Angoras, 89 Indigenous goats and 109 Boer goats; results being depicted in Table 2.

TABLE 2

*Frequency distribution of blood factors in goats*

| Breed | No. of animals | Reagents used | | | | | | | |
|---|---|---|---|---|---|---|---|---|---|
| | | G. 1 | G. 2 | G. 3 | G. 4 | G.5 | G. 6 | G. 7 | G. 8 |
| Angora | 62 | .210 | .435 | .129 | .710 | .694 | .774 | .306 | .500 |
| Indigenous | 89 | .135 | .258 | .427 | .775 | .539 | .663 | .315 | .551 |
| Boer goats | 109 | .349 | .440 | .410 | .697 | .523 | .651 | .413 | .697 |

## DISCUSSION AND CONCLUSIONS

In the hemoglobin types no difference could be found between the three goat breeds. That the beta-globulins are indeed transferrins could be substantiated by the work of WATANABE et al. (1965) who also deduced their mode of inheritance (1966). Their frequencies for $Tf^A$ are somewhat higher than in the material presented here, while SALERNO et al. (1968) confuse the matter by deducing a preponderance of the slower Tf which they call $Tf^A$ and which both WATANABE and SUZUKI and the present authors call $Tf^B$. Two new types both characterized by slower migration, the D type moving slower than the C type, were found, but only in the Angora goat.

Differences exist between Angora and Indigenous goats and also the frequencies given by WATANABE and SUZUKI and SALERNO et al. greatly differ from those presented in this investigation.

The blood grouping reagents could very easily be produced using sheep as recipients. Several fatalities, however, occurred when goat recipients were used. Of the first eight reagents used in this preliminary survey two viz. G.3 and G.4 did not react with any sheep cells and appeared to be goat-specific. SUZUKI and WATANABE (1968) described four blood grouping reagents and EYQUEM et al. (1962) mentioned a variety of goat hemagglutinins without specifying them. In our laboratory, no less than twelve hemolysins have recently been detected in goats following skingrafts (WARD–COX, 1970). These have not been used for blood grouping purposes and will form the basis of a future investigation.

From the blood factor comparison presented here it is evident that the three breeds compared differ greatly, especially with regard to the factors G.1 and G.3. No conclusions could be drawn and it would be unjust to compare these findings with any other goat blood group investigation until any comparison and reference tests are performed.

Regarding the abortion problem in Angora goats it may be concluded that at this stage there is no connection between the results of this investigation and the aborting phenomenon. This does not necessarily imply an exclusion of an immune process, bearing in mind the possible role of cytotoxic factors and, at this stage, the very popular ideas of tissue rejection phenomena. The former have been known to have a bearing on abortions in women (DE VOS, 1970) and the latter is under suspicion from the close relationship that apparently exists between hemolysins and tissue rejection. Thus, the field for research on this interesting problem remains wide open, and further comments would be speculative.

## ACKNOWLEDGEMENTS

We express our thanks to Mr. J. M. VAN DER WESTHUYSEN, Grootfontein for the provision of many of the Angora blood samples.

## REFERENCES

EYQUEM, A., PODLIACHOUK, LUBA and MILLOT, P. (1962), *Ann. N. Y. Acad. Sci.*, **97**, 320.
SALERNO, A., MONTEMURRO, N. and L'AFFLITTO, A. (1970), Proc. XIth Europ. Conf. Anim. Blood Groups., Biochem. Polymorph., Warsaw, 1968, 517–520.
SUZUKI, S. and WATANABE, S. (1970), Proc. XIth Europ. Conf. Anim. Blood Groups., Biochem. Polymorph., Warsaw, 1968, 513–515.
VAN HEERDEN, K. M. (1963), *Onderstepoort J. Vet. Res.*, **30**, 23.
VAN RENSBURG, S. J. (1970), Reproductive physiology and endocrinology of normal and habitually aborting Angora goats. *Thesis*, University of Pretoria, 190 pp.
DE VOS, G. (1970), Personal communication.
WARD-COX, I. S. (1970), Nat. Congr. Soc. Med. Lab. Techn. S. Afr., (in press).
WATANABE, S., NOZAWA, K. and SUZUKI, S. (1965), *Proc. Jap. Acad.*, **41**, 326.
WATANABE, S. and SUZUKI, S. (1966), *Proc. Jap. Acad.*, **42**, 178.
WATANABE, S. and SUZUKI, S. (1967), *Jap. J. Zootech. Sci.*, **38**, 487.

## DISCUSSION

P. MILLOT: Have you any manifestation of hemolytic anaemia or placental anasarca or cytologic abnormalities in the abortive fetuses?

D. R. OSTERHOFF: No satisfactory answer has been found to the abortion problem in Angora goats, therefore we are trying the immunogenetical approach now.

D. O. SCHMID: 1. Why are you using goats for transplantation?
2. Have you done transplantation according to blood-types?

D. R. OSTERHOFF: Goats are used in transplantation experiments because these animals are extremely well suited for this work.

# VII. BLOOD GROUPS AND BIOCHEMICAL POLYMORPHISM IN FISH

*XIIth Europ. Conf. Anim. Blood Groups Biochem. Polymorph., Bp., 1972 (pp. 585—591)*

# BLOOD PROTEINS IN NORTH SEA COD
## *(Gadus morhua L.)*

A. JAMIESON and D. THOMPSON

Ministry of Agriculture, Fisheries and Food, Fisheries Laboratory, Lowestoft, England

## SUMMARY

Cod caught in the North Sea area were tested for transferrin and hemoglobin variants. The results were compared among themselves and with published data. Tests for homogeneity and for genetic balance indicated that most of the cod in the North Sea belong to one race showing only slight genetic differentiation of local stocks. The transferrin data were more uniform than the hemoglobin data.

## INTRODUCTION

The extensive marine range of the cod, *Gadus morhua* L., covers much of the temperate and arctic regions of the North Atlantic, including the North Sea and its adjoining seas. The results from tagging experiments in the North Sea (BEDFORD, 1966) show that very few cod move out of the area. Indeed the young cod are recruited to and remain within limited circuits which overlap in certain seasons, but habitually return to one of the spawning grounds. Cod spawn in the North Sea in the area from the Great Fisher Bank to the Ling Bank, on the Aberdeen Bank or Long Forties, and on the Flamborough grounds between the Dogger Bank and the Yorkshire coast (GRAHAM, 1948). Thus, for practical purposes the adult cod in the North Sea fall loosely into regional stocks without recourse to any genotypic data.

In the North Sea, cod eggs and larvae are transported by the surface currents. These have been described (BÖHNECKE, 1922; LEE and RAMSTER, 1968), but it is difficult to trace the origins of eggs and the fates of larvae after their pelagic phase, particularly since those early stages can be neither tagged nor genotyped. It is possible that larvae from different spawning grounds are mixed together in a common gene pool from which the regional stocks recruit fry at random. Alternatively, the selection and isolation of certain ecotypes could result in deme formation. Where genetic isolation is effective, continuous genotypic observations could be applied, for instance to trace the genetic influence of occasional abundant year-classes.

Recent serological work on cod throughout the North Atlantic has shown that isolation and natural selection separate out genetically diverse regional cod stocks (FRYDENBERG et al., 1965; JAMIESON, 1967a, 1967b, 1970; JAMIESON and JONES, 1967; JAMIESON and JONSSON, 1970; JAMIESON and

TABLE

*Summary of the data obtained from the testing of cod blood*

| Locality | Number of cod | Date | Approximate positions | | Hemoglobins | | |
|---|---|---|---|---|---|---|---|
| | | | Lat o′ | Long o′ | $HbI^1$ | Homozygotes obs. | Homozygotes exp. |
| *North-west Scotland* | | | | | | | |
| Flannan Isles | 52 | June 1967 | 58 20 N | 07 30 W | 0.64 | 36 | 30 |
| Butt of Lewis | 189 | 12 June 1968 | 58 30 N | 07 00 W | 0.61 | 101 | 99 |
| North-west Hebrides | 47 | June 1967 | 58 00 N | 08 00 W | 0.60 | 25 | 24 |
| Stornoway | 29 | 1 December 1965 | 58 00 N | 06 00 W | — | — | — |
| Cape Wrath | 159 | June 1970 | 58 36 N | 05 19 W | 0.63 | 89 | 84 |
| *Northern North Sea* | | | | | | | |
| Hordaland | 216 | March 1962 | 60 20 N | 04 40 E | 0.52 | 99 | 108 |
| Hordaland | 100 | 18 December 1964 | 60 20 N | 04 40 E | — | — | — |
| Shetland | 32 | August 1966 | 60 00 N | 01 00 W | 0.69 | 20 | 18 |
| Fair Isle | 18 | August 1966 | 59 30 N | 01 45 W | 0.58 | 11 | 9 |
| Moray Firth | 29 | 19 January 1970 | 58 00 N | 03 00 W | 0.44 | 19 | 15 |
| Moray Firth | 121 | 1965 | 58 00 N | 03 00 W | 0.58 | 61 | 61 |
| Moray Firth | 61 | April 1966 | 58 00 N | 03 00 W | 0.54 | 36 | → 31 |
| Hirstals | 58 | 1 June 1961 | 57 40 N | 09 00 E | 0.69 | 40 | → 33 |
| Hirstals | 83 | 13 August 1961 | 57 40 N | 09 00 E | 0.63 | 42 | 43 |
| Outer Shoal | 32 | 24 September 1962 | 57 32 N | 05 39 E | 0.58 | 19 | 16 |
| *Middle North Sea* | | | | | | | |
| Aberdeen | 53 | 23 July 1962 | 57 00 N | 02 00 W | 0.65 | 26 | 29 |
| Aberdeen | 27 | 7 December 1965 | 57 00 N | 02 00 W | — | — | — |
| Aberdeen | 188 | 1965 | 57 00 N | 02 00 W | 0.56 | 93 | 95 |
| Aberdeen | 42 | February 1966 | 57 00 N | 02 00 W | 0.54 | 25 | 21 |
| Aberdeen | 60 | September 1967 | 57 00 N | 02 00 W | 0.60 | 32 | 31 |
| Great Fisher Bank | 30 | 20 September 1961 | 56 54 N | 03 40 E | 0.55 | 17 | 15 |
| Hvidesande | 251 | 7 March 1962 | 55 50 N | 07 50 E | 0.63 | 133 | 134 |
| Dogger | 25 | 19 November 1967 | 55 40 N | 03 00 E | — | — | — |
| Farne Islands | 72 | 4 August 1967 | 55 40 N | 01 30 E | 0.56 | 44 | 36 |
| Clay Deep | 16 | 17 March 1966 | 55 00 N | 04 00 W | — | — | — |
| Scarborough | 90 | 29 July 1962 | 54 15 N | 00 20 W | 0.57 | 46 | 46 |
| Flamborough | 98 | 29 February 1963 | 54 10 N | 00 00 | 0.58 | 47 | 50 |
| Flamborough | 69 | 19 May 1967 | 54 10 N | 00 00 | — | — | — |
| Skate Hole | 21 | 19 November 1967 | 54 04 N | 01 50 E | — | — | — |
| Great Silver Pit | 35 | 7 January 1966 | 54 00 N | 02 30 E | — | — | — |
| Great Silver Pit | 25 | 21 November 1967 | 54 00 N | 02 30 E | — | — | — |
| Great Silver Pit | 48 | 29 December 1969 | 54 00 N | 02 30 E | 0.68 | 28 | 28 |
| *Southern North Sea* | | | | | | | |
| Western Mud Hole | 79 | 19 February 1968 | 53 43 N | 03 10 E | 0.74 | 41 | 43 |
| Brown Bank Ground | 33 | 2 February 1970 | 53 25 N | 03 40 E | 0.58 | 23 | 17 |
| Smiths Knoll | 16 | 24 November 1967 | 52 49 N | 02 36 E | — | — | — |
| Winterton Shoal | 38 | 3 December 1969 | 52 45 N | 02 13 E | 0.65 | 20 | 21 |
| Brown Ridges | 75 | 3 January 1970 | 52 35 N | 03 21 E | 0.41 | 56 | → 38 |
| Middle Ground | 23 | 2 February 1966 | 52 30 N | 02 30 E | — | — | — |
| Lowestoft | 98 | November 1966 to April 1967 | 52 27 N | 01 43 E | — | — | — |
| Lowestoft | 155 | November 1967 to April 1968 | 52 27 N | 01 43 E | — | — | — |

1

*samples from the North Sea, English Channel and Outer Hebrides*

| | | Transferrins | | | | | | | |
|---|---|---|---|---|---|---|---|---|---|
| Heterozygotes | | $Tf^B$ | $Tf^C$ | $Tf^D$ | Homozygotes | | Heterozygotes | | References |
| obs. | exp. | | | | obs. | exp. | obs. | exp. | |
| 16 | 22 | — | — | — | — | — | — | — | Wilkins 1970 |
| 87 | 89 | 0.11 | 0.83 | 0.02 | 143 | 133 | 46 | 56 | Jamieson 1970 |
| 22 | 23 | — | — | — | — | — | — | — | Wilkins 1970 |
| — | — | 0.07 | 0.78 | 0.05 | 17 | 18 | 12 | 11 | Jamieson 1970 |
| 70 | 75 | 0.11 | 0.84 | 0.02 | 118 | 113 | 41 | 46 | This paper |
| | | | | | | | | | |
| 117 | 108 | — | — | — | — | — | — | — | Frydenberg *et al.* 1965 |
| — | — | 0.10 | 0.85 | 0.03 | 76 | 73 | 24 | 27 | Møller 1966 |
| 12 | 14 | — | — | — | — | — | — | — | Wilkins 1970 |
| 7 | 9 | — | — | — | — | — | — | — | Wilkins 1970 |
| 10 | 14 | 0.05 | 0.93 | 0.00 | 25 | 25 | 4 | 4 | This paper |
| 60 | 60 | — | — | — | — | — | — | — | Wilkins 1967 |
| 25 ← | 30 | — | — | — | — | — | — | — | Wilkins 1970 |
| 18 ← | 25 | — | — | — | — | — | — | — | Sick 1965b |
| 41 | 39 | — | — | — | — | — | — | — | Sick 1965b |
| 13 | 16 | — | — | — | — | — | — | — | Sick 1965b |
| | | | | | | | | | |
| 27 | 24 | — | — | — | — | — | — | — | Sick 1965b |
| — | — | 0.09 | 0.85 | 0.04 | 21 | 20 | 6 | 7 | Jamieson 1970 |
| 95 | 93 | — | — | — | — | — | — | — | Wilkins 1967 |
| 17 | 21 | — | — | — | — | — | — | — | Wilkins 1970 |
| 28 | 29 | — | — | — | — | — | — | — | Wilkins 1970 |
| 13 | 15 | — | — | — | — | — | — | — | Sick 1965b |
| 118 | 117 | — | — | — | — | — | — | — | Sick 1965b |
| — | — | 0.12 | 0.83 | 0.00 | 22 | 18 | 4 | 8 | Jamieson 1970 |
| 28 | 36 | — | — | — | — | — | — | — | Sick 1965b |
| — | — | 0.09 | 0.88 | 0.00 | 12 | 12 | 4 | 4 | Jamieson 1970 |
| 44 | 44 | — | — | — | — | — | — | — | Sick 1965b |
| 51 | 48 | — | — | — | — | — | — | — | Sick 1965b |
| — | — | 0.05 | 0.93 | 0.01 | 61 | 60 | 8 | 9 | Jamieson 1970 |
| — | — | 0.07 | 0.88 | 0.02 | 17 | 16 | 4 | 5 | Jamieson 1970 |
| — | — | 0.07 | 0.90 | 0.01 | 28 | 28 | 7 | 7 | This paper |
| — | — | 0.04 | 0.92 | 0.04 | 23 | 21 | 2 | 4 | This paper |
| 21 | 21 | 0.11 | 0.86 | 0.00 | 35 | 37 | 13 | 11 | This paper |
| | | | | | | | | | |
| 28 | 27 | 0.12 | 0.84 | 0.04 | 54 | 56 | 25 | 23 | Jamieson 1970 |
| 10 | 16 | 0.05 | 0.94 | 0.00 | 29 | 29 | 4 | 4 | This paper |
| — | — | 0.13 | 0.84 | 0.00 | 11 | 11 | 5 | 5 | This paper |
| 19 | 18 | 0.12 | 0.88 | 0.00 | 33 | 31 | 5 | 7 | This paper |
| 18 ← | 36 | 0.07 | 0.91 | 0.01 | 63 | 62 | 12 | 13 | This paper |
| — | — | 0.07 | 0.83 | 0.02 | 17 | 16 | 6 | 7 | This paper |
| — | — | 0.14 | 0.83 | 0.01 | 68 | 70 | 30 | 28 | Jamieson 1970 |
| | | | | | | | | | |
| — | — | 0.16 | 0.82 | 0.01 | 105 | 109 | 50 | 46 | Jamieson 1970 |

TABLE 1

| Locality | Number of cod | Date | Approximate positions | | Hemoglobins | | |
|---|---|---|---|---|---|---|---|
| | | | Lat o′ | Long o′ | HbI[1] | Homozygotes obs. | Homozygotes exp. |
| *Southern North Sea* | | | | | | | |
| Lowestoft | 172 | November 1968 to April 1969 | 52 27 N | 01 43 E | 0.59 | 90 | 89 |
| Lowestoft | 114 | November 1969 to April 1970 | 52 27 N | 01 43 E | 0.55 | 80 | 58 |
| Brown Ridges | 76 | 30 January 1962 | 52 16 N | 03 10 E | 0.72 | 39 | 45 |
| *English Channel* | | | | | | | |
| Folkestone | 101 | 15 December 1967 | 51 00 N | 01 12 E | — | — | — |
| Folkestone | 56 | 15 December 1968 | 51 00 N | 01 12 E | 0.74 | 30 | 31 |
| Plymouth | 34 | 1 December 1967 | 50 10 N | 04 15 W | — | — | — |
| Plymouth | 33 | 1 December 1968 | 50 10 N | 04 15 W | 0.71 | 14 | 20 |
| Plymouth | 23 | February 1966 | 50 10 N | 04 15 W | 0.50 | 8 | 11 |

OTTERLIND, 1970; MØLLER, 1966a, 1966b, 1968; SICK, 1965a, 1965b; WILKINS, 1967). As part of this overall study the degrees of genetic isolation of the stocks of cod in the North Sea are considered in this paper by presenting and analyzing all of the available genotypic data on blood proteins of cod caught in and about this area.

## MATERIAL

Cod blood samples were collected from batches of cod caught in research vessel trawls and on anglers' lines. Although the batches were collected spasmodically in the absence of any prearranged sampling programme, some repeat samples were obtained in certain places. Some of the material was tested for hemoglobin only and some for transferrin only, but all of the cod bloods arriving at Lowestoft from June 1968 up to the present time (June 1970) were tested for variants at both genetic loci. Each batch of material was tabulated showing its date of collection and a chart reference; some of these are however approximate means. The fishing ground names were taken mainly from the Ordnance Survey Chart number A208/Fishing Grounds.

To make a complete statement of the known information on cod proteins in the North Sea, similar data obtained in other laboratories are quoted and gratefully acknowledged by reference to publications.

(continued)

| Heterozygotes | | Transferrins | | | | | | | |
|---|---|---|---|---|---|---|---|---|---|
| | | Tf$^B$ | Tf$^C$ | Tf$^D$ | Homozygotes | | Heterozygotes | | References |
| obs. | exp. | | | | obs. | exp. | obs. | exp. | |
| 83 | 84 | 0.08 | 0.87 | 0.03 | 141 → | 130 | 30 ← | 41 | This paper |
| 34 ← | 56 | 0.13 | 0.85 | 0.01 | 88 | 84 | 26 | 30 | This paper |
| 37 | 31 | — | — | — | — | — | — | — | Sick 1965b |
| — | — | 0.19 | 0.74 | 0.05 | 58 | 59 | 43 | 42 | This paper |
| 21 | 20 | 0.13 | 0.87 | 0.00 | 47 | 43 | 9 | 13 | This paper |
| — | — | 0.18 | 0.79 | 0.01 | 21 | 22 | 13 | 12 | This paper |
| 20 | 14 | 0.09 | 0.79 | 0.08 | 23 | 20 | 10 | 12 | This paper |
| 15 | 12 | — | — | — | — | — | — | — | Wilkins 1970 |

*Note:* — = "not tested"
obs = "observed"
exp = "expected"

## METHODS

The methods used for collecting the blood and testing and naming the blood proteins are described elsewhere (SICK, 1965a; MØLLER, 1966b; JAMIESON, 1970; JAMIESON and JONSSON, 1970; JAMIESON and OTTERLIND, 1970).

## RESULTS

The transferrin types seen in North Sea cod to date included 1343 examples of the genotype Tf$^C$/Tf$^C$, 336 Tf$^B$/Tf$^C$, 54 Tf$^A$/Tf$^C$, 47 Tf$^C$/Tf$^D$, 36 Tf$^B$/Tf$^B$, 12 Tf$^B$/Tf$^D$ and lesser numbers of a few rare transferrin types. Similarly, this North Sea cod population has yielded 790 examples of hemoglobin genotype HbI$^1$/HbI$^2$, 717 HbI$^1$/HbI$^1$, 309 HbI$^2$/HbI$^2$ and small numbers of rare hemoglobin types.

Tests for genetic equilibrium on these pooled data show significant genetic imbalance ($P < 0.05$ for the transferrin locus and $P < 0.01$ for the hemoglobin locus). In both instances the bias was due to a shortage of heterozygotes. This result indicates some degree of inbreeding or it could be due to non-random sampling. The sampling could be criticized because the catching devices are selective; also, each batch of specimens is taken from a very restricted area.

A search was made for genetic heterogeneity among batches of typed cod and the genetic equilibrium of each batch was tested.

Most of the samples showed approximately 0.60 HbI$^1$ and approximately 0.85 Tf$^C$. One batch caught on the Brown Ridges on 3 January 1970 showed a low proportion of HbI[1] at 0.41 ($P < 0.01$ in a heterogeneity test), but the same sample showed a high frequency of $Tf^C$ at 0.91, which is most characteristic of the North Sea. One sample from Folkestone in December 1967 showed a higher than expected proportion of $Tf^B$ ($P < 0.01$); this did not reappear the following year. Considering the number of batches examined in and about the North Sea area, some spurious observations may be expected. The overall result showed uniformity in allele frequencies. Two unusual frequencies are underlined in Table 1.

Tests for genetic equilibrium in batches of cod showed a limited number of slightly significant departures from expectation, all showing too many homozygotes ($P < 0.05$ in each). As these examples were widely set apart, occurring in the Moray Firth, off the west coast of Denmark and in the southern part of the North Sea, no obvious sub-population pattern emerged. In Table 1 arrows indicate five examples of genetic imbalance.

## DISCUSSION

Cod populations which move between spawning, nursery and feeding grounds are not confined within the geophysical and legal limits of national waters. The cod of the North Sea support an international fishery, and the responsible exploitation of this species depends on a wide and proper appreciation of the genetic sub-structure and identity of its regional stocks.

The blood protein genotypes in cod from the North Sea, sampled in different seasons and places over the past ten years, lend little support to the view that the North Sea cod spawning ground populations are distinct genotypically. This finding does not disprove their isolation as breeding units, but shows that their degree of isolation has not effectively produced genotypic differences on the scale observed in this species in a number of other sea areas. Contrasting the regularity of the unit stock data presented in this paper with the irregularities in those other areas enhances both demonstrations; a new example of a unit stock showing genetic equilibrium is preferable to inventing one.

The two genetic loci were not equally informative. This could be a local effect due to the wide differences in the proportions of the hemoglobin alleles in the adjacent stocks. Independent information at other genetic loci would be helpful in future work. Nevertheless the available data demonstrate the possible application of blood type studies in fisheries management.

## REFERENCES

BEDFORD, B. C. (1966), English cod tagging experiments in the North Sea. International Council for the Exploration of the Sea, C. M. 1966/G9, Gadoid Fish Committee, 9 pp. (mimeo).

BÖHNECKE, G. (1922), Salzgehalt und Strömungen der Nordsee. *Veröff. Inst. Meeresk. Univ. Berl.*, Neue Folge A, Geographisch-naturwiss. Reihe, Heft 10, 1–34.

FRYDENBERG, O., MØLLER, D., NAEVDAL, G. and SICK, K. (1965), Haemoglobin polymorphism in Norwegian cod populations. *Hereditas*, **53**, 257–271.

GRAHAM, M. (1948), Ratoinal Fishing of the Cod of the North Sea. Buckland Lectures for 1939. Edward Arnold and Co., London, 111 pp.

JAMIESON, A. (1967a), New genotypes in cod at Greenland. *Nature*, **215**. 661–662.
JAMIESON, A. (1967b), Genetic diversity in cod. International Council for the Exploration of the Sea, C. M. 1967/F20, Demersal Fish (Northern) Committee, 6 pp. (mimeo).
JAMIESON, A. (1970), Cod transferrins and genetic isolates. Proc. XIth Europ. Conf. Anim. Blood Groups Biochem. Polymorph., Warsaw, 1968, 533–538.
JAMIESON, A. and JONES, B. W. (1967), Two races of cod at Faroe. *Heredity*, **22**, 610–612.
JAMIESON, A. and JÓNSSON, J. (1970), The Greenland component of spawning cod at Iceland. International Council for the Exploration of the Sea, Special Meeting 1969. on The Biochemical and Serological Identifications of Fish Stocks. *Rapp. P.-v Réun. Cons. perm. int. Explor. Mer* (in press).
JAMIESON, A. and OTTERLIND, G. (1970), The use of cod blood protein polymorphisms in the Belt Sea, the Sound and the Baltic Sea. International Council for the Exploration of the Sea, Special Meeting 1969, on The Biochemical and Serological Identification of Fish Stocks. *Rapp. P.-v. Réun. Cons. perm. int. Explor. Mer* (in press).
LEE, A. J. and RAMSTER, J. (1968), The Hydrography of the North Sea. A review of our knowledge in relation to pollution problems. *Helgoländer wiss. Meeresunters.*, **17**. 44–63.
MØLLER, D. (1966a), Genetic differences between cod groups in the Lofoten area. *Nature*, **212**, 824.
MØLLER, D. (1966b), Polymorphism of serum transferrin in cod. *FiskDir. Skr. Ser. HavUnders.*, **14**, 51–60.
MØLLER, D. (1968), Genetic diversity in spawning cod along the Norwegian coast. *Hereditas*, **60**, 1–32.
SICK, K. (1965a), Haemoglobin polymorphism of cod in the Baltic and the Danish Belt Sea. *Hereditas*, **54**, 19–48.
SICK, K. (1965b), Haemoglobin polymorphism of cod in the North Sea and the North Atlantic Ocean. *Hereditas*, **54**, 49–73.
WILKINS, N. P. (1967), Polymorphism of whole blood proteins in the cod (*Gadus morhua* L.). *J. Cons. perm. int. Explor. Mer*, **31**, 77–88, Copenhagen.
WILKINS, N. P. (1970), Haemoglobin polymorphism in cod, whiting and pollack in Scottish waters. International Council for the Exploration of the Sea, Special Meeting 1969, on The Biochemical and Serological Identification of Fish Stocks. *Rapp. P.-v. Réun. Cons. perm. int. Explor. Mer* (in press).

## DISCUSSION

W. DE LIGNY: Did you find variation in the cod in the English Channel? There is some information on cod migrating from there into the North Sea. Did you sample the Channel cod over seasons?

A. JAMIESON: The Channel samples were from Folkstone and Plymouth in two consecutive winters. One sample at Folkstone showed the highest $Tf^B$ value in the data from the region reported here, but this height did not reappear the next winter.

N. P. WILKINS: In addition to Dr. Jamieson's reply to Miss de Ligny's question, I would state that both C. Manwell, and myself, observed a high frequency of abnormal Hb phenotypes in cod from the English Channel near Plymouth, whereas no abnormal Hb phenotypes have been observed in cod from the Southern North Sea.

*XIIth Europ. Conf. Anim. Blood Groups Biochem. Polymorph., Bp., 1972 (pp. 593—595)*

# PRELIMINARY STUDIES ON THE BLOOD OF SARDINELLA FROM THE WEST AFRICAN COAST

J. C. Baron

Office de la Recherche Scientifique et Technique Outre-Mer–Paris (8e) — France

## SUMMARY

**This report gives the preliminary results of blood group investigations on *Sardinella aurita* (C. V.) and *Sardinella eba* (C. V). Human blood and plant extracts were used for erythrocyte antigen detection. Serum protein polymorphism has been studied by starch gel electrophoresis and the transferrin has been identified by Fe 59 autoradiography.**

Two species of Sardinella *(Clupeidae)* have an important commercial value on the West African coast: *Sardinella aurita* (C. V.) and *Sardinella eba* (C. V.). Morphometrical studies indicate that there are several different stocks from Senegal to Congo and we have entered upon serological investigations to identify these subpopulations.

The tropical conditions and the biology of *Clupidae* give special difficulties. The fish must be alive to be bled and the blood doesn't keep very well. The small quantity of blood drawn from the heart doesn't allow extensive absorption techniques.

We tried to detect red cell antigens and biochemical polymorphism in serum, hemoglobin and eye lens. In erythrocyte antigens' investigations we used chiefly human A B O system sera and seed extracts. The interpretation of agglutinations with human sera is very difficult on account of the occurence of a species heteroagglutinin anti-Sardinella. The results obtained with human typing sera absorbed with A and B erythrocytes, and with lecitins from *Phaseolus lunatus* and *Canavalia ensiformis*, indicate the likely occurence of A and B like antigens. No blood group has yet been established. (1) We did not find any natural isoagglutinin or heteroagglutinins against human erythrocytes. Using WA-23 reagent (2) no C antigen has been detected.

The amount of proteins in Sardinella's sera is 46 g/l. These sera have been analysed by means of paper, agar and starch gel electrophoresis. In *S. eba*, with agar medium, 3 principal and 2 secondary lipoproteins have been observed. The electrophoretic pattern reveals on an average 17 components in starch gel with Ashton's buffer. The region No. 8 has been identified to a *transferrin* by autoradiography (3). This transferrin, compared to human sera, has a beta mobility.

(1) Some facts could be directly used. For instance 30% erythrocytes from *S. eba* caught in April 1968 in Abidjan were not agglutinable with human typing sera.

(2) This reagent is a heteroimmune serum produced in the fish *Caulolatilus princeps* by immunization with red cells from the anchovy *Engraulis mordax*. It detects the C antigen on the erythrocytes of Californian Sardine.

(3) The revelation of transferrins by autoradiography has been carried out in the Laboratory of Immunochemistry in the Centre de Transfusion Sanguine de Paris. No polymorphism has been suggested in the transferrins but we expect to find a biochemical polymorphism with the band No. 9 where 4 components A, B, C, D have been detected.

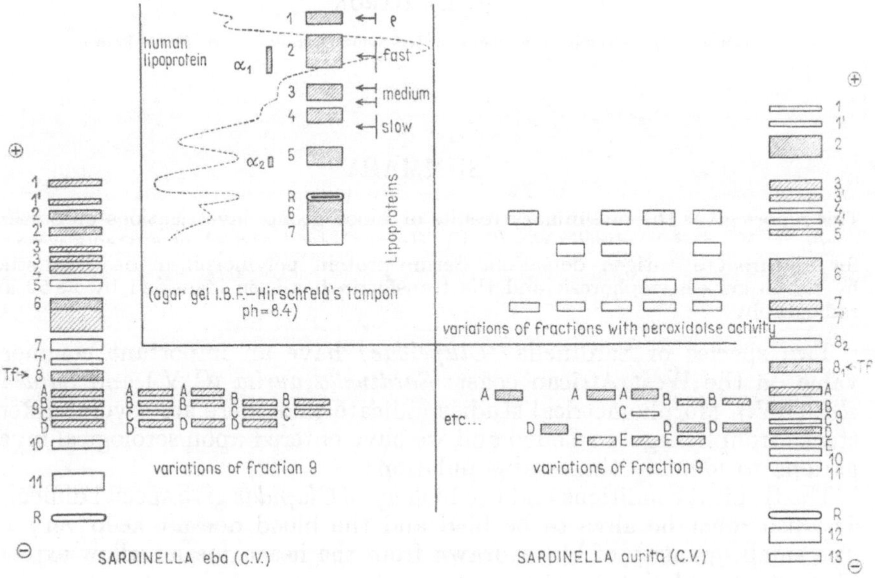
human lipoprotein
fast
medium
slow
Lipoproteins
(agar gel I.B.F.–Hirschfeld's tampon ph=8.4)
variations of fractions with peroxidolse activity
variations of fraction 9
variations of fraction 9
etc...
SARDINELLA eba (C.V.)
SARDINELLA aurita (C.V.)

**Fig. 1. Diagrammatical illustration of electrophoretograms in sera of Sardinelles (13% starch gel – Ashton buffer ph 7.9)**

In *S. aurita* there are on an average 17 components in the serum; this serum can be easily distinguished from the serum of *S. eba* on the basis of the different pattern of the fraction No. 2 (1 band for *S. aurita* and 2 bands for *S. eba*). Variability has been observed with the fractions revealed by benzidin reagent (from 2 to 4 fractions). The transferrin (fraction No. 8) is not polymorphic but seems to have faster migrating subunits. At the level of the fraction No. 9 the bands show different patterns.

No polymorphism has been observed in the hemoglobin of Sardinella. In *S. eba* 2 components giving one precipitation band with hetero-immune sera have been observed.

Preliminary studies on eye lens extracts have been carried out. We got 10 fractions by electrophoresis in agar medium and 7 precipitation bands by immunoelectrophoretic analysis. Eight distinct antigens have been detected by immunodiffusion but they did not permit to separate the different lots of Sardinella.

## REFERENCES

Baron, J. C. (1968), Etude préliminaire sur le sang de deux espèces de sardinelles (Sardinella aurita c.v., Sardinella eba c.v.) Doc. sci. provis. Centre Rech. Océanogr. d'Abidjan. No. 037, 22 p. multigr.

Calaprice, J. R. and Cushing, J. E. (1964), Erythrocyte antigens of california Trouts. California fish and game. **50,** No 3, 152–157.

Creyssel, R., Silberzahn, P., Richard, G. B. and Manuel, Y. (1964), Etude du serum de carpe (Cyprinus carpio) par electrophorèse en gel d'amidon. *Bull. Soc. Chim. biol.*, **46,** 149–159.

Fine, J. M., Drilhon, A., Amouch, P. and Boffa, G. A. (1964), Existence de groupes sériques chez Anguilla vulgaris. Mise en évidence par electrophorèse et autoradiographie de plusieurs types de transferrines. *C. R. Acad. Sci.*, **258,** 753–763.

Jamieson, A. and Jones, W. J. (1967), Two races of Cod at Faroe. *Heredity*, **22,** 610–612.

Krajnovič-Ozretič, M. (1969), Analyses of whole blood proteins in the Adriatic sardine (Sardina pilchardus Walb.). ICES. Special meeting on "The biochemical and serological identification of fish stocks". No 28. Dublin.

de Ligny Wilhelmina, (1969), Serological and biochemical studies on fish populations. *Oceanogr. Mar. Biol. Ann. Rev.*, **7,** 411–513.

Møller, D. and Naevdal, G. (1966), Transferrin polymorphism in fishes, Proc. Xth. Europ. Conf. Anim. Blood Groups Biochem. Polymorph., Paris, 1966, 544 pp.

Odense, P. H., Leung, T. C. and Allen, T. M. (1966), An electrophoretic study of tissue proteins and enzymes of four Canadian cod populations. I.C.E.S. Gadoid Committee G. 14.

Ridgway, G. J. (1966), A complex blood group system in Salmone and trout. Proc. Xth. Europ. Conf. Anim. Blood Groups Biochem. Polymorph. Paris, 1966, 544 pp.

Sindermann, C. J. (1963), Use of plant hemagglutinins in serological studies of clupeoid fishes. *Fish. Bull.*, **63,** 137–141.

Sprague, L. M. and Vrooman, A. M. (1962), A racial analysis of the Pacific Sardine (Sardinops caerulea) based on studies of erythrocyte antigens. *Ann. N. Y. Acad. Sci.*, **97,** 131–138.

Uriel, J., Fine, J. M., Courcon, J. and Le Bourdelles, F. (1957), Contribution à l'étude des protéines et lipoprotéines des sérums d'animaux. *Bull. Soc. Chimie biologique*, XXXIX 1415–1427.

Wieme, R. J. and Kaminski, Marie (1955), Etude de la composition antigénique de l'extrait acqueux du cristallin de boeuf. *Bull. Soc. Chim. Biol.*, **37,** 247.

Wilkins, N. P. (1967), Polymorphism of whole blood proteins in the Cod (Gadus morhua L.). *J. Cons. Int. Explor. Mer.*, **31,** 77–88.

## DISCUSSION

W. de Ligny: You mentioned, that you did not observe polymorphisms in the transferrins of both species of Sardinella studied.

May I ask you how many individuals of each species you analyzed?

I. C. Baron: I analysed 24 S. eba and 58 S. aurita. Each individual serum has been analyzed by autoradiography. It is not impossible that a rare allele would be found after analyzing 1000 individual sera but for population studies such a rare allele would be not useful.

N. P. Wilkins: I agree with Dr. Baron that hemoglobin is not a good genetic marker for identifying populations of *Sardinella*. Indeed, for most, if not all, members of theorder *Isospondylii* (Salmands, Clupeards, etc.) this is so. But it is a good marker for many members of the order *Anacanthinii* (cod, whiting, etc.).

*XIIth Europ. Conf. Anim. Blood Groups Biochem. Polymorph., Bp., 1972 (pp. 597–600)*

# CONNECTIONS BETWEEN ELECTROPHORETIC FRACTIONS OF BLOOD SERUM PROTEINS AND SOME INDICES OF PRODUCTIVITY IN BREAM

A. KIRSIPUU, M. TAMMERT, H. HABERMAN and K. LAUGASTE

**Academy of Sciences of the Estonian SSR, Institute of Zoology and Botany, Limnological Station**

## INTRODUCTION

The aim of our present work is to find out some indices in the blood proteins of the bream which can characterize the productivity of the fish and to ascertain constancy or changeability of these indices at different physiological stages. This paper presents the results of the first stage of our work.

## MATERIAL AND METHODS

About 450 breams from 6 lakes of the Estonian S.S.R. belonging to four drainages were investigated. Most of them were sexually mature. The fish were caught by net in the month of August in the years 1965–1969.

Blood was collected by cardiac puncture, proteins were divided into fractions by electrophoresis on filter paper (serum) and in starch gel (plasma). The paper electrophoresis was carried out in the vertical chamber for 12 hours at a temperature of +5°C and at 400 V. The starch gel electrophoresis was carried out in the horizontal chamber for 12 hours at a temperature of +5°C and current of 4.2 mA/cm². In the starch-gel electrophoretograms the protein and esterase fractions were estimated.

As characteristics of productivity the growth rate, coefficient of condition by Clark, maximum width and lengths of head and trunk were estimated. To ascertain the influence of the physiological state on the blood proteins the fat content between intestines, food found in the intestines, sex, sexual maturity and histochemical composition of the liver were studied. The composition of the liver was studied in 144 breams from the Lake Võrtsjärv: the glycogen, fat and RNA content of liver cells was estimated visually in histological preparations.

## RESULTS

By paper electrophoresis the blood serum proteins of bream were separated into five fractions: $\alpha_1$-, $\alpha_2$-, $\beta$- and $\gamma$-globulins. In many cases $\beta$-globulins were separated into two subfractions. By starch-gel electrophoresis 12–14 fractions were obtained. On the ground of the occurrence of fractions "x", "y" and "z" (presumably transferrins) six genotypes were established (Fig. 1).

The comparison of the electrophoretograms obtained by the different methods showed a weak interdependence only. The frequency of the separation of $\beta$-globulins into two fractions on the strip and, to a certain extent, the percentages of albumins, $\alpha_2$-, $\beta$- and $\gamma$-globulins were connected with

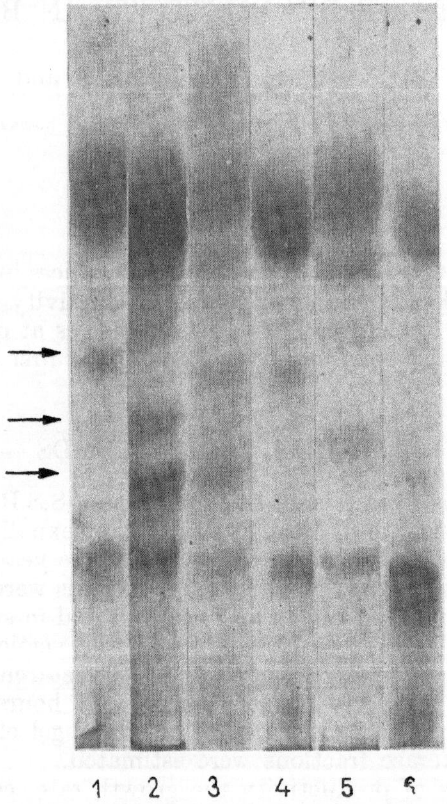
1
2
3
4
5
6

FIG. 1. Starch gel electrophoretograms of blood plasma proteins of bream. Arrows indicate those fractions, which were taken into account by establishing the genotypes: 1 — xy, 2 — yz, 3 — xz, 4 — xx, 5 — yy, 6 — zz

the frequency of the genotypes with alleles y and z (Table 1). The data on the bream from lake Pangodi did not correlate with these relations nor with others described below, evidently because of the entire inbreeding of bream population in this lake.

The frequency of occurrence of the genotypes was different in the lakes investigated. A connection was ascertained between the growth rate (described in Table 1 byt he weight of the 10-year-old individuals) and some mor-

TABLE 1

*Electrophoretic fractions of blood proteins and some indices of productivity in bream from several Estonian lakes*

| Index \ Lake | Veisjärv | Hino | Võrtsjärv | Aheru | Veskijärv | Pangodi |
|---|---|---|---|---|---|---|
| Genotypes with allele y, % | 50 | 45 | 44 | 39 | 36 | 34 |
| Genotypes with allele z, % | 8 | 24 | 29 | 28 | 31 | 23 |
| Genotypes with allele x, % | 42 | 31 | 27 | 33 | 33 | 43 |
| Albumins, % | 26.9 | 25.3 | 23.6 | 21.6 | 23.5 | 23.7 |
| $\alpha_2$-globulins, % | 18.6 | 17.8 | 18.0 | 16.0 | 16.4 | 15.8 |
| $\beta$-globulins, % | 25.5 | 23.7 | 24.4 | 27.6 | 26.1 | 22.8 |
| $\gamma$-globulins, % | 9.9 | 13.3 | 16.2 | 16.0 | 15.5 | 16.3 |
| $\beta + \gamma$-globulins, % | 35.4 | 37.0 | 40.6 | 43.6 | 41.6 | 39.1 |
| Individuals with two $\beta$-fractions, % | 18 | 19 | 24 | 40 | 47 | 16 |
| Weight of 10-year-old individuals, g | 1230 | 760 | 610 | 330 | 140 | 540 |
| Length of head | 20.6 | 21.8 | 22.8 | 23.5 | 23.2 | 22.9 |
| Max. body width | 39.7 | 37.5 | 37.7 | 37.2 | 37.1 | 35.5 |

phometric indices (altitudo maxima, longitudo capitis) which are considered to be indirect characteristics of the individual productivity, on the one hand, and occurrence of the genotypes with allele y, on the other hand. These indices were inversely related with the genotypes with allele z. No connections between protein fractions of starch-gel electrophoresis and physiological state of the fish were ascertained.

Two genotypes were established on the ground of esterases, with one and two fractions respectively. Statistically significant difference between these genotypes was found only in the coefficient of condition and altitudo corporis maxima.

The paper-electrophoretic protein fractions had only weak connection with the indices of productivity. In none of the lakes investigated was found any correlation between percentages of blood protein fractions and growth rate or other productivity indices. However, by comparing the data of several lakes it became evident, that in the lakes where the growth rate of the bream was higher, the mean percentage of albumins was also higher and that of $\beta$- and $\gamma$-globulins was shomewhat lower (Table 1). But correlations were ascertained between the percentages of paper electrophoretic protein fractions and some charachteristics of the physiological state of the fishes. Analyses of the breams from the Lake Võrtsjärv have shown that the maturity index in females had a correlation with albumins ($r = -0.634$) and $\alpha_2$-globulins ($r = 0.636$), fat on the intestines with $\alpha_2$-globulins ($r = -0.260$), fat content in the liver cells with albumins ($r = -0.242$)

glycogen content in the liver cells with albumins (r = 0.324) and RNA content in the liver cells with albumins (r = 0.269), and $\alpha_2$-globulins (r = 0.342).

## DISCUSSION

The materials presented above reveal that the qualitative features of the electrophoretogram — genotype of the transferrins, occurrence of the $\beta$-globulin subfractions — have a constant relation with the metabolism of the fishes. In other words, the constant interrelations of the different kinds of metabolic processes are, of course only in part, determined by these genotypes. Consequently, genotypes, established on the ground of the blood protein fractions, reflect to a certain extent the type of the metabolism and thus, also, the preconditions of individual productivity. Ecological factors, evidently, do not affect genotypes (or their effect is insignificant), but their influence on the realization of the genotypic preconditions is certain. The range of this influence is, to a considerable extent, reflected in the changes of the quantitative relationships of the paper-electrophoretic fractions, which are affected by ecological factors. However, the metabolic rate in a given moment and under given conditions, is an indirect reflection of the type of metabolism and therefore the quantitative relationships of the blood proteins indirectly reflect also the type of metabolism. That is the reason why the individual genetic differences do not appear in the paper electrophoretogram and are detectable as statistical differences of the genofond in different populations only.

## CONCLUSIONS

1) Only a weak interdependence was ascertained between blood protein electrophoretograms of the bream obtained on filter paper and in starch gel. The frequencies of occurrence of two $\beta$-globulin fractions and the percentages of albumin, $\alpha_2$-, $\beta$- and $\gamma$-globulin fractions of paper electrophoretograms were, to some extent, connected with the occurrence of certain fractions in starch-gel electrophoretograms.

2) In the starch-gel electrophoretograms there are three fractions, the frequencies of occurrence of which are in close correlation with the mean individual productivity in the population. No influence of the physiological factors on these fractions was observed.

3) The number of fractions on the paper electrophoretograms and their percentages reflect the mean individual productivity in the population indirectly, because they depend closely upon the physiological state of the fishes.

*XIIth Europ. Conf. Anim. Blood Groups Biochem. Polymorph., Bp. 1972 (pp. 601—606)*

# SOME PROBLEMS IN THE APPLICATION OF BLOOD GROUPING TECHNIQUES TO THE ATLANTIC SALMON (*Salmo salar* L.)

N. P. WILKINS*, G. I. SANGSTER and D. A. CONROY**

Department of Agriculture and Fisheries for Scotland Marine Laboratory, Aberdeen, Great Britain

** Unilever Research Laboratory, Grey Hope Road, Aberdeen, Scotland, Great Britain

## INTRODUCTION

In approaching biochemical and serological studies on wild populations of fish, two major difficulties are immediately apparent: the unavailability of individuals of some species at certain stages of the life cycle due e.g., to migration, and, secondly, the inability to sample repeatedly, single individuals of some species, due to their resistance to establishment in aquaria, or extreme sensitivity to the rigors of handling and sampling. Thus, it is important to establish that the techniques used, and the phenomena observed, in e.g., juvenile individuals, are the same as those used and observed in adult specimens of the same species. This paper presents results of some blood group studies on Atlantic salmon, and discusses some published observations, which indicate problems that may be of general applicability to fish, and that do not appear to have been dealt with in detail elsewhere.

## MATERIAL AND METHODS

Blood was collected by cardiac puncture, with dry Heparin powder as anticoagulant. Agglutination reactions were performed in the usual manner, the degree of agglutination being assessed microscopically.

Blood smears were air dried, and stained with Leishmann stain. Red cell sizes were read microscopically using an eye piece scale, at a magnification at which 1 eye piece scale division = 1.611 $\mu$.

## RESULTS

Natural agglutinating antibodies, which detect individual antigenic differences in salmon erythrocytes, have been observed in the normal unabsorbed serum of individual dogfish (*Squalis acanthias* L.).

Table 1 presents the results obtained when the erythrocytes of a number of salmon taken in Scotland, Canada and West Greenland were tested

* *Present address:* Department of Zoology, University College, Galway, Ireland

TABLE 1

*Agglutination of the erythrocytes of Salmon from Scotland, Canada and the Davis Strait off W. Greenland by the normal serum of a single dogfish* (Serum D. 16) *and the pooled sera of five dogfish* ('Pool')

Reagent : serum D16

| Sample | No. | Degree of agglutination | | | | |
|---|---|---|---|---|---|---|
| | | — | (+) | 1 | 2 | 3 |
| Scotland | 49 | 28 | 17 | 3 | 1 | — |
| Greenland | 71 | 14 | 32 | 18 | 6 | 1 |
| Canada | 47 | 22 | 12 | 13 | — | — |

Reagent : pooled sera

| Sample | No. | — | (+) | 1 | 2 | 3 |
|---|---|---|---|---|---|---|
| Scotland | 49 | 41 | 8 | — | — | — |
| Greenland | 71 | 43 | 26 | 2 | — | — |
| Canada | 47 | 29 | 14 | 4 | — | — |

with various dogfish sera. Differences in the degree of agglutination are evident between the samples from the three areas, suggesting that population differences may exist in the occurrence of the agglutinogens detected with these normal sera.

The observed agglutination was inhibited more-or-less completely by the prior addition of an equal volume of salmon mucus to the dogfish serum, or to the reaction mixture (Table 2).

TABLE 2

*Inhibition of the agglutination of salmon erythrocytes by salmon mucus*

| Cells | Degree of agglutination with: A | B | C |
|---|---|---|---|
| 1 | 2 | — | (+) |
| 2 | 1 | — | (+) |
| 3 | 2 | — | (+) |
| 4 | 1 | — | (+) |
| 5 | 1 | — | (+) |
| 6 | 1 | — | (+) |

A = Dialysed Dogfish serum D24 diluted 1 : 2 with saline
B = Dialysed Dogfish serum D24 diluted 1 : 2 with salmon mucus
C = Dialysed Dogfish serum D24 diluted 1 : 2 with boiled, dialysed salmon mucus

Inhibition occurred irrespective of whether the mucus was obtained from individuals whose cells reacted positively or negatively with the appropriate dogfish serum. Inhibition was also observed when boiled and dialysed

TABLE 3

*Agglutination of salmon erythrocytes by normal dogfish serum D24*

| Cells | A | B | C | D |
|---|---|---|---|---|
| | Degree of agglutination | | | |
| 1 | 2 | — | — | (+) |
| 2 | 2 | — | — | — |
| 3 | 1 | — | — | — |
| 4 | 2 | — | (+) | (+) |

A = Salmon erythrocytes in 1.14% Saline
B = Salmon erythrocytes in Salmon mucus/Saline (1 : 1)
C = Salmon erythrocytes as in B, but washed with 20 vols 1.14% Saline
D = Salmon erythrocytes as in B, but washed with 40 vols 1.14% Saline

mucus was used, suggesting that neither small molecular weight substances, nor heat-labile enzymes, were alone involved. As illustrated in Table 3, inhibition was evident even after washing the cells twice with twenty volumes (each wash) of saline.

The results suggest that the antigenic sites on the erythrocytes are coated or blocked in some manner by a factor, or factors, present in the mucus of all salmon. Similar, but less intensive, inhibition was observed when agglutination was performed with high-titred rabbit immune anti-salmon sera.

The dimensions of the long and short axes of salmon erythrocytes are presented in Table 4 and Fig. 1.

TABLE 4

*Mean long and short axes of the erythrocytes of Scottish salmon of different sizes*

| | No. of cells measured | Long axis | | Short axis | | $\bar{L}/\bar{S}$ |
|---|---|---|---|---|---|---|
| | | Divs. | μ | Divs. | μ | |
| Embryo | | 8.55 | 13.77 | 7.23 | 11.65 | 1.18 |
| Alevin | 423 | 8.84 | 14.25 | 5.81 | 9.37 | 1.52 |
| Fingerling | 273 | 9.18 | 14.80 | 5.40 | 8.70 | 1.70 |
| Parr | 224 | 9.19 | 14.81 | 5.48 | 8.83 | 1.68 |
| Smolt | 155 | 9.45 | 15.23 | 5.35 | 8.62 | 1.77 |
| Adult (Greenland) | 1262 | | 15.48 | | 9.89 | 1.57 |
| Spawning adult | 535 | 9.63 | 15.52 | 5.74 | 9.25 | 1.68 |

The long axis tends to increase, and the short axis to decrease, as the fish grow in length (up to smolt stage), the two-dimensional shape of the cell becoming more elongated. Adult salmon taken in the Davis Strait have a rounder shape (shorter mean long axis, and longer mean short axis) than individuals of similar size taken in brackish water in Scotland in the course of the spawning migration.

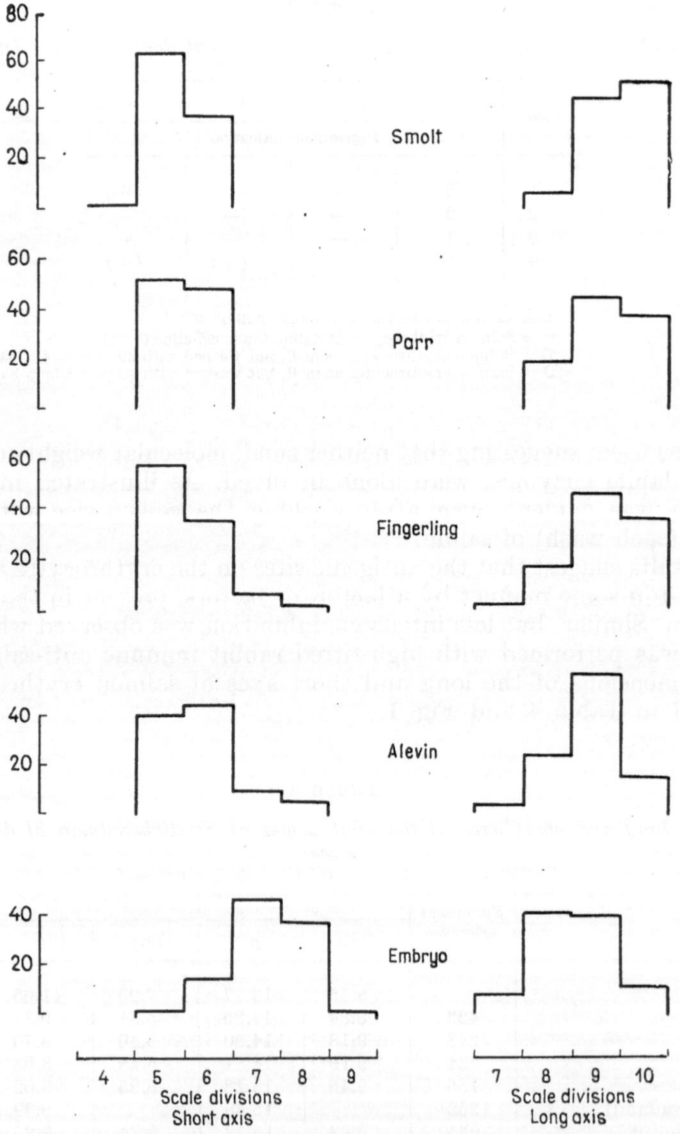

Smolt
Parr
Fingerling
Alevin
Embryo
Scale divisions
Short axis
Scale divisions
Long axis


**Fig. 1.** Histogram of the sizes of the long and short axes of the erythrocytes of Salmon of different sizes

## DISCUSSION

These results illustrate two possible problems in blood group studies on fish: the effect of contamination on the sample i.e. extraneous modification, and secondly, temporal variation in fish tissue i.e. intrinsic variation. In the former case, the observation that mucus contamination may alter the agglutinability of the red cells, as well as presenting an interesting observation in itself, presents an important practical consideration. Severance of the tail (with or without immersion of the severed trunk in Alsever solu tion) has long been one of the standard, routine methods for collecting the blood of fishes, especially from species of which juvenile stages are being sampled, or in which the adults do not grow to a sufficiently large size for cardiac puncture (CUSHING, 1964). Contamination with mucus and other body fluids is difficult to avoid in these circumstances, and the results indicate that repeated washing of the cells may not reverse those changes which may arise through contamination at the time of bleeding.

It is implicit, in the application of blood grouping techniques to the identification of races of any species, that the composition of the red cell population does not vary significantly at different stages of the life cycle, the agglutinogens being detectably present (or absent) at all of the stages normally investigated. Early in our studies it became obvious that the eryhtrocyte population of individual Atlantic salmon was not *biochemically* constant throughout the life cycle: the electrophoretic pattern of the hemoglobins differed in individuals of differing length, although it has not been conclusively shown whether this represents a change in red cell type or simply in cellular metabolism. In either case, the erythrocytes of small and large salmon are demonstrably and reproducibly different in the electrophoretic pattern of their constituent hemoglobins. The preliminary results presented here on erythrocyte size, and two-dimensional shape, suggest that these parameters also may vary in individuals of different size and physiological condition. Similar changes in the size and shape of salmon red cells, occurring in the course of the spawning migration, have been previously reported by NUSENBAUM (1953). Evidence is furthermore available indicating that the agglutination of the erythrocytes of one year old Scottish salmon parr (bled by cardiac puncture) differs from that of two-year-old Scottish salmon smolts, reared under identical conditions (WILKINS, 1970).

Results of a similar nature have been observed in other species of fishes: SANDERS and WRIGHT (1962) presented evidence that the red cells of certain individual brown trout changed during the yearling stage, through a mechanism which rendered them inagglutinable by a specific reagent. VANN (1966) and VANN and CUSHING (1966) quoted by DE LIGNY (1969), noted a higher antigen frequency in large specimens of the Bonito *(Sarda chiliensis)* than in smaller individuals of the same sample. DE LIGNY (1967) noted a correlation between age and AB blood group system in plaice, and FUJINO (1967) found an association between the blood group system and size of fish in Skipjack Tuna *(Katsuwonus pelamis)*. EZELL, SULYA and DODGEN (1969), investigating the behaviour of fish erythrocytes in hypotonic saline solutions, identified two distinct populations of erythrocytes occurring together within certain individuals of the species *Mugil cephalus* and *Archosargus gill*.

In short, it would appear that within individuals of some species of fish the erythrocyte population may be neither homogenous nor temporally constant, and comparisons drawn between the agglutination reactions observed in different populations which differ in age or physiological condition may not always reflect strict population differences.

## REFERENCES

CUSHING, J. E. (1964), The blood groups of marine animals. Advances in Marine Biology, **2,** 85–131. Academic Press, London,

EZELL, G. H., SULYA, LL. and DODGEN, C. L. (1969), Some unusual behaviour of fish erythrocytes in hypotonic saline solutions. *Comp. Biochem. Physiol.*, **30,** 137–147.

FUJINO, K. (1967), Review of subpopulation studies on Skipjack Tuna. *Proc. Conf. west. Ass. St. Game Fish Commnrs.*, **47,** 349–371.

DE LIGNY, WILHELMINA (1967), Physiological aspects of serological and biochemical characteristics in plaice. *Int. Counc. Explor. Sea. C. M.* 1967, F22 (Mimeo) 7 pp.

DE LIGNY, WILHELMINA (1969), Serological and biochemical studies on fish populations. *Oceanogr. Mar. Biol. Ann. Rev.*, **7,** 411–513.

NUSENBAUM, L. M. (1953), Shape of erythrocytes in fish. *Dokl. Akad. Nauk SSSR,* **90,** 889–892.

SANDERS, B. G. and WRIGHT, J. E. (1962), Immunogenetic studies in two trout species of the genus *Salmo. Ann. N. Y. Acad. Sci.*, **97,** 116–130.

VANN, D. C. (1966), *Thesis*, Univ. Calif., Santa Barbara.

VANN, D. C. and CUSHING, J. E. (1966), *Fedn. Proc. Fedn. Am. Socs. exp. Biol.*, **25.** 437.

WILKINS, N. P. (1970), Biochemical and Serological studies on Atlantic Salmon. I.C.E.S. Rapp. et Proc.-Verb. (In press)

# VIII. BLOOD GROUPS AND BIOCHEMICAL POLYMORPHISM IN EXPERIMENTAL ANIMALS

*XIIth Europ. Conf. Anim. Blood Groups Biochem. Polymorph., Bp., 1972 (pp. 609–613)*

# SELECTION AT THE TRANSFERRIN LOCUS IN MICE*

G. C. ASHTON

Department of Genetics, University of Hawaii, Honolulu, Hawaii 96822 U.S.A.

## SUMMARY

**Data on litter size, distribution of litters, sex ratio and segregation were collected from thirteen generations of a laboratory mouse colony segregating for two transferrin alleles, $Trf^a$ and $Trf^b$. There were significant differences between Trfa females and Trfb females, mated to the same Trfa male, in litter size ($P < 0.01$), females producing litters ($P < 0.05$), and females littering within four days of first litter appearing ($P < 0.05$). The mating a♂ × a♀ showed a significant deficiency of a♀ progeny ($P < 0.01$) and from all segregating matings there was a significant deficiency of female progeny with $Trf^a$ ($P < 0.05$). There were significant differences in progeny sex ratio, a♂×b♀ matings producing more male progeny than b♂×b♀ matings ($P < 0.01$). The overall analysis shows $Trf^b$ with a 2.3% advantage over $Trf^a$ under defined conditions. This would lead to elimination of $Trf^a$ from a population and accounts in part for this allele not having been found in natural populations of housemice.**

## INTRODUCTION

The majority of vertebrates show transferrin polymorphism although there are some exceptions. A curious situation occurs in the housemouse. Natural populations do not show transferrin polymorphism (PETRAS, 1967; SELANDER and YANG, 1969), but inbred strains of laboratory mice exhibit one of two alternative alleles (ASHTON and BRADEN, 1961; COHEN and SHREFFLER, 1961). All strains except CBA and strains derived from it have $Trf^b$, while CBA has $Trf^a$. This contrast between natural and laboratory populations caused us to examine this species for evidence of selection at the transferrin locus. A preliminary report was given at the Warsaw meeting in 1968.

## METHODS

The colony was established in May 1966 by single pair matings of six CBA/J males with six C3He/FeJ females and six C3He/FeJ males with six CBA/J females. These two strains, one carrying $Trf^a$ and the other $Trf^b$ were chosen for equality of litter size and mature body size, to minimize maternal effects as far as possible. Data collection commenced with the F3 generation. Full details of the matings will be given elsewhere. In summary

* Supported by PHS research grant No. HD 01831 from the National Institute of Child Health and Human Development.

F3, F6 and F7 were single pair matings producing segregants; F4, F5, F11, F12 and F13 consisted of one male, either Trfa, ab or b mated concurrently in the same cage to a Trfa, ab and b female; and F8, F9 and F10 consisted of one male either Trfa or b, mated concurrently in the same cage to a Trfa and b female. In order to avoid such possible complications as the development of maternal immunological intolerance all our data relate to progeny from virgin females, mated to previously unmated males.

## RESULTS

*Litter size.* In generations F4, F5, F11, F12 and F13, 96 Trfa males were mated in single cages to Trfa, Trfab and Trfb females concurrently as were 100 Trfab males and 111 Trfb males. The Trfa males produced 656 progeny from the Trfa female, 660 from the Trfab females and 697 from Trfb females. The Trfab males produced respectively 759, 734 and 731, and the Trfb males produced 825, 818 and 807. These numbers of progeny do not differ from expectations, but the agreement is misleading due to the relatively small number of progeny from these matings.

If Trfa and Trfb males, mated concurrently to Trfa and Trfb females, only are considered then data from generations F8, F9 and F10 may be added. Then 293 Trfa males produced 2128 offspring from 293 Trfa females and 2260 from Trfb females ($\chi^2 = 3.91$, $P < 0.05$ with no significant heterogeneity between generations). By contrast 310 Trfb males produced 2279 offspring from 310 Trfa females and 2299 from 310 Trfb females ($\chi^2 = 0.09$, N.S.). These data were also analyzed by Duncan's new multiple range test (Steel and Torrie, 1960). The mean litter size of a♂ and b♀ (7.713) differed significantly, $P < 0.01$, from that for a♂×a♀ matings (7.263) and from b♂×a♀ matings (7.352, $P < 0.05$). The mean litter size for b♂×b♀ matings (7.416) did not differ from the others.

*Distribution of litters.* In generations F8–F13 the temporal distribution of litters was recorded. All males and females in an experiment were mated on the same day. The day of appearance of the first litter was called day 1. Differences in means for the numbers of females littering by day 4 was used as a measure of interference with the estrus cycle, comparable to the use of delayed "returns to service" in cattle. The temporal distribution of litters was analyzed as a series of 2×2 contingency tables, comparing Trfa and b females mated to either Trfa or Trfb males. There was no evidence of heterogeneity between generations and only the pooled data are presented for brevity. From 297 cages with Trfa males, only 156 (52.5%) of Trfa females had littered by the fourth day, but 182 (61.3%) of the Trfb female had littered ($P < 0.05$). The comparable values for 298 Trfb males were 173 (58.0%) of Trfa females and 188 (63.1%) of Trfb females (N.S.). Pooling the data gave $P < 0.02$ for the difference between Trfa and Trfb females irrespective of male genotype.

Nine Trfa females and two Trfb females mated to the 297 Trfa males failed to litter ($P < 0.05$). There was no difference between a and b females mated to Trfb males (1 and 2 without litters respectively).

*Segregation data.* Distribution of progeny genotypes from generations F3–F7 and F11–F13 in which segregation was possible were analyzed for deviation from expectation and for heterogeneity between generations. Only one mating type, a♂×ab♀ showed significant effects, and then only for female progeny. The results were a♂×ab♀ gave 414 a♂, 404 ab♂, 338 a♀ and 434 ab♀. The deficiency of a♀ was significant ($P < 0.01$), and consistent between generations.

The gene segregation data were also examined over all segregating matings. The results were as follows:

| | Male Progeny | | Female Progeny | |
|---|---|---|---|---|
| | obs. | exp. | obs. | exp. |
| $Trf^a$ | 4355 | 4367.5 | 4051 | 4149.0 |
| $Trf^b$ | 4221 | 4208.5 | 4131 | 4033.0 |

The difference for female progeny was significant ($P < 0.05$); only 39% of the difference was due to the a♂×ab♀ mating.

*Sex ratios.* Mice were counted and weaned at 21 days and sexed at that time. Only those litters which did not show a loss between birth and weaning were included in the analysis. Sex ratios from those generations with all nine possible mating combinations did not yield significant differences. However, when all generations were considered, and comparisons restricted to matings between homozygotes significant differences were obtained as follows: a♂×a♀ (1.0456), a♂×b♀ (1.1126), b♂×a♀ (1.0265) and b♂×b♀ (0.9063). Trfb♂×Trfb♀ matings produced more males and fewer females compared with a♂×b♀ matings ($P < 0.01$), or with a♂×a♀ matings ($P < 0.05$).

## DISCUSSION

Significant effects were obtained for each of the parameters considered, namely litter size, distribution of litters, segregation and sex ratio. In each case it is possible to interpret the results obtained in such a way that $Trf^a$ appears to be at a disadvantage compared to $Trf^b$.

Considering firstly the segregation data one mating, Trfa♂×Trfab♀, showed a significant deficiency of Trfa female progeny. If the observed and expected distribution of $Trf^a$ and $Trf^b$ genes from all segregating matings are compared there was a significant deficiency of $Trf^a$ genes in the female progeny, but no segregation disturbance in the male progeny. Even with the quite large numbers of progeny available for the analysis, and a difference of 4.8% between the expected distributions of $Trf^a$ and $Trf^b$ genes, the difference was only just significant ($\chi^2 = 4.69$, $P < 0.05$).

The number of Trfa and Trfb females which produced litters when mated to Trfb males showed an advantage for the Trfb females. The superiority, if any, of Trfb females mated to Trfb males, was not discernible statistically. Again, the presence of the $Trf^b$ gene in females confered advantage. As the mouse is a polyestrous animal with seasonal breeding in the wild the number of litters produced during the breeding season would be an important component influencing population gene frequencies.

Litter size differences may be interpreted in the same way. There was no discernible difference in litter size between females mated to Trfb males but there was a highly significant superiority of Trfb females over Trfa females when mated to Trfa males; Trfb females produced an average litter of 7.71, and Trfa females 7.26.

Finally there was a significant effect on sex ratio, such that the most prolific mating, a♂×b♀ produced the most males (52.7%) while the mating b♂×b♀ produced the least (47.5%). The effect of sex ratio on gene frequency has been considered by a number of authors, and most recently by MYERS and KREBS (1970) who have reported an effect of the transferrin locus on sex ratio in *Microtus ochrogaster*. These workers using a computer simulation technique found that increasing the proportion of males present at weaning, i.e. increasing the sex ratio, caused a marked decline in population size after several generations. If this effect also occurs in natural populations of mice then the advantage of greater litter size of a♂×b♀ matings might be offset by the disadvantage of a greater proportion of males produced.

As yet we have insufficient information from the homozygote × heterozygote and double heterozygote matings to permit formulation of a complete model. However the following conclusions may be reached, using data obtained in this study:

1) Trfa females produce smaller litters than b females.

2) Trfa females produce fewer litters than b females, at least when mated to Trfa males.

3) Trfa females produce litters more slowly than b females, as judged by distribution of litters in time.

4) Segregating matings produce fewer $Trf^a$ genes than expected in the female progeny.

5) Matings between a♂ and a♀ produce more male progeny than b♂×b♀ matings; the latter mating would be expected to increase the population size and therefore the frequency of the $Trf^b$ allele.

It is evident that transferrins in the housemouse do not constitute a balanced polymorphism, and that $Trf^b$ is consistently at a selective advantage.

Consequently $Trf^a$ should not be found at polymorphic frequencies in natural populations. In fact it has not yet been reported at all in the wild.

## REFERENCES

ASHTON, G. C. and BRADEN, A. W. H. (1961), Serum β-globulin polymorphism in mice. *Aust. J. Biol. Sci.*, **14**, 248–253.

COHEN, B. L. and SHREFFLER, D. C. (1961), A revised nomenclature for the mouse transferrin locus. *Genet. Res. Camb.*, **2.** 306–308.

MYERS, J. H. and KREBS, C. J. (1970), Sex ratios in open and enclosed vole populations: demographic implications. *Amer. Nat.* (submitted).

PETRAS, M. L. (1967), Studies of natural populations of Mus. I. Biochemical polymorphisms and their bearing on breeding structure. *Evolution*, **21**, 259–274.

SELANDER, R. K. and YANG, S. Y. (1969), Protein polymorphism in wild populations. *Genetics*, **61**, 54.
STEEL, R. G. D. and TORRIE, J. H. (1960), Principles and Procedures of Statistics. McGraw-Hill, New York, p. 107.

## DISCUSSION

J. RENDEL: 1. Interesting to note that there were no segregation disturbances matings ab×a, ab×b. In cattle as you know there have been large deficiencies of heterozygotes in these types of matings. Would you please like to comment on this.

2. What was the age at typing the animals? The reason for this question is that the sex ratio is much influenced by age. It appears that in your data the highest sex ratios were consistently obtained in matings giving 100% heterozygosity. It might be so that the usually lower viability of males is offset by heterozygosity.

3. It would be interesting to study the drift of the gene frequency in a line kept by random mating and being derived from the $F_2$ generation. Has this been done?

G. C. ASHTON: 1. A doubt if there is any homology between what is called $Tf^A$ in cattle and what is called $Trf^a$ in mice.

2. Animals were counted 4 days after birth and sexed at 21 days at weaning. This was strictly adhered to for the reasons you mention. Birth to weaning losses were low and sex ratio at weaning must be close to the sex ratio at birth.

3. This would be well worth doing, but so far we have not done this.

F. K. KRISTJANSSON: In analysing a 3×3 table for litter size it would seem more appropriate to determine whether there are significant differences among mating classes before making comparisons between particular mating classes.

G. C. ASHTON: The analysis was in fact done this way, utilising a multiple linear regression analysis which design variables as suggested by Harvey. A range lost was shown for simple presentation. Analysis by $\chi^2$ is justified because of the experimental design, one male to two or three females.

A. JAMIESON: Please, outline the sources and numbers of wild mouse transferrin data.

G. C. ASHTON: There are published reports by Petras, and by Selandorf from continental United States. Also data are available from Japan, Australia and Hawaii.

F. PIRCHNER: To investigate the effect of the Tf locus upon fitness, male mating behaviour, fertility etc. should be studied for completeness.

G. C. ASHTON: With our limited facilities we considered it preferable to investigate differences between females as a first step. Certainly, for completeness, male behaviour should be investigated.

*XIIth Europ.Conf. Anim. Blood Groups Biochem. Polymorph., Bp., 1972 (pp. 615–620)*

# SEROLOGICAL AND TRANSPLANTATION ANALYSIS OF RECOMBINANT ALLELES AT THE HISTOCOMPATIBILITY-2 LOCUS OF THE MOUSE

P. DÉMANT, JANA BENESOVÁ, JITKA MARTÍNKOVÁ and LIBUŠE OPPLTOVÁ

**Institute of Experimental Biology and Genetics, Czechoslovak Academy of Sciences, Prague, Czechoslovakia**

The histocompatibility-2 (H-2) locus occupies an outstanding position among other mouse H-loci because of the high antigenicity and serological complexity of its products. The product of each H-2 allele consists of a number of serologically distinct antigenic specificities. This serological complexity of H-2 antigens and the possibility to produce large numbers of mice with a desired H-2 genotype make the H-2 locus a suitable tool for the analysis of the mechanism of genetic control in complex immunogenetic systems. Research on the genetic structure of the H-2 locus resulted in detection of a number of recombinant alleles and revealed its genetic complexity (for review see SCHREFFLER, 1967).

One of the interesting features of intra-H-2 recombinations is the tendency of some groups of antigenic specificities to be transferred together in several independent recombinational events. This results in a series of recombinant alleles which all have an identical or apparently identical segment of genetic material. The repeated transmission of the D-end of $H\text{-}2^{d}$ allele to $H\text{-}2^{a}$ and $H\text{-}2^{i}$ and of the K-end of $H\text{-}2^{k}$ allele to $H\text{-}2^{a}$, $H\text{-}2^{h}$, and $H\text{-}2^{m}$ is demonstrated in the Figure 1. The crossing-over between $H\text{-}2^{d}$ and $H\text{-}2^{k}$ which gave rise to the $H\text{-}2^{a}$ allele occurred in the ancestors of the contemporary inbred mouse strains and the recombinational origin of $H\text{-}2^{a}$ was inferred from its serotype (AMOS, GORER and MIKULSKA, 1955; STIMPFLING and RICHARDSON, 1965) and from the existence of the dk-effect (SNELL, 1951). More recently a recombination between $H\text{-}2^{d}$ and $H\text{-}2^{k}$ giving rise to $H\text{-}2^{a}$ was observed by SHREFFLER (1969). Appearance of $H\text{-}2^{h}$ and $H\text{-}2^{i}$ from recombination between $H\text{-}2^{a}$ and $H\text{-}2^{b}$ was described several times. The cross-over origin of $H\text{-}2^{m}$ from $H\text{-}2^{k}$ was suspected long ago (GORER and MIKULSKA, 1959) and on the basis of more detailed serological analysis it was suggested that $H\text{-}2^{m}$ is a recombinant between $H\text{-}2^{k}$ and $H\text{-}2^{q}$ (SNELL, DÉMANT and CHERRY, in press). This suggestion is in agreement with the gene map of $H\text{-}2^{q}$ based on the analysis of $H\text{-}2^{y}$, a recombinant between $H\text{-}2^{a}$ and $H\text{-}2^{q}$ (KLEIN, BEDNÁROVÁ and SCHREFFLER, 1969).

Some antigenic specificities controlled by the D-end of $H\text{-}2^{d}$ allele or by the K-end of the $H\text{-}2^{k}$ allele are present also in products of some H-2 alleles which have not the D-end identical with the D-end of the $H\text{-}2^{d}$

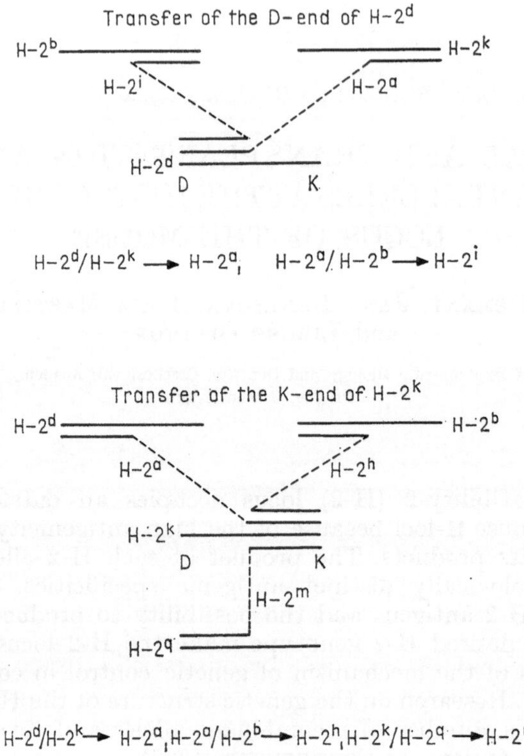

Transfer of the D-end of H-2d
H-2b
H-2k
H-2i
H-2a
H-2d
D
K
H-2d/H-2k → H-2a, H-2a/H-2b → H-2i
Transfer of the K-end of H-2k
H-2d
H-2b
H-2a
H-2h
H-2k
D
K
H-2m
H-2q
H-2d/H-2k → H-2a, H-2a/H-2b → H-2h, H-2k/H-2q → H-2m


FIG. 1

allele or the K-end identical with the K-end of the H-$2^k$ allele. However, these alleles which are called 3-intermediate (DÉMANT, SNELL and CHERRY, in press) or 1-intermediate (SNELL, DÉMANT and CHERRY, in press) alleles, never produce all specificities controlled by the D-end of 3-complete alleles or by the K-end of 1-complete alleles (the 3-complete alleles have the D-end identical with H-$2^d$ and the 1-complete alleles have the K-end identical with H-$2^k$). Moreover, in the products of intermediate alleles the specificities which they share with the complete alleles seem to be present in quantitatively smaller amounts. In quantitative measurements using the $Cr^{51}$ lymphocytotoxic assay the lymphocytes with 1-complete alleles were shown to be 2–10 times superior to lymphocytes with 1-intermediate alleles in their capacity to absorb the antibodies to the K-end of H-$2^k$ (SNELL, DÉMANT and CHERRY, in press). Similarly, the lymphocytes with 3-complete alleles are superior to the lymphocytes with 3-intermediate alleles in their capacity to absorb antibodies to the D-end of H-$2^d$ (DÉMANT, SNELL and CHERRY, in press). An interesting case is represented by the H-$2^o$ allele, a recombinant between H-$2^d$ and H-$2^k$ which has received only the central portion of the K-end of H-$2^k$, i.e. the E-region, the rest being replaced by

TABLE 1

*Hemagglutination activity of anti-3 complex sera*

| ASP | Donor | Recipient | d | a | i | g | o |
|---|---|---|---|---|---|---|---|
| 34 | 5R | 2R | 6.5 | 10.5 | 7 | 0 | 0 |
| 52 | 5R | B10 | 10.5 | 13 | 4 | 0 | 0 |
| 61 | 5R | B10×B10.BR | 5.6 | 8.5 | 4.5 | 0 | 0 |
| 83 | A.SW | B10×ACA | 12.3 | 13[a] | 11 | 0 | 3[a] |
| 86 | 5R | B10×DBA/1 | 14 | 13.5 | 10 | 0 | 4 |
| 89 | B10.A | B10×C3H | 11 | 13 | 10 | 0 | 0 |
| 97 | B10.A | B10 | 14 | 18[a] | 11 | 0 | 0[a] |
| 99 | DBA/1 | B10×C3H | 16 | 17 | 4 | 0 | 1 |
| 207 | A.SW | A.CA×B10.BR | 14 | 15 | 13[a] | 6.6[a] | 5 |
| 211 | B10.A | B10 | 12 | 14[a] | 12 | 1.5 | 5[a] |
| 214 | B10 | B10.BR | 13.3 | 14.1 | 11.8[a] | 12.2[a] | 4.2 |
| 216 | B10 | B10.BR×C3H | 12.5 | 13.2 | 13.5[a] | 8[a] | 0 |
| 225 | B10.AKM | B10×C3H | 2.5 | 10 | 4.2 | 0 | 0 |
| 228 | A.CA | C3H×A.SW | 14 | 13 | 13[a] | 1[a] | 0 |
| 229 | C3H.NB | A.CA×C3H | 13.5 | 12.6 | 13 | 0 | 0 |
| 231 | B10.D2 | C3H.SW×B10.AKM | 14.5 | 14.5 | 13.5 | 2.5 | 0 |
| 233 | B10.D2 | B10×A.CA | 12.5 | 14[a] | 10.5 | 3.5 | 2.2[a] |

[a] — positive reaction with both alleles involved in recombination

the genetic material from the K-end of the $H\text{-}2^d$ allele (SCHREFFLER, AMOS and MARK, 1966). The specificities H-2.1 and H-2.5 received from $H\text{-}2^k$ are present in smaller amount in the product of $H\text{-}2^o$ allele than in $H\text{-}2^k$ and other 1-complete alleles (SNELL, DÉMANT and CHERRY, in press).

Tables 1 and 2 illustrate the serological similarity of antigenic products of 3-complete alleles $H\text{-}2^d$, $H\text{-}2^a$ and $H\text{-}2^i$, and of 1-complete alleles $H\text{-}2^k$, $H\text{-}2^a$, $H\text{-}2^h$ and $H\text{-}2^m$. The antisera listed in these tables react with antigenic specificities of the D-end of $H\text{-}2^d$ and of the K-end of $H\text{-}2^k$ respectively and their reactivity is expressed as the scores of hemagglutination reactions with the red cells. The serological similarity between the 3-complete and 1-complete alleles is not only a qualitative one, i.e. a 3-complete or a 1-complete allele never fails to react when the other 3-complete or 1-complete alleles react positively, but also a quantitative one, reflected in the similarity of the scores of hemagglutination reactions with most of the sera.

The reactions of $H\text{-}2^o$ cells with antisera against the specificities controlled by the K-end of the $H\text{-}2^k$ allele (Table 2) range from zero to the same strength as that observed with the 1-complete alleles. Most close to the reactions of 1-complete alleles are the reactions of the four anti-H-2.5 sera ASP-84, ASP-208 ASP-215 and ASP-238 (ASP is the designation of antisera produced in this laboratory). Similarly strong are the reactions of anti-H-2.1 sera ASP-50, ASP-209, ASP-212 and ASP-237 (the latter is an anti-$H\text{-}2^o$ serum). This is rather surprising because all these anti-1 antisera were prepared against 1-intermediate cells, which are known to exhibit smaller amounts of the antigenic specificity H-2.1 than do 1-complete cells. On the other hand, the anti-1 sera prepared against 1-complete alleles reacted with $H\text{-}2^o$ cells

**Table 2**

*Hemagglutination activity of anti-1 complex sera*

| ASP | Donor | Recipient | k | a | h | m | o | Anti 5 | Other antibodies |
|---|---|---|---|---|---|---|---|---|---|
| 32 | 2R | 5R | 6.5 | 12 | 7 | 4 | 0 | NO | 23 |
| 97[a] | B10.A | B10 | 9.7 | 18 | 10.5 | N.D. | 0[c] | NO | 1, 3, (23, 25, 11) |
| 67[a] | 2R | (B10×B10. D2) | 6.4 | 7.4 | 4.4 | N.D. | 1.3 | NO | 1, (23, 25, 11) |
| 201[a] | WB | B10 | 8.6 | 10.6 | 6.5 | 8 | 1.5 | NO | $1^w$ |
| 83[a] | A.SW | (B10×A.CA) | 13.6 | 13.6 | 14 | 7 | 3.0[c] | NO | 1, 3 |
| 222[a] | B10.BR | (B10×LP.R III) | 12.6 | 9.6 | 11 | 8.4 | 3.2[c] | NO | $1^b$, $3^b$, (23) |
| 202 | HTO | B10 | 5 | 6.1 | 8.5 | 7 | 4.0[c] | NO | 1, 8, 31 |
| 211[a] | B10.A | B10 | 15 | 14 | 15 | 14 | 5.0[c] | NO | 1, 3, (23, 25, 11) |
| 212 | B10.Y | (ACA×5 R) | 12 | 13.2 | 12 | 8.2 | 7.0 | NO | 1 |
| 82 | B10.A | (B10.D2×LP.R III) | 8.0 | 9.9 | 7.8 | N.D. | 2.0 | | $1^b$, ($5^b$, 23) |
| 100 | LP.R III | (B10.D2×C3H.NB) | 9.5 | 10 | 10.3 | N.D. | 2.0 | WEAK | $1^b$, $5^b$, (25, 11) |
| 203 | LP.R III | (B10.D2×DBA/1) | 14.3 | 15.5 | 11.5 | 12 | 2.5 | WEAK | $1^b$, $5^b$, (25) |
| 220[a] | B10.R III | (B10.D2×C3H.NB) | 10.3 | 13 | 13.5 | 11.5 | 2.5 | | $1^b$, ($5^b$, 25, 11) |
| 213 | A.SW | (B10.D2×C3H.NB) | 6.4 | 5.7 | 3.8 | 5.2 | 3.3 | | $1^b$, ($5^b$) |
| 209 | C3H.NB | (B10.D2×A.CA) | 8.6 | 9.4 | 6 | 8.7 | 7.0 | WEAK | 1, 5 |
| 50 | B10.Y | B10.D2 | 12 | 13 | 10 | 9.5 | 9.5 | WEAK | 1, 5 |
| 237 | HTO | BALB/C | 12.6 | 9.8 | 13 | 13 | 10.0 | Present | 1, 5 |
| 54 | B10 | B10.D2 | 9 | 7 | 8.6 | N.D. | N.D. | YES | 5 |
| 84 | 5R | (B10.D2×A.CA) | 14.1 | 15.1 | 11.2 | 9 | 11.0 | YES | 5 |
| 208 | 5R | (B10.D2×LP.R III) | 6.8 | 5.1 | 5.6 | 4.8 | 7.0 | YES | $5^b$ |
| 215 | 5R | (B10.D2×A.CA) | 12.5 | 13.3 | 9.7 | 8.7 | 7.2 | YES | 5 |
| 238 | 5R | (B10.D2×A.CA) | 12 | 15 | 13 | 13 | 10.5 | YES | 5 |

[a] — antiserum reacts strongly with $H\text{-}2^p$ (anti-1)
$1^b$ — blocked anti-1 antibody
$3^b$ — blocked anti-3 antibody
$5^b$ — blocked anti-5 antibody
$1^w$ — a new anti-1 complex antibody
[c] — positive reaction with both $H\text{-}2^d$ and $H\text{-}2^k$
N.D — not done

weakly or negatively in spite of the presence of strong anti-1 antibodies in many of them (Table 2, footnote a). A possible explanation of this observation is that the specificity H-2.1 in 1-intermediate alleles may differ from the specificity H-2.1 in 1-complete alleles also in respect of configuration or spatial arrangement of its determinants on the cell surface. This would make the anti-1 antibodies to 1-intermediate cells to fit better with the H-2.1 receptors on 1-intermediate cells than do the anti-1 antibodies to 1-complete cells.

The availability of a series of H-2 recombinant alleles on an otherwise identical genetic background in congenic resistant strains (SNELL and STIMPFLING, 1966) makes it possible to test the identity of the segments of H-2 genetic material transferred during crossing-over by skin transplantation. The test is performed by grafting the skin of a donor to an $F_1$ hybrid of two inbred strains, each of which shares with the donor strain a segment of the genetic material at one end of the H-2 locus and at least one of them is congenic with the donor strain. If the two segments contributed by the two H-2 alleles in the $F_1$ hybrid recipient comprise the whole genetic material of the H-2 allele of the donor, the grafted skin survives permanently. If one of the segments was transferred by recombination unchanged to several different recombinant alleles, each of these alleles should be able to replace the allele from which it derived this segment of H-2 genetic material in the $F_1$ hybrid recipient, without affecting the permanent graft survival. One series of such tests is shown in Table 3. It demonstrates inability of the B10.A(2R) strain, carrying the 1-complete $H\text{-}2^h$ allele, to provide the expected antigenic complementation of the K-end of other 1-complete alleles. The rejection of $H\text{-}2^a$ skin grafts by congenic $H\text{-}2^d/H\text{-}2^h$ $F_1$ hybrids was observed by Klein (1966). Here the failure of the $H\text{-}2^h$ allele to replace other 1-complete alleles in $F_1$ hybrid recipient is demonstrated in four donor — recipient combinations, in none of which an H-2 or non-H-2 difference should exist. On the other hand, the K-end of $H\text{-}2^h$

TABLE 3

*Deficiency of the K-end of $H\text{-}2^h$ allele in transplantation tests*

| Donor | H-2 | Recipient | H-2 | Incompatibility | | Number of grafts | Rejected | MST±S.E.* |
|---|---|---|---|---|---|---|---|---|
| | | | | H-2 | Non H-2 | | | |
| B10.A | *a* | [B10.A(2R)×Balb/c]$F_1$ | h×d | NO | NO | 10 | 10 | 14.0±0.4 |
| B10.A | *a* | [B10.A(2R)×HTT]$F_1$ | h×t | NO | NO | 20 | 17 | 19.2±3.6 |
| B10.AKM | *m* | [B10.A(2R)×DBA/$_1$]$F_1$ | h×q | NO | NO | 21 | 6 | 49.0±9.8 |
| B10.BR | *k* | [B10.A(2R)×HTO]$F_1$ | h×o | NO | NO | 27 | 19 | 25.8±2.4 |
| B10.A(2R) | *h* | [B10.A×B10]$F_1$ | a×b | NO | NO | 19 | 0 | . |
| B10.A(2R) | *h* | [B10×B10.AKM]$F_1$ | b×m | NO | NO | 20 | 0 | . |
| B10.A(2R) | *h* | [HTG×B10.AKM]$F_1$ | g×m | NO | NO | 20 | 0 | . |
| B10.A(2R) | *h* | [B10.A×HTG]$F_1$ | a×g | NO | NO | 11 | 0 | . |
| B10.A(2R) | *h* | [B10.T×C3H]$F_1$ | b×k | NO | NO | 14 | 0 | . |

* Mean survival time of rejected grafts

allele is readily complemented by the K-end of other 1-complete alleles in the recipient, without impairment of graft survival. One of the likely explanations of this phenomenon is a loss in the K-end of the $H\text{-}2^h$ allele suffered during the recombination between $H\text{-}2^a$ and $H\text{-}2^b$. Our attempts to produce an antiserum against the antigenic specificity supposedly missing in the product of the K-end of $H\text{-}2^h$ have been unsuccessful so far and therefore other explanations of the observed rejections cannot be excluded.

## REFERENCES

Amos, D. B., Gorer, P. A. and Mikulska, Z. B. (1955), *Proc. Roy. Soc.* (London) B, **144,** 369.
David, C. S., Shreffler, D. C. and Stimpfling, J. H. (1969), *Genetics*, **61,** 12.
Démant, P., Snell, G. D. and Cherry, M. (in press).
Gorer, P. A. and Mikulska, Z. B. (1959), *Proc. Roy. Soc.* (London) B, **151,** 57.
Klein, J. (1966), *Fol. biol.* (Praha) **12,** 168.
Klein, J., Bednárová, D. and Shreffler, D. C. (1969), *Genetics*, **61,** 33.
Shreffler, D. C. (1967), Proc. 3rd Int. Congress Human Genet., Hopkins, Baltimore.
Shreffler, D. C., Amos, D. B. and Mark, R. (1966), *Transplantation*, **4,** 300.
Snell, G. D. (1951), *J. Natl. Cancer Inst.*, **11,** 1299.
Snell, G. D. and Stimpfling, J. H. (1966), Biology of Laboratory Mouse, E. L. Green ed., McGraw-Hill, New York.
Snell, G. D., Démant, P. and Cherry, M. (in press).
Stimpfling, J. H. and Richardson, A. (1965), *Genetics*, **51,** 851.

## DISCUSSION

C. Stormont: It was noted on slide No. 2 that mouse strain B10 lacked the antigenic factor D, yet its cells were capable of absorbing anti-D. What is the explanation?

P. Démant: We found the reaction of anti-B sera with $H\text{-}2^b$ cells to be due to anti-35 and anti-36.

C. Stormont: Were there any examples of absorption of antibodies by cells which showed no reactions in the direct tests? I ask this question simply because such phenomena are of regular occurrence in the cattle, sheep and horse tests.

P. Démant: In no case, where the cells failed to react positively, was any absorption activity observed.

*XIIth Europ. Conf. Anim. Blood Groups Biochem. Polymorph., Bp., 1972 (pp. 621—626)*

# FURTHER STUDIES ON HISTOCOMPATIBILITY ANTIGENS AND REPRODUCTIVE PERFORMANCE

MILADA MICKOVÁ and P. IVÁNYI

Institute of Experimental Biology and Genetics, Czechoslovak Academy of Sciences, Prague, Czechoslovakia

## SUMMARY

Further data on male sterility ($W_{st}$) in hybrids obtained from matings of wild males with inbred females are presented. W males with the $W_{st}$ factor were captured in the Prague Zoological Garden. We found no W males with $W_{st}$ factor in other localities. However, hybrid males with the A line were generally less fertile than hybrids with the B10 or C3H line. This is in accordance with the finding that spermatogenesis was impaired to a greater degree in ($A \times W_{st}$) than in ($B10 \times W_{st}$) hybrids. ($BALB/c \times W_{st}$) male hybrids were found to be sterile.

The example of a W male, heterozygous for H-2, captured in the Prague Zoological Garden is described. One of his H-2 alleles (H-2.3+5+) was associated with factor(s) for normal hybrid fertility. His second H-2 allele, H-2 negative for all known antigens was associated with factor(s) for hybrid sterility ($W_{st}$). 4.5% recombination was found between $W_{st}$ and the H-2 locus.

t/+ hybrids from the mating of B10 females with T/t males ($t^6$, $t^{12}$, $t^1$, $t^{w1}$, $t^{w8}$, $t^{w12}$) were found to be within the range of normal fertility.

The aim of our work which was started in 1967 (IVÁNYI and DÉMANT, 1968) was to extend knowledge of the H-2 system, previously studied chiefly on inbred mice to populations of wild mice (W) (house mouse, Mus musculus). In this communication further data are presented on genetic factors associated with the H-2 system and influencing reproductive performance in male mice. Male wild mice captured in different places in the Prague area were routinely tested:

a) for the presence of t alleles;
b) serotyped for H-2 antigens;
c) mated with females from the inbred lines C57B1/10 (B10), A and C3H, other inbred lines and the "$F_1$ hybrids" were tested for fertility.

Fertility of $F_1$ hybrid females was tested by mating them with inbred males of known fertility and the numbers of litters and litter sizes were recorded without attempt to obtain quantitative fertility data. $F_1$ hybrid males were mated to test fertility quantitatively; one male was caged with 5 females from the B10 inbred line for at least two months and the number of offspring obtained during this period was recorded. Male fertility was expressed as the number of offspring per mating unit (OMU); one mating unit corresponded to one male per one female reproductive performance during one month. Thus one male caged with five females for two months was tested for ten mating units. Relative testes weight (TW) (ratio of both testes per total body weight) was also recorded. (For details see IVÁNYI et al., 1969b.)

Our previous findings (IVÁNYI et al., 1969a, b) can be summarized as follows. $F_1$ hybrids from matings of 16 wild (house) male mice (W) to B10, A or C3H inbred females were tested for fertility. While $F_1$ hybrid females were fertile, $F_1$ hybrid males obtained from matings of 11 W males to B10 or A inbred females were all or almost all sterile and $F_1$ hybrid males from matings of the same males to C3H inbred females were fertile. The eleven W males were designated $W_{st}$; the symbol designates a postulated major genetic factor responsible for the observed male sterility. An almost complete arrest of spermatogenesis at the stage of spermatogonia, or primary spermatocytes (i.e. a premeiotic block) were responsible for the observed male sterility. The testes' weight was found to have reduced in sterile or partially fertile hybrids. All 11 $W_{st}$ males were captured in the Prague Zoological Garden in various places and at different time intervals. Five W males produced exclusively fertile $F_1$ hybrid males when mated to females from any of the three inbred strains. The W males producing exclusively fertile $F_1$ hybrids were designated $W_f$. While breeding experiments indicated the involvement of several genetic factors, a major gene associated with the major histocompatibility system H-2 and thus located in the IXth linkage group, seemed to be the primary cause of the observed male sterility. This was found by testing the offspring from [(B10×C3H)♀× ×$W_{st}$ ♂] matings, $F_1$ hybrids of $W_{st}$ males heterozygous for H-2 antigens, and double-cross hybrids. In all experiments, the fertility of hybrid males differed according to the respective H-2 genotype. The impairment of spermatogenesis seems to have resulted from gene interaction of two or more genes; the first of the above experiments indicated that genes located on both IXth chromosomes of the hybrid males were involved. The position of the responsible factors on the linkage map and its relation to the T locus was uncertain. The presence of "typical" t-alleles is not a prerequisite for the observed male sterility. The involvement of the dominant T gene and of H-2 alleles seems to be unlikely. It is hypothetically assumed that this type of male sterility may have been due to t-alleles without the T-modifying effect (IVÁNYI et al., 1969b).

We wish to present here some data which support and extend our previous findings. During 1969 further wild males were captured in the area of the Prague Zoological Garden which obtained the $W_{st}$ factor as described above. On the whole, the fertility of (A×W)$F_1$ hybrids was always relatively more impaired than that of the (B10×W)$F_1$ ones. Wild males captured in other Prague places did not produce completely sterile hybrids with females from the inbred lines tested. However, in some cases the fertility of the hybrids was reduced and (A×W) hybrids were less fertile than (B10×W) hybrids descending from the same W male. (C3H×W)$F_1$ hybrid males were again never sterile.

Testing hybrids obtained from females of further inbred lines it was found that also (BALB/c×$W_{st}$)$F_1$ hybrids were completely sterile. Testing of further inbred lines is in progress.

Since the t alleles, which were absent in our W males from which the sterile hybrids descended are known to interfere with male fertility in mice (DUNN, 1964), we tested the fertility of different (B10×T/t) hybrids.

T/t$^6$, T/t$^{12}$, T/t$^1$, T/t$^{w1}$, T/t$^{w8}$, T/t$^{w12}$ males were mated with B10 females and the fertility of t/+ (i.e. normal tail) male hybrids was investigated. In all experiments, the fertility of these hybrids was normal or even higher than in the B10 parental line (Fig. 1). The significance of the differences observed between the experimental groups remains to be ascertained by further tests.

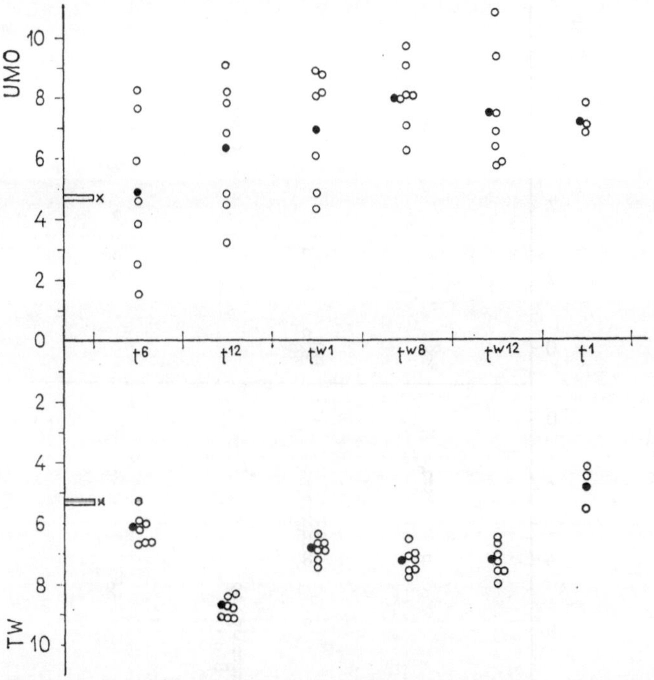
UMO
TW
10
8
6
4
2
0
2
4
6
8
10
t6
t12
tw1
tw8
tw12
t1

FIG. 1. Fertility of t/+ hybrids from the matings of T/t males with B10 females.
○ OMU and TW in individual (B10×T/t) hybrids
• mean values
× "normal control": mean values of OMU and TW in B10 males

In previous studies (IVÁNYI et al., 1969b) a complete sterility of the (A×W$_{st}$) or (B10×W$_{st}$)F$_1$ male hybride was noted and association with H-2 was demonstrated by testing the fertility on segregating populations. We refer here to the example of an H-2 heterozygous W male, whose (B10×W)F$_1$ male hybrid progeny clearly differed in fertility according to their H-2 genotype. W male No. 67 was captured in the Praha Zoological Garden. W ♂ 67 was mated with T/+ females but no tailless progeny was found. When serotyped by the PVP method with a battery of defined anti-H-2 sera reacting with all known H-2 antigens, the W ♂ 67 was typed as H-2.3+5+. Out of 81 (B10×W 67)F$_1$ hybrids 46% were H-2.3+5+ and 54% were H-2.3−5−. [Anti-H2.5 serum, produced in (B10.D2×A.CA) hybrids against B10.A(R5) spleen cells gives a very weak reaction with

H-2.5 produced by $H\text{-}2^b$ and $H\text{-}2^i$ alleles and reacts strongly (titer 1 : 500+++) with H-2.5 produced by $H\text{-}2^a$, $H\text{-}2^k$ or $H\text{-}2^h$ alleles, similarly as with W ♂ 67. Thus $(B10 \times W67)F_1$ hybrids are designated operationally as H-2.5 negative when responding weakly and H-2.5 positive when responding strongly.] Altogether 45 $(B10 \times W67)F_1$ male hybrids were serotyped and tested for fertility. The results are shown in Fig. 2. Twenty-four $F_1$ hybrid males were H-2.3−5− and this $H\text{-}2^{wi}$ allele was designated as $H\text{-}2^{wi-1}$. Twenty-one $F_1$ hybrid males were H-2.3+5+ and this allele was designated as $H\text{-}2^{wi-2}$. Thus all H-2 antigens detectable in the original W male were produced by one of this H-2 alleles, while the products of the second H-2 allele were not detectable by anti-H-2 sera reacting with known H-2 antigens. Such poor H-2 alleles found in W mice ("poor" for known H-2 antigens) were already described in our previous paper (IVÁNYI et al., 1969a).

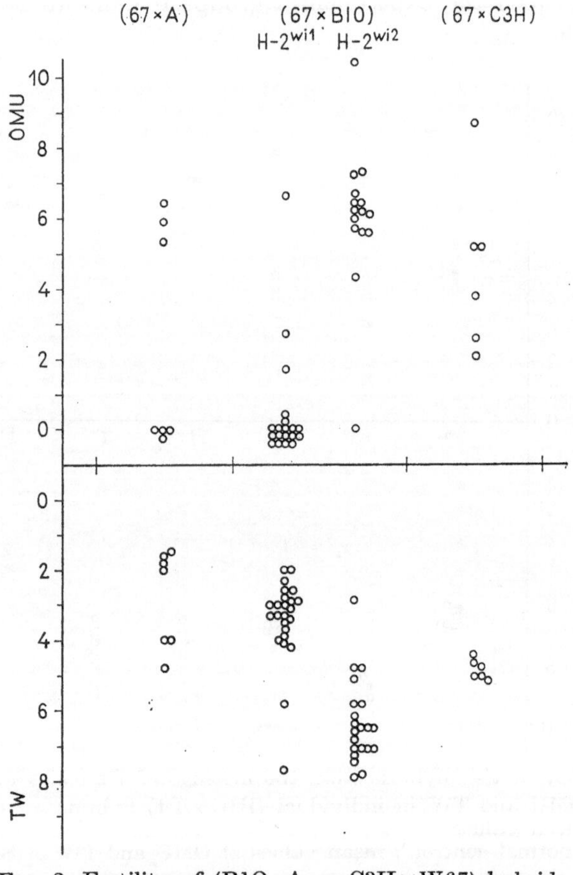

(67×A)
(67×B10)
H-2wi1
H-2wi2
(67×C3H)
OMU
TW
10
8
6
4
2
0
0
2
4
6
8


FIG. 2. Fertility of (B1O, A or C3H×W67) hybrids.
○ OMU and TW in the individual hybrids

Both groups of $F_1$ hybrid males were tested for fertility. Thirty-five animals out of 45 were tested for OMU and TW was measured in each of them. OMU of 15 males with the H-$2^{wi-2}$ allele, with only one exception, varied between 4.3–10.4 with a mean value of 6.0. Twenty males with the H-$2^{wi-1}$ allele had OMU and TW significantly lower ($P < 0.01$) than those with the H-$2^{wi-2}$ allele. Their OMU moved between 2.7 and 0.0 with a mean value 0.4. Out of 18 males with the H-$2^{wi-1}$ allele 14 were sterile (OMU 0.0) while the remaining 6 males had OMU values of 0.1, 0.2, 0.4, 1.7, 2.7 and 6.6. The first four males from these exceptions did not produce offspring until the end of the mating period, when some embryos were found in the females at biopsy. The number of embryos found was regularly lower than the litter size observed in the fertile sibs of the sterile males, i.e. 4, 1, 4, 2, 3, 1, 1, 11

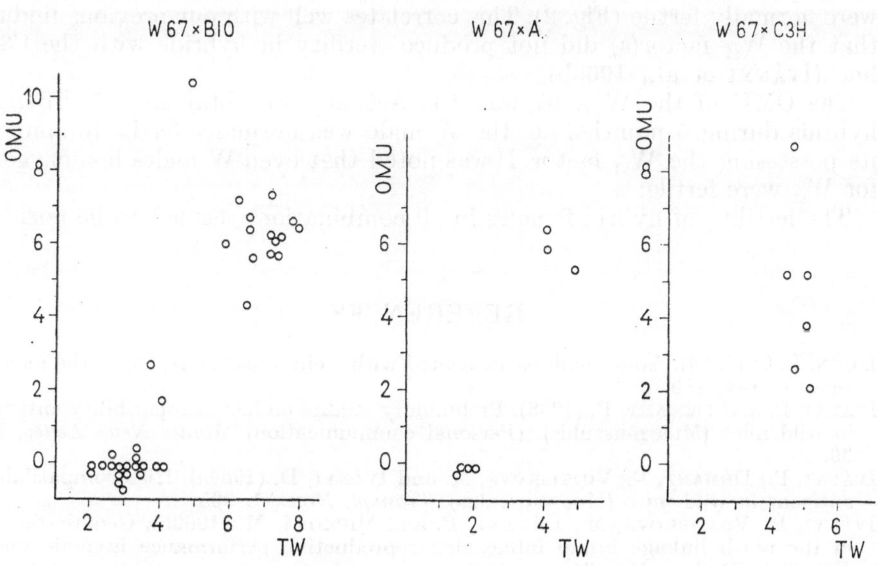
W 67×B10
W 67×A
W 67×C3H
OMU
TW
10
8
6
4
2
0
2
4
6
8

FIG. 3. Correlation between OMU and TW in the (B1O, A or C3H×W67) hybrid males

embryos were found in the individual pregnant females. Further studies are needed to understand how some practically sterile hybrids (zero values of OMU after 10 mating units and low TW) can produce a few embryos and of what the final outcome of such pregnancies will be like.

The correlation between the H-2 types and TW was also highly significant ($P < 0.01$). Only one out of the 24 H-$2^{wi-2}$ hybrids was sterile and had a low TW. The mean value of TW in this group was 6.3 (7.9–4.8, or 2.9 for the sterile male). The mean value of TW in the hybrids with the H-$2^{wi-1}$ allele was 3.4 (2.0–4.1, or 7.7 for the fertile male). The TW of those males which produced a few embryos or newborns (OMU 0.1–1.7) was in average not higher than that of the whole group with the H-$2^{wi-1}$ allele. Figure 3 shows the striking correlation between OMU and TW.

The two exceptions from 45 males indicate a 4.5% recombination between the genetic factor ($W_{st}$) responsible for male sterility and the H-2 locus.

Only 7 (A×W67)$F_1$ hybrid males were tested. Four were sterile with very low TW and three were normally or almost normally fertile (Fig. 2). Serotyping of the differences between H-$2^{wi-1}$ and H-$2^{wi-2}$ was not possible, because A(H-$2^a$) mice were strongly positive for both H-2.3 and H-2.5. However, the subdivision into two subgroups indicates that the two H-$2^{wi}$ alleles behaved differently also in these hybrids. The TW values of the sterile (A×W67)$F_1$ hybrids were very low, lower than those of the sterile (B10×W67) hybrids. This is in accordance with our previous finding that the disturbance of spermatogenesis in (A×$W_{st}$) hybrids was relatively more distinct than in (B10×$W_{st}$) ones.

Six (C3H×W67)$F_1$ hybrid males were tested for OMU and TW and 1.1 were normally fertile (Fig. 2). This correlates well with our previous finding that the $W_{st}$ factor(s) did not produce sterility in hybrids with the C3H line (Iványi et al., 1969b).

The OMU of the W ♂ 67 was 7.1. Actually we obtained 167 different hybrids during 5 months, i.e. the W male was normally fertile in spite of its possessing the $W_{st}$ factor. It was noted that even W males homozygous for $W_{st}$ were fertile.

The fertility of hybrid females in all combinations seemed to be normal.

## REFERENCES

Dunn, L. C. (1964), Abnormalities associated with a chromosome region in the mouse. *Science*, **144**, 260.

Iványi, P. and Démant, P. (1968), Preliminary studies on histocompatibility antigens in wild mice (Mus musculus). (Personal communication). *Mouse News Letter*, **38**, 36.

Iványi, P., Démant, P., Vojtisková, M. and Iványi, D. (1969a), Histocompatibility antigens in wild mice (Mus musculus). *Transpl. Proc.*, **1**, 365.

Iványi, P., Vojtišková, M., Démant, P. and Micková, M. (1969b), Genetic factors in the ninth linkage group influencing reproductive performance in male mice. *Fol. biol.* (Praha), **15**, 401.

*XIIthEurop. Conf. Anim. Blood Groups Biochem. Polymorph., Bp., 1972 (pp. 627—630)*

# TEN ALLELES OF THE RtH-1 SYSTEM IN 34 INBRED STRAINS AND 2 RANDOM BRED POPULATIONS OF LABORATORY RATS*

O. Štark, V. Křen and E. Günther**

Department of Biology, Medical Faculty, Charles University Prague, Czechoslovakia
** Max-Planck-Institut für Immunbiologie, Freiburg, GFR

The findings on the main histocompatibility system of the rat (RtH-1 or Ag-B) and its role in some alloimmune reactions were reported earlier (Štark et al., 1966; Křen and Štark, 1968). Two years ago 9 different RtH-1 (H-1) alleles were described in 21 inbred strains (Štark et al.,1968). In the course of serological typing of further 13 inbred and 2 random bred strains, using standard polyvalent and monospecific anti-H-1 reagents and absorbed newly produced antisera, only one new H-1 allele was identified in HW rats originating at the University of Saarland in Homburg.

Erythrocyte typing of HW rats had surprising results as all reactions were completely negative. Only after repeated immunization of LEW and LEW.AVN rats by HW cells were obtained some polyvalent sera, which were absorbed to obtain monospecific sera. The antigen of HW rats detected in this way was found to segregate alternately with antigens 5 of the LEW strain and 1 of the DA strain in corresponding interstrain hybrids. Thus the newly detected antigen of HW rats was included into the H-1 table and denoted by the number 22 (Table 1). The composition of the antigenic product of the tenth allele $H\text{-}1^h$ has not yet been defined in full detail, but negative reactions with type sera showed that it did not contain antigens 1, 3, 5, 6, 10, 11, 12 and 19. The H-1 table will be corrected after having completed the analysis of serological relations among newly produced congenic lines.

The incidence of 10 H-1 alleles in the 34 inbred strains is highly disproportional (Table 1). Most of the newly identified strains (PD is referred by Křen et al., in this issue) belong to the 3 largest groups: $H\text{-}1^d$ allele being present in 7, $H\text{-}1^l$ in 8 and $H\text{-}1^w$ in 9 strains. This may be important, suggesting not only the possible common origin of the above strains but also a strong selection value of the 3 alleles specified in the foregoing. The presence of the $H\text{-}1^d$ allele in 7 out of 10 BD strains is not surprising when the common ancestry of all these strains is considered. But only 1 out of 3 descendent strains still carries the $H\text{-}1^d$ allele (BD I), whereas the other 2 carry the $H\text{-}1^w$ (BD II) and $H\text{-}1^l$ (BD III) alleles. There are some other discrepancies in both groups carrying $H\text{-}1^l$ and $H\text{-}1^w$ alleles, evidently

* This work was partly supported by the Deutsche Forschungsgemeinschaft.

TABLE

*Composition of antigenic products of ten RtH-1 alleles and their occurrence*

| RtH-1 allele | Individual antigenic specificities | | | | | | | | | | | | | | | | | | | | | |
|---|---|---|---|---|---|---|---|---|---|---|---|---|---|---|---|---|---|---|---|---|---|---|
| | 1 | 2 | 3 | 4 | 5 | 6 | 7 | 8 | 9 | 10 | 11 | 12 | 13 | 14 | 15 | 16 | 17 | 18 | 19 | 20 | 21 | 22 |
| H-1$^{a}$ | 1 | 2 | — | 4 | — | — | 7 | — | — | — | — | — | 13 | 14 | — | — | 17 | — | — | 20 | 21 | — |
| H-1$^{b}$ | — | 2 | 3 | — | — | — | — | 8 | 9 | — | — | — | 13 | — | — | — | — | 18 | — | — | — | — |
| H-1$^{c}$ | — | — | — | — | — | — | — | — | 9 | 10 | — | — | 13 | — | 15 | — | 17 | — | — | — | — | — |
| H-1$^{d}$ | — | — | — | 4 | — | — | — | 8 | 9 | — | — | 12 | 13 | 14 | — | 16 | — | 18 | — | 20 | 21 | — |
| H-1$^{e}$ | — | 2 | — | 4 | — | — | — | — | — | — | — | — | 13 | 14 | — | — | 17 | — | — | — | 21 | — |
| H-1$^{f}$ | — | — | — | 4 | — | — | — | 8 | — | — | — | — | 13 | — | — | — | — | 18 | 19 | 20 | 21 | — |
| H-1$^{h}$ | — | — | — | | — | — | | | | — | — | — | | | | | | | — | | | 22 |
| H-1$^{l}$ | — | — | — | — | 5 | — | 7 | 8 | — | — | — | — | — | — | — | 16 | 17 | — | — | — | — | — |
| H-1$^{n}$ | — | — | — | — | — | — | — | — | 9 | — | 11 | — | 13 | — | 15 | — | — | — | — | — | — | — |
| H-1$^{w}$ | — | — | — | 4 | — | 6 | — | — | 9 | — | — | — | — | — | — | — | 17 | — | — | — | — | — |

Symbols in brackets indicate the origin of rat strains

characterizing some American strains from the Wistar Institute. Together with American strains there are some other strains present (BS, HS, BDE, LEP) relating to various domestic or domesticated wild rats.

All the 3 most frequently occurring alleles were found in two SPF strains from the Zentralinstitut für Versuchstierzucht in Hannover (Table 2). Both strains have been bred as closed colonies, each of them having about 800 breeding couples selected at random according to the parents' fertility. Both samples were evaluated statistically using the $\chi^2$ test, with the fol-

TABLE 2

*Distribution of H-1 alleles in 2 random bred strains of SPF rats*

| Sprague dawley nih/han | | | | | |
|---|---|---|---|---|---|
| Sex | H-1$^{l}$/H-1$^{l}$ | H-1$^{l}$/H-1$^{w}$ | H-1Z/H-1$^{w}$ | Total | Gene frequency |
| ♀ | 65 | 45 | 10 | 120 | H-1$^{l}$ = 0.7239 |
| ♂ | 57 | 44 | 9 | 110 | H-1$^{w}$ = 0.2761 |
| Total | 122 | 89 | 19 | 230 | ± 0.0209 |
| f= | 0.5304 | 0.3870 | 0.0826 | | |

| Wistar af/han | | | | | |
|---|---|---|---|---|---|
| Sex | H-1$^{d}$/H-1$^{d}$ | H-1$^{d}$/H-1$^{w}$ | H-1$^{w}$/H-1$^{w}$ | Total | Gene frequency |
| ♀ | 50 | 39 | 10 | 99 | H-1$^{d}$ = 0.7186 |
| ♂ | 56 | 35 | 9 | 100 | H-1$^{w}$ = 0.2814 |
| Total | 106 | 74 | 19 | 199 | ± 0.0225 |
| f= | 0.5327 | 0.3719 | 0.0954 | | |

1

*in 34 inbred strains of rats*

| Type strains | Inbred strains tested | | Congenic strains (CS) |
|---|---|---|---|
| | previously | newly | |
| AVN (CS) | DA (USA) | | LEW.AVN |
| BP (CS) | | | LEW.BP |
| CAP (POL) | Y 59 (Y) | PD (CS) | |
| BD V (D) | BD X (D) | BD I, IV, VI, VIII, IX (D) | LEW.BD V |
| BD VII (D) | | | |
| AS 2 (NZ) | | | |
| LEW (USA) | AS, BS, HS (NZ), AGA (Y), CDF (D) | HW (D), BD III (D), F 344 (USA) | |
| BN (USA) | | | LEW.BN |
| WP (CS) | WR, LEP (CS), VM (Y), BD II (D) | WAG (GB), R (N), BDE, E 3 (D) | LEW.WP |

lowing interesting results: 1. Each strain carries only 2 H-1 alleles. 2. The H-$1^w$ allele common in both strains shows the same gene frequency in both populations. 3. Both populations are in a genetic equilibrium corresponding to Hardy–Weinberg's law (0.9 p 0.9 for Sprague Dawley, and 0.5 p 0.6 for Wistar). Further examination of both populations 2–3 generations later will bring about more information.

Our finding of only 10 H-1 alleles in 36 rat strains of European, American and New Zealandian origin corresponds to the analogous data of Palm, (1970), who found 6 different Ag-B alleles in 18 American strains. However, comparison in more detail is impossible, for Palm published neither the names of all strains tested nor differences of Ag-B antigens detected. We really suppose the situation to be mature for Rt H-1 (Ag-B) reference testing and for unifying the nomenclature of the main histocompatibility system of the rat.

## ACKNOWLEDGEMENTS

Authors are indebted to Dr. F. Soukup (Prague) for statistical evaluation and to Mrs. J. Radecker (Freiburg) and Mr. L. Vojcik (Prague) for their excellent technical assistance.

## REFERENCES

Křen, V. and Štark, O. (1970), Tolerance induction to H-1 antigenic differences in the rat. Proc. XIth Europ. Conf. Anim. Blood Groups Biochem. Polymorph., Warsaw, 1968, 561–566.

Palm, Joy (1970), "Ontogeny" of the major histocompatibility locus in rats — a problem in nomenclature. *Transplantation*, **9**, 161–163.

Štark, O., Křen, V. and Frenzl, B. (1966), Histocompatibility locus in the rat. Proc. Xth Europ. Conf. Anim. Blood Groups Biochem. Polymorph. Paris, 1966, 501–506.

Štark, O., Křen, V. and Frenzl, B. (1970), Further analysis of the Histocompatibility-1 system of the rat. Proc. XIth Europ. Conf. Anim. Blood Groups Biochem. Polymorph., 1968, Warsaw, 551–555.

## DISCUSSION

P. Iványi: Dr. Štark, could you give more data about the past history of the rat strains which still segregate in the Rt H-1 system? Is this situation comparable to the persisting segregation of B alleles found in highly inbred chicken lines?

O. Štark: Both strains originated from two small breeding nuclei imported 2–3 years ago; the strains are maintained by random-breeding. The original H-1 characteristic is not known. The trend of the population development will be seen only after having tested both populations 2–3 generations later.

*XIIth Europ. Conf. Anim. Blood Groups Biochem. Polymorph., Bp., 1972 (pp. 631–635)*

# SIMPLIFIED ALLOGENIC SYSTEMS PROVIDED BY CONGENIC RESISTANT LINES OF RATS AND THEIR PARENTAL STRAINS*

O. ŠTARK and E. GÜNTHER**

Department of Biology, Medical Faculty, Prague, Czechoslovakia
Max-Planck-Institut für Immunbiologie, Freiburg, GFR

## INTRODUCTION

Differences in the main histocompatibility system of the rat (RtH-1 or-Ag-B) are considered as decisive not only for serological but also for cell-mediated alloimmune interactions. They may be studied in a more simplified and pure system provided by congenic resistant (CR) lines representing so-called biologically isolated H-1 alleles (ŠTARK and KŘEN, 1969). In our experiments such strains were used for studying the antigenicity or stimulatory capacity of the antigenic products of four H-1 alleles. An in vivo and in vitro system was used, i.e. tumour grafting and mixed lymphocyte culture test, and the results were correlated to the humoral immune response following skin grafting. The following strains were involved: **LEW.AVN, LEW.BD V, LEW.BN as CR lines and LEW as one of the parental strains. Figure 1 shows that the H-1$^a$, H-1$^d$ and H-1$^n$ alleles were**

| Parental strains | | Congenic resistant lines | | |
|---|---|---|---|---|
| Donor of genetic background | Donor of H-1 allele | Symbol | H-1 allele | Its antigenic product (=individual antigens) |
| LEWIS | × AVN ----- | L.AVN | H-1$^a$ | 1 2 . 4 . . 7 8 . . . . 13 14 . . 17 . . 20 21 |
| | × BDV ----- | L.BDV | H-1$^d$ | . . . 4 . . 8 9 . . 12 13 14 . 16 . 18 . 20 21 |
| | × BN ----- | L.BN | H-1$^n$ | . . . . . . . . 9 . 11 . 13 . 15 . . . . . |
| | | LEW | H-1$^l$ | . . . . 5 . 7 8 . . . . . . . 16 17 . . . |

FIG. 1. Scheme of the congenic lines used, their parental strains, the respective H-1 alleles and their antigenic products

* This work was partly supported by the Deutsche Forschungsgemeinschaft (DFG).
** Scholarship of the DFG.

transferred from the inbred strains AVN, BD V and BN to the common genetic background of the LEW inbred strain. A simple back-cross system was used, selection was done serologically by eliminating the H-1[l] allele.

## EXPERIMENTS

Previously published data about hemagglutinin and cytotoxin titres 11 and 15 days after skin grafting (ŠTARK et al., 1970) are shown schematically n Fig. 2. The LEW skin elicited the highest antibody response in the

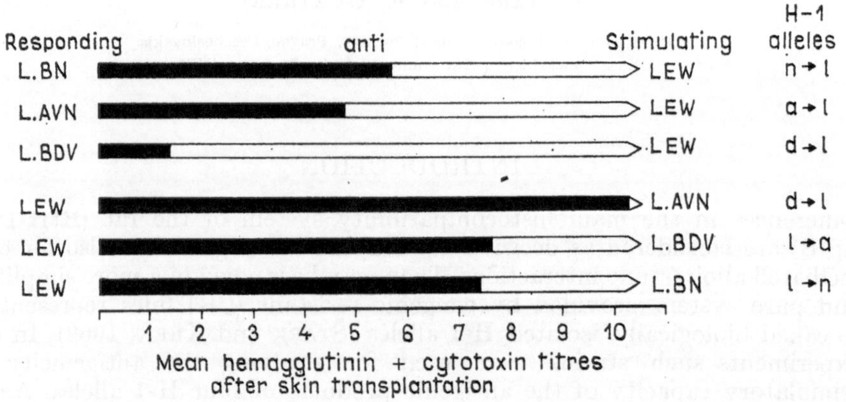
Responding
anti
Stimulating
H-1 alleles
L.BN
LEW
n→l
L.AVN
LEW
a→l
L.BDV
LEW
d→l
LEW
L.AVN
d→l
LEW
L.BDV
l→a
LEW
L.BN
l→n
1 2 3 4 5 6 7 8 9 10
Mean hemagglutinin + cytotoxin titres after skin transplantation

FIG. 2. Graph showing the mean values of hemagglutinin and cytotoxin titres after mutual first graft exchange

LEW.BN hosts, and the lowest one in the LEW.BD V. Vice versa, LEW reacted most strongly against LEW.AVN.

For tumour grafting a stable cell-line of a spontaneous mammary carcinoma of LEW origin (CaM-LEW) was used. This tumour was shown to possess all the H-1 antigens of the syngenic strain even after a long-term cultivation in vitro (KŘENOVÁ et al., 1969).

In the first part of these experiments a wide range of CaM-LEW cell-doses were injected intraperitoneally and deaths of hosts were evaluated. Each group consisted of at least 6 female and male animals aged 6–9 weeks, on a total 126 animals. Figure 3 shows that CaM-LEW tumour grew across the H-1 barrier but in a different manner. While more than 50% of the syngenic LEW animals died already after the injection of $2 \times 10^3$ cells, the $LD_{50}$ for LEW.BD V was $1 \times 10^5$, for LEW.AVN $1 \times 10^6$ (values taken by probit analysis), in LEW.BN $5 \times 10^7$ failed to kill the hosts.

In the second series a constant dose of $2 \times 10^6$ CaM-LEW cells was injected subcutaneously into 12-week-old animals. Tumour growth was assessed as the product of 3 diameters. The same differences were found (Fig. 4): in the syngenic hosts the tumour grew steadily, but in the allogenic environment its volume was less and regression could be observed first in LEW.BN, next in LEW.AVN.

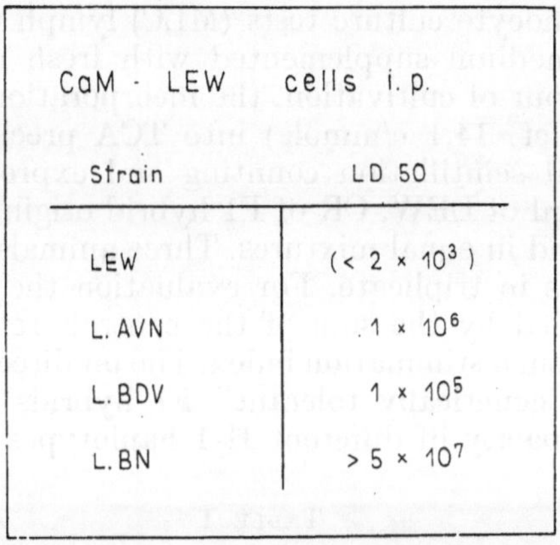
CaM - LEW cells i. p.
Strain
LD 50
LEW
(< 2 × 10³ )
L. AVN
1 × 10⁶
L. BDV
1 × 10⁵
L. BN
> 5 × 10⁷

FIG. 3. $LD_{50}$ of CaM-LEW cells injected intraperitoneally into syngenic and allogenic hosts

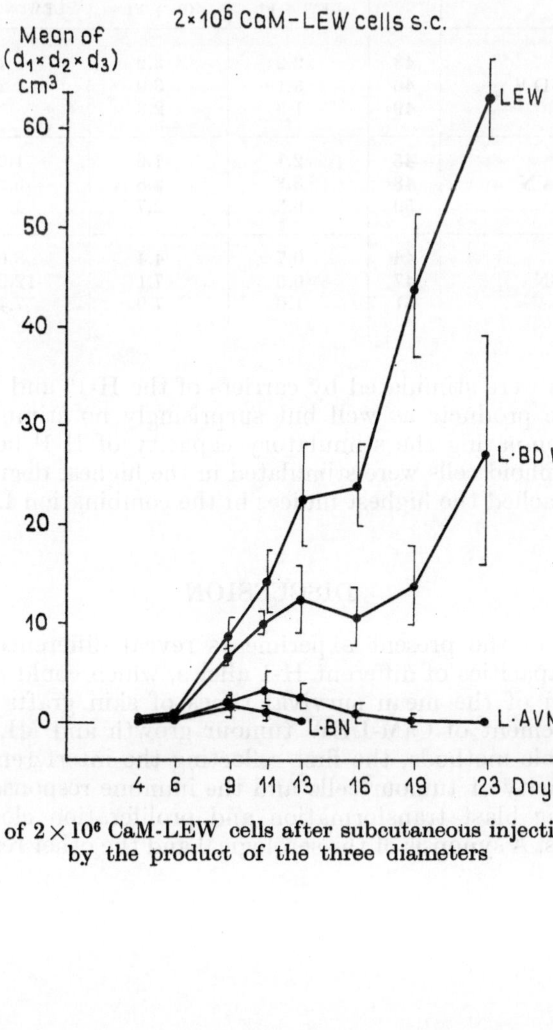
2×10⁶ CaM-LEW cells s.c.
Mean of
(d₁×d₂×d₃)
cm³
60
50
40
30
20
10
0
LEW
L·BDV
L·BN
L·AVN
4
6
9
11
13
16
19
23 Days

FIG. 4. Growth of $2 \times 10^6$ CaM-LEW cells after subcutaneous injection, demonstrated by the product of the three diameters

For mixed lymphocyte culture tests (MLC) lymph node cells were cultivated in Eagle's medium supplemented with fresh rat serum. From the 42nd to the 48th hour of cultivation, the incorporation of added $^3$H-thymidine (2 $\mu$c, spec. act. 14.1 c/mmole) into TCA precipitable material was measured by liquid scintillation counting and expressed in c.p.m. Eight million cells per 2 ml of LEW, CR of F1 hybrid origin were cultivated each alone as controls and in equal mixtures. Three animals of each combination were tested, always in triplicate. For evaluation the counts of the mixed cultures were divided by the sum of the controls referred to an identical cell number, to obtain a stimulation index. The unidirectional MLC by means of the respective "genetically tolerant" F1 hybrids gave information on the stimulatory capacity of different H-1 haplotypes (Table 1). The LEW

TABLE 1

*Indices of mixed lymphocyte culture reactions on the second day*

| Congenic strain | Exp. No. | Strain combinations | | |
|---|---|---|---|---|
| | | LEW + F1 | CR + F1 | LEW + CR |
| L.BD V | 43 | 2.2 | 3.2 | 6.4 |
| | 46 | 3.1 | 3.0 | 1.2 |
| | 49 | 1.3 | 2.3 | |
| L.AVN | 45 | 2.1 | 1.6 | 1.9 |
| | 48 | 3.3 | 2.8 | 5.2 |
| | 50 | 6.3 | 2.7 | 4.3 |
| L.BN | 44 | 0,7 | 4.4 | 8.6 |
| | 47 | 0.5 | 7.1 | 12.9 |
| | 51 | 1.0 | 7.9 | 7.4 |

lymphoid cells were stimulated by carriers of the H-1$^d$ and by those of the H-1$^a$ antigenic products as well but surprisingly no stimulation by H-1$^n$ was found. Comparing the stimulatory capacity of H-1$^l$ bearing F1 cells, LEW.BN lymphoid cells were stimulated in the highest degree. The mutual stimulation reached the highest indices in the combination LEW-LEW.BN.

## DISCUSSION

The results of the present experiments reveal differences between the stimulatory capacities of different H-1 alleles, which could not be detected by comparison of the mean survival times of skin grafts (ŠTARK et al., 1970). Measurement of CaM-LEW tumour growth and MLC test seem to be more suitable methods, the first reflecting the interference of the proliferative capacity of tumour cells and the immune response to them, the second showing blast transformation and proliferation closely related to H-1 differences. A synopsis of the serological and the other results described including the still preliminary values of second-day MLC, leads to the same or to an at least not contradictory range of the stimulatory capacities.

As the antigenic product of each allele is known to represent a certain combination of individual antigens (Fig. 1), the stimulatory capacity may be supposed to be related with them. The results favour this concept of correlation: in the case of LEW.BN anti-LEW 5, individual antigens are stimulatory, in LEW.AVN and LEW.BD V anti-LEW only 3, but different, do so. In the reverse combinations, 8 individual antigens of LEW.BD V, 7 of LEW.AVN and only 4 of LEW.BN origin are alien to LEW lymphoid cells. If this concept, which is discussed also for mice and for men is not too simple or even wrong, the existing inconsistencies may be explained either by differences between the antigenicity of the individual antigens or by different arrangements on the cell-surface and, on the other hand, by the use of conventionally maintained animals which are exposed to many other antigenic stimuli during the experiments.

## REFERENCES

Křenová, D., Křen, V. and Štark, O. (1969), Properties of spontaneous rat mammary carcinoma CaM-LEW. II. Karyological characteristics and antigenic properties of CaM-LEW tumour, long term passaged in vivo and in vitro. *Neoplasma*, **16**, 523–530.

Štark, O. and Křen, V. (1969), Five congenic resistant lines of rats differing at the RtH-1 locus. *Transplantation*, **8**, 200–203.

Štark, O., Křenová, D., Křen, V. and Frenzl, B. (1970), Immunological relationship between five congenic lines of rats and their parental strains. *Fol. biol.*, Praha, **16**, 1–11.

## DISCUSSION

P. Iványi: We previously held the view that the number of H-2 differences is important for the values in MLC. We know today that this standpoint was premature and wrong. Rychliková et al. (Folia biol., 1970) published evidence that the products of the H-2 K-end are greatly superior to the products of the H-2 D-end regardless of the number of antigenic differences involved.

E. Günther: Concerning the H-1 region of the rat — as far as I know — still nothing is known about recombination or inner structure of this region. Perhaps the lack of stimulation in the 3 tests of the combinations LEW + (L.BN × LEW) may be interesting in this context.

P. Démant: In the H-2 system the individual specificities are exceptionally strong in stimulating antibody response. Were not the cases in which you observed the best antibody response, due to the incompatibility in individual specificity?

O. Štark: It is not yet possible to decide whether the different immunogeneity of H-1 complex antigens is due to the quantity of the composing antigens or whether it depends more upon some "strong" individual specificities. The last explanation would better agree with what is known for H-2 of the mouse.

*XIIth Europ. Conf. Anim. Blood Groups Biochem. Polymorph., Bp., 1972 (pp. 637–639)*

# FACTORS INFLUENCING RUNT SYMDROME AND TOLERANCE INDUCTION IN RATS

V. KŘEN, DRAHOMÍRA KŘENOVÁ and O. ŠTARK

Department of Biology, Medical Faculty, Charles University, Prague, Czechoslovakia

As it was shown previously (KŘEN and ŠTARK, 1968) skin graft tolerance following neonatal spleen cell administration may be induced in H-1 compatible combinations only. On the other hand, there was considerable variability of runt syndrome (r.s.) incidence in both H-1 compatible and incompatible combinations. Production of congenic resistant (CR) lines of rats (ŠTARK and KŘEN, 1969) differing only at the H-1 locus and identical for their LEW genetic background permitted a more precise study of the conditions of r.s. and tolerance development. For this purpose three types of interstrain combinations were studied involving (1) H-1 difference only (LEW-CR), (2), H-1 + multiple non H-1 differences in the combination of LEW and fully allogenic strains from which CR lines were derived and (3) multiple non H-1 differences represented only by the combination of LEW.CR and H-1 compatible allogenic strains. Neonatal intravenous spleen cell administration was used throughout the whole work.

1) Comparing the r.s. inducing capacity of fully allogenic and congenic spleen cells in newborn LEW recipients, the increased efficacy of congenic spleen cells is apparent. The syngenic background in CR-LEW combinations revealed the importance of H-1 incompatibility for GvHR development even in fully allogenic combinations (BN-LEW, WP-LEW) with very low r.s. incidence. A less pronounced increase of fatal r.s. incidence was found in LEW.BP-LEW, as compared with BP-LEW for high efficacy of BP spleen cells itself.

2) There was some variability of r.s. incidence in the combinations (LEW-CR) confined to the H-1 difference only. The mortality rate in individual LEW-CR combinations was correlated with the number of the H-1 antigenic specificities involved (Table 1).

Our finding is in agreement with that of EICHWALD et al. (1969) who found in mice no correlation between splenic enlargement and the number of H-2 antigenic differences. On the other hand, there was a certain relationship between the strength of antibody and GvH reactions against target antigens. The strongest GvH reaction was observed in the LEW-LEW.BP group with the weakest antibody reaction following the first set skin graft. In contrast, the GvH reaction was weaker in the LEW-LEW.AVN combination, showing the strongest antibody response. In

TABLE 1

*Runt syndrome incidence and H-1 differences*

| LEW H-1 | —CR Allele | H-1 Total number | Antigenic difference Individual specificities | Antibody response* | R. s. Died/total | Cell dose × $10^6$ | Deaths % |
|---|---|---|---|---|---|---|---|
| combination | | | | | | | |
| a | l | 3 | 5, 8, 16 | 4.8 | 10/11 | 10 | 90.9 |
| l | a | 7 | 1, 2, 4, 13, 14, 20, 21 | 10.3 | 12/16 | 10 | 79.3 |
| n | l | 5 | 5, 7, 8, 16, 17 | 5.7 | 5/5 | 10 | 100.0 |
| l | n | 4 | 9, 11, 13, 15 | 7.4 | 2/6 | 10 | 33.0 |
| w | l | 4 | 5, 7 8, 16 | 4.6 | 10/10 | 10 | 100.0 |
| l | w | 3 | 4, 6, 9 | 5.6 | 5/9 | 10 | 55.5 |
| b | l | 4 | 5, 7, 16, 17 | 5.1 | 10/10 | 10 | 100.0 |
| | | | | | 10/12 | 5 | 83.3 |
| l | b | 5 | 2, 3, 9, 13, 18 | 2.7 | 19/19 | 10 | 100.0 |
| | | | | | 21/21 | 5 | 100.0 |

* mean values of antibody response after ŠTARK et al. 1970.

general, in all the CR-LEW mutual combinations it appeared more or less clearly that GvHR was stronger in the combinations with weaker antibody reaction. This seems to remind of the recent finding by RYCHLÍKOVÁ et al. (1970) in mice that in lymphocyte transformation in mixed cultures the D end of the H-2 locus did not provide significant stimulation despite of the strong antigenicity of its products.

3) In H-1 compatible, non H-1 different combinations no fatal r.s. occurred (5 groups with 52 animals tested) with the exception of BP-LEW.BP group in which 6/15 recipients of $30 \times 10^6$ BP spleen cells died.

4) Skin graft tolerance was not induced in any survivor of H-1 or H-1 + non H-1 incompatible spleen cell inoculation and all grafts were rejected within 15 days. On the contrary, the recipients of multiple non H-1 different spleen cell inocula were found to be highly tolerant, showing survival of the majority of the skin grafts over 50 days. In this respect, the BP strain was an exception again and no skin graft in LEW.BP recipients survived beyond the 50th day, the majority of them having been rejected within 20 days.

5) The exceptional position of the BP strain in rat immunological relationships, especially in GvHR resistance, is connected chiefly with the non H-1 factors, as shown by the high sensitivity of LEW.BP newborns to GvHR, mediated by LEW spleen cells. We started experiments on various BP hybrids in order to study the presumed "factors of the GvHR resistance" of the BP strain (Table 2). These factors seem to be of dominant nature ensuring 100% resistance of LEW.BP×BP hybrids towards LEW spleen cells in twice the dose which kills 100% of LEW.BP newborns. Nearly 50% resistance was found in $(BP \times LEW)F_1$ newborns which being $H\text{-}1^b/H\text{-}1^l$ moreover, genetically are active to LEW spleen cells. Segregation of the resistance to LEW spleen cells was observed among Bc hybrids (LEW.BP×BP)×LEW.BP. The mortality rate was reverse at the two

TABLE 2

*Genetics of the GvHR resistance of the BP strain*

| LEW spleen cells | | Number | | |
|---|---|---|---|---|
| into | dose $\times 10^6$ | injected | Died | % |
| BP | 5 | 6 | 0 | 0 |
| | 30 (iP) | 8 | 0 | 0 |
| BP×LEW | 10 | 15 | 8 | 53.3 |
| LEW.BP | 5 | 21 | 21 | 100.0 |
| BP×LEW.BP | 10—12 | 29 | 0 | 0 |
| (BP×LEW.BP)× | 5 | 30 | 5 | 16.6 |
| ×LEW.BP | 10 | 52 | 42 | 80.7 |

cell dose levels, being 42 : 10 for $10\times10^6$ and 5 : 25 for $5\times10^6$ LEW spleen cells. These summarized results seem to fit in with the hypothesis of dihybrid segregation where two dominant factors are needed for the resistance to $10\times10^6$ ($\chi^2 = 0.923$, $P < 0.3$) and one of these factors is sufficient to ensure resistance to $5\times10^6$ cell dose ($\chi^2 = 1.100$, $P < 0.2$).

The mortality rate depended, however, on the type of females producing the treated newborns. It was higher in the offspring of LEW.BP females than in those of LEW.BP×BP ones heterozygous for BP genetic background; this difference was more pronounced at the $5\times10^6$ dose level. This seems to indicate that in addition to other factors, those of the maternal environment may play a role in the resistance to GvH reaction.

## REFERENCES

EICHWALD, E. J., HART, E. A. and EICHWALD, B. (1969), Genetic aspects of the graft versus-host reaction in mice. *Fol. biol.* Praha, **15**, 254.

KŘEN, V. and ŠTARK, O. (1970), Tolerance induction to H-1 antigenic differences in the rat. Proc. XIth Europ. Conf. Anim. Blood Groups Biochem. Polymorph., Warsaw, 1968, 561–566.

RYCHLÍKOVÁ, M., DÉMANT, P. and IVÁNYI, P. (1970), The predominant role of the K-end of the H-2 locus in lymphocyte transformation in mixed cultures. *Fol. Biol.* Praha, **16**, 218.

ŠTARK, O. and KŘEN, V. (1969), Five congenic resistant lines of rats differing at the Rt H-1 locus. *Transplantation*, **8**, 200.

ŠTARK, O., KŘENOVÁ, D., KŘEN, V. and FRENZL, B. (1970), Immunological relationships between five congenic lines of rats and their parental strains. *Fol. Biol.* Praha, **16**, 1.

*XIIth Europ. Conf. Anim. Blood Groups Biochem. Polymorph., Bp., 1972 (pp. 641–644)*

# GENETICS OF THE POLYDACTYLY IN RATS AND INDEPENDENT SEGREGATION OF POLYDACTYLOUS AND Rt H-1 ALLELES

V. KŘEN, DRAHOMÍRA KŘENOVÁ, MILOSLAVA KRŠIAKOVÁ and B. FRENZL

Department of Biology, Medical Faculty, Charles University, Prague, Czechoslovakia

## INTRODUCTION

Extensive genetic studies are required to determine the Rt H-1 locus position in some of the rat linkage groups. The analysis of the relationship of Rt H-1 locus with coat colour gene has so far had (PALM, 1963; ŠTARK et al., 1969) negative results. Because of the limited number of usable mutants (ROBINSON, 1965) it is in the interest of rat genetics to search for them systematically and collect carefully the carrier strains of the mutants found. The occurence of polydactyly in a close colony of random bred rats was, therefore, utilized to establish a polydactylous strain (PD); on such rats a genetic study was performed the results of which are reported in this communication.

## MATERIAL AND METHODS

Polydactylous rats were found in rat population* originating probably from Wistar stock imported to Czechoslovakia after 1945 from Denmark and bred for several years in Konárovice farm. Ten years ago a small breeding stock was established from 15 rats of this origin (10 females and 5 males). Rats were bred with the exclusion of brother-sister mating. Besides polydactyly, another anomaly of tibial bone, not previously investigated, was found in this population. LEW rats were used for hybridization experiments and for anti-PD serum production. CAP rats were used for mutual skin grafting with PD rats. The transplantation technique and the serological methods used (immunization, absorption, hemagglutination and cytotoxic test) were those previously described by ŠTARK et al. (1968). For cross-reaction analysis, a panel of H-1 different RBC was used from the following rat strains: AVN, BP, CAP, BD V, BD VII, AS 2, LEW, BN, WP and congenic resistant lines LEW.AVN, LEW.BP, LEW.BN, LEW.BD V, LEW.WP and LEW.AS 2.

* Polydactylous rats were kindly supplied by Dr. Czabanová from the 2nd Institute of Morbid Anatomy, Medical Faculty, Charles University, Prague.

## RESULTS AND DISCUSSION

*1. The characteristics and genetics of polydactyly.* Six, rarely seven digits were on the hind feet of the affected rats. According to X-ray photograph compared with the skeleton of normal rat, the first digit seemed to be the additional one – thus representing a realization of the prehallucial ray. All offsprings of polydactylous parents were polydactylous as well; some of them, however, exhibited syndactyly of the 5th and 6th digits, or a poorly developed 6th digit, indicating the varied manifestation of the polydactylous trait. The genetical relationship of polydactyly was studied on the hybrids LEW×PD as shown in Table 1. All the $F_1$ hybrids (PD×LEW) were normodactylous (ND), in $F_2$ and Bc hybrids the polydactylous trait segregated in a ratio closely fitting with the simple Mendelian ratios (1 : 3 and 1 : 1 resp.). The results indicated determination of polydactyly by a pair of recessive genes.

*2. The Rt H-1, antigenicity of Pd rats.* For the determination of the Rt H-1 antigenicity of PD rats, a panel of anti-H-1 antisera was used. In this test, PD erythrocytes gave similar hemagglutination reactions as CAP ones. Further on, the probable H-1 identity of PD and CAP rats was indicated by the comparison of cross reactions of LEW anti-CAP and LEW anti-PD antisera together with the results of the absorption of LEW anti-PD antiserum with buffy coats of WP, BN and CAP strains. No production of hemagglutinating and cytotoxic antibodies was ascertained following mutual skin transplantation between PD and CAP strains, in spite of complete graft destruction in 15–16 days. Following the 2nd transplantation, only weak hemagglutinating antibodies, PD anti-CAP were found, the occurrence of which was explicable by eventual differences of non-H-1 in B-1 system. The segregation of H-1$^c$ and H-1$^e$ antigens in $F_2$ (LEW×PD) and Bc (LEW×PD)×PD hybrid populations (see Tables 2 and 3) was tested by means of LEW anti-PD and LEW.BD V anti-LEW antisera. The ratios found either fit with, or do not differ significantly from simple Mendelian ratios 1 : 2 : 1,

TABLE 1
*Genetical relationship of polydactyly*

| | Type of mating | Progeny PD | ND | Total | Number of litters |
|---|---|---|---|---|---|
| | PD×PD | 104 | 0 | 104 | 15 |
| $F_1$ | PD×LEW | | | | |
| | f m | 0 | 22 | 22 | 2 |
| | m f | 0 | 20 | 20 | 2 |
| | Total | 0 | 42 | 42 | 4 |
| Bc | (PD×LEW)×PD | | | | |
| | f m | 38 | 34 | 72 | 6 |
| | m f | 42 | 48 | 90 | 10 |
| | Total | 80 | 82 | 162 | 16 |
| $F_2$ | (PD×LEW) | 36 | 102 | 138 | 12 |

TABLE 2

*Segregation of polydactyly and H-1 antigens in (PD×LEW) $F_2$ hybrids*

| | H-1 Antigens | | | | | | |
|---|---|---|---|---|---|---|---|
| | c | 1 + c | 1 | Total | d. f. | $\chi^2$ | $p$ |
| ND | 18 | 41 | 23 | 82 | 5 | 1.0535 | $> 0.975$ |
| PD | 8 | 16 | 7 | 31 | | | $< 0.950$ |
| Total | 26 | 57 | 30 | 113 | | | |

1 : 1. This finding together with cross-reaction and absorption analyses showed antibodies to H-$1^c$ antigenic specificities to have been active in the LEW anti-PD antiserum. The erythrocytes of all PD rats from the original population were agglutinated by LEW anti-PD serum, indicating the H-$1^c$ homogeneity of PD rats. A considerable degree of isohistogeneity also in further non-H-1 systems, was suggested by the long-term survival of skin grafts exchanged among PD rats.

3. *Attempts to determine the relationship between polydactyly and Rt H-1 antigenicity.* A possible association of the polydactylous trait with H-$1^c$ antigenicity was examined on $F_2$ (LEW×PD) and Bc (LEW×PD)×PD hybrid populations (Tables 2 and 3).

As both traits were unrelated to sex, results were summarized for males and females. No close association between the two traits was found in either hybrid group and the results did not differ significantly from those expected for independent segregation. We believe, however, that the existence of a weak linkage cannot be excluded considering the segregation ratios in Bc hybrids with $\chi^2$ values relatively high ($\chi^2_L = 1.7193$, $0.2 > p > 0.1$) Moreover, the numbers of Bc hybrids were not sufficient for the testing of weak linkage; thus further studies remain to be carried out to resolve the problem.

TABLE 3

*Segregation of polydactyly and H-1 antigens in Bc (PD×LEW)×PD hybrids*

| | H-1 antigens | | | | |
|---|---|---|---|---|---|
| | 1 + c | c | Total | | |
| ND | 39 | 23 | 62 | $\chi^2_{PD} = 0.8772$ | $0.4 > p > 0.3$ |
| PD | 27 | 25 | 52 | $\chi^2_{H-1} = 2.8421$ | $0.1 > p > 0.05$ |
| Total | 66 | 48 | 114 | $\chi^2_L = 1.7193$ | $0.2 > p > 0.1$ |
| | | | | $\chi^2_\Sigma = 5.4386$ | $0.1 > p > 0.05$ |

## ACKNOWLEDGEMENTS

The authors are indebted to Dr KUDOVÁ and Dr SOUKUP for statistical evaluation of the results.

## REFERENCES

PALM, J. (1963), Histocompatibility and linkage relationships of loci determining isoantigens of the rat, Proc. XIth Inter. Congress on Genetics, **1**, 198.
ROBINSON, R. (1965), Genetics of the Norway Rat, London.
ŠTARK, O., FRENZL, B. and KŘEN, V. (1968), Erythrocyte and transplantation antigens in inbred strains of rats. VII. H-1 alleles of the LEP, CAP, BN, BD V, BD VII and BD X strains, *Fol. Biol.* Praha, **14.** 169.
ŠTARK, O., KŘEN, V., FRENZL, B. and KRŠIAKOVÁ, M. (1969), Independent segregation of the Rt H-1 alleles and coat colour genes in the rat, *Fol. Biol.* Praha, **15,** 470.

*XIIth Europ. Conf. Anim. Blood Groups Biochem. Polymorph., Bp., 1972 (pp. 645 – 647)*

# THE ANALYSIS OF ERYTHROCYTE ANTIGEN (B-1) OF THE RAT GENETICALLY INDEPENDENT FROM THE Rt H-1 LOCUS AND THE SEROLOGIC PRODUCTION OF B-1 NEGATIVE CONGENIC LINE

B. Frenzl, V. Křen and O. Štark

Department of Biology, Medical Faculty, Charles University, Prague, Czechoslovakia

## SUMMARY

The complex character of the erythrocyte antigen B-1 is suggested on the basis of the absorption analysis of the anti-B-1 antiserum. For the purpose of detailed analysis of B-1 antigen the serological production of two congenic strains was started by transferring the B-1 negativity from WP strain into LEW.WP and BP background.

The blood group antigen (B-1) was originally detected in the combination of two rat strains, BP (B-1 positive) and WP (B-1 negative), in the authors' laboratory. The B-1 positivity was determined by dominant factors and its segregation in interstrain hybrids (WP×BP) fitted in with the simple Mendelian ratios (Frenzl et al., 1960a). Later on, it was proven that B-1 antigen segregated independently on the Rt H-1 system antigens and had no detectable histocompatible effect (Štark et al., 1967a, 1967b). With regard to the role of B-1 antigen in the induction of experimental fetal erythroblastosis (Frenzl et al., 1960b) attempts were made to produce B-1 positive congenic lines on WP genetic background, which remained to be the only B-1 negative strain in the whole panel of the strains tested. But these attempts failed; since after the 20th generation of WP inbreeding the mortality of newborns continued to increase which made the further development of this strain as well as of congenic lines on its background impossible.

The production of B-1 negative strains was, therefore, started in 1968 by transferring the "B-1 negativity" of the WP strain into the genetic background of LEW.WP and BP strains. Because of the recessive nature of B-1 — the M system of mating (cross-intercross) was used according to Snell and Bunker (1965). The differentiation of B-1+ and B-1− individuals segregating in the intercross generations was made by using anti B-1 antiserum obtained from (WP×BP) $F_2$ H-$1^b$/H-$1^b$ B-1 negative animals after immunization with the whole blood BP. This type of antiserum agglutinated only B-1+ RBC, thus distinguishing B-1+ and B-1− hybrids by the Mendelian segregation ratios (Table 1).

## PRODUCTION OF B-1 NEGATIVE CONGENIC LINES

The course of the production of B-1 negative congenic line LEW.WP is documented in Table 2.

TABLE 1

*Segregation of B-1 antigen intercross hybrids*

| Intercross hybrids | B-1 + | B-1 − |
|---|---|---|
| BP×WP (G1) | 18 | 7 |
| LEW.WP×WP (G2, G4, G6) | 93 | 34 |
| Total | 111 | 41 |

A considerable mortality of newborns was ascertained in the first intercross generation which was obviously not associated with the B-1 negativity as shown by the shift of segregation ratios (39+ : 16−) toward the prevalence of B-1 negative animals. In the following cross-generation, mortality of newborns occurred only among the offspring of B-1 negative intercross females with LEW.WP males and not vice versa. It seems, therefore, obvious that factors transferred from WP background and functioning in females were the cause of the newborns' mortality.

Similarly, a considerable newborn mortality was the complication of BP.B-1− congenic line production. It was, moreover, necessary to produce a sufficient number of B-1− animals homozygous in the H-1$^b$ allele. For that reason several additional intercross matings were performed and B-1−, H-1$^b$/H-1$^b$ animals were selected so as to ensure both anti B-1+ antiserum and BP.B-1− congenic line production.

## THE COMPLEXITY OF B-1 ANTIGEN (TABLE 3)

The above antiserum, enabling distinction of B-1 positivity and negativity, was used for the study of the supposed complexity of the B-1 antigen. This antiserum, giving readily detectable cross-reactions with many strain-different erythrocytes, was subjected to absorption analysis with RBC from 5 rat strains, AVN, BN, CAP, LEW and WAG. The results are shown in Table 3. None of the type RBCs used absorbed all antibodies leaving besides the positive reaction with BP erythrocytes at least one more cross-reaction. LEW erythrocytes were found to absorbe a minimum of the antibody spectrum and, on the other hand, the positive cross-reaction of the

TABLE 2

*LEW.WP×WP B-1 − strain production (cross-intercross)*

| Parents ♀ ♂ | Progeny generation | Number of litters | Born | Survived | B-1 + | B-1 − |
|---|---|---|---|---|---|---|
| LEW.WP×WP | G1 | 3 | | 14* | | |
| G1×G1 | G2 | 13 | 130 | 55 | 39 | 16 |
| G2×LEW.WP | G3 | 4 | 44 | 14 | | |
| LEW.WP×G2 | G3 | 2 | 15 | 15 | | |
| G3×G3 | G4 | 11 | 95 | 57 | 43 | 14 |
| G4×LEW.WP | G5 | 4 | 31 | 28 | | |
| G5×G5 | G6 | 2 | 20 | 15 | 11 | 4 |

* Litters reduced artificially, no mortality observed

serum with LEW disappeared, owing to absorption with all other type RBCs. In accordance with this finding, also the cross-reaction of all B-1+ (LEW.WP×WP) intercross hybrids disappeared. These findings enabled a more precise formulation of the hypothesis on the complex character of the B-1 antigen of the BP strain, for at least 4 antigenic specificities. There were only three negative cross-reactions which were at variance with our "4 specificity hypothesis", two of which (with CAP RBC) were explicable by the low agglutinability of CAP erythrocytes, noted in the H-1 antigen testing as well. Regardless of these discrepancies and of the possible inaccuracy of the hypothesis, the changes of which may be envisaged in future work, the absorption analysis indicated the B-1 to be a second complex immunogenetic system of the rat. The definite verification of the hypothesis advanced requires a thorough and precise analysis, depending primarily on the production of B-1 negative congenic lines.

TABLE 3

*Cross-reaction and absorption analysis of anti-B-1 antiserum*

| RBC | Antiserum adsorbed with RBC | | | | | | |
|---|---|---|---|---|---|---|---|
| | Not abs. | AVN | BN | CAP | LEW | WAG | Supposed specificities |
| WP | — | — | — | — | — | — | 0 |
| BP | + | + | + | + | + | + | 1, 2, 3, 4 |
| WAG | + | + | + | + | + | — | 1, 2, 3 |
| AVN | + | — | + | + | + | [−] | 2, 3, 4 |
| BN | + | — | — | — | + | + | 3, 4 |
| CAP | + | — | — | — | [−] | [−] | 3, 4 |
| LEW | + | — | — | — | — | — | 3 |
| LEW.WP×WP B-1 + | + | — | — | — | — | — | 3 |
| B-1 — | — | — | — | — | — | — | 0 |

## REFERENCES

FRENZL, B., KŘEN, V. and ŠTARK, O. (1960), Attempt to determine blood groups in rats, *Fol. biol.* Praha, **6**, 121–126.

FRENZL, B., KŘEN, V., ŠTARK, O., SMETANA, K. and KRAUS, R. (1960), Experimental erythroblastosis in rats, *Fol. biol.* Praha, **6**, 135–144.

SNELL, G. D. and BUNKER, H. P. (1965), Histocompatibility genes of mice. V. Five new histocompatibility loci identified by congenic resistant lines on a C57BL/10 background, *Transplantation*, **3**, 235.

ŠTARK, O. and KŘEN, V. (1967), Erythrocyte and transplantation antigens in inbred strains of rats. II. Antigens of the AVN strain, *Fol. biol.* Praha, **13**, 93–99.

ŠTARK, O. and KŘEN, V. (1967), Erythrocyte and transplantation antigens in inbred strains of rats. III. Antigens of the BP strain, *Fol. biol.* Praha, **13**, 299–305.

*XIIth Europ. Conf. Anim. Blood Groups Biochem. Polymorph., Bp., 1972 (pp. 649–651)*

# GENETIC COMPARISON OF FOUR GEOGRAPHIC ISOLATES OF THE MOLE RAT (*Spalax ehrenbergi*)

C. R. SHAW

**Department of Biology, The University of Texas, M. D. Anderson Hospital and Tumour Institute Houston, Texas**

The standard technique of the geneticist, the experimental cross, has long enabled us to analyze the genetic differences, for example, between two strains of mice or between two strains of whales. But, as Professor JAMES CROW recently pointed out, "A far more challenging question is that of genetic differences between a mouse and a whale." He goes on further "Geneticists have long wondered whether genes in distantly related species, such as the mouse and whale, are mostly alike or mostly different."

The methods are now available for answering this question. These methods are the tools of the young but rapidly burgeoning field of molecular genetics. Ideally, we would like to make direct comparisons of homologous segments of the DNA. For the moment, however, we must be content mainly with comparisons of the products of the genes, the polypeptides. Analysis of the primary structure, that is, the amino acid sequence, of proteins is now possible, but it is not yet feasible to study large numbers of different proteins in this way. In our laboratory, we do comparisons of proteins by gel electrophoresis. This is a sensitive method for demonstrating structural differences, as these are reflected in altered electrophoretic mobility. Here, it is feasible to study relatively large numbers of proteins from large numbers of organisms.

The zymogram method involves electrophoretic separation of proteins in crude tissue extracts. Specific enzyme stains then permit detection of homologous proteins from different organisms. Observing the positions in starch gel after electrophoresis, of malate dehydrogenase from several animals, we can see that some have identical positions, others different.

By studying a number of enzymes in two different species, we can express the degree of genetic relatedness by simply counting the number of molecules which are the same and the number which are different and expressing the ratio of the same divided by the total. For example, if we compare several enzymes in three species of the bat genus, *Lasurius*, and we score these results by doing pair comparisons, i.e. comparing *seminolus* and *cinereus*, G6PD is different, 6PGD is different, LDH is the same, $\alpha$-glycerophosphate dehydrogenase is different, and leucine aminopeptidase is the same. Continuing through various other enzymes for the same species, we derived the results, and the amount of genetic relatedness between each pair of these three species is indicated.

A number of studies on a wide variety of species of other organisms including bacteria, fresh water trout, a family of salt water bony fishes, *Sciaenidae*, and the slime molds, *Myxomycetes*, all demonstrate that roughly this same order of genetic relatedness exists among species in the same genus. Thus, to return to Dr. CROW's original question, we have not actually compared the genes of a mouse and a whale (although we would be happy to do so if somebody would supply the whale), but I would hazard an educated guess that almost all of their genes would be different.

Now, it must be pointed out that in all of these studies, relatively small numbers of organisms have been examined, and the matter of polymorphism within a species was ignored. This was feasible in these small numbers simply because very little polymorphism was encountered, and the predominant form of each enzyme was utilized. Polymorphism of course complicates the picture, but for accurate genetic comparison of any two groups of organisms, polymorphism must be taken into account. The remainder of this paper deals with a study on the Israeli mole rat, *Spalax ehrenbergi*, in which statistically significant numbers were studied, and where some polymorphisms were encountered.

*Spalax* occurs as four geographic isolates in Israel. It has long been uncertain as to whether these four groups should be considered subspecies. The groups are partially interfertile. They are morphologically indistinguishable. NEVO has recently found that each of the groups has a different chromosome number, the numbers being 52, 54, 58 and 60. A few hybrids have been found with intermediate numbers. Total numbers of chromosome arms are all the same.

## METHODS

By the method of starch gel electrophoresis, we have studied thirteen different enzyme systems in the four *Spalax* populations. These thirteen enzymes represent seventeen genetic loci, the four extra loci resulting from the fact that several of the enzymes, such as lactate dehydrogenase, occur as isozymes and are controlled by more than one gene locus. The four populations have been arbitrarily designated A, B, C and D, and the total numbers studied from each were 138, 60, 74 and 115 respectively.

Most of enzymes were studied in extracts of liver, although in several cases either kidney or lung extracts were used, for example, lactate dehydrogenase was studied in lung, inasmuch as liver contains only one of the LDH isozymes, the LDH-5.

## RESULTS

Of the seventeen loci studied, twelve showed no variation. That is, they were the same both within and between populations. Five loci showed polymorphisms. In all cases, the polymorphisms consisted only of two different forms. The two forms were always arbitrarily called A and B, the A in each case being the more anodal on the electrophoretic gel. Not all populations were polymorphic for all of these loci. In fact, in no case is a locus polymorphic in all the four populations. On the other hand, when the groups are compared in pairs, one finds that in all but one case, the pairs of groups share at least one of the polymorphisms. For example, groups B and C are both polymorphic for G6PD, groups C and D are both polymorphic for G6PD and 6PGD. The one exception is the pair A and B, where none of the polymorphisms in one group occurs in the other group.

These results are strongly suggestive that either the four geographic isolates are presently exchanging genes, or that they did so until rather recently. If they are incipient subspecies, they have not yet been diverging long enough to have achieved genetic uniqueness, as have the separate species of other organisms described earlier. The fact of their different chromosome numbers would indicate that they have evolved a chromosomal isolating mechanism, and one may expect that they will diverge into genetically distinguishable groups in the future.

Incidentally, the frequency of polymorphism in these populations, five loci out of seventeen, i.e. 29.4%, agrees well with the figure of 30% reported in natural populations of house mouse (SELANDER and YANG, 1969), and of 36% in *Drosophila pseudoobscura* (LEWONTIN and HUBBY, 1966). It is expected that larger samples would increase this frequency of polymorphism to some extent.

Beyond simply demonstrating the genetic similarity of these four populations, we have undertaken a quantitative determination of the amount of genetic similarity (or proximity). The frequencies of alleles at each locus were counted, and then comparisons between each pair of the four populations were made by a formula. The results were computed by hand ($n = 17$), but the method could easily be adapted to computerization, for larger numbers and more populations. The results express genetic distance between each pair in a dendrogram.

As expected, the distances are all very close in these four groups. The method should be equally useful in comparing more distantly related species, as in the case of the bat species previously noted. We are currently collecting larger numbers of these species or analysis by this method. The method of course has the advantage over the one described earlier for the bat comparisons, as it is able to deal with the problem of intraspecific polymorphism.

## REFERENCES

LEWONTIN and HUBBY (1966), *Genetics* **54.** 595.
SELANDER and YOUNG (1969), *Genetics* **63.** 653.

## DISCUSSION

C. STORMONT: I am wondering if you are familiar with reports by WARHMAN on the moles which apparently like Spalax, form a series of incipient species which apparently differ only in chromosome numbers?

Another series of alleles in which many of the isozymes have identical electrophoretic mobilities but differ physiologically and/or biochemically is the G-6PD series in man.

C. R. SHAW: I don't remember the results on the Israeli moles.

*XIIth Europ. Conf. Anim. Blood Groups Biochem. Polymorph., Bp., 1972 (pp. 653—655)*

# A THIRD ALLELE IN THE Es-3 SYSTEM OF RABBIT BLOOD ESTERASES

YOSHIKO SUZUKI and C. STORMONT

Serology Laboratory, Department of Veterinary Microbiology, University of California, Davis, California

Studies of rabbit blood by means of the zymogram method have led to the elucidation of three genetic systems of esterases. The first one, now named Es-1, was described by GRUNDER, SARTORE and STORMONT in 1965. In 1970 SCHIFF and STORMONT described two additional systems: Es-2, a system of platelet esterases and Es-3, another system of red cell esterases. Each of these systems comprise three phenotypes designated A, AB and B, which are under the control of a pair of co-dominant alleles.

In connection with our studies on the serum esterases of rabbits (STORMONT and SUZUKI, 1970) the red cells of each rabbit studied were, whenever possible, phenotyped for the three esterase systems. Red cell samples were prepared and phenotyped according to the method of SCHIFF and STORMONT (1970). Although 537 individuals were phenotyped, 200 were members of families for which pedigree information was still unavailable at this writing; therefore, no mention is made of them in the tabular material.

Classification into known phenotypes proceeded normally for systems Es-1 and Es-2, but for system Es-3 a new phenotype was recognized in samples from seven individuals. This phenotype was designated AC, the C signifying a pattern of zones migrating more slowly than B. In Fig. 1 is a photograph of a gel slice showing the representative phenotypes of system Es-3.

In finding a new phenotype such as AC an assumption can be made that C zones represent the expression of a new allele Es-$3^C$. In our case there was supporting evidence for the existence of allele Es-$3^C$ because five of the seven individuals of type AC were members of a single family (see Table 1). A mating of a male of type A with a female of type AC resulted in a litter of eight offspring, four of type A and four of type AC. Thus we can now regard Es-3 as a multiple allelic system in which there are three alleles: Es-$3^A$, Es-$3^B$ and Es-$3^C$.

Estimates of the gene frequencies in six populations of rabbits are shown in Table 2. The frequency of Es-$3^B$, ranging from 0.03 to 0.26, is lower than that of Es-$3^A$ for all groups. The very low figure of 0.03 for the SSI population may be due merely to a distortion resulting from the small sampling number.

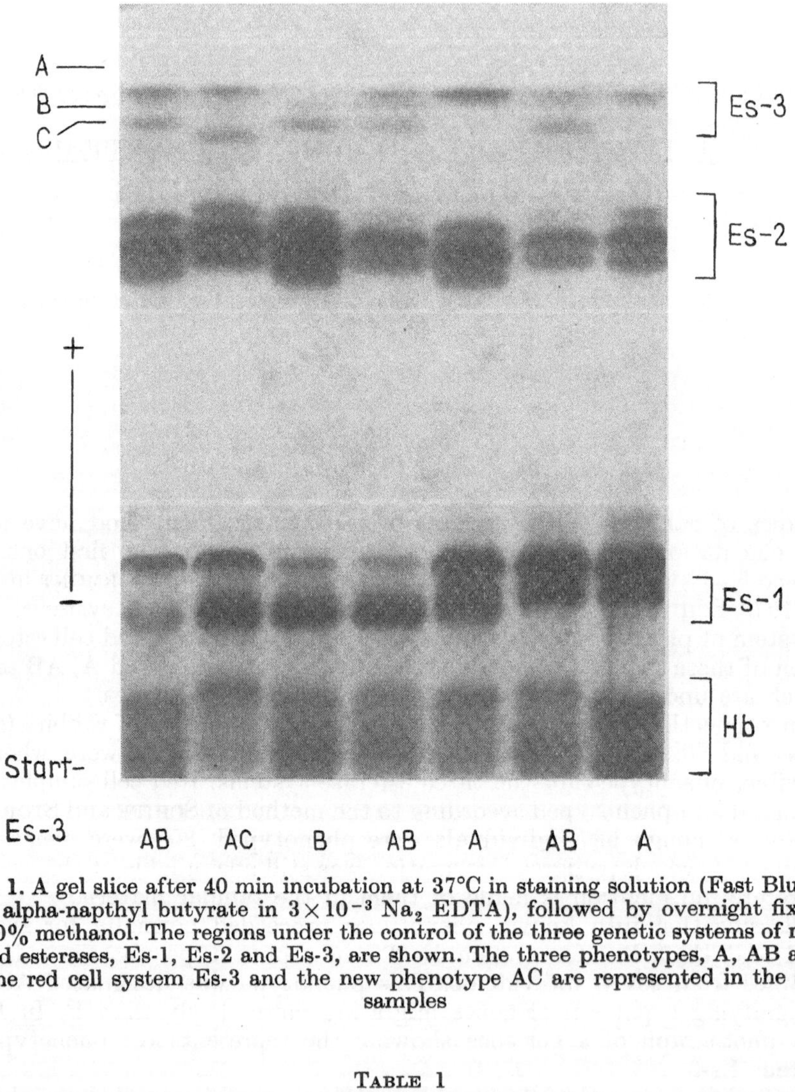
A
B
C
Es-3
Es-2
+
−
Es-1
Hb
Start-
Es-3
AB AC B AB A AB A

FIG. 1. A gel slice after 40 min incubation at 37°C in staining solution (Fast Blue BB and alpha-napthyl butyrate in $3 \times 10^{-3}$ $Na_2$ EDTA), followed by overnight fixation in 50% methanol. The regions under the control of the three genetic systems of rabbit blood esterases, Es-1, Es-2 and Es-3, are shown. The three phenotypes, A, AB and B of the red cell system Es-3 and the new phenotype AC are represented in the seven samples

TABLE 1

*Additional data on the inheritance of esterase phenotypes in the Es-3 system of rabbit red cells*

| Number of litters | Matings | Number of offspring of phenotype | | |
|---|---|---|---|---|
| | | A | AB | AC |
| 25 | A×A | 125 | 0 | 0 |
| 7 | A×AB | 15 | 16 | 0 |
| 1 | A×AC | 4 | 0 | 4 |

TABLE 2

*Estimates of gene frequencies (Parents or adults only)*

| Source | Es-3[A] | Es-3[B] | Es-3[C] | Number tested |
|---|---|---|---|---|
| *Davis | 0.83 | 0.17 | 0.00 | 101 |
| *Windsor | 0.81 | 0.19 | 0.00 | 43 |
| *Los Gatos | 0.83 | 0.17 | 0.00 | 54 |
| SSI | 0.95 | 0.03 | 0.02 | 29 |
| **COH | 0.86 | 0.14 | 0.00 | 97 |
| Saratoga | 0.73 | 0.26 | 0.01 | 47 |

* From SCHIFF and STORMONT (1970)
** All were of New Zealand white breed except for "COH" which was a panmictic population maintained at Duarte, Calif.

As might be expected, certain relationships have been observed among three of the four known systems of esterases in rabbits. SCHIFF and STORMONT (1970) have found a linkage relationship between Es-1 and Es-2. We have found a close association between the serum esterases and Es-2 (unpublished). It is, therefore, interesting to note that Es-3 alone appears to be independent of the other esterase systems.

## REFERENCES

GRUNDER, A. A., SARTORE, G. and STORMONT, C. (1965), Genetic variation in red cell esterases of rabbits. *Genetics*, **52**, 1345–1353.

SCHIFF, R. and STORMONT, C. (1970), The biochemical genetics of rabbit blood esterases: Two new esterase loci. *Biochem. Genet.*, **4**, 11–24.

STORMONT, C. and SUZUKI, Y. (1970), Atropinesterase and cocainesterase of rabbit serum: Localization of the enzyme activity in isozymes. *Science*, **167**, 200–202.

*XIIth Europ. Conf. Anim. Blood Groups Biochem. Polymorph., Bp., 1972 (pp. 657–660)*

# HEMOGLOBIN POLYMORPHISM IN RABBITS

S. J. L. Ramos, R. A. Mansilla and H. P. Blazquez

**Patronato de Biología Animal, Department of Genetics, Madrid, Spain**

## SUMMARY

**Five hundred random blood samples from two Spanish rabbit breeds, White Giant and Brown Giant, were studied for their hemoglobin variants by the paper and cellulose acetate electrophoretic techniques.**

**The hemoglobin solution was prepared according to the method described by Naik and Sukumaran (1966).**

**The random samples showed two hemoglobin variants: — Hb-I and Hb-II, the Hb-I showing a higher migration velocity than the Hb-II.**

**Examinations are in progress to find out the possible means of inheritance of the hemoglobin variants.**

## INTRODUCTION

In the last 18 years, extensive research was conducted on hemoglobin in humans and in numerous animal species, both vertebrate and invertebrate, with the aim to investigate normal and abnormal variants, as well as the different types of hemoglobin chains. Molecular variations were confirmed particularly in beta chains, and some interesting characteristics were determined, such as oxygen binding capacity, different speeds of electrophoretic migration, etc.

The first studies on these subjects were reviewed by Huisman in 1959. Also of great interest are the reports of Gratzer and Allison (1960) and more recently, of Manwell (1964), and Antonini (1965). Nevertheless, literary data have been scanty on the electrophoretic behavior of rabbit hemoglobin. Other authors described the existence of a single variant in adult animals; in this sense, rabbit hemoglobin can be compared to that of other species such as the donkey, camel, pig, dog or cat.

In Spain the first studies on the electrophoretic mobility of the hemoglobin of some animal species are presently in progress. In the Genetics Department of the "Patronato de Biología Animal", work was started in 1969, and now we report the first results related to two breeds of rabbits very common in the country.

## MATERIAL AND METHODS

Citrated blood samples taken under sterile conditions from the marginal ear vein of 500 adult rabbits, both does and bucks, of the Spanish White Giant and Spanish Brown Giant breeds were examined.

The hemoglobin solution was prepared according to the hemolyzation technique suggested by NAIK and SUKUMARAN (1966). The solution was adjusted to 10% with distilled water, toluene was added and the material was centrifuged at 2500 r.p.m. for one hour, to preserve it as carboxy-hemoglobin. The paper electrophoresis technique described by SMITH and CONLEY in 1963, was used with a veronal buffer of pH 8.6.

The electrophoretic separation using cellulose acetate was carried out in an Elphor-Integraph apparatus for 25 minutes at 200 V and 20°C, with a veronal buffer of 0.05 M and pH 8.6.

## RESULTS AND CONCLUSIONS

The specimens analyzed showed two hemoglobin variants called Hb-I and Hb-II, the former (Hb-I) having higher migration velocity than the latter (Hb-II) (Figs. 1 and 2).

No significant difference was found in hemoglobin variants between breeds or sexes, as can be seen from Table 1.

The percentage of samples with the Hb-I variant was superior to that with Hb-II.

No specimen showed a mixture of both hemoglobins, which can be interpreted in two possible ways: either each variant is determined by genes that are not codominant, or the samples were not suitable for finding the third variant (Hb-I + Hb-II). The latter possibility was disregarded,

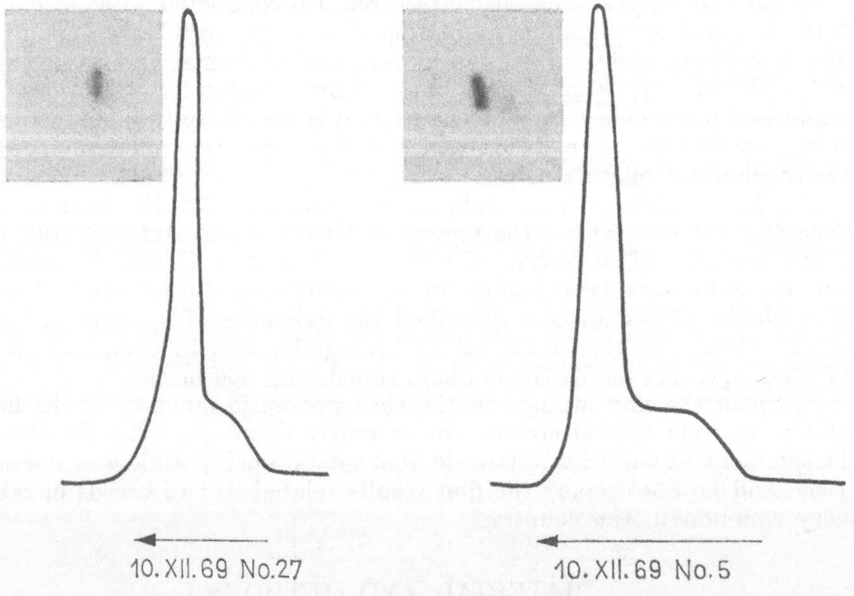
10. XII. 69 No. 27
10. XII. 69 No. 5

FIG. 1. Hemoglobin variants in rabbits (paper electrophoresis, veronal buffer pH 8.6)

TABLE 1

*Types of hemoglobin in different rabbit specimens*

| Breed | Sex | Hb-I | Hb-II | No. of animals |
|---|---|---|---|---|
| S. White | Bucks | 136 | 4 | 140 |
| Giant | Does | 118 | 5 | 123 |
| S. Brown | Bucks | 122 | 3 | 125 |
| Giant | Does | 109 | 3 | 112 |
| Total | | 485 | 15 | 500 |

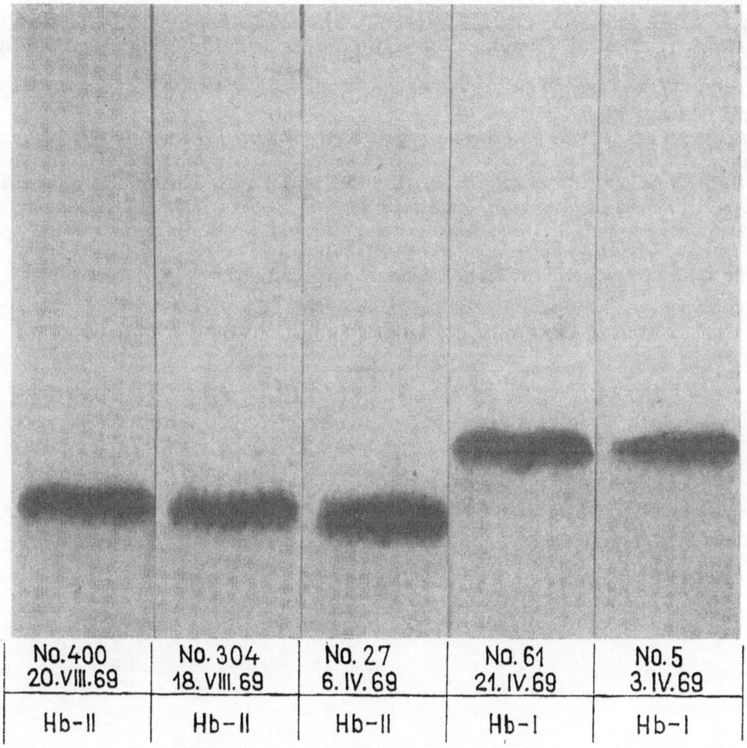
No.400
20.VIII.69
Hb-II
No. 304
18. VIII. 69
Hb-II
No. 27
6. IV. 69
Hb-II
No. 61
21. IV. 69
Hb-I
No. 5
3. IV. 69
Hb-I

FIG. 2. Hemoglobin variants in rabbits: Hb-I (right) and Hb-II (left) (cellulose acetate electrophoresis, veronal buffer pH 8.6)

because the population studied was not obtained with a high degree of homozygosis.

On the other hand, a study within the families of each breed seemed to indicate a typical Mendelian inheritance with dominance of the Hb-I fraction over the Hb-II, although we believe the data obtained were insufficient to permit a definitive conclusion.

Possible variants in fetal and cord blood have not been studied.

We wish to emphasize the difference between our findings and those of other authors, who have described only a single component in rabbit hemoglobin.

Both electrophoretic methods employed gave similar results.

## REFERENCES

ANTONINI, E. (1965), *Physiol. Rev.*, **45**, 123–170.

BAGLIONE, C. (1963), In *Molecular Genetics*, ed. J. H. Taylor Acad. Press, New York, 405.

GRATZER, W. B. and ALLISON, A. C. (1960), *Biological Reviews*, **35**, 459–506.

HUISMAN, T. H. J. et al. (1959), Symposium on abnormal haemoglobins, Istanbul 425, Blackwell, Oxford.

HUISMAN, T. H. J.(1966), Proc. Xth Europ. Conf. Anim. Blood Groups Biochem. Polymorph., Paris, 1966.

MANWELL, C. (1964), Oxygen in the Animal Organism. Proceedings. Pergamon Press, London.

NAIK, S. N. and SUKUMARAN, P. K. (1966), Proc. Xth Europ. Conf. Anim. Blood Groups Biochem. Polymorph., Paris, 1966.

PERUTZ, M. F. et al (1959), *J. Molec. Biol.*, **1**, 402.

RIGGS, A. (1965), *Physiol. Rev.*, **45**, 619–670.

THOMPSON, J. S. (1968), *Genética Médica*. Ed. Salvat.

*XIIth Europ. Conf. Anim. Blood Groups Biochem. Polymorph., Bp., 1972 (pp. 661–665)*

# THE Ab9 ALLELE OF RABBIT IMMUNOGLOBULINS. A GENE CAUSING IMPAIRMENT OF THE IMMUNE POTENTIAL

R. M. Tosi, A. L. Luzzati and A. O. Carbonara

**Istituto di Genetica Medica, Università di Torino, Italy**

Several genetic markers are known on the polipeptide chains of rabbit immunoglobulins. Four allelic variants (Ab4, Ab5, Ab6 and Ab9) of the gene controlling the light chain of kappa type have been discovered; these variants are recognized by immunological tests, making use of alloantisera and are referred to as b locus allotypes.

The gene frequencies of Ab4, Ab5 and Ab6 respectively are 85, 11 and 3.5% thus they can be considered in polymorphic equilibrium. On the contrary, the Ab9 allotype (Dubiski and Muller, 1967; Carbonara and Mancini, 1968; Mancini et al., 1969; Carbonara et al., 1969), which is the object of the present communication, has a very low gene frequency (about 0.2%), so that it could be considered as a rare mutation conferring a low survival value. A reasonable working hypothesis was that the Ab9 gene is associated with the impairment of some immunological functions. If this is true, the importance of Ab9 appears obvious: like in bacterial genetics a rare mutant can give a clearer insight into the normal function, so in this case informations on the physiology of the immune response can possibly be gained.

The first experimental approach to this problem was to test the phenotypic expression of the different alleles, i.e. the amount of Ig molecules carrying each allotypic specificity in the serum of heterozygous rabbits. Table 1 shows clearly that in any genotypic combination Ab9 has a lower

TABLE 1

*Relative ratios of allelic Ig molecules in rabbits heterozygous at the b locus*

| Rabbits *n* | Allo-type | % Ab4 | % Ab5 | % Ab6 | % Ab9 |
|---|---|---|---|---|---|
| 10 | 4/5 | 64.5 | 35.5 | — | — |
| 10 | 4/6 | 59.3 | — | 40.7 | — |
| 14 | 4/9 | 83.6 | — | — | 16.4 |
| 10 | 5/9 | — | 73.3 | — | 26.7 |
| 2 | 6/9 | — | — | 76.2 | 23.8 |

Quantitative determinations were performed by radial immunodiffusion, using specific anti-allotype sera.

phenotypic expression than the other alleles. Very similar ratios were found at the cellular level when in heterozygous rabbits the plasmacells containing one or the other allelic allotype were counted. Therefore a reasonable conclusion is that Ab9 possesses, in comparison to the other allotype, a lower probability of being activated during cellular differentiation. In heterozygous rabbits the other allele compensates for the low phenotypic expression of Ab9. On the other hand, in rabbits which are homozygous for Ab9, as shown in Table 2, the serum level of Ig carrying L chains of lambda type (so-called $b^-$ molecules) appears to be considerably higher,

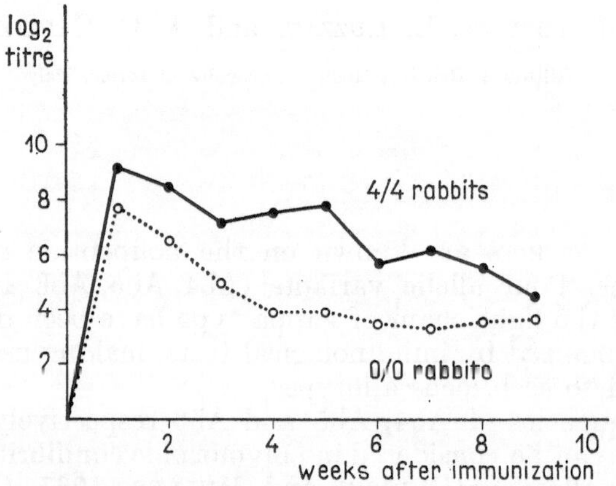
log2
titre
10
8
6
4
2
4/4 rabbits
0/0 rabbits
2
4
6
8
10
weeks after immunization

**Fig. 1.** Response of Ab4 homozygous (continuous line) and Ab9 homozygous rabbits (broken line) to *Salmonella enteridis* endotoxin (1 microgram injected intramuscularly, emulsified with Freund complete adjuvant). Results are expressed as agglutinating titers using sheep red blood cells coated with endotoxin. Each point represents the geometric mean of the titers obtained in four animals

especially when compared with homozygous Ab4 rabbits. Also in this case a compensatory mechanism is a likely explanation. The data presented in Table 2 also show that homozygous Ab9 rabbits have a lower serum Ig

TABLE 2

*Proportion of Ig molecules carrying light chains of lambda type ($b^-$ molecules)*

| Rabbits *n* | Allo-type | Total Ig mg/ml | $b^+$ Ig (kappa) mg/ml | $b^-$ Ig (lambda) mg/ml | $b^-$ Ig (lambda) % |
|---|---|---|---|---|---|
| 19 | 4/4 | 10.20 | 9.50 | 0.70 | 7 |
| 18 | 5/5 | 8.30 | 6.20 | 1.80 | 21 |
| 12 | 6/6 | 7.80 | 6.10 | 1.70 | 22 |
| 12 | 9/9 | 6.80 | 4.60 | 2.10 | 30 |
| t. 4/4 v. | 9/9 | 16.6 | 24.2 | 4.85 | |
| P | | <0.001 | <0.001 | <0.001 | |

Total Ig were determined by a goat anti rabbit Ig serum; $b^+$ molecules by the specific anti allotype sera. The amount of $b^-$ molecules (lambda chains) was calculated as difference between the two above measurements.

level. This finding was more firmly established by extending the assay to a larger animal sample. 36 Ab9 homozygous and 36 Ab4 homozygous rabbits have been tested, by radial diffusion method (MANCINI, CARBONARA and HEREMANS, 1965), for *IgG* level (using an anti Fc serum). The average amount is for Ab9 rabbits $4.53 \pm 1.72$ mg/ml, for Ab4 rabbits $7.94 \pm 2.18$ mg/ml. The difference is significant ($P < .01$).

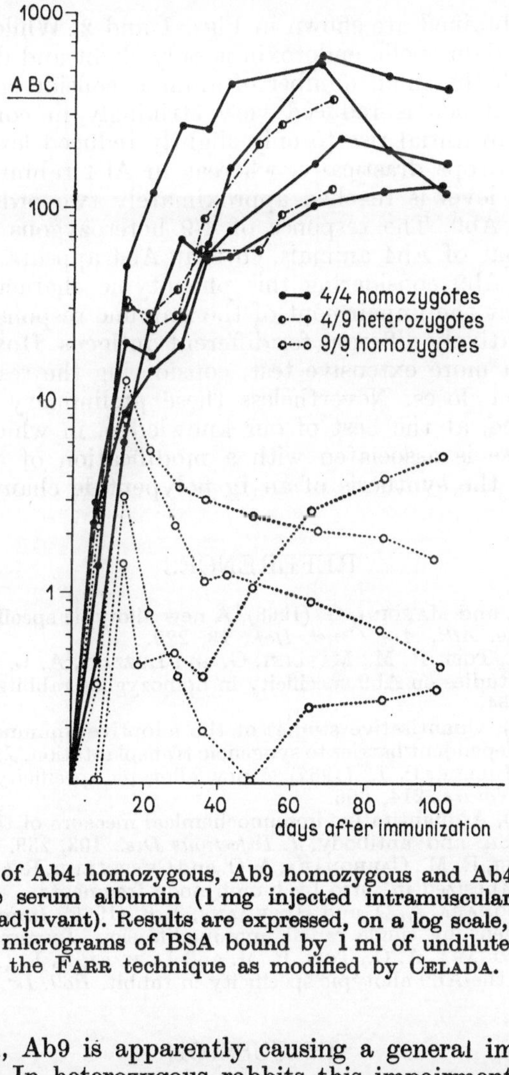
ABC
1000
100
10
1
4/4 homozygotes
4/9 heterozygotes
9/9 homozygotes
20
40
60
80
100
days after immunization

FIG. 2. Response of Ab4 homozygous, Ab9 homozygous and Ab4/Ab9 heterozygous rabbits to bovine serum albumin (1 mg injected intramuscularly emulsified with Freund complete adjuvant). Results are expressed, on a log scale, as antigen binding capacity (ABC = micrograms of BSA bound by 1 ml of undiluted antiserum) using the FARR technique as modified by CELADA.

In conclusion, Ab9 is apparently causing a general impairment in the synthesis of Ig. In heterozygous rabbits this impairment is compensated by the other allele. In homozygotes, despite the compensation provided by the gene(s) controlling light chains of another type, the Ig level is significantly decreased.

Having thus established the lowered expression of the Ab9 gene from the point of view of the overall Ig production, it was of interest to test whether comparable deficiencies could be demonstrated in the immune response to specific antigens. In a preliminary survey, two antigens of different chemical consitution were considered: a protein antigen, Bovine Serum Albumin (BSA) and a polisaccharide, purified endotoxin from *Salmonella enteridis*.

The results obtained are shown in Figs. 1 and 2. While the difference in the response to Salmonella endotoxin is only slight and does not reach the significance with the small number of animals considered, the response to BSA of Ab9 animals is reduced very strikingly in comparison to Ab4 animals. After an initial rise to only slightly reduced levels, the antibody of Ab9 rabbits drops drastically, whereas in Ab4 rabbits keeps going up until a plateau level is reached approximately two orders of magnitude higher than in Ab9. The response of 4/9 heterozygous rabbits is superimposable to that of Ab4 animals, so that Ab4 appears to be completely dominant over Ab9 considering this phenotypic character. These results may suggest that the impairment of the immune response caused by Ab9 may be quantitatively different for different antigens. However such a conclusion awaits a more extensive test, considering the response to a wider range of antigen doses. Nevertheless these preliminary results represent the first instance, at the best of our knowledge, in which a defect of an immune response is associated with a modification of a structural gene participating in the synthesis of an Ig polypeptide chain.

## REFERENCES


CARBONARA, A. O. and MANCINI, G. (1968), A new allotypic specificity (A9) of rabbit immunoglobulins. *Atti, Ass. Genet. Ital.*, **13**, 229.

CARBONARA, A. O., TOSI, R. M., MANCINI, G. and LUZZATI, A. L. (1969), Further immunochemical studies on Ab9 specificity in homozygous rabbits. *Boll. Ist. Sieroter. Milanese*, **48**, 154.

CELADA, F. (1966), Quantitative studies of the adoptive immunological memory in mice 1. An age-dependent barrier to syngeneic transplantation. *J. Expt. Med.*, **124**, 1.

DUBISKI, S. and MULLER, P. J. (1967), A new allotypic specificity (A9) of rabbit immunoglobulin. *Nature*, **214**, 696.

FARR, R. S. (1958), A quantitative immunochemical measure of the primary interaction between BSA and antibody. *J. Infectious Dis.*, **103**, 239.

LUZZATI, A. L., TOSI, R. M., CARBONARA, A. O. and CEPPELLINI, R. (1968), Homogeneity of antibody synthesized in vitro by lymph node fragments.

MANCINI, G., CARBONARA, A. O. and HEREMANS, J. F. (1965), Immunochemical quantitation of antigens by single radial immunodiffusion. *Immunochemistry*, **2**, 235.

MANCINI, G., CARBONARA, A. O., TOSI, R. M. and LUZZATI, A. L. (1969), An immunogenetic study of the Ab9 allotypic specificity in rabbit. *Boll. Ist. Sieroter. Milanese* **48**, 142.


## DISCUSSION

A. PERRAMON: What about lymphocytemia in rabbits homozygous 9/9 and in normal ones?

R. M. TOSI: We have told that the amount of lymphocytes in rabbit's blood shows very relevant variations both between different animals and in single animals in different physiologic conditions. — Therefore we were discouraged in going through this determinations. — However, it is certainly our intention to continue this research at the cellular, rather than at the molecular level.

W. E. Briles: What was the genetic origin of the allele 9?

R. M. Tosi: All our 9/9 rabbits were obtained through intensive breeding from two original rabbits. One was found by an extensive screening of rabbits from dealers in the Torino area. The other was obtained from a dealer in Germany. It is quite interesting that in the rabbits sold by this single dealer, the gene 9 were represented in a predominant proportion, in comparison with the other alleles.

*XIIth Europ. Conf. Anim. Blood Groups Biochem. Polymorph., Bp., 1972 (pp. 667–670)*

# THE GROUP *a* ALLOTYPES OF RABBIT IMMUNOGLOBULINS: A GENETIC PUZZLE

SIMONETTA LANDUCCI-TOSI

Osservatorio di Genetica Animale, Torino, Italy

The basic structure of immunoglobulins (Ig) is that of a tetramer composed of two "light" polypeptide chains, which can be either of type $\varkappa$ or $\lambda$, and of two "heavy" chains which can be of different classes ($\gamma$, $\mu$, $\alpha$, etc.). The heavy chains of IgG and IgM consist of about 400 aminoacids and are respectively of class $\gamma$ and $\mu$. The Ig molecule is symmetric, i.e. the two light chains are identical, as are identical between each other the two heavy chains. Limiting ourselves to the heavy chain polymorphisms, several groups of allelic markers (or allotypes) are known, each specific for a given class of heavy chain. For instance, e group allotypes (14 and 15) are present on $\gamma$ chains only, *MS* group on $\mu$ chains, *f* group on $\alpha$ chains.

Furthermore, another group is known, the group *a*, which comprises three allotypes, 1, 2 and 3 behaving as alleles. The peculiarity of these allotypes resides in their being present, at the molecular level, on all classes of heavy chains. Since these allotypes behave as single Mendelian units, it must be concluded that one locus controls not only one but at least three polypeptide chains. This is a direct contradiction of the principle "one gene — one polypeptide chain". Several lines of evidence, both immunological (TODD, 1963; LICHTER, 1967; PERNIS et al., 1968) and chemical (WILKINSON, 1969a; WILKINSON, 1969b), exist that strongly support this fact, i.e. the presence of *a* locus allotypes on the different Ig classes. However, in view of the general importance of this matter, it was thought worthwhile to reconsider it, following these experimental lines:

1) Test the above hypothesis with a direct and specific immunological reaction, by using insoluble antisera and labelled Ig for the study of the allotypes 1 and 2 on IgG and IgM. Details of this procedure are more throughly described by LANDUCCI-TOSI et al. (1970).

2) The different *a* group allotypes possess different expressivity. For instance, in 1/2 heterozygous rabbits the amount of IgG possessing allotype 1 is about three times as that of IgG carrying allotype 2. I have tried to test whether the same quantitative relationship between allotype 1 and allotype 2 is maintained in the IgM isolated from the same rabbits. This is expected if the same group *a* allotypes possessing the same expressivity are really present both on IgG and on IgM.

3) In some experimental conditions, e.g. injecting soon after the birth an anti-allotype serum, it is possible to obtain an "allotype suppression".

For instance if a heterozygous 1/2 rabbit is injected with anti-1 antibodies, the expressivity of the allele 1 is substancially decreased and the ratio between the two allotypes is completely reversed. If the same allotypes are really present on IgM, these should be affected by allotype suppression qualitatively in the same direction and quantitatively to the same extent.

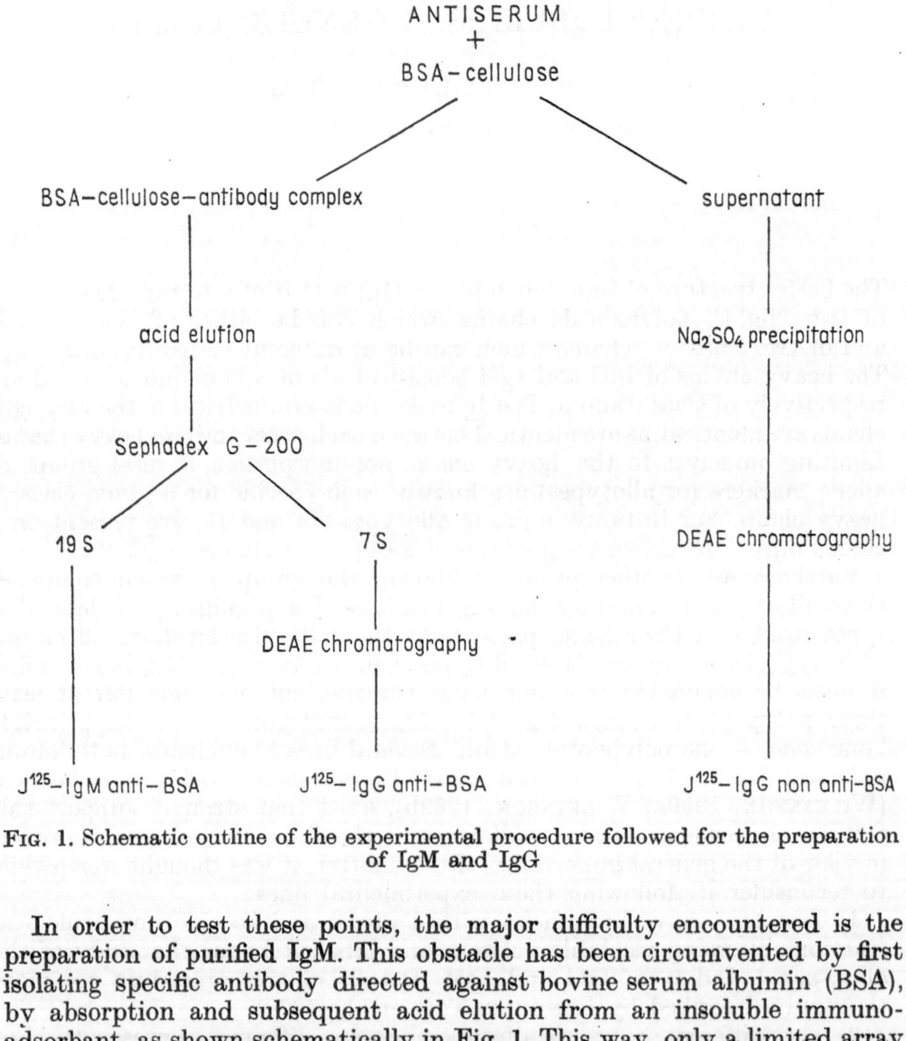
ANTISERUM
+
BSA-cellulose
BSA-cellulose-antibody complex
supernatant
acid elution
$Na_2SO_4$ precipitation
Sephadex G-200
19 S
7 S
DEAE chromatography
DEAE chromatography
$J^{125}$-IgM anti-BSA
$J^{125}$-IgG anti-BSA
$J^{125}$-IgG non anti-BSA

FIG. 1. Schematic outline of the experimental procedure followed for the preparation of IgM and IgG

In order to test these points, the major difficulty encountered is the preparation of purified IgM. This obstacle has been circumvented by first isolating specific antibody directed against bovine serum albumin (BSA), by absorption and subsequent acid elution from an insoluble immunoadsorbant, as shown schematically in Fig. 1. This way, only a limited array of IgM molecules, i.e. those possessing anti-BSA activity, is analyzed, but the purification procedure becomes much simpler.

The results of the binding experiments, using radioiodinated Ig preparations and insoluble anti-allotype sera, are shown in Fig. 2 and summarized in Table 1. The following conclusions can be drawn from them:

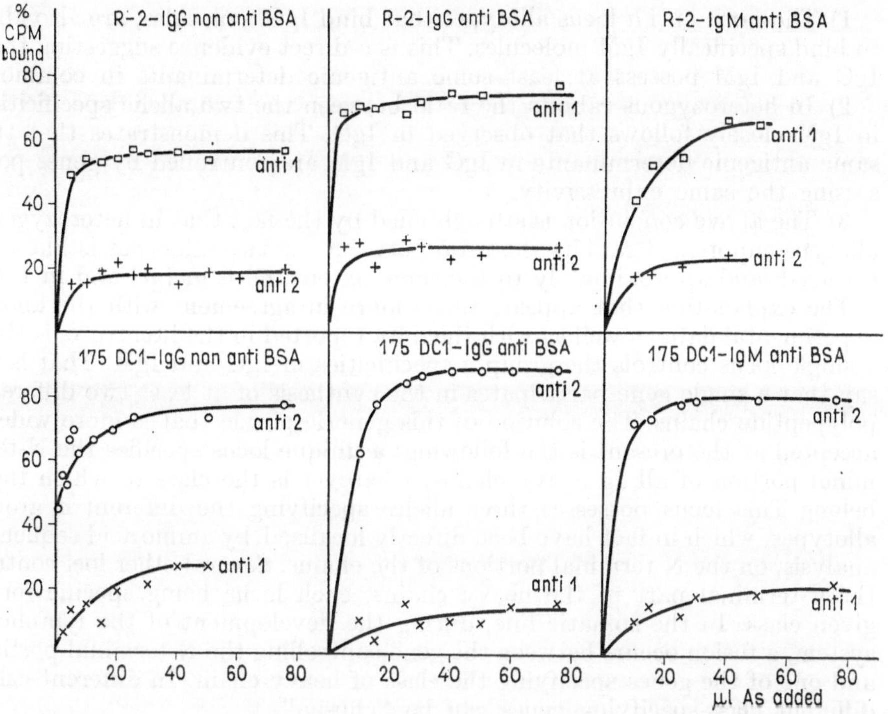
% CPM bound
R-2-IgG non anti BSA
R-2-IgG anti BSA
R-2-IgM anti BSA
175 DC1-IgG non anti BSA
175 DC1-IgG anti BSA
175 DC1-IgM anti BSA
anti 1
anti 2
µl As added

FIG. 2. Binding curves of $I^{125}$-labelled Ig preparations using insoluble antiallotype sera. R-2 is a normal rabbit, 1/2 heterozygous. 175 DC-1 is a rabbit with the same genotype, but allotype-suppressed for 1

TABLE 1

*Proportion of molecules carrying the allotypic specificities A1 and A2 in heterozygous rabbits*

| | % A1 | % A2 | Total | % $\frac{A1}{A1 + A2}$ |
|---|---|---|---|---|
| **Rabbit R-2 (normal)** | | | | |
| IgG non anti-BSA | 61.1 | 20.0 | 81.1 | 75.3 |
| IgG anti-BSA | 73.2 | 22.9 | 96.1 | 76.2 |
| IgM anti-BSA | 63.6 | 31.6 | 95.2 | 66.8 |
| **Rabbit R-1 (normal)** | | | | |
| IgG non anti-BSA | 64.9 | 18.2 | 83.1 | 78.1 |
| IgG anti-BSA | 66.2 | 31.2 | 97.4 | 68.0 |
| IgM anti-BSA | 55.4 | 27.0 | 82.4 | 67.2 |
| **Rabbit 175 DC1 (A1-suppressed)** | | | | |
| IgG non anti-BSA | 26.2 | 72.8 | 99.0 | 26.4 |
| IgG anti-BSA | 12.0 | 87.8 | 99.8 | 12.0 |
| IgM anti-BSA | 19.5 | 79.0 | 98.5 | 19.8 |

1) The same anti *a* locus allotypes that bind IgG molecules, are also able to bind specifically IgM molecules. This is a direct evidence suggesting that IgG and IgM possess at least some antigenic determinants in common.

2) In heterozygous rabbits the ratio between the two allelic specificities in IgM closely follows that observed in IgG. This demonstrates that the same antigenic determinants in IgG and IgM are controlled by genes possessing the same expressivity.

3) The above conclusion is strenghtened by the fact that in heterozygous allotype suppressed rabbits the ratio between the two allotypes is likewise reversed, and approximately to the same extent, both in IgG and in IgM.

The explanation that appears to be more in agreement with the above experimental data as well as with the data reported in the literature, is that a single locus controls the group *a* specificities in IgG and IgM. That is to say that a single gene participates in the synthesis of at least two different polypeptide chains. The solution of this genetic puzzle that is more widely accepted at the present is the following: a unique locus specifies the N-terminal portion of all Ig heavy chains, whatever is the class to which they belong This locus possesses three alleles specifying the different *a* group allotypes, which in fact have been directly localized, by amino acid sequence analysis, on the N-terminal portions of the chains. Several other loci control the C-terminal part of the heavy chains, each locus being specific for a given class. In the somatic line, during the development of the lymphoid system, a fusion occurs between the gene controlling the N-terminal portion and one of the genes specifying the class of heavy chain. In different cells, different class-specifying genes can be "chosen".

Although this hypothesis may look highly implausible, I think it should be remembered that striking evidences in favour of it are accumulating, especially through amino-acid sequence analysis of human Ig (Press and Hogg, 1969). On the other hand, the immunological function represents such a complex and highly organized biological phenomenon that it should not be surprising to find some peculiar mechanism, as the above described genic fusion, not encountered in other biological systems.

## REFERENCES

Landucci-Tosi, S., Mage, R. G. and Dubiski, S. (1970), Distribution of allotypic specificities A1, A2, A14 and A15 among IgG molecules, *J. Immun.*, **104**, 641.

Lichter, E. A. (1967), Rabbit A and M Immunoglobulins with allotypic specificities controlled by the *a* locus. *J. Immun.*, **98**, 139.

Pernis, B., Torrigiani, G., Amante, L., Kelus, A. S. and Cebra, J. J. (1968), Identical allotypic markers of heavy polypeptide chains present in different Immunoglobulin classes. *Immunology*, **14**, 445.

Press, E. M. and Hogg, N. M. (1969), Comparative study of two IgG Fd fragments, *Nature*, **223**, 807.

Todd, C. W. (1963), Allotypy in rabbit 19S protein, *Biochem. Biophys. Res. Comm.*, **11**, 170.

Wilkinson, J. M. (1969a), Variation in the N-terminal sequence of heavy chains of IgG from rabbits of different allotype, *Biochem. J.*, **112**, 173.

Wilkinson, J. M. (1969b), H chains of IgA from rabbits of different allotype composition and N-terminal sequence, *Nature*, **223**, 616.

# IX. BLOOD GROUPS AND BIOCHEMICAL POLYMORPHISM IN OTHER SPECIES

*XIIth Europ. Conf. Anim. Blood Groups Biochem. Polymorph., Bp., 1972 (pp. 673—678)*

# FURTHER IMMUNOGENETIC INVESTIGATIONS OF BREEDING FOXES

H. Balbierz and Maria Nikołajczuk

Department of Immunopathology, Faculty of Veterinary Science, Wrocław, Poland

## INTRODUCTION

Fox breeding admits mating of one female with two or more males during copulation season. This presents particularly interesting opportunities for examining the genetic determinants of the offspring which may derive from a single father or else some cubs derive from one dog-fox and some from another.

These considerations initiated us to undertake research on a group of foxes kept on a small private breeder farm.

Our earlier investigations (Kaminski and Balbierz, 1965; Kaminski et al., 1966) of the serum proteins of silver, blue, and platinum foxes indicated a race differentiation which came into display by variations of quantity and intensity of protein bands in certain zones. This paper is a report of complementary investigations of haptoglobin and acid erythrocytic phosphatase systems.

Separation of serum was effected in a system of buffers according to Gahne (1962) and in a borate system according to Poulik (1957). Separation of serum in buffer according to Gahne provides more distinct segregation of the beta globulin zone from the rest of protein fractions.

Haptoglobin was tested with phosphate buffer according to Laurell (1959) and with borate buffer according to Prokop et al. (1963), as described by Buschmann and Schmid (1968).

Acid erythrocytic phosphatases were separated according to the methods described by Hopkinson et al. (1963) for acid erythrocyte phosphatases determination in man.

## RESULTS AND DISCUSSION

In the group under investigation, five dog-foxes (fathers) were mated with six vixens. Blood samples taken from the vixens' issue, i.e., twenty-three cubs, were examined. In addition, a group of twenty-eight foxes selected at random and examined in the previous year, were included for the interpretation of the beta-globulin zone. The interpretation of the beta-globulin system was based on earlier results (Kaminski and Balbierz, 1965; Kaminski et al., 1966) and on the pattern of two-dimensional electrophoresis, the first dimension being the separation on chromatographic paper in barbituric buffer of pH 8.6; i.s. 0.06; while the second dimension separation in starch gel, using the buffer system of GAHNE (Fig. 1).

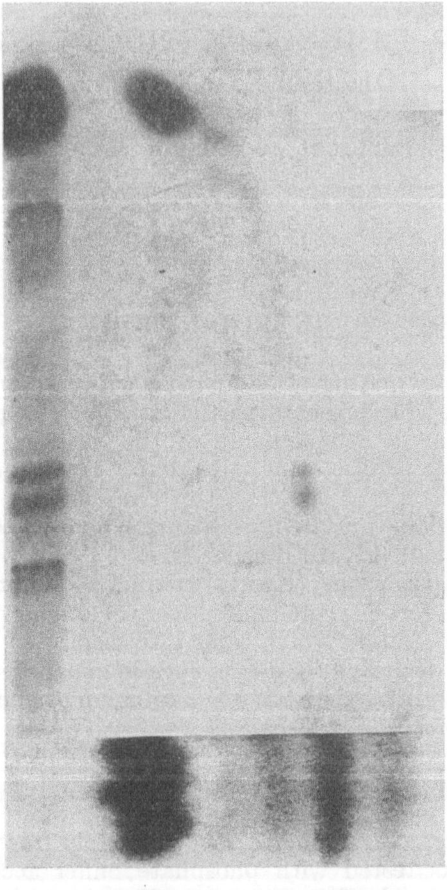

**FIG. 1. Two-dimensional separation of blue-fox serum. First dimension: paper electrophoresis. Second dimension: electrophoresis in starch gel (See explanation in the text)**

Two-dimensional separation facilitated the orientation of the zones of fox-serum protein fractions and also made possible to refer the results to data by others. After electrophoretic separation, the number of protein bands in the beta-globulin zone amounted to 2, 3 or 4 strips (Fig. 2), the front strip being, as a rule, the faintest, others displaying varying intensities of colour. The comparison of patterns according to family data — six mothers, each of them mated with two dog-foxes, and twenty-three cubs — reveals a correspondence between parents and offspring provided that all animals examined had two genes, and that each gene gave a two-strip system, one with a faster migration (Type A), the other with a slower one (Type B), the front band of the slow system having a migration velocity corresponding to a slow band of the faster system. Fast monozygous

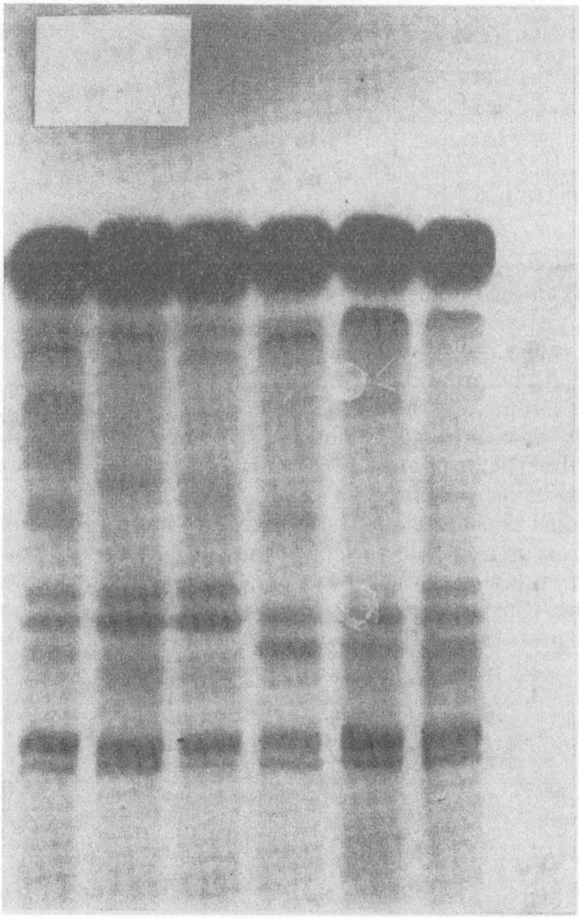

FIG. 2. Separation in starch gel of blue-fox sera

system (Table 1) which had been discovered in the group of twenty-eight foxes examined in the previous year (Fig. 2), was absent in the family group. Among the dog-foxes that were mated, one displayed a four-strip system in the beta-globulin zone, yet no such system was found either in the offspring of the examined families or in the group of sera examined in the previous year.

This considerable mutability of frequencies of certain systems in the fox beta-globulin zone had already been noted in our earlier paper (KAMINSKI

TABLE 1

*Results of examination of a group of animals chosen at random*

| Year | Animals | Types of beta-globulins | | | Types of acid erythroc. phosphatases | | |
|---|---|---|---|---|---|---|---|
| | | A | B | AB | a | b | ab |
| | Dog-foxes | 0 | 3 | 1 + 1 BX | 1 | 1 | 3 |
| 1968 | Vixens | 0 | 3 | 3 | 3 | 0 | 3 |
| | Offspring | 0 | 13 | 10 | 12 | 0 | 11 |
| 1967 | 28 individuals | 4 | 10 | 14 | | not tested | |

et al., 1966) describing findings in the same fox-breeding farm for two consecutive years.

Apart from the question of the actual number of Tf genes in blue foxes, we take it that the animals we had examined, had two genes yielding three phenotypic systems. As shown in Table 2, the group investigated according to the family data is distinguished — by the absence of a homozygous A system and by the predominance of B system, which corresponds to the parents' copulation patterns.

Haptoglobins separated in the above two buffer systems showed, after addition of fox hemoglobin to the serum, insignificant variations of Hp-Hb migration complex without, however, separating into zones within the complex. The comparison of the patterns in families does not justify a genetic interpretation of the patterns. The differences observed (Fig. 3a

TABLE 2

*Results of family examinations*

| Type of beta-globulins and of acid erythroc. phosphatases* | | |
|---|---|---|
| Kind of mating | Offspring | |
| ♀ B(a) × ♂ B(ab) ♂ AB(ab) | 1—B | 1 (ab) |
| ♀ B(ab) × ♂ AB(ab) ♂ AB(ab) | 5—B<br>1—AB | 3 (a)<br>3 (ab) |
| ♀ B(a) × ♂ B(a) ♂BX (b) | 2—B | 1(a)<br>1(ab) |
| ♀ AB(ab) × ♂ B(ab) ♂ BX (b) | 3—B<br>2—AB | 1 (a)<br>4 (ab) |
| ♀ AB(ab) × ♂ B(ab) ♂ B(a) | 1—B<br>5—AB | 4 (a)<br>2 (ab) |
| ♀ AB(a) × ♂ B(ab) ♂ B(a) | 1—B<br>2—AB | 3 (a) |

* In brackets.

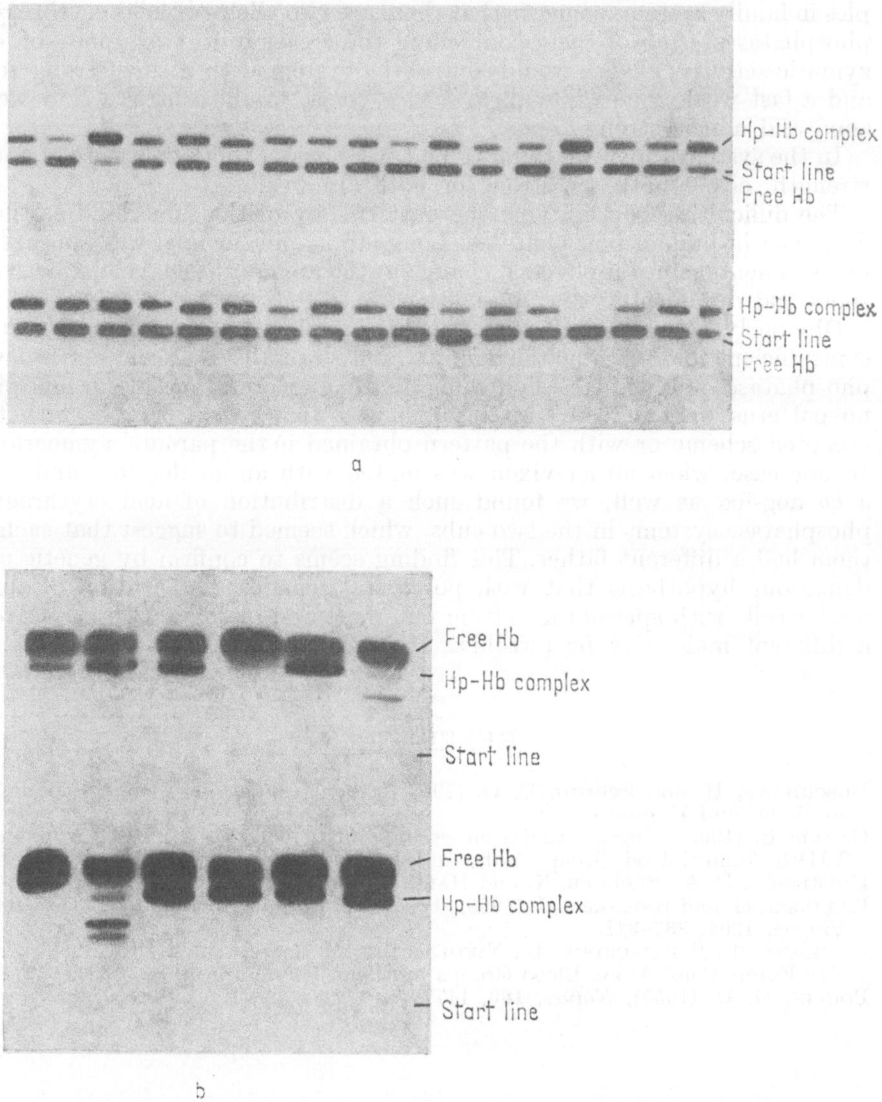
Hp-Hb complex
Start line
Free Hb
Hp-Hb complex
Start line
Free Hb
a
Free Hb
Hp-Hb complex
Start line
Free Hb
Hp-Hb complex
Start line
b

**Fig. 3.** Separation of haptoglobins in starch gel: a — in buffer according to Laurell, b — in buffer according to Prokop et al. x/ human serum

and 3b) seem to indicate that in certain serum samples haptoglobin levels are unequal or that haptoglobin displays unequal ability to bind hemoglobin.

More appropriate conditions of separation must apparently be found to demonstrate Hp differentiation in foxes.

To separate acid erythrocyte phosphatases, purified hemolysate shaken with toluene were used. The comparison of zymograms of forty-three samples in family systems suggests that there are two alleles of acid erythrocyte phosphatases, each of them controlling the creation of two zones of enzymatic activity; allele *a* would control the system with a slow strong zone and a fast weak zone while allele *b*, vice versa, would control a slow weak zone and a fast strong zone.

In the system which we think to be heterozygous, both zones are of equal strength; either both are strong or both are weak.

The difficulties in classifying the patterns are of the same kind as those described in human phosphatases. Long storage of hemolysates appears to be detrimental since it effects a change in the amount of enzymatic activity spots and diminishing their strength.

Obviously, a small number of animals, though arranged in family groups, is insufficient to determine the definite number of alleles of acid erythrocyte phosphatases in blue foxes. Following family interpretations of our material, no patterns were detected in the offspring, that would comply with the accepted scheme or with the pattern obtained in the parents' zymograms. In one case, when an *aa* vixen was mated with an *aa* dog-fox, and with a *bb* dog-fox as well, we found such a distribution of acid erythrocyte phosphatase systems in the two cubs, which seemed to suggest that each of them had a different father. This finding seems to confirm by genetic evidence our hypothesis that with polyfoetal females, fertilization of some ovular cells with spermatozoa from one male and of some with those from a different male may be possible.

## REFERENCES


BUSCHMANN, H. and SCHMID, D. O. (1968), Serumgruppen bei Tieren. Paul Parey in Berlin and Hamburg.

GAHNE, B. (1962), Recent studies on serum protein polymorphism in cattle. Proc. VIIIth Anim. Blood Group Conf., in Europe, Ljubljana, (Mimeo).

HOPKINSON, D. A., SPENCER, N. and HARRIS, H. (1963), *Nature*, **199.** 969.

KAMINSKI, M. and BALBIERZ, H. (1965), Proc. 9th Europ. Conf. Anim. Blood Group, Prague, 1964, 337–341.

KAMINSKI, M., PODLIACHOUK, L., NIKOŁAJCZUK, M. and BALBIERZ, H. (1966), Proc. Xth Europ. Conf. Anim. Blood Groups Biochem. Polymorph. Paris, 1966, 315–318.

POULIK, M. D (1957), *Nature*, **180,** 1477.

*XIIth Europ. Conf. Anim. Blood Groups Biochem. Polymorph., Bp., 1972 (pp. 679—682)*

# CONTRIBUTION ON THE STUDY OF THE A-B-O BLOOD GROUPS IN BABOONS

## ONTOGENETIC INVESTIGATION OF THE DEVELOPMENT OF HEMAGGLUTININS

LUBA PODLIACHOUK and P. DUBOUCH

Institut Pasteur, Paris and Institut de Recherches scientifiques sur le Cancer, Villejuif, France

## INTRODUCTION

Baboons do not contain A-B-O blood groups in their red cells, but they do contain corresponding substances in their saliva (MOOR-JANKOWSKI et al., 1964; MOOR-JANKOWSKI et al., 1965).

Moreover the baboon serum contains hemagglutinins specific of human A and B factors. Their presence agrees with the rule of Landsteiner: animals secreting A and H substance contain anti-B, those secreting B and H substance contain anti-A.

Therefore to determine the blood group of the animal, it is necessary to examine both: serum and saliva.

It must be pointed out that one may occasionally detect in the serum of an animal the antibodies of the same type as its blood group. Such agglutinins are not inhibited by the saliva of human secretors or that of a baboon of corresponding blood group. The antigens A and B possess probably two different specificities: one present in the human saliva and in the red cells (designated $A^s$ or $B^s$), the other present in red cells only (designated $A^c$ or $B^c$). The baboon possesses the $A^s$ or $B^s$ specificity, but not the $A^c$ or $B^c$. His serum can therefore contain anti-$A^c$ or anti-$B^c$ (MOOR-JANKOWSKI et al., 1964; MOOR-JANKOWSKI et al., 1965).

We have previously investigated 139 baboons (PODLIACHOUK and DUBOUCH, 1970) from a baboon colony of the Primatology Division of the Institut de Recherches scientifiques sur le Cancer of Villejuif (near Paris), founded in 1964 (DUBOUCH et al., 1967; DUBOUCH and CAUBEL-KHALADI, 1969).

In sera of newborn baboons no anti-A or anti-B agglutinins were detected. To corroborate the blood group deduced by the study of the saliva, we had to wait several months before the anti-A or anti-B appeared. That is why we consider so important the knowledge of the ontogenesis of hemagglutinins in baboons.

## MATERIAL AND METHODS

In this study 145 baboons belonging to the Papio species: *Papio Cynocephalus Linné* and *Papio Anubis Cuvier* have been investigated. 32 belonged to the group A, 50 to group B and 63 to group AB.

The titer of the anti-A and anti-B has been determined by titration of the serum (previously absorbed by a bottom of human group O red cells) with regard to human cells of group A and B.

16 baboons (4 belonging to group A and 12 to group B) have been investigated during several months from the very day of their birth until several weeks old. Now the majority of them has reached the age of 5 to 10 months; the youngest is 3 months old, a few more than one year old.

## RESULTS

Table 1 designates the titer of the anti-A and anti-B in various animals during the first 12 months of their life. No agglutinin could be detected in the serum before 3 or 4 months of age. When they appear, their titer is very low (1/1 to 1/2) and increases gradually. Still it does not generally exceed 1/32 at the age of one year. The evolution of anti-B sera seems to be slower than that of anti-A.

TABLE 1

*Hemagglutinin titers in various baboons during the first 12 months of age*

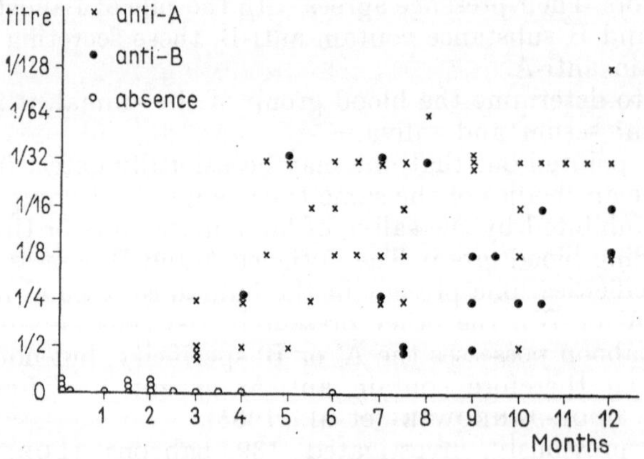
titre
× anti-A
• anti-B
o absence
1/128
1/64
1/32
1/16
1/8
1/4
1/2
0
1 2 3 4 5 6 7 8 9 10 11 12
Months

TABLE 2

*Hemagglutinin distributions in 73 baboons more than one year old*

| Titer | 8 | 16 | 32 | 64 | 128 | 256 | 512 |
|---|---|---|---|---|---|---|---|
| Anti-A No. | 3 | 6 | 16 | 5 | 8 | 2 | 3 |
| % | 7 | 14 | 37.2 | 11.6 | 18.6 | 4.6 | 7 |
| Anti-B No. | 3 | 4 | 9 | 6 | 6 | 1 | 1 |
| % | 10 | 13.4 | 30 | 20 | 20 | 3.3 | 3.3 |

In some cases a slight decline of the titer has been observed in the winter season. Variations of the natural antibody titer with the season, in men and in various animals, have been reported by numerous investigators.

In baboons more than one year old (30 of group A and 43 B) the titer of anti-A and anti-B varies from 1/8 to 1/512 (Table 2). Their assessment in percentage is presented in Fig. 1. The most frequent titer of hemagglutinins is 1/32.

During this study, antibodies that were not conformable to the rule of Landsteiner were observed 4 times: anti-A in 3 A secretors (titers: 1/8, 1/16, 1/16) and anti-B in one B secretor (titer 1/64).

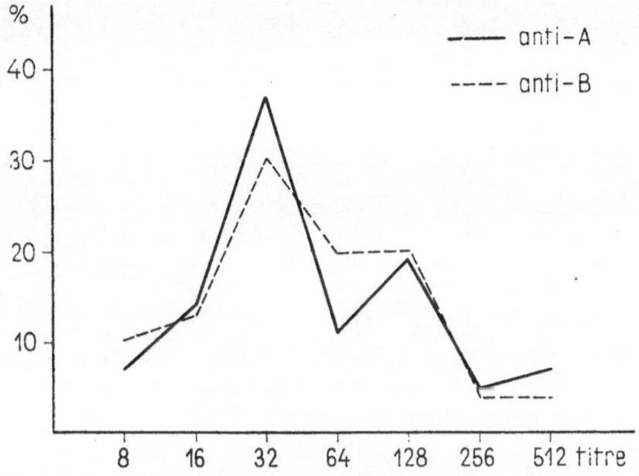
%
40
30
20
10
anti-A
anti-B
8
16
32
64
128
256
512 titre

FIG. 1. Hemagglutinin distribution (percentage) in 73 baboons more than one year old

Both types of agglutinins are 2-mercaptoethanol sensitive and thermolabile after 10-minute heating at 70°C (which suggests presence of IgM globulins).

## CONCLUSION

The age of appearance of anti-A and anti-B in the serum of baboons (3 or 4 months) is more or less the same as that of men (3–6 months). To conclude, we think that it is unwise to establish the blood group of a baboon before the age of 4 months.

## ACKNOWLEDGEMENTS

The authors wish to express their gratitude to Mrs. R. BEAUD for skilld technical assistance.

## REFERENCES

DUBOUCH, P., CAUBEL-KHALADI, M. and ASSELIN, H. G. (1967), Etablissement d'un élevage de babouins destinés à la cancérologie et résultats obtenus. *Ann. Inst. Pasteur,* **112**, 195.

DUBOUCH, P. and CAUBEL-KHALADI, M. (1969), Experiences in laboratory breeding of baboons. *Ann. N. Y. Acad. Sci.*, **162**, 278–281.

MOOR-JANKOWSKI, J., WIENER, A. S. and GORDON, E. B. (1964), Blood groups of apes and monkeys. I. The A-B-O blood groups in baboons. *Transfusion*, **4**, 92.

MOOR-JANKOWSKI, J., HUSER, H. J., WIENER, A. S., KALTER, S. S., PALLOTA, A. J. and GUTHRIE, C. B. (1965), Hematology, blood groups, serum isoantigens and preservation of blood of the baboons. The baboon in medical research. Proc. Ist. Intern. Symposium Univ. of Texas Press, Austin. p. 363.

PODLIACHOUK. LUBA and DUBOUCH, P. (1970), The A-B-O blood groups in a baboon colony in France. *Vox Sang.*, (in press).

*XIIth Europ. Conf. Anim. BloodGroups Biochem. Polymorph., Bp., 1972 (pp. 683–686)*

# NATURAL VARIATION OF THE BLOOD PROTEIN TYPES OF THE AFRICAN ELEPHANT *(Loxodonta africana)*

D. R. OSTERHOFF, E. YOUNG and I. S. WARD-COX

Department of Zootechnology, Faculty of Veterinary Science, University of Pretoria, Onderstepoort, and Veterinary Investigation Centre, Kruger National Park, Rep. of South Africa

## SUMMARY

Concurrent with surveys done previously on Rhinoceros and Buffalo blood samples, similar tests were performed on samples from an African Elephant herd in the Kruger National Park. Contrary to expectations, a considerable variation was found to exist within the various serum fractions, a phenomenon quite unknown as yet amongst African Buffalo. The transferrin type variations also excelled those found previously in the populations of White and Black Rhinoceros.

Genetic markers could be of considerable help to the Nature Conservation Officers in the vast expanses of game parks in Africa, particularly in view of the recently introduced culling programs. The role of the geneticist in the preservation of species unknowingly becoming extinct and the culling of those in a boom stage of proliferation is discussed.

## INTRODUCTION

In a previous study of the African buffalo (OSTERHOFF et al., 1970), it could be concluded that these animals were genetically homogeneous with regard to the serum types analysed by current techniques. The homogeneity within two herds of buffalo was coupled to a very high degree of inbreeding resulting in certain morphological abnormalities. A related study was carried out by OSTERHOFF and KEEP (1970) on the Black and White Rhinoceros of Zululand. Here, a related problem was investigated because the Black variety, which is actually threatened with extinction, displays a remarkable degree of genetic uniformity regarding serum types as compared with his White brother.

A similar study in African elephant was undertaken to estimate the degree of homogeneity and to study possible migration routes of different herds.

## MATERIAL AND METHODS

Progressing from north to south blood samples were obtained from animals in the following areas in the Kruger National Park: Mahlangene (7), Letaba (8) Satara (6), Nwanedzi (10), Tshokwane (20), Skukuza (16) and Crocodile Bridge (17). All 84 specimens were subjected to serum protein analyses and 30 of these to hemoglobin analyses, using the classical technique of starch gel zone electrophoresis. The samples were obtained from animals either being immobilized in the identification and marking programme or from animals being actually shot in areas of the Park where an "elephant problem" exists, i.e. where vast areas have been stripped of tree vegetation and established water-holes have been converted into mud pools by elephants (PIENAAR et al., 1966; VAN WYK and FAIRALL, 1969).

The actual electrophoretic typing had to be done in the Veterinary Investigation Centre in the Kruger National Park, since it was not allowed to take samples out of the reserve due to the risk of spreading foot and mouth disease.

## RESULTS

In spite of the fact that no iron-binding experiments could be performed, it is assumed that the beta-globulin bands depicted in Fig. 1 are in fact the transferrin bands. The designation of the different types are given tentatively.

Albumin typing was also performed but no variation could be established with current techniques. Amylase typing proved to be difficult since the concentration of this enzyme was extremely high, and the interesting variation could not be clearly recorded.

Unfortunately, the hemoglobin typing could be performed on only 30 blood specimens, the two types found are illustrated in Fig. 2.

The slower moving double band of the hemoglobins was only noticed in the animals from Nwanedzi, not a single animal was found showing the heterozygous pattern.

In Tables 1 and 2 the different transferrin and hemoglobin types are shown, grouped according the regions of origin.

TABLE 1

*Transferrin phenotype distribution in African elephant from the Kruger National Park*

| Region | Transferrin phenotypes | | | Total |
|---|---|---|---|---|
| | FF | FS | SS | |
| Mahlangene | — | — | 7 | 7 |
| Letaba | — | 2 | 6 | 8 |
| Satara | — | 5 | 1 | 6 |
| Nwanedzi | — | 2 | 8 | 10 |
| Tshokwane | 2 | 3 | 15 | 20 |
| Skukuza | 2 | 8 | 6 | 16 |
| Crocodile Bridge | — | 5 | 12 | 17 |
| Total | 4 | 25 | 55 | 84 |

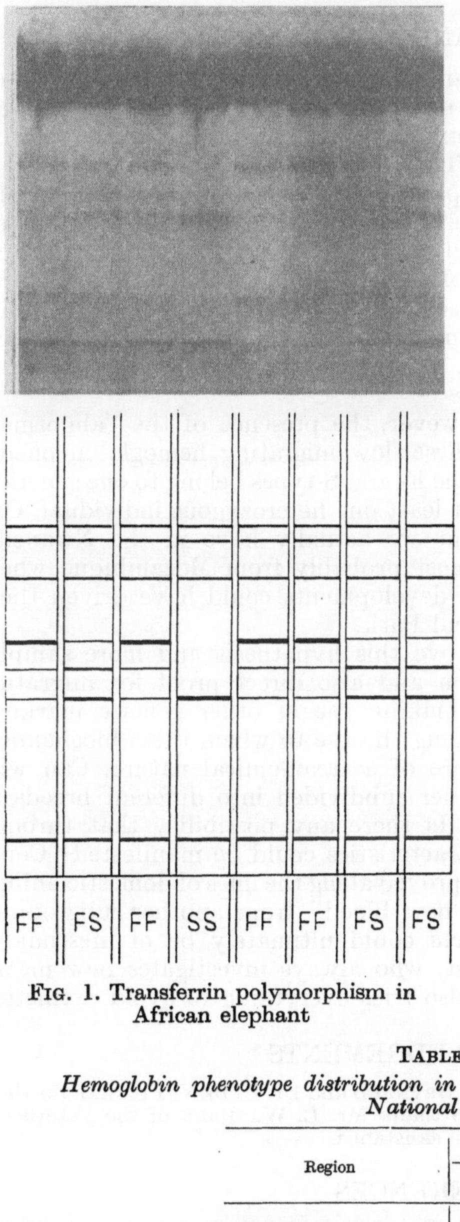
FF
FS
FF
SS
FF
FF
FS
FS

FIG. 1. Transferrin polymorphism in African elephant

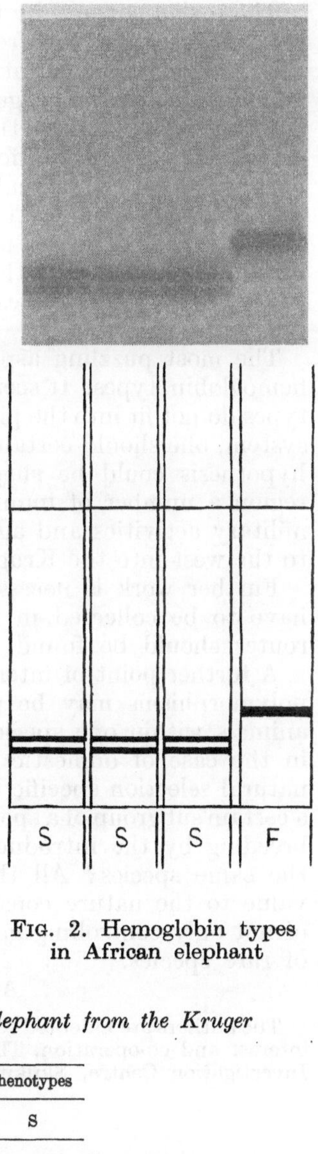
S
S
S
F

FIG. 2. Hemoglobin types in African elephant

TABLE 2

*Hemoglobin phenotype distribution in African elephant from the Kruger National Park*

| Region | Hemoglobin phenotypes | |
|---|---|---|
| | F | S |
| Nwanedzi | 6 | 11 |
| Mahlangene | 5 | — |
| Satara | 2 | — |
| Letaba | 6 | — |
| Total | 19 | 11 |

## DISCUSSION AND CONCLUSIONS

The region covered by the elephants included in the present study represents an area 240 km from north to south and 50 km from east to west with the greatest concentration of animals in the north, due to the relative abudance of natural vegetation (PIENAAR et al., 1966).

From available data there is apparently no admixture of taxonomically and phylogenetically different herds. From the results in Table 1, however, it appears that the animals are genetically very similar and certainly belong to the same gene pool. The overall observed numbers are in complete agreement with the expected number according to the Hardy–Weinberg-equilibrium. One might state that the degree of inbreeding seems to be comparatively lower than in African buffalo, although not as low as that of the Zebra (OSTERHOFF, 1966).

The most puzzling aspect is, however, the presence of the "abnormal" hemoglobin types. It seems that these slow migrating hemoglobin phenotypes do not fit into the picture. If the F- and S-types belong to one genetical system, one should certainly find at least one heterozygous individual. One hypothesis could be suggested that we actually have in the Nwanedzi region a number of immigrants, most probably from Mozambique where military activities and agricultural developments could have driven them to the west into the Kruger National Park.

Further work is necessary to prove this hypothesis and more samples have to be collected in that region and also direct proof for migration routes should be found, also by making use of other genetic markers.

A further point of interest regarding the use to which these biochemical polymorphisms may be put is more of a taxonomical nature. Can wild animals within one species be further subdivided into different breeds as in the case of domestic animals? Is there any possibility that through natural selection specific breed characteristics could be manifested? Could a certain subgroup of a species be improved along the lines of domestic animal breeding by the introduction of "new blood" from another subgroup of the same species? All these aspects could ultimately be of inestimable value to the nature conservationist, who always investigates new means of not only controlling game but also tries everything to avoid extinction of rare species.

## ACKNOWLEDGEMENTS

The authors are indebted to Mr. A. M. BRYNARD and Dr. U. DE V. PIENAAR for their interest and co-operation. They wish to thank Mr. L. WAGENER of the Veterinary Investigation Centre, Skukuza, for his assistance.

## REFERENCES

OSTERHOFF, D. R., YOUNG, E. and WARD-COX, I. S. (1970), *Jl S. Afr. vet. med.*, **41**, 33.
OSTERHOFF, D. R. and KEEP, M. R. (1970), *The Lammergeyer*, **11**, 50.
OSTERHOFF, D. R. (1966), Proc. Xth Europ. Conf. Anim. Blood Groups Biochem. Polymorph. Paris, 1966, 345.
PIENAAR, U. DE V., VAN NIEKERK, J. W., YOUNG, E. VAN WYK, P. and FAIRALL, N. (1966), *Koedoe*, **9**, 108.
PIENAAR, U. DE V., VAN WYK, P. and FAIRALL, N. (1966), *Koedoe*, **9**. 40.
VAN WYK, P. and FAIRALL, N. (1969), *Koedoe*, **12**, 57.